6/16

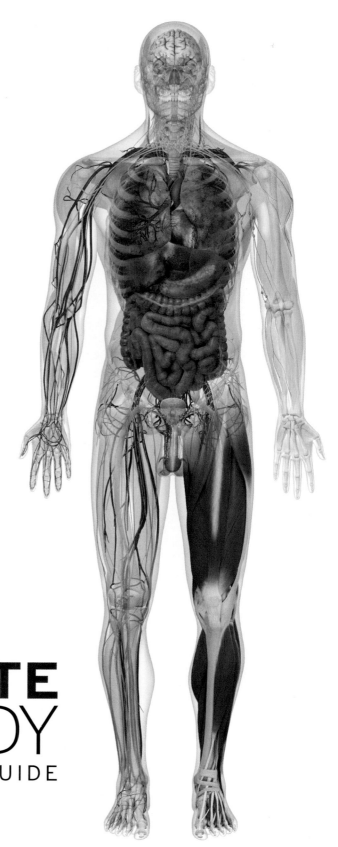

THE COMPLETE HUMANBODY
THE **DEFINITIVE** VISUAL GUIDE

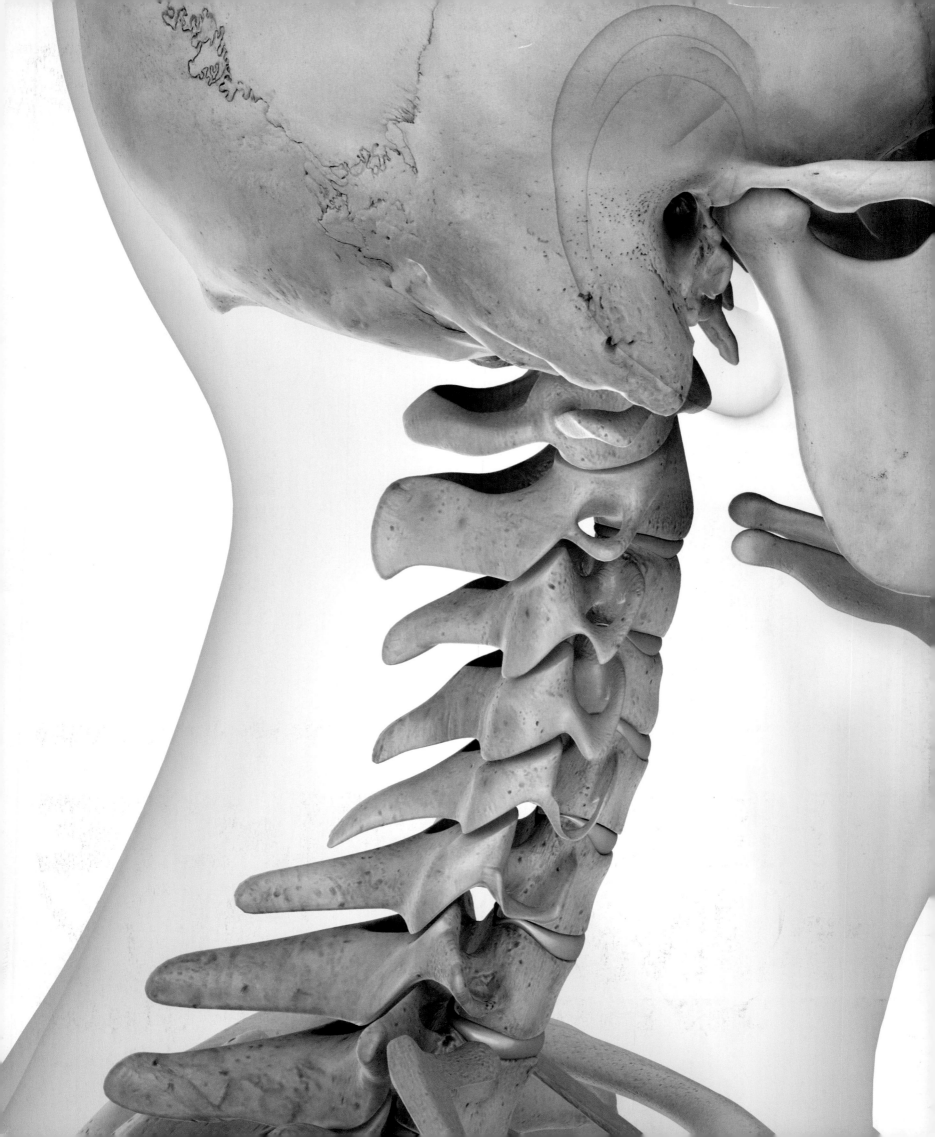

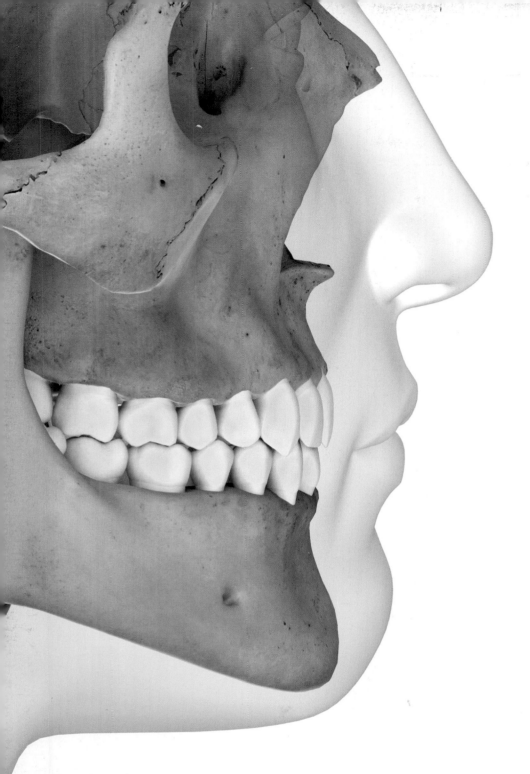

PROFESSOR ALICE ROBERTS

THE COMPLETE HUMANBODY
THE **DEFINITIVE** VISUAL GUIDE

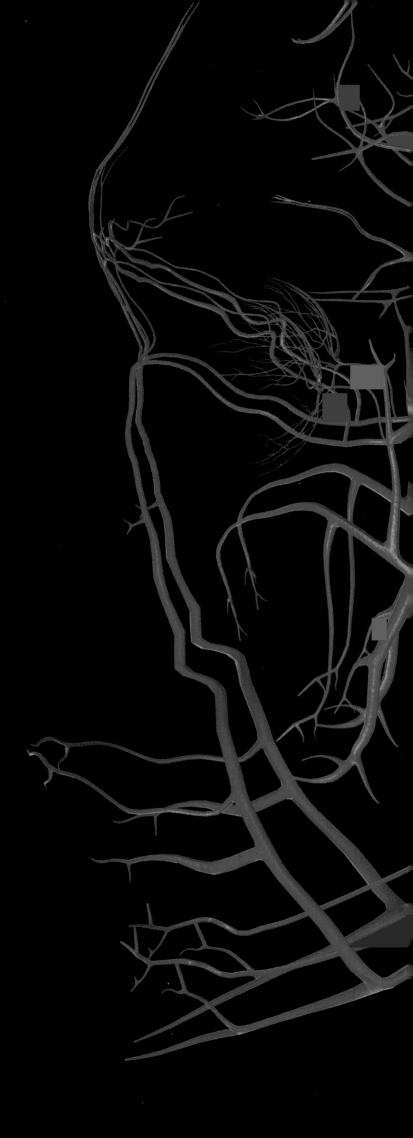

 Penguin Random House

Senior Art Editor
Ina Stradins
Project Art Editors
Alison Gardner, Yen Mai Tsang, Francis Wong
Designers
Sonia Barbate, Clare Joyce,
Helen McTeer, Simon Murrell, Steve Knowlden
Design Assistants
Riccie Janus, Alex Lloyd,
Fiona Macdonald, Rebecca Tennant
Pre-production producers
Phil Sergeant, Robert Dunn
Creative Technical Support
Adam Brackenbury
Jacket Designer
Mark Cavanagh
Managing Art Editor
Michelle Baxter, Michael Duffy
Art Director
Philip Ormerod

Senior Editors
Angeles Gavira, Janet Mohun, Wendy Horobin
Project Editors
Joanna Edwards, Nicola Hodgson, Ruth O'Rourke-
Jones, Nikki Sims, David Summers
Editors
Martha Evatt, Salima Hirani, Steve Setford
Editorial Assistant
Elizabeth Munsey
US Editors
Jill Hamilton, Jane Perlmutter, Margaret Parrish
Indexer
Hilary Bird
Production Controllers
Inderjit Bhullar, Mary Slater
Picture Researcher
Liz Moore
Managing Editor
Angeles Gavira
Publisher
Liz Wheeler
Reference Publisher
Jonathan Metcalf

Illustrators
Medi-Mation (Creative Director: Rajeev Doshi)
Medi-Mation
Medical & Scientific Visualization

Antbits Ltd (Richard Tibbitts)
Deborah Maizels
Dotnamestudios (Andrew Kerr)

Editor-in-Chief Professor Alice Roberts

Authors

THE INTEGRATED BODY
Linda Geddes

ANATOMY
Professor Alice Roberts

HOW THE BODY WORKS
HAIR, NAILS, AND SKIN: Richard Walker
MUSCULOSKELETAL: Richard Walker
NERVOUS SYSTEM: Steve Parker
RESPIRATORY SYSTEM: Dr. Justine Davies
CARDIOVASCULAR SYSTEM: Dr. Justine Davies
LYMPHATIC AND IMMUNE SYSTEM:
Daniel Price
DIGESTIVE SYSTEM: Richard Walker
URINARY SYSTEM: Dr. Sheena Meredith
REPRODUCTIVE SYSTEM: Dr. Gillian Jenkins
ENDOCRINE SYSTEM: Dr. Mimi Chen,
Andrea Bagg

LIFE CYCLE
Authors: Dr. Gillian Jenkins, Dr. Sheena Meredith
Consultant: Professor Mark Hanson

DISEASES AND DISORDERS
Authors: Dr. Fintan Coyle (allergies, blood,
digestive, hair and nails, respiratory, skin)
Dr. Gillian Jenkins (cardiovascular, endocrine,
infertility, reproductive, STDs, urinary)
Dr. Mary Selby (cancer, eye and ear, infectious
disease, inherited disease, nervous system,
mental health, musculoskeletal)
Consultants: Cordelia T Grimm, MD,
Dr. Rob Hicks

Consultants

THE INTEGRATED BODY
Professor Mark Hanson, Southampton
General Hospital

ANATOMY
Professor Harold Ellis, King's College, London
Professor Susan Standring, King's College London

HOW THE BODY WORKS
HAIR, NAILS, AND SKIN:
Professor David Gawkrodger, Royal Hallamshire
Hospital, Sheffield

MUSCULOSKELETAL SYSTEM:
Dr. Christopher Smith, King's College London
Dr. James Barnes, Bristol Royal Hospital for Children

NERVOUS SYSTEM:
Dr. Adrian Pini, King's College London

RESPIRATORY SYSTEM:
Dr. Cedric Demaine, King's College London

CARDIOVASCULAR SYSTEM:
Dr. Cedric Demaine, King's College London

IMMUNE AND LYMPHATIC SYSTEM:
Dr. Lindsay Nicholson, University of Bristol

DIGESTIVE SYSTEM:
Dr. Richard Naftalin, King's College London

URINARY SYSTEM:
Dr. Richard Naftalin, King's College London

REPRODUCTIVE SYSTEM:
Dr. Cedric Demaine, King's College London

ENDOCRINE SYSTEM:
Professor Gareth Williams, University of Bristol
Dr. Mimi Chen, Royal United Hospitals NHS
Foundation Trust, Bristol

Researchers: Christoper Rao,
Kathie Wong, Imperial College, London

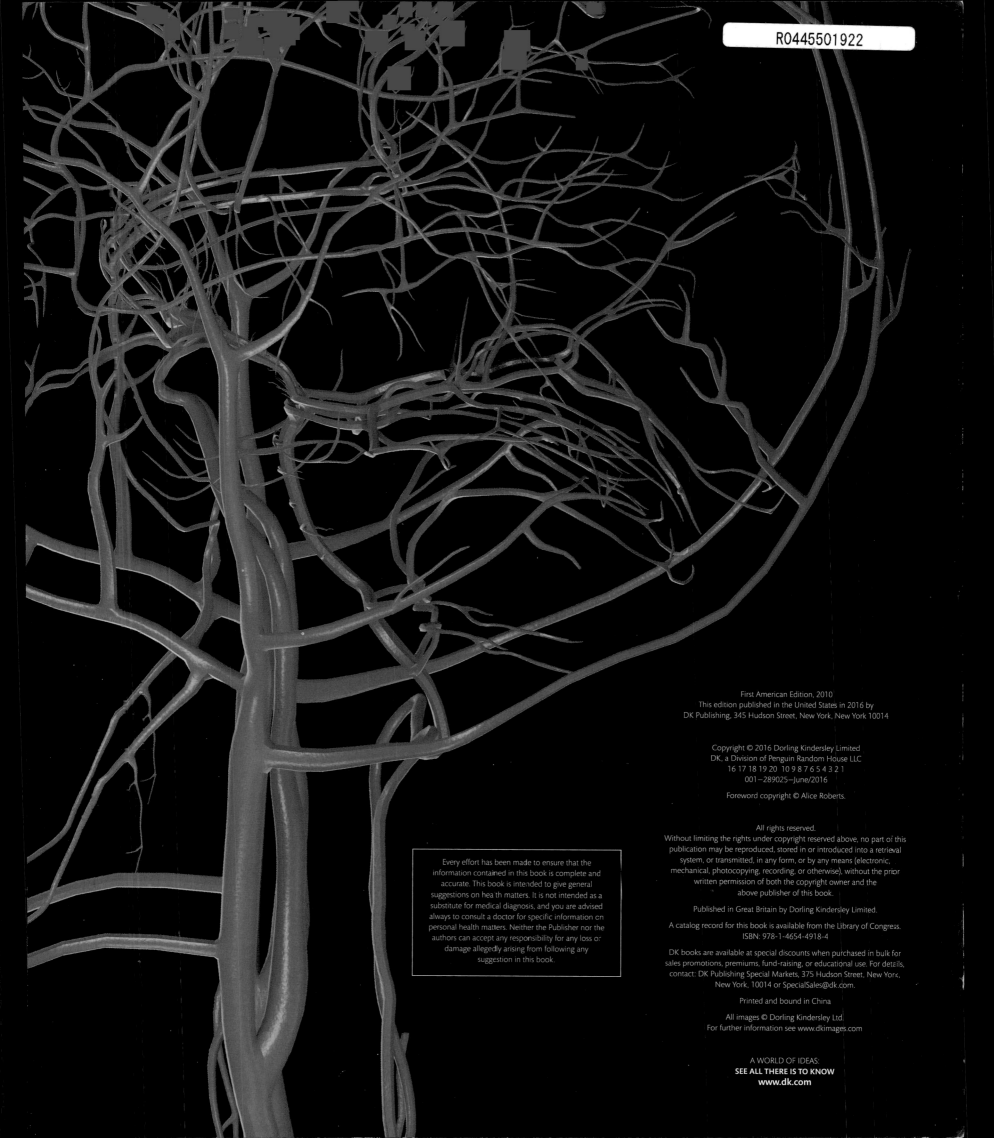

First American Edition, 2010
This edition published in the United States in 2016 by
DK Publishing, 345 Hudson Street, New York, New York 10014

Copyright © 2016 Dorling Kindersley Limited
DK, a Division of Penguin Random House LLC
16 17 18 19 20 10 9 8 7 6 5 4 3 2 1
001—289025—June/2016

Foreword copyright © Alice Roberts.

Published in Great Britain by Dorling Kindersley Limited.

A catalog record for this book is available from the Library of Congress.
ISBN: 978-1-4654-4918-4

DK books are available at special discounts when purchased in bulk for
sales promotions, premiums, fund-raising, or educational use. For details,
contact: DK Publishing Special Markets, 375 Hudson Street, New York,
New York, 10014 or SpecialSales@dk.com.

Printed and bound in China

All images © Dorling Kindersley Ltd.
For further information see www.dkimages.com

A WORLD OF IDEAS:
SEE ALL THERE IS TO KNOW
www.dk.com

CONTENTS

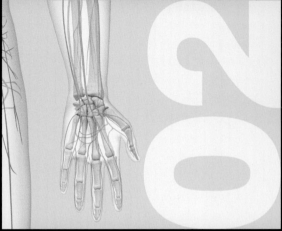
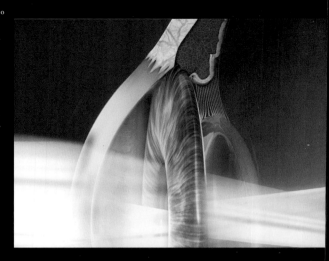

010
THE INTEGRATED BODY

028
ANATOMY

288
HOW THE BODY WORKS

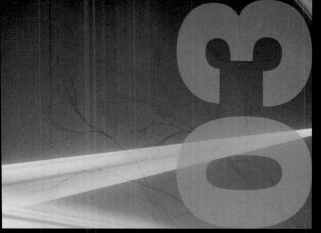
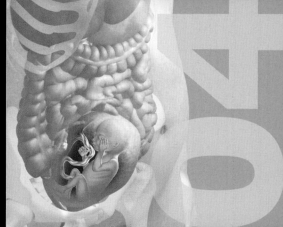
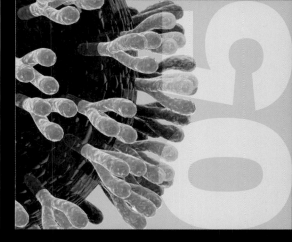

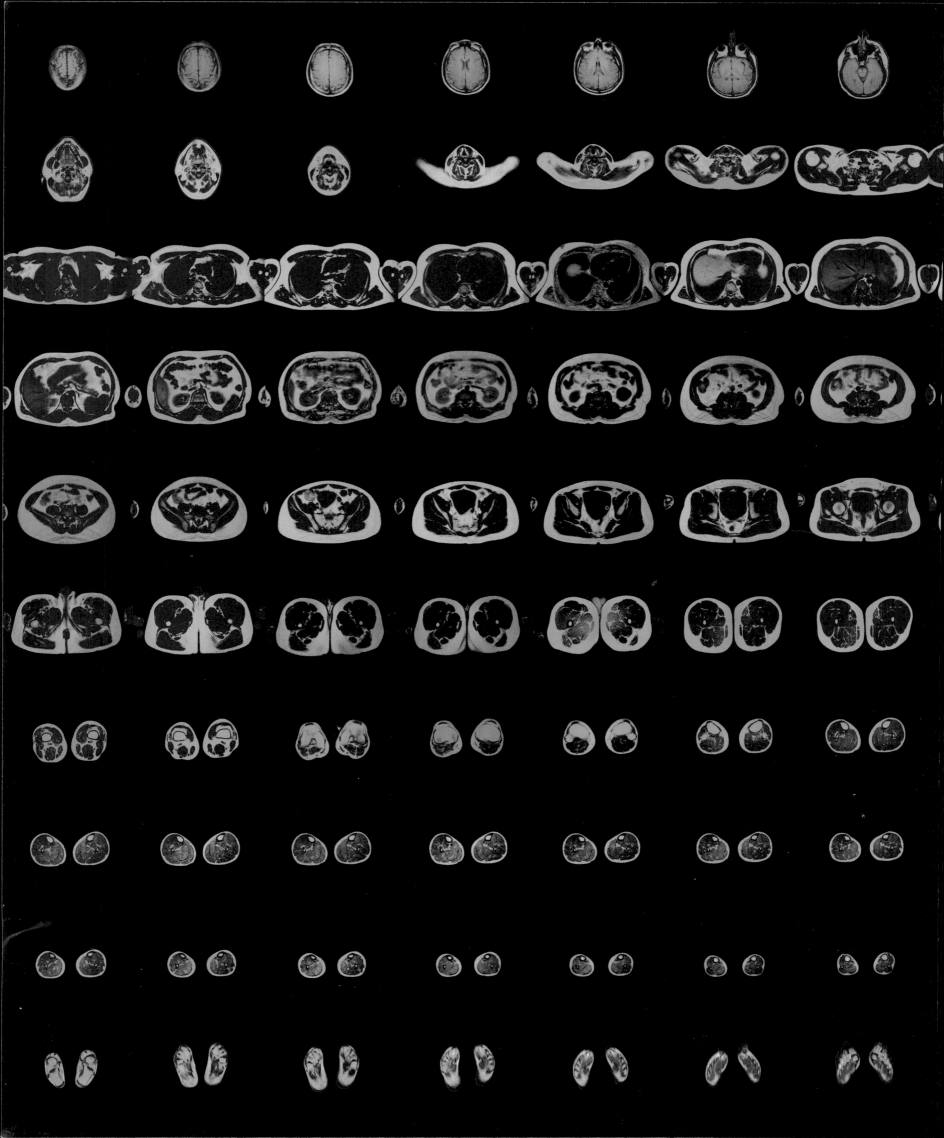

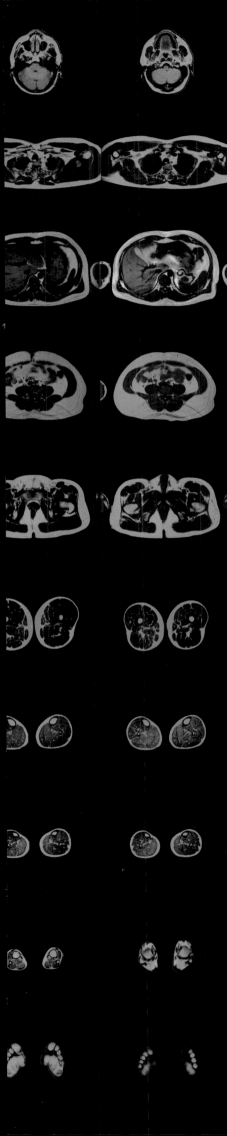

FOREWORD

The study of the human body has an extremely long history. The Edwin Smith papyrus, dating to around 1600 BCE, is the earliest known medical document. It's a sort of early surgical textbook, listing various afflictions and ways of treating them. Even if those are treatments that we wouldn't necessarily recommend today, the papyrus shows us that the ancient Egyptians had some knowledge of the internal structure of the body—they knew about the brain, heart, liver, and kidneys, even if they didn't understand how these organs functioned.

Historically, finding out about the structure of the human body involved dissection; the word "anatomy" literally means "to cut up." After all, when you're trying to find out how a machine works, it's not particularly helpful just to look at the outside of it and try to imagine the machinery inside. I remember a physics practical at school, when we were tasked with finding out how a toaster worked. We found out by taking it apart—although I must admit that we miserably failed to put it back together again (so it's probably a good thing that I ended up as an anatomist rather than a surgeon). Most medical schools still have dissection rooms, where medical students can learn about the structure of the body in a practical, hands-on way. Being able to

learn in this way is a great privilege and depends entirely on the generosity of people who bequeath their bodies to medical science. But in addition to dissection, we now have other techniques with which to explore the structure of the human body: cutting it up virtually using X-rays, computed tomography (CT) and magnetic resonance imaging (MRI), or studying the minute detail of its architecture using electron microscopy.

The first section of this book is an atlas of human anatomy. The body is like a very complicated jigsaw, with organs packed closely together and nestled into cavities, with nerves and vessels twisting around each other, branching inside organs, or piercing through muscles. It can be very hard to appreciate the way that all these elements are organized, but the illustrators have been able to strip down and present the anatomy in a way that is not really possible in the dissection room—showing the bones, muscles, blood vessels, nerves, and organs of the body in turn.

Of course, this isn't an inanimate sculpture, but a working machine. The function of the body becomes the main theme of the second part of the book, as we focus on physiology. Many of us only start to think about how the human body is constructed, and how it works, when something goes wrong with it. The final section looks at some of the problems that interfere with the smooth running of our bodies.

This book—which is a bit like a user's manual—should be of interest to anyone, young or old, who inhabits a human body.

PROFESSOR ALICE ROBERTS

The body piece by piece
A series of magnetic resonance imaging (MRI) scans show horizontal slices through the body, starting with the head and working downward, through the thorax and upper limbs, to the lower limbs, and finally the feet.

the integrated body

The human body comprises trillions of cells, each one a complex unit with intricate workings in itself. Cells are the building blocks of tissues, organs, and eventually, the integrated body systems that all interact—allowing us to function and survive.

010
THE INTEGRATED BODY

HUMAN EVOLUTION

Who are we? Where are we from? We can attempt to answer these questions by studying human evolution. Evolution provides a context for understanding the structure and function of our bodies, and even how we behave and think.

ANCIENT ORIGINS

In placing our species within the animal kingdom, it is clear that we are primates—mammals with large brains compared to other mammals, good eyesight, and, usually, opposable thumbs. Primates diverged, or branched off, from other mammal groups on the evolutionary tree at least 65 million years ago, and possibly as far back as 85 million years ago (see below).

Within the primates, we share with a clutch of other species—the apes—a range of anatomical features: a large body with a chest that is flattened front-to-back; shoulder blades on the back of the chest, supported by long collarbones; arms and hands designed for swinging from branches; and the lack of a tail.

The earliest apes emerged in East Africa at least 20 million years ago, and for the following 15 million years a profusion of ape species existed across Africa, Asia, and Europe. The picture today is very different: humans represent one populous, globally distributed species, contrasting with very small populations of other apes, which are threatened with habitat loss and extinction.

Braincase is slightly larger than in monkeylike species

Face is flatter than in monkeylike species

Robust, apelike jaw

Possible ancestor
Proconsul lived in Africa 27–17 million years ago. Although it has some more primitive primate characteristics, it may be an early ape and even a common ancestor of living apes, including humans.

UNUSUAL PRIMATE

From bush babies to bonobos, lorises and lemurs, to gibbons and gorillas, primates are a diverse bunch of animals, bound together by a common ancestral heritage (see below) and a penchant for living in trees. Humans are unusual primates, having developed a new way of getting around—on two legs, on the ground. However, we still share many characteristics with the other members of the wider primate family tree: five digits on our hands and feet; opposable thumbs, which can be brought into contact with the tips of the fingers (other primates have opposable big toes as well); large, forward-facing eyes, which allow good depth perception; nails rather than claws on our fingers and toes; year-round breeding and long gestation periods, with only one or two offspring produced per pregnancy; and flexible behavior with a strong emphasis on learning.

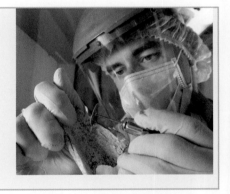

SCIENCE
DATING SPECIES DIVERGENCE

Historically, figuring out evolutionary relationships between living species depended on comparing their anatomy and behavior. Recently, scientists began to compare species' proteins and DNA, using differences in these molecules to construct family trees. Assuming a uniform rate of change, and calibrating the tree using dates from fossils, the dates of divergence of each branch or lineage can be calculated.

Primate family tree
This diagram explains the evolutionary relationships between living primates. It shows how humans are most closely related to chimpanzees, and that apes are more closely related to Old World monkeys (including baboons) than New World monkeys (including squirrel monkeys). All monkeys and apes are shown to be more closely related to each other than to prosimians (including lemurs and bush babies).

MILLIONS OF YEARS AGO

80 | 70 | 60 | 50 | 40 | 30 | 20 | 10 | 0

Human | Chimpanzee | Gorilla | Orangutan | Gibbon | Baboon | Macaque | Vervet | Squirrel monkey | Marmoset | Titi monkey | Mouse lemur | Lemur | Bush baby

GREAT APE

Although we might like to think of ourselves as separate from other apes, our anatomy and genetic makeup places us firmly in that group. Classically, the apes have been divided into two families: lesser apes (gibbons and siamangs) and great apes (orangutans, gorillas, and chimpanzees), with humans and their ancestors placed in a separate family hominids. But, since genetic studies have shown such a close relationship between the African apes and humans, it makes more sense to group humans, chimpanzees, and gorillas together as hominids. Humans and their ancestors are then known as hominins.

Not only that, but humans are genetically closer to chimpanzees than either humans or chimpanzees are to gorillas. It's not surprising that humans have been called the "third chimpanzee."

Human skull
The skull in humans is dominated by a massive braincase, with a volume of 1,100–1,700 cubic centimeters (cc). Its teeth, jaws, and areas of attachment for chewing muscles are small in comparison with other apes. The brow ridges over the eye sockets are subtle and the face is relatively flat.

Chimpanzee skull
Chimpanzees have a relatively small, rounded braincase, accommodating a brain of 300–500 cubic centimeters in volume. The face is relatively large, with a fairly prominent brow ridge and jaws that project forward.

Gorilla skull
The occipital torus is high on the skull, with a large area for the attachment of strong neck muscles below it. The male gorilla has a massive brow ridge and a large sagittal crest for the attachment of strong jaw muscles. The size of the braincase is 350–700 cubic centimeters.

Orangutan skull
Like the chimpanzee, the orangutan has a relatively small braincase, with a volume of 300–500 cubic centimeters, and a large face. The skull is extremely prognathic, with strongly projecting jaws. The brow ridge is much smaller than in gorillas or chimpanzees.

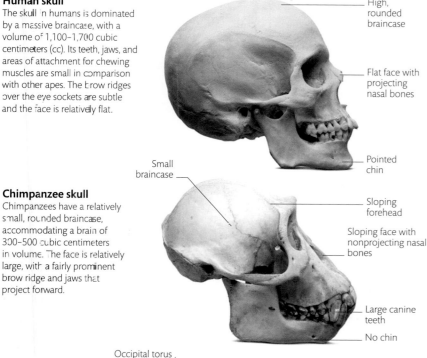

High, rounded braincase

Flat face with projecting nasal bones

Small braincase

Pointed chin

Sloping forehead

Sloping face with nonprojecting nasal bones

Large canine teeth

No chin

Occipital torus

Large sagittal crest

Massive brow ridge

Flat forehead

Long, sloping face

Large, projecting jaw, but no chin

Small braincase

Small brow ridge

Strongly projecting jaws

OUR CLOSEST RELATIVE

Science has shown that humans and chimpanzees shared a common ancestor some 5–8 million years ago. Comparing ourselves with our closest relative gives us an opportunity to identify the unique features that make us human.

Humans have developed two major defining characteristics—upright walking on two legs, and large brains—but there are many other differences between us and chimpanzees. The human population is huge and globally distributed, but we are, in fact, less genetically diverse than chimpanzees, probably because our species is much younger. Reproduction is quite similar, although human females reach puberty later, and also live for a long time after menopause. Humans live up to 80 years, while chimpanzees may live up to 40 or 50 years in the wild. Chimpanzees live in large, hierarchical social groups, with relationships strengthened by social grooming; humans have even more complex social organization. Furthermore, although chimpanzees can be taught to use sign language, humans are uniquely adept at communicating thoughts and ideas through complex language systems.

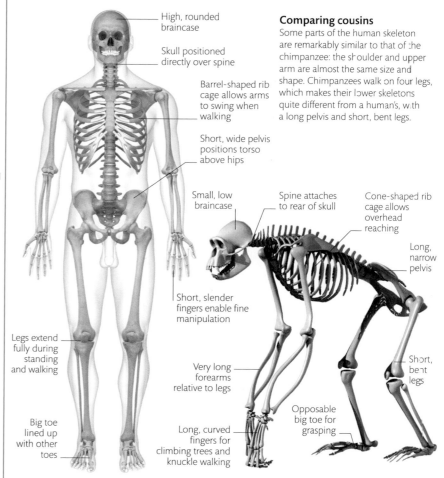

High, rounded braincase

Skull positioned directly over spine

Barrel-shaped rib cage allows arms to swing when walking

Short, wide pelvis positions torso above hips

Small, low braincase

Spine attaches to rear of skull

Cone-shaped rib cage allows overhead reaching

Long, narrow pelvis

Short, slender fingers enable fine manipulation

Legs extend fully during standing and walking

Very long forearms relative to legs

Short, bent legs

Big toe lined up with other toes

Long, curved fingers for climbing trees and knuckle walking

Opposable big toe for grasping

Comparing cousins
Some parts of the human skeleton are remarkably similar to that of the chimpanzee: the shoulder and upper arm are almost the same size and shape. Chimpanzees walk on four legs, which makes their lower skeletons quite different from a human's, with a long pelvis and short, bent legs.

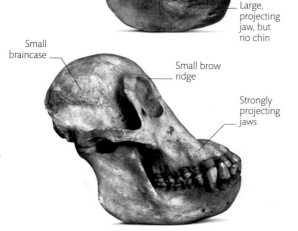

Dependent young
A human baby is born at an earlier stage of brain development than a chimpanzee baby, and is more helpless and dependent on caregivers. Even so, the human baby's head is relatively large at birth, making for a longer and more difficult delivery.

HUMAN ANCESTORS

Humans and their ancestors are known as hominins. The hominin fossil record begins in East Africa, with many finds from the Rift Valley. Early species walked upright, but large brains and tool-making came along later, with the appearance of our own genus, *Homo*.

THE FOSSIL RECORD

In the last two decades, exciting discoveries have pushed back the dates of the earliest hominin ancestors, and provoked controversy over when humans first left Africa.

Fossils of a few possible early hominins have been found in East and Central Africa, dating to more than 5 million years ago. The oldest of these is *Sahelanthropus tchadensis*, which, from the position of the foramen magnum (the large hole where the spinal cord exits) on its fossil skull, appears to have stood upright on two legs. Fossilized limb bones of *Ardipithecus ramidus* suggest that it clambered around in trees as well as being able to walk on two legs on the ground. From 4.5 million years ago, a range of fossil species known collectively as australopithecines emerged. These hominins were well adapted to upright walking, but did not have the long legs and large brains of the *Homo* genus.

Until recently, it was thought that *Homo erectus* was the first hominin to leave Africa, and its fossils are found as far east as China. However, discoveries of small hominins in Indonesia suggest that there may have been an earlier expansion out of Africa.

We are the only hominin species on the planet today, but this is unusual: for most of human evolutionary history, there have been several species overlapping with each other.

PRESENT DAY
0
1 MYA
2 MYA
3 MYA
4 MYA
5 MYA
6 MYA
7 MYA

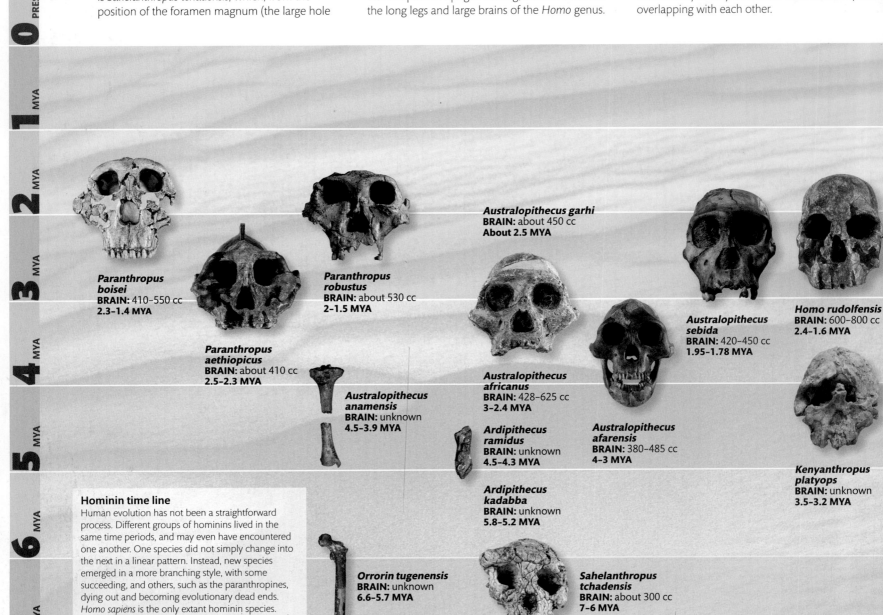

Australopithecus garhi
BRAIN: about 450 cc
About 2.5 MYA

Paranthropus boisei
BRAIN: 410–550 cc
2.3–1.4 MYA

Paranthropus robustus
BRAIN: about 530 cc
2–1.5 MYA

Australopithecus sebida
BRAIN: 420–450 cc
1.95–1.78 MYA

Homo rudolfensis
BRAIN: 600–800 cc
2.4–1.6 MYA

Paranthropus aethiopicus
BRAIN: about 410 cc
2.5–2.3 MYA

Australopithecus anamensis
BRAIN: unknown
4.5–3.9 MYA

Australopithecus africanus
BRAIN: 428–625 cc
3–2.4 MYA

Ardipithecus ramidus
BRAIN: unknown
4.5–4.3 MYA

Australopithecus afarensis
BRAIN: 380–485 cc
4–3 MYA

Kenyanthropus platyops
BRAIN: unknown
3.5–3.2 MYA

Ardipithecus kadabba
BRAIN: unknown
5.8–5.2 MYA

Hominin time line
Human evolution has not been a straightforward process. Different groups of hominins lived in the same time periods, and may even have encountered one another. One species did not simply change into the next in a linear pattern. Instead, new species emerged in a more branching style, with some succeeding, and others, such as the paranthropines, dying out and becoming evolutionary dead ends. *Homo sapiens* is the only extant hominin species.

Orrorin tugenensis
BRAIN: unknown
6.6–5.7 MYA

Sahelanthropus tchadensis
BRAIN: about 300 cc
7–6 MYA

MODERN HUMANS

From around 600,000 years ago, a species called *Homo heidelbergensis* existed in Africa and Europe. This ancestral species may have evolved into Neanderthals (*Homo neanderthalensis*) in Europe, about 400,000 years ago and anatomically modern humans (*Homo sapiens*) in Africa, around 200,000 years ago. Although it is difficult to draw a line between the later fossils of *Homo heidelbergensis* and the earliest fossils of *Homo sapiens*, the rounded cranium of Omo II, discovered by the renowned Kenyan paleoanthropologist Richard Leakey and his team in southern Ethiopia, and now dated to around 195,000 years ago, is accepted by many to be the earliest fossil of a modern human (see below).

The fossil, archaeological, and climatic evidence suggests that modern humans expanded out of Africa between 50,000 and 80,000 years ago. People spread out of Africa along the rim of the Indian Ocean to Australia, and northward, into Europe, northeast Asia, and later, into the Americas.

Modern behavior
This piece of ocher found at Pinnacle Point, South Africa, suggests that humans were using pigment more than 160,000 years ago.

EXTINCT COUSINS

Neanderthals lived in Europe for hundreds of thousands of years before modern humans arrived on the scene some 40,000 years ago. The last known evidence of Neanderthals is from Gibraltar, around 25,000 years ago. The question of whether Neanderthals and modern humans met and interacted is hotly debated. There are a few fossils that some anthropologists believe show features of both species, leading to the controversial suggestion that modern humans and Neanderthals interbred with each other. Analysis of DNA from Neanderthal fossils has not shown any genetic evidence for interbreeding.

Varied diets
Archaeological evidence from Gibraltar suggests that, like humans, Neanderthals were eating a varied diet including shellfish, small animals and birds, and possibly even dolphins.

Homo habilis
BRAIN: 500–550 cc
2.4–1.4 MYA

Homo ergaster
BRAIN: 600–910 cc
1.9–1.5 MYA

Homo erectus
BRAIN: 750–1,300 cc
1.8 MYA–30,000 YA

Homo floresiensis
BRAIN: about 400 cc
95,000–12,000 YA

Homo antecessor
BRAIN: about 1,000 cc
780,000–500,000 YA

Homo heidelbergensis
BRAIN: 1,100–1,400 cc
600,000–100,000 YA

Homo neanderthalensis
BRAIN: about 1,412 cc
400,000–28,000 YA

Homo sapiens
BRAIN: 1,000–2,000 cc
200,000 YA–present

OUR OLDEST REMAINS

In 1967, a team led by the paleoanthropologist Richard Leakey discovered fossils of our own species in the dunelike hills of the Kibish formation near the Omo River in Ethiopia (shown here). The fossils were found sandwiched between layers of ancient volcanic rock. In 2005, scientists applied new dating techniques to these volcanic layers, and pushed back the date of the fossils to around 195,000 years old. This makes them the oldest known remains of *Homo sapiens* in the world.

HUMAN GENETIC FORMULA

DNA (deoxyribonucleic acid) is the blueprint for all life, from the humblest yeast to the human being. It provides a set of instructions on how to assemble the many thousands of different proteins that make us who we are. It also tightly regulates this assembly, ensuring that it does not run out of control.

THE MOLECULE OF LIFE

Although we all look different, the basic structure of our DNA is identical. It consists of chemical building blocks called bases, or nucleotides. What varies between individuals is the precise order in which these bases are connected into pairs. When base pairs are strung together they can form functional units called genes, which "spell out" the instructions for making a protein. Each gene encodes a single protein, although some complex proteins are encoded by more than one gene. Proteins have a wide range of vital functions in the body. They form structures such as skin or hair, they carry signals around the body, and they fight off infectious agents such as bacteria. Proteins also make up cells, the basic units of the body, and perform the thousands of basic biochemical processes needed to sustain life. However, only about 1.5 percent of our DNA encodes genes. The rest consists of regulatory sequences, structural DNA, or has no obvious purpose—so-called junk DNA.

DNA micrograph
Although DNA is extremely small, its structure can be observed by using a scanning tunneling microscope, which has magnified this image around two million times.

DNA backbone
Formed of alternating units of phosphate and a sugar called deoxyribose

DNA double helix
In the vast majority of organisms, including humans, long strands of DNA twist around each other to form a right-handed spiral structure called a double helix. The helix consists of a sugar (deoxyribose) and phosphate backbone and complementary base pairs that stick together in the middle. Each twist of the helix contains around ten base pairs.

Guanine
Cytosine

Thymine
Adenine

BASE PAIRS

DNA consists of building blocks called bases. There are four types: adenine (A), thymine (T), cytosine (C), and guanine (G). Each base is attached to a phosphate group and a deoxyribose sugar ring to form a nucleotide. In humans, bases pair up to form a double-stranded helix in which adenine pairs with thymine, and cytosine with guanine. The two strands are "complementary" to each other. Even if they are unwound and unzipped, they can realign and rejoin.

Forming bonds
The two strands of the double helix join by forming hydrogen bonds. When guanine binds with cytosine, three bonds are formed, and when adenine binds with thymine, they form two.

Three bonds join C and G

Phosphate

C — G

T — A

G — C

A — T

Sugar

Two bonds join A and T

GENES

A gene is a unit of DNA needed to make a protein. Genes range in size from just a few hundred to millions of base pairs. They control our development, but are also switched on and off in response to environmental factors. For example, when an immune cell encounters a bacterium, genes are switched on that produce antibodies to destroy it. Gene expression is regulated by proteins that bind to regulatory sequences within each gene. Genes contain regions that are translated into protein (exons) and noncoding regions (introns).

Eye color
The genetics of eye color are incredibly complex, and many different genes are involved.

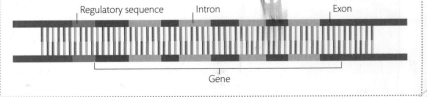

Regulatory sequence　　Intron　　Exon

Gene

PACKAGING DNA

The human genome is composed of approximately 3 billion bases of DNA—about 5½ ft (2 m) of DNA in every cell if it was stretched from end to end. Therefore, our DNA must be packaged up in order to fit inside each tiny cell. DNA is concentrated into dense structures called chromosomes, and each cell contains 23 pairs of chromosomes (46 in total)—one set from the mother and another set from the father. To package up DNA, the double helix must first be coiled around histone proteins, forming a structure that looks a little like beads on a string. These histone "beads" then wind up and lock together into densely coiled "chromatin," which, when a cell prepares to divide, further winds back on itself into tightly coiled chromosomes.

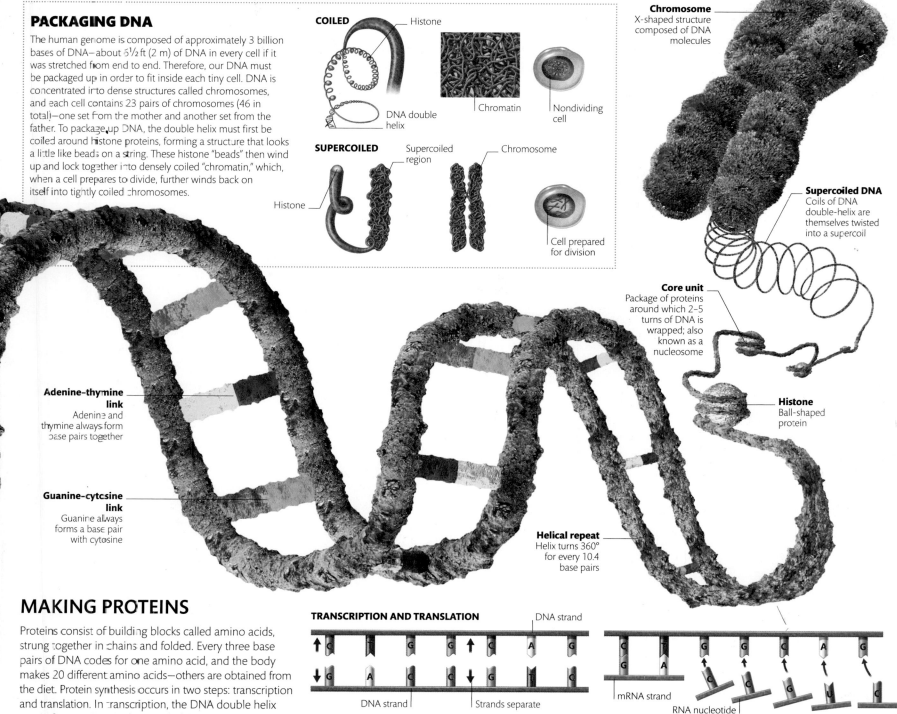

COILED

Histone

DNA double helix

Chromatin

Nondividing cell

SUPERCOILED

Supercoiled region

Chromosome

Histone

Cell prepared for division

Chromosome
X-shaped structure composed of DNA molecules

Supercoiled DNA
Coils of DNA double-helix are themselves twisted into a supercoil

Core unit
Package of proteins around which 2–5 turns of DNA is wrapped; also known as a nucleosome

Histone
Ball-shaped protein

Adenine-thymine link
Adenine and thymine always form base pairs together

Guanine-cytosine link
Guanine always forms a base pair with cytosine

Helical repeat
Helix turns 360° for every 10.4 base pairs

MAKING PROTEINS

Proteins consist of building blocks called amino acids, strung together in chains and folded. Every three base pairs of DNA codes for one amino acid, and the body makes 20 different amino acids—others are obtained from the diet. Protein synthesis occurs in two steps: transcription and translation. In transcription, the DNA double helix unwinds, exposing single-stranded DNA. Complementary sequences of a related molecule called RNA (ribonucleic acid) then create a copy of the DNA sequence that can be translated into protein. This "messenger RNA" travels to ribosomes, where it is translated into strings of amino acids. These are then folded into the 3-D structure of a protein.

Cell nucleus
DNA is found in a structure at the center of the cell called the nucleus. The first stage of protein synthesis takes place here.

TRANSCRIPTION AND TRANSLATION

DNA strand

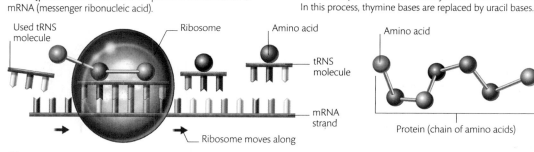

DNA strand

Strands separate

mRNA strand

RNA nucleotide

1 Inside the nucleus of the cell, the DNA strands temporarily separate. One will act as a template for the formation of mRNA (messenger ribonucleic acid).

2 RNA nucleotides with correctly corresponding bases lock onto the exposed DNA bases and join to form a strand of mRNA. In this process, thymine bases are replaced by uracil bases.

Used tRNS molecule

Ribosome

Amino acid

Amino acid

tRNS molecule

mRNA strand

Ribosome moves along

Protein (chain of amino acids)

3 The mRNA strand attaches to a ribosome, which passes along the strand. Within the ribosome, individual tRNS (transfer ribonucleic acid) molecules, each carrying an amino acid, slot onto the mRNA.

4 As the ribosome moves along the mRNA, it produces a specific sequence of amino acids, which combine to form a particular protein.

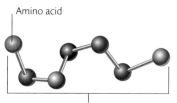

THE HUMAN GENOME

Different organisms contain different genes, but a surprisingly large proportion of genes are shared between organisms. For example, roughly half of the genes found in humans are also found in bananas. However, it would not be possible to substitute the banana version of a gene for a human one because variations in the order of the base pairs within each gene also distinguish us. Humans all possess more or less the same genes, but many of the differences between individuals can be explained by subtle variations within each gene. The extent of these variations is smaller than between humans and animals, and smaller still than the differences between humans and plants. In humans, DNA differs by only around 0.2 percent, while human DNA differs from chimpanzee DNA by around 5 percent.

Human genes are divided unevenly between 23 pairs of chromosomes, and each chromosome consists of gene-rich and gene-poor sections. When chromosomes are stained, differences in these regions show up as light and dark bands, giving chromosomes a striped appearance. We still don't know exactly how many protein-coding genes there are in the human genome, but researchers currently estimate between 20,000 and 25,000.

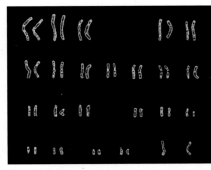

Karyotype
This is an organized profile of the chromosomes in someone's cells, arranged by size. Studying someone's karyotype enables doctors to determine whether any chromosomes are missing or abnormal.

GENETIC PROFILING

Apart from subtle genetic variations, humans also vary in their noncoding DNA. This so-called junk DNA accounts for vast tracts of our genetic material, and we still have little understanding of what it does. However, that does not make it useless. Forensic scientists look at variations in noncoding DNA to match criminal suspects to crime scenes. To do this, they analyze short, repeating sequences of DNA within noncoding regions, called short-tandem-repeats (STRs). The precise number of repeats is highly variable between individuals. In one method, forensic scientists compare ten of these repeating regions, chopping them up and then separating them on the basis of their size to generate a series of bands called a DNA profile or fingerprint.

Shared characteristics
Genetic profiling can also be used to prove family relationships. Here, two children are shown to share bands with each parent, proving they are related.

Chromosome banding
Each chromosome has two arms, and staining reveals that these are divided into bands. Each band is numbered, making it possible to locate a specific gene if you know its address. These are the bandings on chromosome 7.

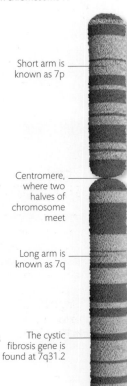

Short arm is known as 7p

Centromere, where two halves of chromosome meet

Long arm is known as 7q

The cystic fibrosis gene is found at 7q31.2

Chromosome complement
The human genome is stored on 23 pairs of chromosomes—46 in total. Of these, 22 pairs store general genetic information and are called autosomes, while the remaining pair determines whether you are male or female. There are two types of sex chromosome: X and Y. Men have one X and one Y, while women have two X chromosomes.

There is **no known function** for **97 percent** of the DNA in the human genome—sometimes known as **junk DNA**.

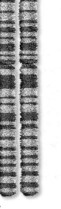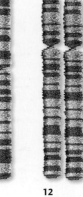

	1	2	3	4	5	6	7	8	9	10	11	12
Number of genes:	4,234	3,078	3,723	542	737	2,277	4,171	1,400	1,931	1,776	546	1,698
Associations and conditions:	Alzheimer's disease; Parkinson's disease; glaucoma; prostate cancer; brain size	Color blindness; red hair; breast cancer; Crohn's disease; amyotrophic lateral sclerosis (ALS); high cholesterol	Deafness; autism; cataracts; susceptibility to HIV infection; diabetes; Charcot-Marie-Tooth disease	Blood vessel growth; immune system genes; bladder cancer; Huntington's disease; deafness; hemophilia; Parkinson's disease	DNA repair; nicotine addiction; Parkinson's disease; Cri du Chat syndrome; breast cancer; Crohn's disease	Cannabis receptor; cartilage strength; immune system genes; epilepsy; type 1 diabetes; rheumatoid arthritis	Pain perception; muscle, tendon and bone formation; cystic fibrosis; schizophrenia; Williams syndrome; deafness; type 2 diabetes	Brain development and function; cleft lip and palate; schizophrenia; Werner syndrome	Blood group; albinism; bladder cancer; porphyria	Inflammation; DNA repair; breast cancer; Usher's syndrome	Sense of smell; hemoglobin production; autism; albinism; sickle-cell anemia; breast cancer; bladder cancer	Cartilage and muscle strength; narcolepsy; stuttering; Parkinson's disease

THE SUM OF ONE'S GENES

At the simplest level, each gene encodes a protein, and each protein results in a distinct trait or phenotype. In humans, this is best illustrated by inherited diseases like cystic fibrosis. Here, a mutation in the CFTR gene, which makes a protein found in mucus, sweat, and digestive juices, results in the accumulation of thick mucus in the lungs, leaving carriers of the defective gene more susceptible to lung infections. If we know what a specific gene looks like in a healthy person, and how it looks if it has gone wrong, it may be possible to devise a genetic test to find out whether someone is at risk of disease. For example, mutations in a gene called BRCA1 can predict if a woman is at high risk of developing one form of breast cancer. However, many traits—such as height or hair color—are influenced by several genes working together. And genes are only part of the equation. In the case of personality or lifespan, multiple genes interact with environmental factors, such as upbringing and diet, to shape who we are and who we will become (see p.410).

Human diversity

Although all humans carry more or less the same genes in terms of the proteins they manufacture, the vast number of possible combinations of genes, and the ways they are expressed, explains the huge diversity in the human body across the world's population.

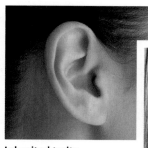

Inherited traits

Humans possess two copies of each gene, but not all genes are equal. Dominant genes show their effect even if there is only one in a pair, while recessive genes need two copies (see p.411). Free-hanging earlobes are caused by the dominant form of a gene, while attached earlobes are recessive.

(see p.410).
(see p.411)

BREAKTHROUGHS
GENETIC ENGINEERING

This form of gene manipulation enables us to substitute a defective gene with a functional one, or introduce new genes. Glow-in-the-dark mice were created by introducing a jellyfish gene that encodes a fluorescent protein into the mouse genome. Finding safe ways of delivering replacement genes to the correct cells in humans could lead to cures for many types of inherited diseases—so-called gene therapy.

13
Number of genes: 925
Associations and conditions:
LSD receptor; breast cancer (BRCA2 gene); bladder cancer; deafness; Wilson's disease

14
Number of genes: 1,887
Associations and conditions:
Antibody production; Alzheimer's disease; amyotrophic latera sclerosis (ALS); muscular dystrophy

15
Number of genes: 1,377
Associations and conditions:
Eye color; skin color; Angelman syndrome; breast cancer; Tay-Sachs disease; Marfan syndrome

16
Number of genes: 1,561
Associations and conditions:
Red hair; obesity; Crohn's disease; breast cancer; trisomy 16 (most common chromosomal cause of miscarriage)

17
Number of genes: 2,417
Associations and conditions:
Connective tissue function; early onset breast cancer (BRCA1); brittle bone disease; bladder cancer

18
Number of genes: 756
Associations and conditions:
Edward's syndrome; Paget's disease; porphyria; selective mutism

19
Number of genes: 1,984
Associations and conditions:
Cognition; Alzheimer's disease; cardiovascular disease; high cholesterol; hereditary stroke

20
Number of genes: 1,019
Associations and conditions:
Celiac disease; type 1 diabetes; prion diseases

21
Number of genes: 595
Associations and conditions:
Down syndrome; Alzheimer's disease; amyotrophic lateral sclerosis (ALS); deafness

22
Number of genes: 1,841
Associations and conditions:
Antibody production; breast cancer; schizophrenia; amyotrophic lateral sclerosis (ALS)

X
Number of genes: 1,860
Associations and conditions:
Breast cancer; color blindness; hemophilia; fragile X syndrome; Turner syndrome; Klinefelter's syndrome

Y
Number of genes: 454
Associations and conditions:
Male fertility and testicular development

THE CELL

It is hard to comprehend what 75 trillion cells looks like, but observing yourself in a mirror would be a good start. That is how many cells exist in the average human body—and we replace millions of these cells every single day.

CELL ANATOMY

The cell is the basic functional unit of the human body. Cells are extremely small, typically only about 0.01 mm across—even our largest cells are no bigger than the width of a human hair. They are also immensely versatile: some can form sheets like those in your skin or lining your mouth, while others can store or generate energy, such as fat and muscle cells. Despite their amazing diversity, there are certain features that all cells have in common, including an outer membrane, a control center called a nucleus, and tiny powerhouses called mitochondria.

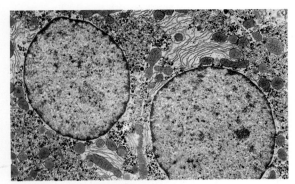

Liver cell
These cells make protein, cholesterol, and bile, and detoxify and modify substances from the blood. This requires lots of energy, so liver cells are packed with mitochondria (orange).

CELL METABOLISM

When individual cells break down nutrients to generate energy for building new proteins or nucleic acids, it is known as cell metabolism. Cells use a variety of fuels to generate energy, but the most common one is glucose, which is transformed into adenosine triphosphate (ATP). This takes place in structures called mitochondria through a process called cellular respiration: enzymes within the mitochondria react with oxygen and glucose to produce ATP, carbon dioxide, and water. Energy is released when ATP is converted into adenoside diphosphate (ADP) via the loss of a phosphate group.

Mitochondrion
While the number of mitochondria varies between different cells, all have the same basic structure: an outer membrane and a highly folded inner membrane, where the production of energy actually takes place.

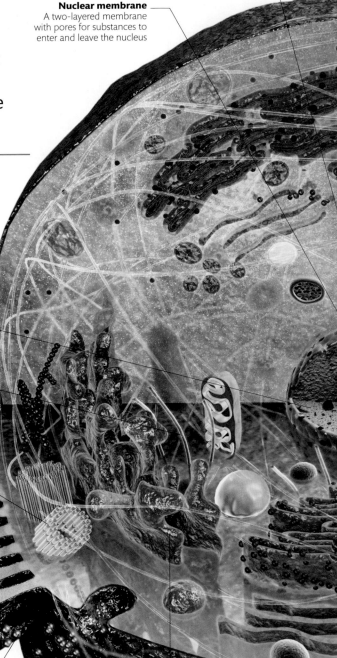

Generic cell
At a cell's heart is the nucleus, where the genetic material is stored and the first stages of protein synthesis occur. Cells also contain other structures for assembling proteins, including ribosomes, the endoplasmic reticulum, and Golgi apparatus. The mitochondria provide the cell with energy.

Nucleolus
The region at the center of the nucleus; plays a vital role in ribosome production

Nucleus
The cell's control center, containing chromatin and most of the cell's DNA

Nuclear membrane
A two-layered membrane with pores for substances to enter and leave the nucleus

Nucleoplasm
Fluid within the nucleus, in which nucleolus and chromosomes float

Microtubules
Part of cell's cytoskeleton, these aid movement of substances through the watery cytoplasm

Centriole
Composed of two cylinders of tubules; essential to cell reproduction

Microvilli
These projections increase the cell's surface area, aiding absorption of nutrients

Released secretions
Secretions are released from the cell by exytosis, when a vesicle merges with the cell membrane and releases its contents

Secretory vesicle
Sac containing various substances, such as enzymes, that are produced by the cell and secreted at the cell membrane

Golgi complex
A structure that processes and repackages proteins produced in the rough endoplasmic reticulum for release at the cell membrane

Lysosome
Produces powerful enzymes that aid in digestion and excretion of substances and worn-out organelles

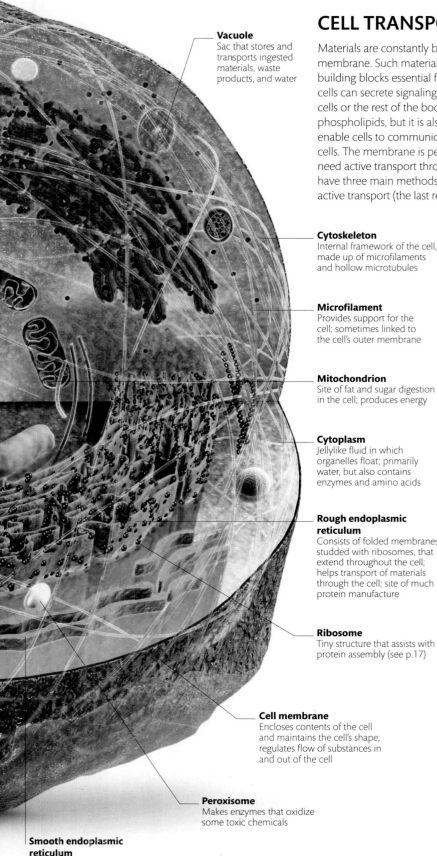

Vacuole
Sac that stores and transports ingested materials, waste products, and water

Cytoskeleton
Internal framework of the cell, made up of microfilaments and hollow microtubules

Microfilament
Provides support for the cell; sometimes linked to the cell's outer membrane

Mitochondrion
Site of fat and sugar digestion in the cell; produces energy

Cytoplasm
Jellylike fluid in which organelles float; primarily water, but also contains enzymes and amino acids

Rough endoplasmic reticulum
Consists of folded membranes, studded with ribosomes, that extend throughout the cell; helps transport of materials through the cell; site of much protein manufacture

Ribosome
Tiny structure that assists with protein assembly (see p.17)

Cell membrane
Encloses contents of the cell and maintains the cell's shape; regulates flow of substances in and out of the cell

Peroxisome
Makes enzymes that oxidize some toxic chemicals

Smooth endoplasmic reticulum
Network of tubes and flat, curved sacs that helps to transport materials through the cell; site of calcium storage; main location of fat metabolism

CELL TRANSPORT

Materials are constantly being transported in and out of the cell via the cell membrane. Such materials could include fuel for generating energy, or building blocks essential for protein assembly, such as amino acids. Some cells can secrete signaling molecules to communicate with neighboring cells or the rest of the body. The cell membrane is largely composed of phospholipids, but it is also studded with proteins that facilitate transport, enable cells to communicate with one another, and identify a cell to other cells. The membrane is permeable to some molecules, but other molecules need active transport through special channels in the membrane. Cells have three main methods of transport: diffusion, facilitated diffusion, and active transport (the last requires energy).

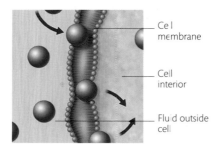

Cell membrane
Cell interior
Fluid outside cell

Diffusion
Molecules passively cross the membrane from areas of high to low concentration. Water and oxygen both cross by diffusion.

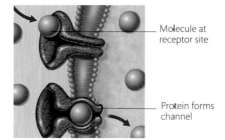

Carrier protein
Cell interior

Molecule at receptor site
Protein forms channel

Facilitated diffusion
A carrier protein, or protein pore, binds with a molecule outside the cell, then changes shape and ejects the molecule into the cell.

Active transport
Molecules bind to a receptor site on the cell membrane, triggering a protein, which changes into a channel that molecules travel through.

MAKING NEW BODY CELLS

Some cells are constantly replacing themselves; others last a lifetime. While the cells lining the mouth are replaced every couple of days, some of the nerve cells in the brain have been there since before birth. Stem cells are specialized cells that are constantly dividing and giving rise to new cells, such as blood cells, immune cells, or fat cells. Cell division requires that a cell's DNA is accurately copied and then shared equally between two "daughter" cells, by a process called mitosis. The chromosomes are first replicated before being pulled to opposite ends of the cell. The cell then divides to produce two daughter cells, with the cytoplasm and organelles being shared between the two cells.

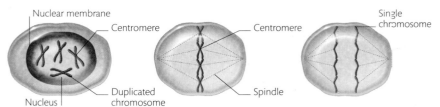

Nuclear membrane
Centromere
Centromere
Single chromosome
Nucleus
Duplicated chromosome
Spindle

1 Preparation
The cell produces proteins and new organelles, and duplicates its DNA. The DNA condenses into X-shaped chromosomes.

2 Alignment
The chromosomes line up along a network of filaments called the spindle. This is linked to a larger network called the cytoskeleton.

3 Separation
The chromosomes are pulled apart and move to opposite ends of the cell. Each end has an identical set of chromosomes.

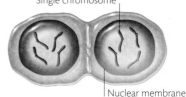

Single chromosome
Nuclear membrane

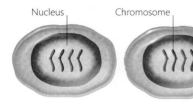

Nucleus
Chromosome

4 Splitting
The cell now splits in two, with the cytoplasm, cell membrane, and remaining organelles being shared roughly equally between the two daughter cells.

5 Offspring
Each daughter cell contains a complete copy of the DNA from the parent cell; this enables it to continue growing, and eventually divide itself.

CELLS AND TISSUES

Cells are the building blocks from which the human body is made. Some cells work alone—such as red blood cells, which carry oxygen around the body, or sperm, which fertilize egg cells—but many are organized into tissues, where cells with different functions join forces to accomplish one or more specific tasks.

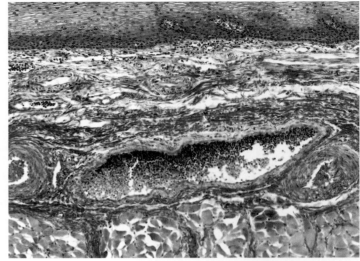

Integrated tissues
This section through the wall of the esophagus shows a combination of different tissues: lining epithelium (pink, top); collagen connective tissue (blue); blood vessels (circular); skeletal muscle fibers (purple, bottom).

CELL TYPES

There are more than 200 different types of cell in the body, each type specially adapted to its own particular function. Every cell contains the same genetic information, but not all of the genes are "switched on" in every cell. It is this pattern of gene expression that dictates what the cell looks like, how it behaves, and what role it performs in the body. A cell's fate is largely determined before birth, influenced by its position in the body and the cocktail of chemical messengers that it is exposed to in that environment. Early during development, stem cells begin to differentiate into three layers of more specialized cells called the ectoderm, endoderm, and mesoderm. Cells of the ectoderm will form the skin and nails, the epithelial lining of the nose, mouth, and anus, the eyes, and the brain and spinal cord. Cells of the endoderm become the inner linings of the digestive tract, the respiratory linings, and glandular organs including the liver and pancreas. Mesoderm cells develop into the muscles, circulatory system, and the excretory system, including the kidneys.

SCIENCE
STEM CELLS

A few days after fertilization, an embryo consists of a ball of "embryonic stem cells" (ESCs). These cells have the potential to become any type of cell in the body. Scientists are trying to harness this property to grow replacement body parts. As the embryo grows, the stem cells become increasingly restricted in their potential. By the time we are born most of our cells are fully differentiated, but a small number of adult stem cells remain in parts of the body, including in bone marrow. While not as universal in their potential as ESCs, they do have some flexibility in terms of what they can become. Scientists believe that these cells could also be used to help cure disease.

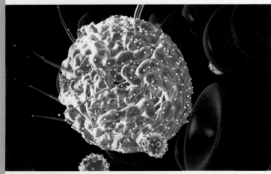

Adult stem cells
Adult stem cells, such as the large white cell in this image, are present in bone marrow, where they multiply and produce millions of blood cells, including red blood cells, also seen here.

200

The number of different **types of cell** in the human body. Most are **organized in groups** to form tissues.

Red blood cells
Unlike all other human cells, red blood cells lack a nucleus and most organelles. Instead, they are packed with an oxygen-carrying protein called hemoglobin, which gives blood its red color. Red blood cells develop in the bone marrow and circulate for around 120 days, before being broken down and recycled.

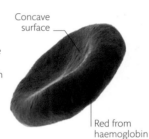

Concave surface

Red from haemoglobin

Adipose (fat) cells
These cells are highly adapted for the storage of fat, and the bulk of their interior is taken up by a large droplet of semiliquid fat. When we gain weight, our adipose cells swell up and fill with even more fat, though eventually they also start to increase in number.

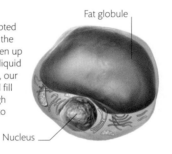

Fat globule

Nucleus

Sperm cells
Sperm are male reproductive cells with a tail that enables them to swim up the female reproductive tract and fertilize an egg. Sperm contain just 23 chromosomes; in fertilization, these pair up with an egg's 23 chromosomes to create an embryo with the normal 46 chromosomes per cell.

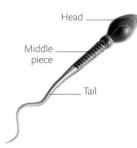

Head

Middle piece

Tail

Photoreceptor cells
These occur at the back of the eye. They contain a light-sensitive pigment and generate electrical signals when struck by light, enabling us to see. There are two main photoreceptor types: rods (below) see in black and white, and work well in low light; cones work better in bright light, and are able to detect colors.

Nucleus

Pigment-containing part

Epithelial cells
These cells are barrier cells lining the cavities and surfaces of the body. They include skin cells and the cells lining the lungs and reproductive tracts. Some epithelial cells have fingerlike projections called "cilia" that can waft eggs down the fallopian tubes, or push mucus out of the lungs, for example.

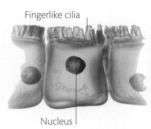

Fingerlike cilia

Nucleus

Nerve cells
These electrically excitable cells transmit electrical signals, or "action potentials," down an extended stem called an axon. Found throughout the body, they enable you to move and feel sensations such as pain. They communicate with each other across connections called synapses.

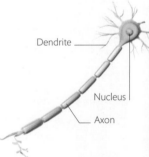

Dendrite

Nucleus

Axon

Ovum (egg) cells
One of the largest cells in the body, a human egg is still only just visible to the naked eye. Eggs are the female reproductive cells and, like sperm, they contain just 23 chromosomes. Every woman is born with a finite number of eggs, which decreases as she ages.

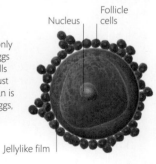

Follicle cells

Nucleus

Jellylike film

Smooth muscle cells
One of three types of muscle cell, smooth muscle cells are spindle-shaped cells found in the arteries and the digestive tract that produce long, wavelike contractions. To do this, they are packed with contractile filaments, and large numbers of mitochondria that supply the energy they need.

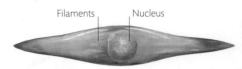

Filaments

Nucleus

TISSUE TYPES

Cells often group together with their own kind to form tissues that perform a specific function. However, not all cells within a tissue are necessarily identical. The four main types of tissue in the human body are muscle, connective tissue, nervous tissue, and epithelial tissue. Within these groups, different forms of these tissues can have very different appearances and functions. For example, blood, bone, and cartilage are all types of connective tissue, but so are fat layers, tendons, ligaments, and the fibrous tissue that holds organs and epithelial layers in place. Organs such as the heart and lungs are composed of several different kinds of tissue.

Skeletal muscle
This tissue performs voluntary movements of the limbs. Unlike smooth muscle, skeletal muscle cells are arranged into bundles of fibers, which connect to bones via tendons. They are packed with highly organized filaments that slide over one another to produce contractions.

MUSCLE FIBERS

Smooth muscle
Able to contract in long, wavelike motions without conscious thought, smooth muscle is found in sheets on the walls of the blood vessels, stomach, intestines, and bladder. It is vital for maintaining blood pressure and for pushing food through the digestive system.

SMALL INTESTINE

Spongy bone
Bone cells secrete a hard material that makes bones strong and brittle. Spongy bone is found in the center of bones, and is softer and weaker than the compact bone. The latticelike spaces in spongy bone are filled with bone marrow or connective tissue.

END OF THE FEMUR

Cartilage
This stiff, rubbery connective tissue is composed of cells called chondrocytes embedded in a matrix of gel-like material, which the cells secrete. Cartilage is found in the joints between bones, and in the ear and nose. The high water content of cartilage makes it tough but flexible.

NOSE CARTILAGE

Loose connective tissue
This type of tissue also contains cells called fibroblasts, but the fibers they secrete are loosely organized and run in random directions, making the tissue quite pliable. Loose connective tissue holds organs in place, and provides cushioning and support.

DERMAL TISSUE

Dense connective tissue
This contains fibroblast cells, which secrete the fibrous protein called type 1 collagen. The fibers are organized into a regular parallel pattern, making the tissue very strong. Dense connective tissue occurs in the base layer of skin, and forms structures such as ligaments and tendons.

KNEE LIGAMENTS

Adipose tissue
A type of connective tissue, adipose tissue is composed of fat cells called adipocytes, as well as some fibroblast cells, immune cells, and blood vessels. Its main function is to act as an energy store, and to cushion, protect, and insulate the body.

SUBCUTANEOUS FAT

Epithelial tissue
This tissue forms a covering or lining for internal and external body surfaces. Some epithelial tissues can secrete substances such as digestive enzymes; others can absorb substances like food or water.

STOMACH WALL

Nerve tissue
This forms the brain, spinal cord, and the nerves that control movement, transmit sensation, and regulate many body functions. It is mainly made up of networks of nerve cells (see opposite).

UPPER SPINAL CORD

BODY COMPOSITION

If the 75 trillion cells that make up the human body led an isolated, anarchic existence, it would be no more than a shapeless mass. Instead, those cells are precisely organized, taking their place within the hierarchical structure that is a fully functioning human being.

LEVELS OF ORGANIZATION

The overall organization of the human body can be visualized in the form of a hierarchy of levels, as shown below. At its lowest level are the body's basic chemical constituents. As the hierarchy ascends, the number of components in each of its levels—cells, tissues, organs, and systems—decreases progressively, culminating in a single organism at its apex.

More than 20 chemical elements are found in the body, with just four—oxygen, carbon, hydrogen, and nitrogen—comprising around 96 percent of body mass. Each element is composed of atoms, the tiny building blocks of matter, of which there are quadrillions in the body. Atoms of different elements generally combine with others to form molecules such as water (hydrogen and oxygen atoms), and the many organic molecules, including proteins and DNA. These organic molecules are constructed around a "skeleton" of linked carbon atoms.

Cells are the smallest of all living units. They are created from chemical molecules, which shape their outer covering and inner structures, and drive the metabolic reactions that keep them alive. There are more than 200 types of cell in the human body, each adapted to carry out a specific role, but not in isolation (see p.22). Groups of similar cells with the same function form and cooperate within communities called tissues. The body's four basic tissue types are epithelial, which covers surfaces and lines cavities; connective, which supports and protects body structures; muscular, which creates movement; and nervous, which facilitates rapid internal communication (see p.23).

Organs, such as the liver, brain, and heart are discrete structures built from at least two types of tissue. Each has a specialized role or roles that no other organ can perform. Where organs collectively have a common purpose, they are linked together within a system, such as the cardiovascular system, which transports oxygen and nutrients around the body, and which is overviewed here. Integrated and interdependent, the body's systems combine to produce a complete human (see pp.26–27).

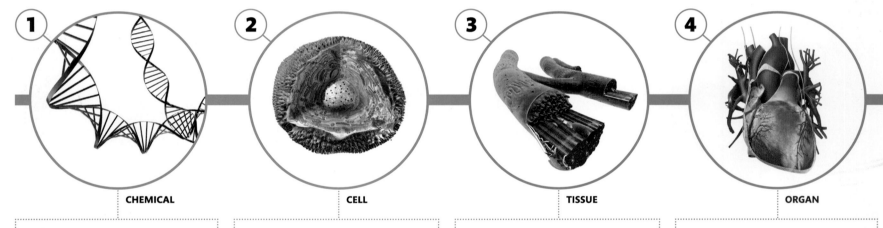

1 CHEMICAL **2** CELL **3** TISSUE **4** ORGAN

CHEMICALS

Key among the chemicals inside all cells is DNA (see pp.16–17). Its long molecules resemble twisted ladders, their "rungs" made from bases that provide the instructions for making proteins. These, in turn, perform many roles, from building cells to controlling chemical reactions.

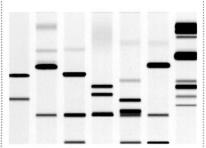

DNA sequencing
The bases of DNA can be isolated and separated by scientists. Such sequencing allows them to "read" the instructions coded within the molecules.

CELLS

While cells may differ in size and shape according to their function (see p.22), all possess the same basic features: an outer boundary membrane; organelles, floating within a jellylike cytoplasm; and a nucleus, which contains DNA (see pp.20–21). Cells are the body's most basic living components.

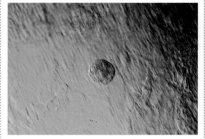

Stem cells
These unspecialized cells have the unique ability to differentiate, or develop, into a wide range of specialized tissue cells such as muscle, brain, or blood cells.

HEART TISSUE

One of three types of muscle tissue, cardiac muscle is found only in the walls of the heart. Its constituent cells contract together to make the heart squeeze and pump, and, working as a network, conduct the signals that ensure that the pumping is precisely coordinated.

Muscle fibers
The cells, or fibers, in cardiac tissue are long and cylindrical and have branches that form junctions with other cells to create an interconnected network.

HEART

Like other organs, the heart is made of several types of tissue, including cardiac muscle tissue. Among the other types present are connective tissues, which protect the heart and hold the other tissues together, and epithelial tissues, which line its chambers and cover its valves.

Complex structure
The heart has a complex structure. Internally it has four chambers through which blood is pumped by its muscular walls. It is connected to a vast network of veins and arteries.

75 trillion

The total **number of cells** that make up the average **human body**.

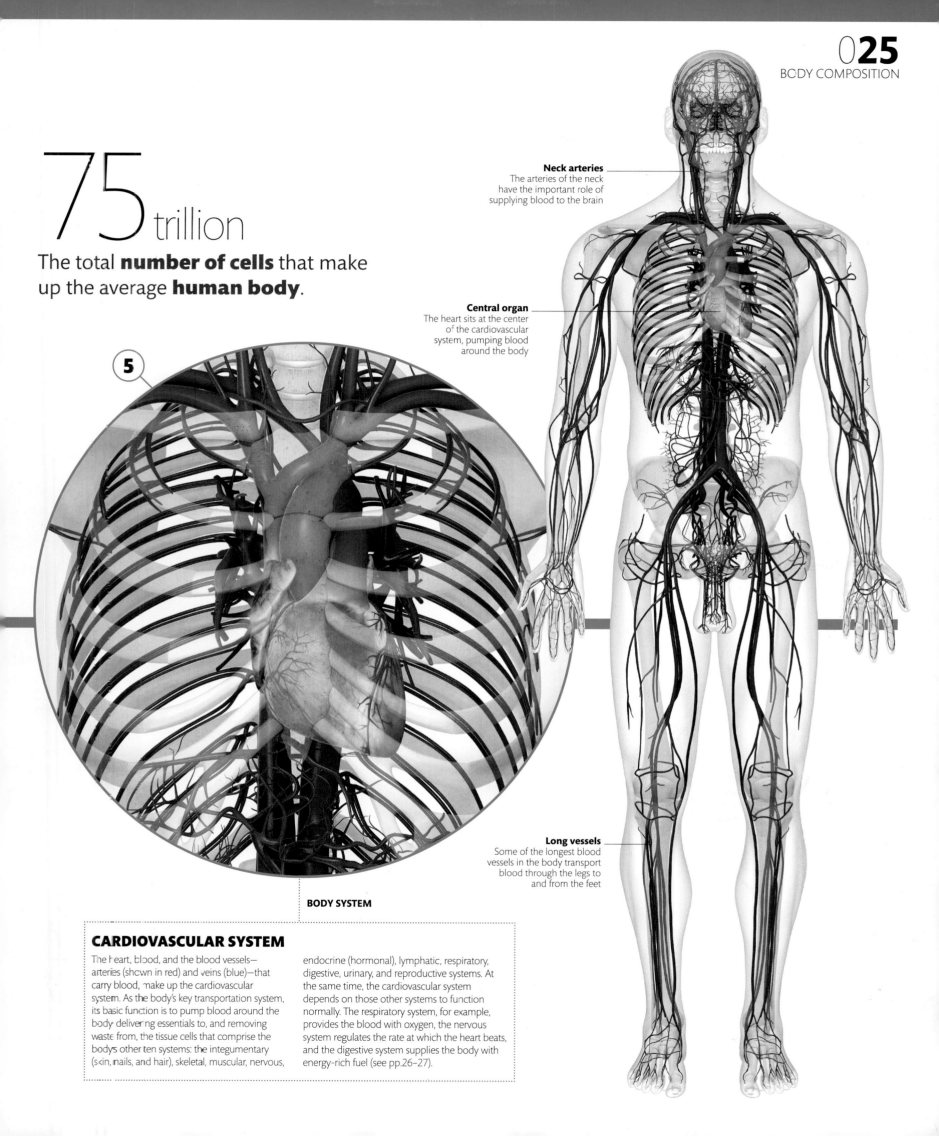

Neck arteries
The arteries of the neck have the important role of supplying blood to the brain

Central organ
The heart sits at the center of the cardiovascular system, pumping blood around the body

Long vessels
Some of the longest blood vessels in the body transport blood through the legs to and from the feet

BODY SYSTEM

CARDIOVASCULAR SYSTEM

The heart, blood, and the blood vessels—arteries (shown in red) and veins (blue)—that carry blood, make up the cardiovascular system. As the body's key transportation system, its basic function is to pump blood around the body, delivering essentials to, and removing waste from, the tissue cells that comprise the body's other ten systems: the integumentary (skin, nails, and hair), skeletal, muscular, nervous, endocrine (hormonal), lymphatic, respiratory, digestive, urinary, and reproductive systems. At the same time, the cardiovascular system depends on those other systems to function normally. The respiratory system, for example, provides the blood with oxygen, the nervous system regulates the rate at which the heart beats, and the digestive system supplies the body with energy-rich fuel (see pp.26–27).

BODY SYSTEMS

The human body can do many different things. It can digest food, think, move, even reproduce and create new life. Each of these tasks is performed by a different body system—a group of organs and tissues working together to complete that task. However, good health and body efficiency rely on the different body systems working together in harmony.

SYSTEM INTERACTION

Think about what your body is doing right now. You are breathing, your heart is beating, and your blood pressure is under control. You are also conscious and alert. If you were to start running, specialized cells called chemoreceptors would detect a change in your body's metabolic requirements and signal to the brain to release adrenaline. This would in turn signal to the heart to beat faster, boosting blood circulation and enabling more oxygen to reach the muscles. After a while, cells in the hypothalamus might detect an increase in body temperature and send a signal to the skin to produce sweat, which would evaporate and cool you down.

The individual systems of the body are linked together by a vast network of positive and negative feedback loops. These use signaling molecules such as hormones and electrical impulses from nerves to communicate and maintain a state of equilibrium. Here, the basic components and functions of each system are described, and examples of system interactions are examined.

ENDOCRINE SYSTEM

Like the nervous system, the endocrine system communicates messages between the rest of the body's systems, enabling them to be closely monitored and controlled. It uses chemical messengers called hormones, which are usually secreted into the blood from specialized glands.

LYMPHATIC SYSTEM

The lymphatic system is composed of a network of vessels and nodes, which drain fluid from blood capillaries and return it to the veins. Its main functions are to maintain fluid balance within the cardiovascular system and to distribute immune cells from the immune system around the body. Movement of lymph fluid relies on the contraction and relaxation of smooth muscles within the muscular system.

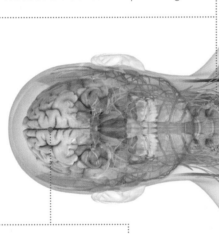

CONTROLLING THE HEART

Working together, nerves of the sympathetic and parasympathetic nervous systems regulate the heart and cardiac output (see p.353). Sympathetic nerves release chemicals that increase heart rate and the force of cardiac muscle contractions. The vagus nerve, from the parasympathetic system, releases a chemical that slows the heart rate and reduces cardiac output.

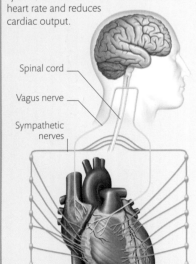

Spinal cord

Vagus nerve

Sympathetic nerves

NERVOUS SYSTEM

The brain, spinal cord, and nerves work together to collect, process, and disseminate information from the body's internal and external environments. The nervous system communicates through networks of nerve cells, which connect with every other body system. The brain controls and monitors all of these systems to make sure that they are performing normally and receiving everything they need.

RESPIRATORY SYSTEM

Every cell in the body needs oxygen and must get rid of the waste product carbon dioxide in order to function—regardless of which body system it belongs to. The respiratory system allows this to happen by breathing air into the lungs, where the passive exchange of these molecules occurs between the air and blood. The cardiovascular system transports oxygen and carbon dioxide between the cells and the lungs.

BREATHING IN AND OUT

The mechanics of breathing rely upon an interaction between the respiratory and muscular systems. Together with three accessory muscles, the intercostal muscles and the diaphragm contract to increase the volume of the chest cavity (see pp.342–43). This forces air down into the lungs. A different set of muscles is used during forced exhalation. These rapidly shrink the chest cavity, forcing air out of the lungs.

Accessory and intercostal muscles

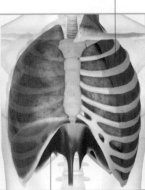

Diaphragm

DIGESTIVE SYSTEM

In addition to oxygen every cell needs energy in order to function. The digestive system processes and breaks down the food we eat so that a variety of nutrients can be absorbed from the intestines into the circulatory system. These are then delivered to the cells of every body system in order to provide them with energy.

MUSCULAR SYSTEM

The muscular system is made up of three types of muscle: skeletal, smooth, and cardiac. It is responsible for generating movement—both of the limbs and within the other body systems. For example, smooth muscle aids the digestive system by helping to propel food down the esophagus and through the stomach, intestines, and rectum. And the respiratory system could not function without the muscles of the thorax contracting to fill the lungs with air (see opposite).

SKELETAL SYSTEM

This system uses bones, cartilage, ligaments, and tendons to provide the body with structural support and protection. It encases much of the nervous system within a protective skull and vertebrae, and the vital organs of the respiratory and circulatory systems within the rib cage. The skeletal system also supports the circulatory and immune systems by manufacturing red and white blood cells.

CIRCULATING BLOOD

The veins of the cardiovascular system rely on the direct action of skeletal muscles to transport deoxygenated blood from the body's extremities back to the heart (see p.355). As shown here, in the muscles and veins of the lower leg, muscle contractions compress nearby veins, forcing the blood upward. When the muscles relax, the one-way valves within the veins prevent the blood from flowing back down, and the vein fills up with blood from below. The same process is used by the lymphatic system as muscle contractions aid the transportation of lymph through lymph vessels (see p.358).

Blood forced upward

Contracting muscle

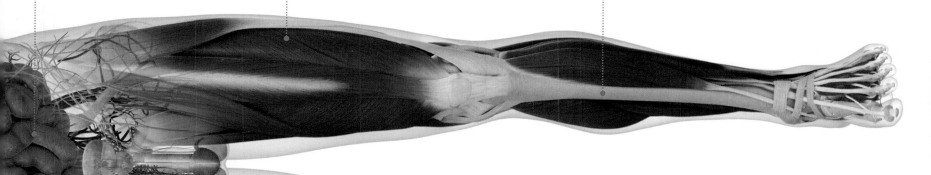

REPRODUCTIVE SYSTEM

Although the reproductive system is not essential for maintaining life, it is needed to propagate it. Both the testes of the male and the ovaries of the female produce gametes in the form of sperm and eggs, which fuse to create an embryo. The testes and ovaries also produce hormones including estrogen and testosterone, so also form part of the endocrine system.

MAKING URINE

The kidney is the site of a key interaction between the urinary and cardiovascular systems (see p.381). Urine is produced as nephrons, the kidney's functional units, filter the blood. Within each nephron, blood is forced through a glomerulus (cluster of capillaries) and filtered by its sievelike membranes. The filtrate passes through a series of tubules through which some glucose, salts, and water are reabsorbed into the blood stream. What remains, including urea and waste products, is excreted as urine.

CARDIOVASCULAR SYSTEM

The cardiovascular system uses blood to carry oxygen from the respiratory system and nutrients from the digestive system to cells of all the body's systems. It also removes products from these cells. At the center of the cardiovascular system lies the muscular heart, which pumps the blood through the blood vessels.

URINARY SYSTEM

The urinary system filters and removes many of the waste products generated by the other body systems, such as the digestive system. It does this by filtering blood through the kidneys and producing urine, which is collected in the bladder and then excreted through the urethra (see right). The kidneys also help maintain blood pressure within the cardiovascular system by ensuring that the correct amount of water is reabsorbed by the blood.

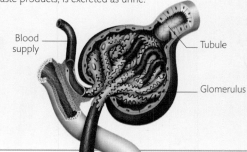

Blood supply

Tubule

Glomerulus

anatomy

The human body is a "living machine" with many complex working parts. To understand how the body functions it is vital to know how it is assembled. Advances in technology allow us to strip back the outer layers and reveal the wonders inside.

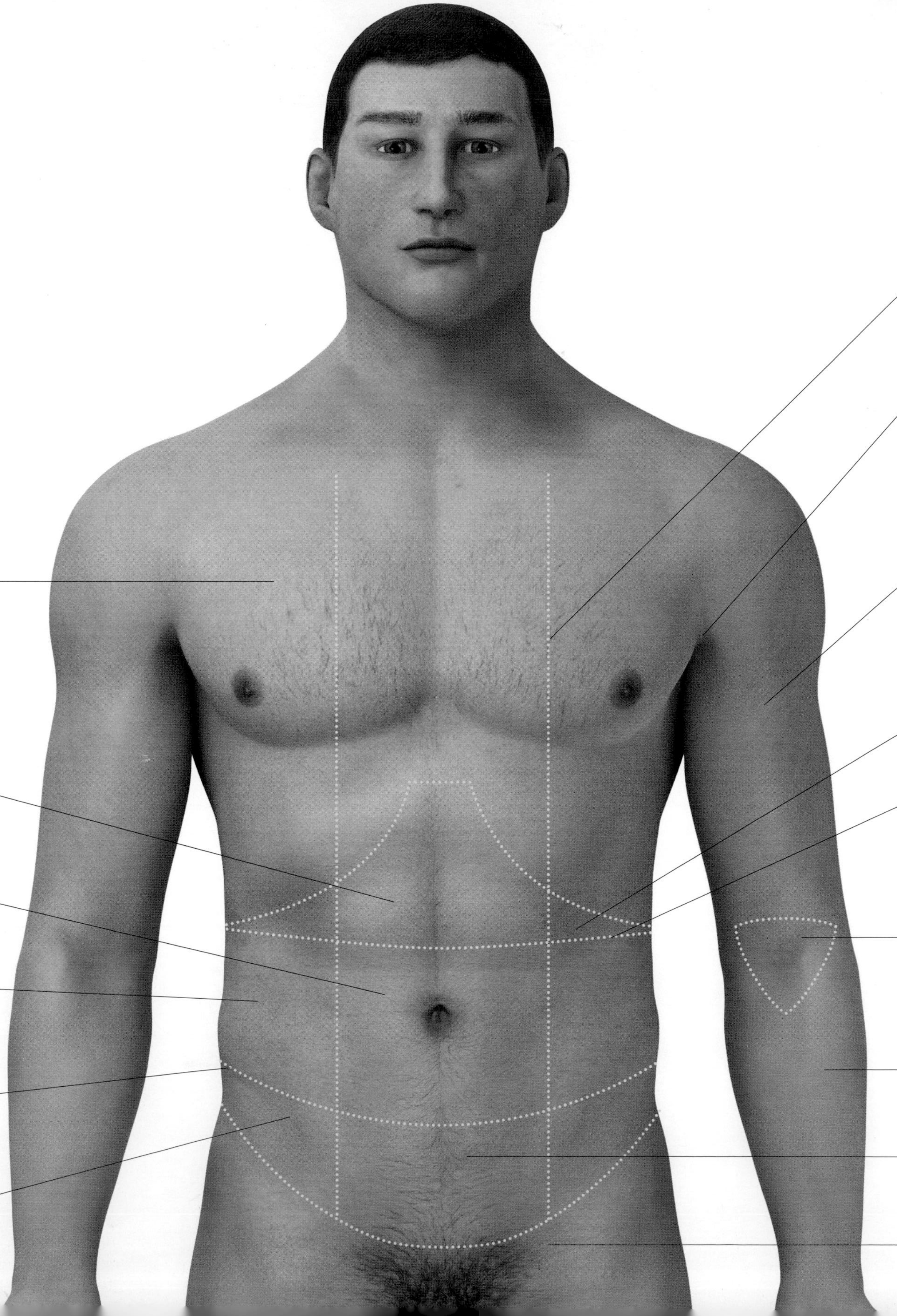

Midclavicular line
A vertical line running down from the midpoint of each clavicle

Axilla
Loosely, the armpit; more precisely, the pyramid-shaped part of the body between the upper arm and the side of the thorax. Floored by the skin of the armpit, it reaches up to the level of the clavicle, top of the scapula, and first rib

Anterior surface of arm
'Anterior' means front, and always refers to the body when it is in the "anatomical position" shown here. Strictly speaking, "arm" only relates to the part of the upper limb between the shoulder and the elbow

Hypochondrial region
The abdominal region under the ribs on each side

Transpyloric plane
Horizontal plane joining the tips of the ninth costal cartilages, at the margins of the ribcage; also level with the first lumbar vertebra and the pylorus of the stomach

Cubital fossa
Triangular area anterior to (in front of) the elbow, bounded above by a line between the bony epicondyles of the humerus on each side, and framed below by the pronator teres and brachioradialis muscles

Anterior surface of forearm
Anatomically—and colloquially—the forearm is the part of the body between the elbow and the wrist

Suprapubic region
The part of the abdomen that lies just above the pubic bones of the pelvis

Inguinal region
Refers to the groin area, where the thigh meets the trunk

Pectoral region
The chest; sometimes used to refer to just the upper chest, where the pectoral muscles lie

Epigastric region
Area of the abdominal wall above the transpyloric plane, and framed by the diverging margins of the ribcage

Umbilical region
Central region of the abdomen, around the umbilicus (navel)

Lumbar region
Refers to the sides of the abdominal wall, between the transpyloric and intertubercular planes

Intertubercular plane
This plane passes through the iliac tubercles—bony landmarks on the pelvis—and lies at the level of the fifth lumbar vertebra

Iliac region
The area below the intertubercular plane and lateral to (to the side of) the midclavicular line; may also be referred to as the "iliac fossa"

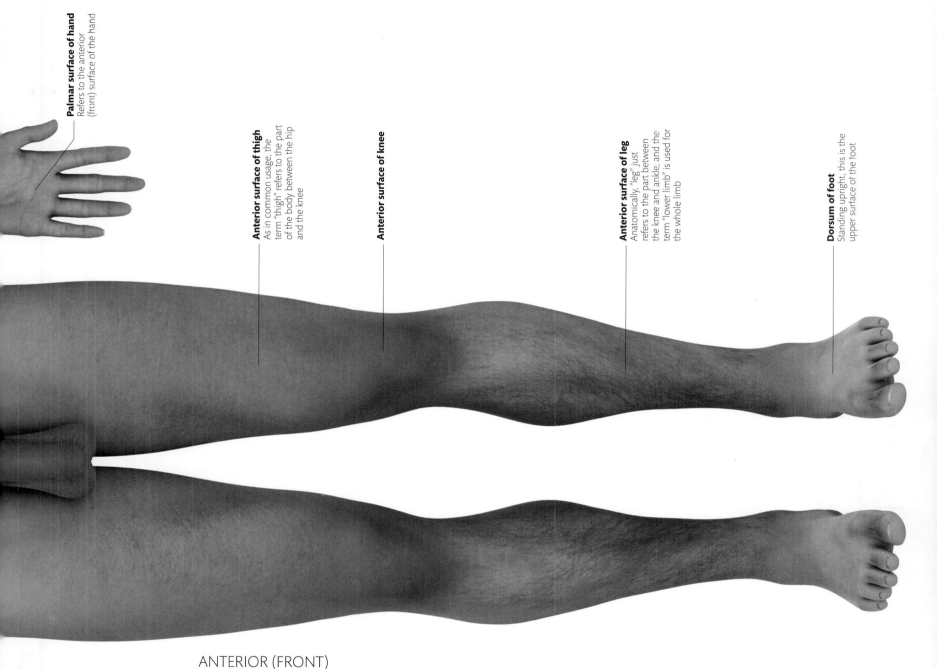

Palmar surface of hand
Refers to the anterior (front) surface of the hand

Anterior surface of thigh
As in common usage, the term "thigh" refers to the part of the body between the hip and the knee

Anterior surface of knee

Anterior surface of leg
Anatomically, "leg" just refers to the part between the knee and ankle, and the term "lower limb" is used for the whole limb

Dorsum of foot
Standing upright, this is the upper surface of the foot

ANTERIOR (FRONT)

ANATOMICAL TERMINOLOGY

Anatomical language allows us to describe the structure of the body clearly and precisely. It is useful to be able to describe areas and parts, as well as the planes and lines used to map out the body, in much more accurate and detailed terms than would be possible colloquially. Rather than recording that a patient had a tender area "somewhere on the left side of the belly," a doctor can be more precise and say that the patient's painful area was "the left lumbar region," and other doctors will know exactly what is meant.

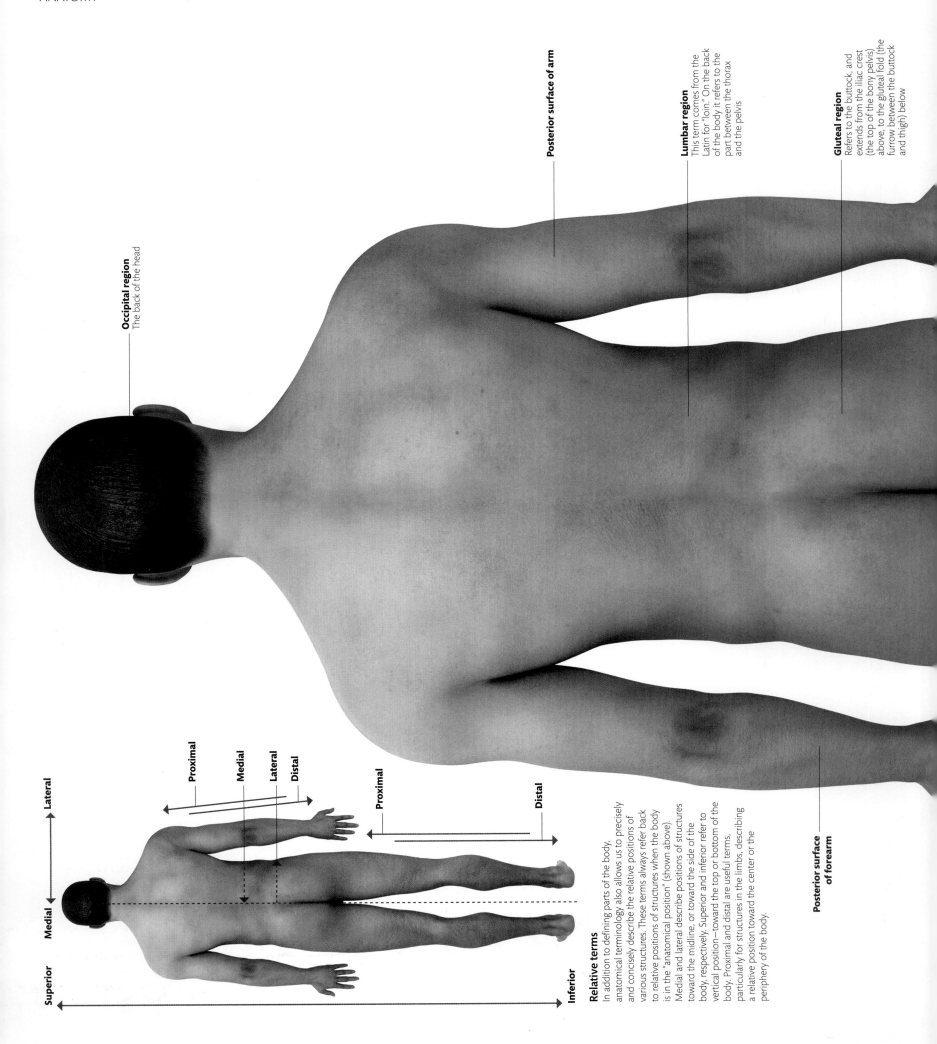

Occipital region
The back of the head

Posterior surface of arm

Lumbar region
This term comes from the Latin for "loin." On the back of the body it refers to the part between the thorax and the pelvis

Gluteal region
Refers to the buttock, and extends from the iliac crest (the top of the bony pelvis) above, to the gluteal fold (the furrow between the buttock and thigh) below

Posterior surface of forearm

Lateral

Medial

Proximal

Medial

Lateral

Distal

Proximal

Distal

Superior

Medial

Inferior

Relative terms

In addition to defining parts of the body, anatomical terminology also allows us to precisely and concisely describe the relative positions of various structures. These terms always refer back to relative positions of structures when the body is in the "anatomical position" (shown above). Medial and lateral describe positions of structures toward the midline, or toward the side of the body, respectively. Superior and inferior refer to vertical position—toward the top or bottom of the body. Proximal and distal are useful terms, particularly for structures in the limbs, describing a relative position toward the center or the periphery of the body.

0**33**
ANATOMICAL TERMINOLOGY

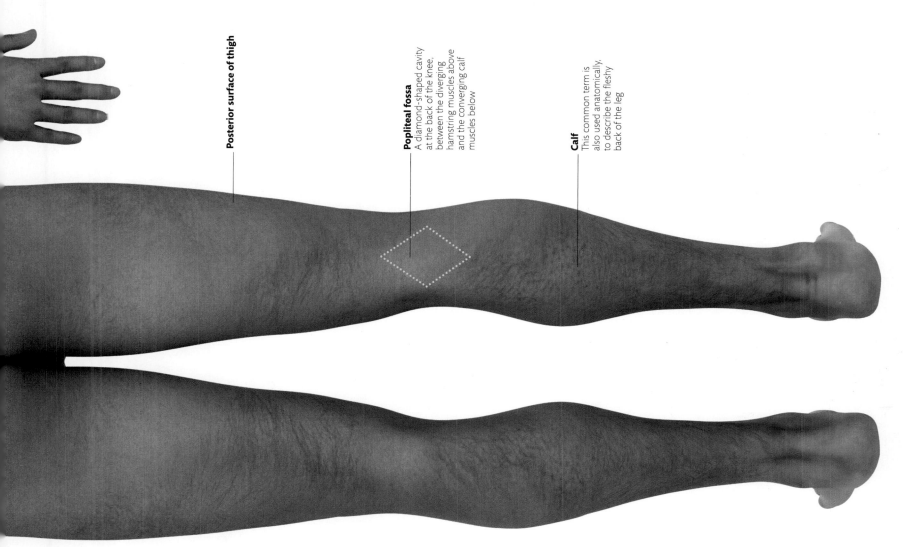

Posterior surface of thigh

Popliteal fossa
A diamond-shaped cavity at the back of the knee, between the diverging hamstring muscles above and the converging calf muscles below

Calf
This common term is also used anatomically, to describe the fleshy back of the leg

POSTERIOR (BACK)

Dorsum of hand
The back of the hand

ANATOMICAL TERMINOLOGY

The illustration shows some of the terms used for the broader regions of the back of the body, and those used to describe relative position. Where our everyday language may have names for larger structures—such as the shoulder or hip—it soon runs out when it comes to finer detail. So anatomists have created names for specific structures, usually derived from Latin or Greek. The pages that follow show the detailed structure of the head and neck, thorax, abdomen, and limbs. The anatomical language is there to illuminate rather than confuse. Some of the terms may seem unfamiliar and even unnecessary at first, but they enable precise description and clear communication.

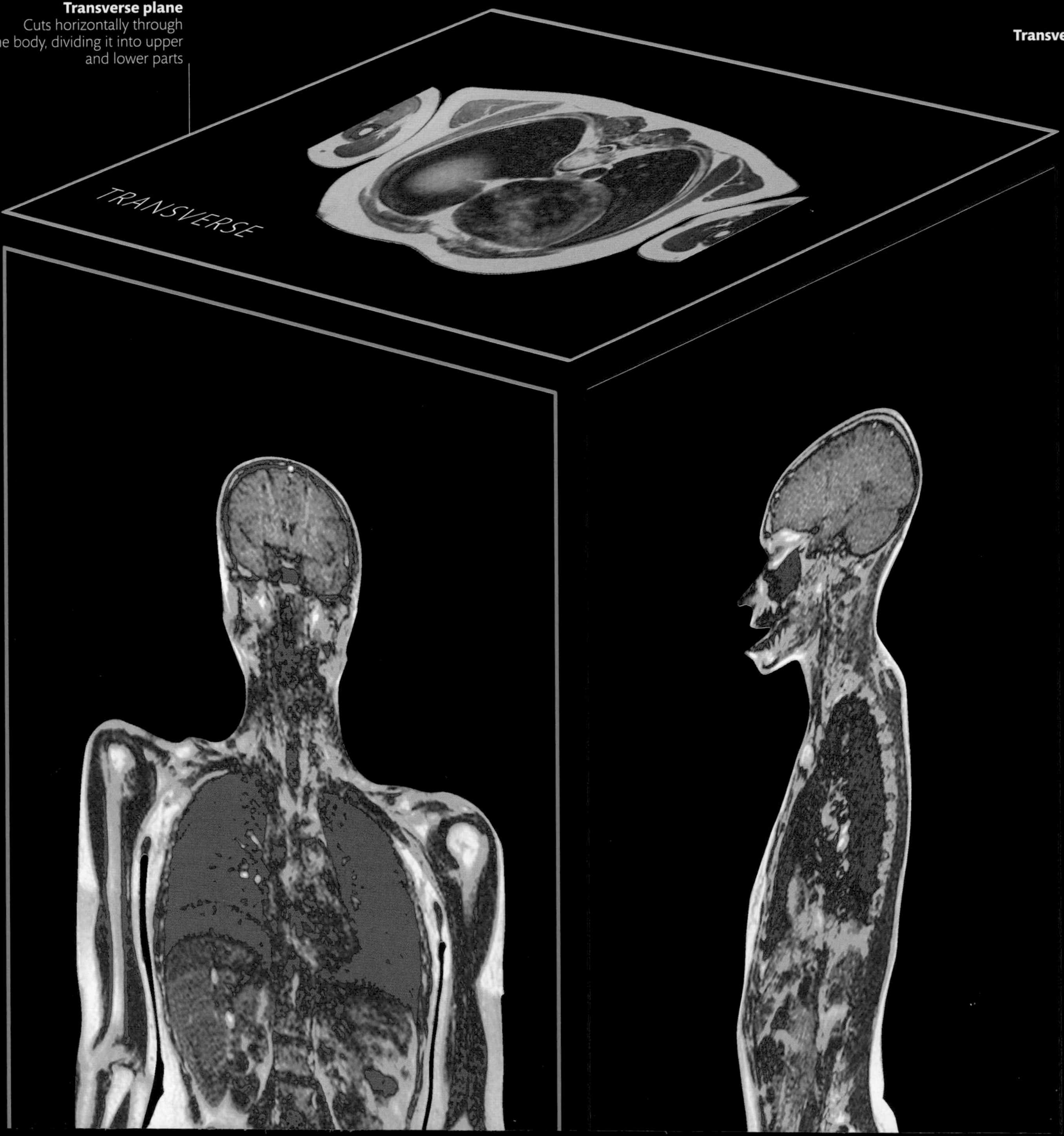

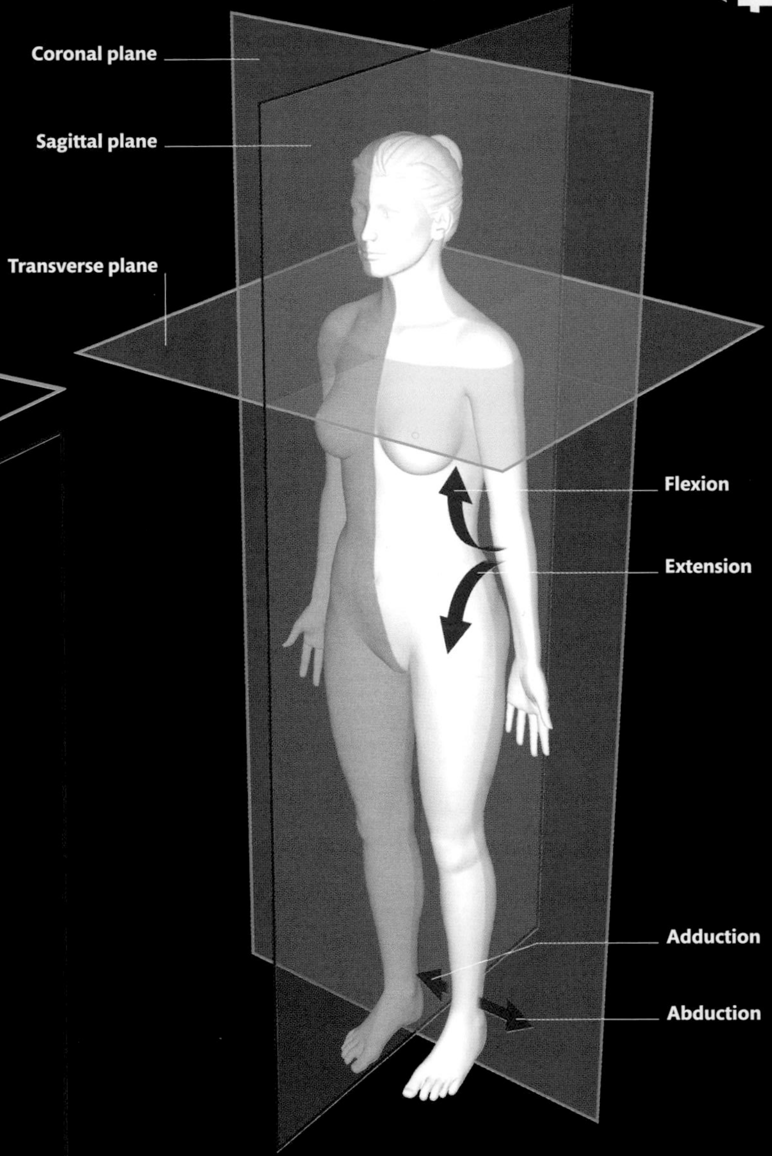

Flexion

Extension

Adduction

Abduction

Coronal plane

Sagittal plane

Transverse plane

Anatomical terms for movement

The diagram above shows the three planes—sagittal, coronal, and transverse—cutting through a body, and to the left are examples of real MRI scans demonstrating views along those planes. The above image also illustrates some medical terms that are used to describe certain movements of body parts: flexion decreases the angle of a joint, such as the elbow, while extension increases it; adduction draws a limb closer to the sagittal plane, while abduction moves it farther away from that plane.

Transverse plane
Cuts horizontally through the body, dividing it into upper and lower parts

TRANSVERSE

PLANES AND MOVEMENT

Sometimes it s easier to appreciate and understand anatomy by dividing the three-d mensional body up into two-d mensional slices. Computed tomography (CT) and magnetic resonance imaging (MRI) scans are examples of medical imaging techniques that show the body in slices or sections. The orientation of these slices or sections are described as sagittal, coronal, or transverse—as shown in these images. Precise anatomical terms are also used to define the absolute and relative positions of structures within the body (see pp.30–33), and to describe movements of joints, such as abduction, adduction, flexion, and extension (see left). Some joints, such as the shoulder and hip joint, also allow rotation of a limb along its axis. A special type of rotation between the forearm bones allows the palm to be moved from a forward or upward-facing position (supination) to a backward or downward-facing position (pronation).

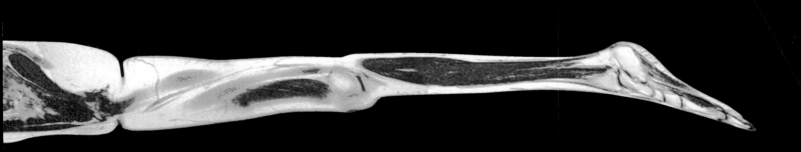

SAGITTAL

Sagittal plane
Cuts vertically down the body, through or parallel to the sternum

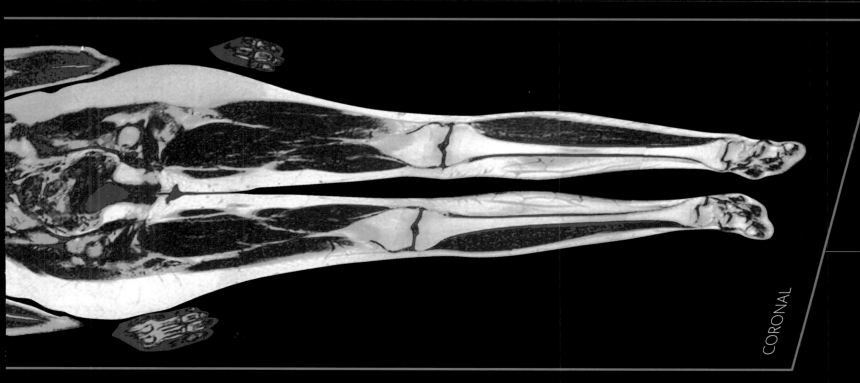

CORONAL

Coronal plane
Cuts vertically down the body, through or parallel to the shoulders

SKIN, HAIR, AND NAILS

· Skin, hair, and nail structure pp.38–39

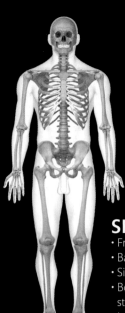

SKELETAL

· Front pp.40–41
· Back pp.42–43
· Side pp.44–45
· Bone and cartilage structure pp.46–47
· Joint and ligament structure pp.48–49

MUSCULAR

· Front (superficial on right side of body; deep on left side) pp.50–51
· Back (superficial on right side of body; deep on left side) pp.52–53
· Side pp.54–55
· Muscle attachments pp.56–57
· Muscle structure pp.58–59

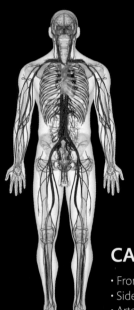

CARDIOVASCULAR

· Front pp.68–69
· Side pp.70–71
· Artery, vein, capillary structure pp.72–73

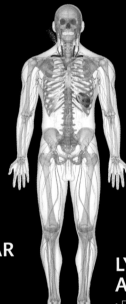

LYMPHATIC AND IMMUNE

· Front pp.74–75
· Side pp.76–77

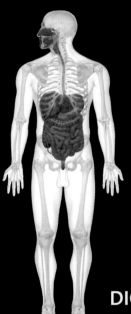

DIGESTIVE

· Front pp.78–79

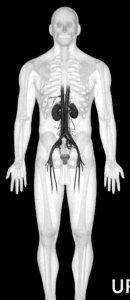

URINARY

· Front (male main; female inset) pp.80–81

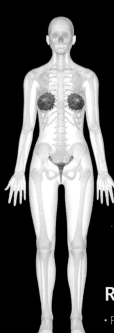

REPRODUCTIVE

· Front (female main; male inset) pp.82–83

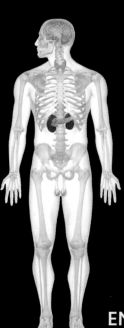

ENDOCRINE

· Front pp.84–85

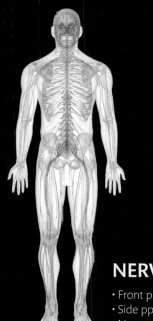

NERVOUS

- Front pp.60–61
- Side pp.62–63
- Nerve structure pp.64–65

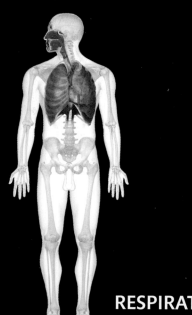

RESPIRATORY

- Front pp.66–67

The body has 11 main body systems. None of these works in isolation, for example the endocrine and nervous systems work together closely, as do the respiratory and cardiovascular systems. However, in order to understand how the body is put together, it helps to break it down system by system. In this part of the **Anatomy** chapter, an overview of the basic anatomy of each of the 11 systems is given before being broken down into more detail in the **Anatomy Atlas**.

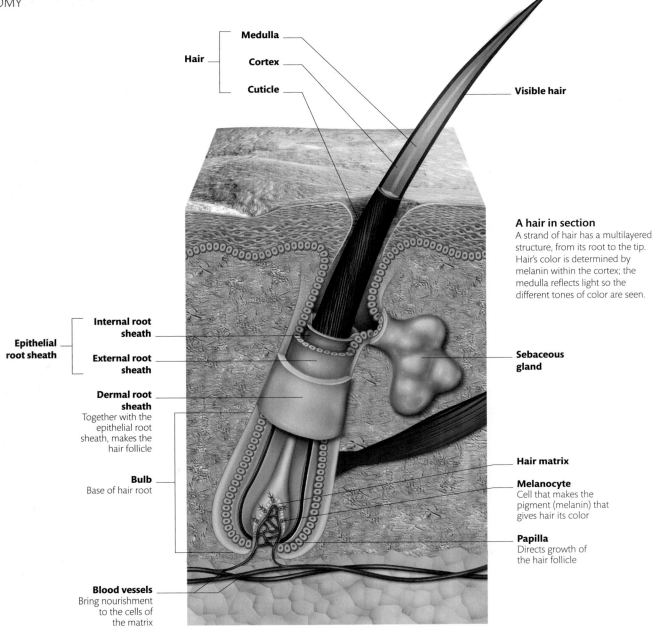

Hair
- Medulla
- Cortex
- Cuticle

Visible hair

A hair in section
A strand of hair has a multilayered structure, from its root to the tip. Hair's color is determined by melanin within the cortex; the medulla reflects light so the different tones of color are seen.

Epithelial root sheath
- **Internal root sheath**
- **External root sheath**

Dermal root sheath
Together with the epithelial root sheath, makes the hair follicle

Sebaceous gland

Bulb
Base of hair root

Hair matrix

Melanocyte
Cell that makes the pigment (melanin) that gives hair its color

Papilla
Directs growth of the hair follicle

Blood vessels
Bring nourishment to the cells of the matrix

SECTION THROUGH A HAIR

SKIN, HAIR, AND NAIL STRUCTURE

The skin is our largest organ, weighing about 9 lb (4 kg) and covering an area of about 21 square feet (2 square meters). It forms a tough, waterproof layer, which protects us from the elements. However, it offers much more than protection: the skin lets us appreciate the texture and temperature of our environment; it regulates body temperature; it allows excretion in sweat, communication through blushing, gripping due to ridges on our fingertips, and vitamin D production in sunlight.

Thick head hairs and fine body hairs help to keep us warm and dry. All visible hair is in fact dead; hairs are only alive at their root. Constantly growing and self-repairing, nails protect fingers and toes but also enhance their sensitivity.

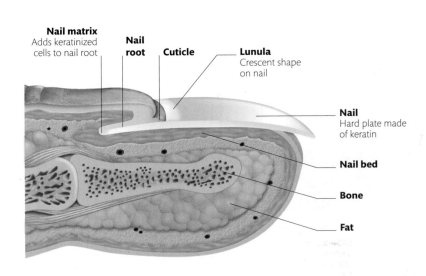

Nail matrix
Adds keratinized cells to nail root

Nail root

Cuticle

Lunula
Crescent shape on nail

Nail
Hard plate made of keratin

Nail bed

Bone

Fat

SECTION THROUGH A NAIL

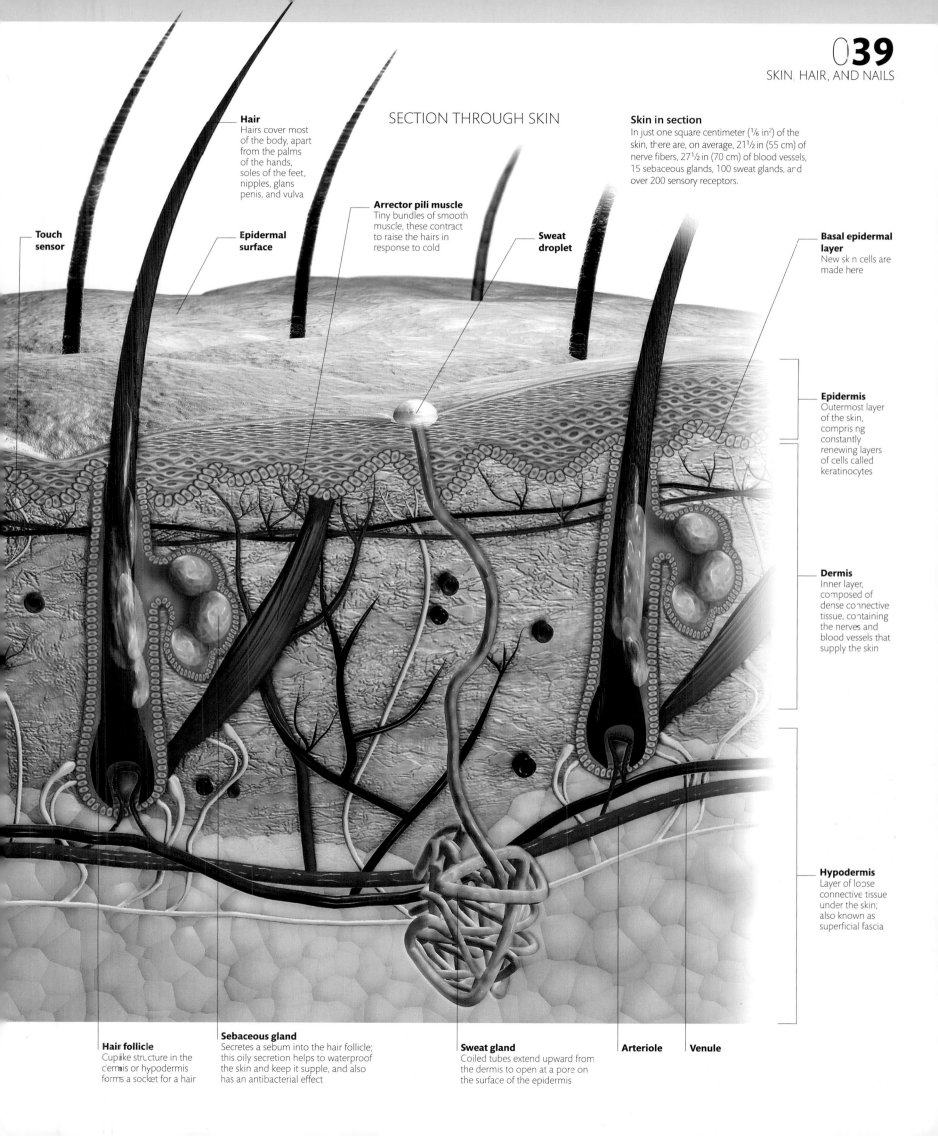

SECTION THROUGH SKIN

Hair
Hairs cover most of the body, apart from the palms of the hands, soles of the feet, nipples, glans penis, and vulva

Arrector pili muscle
Tiny bundles of smooth muscle, these contract to raise the hairs in response to cold

Skin in section
In just one square centimeter (⅙ in²) of the skin, there are, on average, 21½ in (55 cm) of nerve fibers, 27½ in (70 cm) of blood vessels, 15 sebaceous glands, 100 sweat glands, and over 200 sensory receptors.

Touch sensor

Epidermal surface

Sweat droplet

Basal epidermal layer
New skin cells are made here

Epidermis
Outermost layer of the skin, comprising constantly renewing layers of cells called keratinocytes

Dermis
Inner layer, composed of dense connective tissue, containing the nerves and blood vessels that supply the skin

Hypodermis
Layer of loose connective tissue under the skin; also known as superficial fascia

Hair follicle
Cuplike structure in the dermis or hypodermis forms a socket for a hair

Sebaceous gland
Secretes a sebum into the hair follicle; this oily secretion helps to waterproof the skin and keep it supple, and also has an antibacterial effect

Sweat gland
Coiled tubes extend upward from the dermis to open at a pore on the surface of the epidermis

Arteriole

Venule

Clavicle
Traces a sinuous curve at the base of the neck; it acts as a strut supporting the shoulder

Scapula
Connects the arm to the trunk, and provides a secure but mobile anchor for the arm, allowing the shoulders to be retracted backward, protracted forward, and elevated

Humerus

Ulna
Wide at its proximal end, where it articulates with the humerus at the elbow, this bone tapers down to a pointed styloid process near the wrist

Radius
Forearm bone; it can rotate around the ulna to alter the orientation of the hand

Carpals
Eight small bones in the base of the hand. Two articulate with the radius to form the wrist joint

Cranium
Contains and protects the brain and the organs of special sense—the eyes, ears, and nose—and provides the supporting framework of the face

Mandible
A single bone, the jaw contains the lower teeth and provides attachment for the chewing muscles

ANTERIOR (FRONT)

Vertebral column
Comprises stacked vertebrae and forms a strong, flexible backbone for the skeleton

Manubrium

Gladiolus

Xiphoid process

Sternum
Breastbone; made up of the manubrium, the body (gladiolus), and the xiphoid process. Anchors the upper seven costal cartilages

Costal cartilages
Attach the upper ribs to the sternum, and lower ribs to each other, and give the ribcage flexibility

Ribs
Twelve pairs of curving bones form the ribcage

Pelvis
Oddly shaped bone also called the innominate bone ("bone without a name")

Sacrum
Formed from five fused vertebrae; it provides a strong connection between the pelvis and the spine

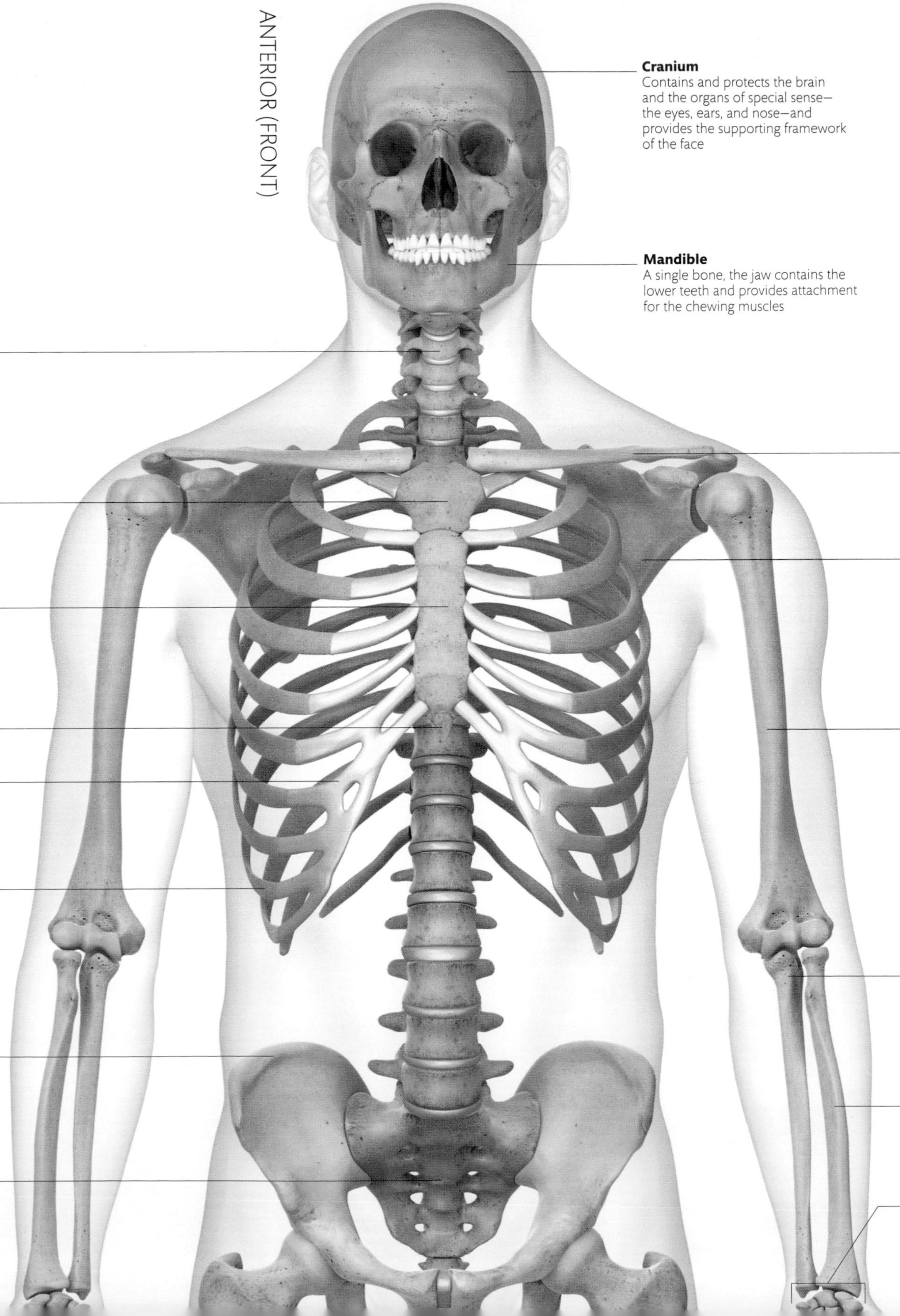

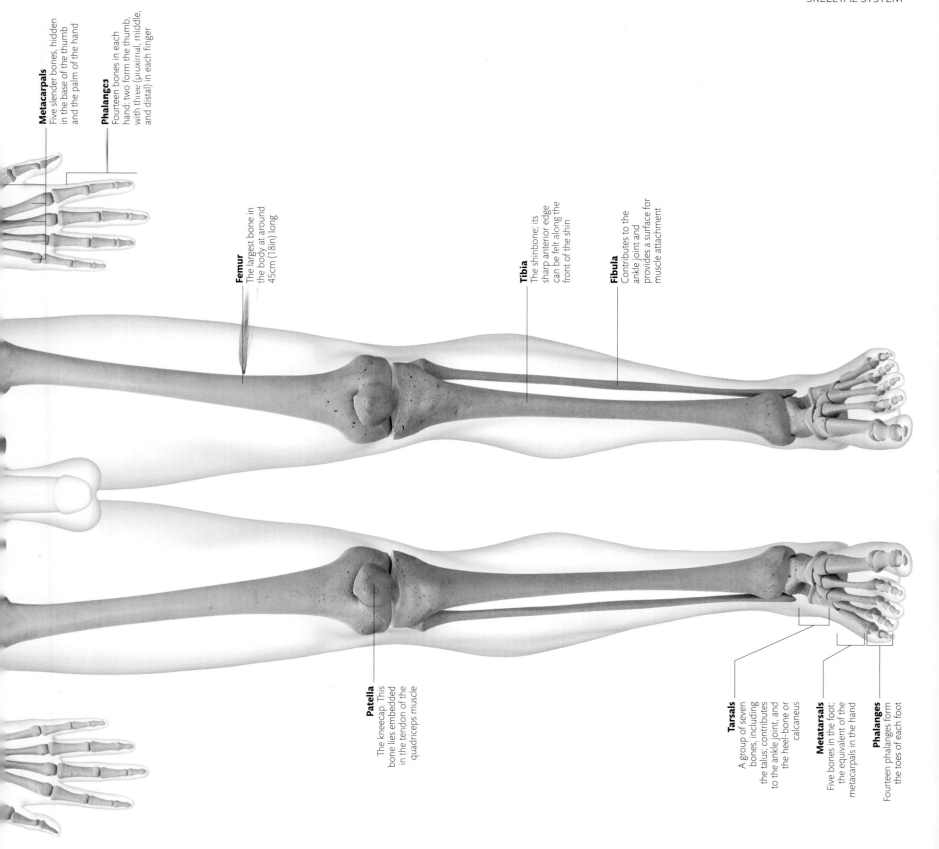

Metacarpals
Five slender bones, hidden in the base of the thumb and the palm of the hand

Phalanges
Fourteen bones in each hand; two form the thumb, with three (proximal, middle, and distal) in each finger

Femur
The largest bone in the body at around 45cm (18in) long

Tibia
The shinbone; its sharp anterior edge can be felt along the front of the shin

Fibula
Contributes to the ankle joint and provides a surface for muscle attachment

Patella
The kneecap. This bone lies embedded in the tendon of the quadriceps muscle

Tarsals
A group of seven bones, including the talus; contributes to the ankle joint, and the heel-bone or calcaneus

Metatarsals
Five bones in the foot; the equivalent of the metacarpals in the hand

Phalanges
Fourteen phalanges form the toes of each foot

SKELETAL
SYSTEM

The skeleton gives the body its shape, supports the weight of all our other tissues, provides attachment for muscles, and forms a system of linked levers that the muscles can move. The skeleton also plays an important role in protecting delicate organs and tissues, such as the brain within the skull, the spinal cord within the protective arches of the vertebrae, and the heart and lungs within the ribcage.

The human skeleton differs between the sexes. This is most obvious in the pelvis, which must form the birth canal in a woman; the pelvis of a woman is usually wider than that of a man. The skull also varies: men tend to have a larger brow and more prominent areas for muscle attachment on the back of the head. The entire skeleton tends to be larger and more robust in a man.

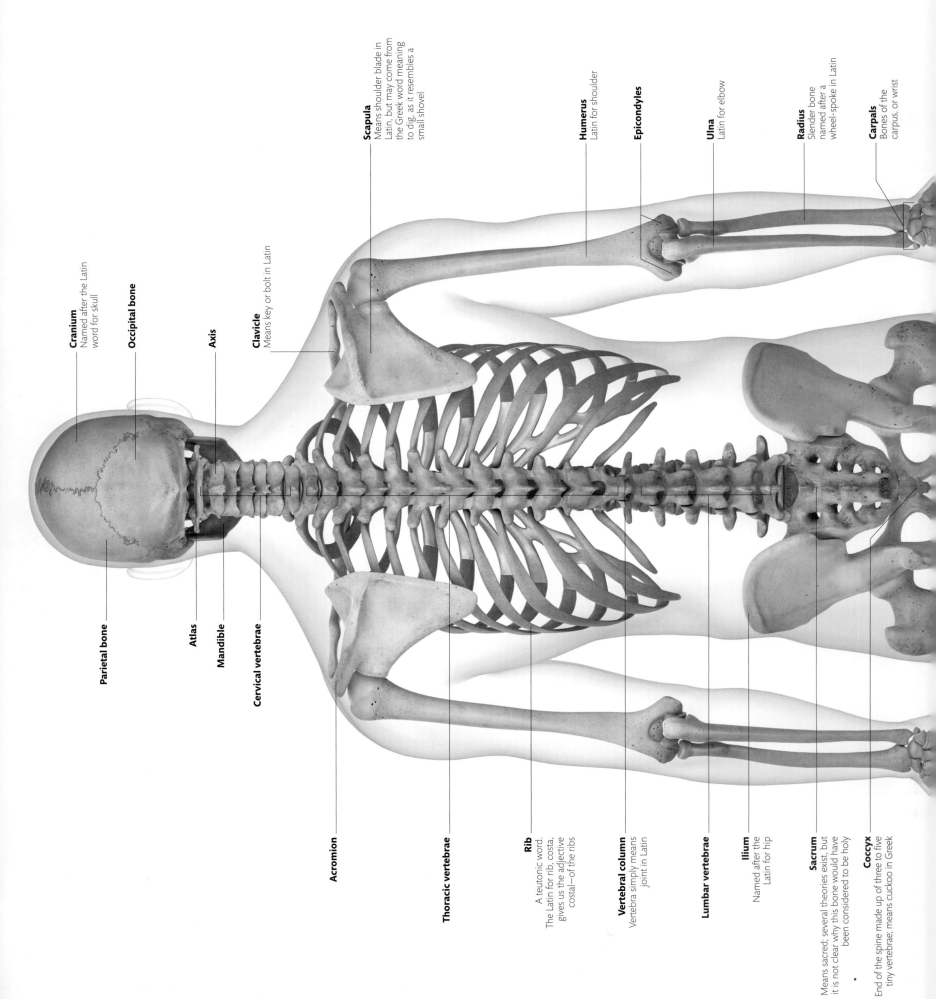

Scapula
Means shoulder blade in Latin, but may come from the Greek word meaning to dig, as it resembles a small shovel

Humerus
Latin for shoulder

Epicondyles

Ulna
Latin for elbow

Radius
Slender bone named after a wheel-spoke in Latin

Carpals
Bones of the carpus, or wrist

Cranium
Named after the Latin word for skull

Occipital bone

Axis

Clavicle
Means key or bolt in Latin

Parietal bone

Atlas

Mandible

Cervical vertebrae

Acromion

Thoracic vertebrae

Rib
A teutonic word. The Latin for rib, costa, gives us the adjective costal—of the ribs

Vertebral column
Vertebra simply means joint in Latin

Lumbar vertebrae

Ilium
Named after the Latin for hip

Sacrum
Means sacred; several theories exist, but it is not clear why this bone would have been considered to be holy

Coccyx
End of the spine made up of three to five tiny vertebrae; means cuckoo in Greek

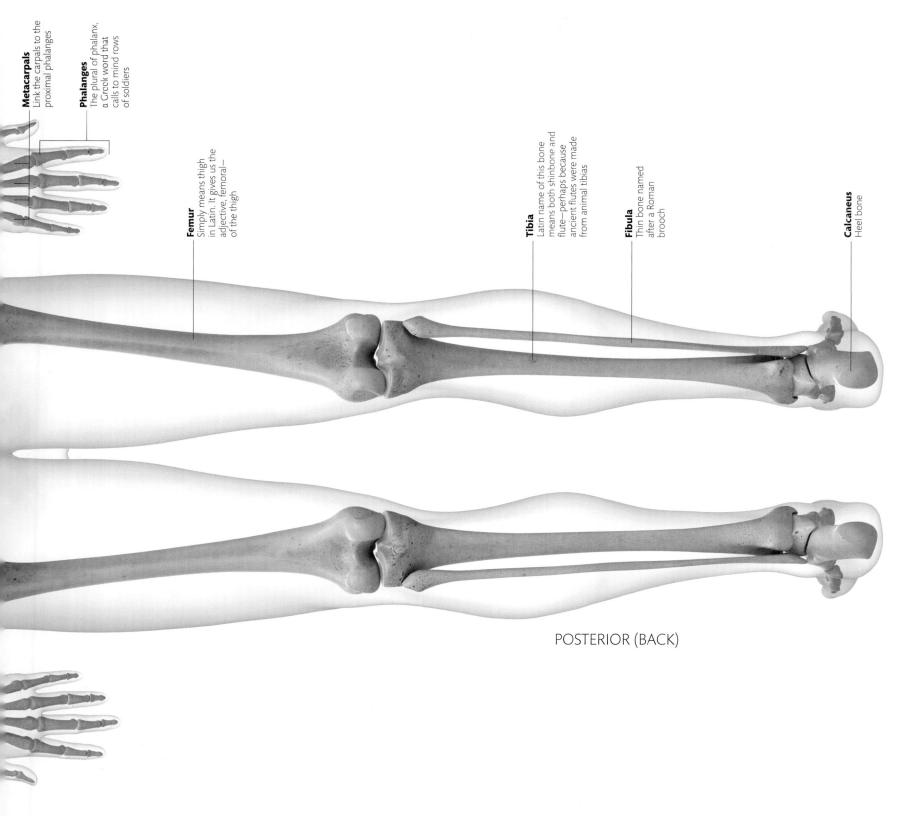

Metacarpals
Link the carpals to the proximal phalanges

Phalanges
The plural of phalanx, a Greek word that calls to mind rows of soldiers

Femur
Simply means thigh in Latin. It gives us the adjective, femoral—of the thigh

Tibia
Latin name of this bone means both shinbone and flute—perhaps because ancient flutes were made from animal tibias

Fibula
Thin bone named after a Roman brooch

Calcaneus
Heel bone

POSTERIOR (BACK)

SKELETAL
SYSTEM

It is important to remember that bone is a living, dynamic tissue that constantly restructures itself in response to mechanical changes. We are familiar with the idea that if we work out at the gym our muscles develop in response—we can see the effects. But deep under the skin, our bones also respond to the change by slightly altering their architecture. Bones are full of blood vessels, and bleed when broken. Arteries enter bones through small holes in the surface, visible to the naked eye, called nutrient foramina. The surface, or periosteum, of a bone is supplied with sensory nerves, so it's not surprising that when we damage a bone it produces a lot of pain.

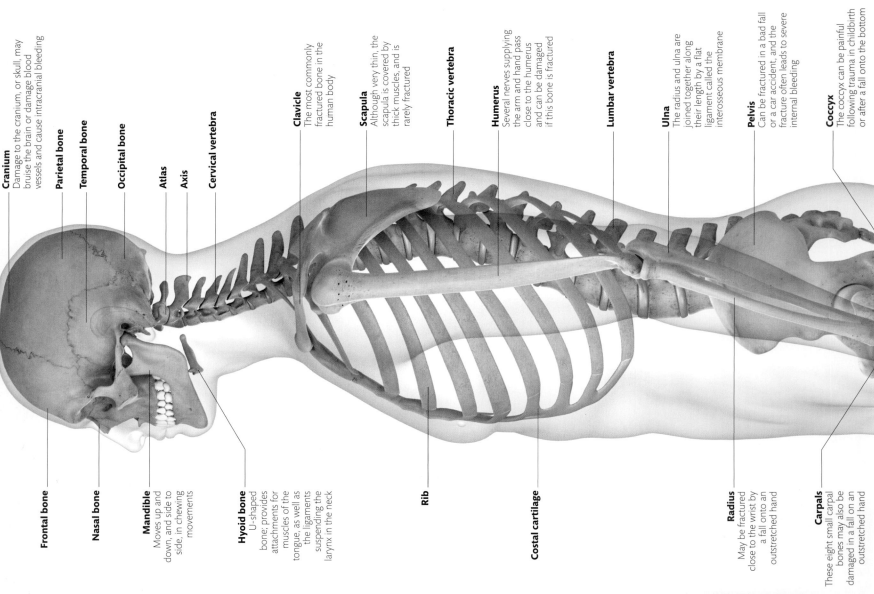

Cranium
Damage to the cranium, or skull, may bruise the brain or damage blood vessels and cause intracranial bleeding

Parietal bone

Temporal bone

Occipital bone

Atlas

Axis

Cervical vertebra

Clavicle
The most commonly fractured bone in the human body

Scapula
Although very thin, the scapula is covered by thick muscles, and is rarely fractured

Thoracic vertebra

Humerus
Several nerves supplying the arm and hand pass close to the humerus and can be damaged if this bone is fractured

Lumbar vertebra

Ulna
The radius and ulna are joined together along their length by a flat ligament called the interosseous membrane

Pelvis
Can be fractured in a bad fall or a car accident, and the fracture often leads to severe internal bleeding

Coccyx
The coccyx can be painful following trauma in childbirth or after a fall onto the bottom

Frontal bone

Nasal bone

Mandible
Moves up and down, and side to side, in chewing movements

Hyoid bone
U-shaped bone; provides attachments for muscles of the tongue, as well as the ligaments suspending the larynx in the neck

Rib

Costal cartilage

Radius
May be fractured close to the wrist by a fall onto an outstretched hand

Carpals
These eight small carpal bones may also be damaged in a fall on an outstretched hand

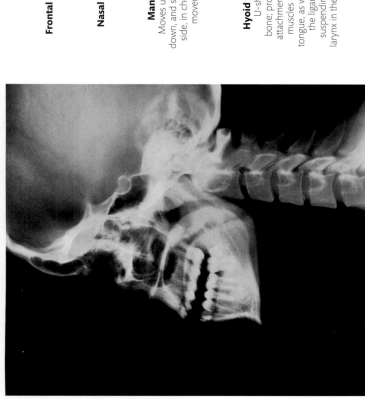

Lateral radiograph of a skull and cervical spine
On radiographs—images produced using X-rays—bone appears bright, while air spaces are dark. The part of the skull just above the spine looks very bright here—this is the extremely dense petrous, or "stony", part of the temporal bone.

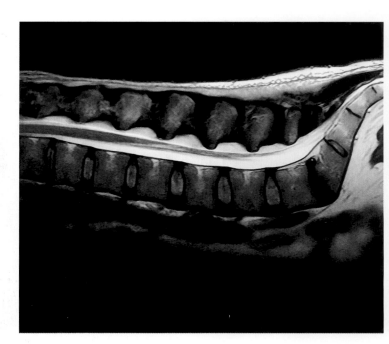

MRI of a lumbar spine
Protected within the vertebral column, the tapering tail end of the spinal cord can be seen, in blue. The fluid and fat around the cord appears white.

SKELETAL
SYSTEM

After teeth, bone is the hardest material in the human body. Bone mineral—made of calcium and phosphate salts—gives bone its hardness and rigidity. It also acts as the body's calcium store: if the level of calcium in the blood drops, calcium will be freed from the bones. Cartilage is another component of the skeleton. Many bones develop as cartilage "models" in the embryo, and later ossify, or turn to bone. But cartilage persists into adulthood at certain sites, such as at the surfaces of joints and as the costal cartilages that join the ribs to the sternum. Cartilage is not as hard as bone, but it has other useful properties. The costal cartilages give the rib cage some flexibility, and the cartilage lining the surface of joints resists compression well and provides a smooth, low-friction surface.

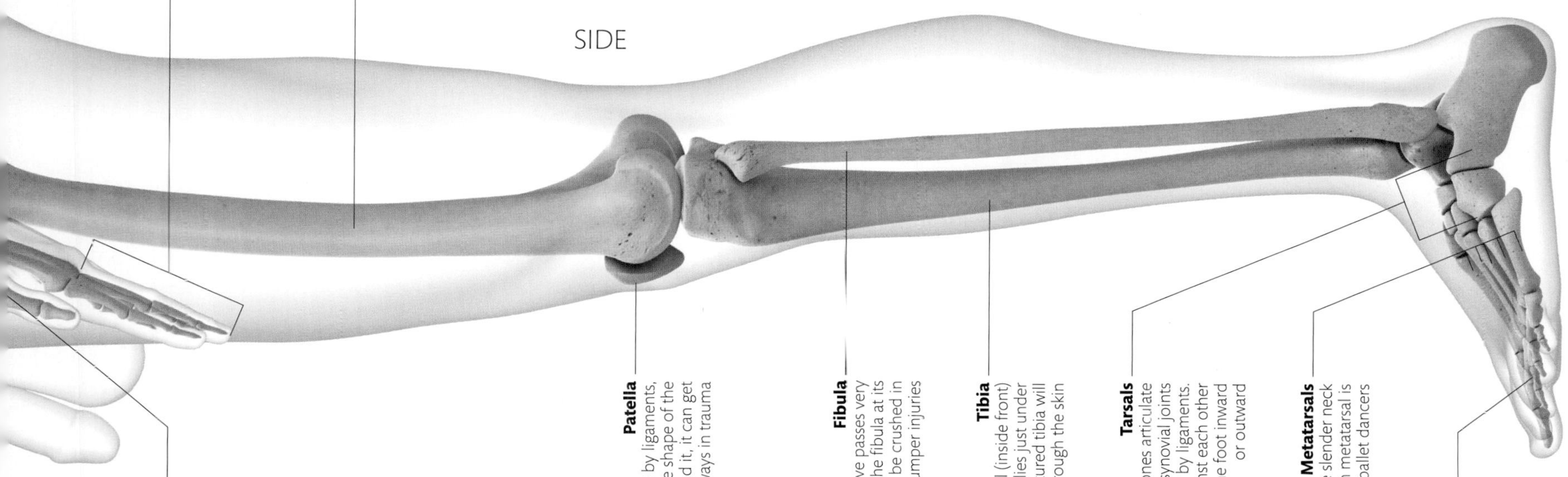

SIDE

Phalanges
Fingers tend to stick out and get knocked, twisted, and crushed. If the finger swells up and is very painful, a phalanx may have been fractured

Femur
Large arteries pass close to this bone, and fractures can lead to considerable bleeding

Metacarpal
The first metacarpal is the key to our opposable thumbs: it is very mobile and can be brought across the palm, bringing the thumb into a position where it can touch the other fingers

Patella
Usually held in place by ligaments, muscles, and the shape of the femur behind it, it can get dislocated sideways in trauma

Fibula
An important nerve passes very close to the neck of the fibula at its upper end, and can be crushed in car bumper injuries

Tibia
The anteromedial (inside front) surface of the tibia lies just under the skin, and a fractured tibia will often stick out through the skin

Tarsals
The seven tarsal bones articulate with each other with synovial joints and are held together by ligaments. They can twist against each other to move the sole of the foot inward or outward

Metatarsals
Fracture of the slender neck of the fifth metatarsal is common in ballet dancers

Phalanx
African apes have opposable great toes, somewhat like our thumbs. This opposability was lost during human evolution because we use our feet more as platforms to stand, run, and walk on—rather than to grasp things

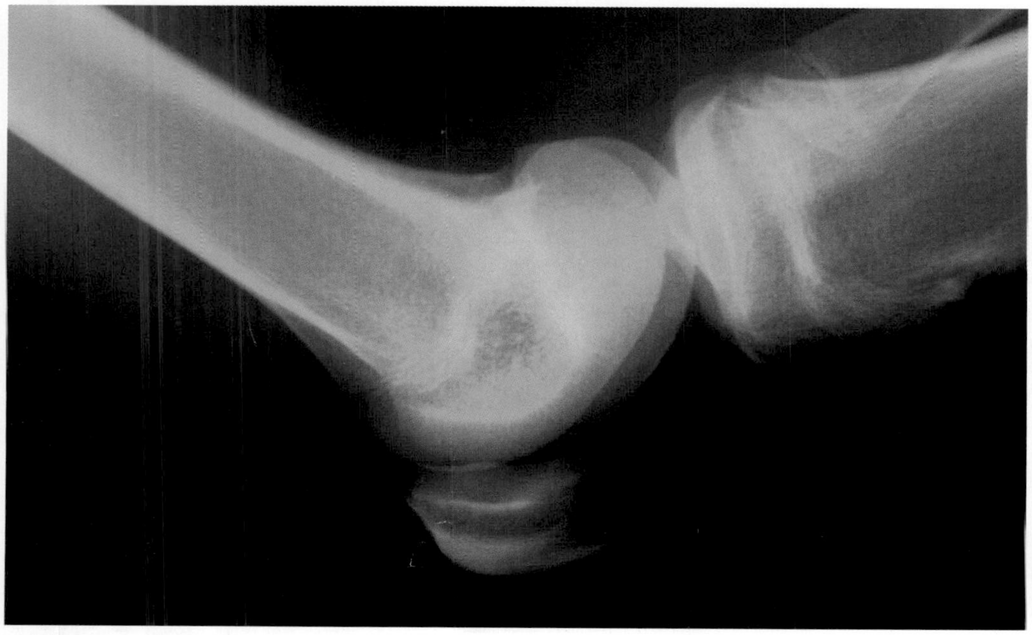

Lateral radiograph of a knee
The knee is half flexed here, showing how the curved condyles of the femur rotate on the tibia below. The patella is embedded in the quadriceps tendon (invisible on a radiograph), which runs over the front of the knee.

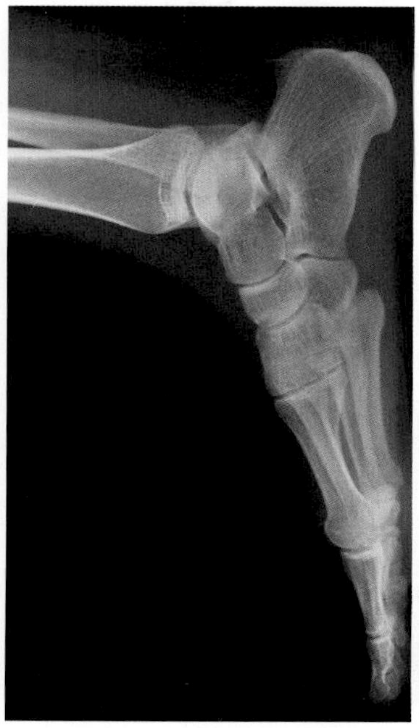

Lateral radiograph of a foot
The hinge joint that forms the ankle can be clearly seen here—between the tibia and fibula of the leg and the uppermost tarsal bone, the talus. The bones of the foot can be seen to form an arch, which is supported by tendons and ligaments.

LONG BONE

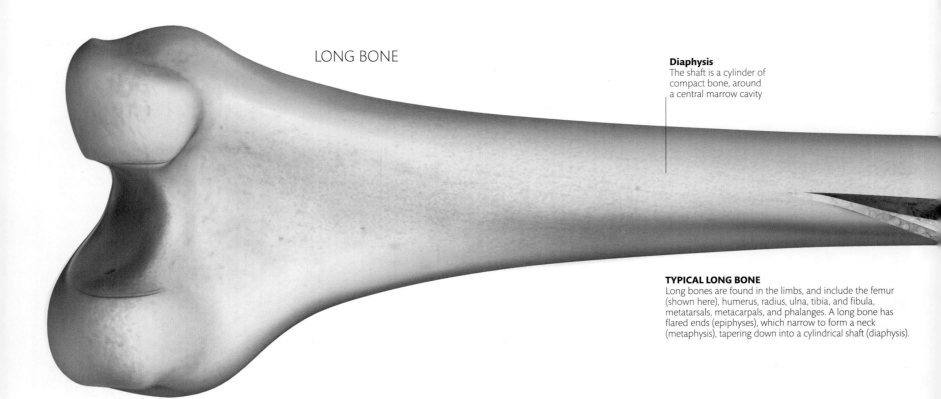

Diaphysis
The shaft is a cylinder of compact bone, around a central marrow cavity

TYPICAL LONG BONE
Long bones are found in the limbs, and include the femur (shown here), humerus, radius, ulna, tibia, and fibula, metatarsals, metacarpals, and phalanges. A long bone has flared ends (epiphyses), which narrow to form a neck (metaphysis), tapering down into a cylindrical shaft (diaphysis).

TRANSVERSE

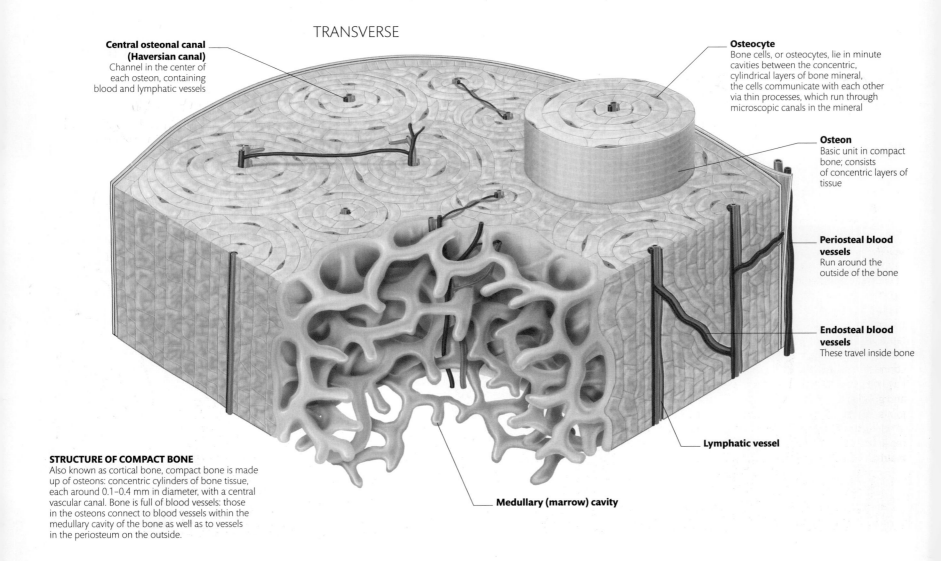

Central osteonal canal (Haversian canal)
Channel in the center of each osteon, containing blood and lymphatic vessels

Osteocyte
Bone cells, or osteocytes, lie in minute cavities between the concentric, cylindrical layers of bone mineral, the cells communicate with each other via thin processes, which run through microscopic canals in the mineral

Osteon
Basic unit in compact bone; consists of concentric layers of tissue

Periosteal blood vessels
Run around the outside of the bone

Endosteal blood vessels
These travel inside bone

Lymphatic vessel

Medullary (marrow) cavity

STRUCTURE OF COMPACT BONE
Also known as cortical bone, compact bone is made up of osteons: concentric cylinders of bone tissue, each around 0.1–0.4 mm in diameter, with a central vascular canal. Bone is full of blood vessels: those in the osteons connect to blood vessels within the medullary cavity of the bone as well as to vessels in the periosteum on the outside.

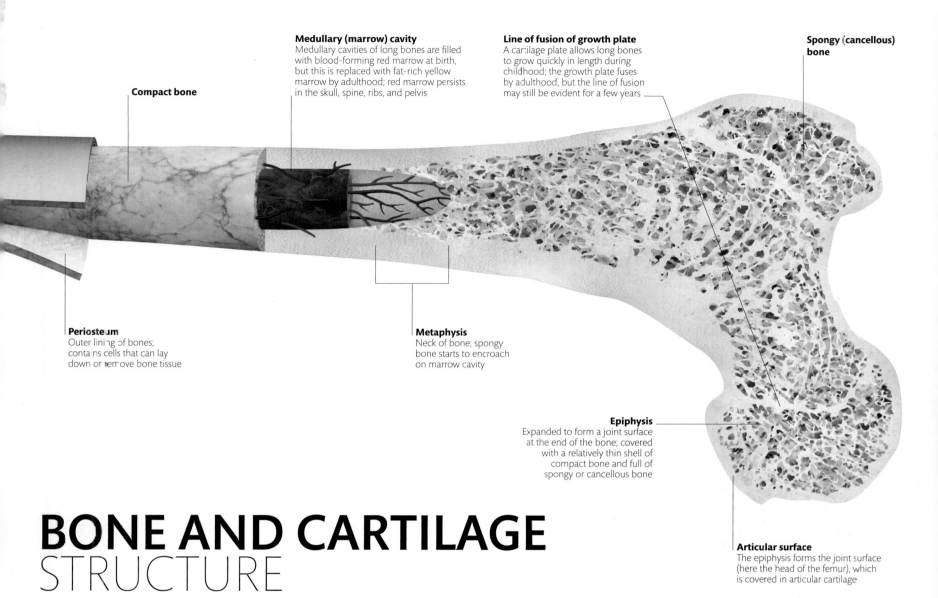

Compact bone

Medullary (marrow) cavity
Medullary cavities of long bones are filled with blood-forming red marrow at birth, but this is replaced with fat-rich yellow marrow by adulthood; red marrow persists in the skull, spine, ribs, and pelvis

Line of fusion of growth plate
A cartilage plate allows long bones to grow quickly in length during childhood; the growth plate fuses by adulthood, but the line of fusion may still be evident for a few years

Spongy (cancellous) bone

Periosteum
Outer lining of bones; contains cells that can lay down or remove bone tissue

Metaphysis
Neck of bone; spongy bone starts to encroach on marrow cavity

Epiphysis
Expanded to form a joint surface at the end of the bone; covered with a relatively thin shell of compact bone and full of spongy or cancellous bone

Articular surface
The epiphysis forms the joint surface (here the head of the femur), which is covered in articular cartilage

BONE AND CARTILAGE
STRUCTURE

The adult skeleton is mainly made of bone, with just a little cartilage in some places—such as the costal cartilages which complete the ribs. Most of the human skeleton develops first as cartilage, which is later replaced by bone (see pp.300–01). At just 8 weeks, a fetus already has cartilage models of almost all the components of the skeleton, some of which are just starting to transform into bone. This transformation continues during fetal development and throughout childhood. But there are still cartilage plates near the end of the bones in an adolescent's skeleton, enabling rapid growth. When growth is finally complete, those plates close and become bone. Bone and cartilage are both connective tissues, with cells embedded in a matrix, but they have different properties. Cartilage is a stiff but flexible tissue and good at load bearing, which is why it is involved in joints. But it has virtually no blood vessels and is very bad at self repair. In contrast, bone is full of blood vessels and repairs very well. Bone cells are embedded in a mineralized matrix, creating an extremely hard, strong tissue.

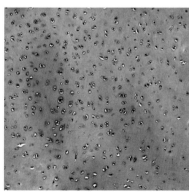

CARTILAGE
This tissue is made up of specialized cells called chondrocytes (seen clearly here) contained within a gel-like matrix embedded with fibers, including collagen and elastin. The different types of cartilage include hyaline, elastic, and fibrocartilage, which differ in the proportion of these constituents.

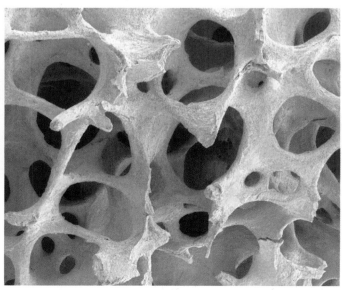

SPONGY BONE
Also known as cancellous bone, this is found in the epiphyses of long bones, and completely fills bones such as the vertebrae, carpals, and tarsals. It is made of minute interlinking struts or trabeculae (seen in this magnified image), giving it a spongy appearance, with bone marrow occupying the spaces between the trabeculae.

JOINT AND LIGAMENT
STRUCTURE

During development of the embryo, the connective tissue between developing bones forms joints—either remaining solid, creating a fibrous or cartilaginous joint, or creating cavities, to form a synovial joint. Fibrous joints are linked by microscopic fibers of collagen. They include the sutures of the skull, the teeth sockets (gomphoses), and the lower joint between the tibia and fibula. Cartilaginous joints include the junctions between ribs and costal cartilages, joints between the components of the sternum, and the pubic symphysis. The intervertebral disks are also specialized cartilaginous joints. Synovial joints contain lubricating fluid, and the joint surfaces are lined with cartilage to reduce friction. They tend to be very mobile joints (see pp.302–03).

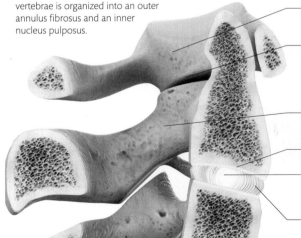

Fibula

Tibia

Inferior tibiofibular joint
The bones are united here by a ligament, whereas the superior tibiofibular joint is synovial

Syndesmosis
From the Greek for joined together; the lower ends of the tibia and fibula are firmly bound together by fibrous tissue. The interosseous membranes of the forearm and lower leg could also be described as syndesmoses.

ANKLE

FIBROUS JOINTS

Gomphosis
This name comes from the Greek word for bolted together. The fibrous tissue of the periodontal ligament connects the cement of the tooth to the bone of the socket.

Alveolar bone
Bone of the maxilla or mandible forming the tooth socket (alveolus)

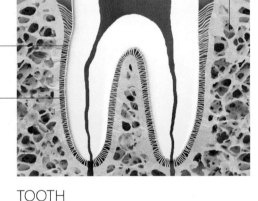

Cement
Covers the roots of the tooth

Periodontal ligament
Dense connective tissue anchoring the tooth in the socket

TOOTH

Suture
These joints exist between flat bones of the skull. They are flexible in the skull of a newborn baby, and allow growth of the skull throughout childhood. The sutures in the adult skull are interlocking, practically immovable joints, and eventually fuse completely in later adulthood.

Uniting layer

Bone

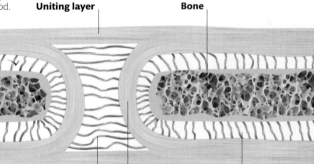

SKULL

Middle layer

Capsular layer

Cambial layer

CARTILAGINOUS JOINTS

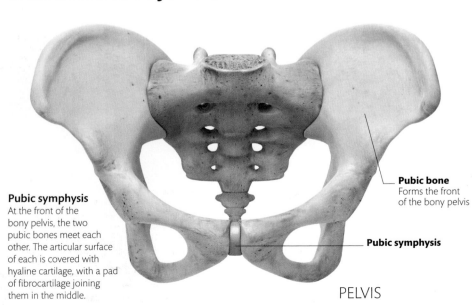

Pubic symphysis
At the front of the bony pelvis, the two pubic bones meet each other. The articular surface of each is covered with hyaline cartilage, with a pad of fibrocartilage joining them in the middle.

Pubic bone
Forms the front of the bony pelvis

Pubic symphysis

PELVIS

Intervertebral disk
The fibrocartilage pad or disk between vertebrae is organized into an outer annulus fibrosus and an inner nucleus pulposus.

Atlas (first cervical vertebra)

Zygapophyseal joint
Small synovial joints between the neural arches at the back of the spine

Axis (second cervical vertebra)

Hyaline cartilage

Nucleus pulposus
Inner, gel-like center of the disk

Annulus fibrosus
Outer, fibrous ring of the disk

SPINE

SYNOVIAL JOINTS

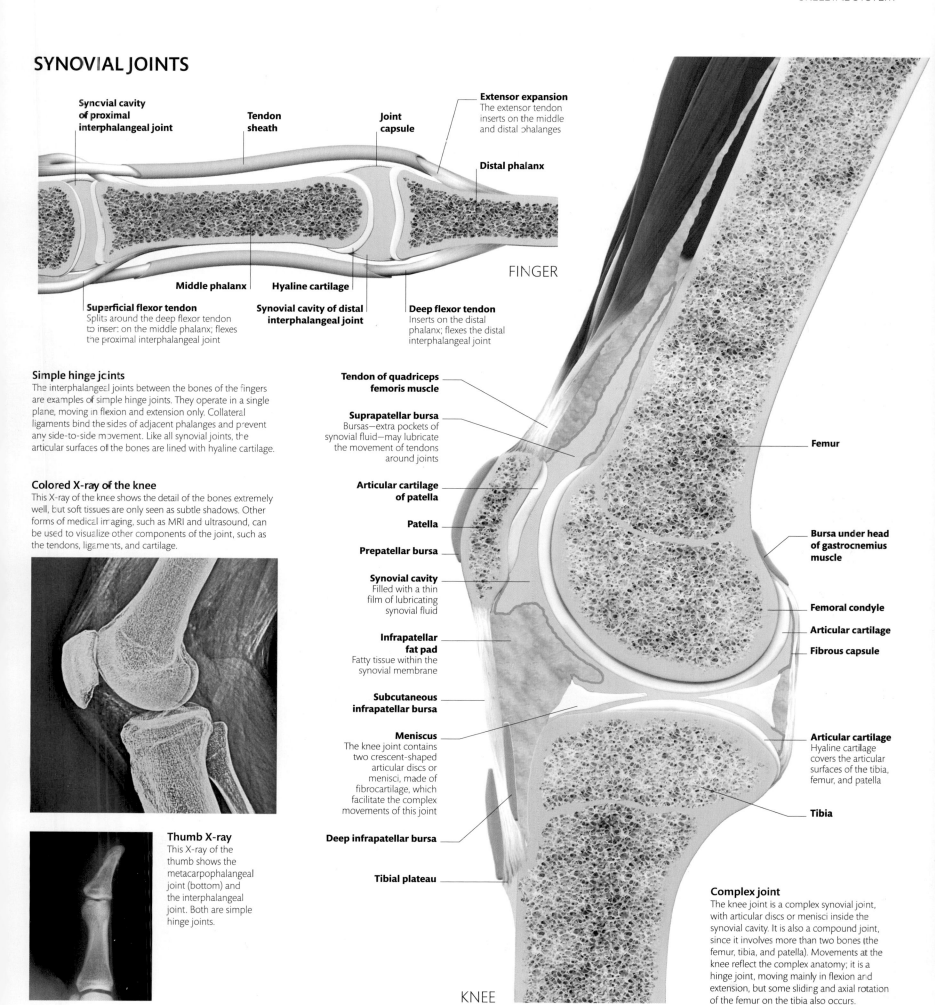

Syncvial cavity of proximal interphalangeal joint

Tendon sheath

Joint capsule

Extensor expansion
The extensor tendon inserts on the middle and distal phalanges

Distal phalanx

FINGER

Middle phalanx

Hyaline cartilage

Superficial flexor tendon
Splits around the deep flexor tendon to insert on the middle phalanx; flexes the proximal interphalangeal joint

Synovial cavity of distal interphalangeal joint

Deep flexor tendon
Inserts on the distal phalanx; flexes the distal interphalangeal joint

Simple hinge joints
The interphalangeal joints between the bones of the fingers are examples of simple hinge joints. They operate in a single plane, moving in flexion and extension only. Collateral ligaments bind the sides of adjacent phalanges and prevent any side-to-side movement. Like all synovial joints, the articular surfaces of the bones are lined with hyaline cartilage.

Colored X-ray of the knee
This X-ray of the knee shows the detail of the bones extremely well, but soft tissues are only seen as subtle shadows. Other forms of medical imaging, such as MRI and ultrasound, can be used to visualize other components of the joint, such as the tendons, ligaments, and cartilage.

Thumb X-ray
This X-ray of the thumb shows the metacarpophalangeal joint (bottom) and the interphalangeal joint. Both are simple hinge joints.

Tendon of quadriceps femoris muscle

Suprapatellar bursa
Bursas—extra pockets of synovial fluid—may lubricate the movement of tendons around joints

Articular cartilage of patella

Patella

Prepatellar bursa

Synovial cavity
Filled with a thin film of lubricating synovial fluid

Infrapatellar fat pad
Fatty tissue within the synovial membrane

Subcutaneous infrapatellar bursa

Meniscus
The knee joint contains two crescent-shaped articular discs or menisci, made of fibrocartilage, which facilitate the complex movements of this joint

Deep infrapatellar bursa

Tibial plateau

Femur

Bursa under head of gastrocnemius muscle

Femoral condyle

Articular cartilage

Fibrous capsule

Articular cartilage
Hyaline cartilage covers the articular surfaces of the tibia, femur, and patella

Tibia

KNEE

Complex joint
The knee joint is a complex synovial joint, with articular discs or menisci inside the synovial cavity. It is also a compound joint, since it involves more than two bones (the femur, tibia, and patella). Movements at the knee reflect the complex anatomy; it is a hinge joint, moving mainly in flexion and extension, but some sliding and axial rotation of the femur on the tibia also occurs.

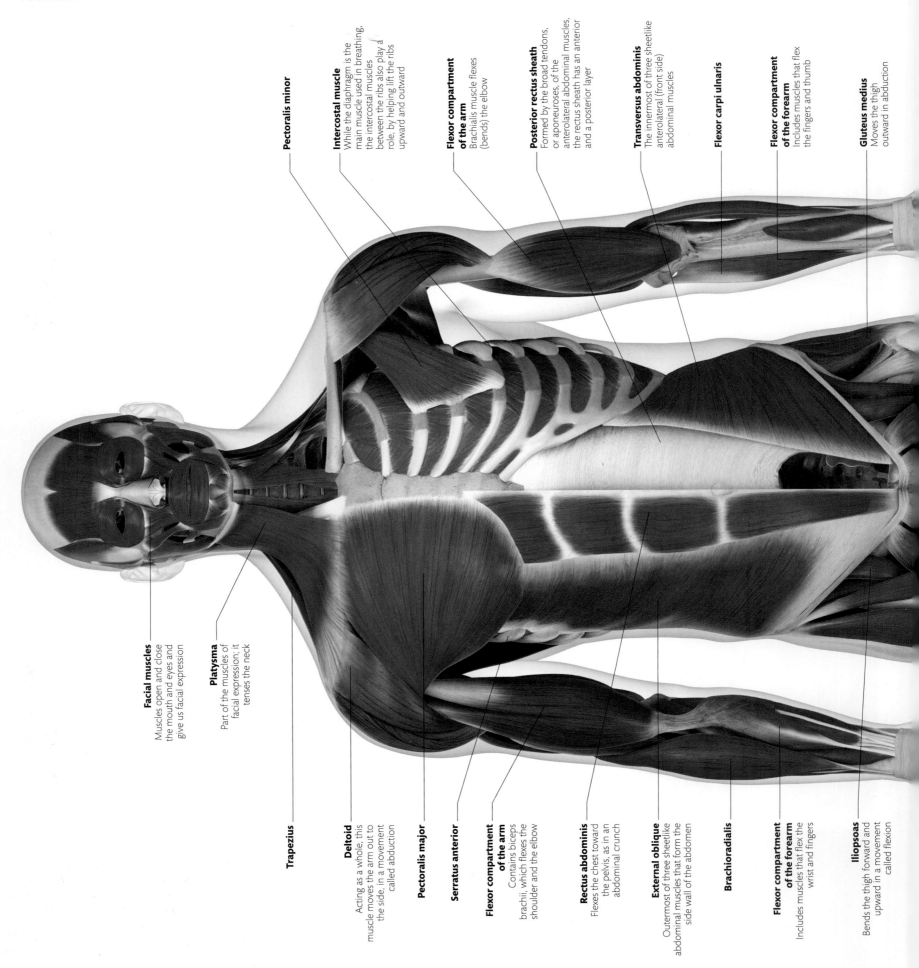

Pectoralis minor

Intercostal muscle
While the diaphragm is the main muscle used in breathing, the intercostal muscles between the ribs also play a role, by helping lift the ribs upward and outward

Flexor compartment of the arm
Brachialis muscle flexes (bends) the elbow

Posterior rectus sheath
Formed by the broad tendons, or aponeuroses, of the anterolateral abdominal muscles, the rectus sheath has an anterior and a posterior layer

Transversus abdominis
The innermost of three sheetlike anterolateral (front side) abdominal muscles

Flexor carpi ulnaris

Flexor compartment of the forearm
Includes muscles that flex the fingers and thumb

Gluteus medius
Moves the thigh outward in abduction

Facial muscles
Muscles open and close the mouth and eyes and give us facial expression

Platysma
Part of the muscles of facial expression; it tenses the neck

Trapezius

Deltoid
Acting as a whole, this muscle moves the arm out to the side, in a movement called abduction

Pectoralis major

Serratus anterior

Flexor compartment of the arm
Contains biceps brachii, which flexes the shoulder and the elbow

Rectus abdominis
Flexes the chest toward the pelvis, as in an abdominal crunch

External oblique
Outermost of three sheetlike abdominal muscles that form the side wall of the abdomen

Brachioradialis

Flexor compartment of the forearm
Includes muscles that flex the wrist and fingers

Iliopsoas
Bends the thigh forward and upward in a movement called flexion

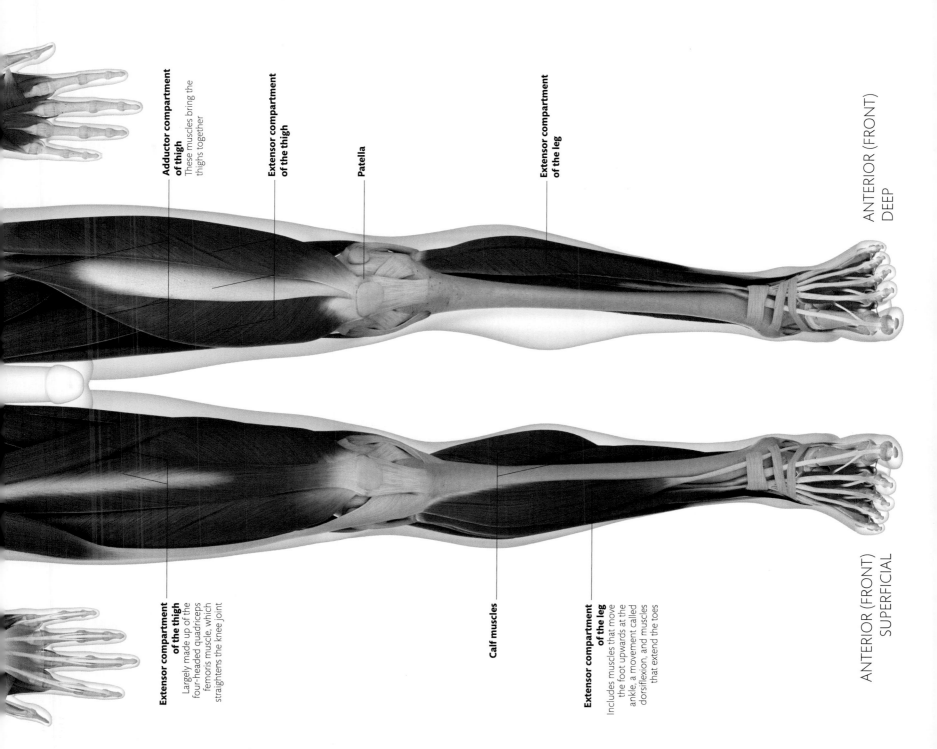

Adductor compartment of thigh
These muscles bring the thighs together

Extensor compartment of the thigh

Patella

Extensor compartment of the leg

ANTERIOR (FRONT)
DEEP

Extensor compartment of the thigh
Largely made up of the four-headed quadriceps femoris muscle, which straightens the knee joint

Calf muscles

Extensor compartment of the leg
Includes muscles that move the foot upwards at the ankle, a movement called dorsiflexion, and muscles that extend the toes

ANTERIOR (FRONT)
SUPERFICIAL

MUSCULAR
SYSTEM

Muscles attach to the skeleton by means of tendons, aponeuroses (flat, sheetlike tendons), and bands of connective tissue called fascia. Muscles are well supplied with blood vessels and appear reddish; tendons have a sparse vascular supply and look white. The "action" of a muscle refers to the movement it produces as it contracts. Muscle action has been investigated both by observing living people and by dissection of cadavers to pinpoint the precise attachments of muscles. Electromyography (EMG)—using electrodes to detect the electrical activity that accompanies muscle contraction—has proved invaluable in revealing which muscles act to produce a specific movement.

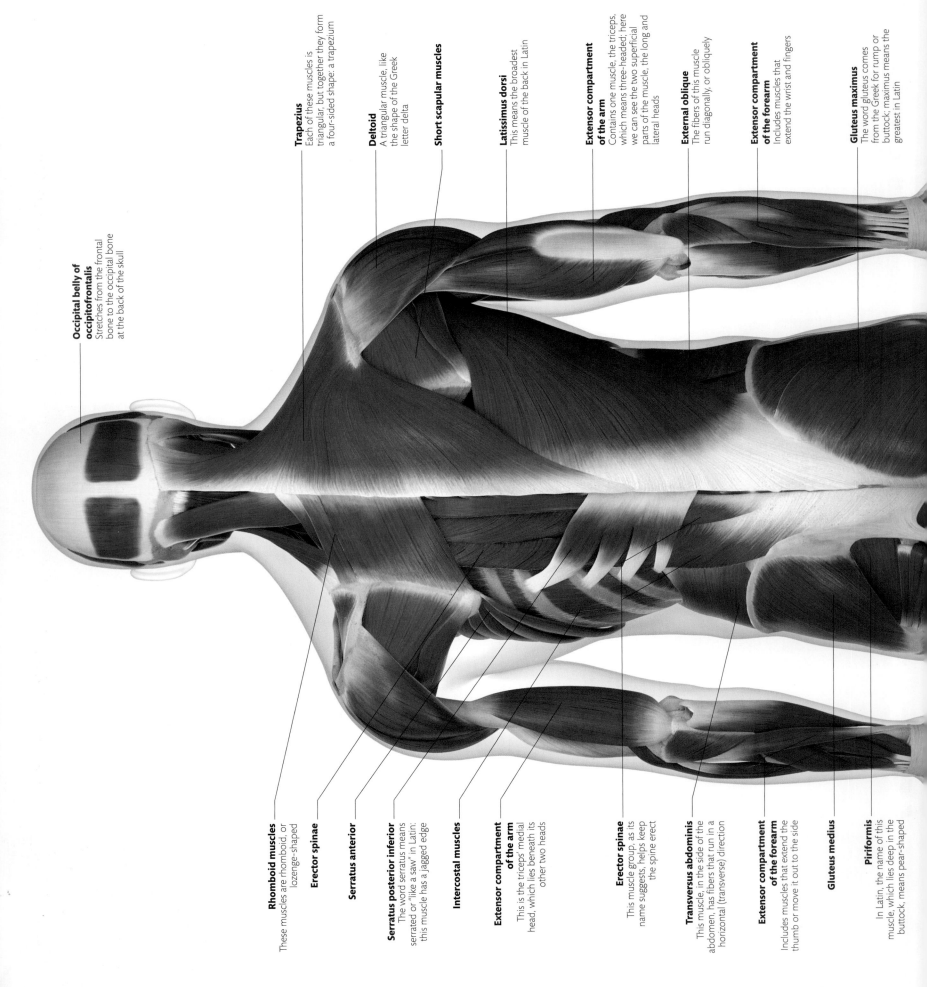

Occipital belly of occipitofrontalis
Stretches from the frontal bone to the occipital bone at the back of the skull

Trapezius
Each of these muscles is triangular, but together they form a four-sided shape: a trapezium

Deltoid
A triangular muscle, like the shape of the Greek letter delta

Short scapular muscles

Latissimus dorsi
This means the broadest muscle of the back in Latin

Extensor compartment of the arm
Contains one muscle, the triceps, which means three-headed; here we can see the two superficial parts of the muscle, the long and lateral heads

External oblique
The fibers of this muscle run diagonally, or obliquely

Extensor compartment of the forearm
Includes muscles that extend the wrist and fingers

Gluteus maximus
The word gluteus comes from the Greek for rump or buttock; maximus means the greatest in Latin

Rhomboid muscles
These muscles are rhomboid, or lozenge-shaped

Erector spinae

Serratus anterior

Serratus posterior inferior
The word serratus means serrated or "like a saw" in Latin: this muscle has a jagged edge

Intercostal muscles

Extensor compartment of the arm
This is the triceps' medial head, which lies beneath its other two heads

Erector spinae
This muscle group, as its name suggests, helps keep the spine erect

Transversus abdominis
This muscle, in the side of the abdomen, has fibers that run in a horizontal (transverse) direction

Extensor compartment of the forearm
Includes muscles that extend the thumb or move it out to the side

Gluteus medius

Piriformis
In Latin, the name of this muscle, which lies deep in the buttock, means pear-shaped

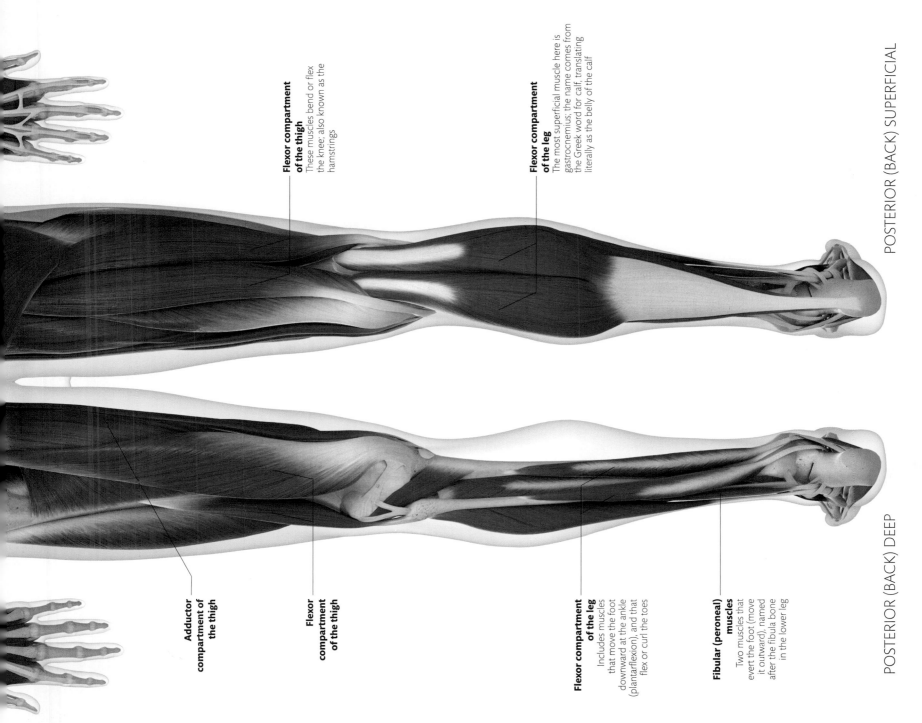

POSTERIOR (BACK) SUPERFICIAL

Flexor compartment of the thigh
These muscles bend or flex the knee; also known as the hamstrings

Flexor compartment of the leg
The most superficial muscle here is gastrocnemius; the name comes from the Greek word for calf, translating literally as the belly of the calf

POSTERIOR (BACK) DEEP

Adductor compartment of the thigh

Flexor compartment of the thigh

Flexor compartment of the leg
Includes muscles that move the foot downward at the ankle (plantarflexion), and that flex or curl the toes

Fibular (peroneal) muscles
Two muscles that evert the foot (move it outward), named after the fibula bone in the lower leg

MUSCULAR
SYSTEM

Most muscle names are derived from Latin or Greek. They can refer to a muscle's shape, size, attachments, number of heads, position or depth in the body, or the action it produces when it contracts. Names that end in -oid refer to the shape of the muscle. Deltoid, for example, means triangle-shaped, and rhomboid means diamond-shaped. Many muscles have two-part names. These names often refer to both a characteristic of the muscle and the muscle's position in the body. Rectus abdominis, for example, means straight [muscle] of the abdomen, and biceps brachii means two-headed [muscle] of the arm. Some muscles have names that describe their action, such as flexor digitorum, which simply means flexor of the fingers.

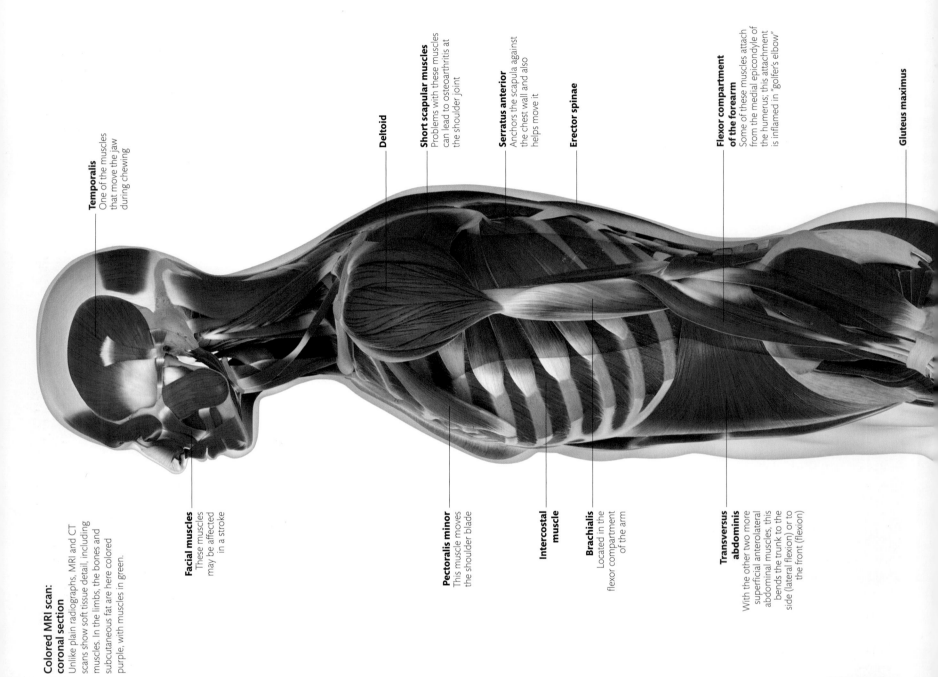

Colored MRI scan: coronal section
Unlike plain radiographs, MRI and CT scans show soft tissue detail, including muscles. In the limbs, the bones and subcutaneous fat are here colored purple, with muscles in green.

Temporalis
One of the muscles that move the jaw during chewing

Deltoid

Short scapular muscles
Problems with these muscles can lead to osteoarthritis at the shoulder joint

Serratus anterior
Anchors the scapula against the chest wall and also helps move it

Erector spinae

Flexor compartment of the forearm
Some of these muscles attach from the medial epicondyle of the humerus; this attachment is inflamed in "golfer's elbow"

Gluteus maximus

Facial muscles
These muscles may be affected in a stroke

Pectoralis minor
This muscle moves the shoulder blade

Intercostal muscle

Brachialis
Located in the flexor compartment of the arm

Transversus abdominis
With the other two more superficial anterolateral abdominal muscles, this bends the trunk to the side (lateral flexion) or to the front (flexion)

MUSCULAR
SYSTEM

The force produced by muscles of different shapes varies. Long, thin muscles tend to contract a lot but exert low forces. Muscles with many fibers attaching to a tendon at an angle, such as the deltoid, shorten less during contraction but produce greater forces. Although the shape of muscles varies, there is a general rule that the force generated by the contracting muscle fibers will be directed along the line of the tendon. Muscle fibers will enlarge in response to intense exercise. Conversely, if muscles are unused for just a few months, they start to waste away. Consequently, physical activity is very important in maintaining muscle bulk.

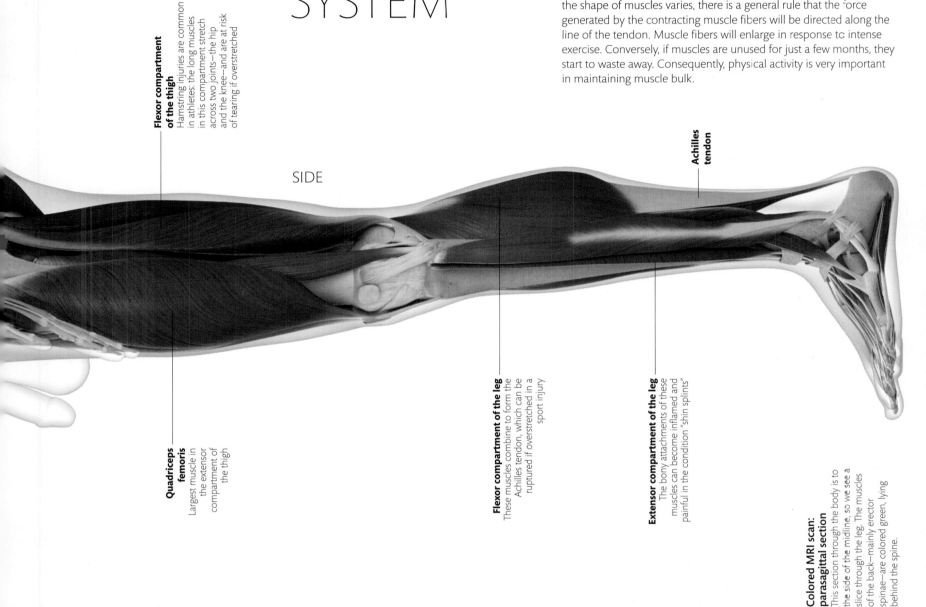

SIDE

Flexor compartment of the thigh
Hamstring injuries are common in athletes: the long muscles in this compartment stretch across two joints—the hip and the knee—and are at risk of tearing if overstretched

Achilles tendon

Quadriceps femoris
Largest muscle in the extensor compartment of the thigh

Flexor compartment of the leg
These muscles combine to form the Achilles tendon, which can be ruptured if overstretched in a sport injury

Extensor compartment of the leg
The bony attachments of these muscles can become inflamed and painful in the condition "shin splints"

Colored MRI scan: parasagittal section
This section through the body is to the side of the midline, so we see a slice through the leg. The muscles of the back—mainly erector spinae—are colored green, lying behind the spine.

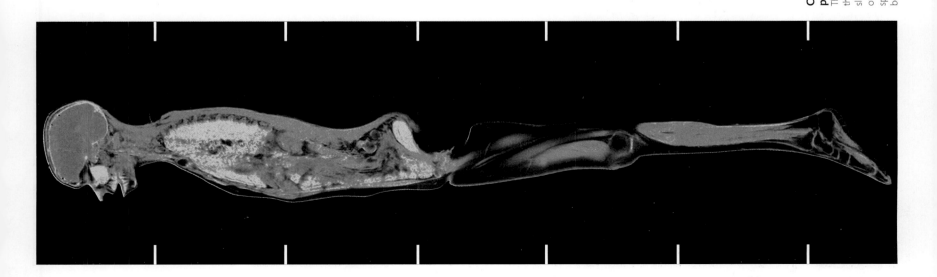

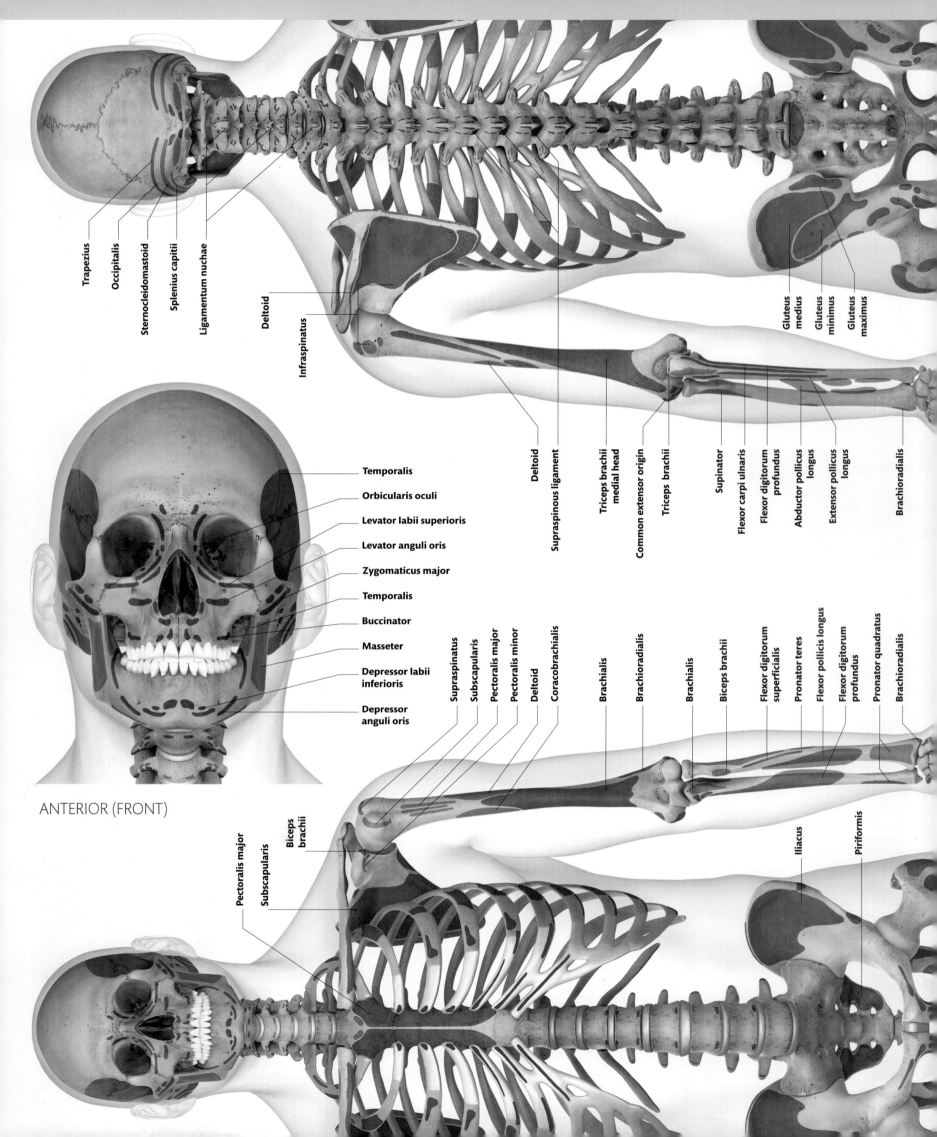

Trapezius
Occipitalis
Sternocleidomastoid
Splenius capitii
Ligamentum nuchae

Deltoid

Infraspinatus

Deltoid

Supraspinous ligament

Triceps brachii medial head

Common extensor origin

Triceps brachii

Supinator

Flexor carpi ulnaris

Flexor digitorum profundus

Abductor pollicus longus

Extensor pollicus longus

Brachioradialis

Gluteus medius
Gluteus minimus
Gluteus maximus

Temporalis
Orbicularis oculi
Levator labii superioris
Levator anguli oris
Zygomaticus major
Temporalis
Buccinator
Masseter
Depressor labii inferioris
Depressor anguli oris

ANTERIOR (FRONT)

Supraspinatus
Subscapularis
Pectoralis major
Pectoralis minor
Deltoid
Coracobrachialis

Brachialis

Brachioradialis

Brachialis

Biceps brachii

Flexor digitorum superficialis

Pronator teres

Flexor pollicis longus

Flexor digitorum profundus

Pronator quadratus

Brachioradialis

Pectoralis major
Subscapularis

Biceps brachii

Iliacus

Piriformis

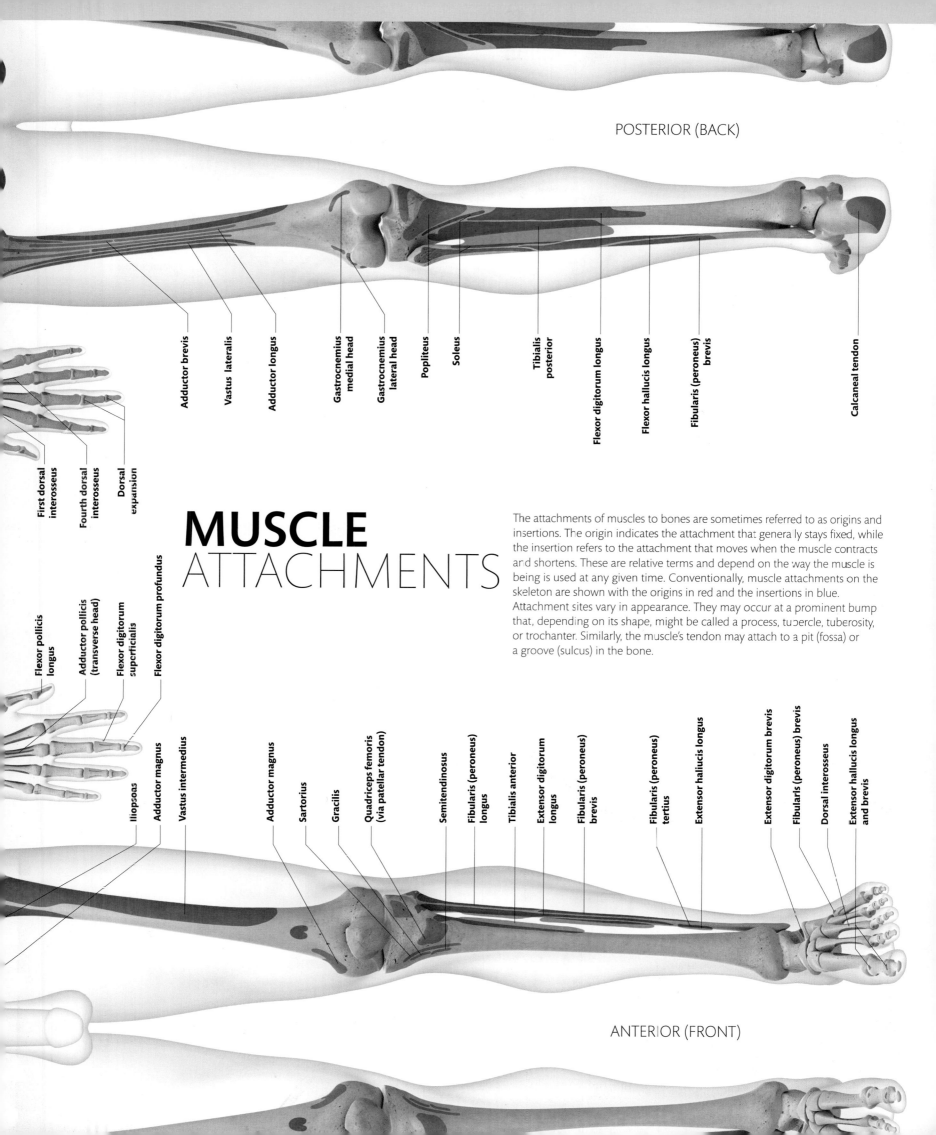

First dorsal interosseus

Fourth dorsal interosseus

Dorsal expansion

Adductor brevis

Vastus lateralis

Adductor longus

Gastrocnemius medial head

Gastrocnemius lateral head

Popliteus

Soleus

Tibialis posterior

Flexor digitorum longus

Flexor hallucis longus

Fibularis (peroneus) brevis

Calcaneal tendon

MUSCLE
ATTACHMENTS

The attachments of muscles to bones are sometimes referred to as origins and insertions. The origin indicates the attachment that generally stays fixed, while the insertion refers to the attachment that moves when the muscle contracts and shortens. These are relative terms and depend on the way the muscle is being is used at any given time. Conventionally, muscle attachments on the skeleton are shown with the origins in red and the insertions in blue. Attachment sites vary in appearance. They may occur at a prominent bump that, depending on its shape, might be called a process, tubercle, tuberosity, or trochanter. Similarly, the muscle's tendon may attach to a pit (fossa) or a groove (sulcus) in the bone.

Flexor pollicis longus

Adductor pollicis (transverse head)

Flexor digitorum superficialis

Flexor digitorum profundus

Iliopsoas

Adductor magnus

Vastus intermedius

Adductor magnus

Sartorius

Gracilis

Quadriceps femoris (via patellar tendon)

Semitendinosus

Fibularis (peroneus) longus

Tibialis anterior

Extensor digitorum longus

Fibularis (peroneus) brevis

Fibularis (peroneus) tertius

Extensor hallucis longus

Extensor digitorum brevis

Fibularis (peroneus) brevis

Dorsal interosseus

Extensor hallucis longus and brevis

ANTERIOR (FRONT)

Perimysium

Fascicle
A bundle of muscle fibers, packed in connective tissue called endomysium and contained in a sheath of perimysium

Epimysium

Parallel bundles
Skeletal muscle includes familiar muscles such as biceps or quadriceps. It is composed of parallel bundles of muscle fibers, which are conglomerations of many cells. Skeletal muscles are supplied by somatic motor nerves, which are part of the peripheral nervous system (see p.310) and are generally under conscious control.

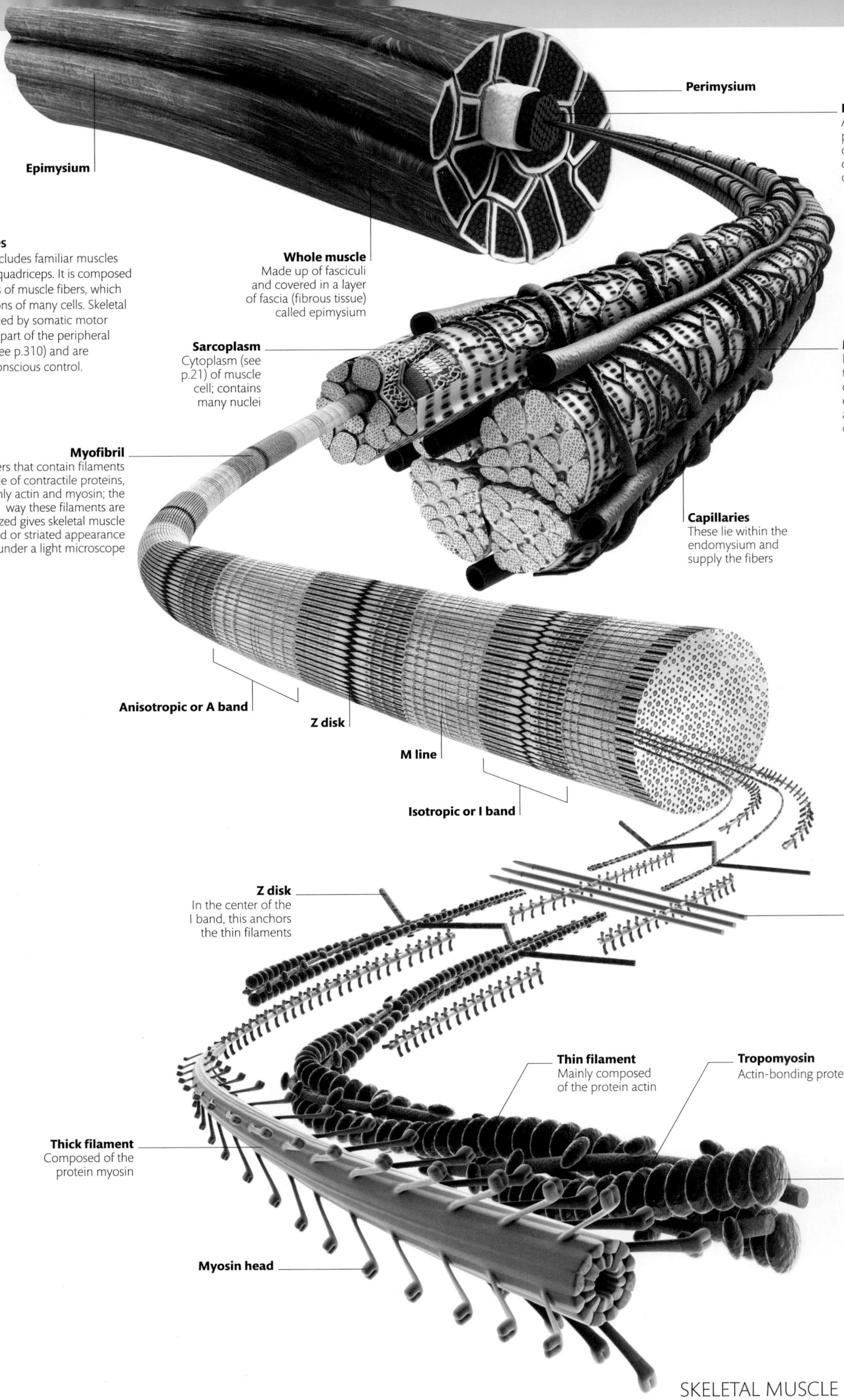

Whole muscle
Made up of fasciculi and covered in a layer of fascia (fibrous tissue) called epimysium

Sarcoplasm
Cytoplasm (see p.21) of muscle cell; contains many nuclei

Muscle fiber
Formed by many cells merged together, and therefore containing many nuclei, these cylindrical units range from a few millimeters to several centimeters in length

Myofibril
Fibers that contain filaments made of contractile proteins, mainly actin and myosin; the way these filaments are organized gives skeletal muscle a striped or striated appearance under a light microscope

Capillaries
These lie within the endomysium and supply the fibers

Anisotropic or A band

Z disk

M line

Isotropic or I band

Z disk
In the center of the I band, this anchors the thin filaments

M line
In the center of the A band, this connects the thick filaments

Thin filament
Mainly composed of the protein actin

Tropomyosin
Actin-bonding protein

Thick filament
Composed of the protein myosin

Actin

Myosin head

SKELETAL MUSCLE

MUSCLE
STRUCTURE

Muscle cells possess a special ability to contract. Also called myocytes, muscle cells are packed full of the long, filamentous proteins actin and myosin, which ratchet past each other to change the length of the cell itself (see p.304). There are three main types of muscle in the human body: skeletal or voluntary muscle, cardiac muscle, and smooth or involuntary muscle. Each of these has a distinctive microscopic structure. Skeletal muscle also varies in its overall shape and structure, depending on its function.

SMOOTH MUSCLE

Smooth muscle cell
These spindle-shaped cells contain actin and myosin; unlike in skeletal and cardiac muscle, the proteins are not lined up, so smooth muscle does not appear striated

CARDIAC MUSCLE

Intercalated disk
These elaborate junctions firmly bind cardiac muscle cells together

Cell nucleus

Cardiac muscle cell

Mitochondrion
Muscle cells are packed with energy-producing mitochondria

Myofibril
The myofibrils of cardiac muscle are organized in a similar way to those in skeletal muscle, giving a striated appearance under a light microscope

Heart muscle
Also called myocardium, cardiac muscle is only found in the heart. It exists as a network of interconnected fibers, and it spontaneously, rhythmically contracts. Autonomic nerves can increase or reduce the rate of contraction, matching the heart's output to the body's needs.

Mitochondrion

Intermediate filament

Dense body

Cell nucleus
Lies in the centre of the cell

Actin filament

Myosin filament

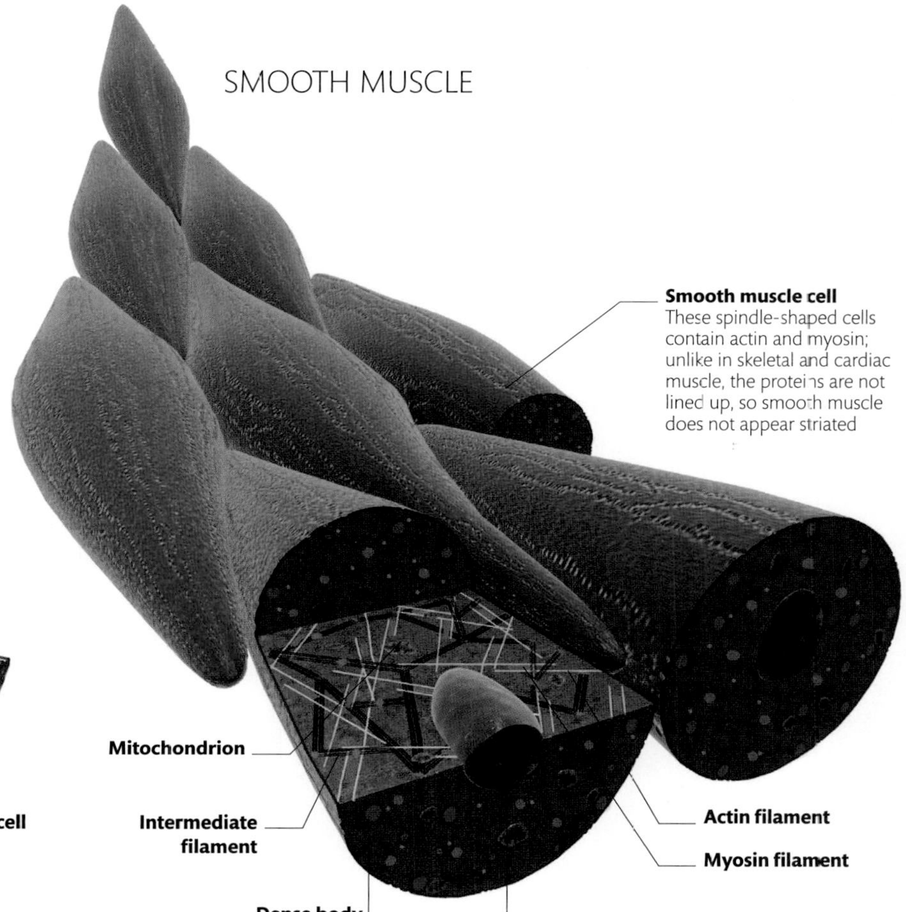

Tapering cells
This type of muscle is made of individual, tapering cells and is supplied by autonomic motor nerves, which control the operation of body systems, usually at a subconscious level. It is found in the organs of the body, particularly in the walls of tubes such as the gut, blood vessels, and the respiratory tract.

MUSCLE SHAPES

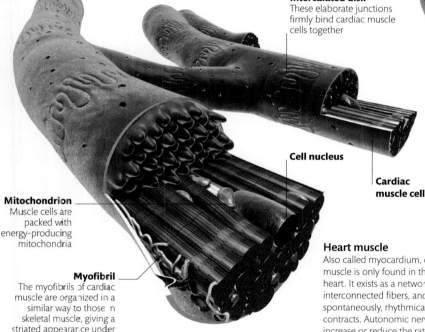
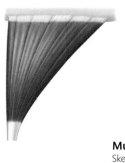

UNIPENNATE

BIPENNATE

MULTIPENNATE

STRAP

TRIANGULAR

QUADRATE

CIRCULAR OR SPHINCTERIC

Muscular variation
Skeletal muscles vary hugely in size and shape. In some, such as strap or quadrate muscles, the muscle fibers are parallel with the direction of pull. In others, the fibers are obliquely oriented— as in triangular or pennate (featherlike) muscles.

FUSIFORM

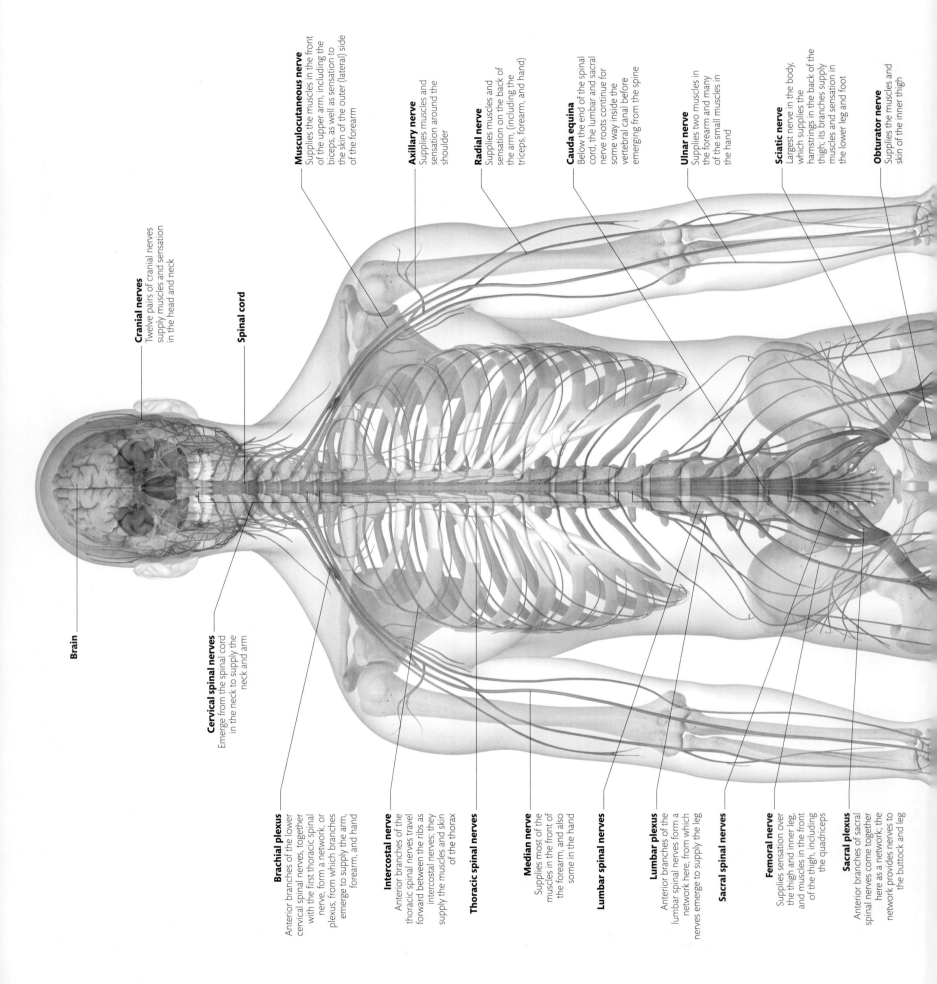

Musculocutaneous nerve
Supplies the muscles in the front of the upper arm, including the biceps, as well as sensation to the skin of the outer (lateral) side of the forearm

Axillary nerve
Supplies muscles and sensation around the shoulder

Radial nerve
Supplies muscles and sensation on the back of the arm, (including the triceps, forearm, and hand)

Cauda equina
Below the end of the spinal cord, the lumbar and sacral nerve roots continue for some way inside the vertebral canal before emerging from the spine

Ulnar nerve
Supplies two muscles in the forearm and many of the small muscles in the hand

Sciatic nerve
Largest nerve in the body, which supplies the hamstrings in the back of the thigh; its branches supply muscles and sensation in the lower leg and foot

Obturator nerve
Supplies the muscles and skin of the inner thigh

Cranial nerves
Twelve pairs of cranial nerves supply muscles and sensation in the head and neck

Spinal cord

Brain

Cervical spinal nerves
Emerge from the spinal cord in the neck to supply the neck and arm

Brachial plexus
Anterior branches of the lower cervical spinal nerves, together with the first thoracic spinal nerve, form a network, or plexus, from which branches emerge to supply the arm, forearm, and hand

Intercostal nerve
Anterior branches of the thoracic spinal nerves travel forward between the ribs as intercostal nerves; they supply the muscles and skin of the thorax

Thoracic spinal nerves

Median nerve
Supplies most of the muscles in the front of the forearm, and also some in the hand

Lumbar spinal nerves

Lumbar plexus
Anterior branches of the lumbar spinal nerves form a network here, from which nerves emerge to supply the leg

Sacral spinal nerves

Femoral nerve
Supplies sensation over the thigh and inner leg, and muscles in the front of the thigh, including the quadriceps

Sacral plexus
Anterior branches of sacral spinal nerves come together here as a network; the network provides nerves to the buttock and leg

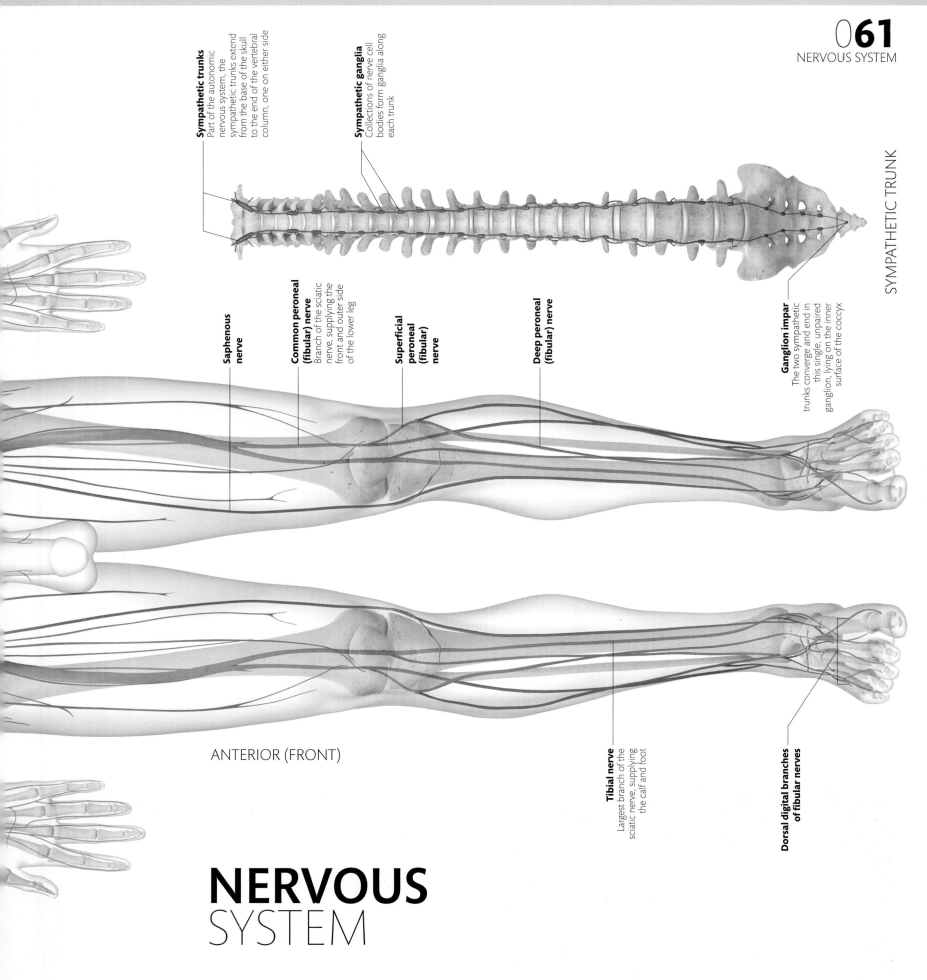

Sympathetic trunks Part of the autonomic nervous system, the sympathetic trunks extend from the base of the skull to the end of the vertebral column, one on either side

Sympathetic ganglia Collections of nerve cell bodies form ganglia along each trunk

SYMPATHETIC TRUNK

Saphenous nerve

Common peroneal (fibular) nerve Branch of the sciatic nerve, supplying the front and outer side of the lower leg

Superficial peroneal (fibular) nerve

Deep peroneal (fibular) nerve

Ganglion impar The two sympathetic trunks converge and end in this single, unpaired ganglion, lying on the inner surface of the coccyx

ANTERIOR (FRONT)

Tibial nerve Largest branch of the sciatic nerve, supplying the calf and foot

Dorsal digital branches of fibular nerves

NERVOUS
SYSTEM

The nervous system contains billions of intercommunicating nerve cells, or neurons. It can be broadly divided into the central nervous system (brain and spinal cord) and the peripheral nervous system (cranial and spinal nerves and their branches). The brain and spinal cord are protected by the skull and vertebral column respectively. Cranial nerves exit through holes in the skull to supply the head and neck; spinal nerves leave via gaps between vertebrae to supply the rest of the body. You can also divide the nervous system by function. The part that deals more with the way we sense and interact with our surroundings is called the somatic nervous system. The part involved with sensing and controlling our internal environments—affecting glands or heart rate, for example—is the autonomic nervous system.

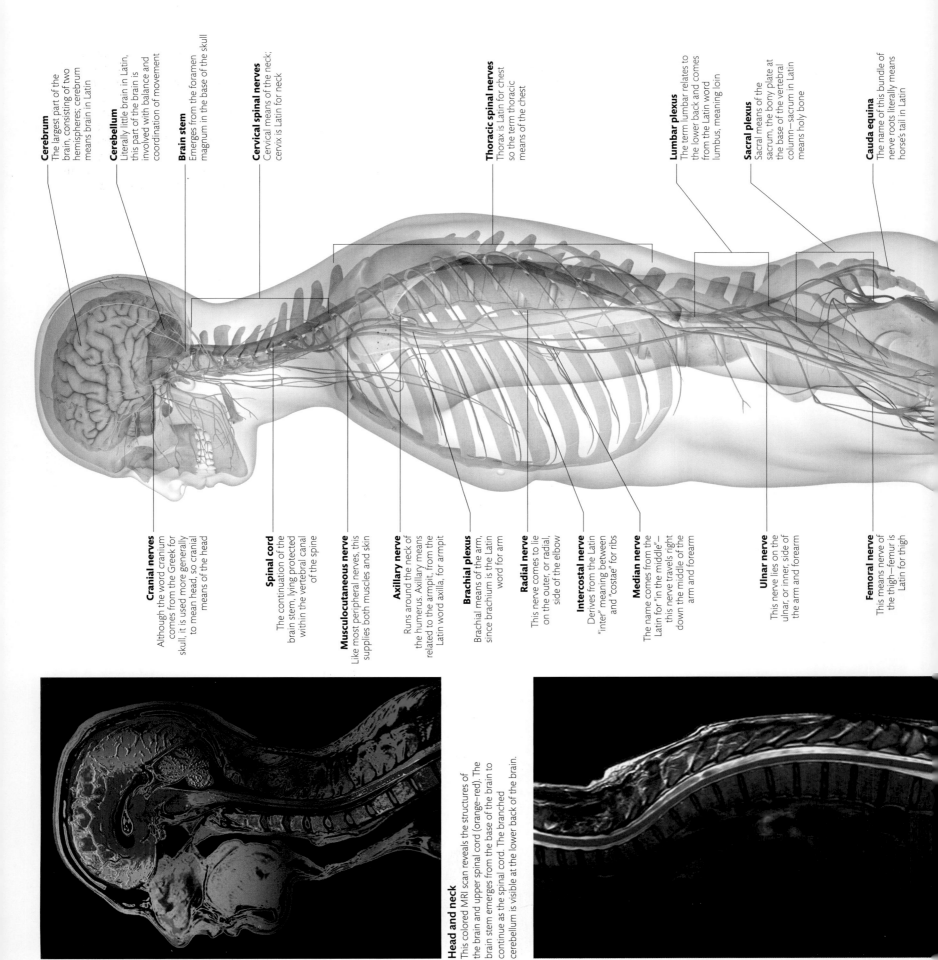

Cerebrum
The largest part of the brain, consisting of two hemispheres; cerebrum means brain in Latin

Cerebellum
Literally little brain in Latin, this part of the brain is involved with balance and coordination of movement

Brain stem
Emerges from the foramen magnum in the base of the skull

Cervical spinal nerves
Cervical means of the neck; cervix is Latin for neck

Thoracic spinal nerves
Thorax is Latin for chest so the term thoracic means of the chest

Lumbar plexus
The term lumbar relates to the lower back and comes from the Latin word lumbus, meaning loin

Sacral plexus
Sacral means of the sacrum, the bony plate at the base of the vertebral column—sacrum in Latin means holy bone

Cauda equina
The name of this bundle of nerve roots literally means horse's tail in Latin

Cranial nerves
Although the word cranium comes from the Greek for skull, it is used more generally to mean head, so cranial means of the head

Spinal cord
The continuation of the brain stem, lying protected within the vertebral canal of the spine

Musculocutaneous nerve
Like most peripheral nerves, this supplies both muscles and skin

Axillary nerve
Runs around the neck of the humerus. Axillary means related to the armpit, from the Latin word axilla, for armpit

Brachial plexus
Brachial means of the arm, since brachium is the Latin word for arm

Radial nerve
This nerve comes to lie on the outer, or radial, side of the elbow

Intercostal nerve
Derives from the Latin "inter" meaning between and "costae" for ribs

Median nerve
The name comes from the Latin for "in the middle"— this nerve travels right down the middle of the arm and forearm

Ulnar nerve
This nerve lies on the ulnar, or inner, side of the arm and forearm

Femoral nerve
This means nerve of the thigh—femur is Latin for thigh

Head and neck
This colored MRI scan reveals the structures of the brain and upper spinal cord (orange–red). The brain stem emerges from the base of the brain to continue as the spinal cord. The branched cerebellum is visible at the lower back of the brain.

NERVOUS SYSTEM

Twelve cranial nerves emerge from the brain and brain stem to supply structures in the head and neck, including the eyes, ears, nose, and mouth. Thirty one pairs of spinal nerves sprout from the spinal cord, with eight cervical, twelve thoracic, five lumbar, five sacral, and one coccygeal on each side. These nerves branch to supply tissues behind and in front of the vertebral column. In the cervical, lumbar, and sacral regions, nerves join together to form networks, or "plexuses," before branching again to supply the limbs. Most peripheral nerves contain both nerve fibers that carry messages out to muscles, and fibers that convey sensory information back to the central nervous system.

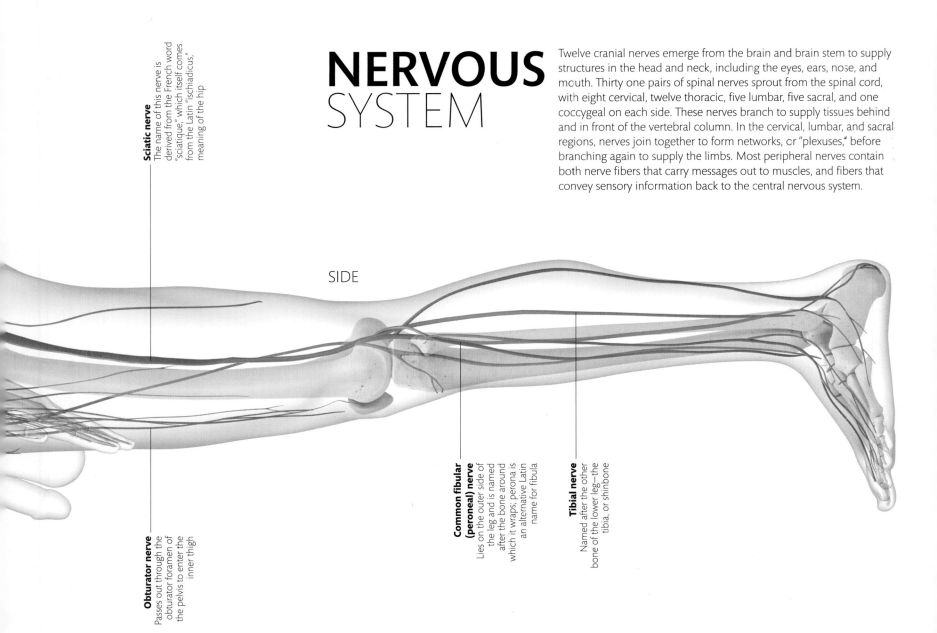

SIDE

Sciatic nerve
The name of this nerve is derived from the French word "sciatique," which itself comes from the Latin "ischiadicus," meaning of the hip

Obturator nerve
Passes out through the obturator foramen of the pelvis to enter the inner thigh

Common fibular (peroneal) nerve
Lies on the outer side of the leg and is named after the bone around which it wraps; perona is an alternative Latin name for fibula

Tibial nerve
Named after the other bone of the lower leg—the tibia, or shinbone

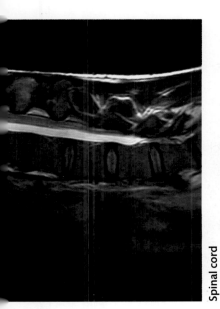

Spinal cord
The protective vertebrae surrounding the spinal cord appear as blue blocks in this MRI of the spine. The spinal cord is shown as a dark blue column lying within the pale blue sheath of the dura mater. Toward the lower right is the cauda equina.

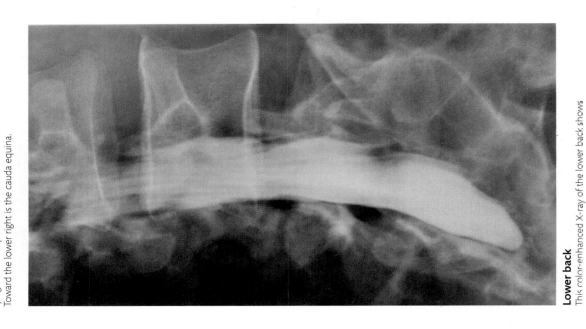

Lower back
This color-enhanced X-ray of the lower back shows the dural sac (white), which sheaths the spinal cord and its emerging nerves. The column of vertebrae (orange) ends in the sacrum, which connects the vertebrae to the pelvis.

NEURON

NERVE STRUCTURE

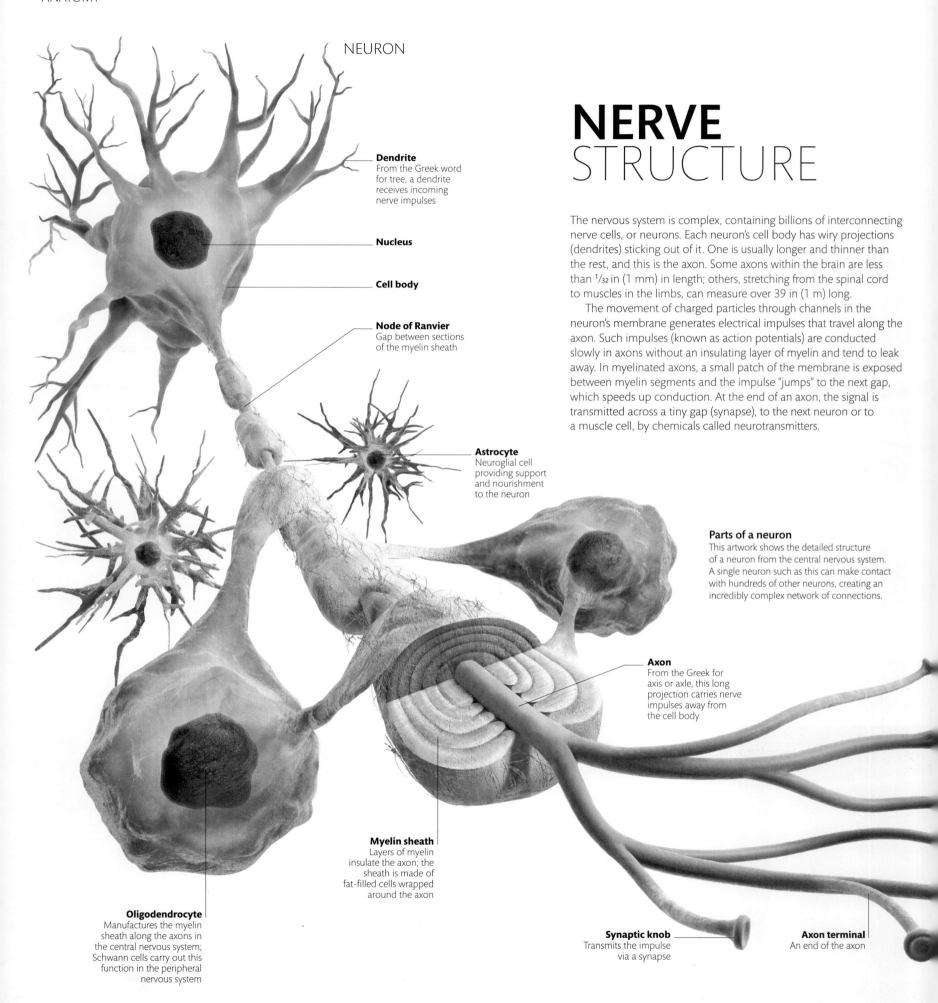

Dendrite
From the Greek word
for tree, a dendrite
receives incoming
nerve impulses

Nucleus

Cell body

Node of Ranvier
Gap between sections
of the myelin sheath

Astrocyte
Neuroglial cell
providing support
and nourishment
to the neuron

Oligodendrocyte
Manufactures the myelin
sheath along the axons in
the central nervous system;
Schwann cells carry out this
function in the peripheral
nervous system

Myelin sheath
Layers of myelin
insulate the axon; the
sheath is made of
fat-filled cells wrapped
around the axon

Synaptic knob
Transmits the impulse
via a synapse

Axon terminal
An end of the axon

The nervous system is complex, containing billions of interconnecting nerve cells, or neurons. Each neuron's cell body has wiry projections (dendrites) sticking out of it. One is usually longer and thinner than the rest, and this is the axon. Some axons within the brain are less than $1/32$ in (1 mm) in length; others, stretching from the spinal cord to muscles in the limbs, can measure over 39 in (1 m) long.

The movement of charged particles through channels in the neuron's membrane generates electrical impulses that travel along the axon. Such impulses (known as action potentials) are conducted slowly in axons without an insulating layer of myelin and tend to leak away. In myelinated axons, a small patch of the membrane is exposed between myelin segments and the impulse "jumps" to the next gap, which speeds up conduction. At the end of an axon, the signal is transmitted across a tiny gap (synapse), to the next neuron or to a muscle cell, by chemicals called neurotransmitters.

Parts of a neuron
This artwork shows the detailed structure of a neuron from the central nervous system. A single neuron such as this can make contact with hundreds of other neurons, creating an incredibly complex network of connections.

Axon
From the Greek for
axis or axle, this long
projection carries nerve
impulses away from
the cell body

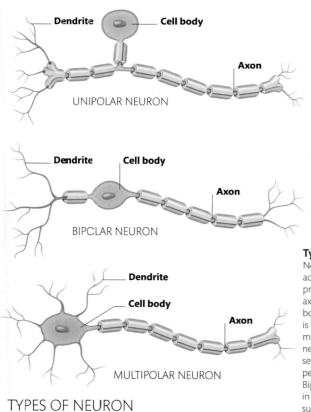

TYPES OF NEURON

Dendrite — Cell body

Axon

UNIPOLAR NEURON

Dendrite — Cell body

Axon

BIPOLAR NEURON

Dendrite

Cell body

Axon

MULTIPOLAR NEURON

Types of neuron
Neurons can be classified according to how many projections (dendrites and axons) extend from the cell body. The most common is multipolar, with three or more projections. Unipolar neurons lie mainly in the sensory nerves of the peripheral nervous system. Bipolar neurons are found in only a few locations, such as the eye's retina.

PERIPHERAL NERVE

Axon

Myelin sheath

Nerve fiber

Endoneurium
Layer of delicate connective tissue around the myelin sheath

Nerve fascicle
Bundle or group of nerve fibers

Perineurium
Sheathlike wrapping for a fascicle

Blood vessels

Epineurium
Strong protective outer covering for the whole nerve

Nerve structure
Peripheral nerves comprise bundles of bundles of nerve fibers. Axons are wrapped in a layer of packing tissue called endoneurium. Small bundles of these nerve fibers are packaged in perineurium to form fascicles, and several fascicles are bundled within epineurium to form the nerve.

Structure of the spinal cord
Like the brain, the spinal cord contains grey matter (mostly neuron cell bodies) and white matter (axons), and is covered in the same three layers of meninges: dura mater, arachnoid, and pia mater (see p.115).

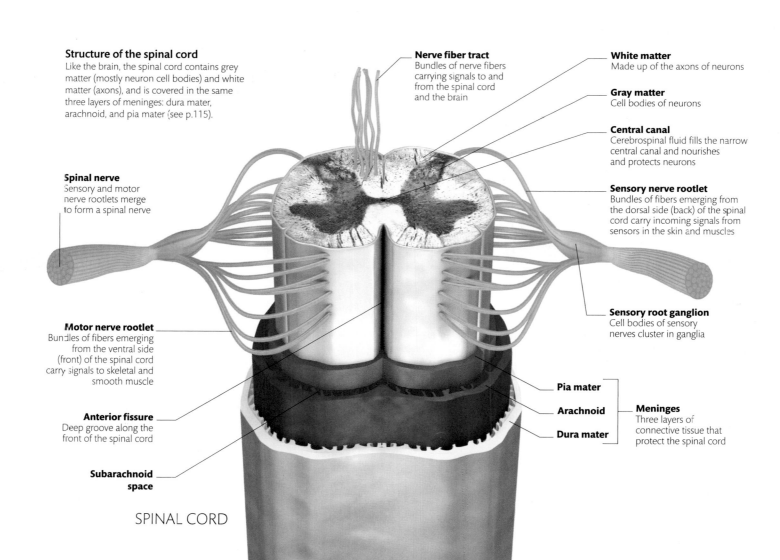

Nerve fiber tract
Bundles of nerve fibers carrying signals to and from the spinal cord and the brain

White matter
Made up of the axons of neurons

Gray matter
Cell bodies of neurons

Central canal
Cerebrospinal fluid fills the narrow central canal and nourishes and protects neurons

Sensory nerve rootlet
Bundles of fibers emerging from the dorsal side (back) of the spinal cord carry incoming signals from sensors in the skin and muscles

Sensory root ganglion
Cell bodies of sensory nerves cluster in ganglia

Spinal nerve
Sensory and motor nerve rootlets merge to form a spinal nerve

Motor nerve rootlet
Bundles of fibers emerging from the ventral side (front) of the spinal cord carry signals to skeletal and smooth muscle

Anterior fissure
Deep groove along the front of the spinal cord

Subarachnoid space

Pia mater

Arachnoid

Dura mater

Meninges
Three layers of connective tissue that protect the spinal cord

SPINAL CORD

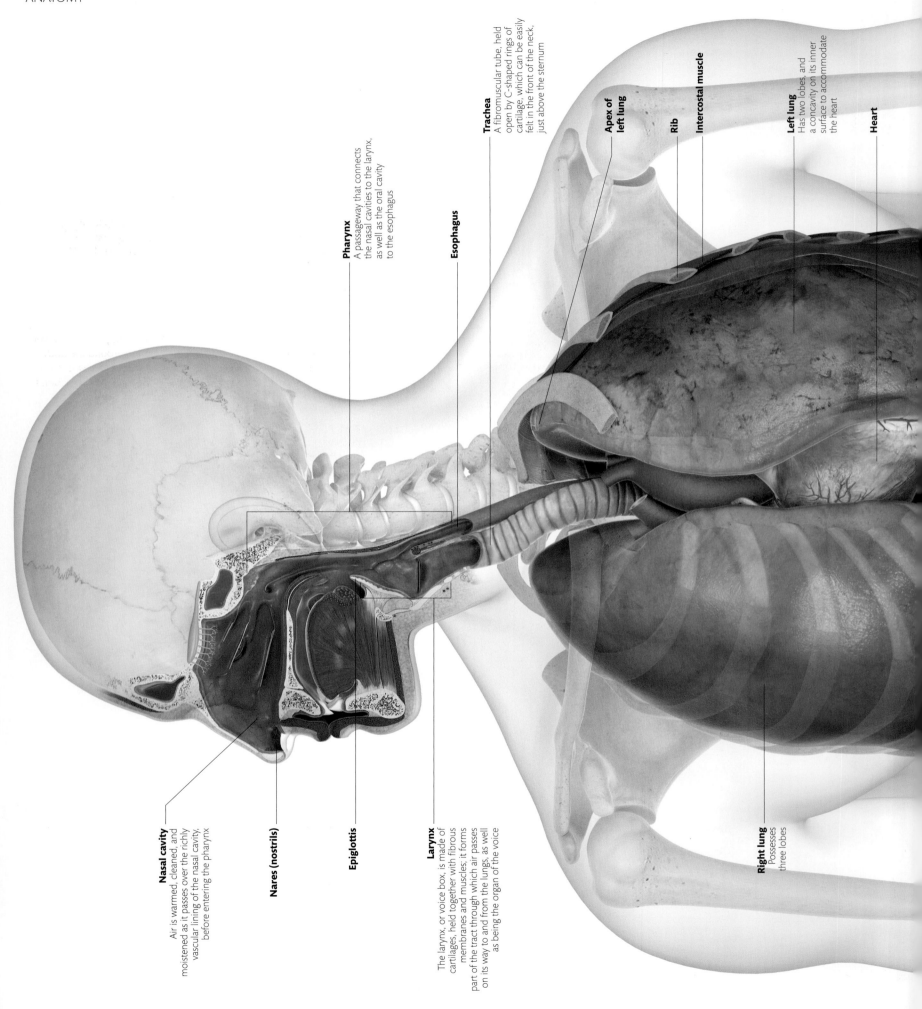

Pharynx
A passageway that connects the nasal cavities to the larynx, as well as the oral cavity to the esophagus

Esophagus

Trachea
A fibromuscular tube, held open by C-shaped rings of cartilage, which can be easily felt in the front of the neck, just above the sternum

Apex of left lung

Rib

Intercostal muscle

Left lung
Has two lobes, and a concavity on its inner surface to accommodate the heart

Heart

Nasal cavity
Air is warmed, cleaned, and moistened as it passes over the richly vascular lining of the nasal cavity, before entering the pharynx

Nares (nostrils)

Epiglottis

Larynx
The larynx, or voice box, is made of cartilages, held together with fibrous membranes and muscles; it forms part of the tract through which air passes on its way to and from the lungs, as well as being the organ of the voice

Right lung
Possesses three lobes

Visceral pleura
This membrane covers the surface of the lungs themselves

Pleural cavity
Potential space between the parietal and visceral layers of the pleura, containing a thin film of pleural fluid that lubricates the lungs as they move within the chest

Parietal pleura
Membrane that lines the inner surface of the chest wall

Diaphragm
Main muscle of breathing, supplied by the phrenic nerve; the diaphragm flattens as it contracts, increasing the volume of the thorax, producing a drop in pressure inside the lungs which draws breath into them

RESPIRATORY
SYSTEM

Every cell in the human body needs to get hold of oxygen, and to get rid of carbon dioxide. These gases are transported around the body in the blood, but the actual transfer of gases between the air and the blood occurs in the lungs. The lungs have extremely thin membranes that allow the gases to pass across easily. But air also needs to be regularly drawn in and out of the lungs, to expel the building carbon dioxide and to bring in fresh oxygen, and this is brought about by respiration—commonly called breathing. The respiratory system includes the airways on the way to the lungs: the nasal cavities, parts of the pharynx, the larynx, the trachea, and the bronchi (see p.153).

ANTERIOR (FRONT)

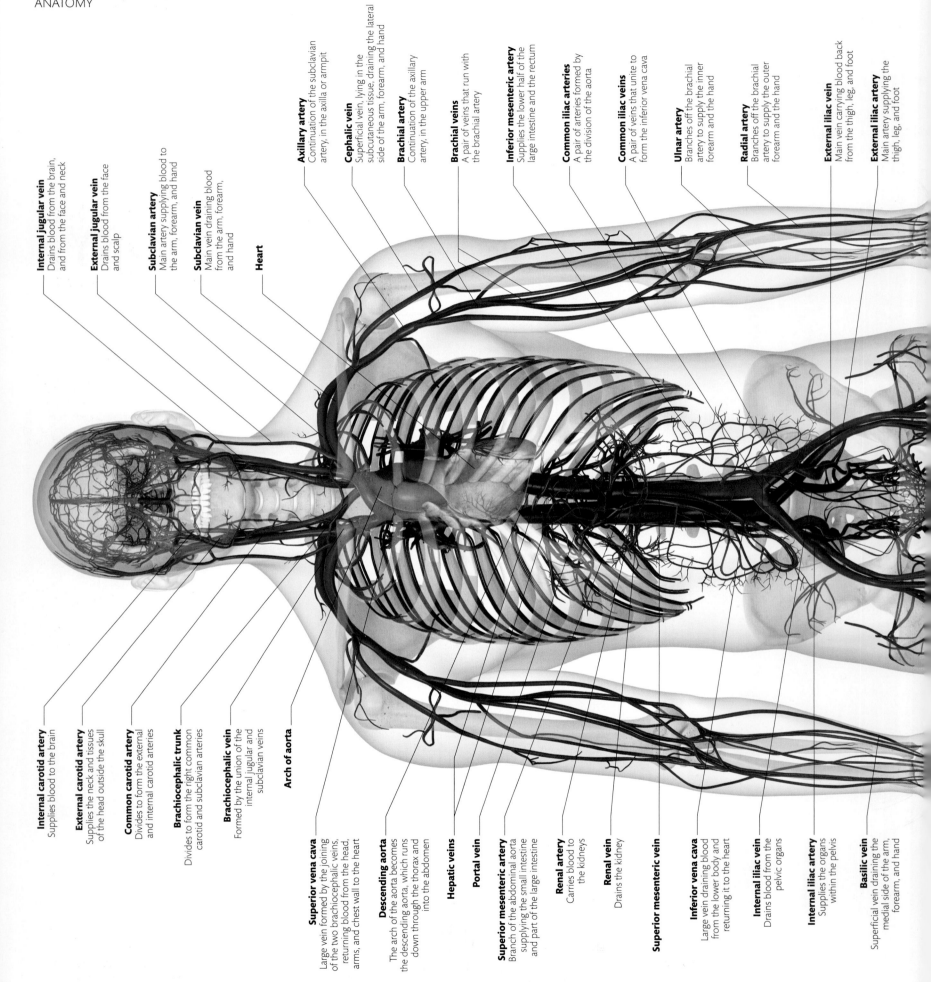

Internal jugular vein
Drains blood from the brain,
and from the face and neck

External jugular vein
Drains blood from the face
and scalp

Subclavian artery
Main artery supplying blood to
the arm, forearm, and hand

Subclavian vein
Main vein draining blood
from the arm, forearm,
and hand

Heart

Axillary artery
Continuation of the subclavian
artery, in the axilla or armpit

Cephalic vein
Superficial vein, lying in the
subcutaneous tissue, draining the lateral
side of the arm, forearm, and hand

Brachial artery
Continuation of the axillary
artery, in the upper arm

Brachial veins
A pair of veins that run with
the brachial artery

Inferior mesenteric artery
Supplies the lower half of the
large intestine and the rectum

Common iliac arteries
A pair of arteries formed by
the division of the aorta

Common iliac veins
A pair of veins that unite to
form the inferior vena cava

Ulnar artery
Branches off the brachial
artery to supply the inner
forearm and the hand

Radial artery
Branches off the brachial
artery to supply the outer
forearm and the hand

External iliac vein
Main vein carrying blood back
from the thigh, leg, and foot

External iliac artery
Main artery supplying the
thigh, leg, and foot

Internal carotid artery
Supplies blood to the brain

External carotid artery
Supplies the neck and tissues
of the head outside the skull

Common carotid artery
Divides to form the external
and internal carotid arteries

Brachiocephalic trunk
Divides to form the right common
carotid and subclavian arteries

Brachiocephalic vein
Formed by the union of the
internal jugular and
subclavian veins

Arch of aorta

Superior vena cava
Large vein formed by the joining
of the two brachiocephalic veins,
returning blood from the head,
arms, and chest wall to the heart

Descending aorta
The arch of the aorta becomes
the descending aorta, which runs
down through the thorax and
into the abdomen

Hepatic veins

Portal vein

Superior mesenteric artery
Branch of the abdominal aorta
supplying the small intestine
and part of the large intestine

Renal artery
Carries blood to
the kidneys

Renal vein
Drains the kidney

Superior mesenteric vein

Inferior vena cava
Large vein draining blood
from the lower body and
returning it to the heart

Internal iliac vein
Drains blood from the
pelvic organs

Internal iliac artery
Supplies the organs
within the pelvis

Basilic vein
Superficial vein draining the
medial side of the arm,
forearm, and hand

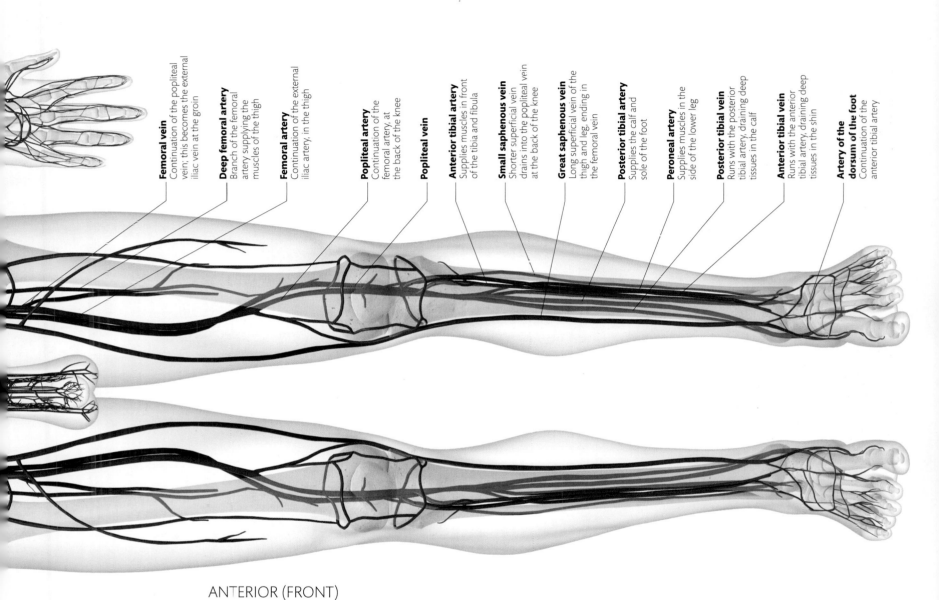

Femoral vein
Continuation of the popliteal vein; this becomes the external iliac vein at the groin

Deep femoral artery
Branch of the femoral artery supplying the muscles of the thigh

Femoral artery
Continuation of the external iliac artery, in the thigh

Popliteal artery
Continuation of the femoral artery, at the back of the knee

Popliteal vein

Anterior tibial artery
Supplies muscles in front of the tibia and fibula

Small saphenous vein
Shorter superficial vein drains into the popliteal vein at the back of the knee

Great saphenous vein
Long superficial vein of the thigh and leg, ending in the femoral vein

Posterior tibial artery
Supplies the calf and sole of the foot

Peroneal artery
Supplies muscles in the side of the lower leg

Posterior tibial vein
Runs with the posterior tibial artery, draining deep tissues in the calf

Anterior tibial vein
Runs with the anterior tibial artery, draining deep tissues in the shin

Artery of the dorsum of the foot
Continuation of the anterior tibial artery

ANTERIOR (FRONT)

CARDIOVASCULAR
SYSTEM

The heart and blood vessels deliver useful substances—oxygen from the lungs, nutrients from the gut, white blood cells to protect against infection, and hormones—to the tissues of the body. The blood also removes waste products and takes them to other organs—mainly the liver and kidneys—for excretion. The heart is a muscular pump that contracts to push blood through the body's network of vessels. Arteries are vessels that carry blood away from the heart; veins take blood back to it. Arteries branch into smaller and smaller vessels, eventually leading to capillaries. Tiny vessels taking blood away from capillary networks join up, like the tributaries of a river, to form veins.

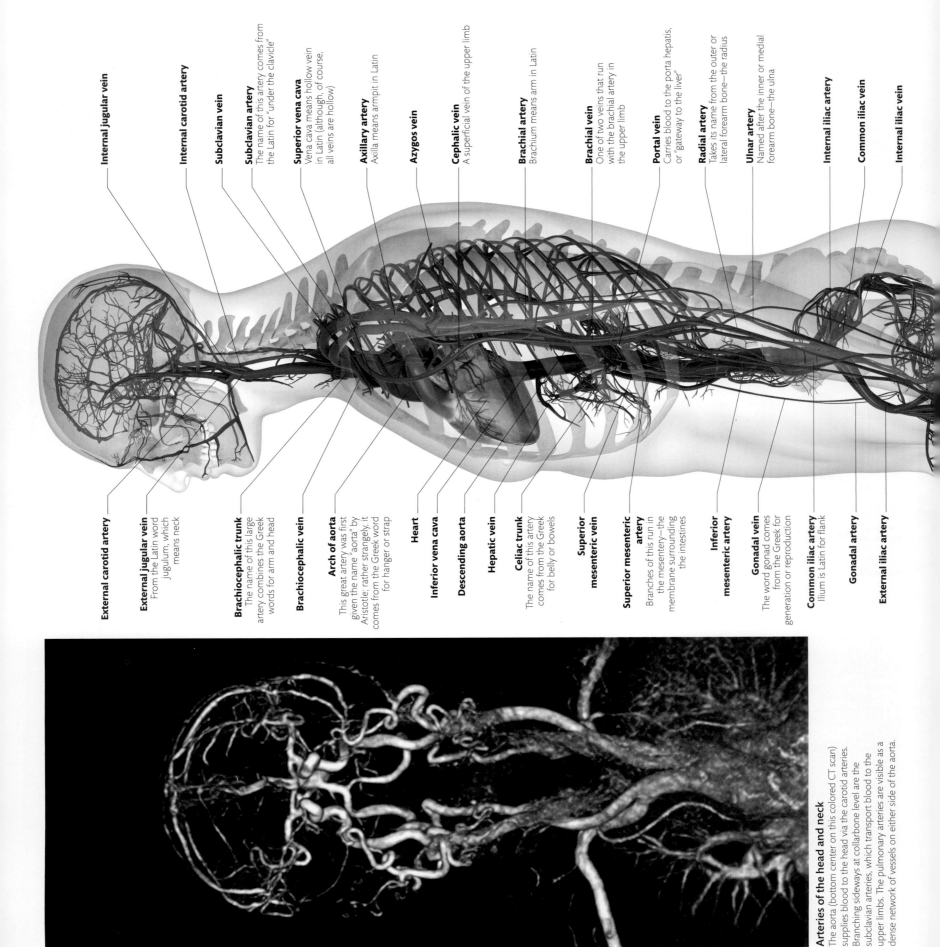

Internal jugular vein

Internal carotid artery

Subclavian vein

Subclavian artery
The name of this artery comes from the Latin for "under the clavicle"

Superior vena cava
Vena cava means hollow vein in Latin (although, of course, all veins are hollow)

Axillary artery
Axilla means armpit in Latin

Azygos vein

Cephalic vein
A superficial vein of the upper limb

Brachial artery
Brachium means arm in Latin

Brachial vein
One of two veins that run with the brachial artery in the upper limb

Portal vein
Carries blood to the porta hepatis, or "gateway to the liver"

Radial artery
Takes its name from the outer or lateral forearm bone—the radius

Ulnar artery
Named after the inner or medial forearm bone—the ulna

Internal iliac artery

Common iliac vein

Internal iliac vein

External carotid artery

External jugular vein
From the Latin word jugulum, which means neck

Brachiocephalic trunk
The name of this large artery combines the Greek words for arm and head

Brachiocephalic vein

Arch of aorta
This great artery was first given the name "aorta" by Aristotle; rather strangely, it comes from the Greek word for hanger or strap

Heart

Inferior vena cava

Descending aorta

Hepatic vein

Celiac trunk
The name of this artery comes from the Greek for belly or bowels

Superior mesenteric vein

Superior mesenteric artery
Branches of this run in the mesentery—the membrane surrounding the intestines

Inferior mesenteric artery

Gonadal vein
The word gonad comes from the Greek for generation or reproduction

Common iliac artery
Ilium is Latin for flank

Gonadal artery

External iliac artery

Arteries of the head and neck
The aorta (bottom center on this colored CT scan) supplies blood to the head via the carotid arteries. Branching sideways at collarbone level are the subclavian arteries, which transport blood to the upper limbs. The pulmonary arteries are visible as a dense network of vessels on either side of the aorta.

CARDIOVASCULAR
SYSTEM

The circulation can be divided in two: the pulmonary circulation carries blood pumped by the right side of the heart to the lungs, and the systemic circulation carries blood pumped by the more powerful left side of the heart to the rest of the body. The pressure in the pulmonary circulation is relatively low, to prevent fluid being forced out of capillaries into the alveoli of the lungs. The pressure in the systemic circulation (which is what is measured with a blood-pressure cuff on the arm) is much higher, easily enough to push blood all the way up to your brain, into all your other organs, and out to your fingers and toes.

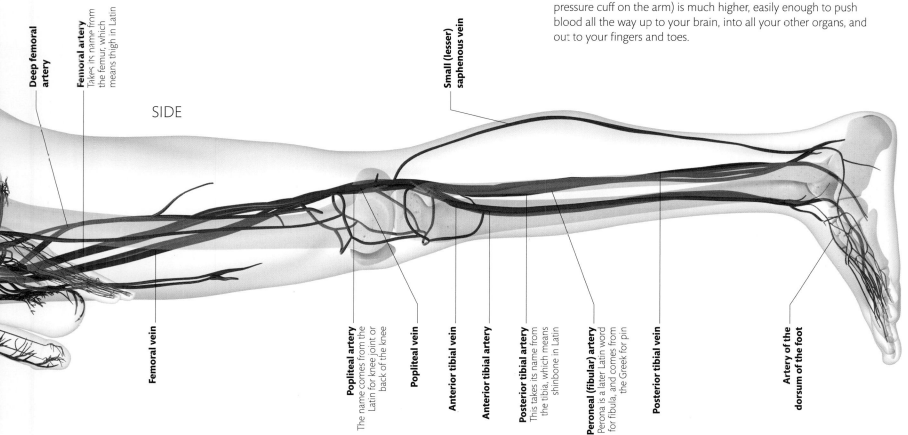

SIDE

Deep femoral artery

Femoral artery
Takes its name from the femur, which means thigh in Latin

Small (lesser) saphenous vein

Femoral vein

Popliteal artery
The name comes from the Latin for knee joint or back of the knee

Popliteal vein

Anterior tibial vein

Anterior tibial artery

Posterior tibial artery
This takes its name from the tibia, which means shinbone in Latin

Peroneal (fibular) artery
Perona is a later Latin word for fibula, and comes from the Greek for pin

Posterior tibial vein

Artery of the dorsum of the foot

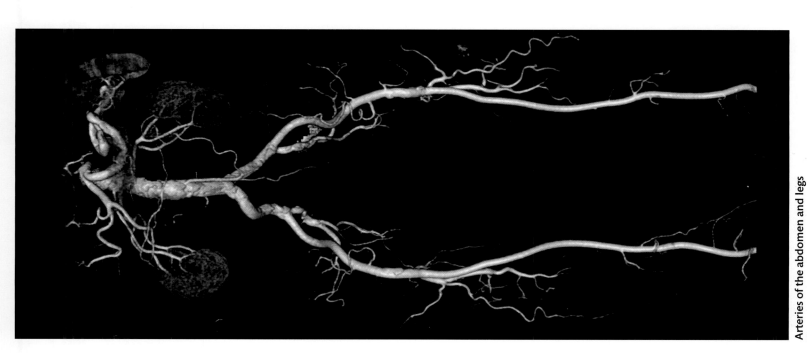

Arteries of the abdomen and legs
This color-enhanced CT angiogram shows the abdominal aorta and the arteries of the legs. Also visible are the kidneys and spleen. The large artery traveling through each thigh is the femoral artery; this becomes the popliteal artery behind the knee and branches into the tibial arteries in the lower leg.

Tunica adventitia
The outermost coat, composed of connective tissue and elastic fibers

Tunica media
Consists mainly of smooth muscle; this is the thickest layer in an artery

Internal elastic lamina
Prominent in large arteries, including the aorta and its main branches; the layer between the tunica media and tunica intima

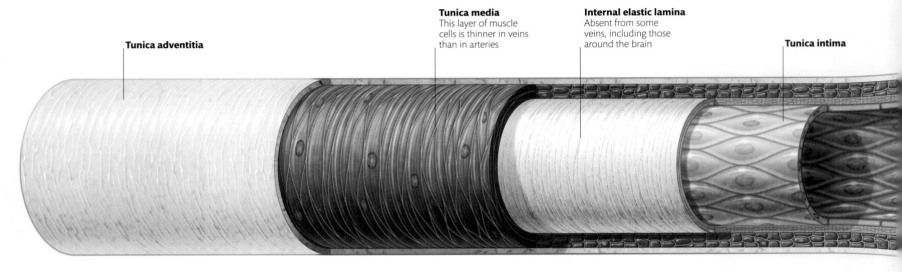

ARTERY

Tunica adventitia

Tunica media
This layer of muscle cells is thinner in veins than in arteries

Internal elastic lamina
Absent from some veins, including those around the brain

Tunica intima

VEIN

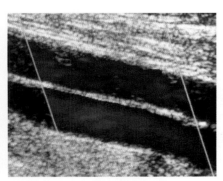

Color doppler
A doppler ultrasound probe can detect the difference between blood flowing to and from the detector. This scan shows the blood that flows in an artery in the leg as red, and the blood in the vein as blue.

Endothelium
A single layer of flattened cells that forms the thin wall of capillaries

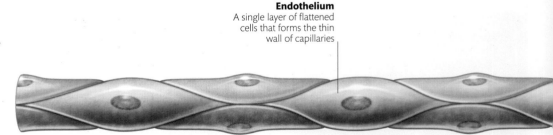

CAPILLARY

ARTERY, VEIN, CAPILLARY
STRUCTURE

The cardiovascular system consists of the heart, blood, and blood vessels—comprising arteries, arterioles, capillaries, venules, and veins.

The heart contracts to keep the blood continually moving through a vast network of blood vessels. Arteries carry blood away from the heart to organs and tissues, whereas veins carry blood back to the heart. Both arteries and veins have walls made up of three main layers: the innermost lining or tunica intima, the middle tunica media, and the outer wrapping, or tunica adventitia. While the tunica media is a thick layer in arteries, it is very thin in veins, and completely absent from capillaries, the walls of which comprise just a single layer of endothelial cells.

The cardiovascular system carries oxygen from the lungs, nutrients from the gut, hormones, and white blood cells for the body's defense system. It also picks up waste from all body tissues and carries it to the appropriate organs for excretion.

Tunica intima
The innermost lining of an artery; made up of a single layer of flattened cells, also known as the endothelium

Artery cross section
Arteries range from less than ⅟₂₅ in (1 mm) to up to 1¼ in (3 cm) in diameter

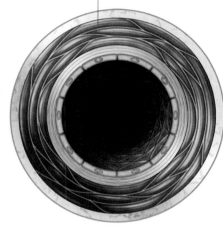

Artery

The largest arteries of the body contain a good proportion of elastic tissue within the internal elastic lamina and tunica media layers. The thick walls and elastic nature of arteries mean they can withstand the high pressure that occurs when the heart contracts and also keep blood flowing between heartbeats. There is less elastic tissue in smaller, muscular arteries, and even less in the smallest arteries, or arterioles.

Valve
Allows blood to flow only toward the heart

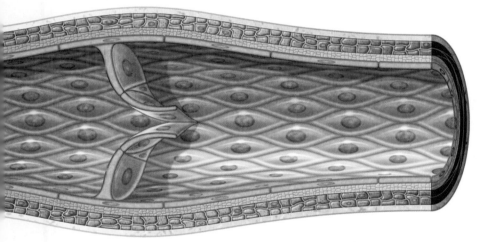

Vein cross section
The largest veins measure up to 1¼ in (3 cm) in diameter

Vein

Veins have much thinner walls than arteries and contain proportionately less muscle and more connective and elastic tissue. Capillaries converge to form tiny veins, or venules, which then join up to form larger veins. Most veins contain simple, pocketlike valves to keep blood flowing in the right direction.

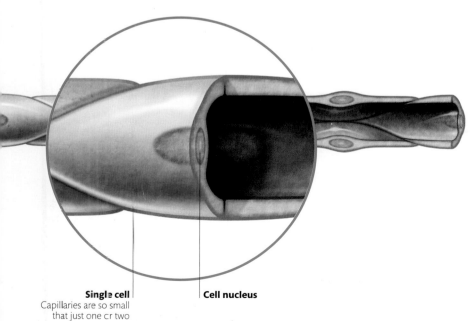

Capillary cross section
Capillaries measure just ⅟₂,₅₀₀ in (0.01 mm) in diameter—this capillary is not shown to scale with the other vessels

Capillary

The walls of a capillary are extremely thin, formed by just a single layer of flattened cells. This allows substances to transfer between the blood inside the capillary and the surrounding tissue. Some capillaries have pores, or fenestrations, to make the exchange of substances even easier.

Single cell
Capillaries are so small that just one or two cells wrap around their diameter

Cell nucleus

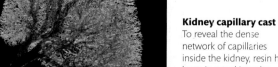

Kidney capillary cast

To reveal the dense network of capillaries inside the kidney, resin has been injected into the renal artery and allowed to set. The tissue of the organ has dissolved away.

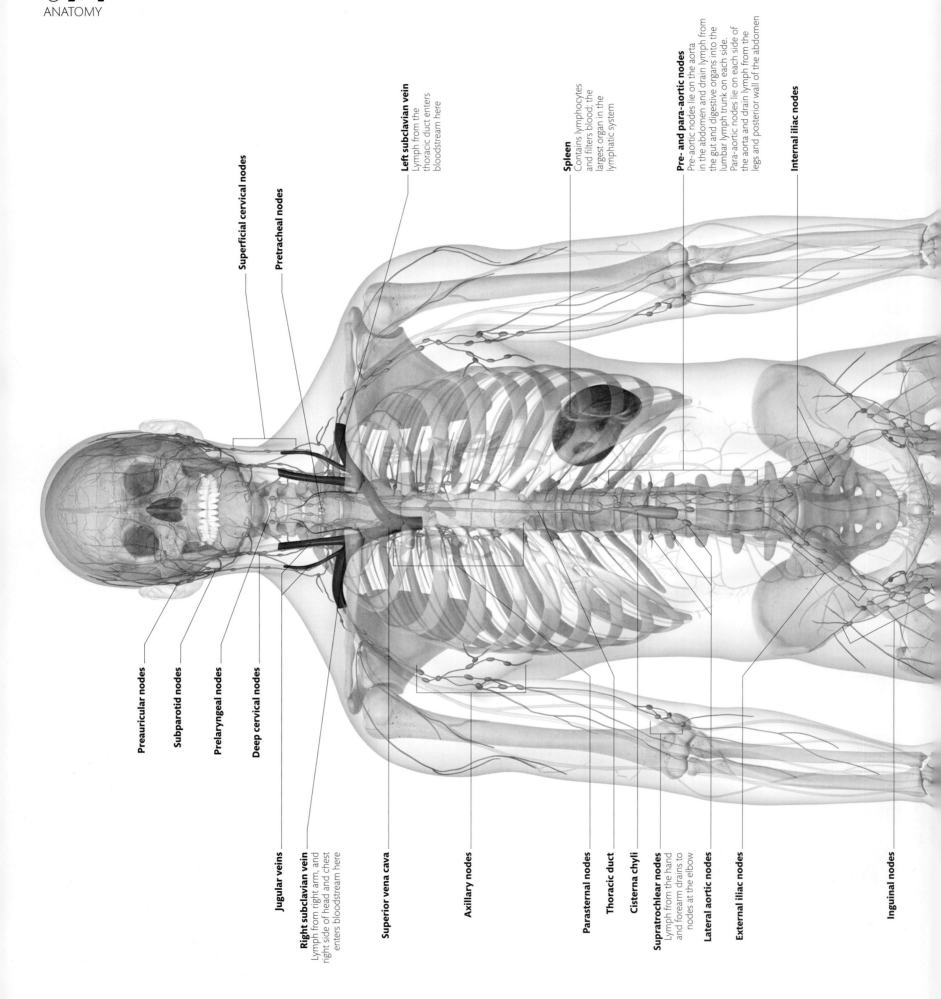

Superficial cervical nodes

Pretracheal nodes

Left subclavian vein
Lymph from the thoracic duct enters bloodstream here

Spleen
Contains lymphocytes and filters blood; the largest organ in the lymphatic system

Pre- and para-aortic nodes
Pre-aortic nodes lie on the aorta in the abdomen and drain lymph from the gut and digestive organs into the lumbar lymph trunk on each side. Para-aortic nodes lie on each side of the aorta and drain lymph from the legs and posterior wall of the abdomen

Internal iliac nodes

Preauricular nodes

Subparotid nodes

Prelaryngeal nodes

Deep cervical nodes

Jugular veins

Right subclavian vein
Lymph from right arm, and right side of head and chest enters bloodstream here

Superior vena cava

Axillary nodes

Parasternal nodes

Thoracic duct

Cisterna chyli

Supratrochlear nodes
Lymph from the hand and forearm drains to nodes at the elbow

Lateral aortic nodes

External iliac nodes

Inguinal nodes

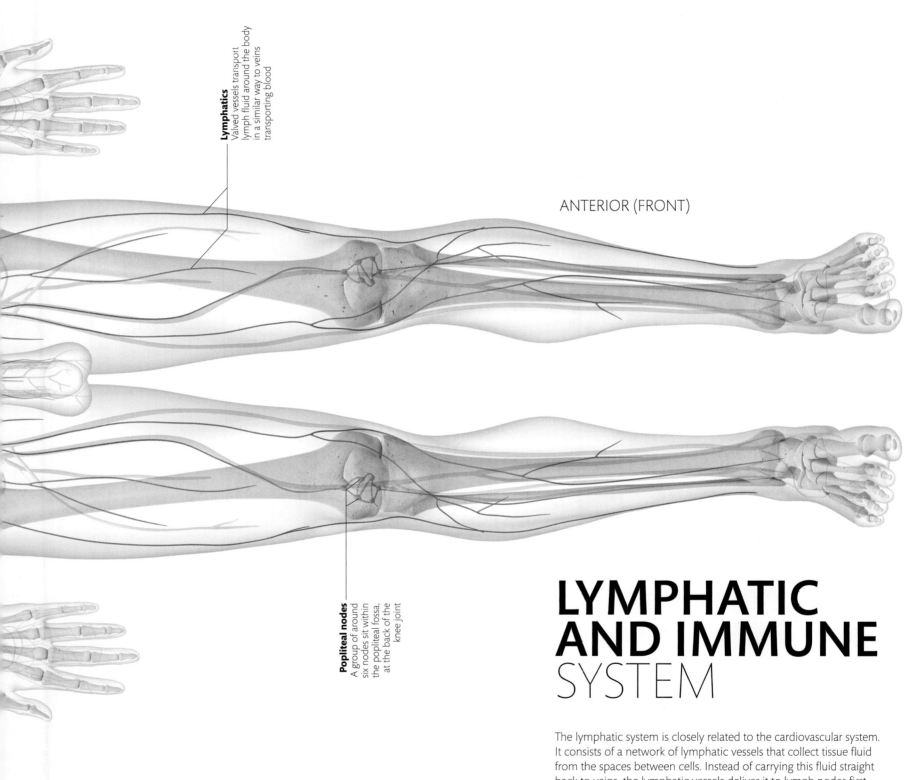

Lymphatics
Valved vessels transport lymph fluid around the body in a similar way to veins transporting blood

ANTERIOR (FRONT)

Popliteal nodes
A group of around six nodes sit within the popliteal fossa, at the back of the knee joint

LYMPHATIC AND IMMUNE SYSTEM

The lymphatic system is closely related to the cardiovascular system. It consists of a network of lymphatic vessels that collect tissue fluid from the spaces between cells. Instead of carrying this fluid straight back to veins, the lymphatic vessels deliver it to lymph nodes first. These nodes, like the tonsils, spleen, and thymus, are "lymphoid tissues," meaning that they all contain immune cells known as lymphocytes. The nodes are therefore part of the immune system. There are also patches of lymphoid tissue in the walls of the bronchi and the gut. The spleen, which lies tucked up under the ribs on the left side of the abdomen, has two important roles: it is a lymphoid organ, and it also removes old red blood cells from the circulation.

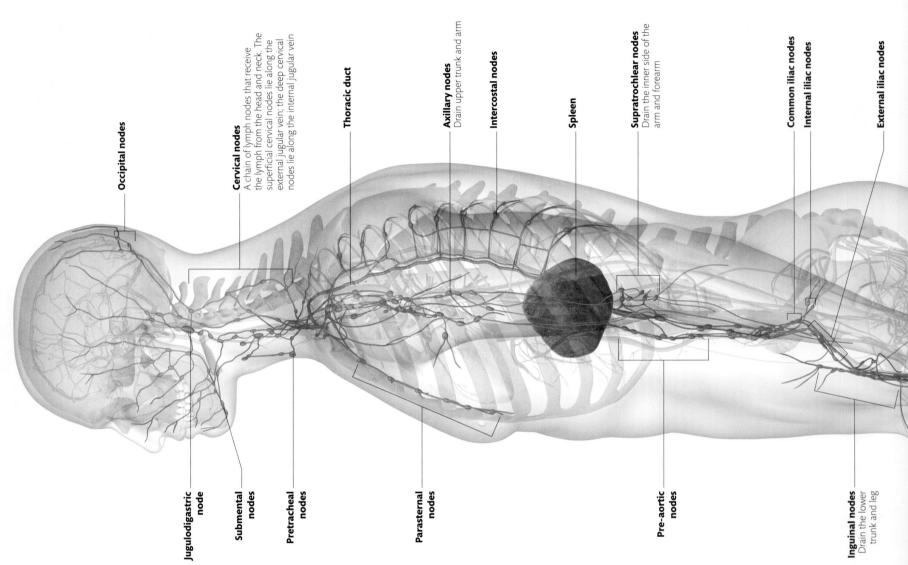

Occipital nodes

Cervical nodes
A chain of lymph nodes that receive the lymph from the head and neck. The superficial cervical nodes lie along the external jugular vein; the deep cervical nodes lie along the internal jugular vein

Thoracic duct

Axillary nodes
Drain upper trunk and arm

Intercostal nodes

Spleen

Supratrochlear nodes
Drain the inner side of the arm and forearm

Common iliac nodes

Internal iliac nodes

External iliac nodes

Jugulodigastric node

Submental nodes

Pretracheal nodes

Parasternal nodes

Pre-aortic nodes

Inguinal nodes
Drain the lower trunk and leg

Lymph node
There are around 450 lymph nodes in the adult body. Lymph nodes vary in size from ¹⁄₃₂ in (1 mm) to over 1 in (2 cm) in length and tend to be oval. Several lymphatic vessels bring lymph to the node, and a single vessel carries it away.

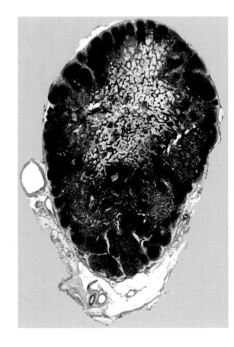

Cross section of a lymph node
Lymph nodes possess a capsule (stained pink in this section), an outer cortex packed full of lymphocytes (deep purple), and an inner medulla made up of lymphatic channels (blue).

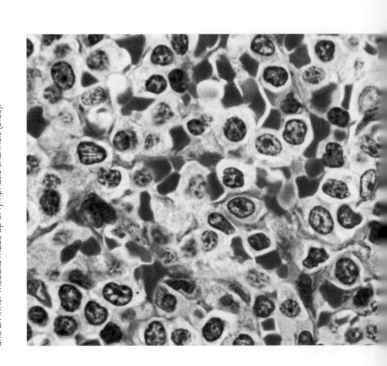

LYMPHATIC AND IMMUNE SYSTEM

The immune system is the body's defense mechanism against external and internal threats. Skin forms a physical barrier to infection, and the antibacterial sebum secreted onto it is a chemical barrier. There are also important immune molecules, including antibodies, and a great range of immune cells, including lymphocytes, that are all made in the bone marrow. Some lymphocytes mature in the bone marrow, whereas others move to the thymus to develop. The thymus is a large gland, low in the neck in children (see p.163), which largely disappears in adulthood. Mature lymphocytes take up residence in the lymph nodes, where they check incoming tissue fluid for potential invaders.

SIDE

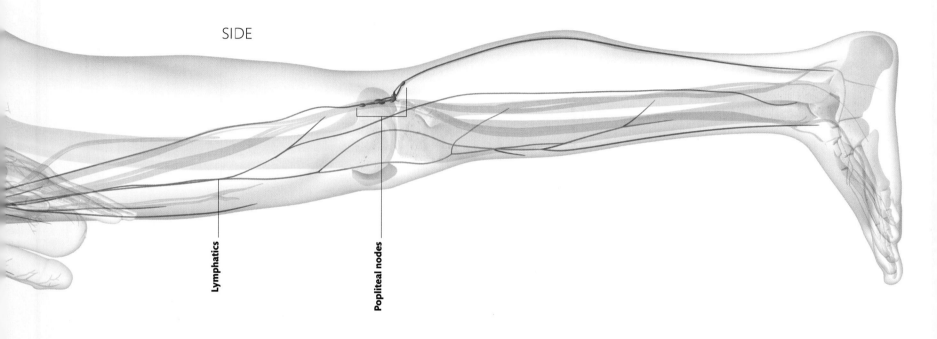

Lymphatics

Popliteal nodes

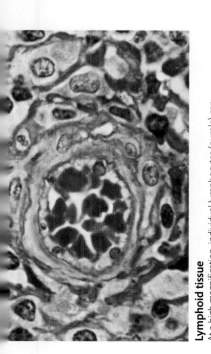

Lymphoid tissue
At a high magnification, individual lymphocytes (purple) can be seen in a section of lymphoid tissue. The blue circle in the image is an arteriole, packed full of blood cells (stained pink).

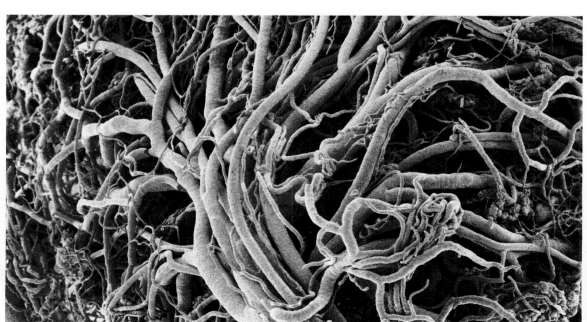

Blood vessels of lymph node
This image, produced using a scanning electron microscope, shows a resin cast of the dense network of tiny blood vessels inside a lymph node.

DIGESTIVE SYSTEM

The digestive system comprises the organs that enable us to take in food, break it down physically and chemically, extract useful nutrients from it, and excrete what we don't need. This process begins in the mouth, where the teeth, tongue, and saliva work together to form a food into a moist ball that can be swallowed. The mouth, pharynx, stomach, intestines, rectum, and anal canal form a long tube that is referred to as the digestive tract. It usually takes between one and two days for ingested food to travel all the way from the mouth to the anus. Other organs—including the salivary glands, liver, gallbladder, and pancreas—complete the digestive system.

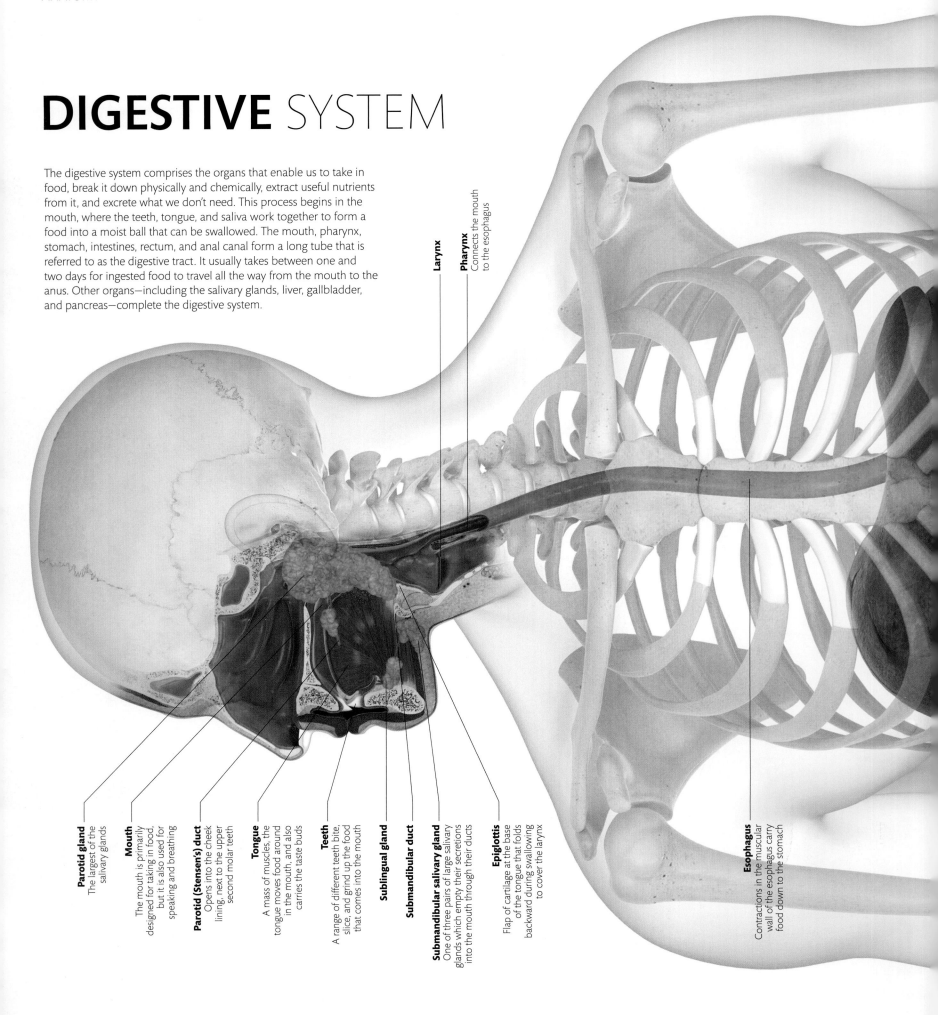

Larynx

Pharynx
Connects the mouth to the esophagus

Parotid gland
The largest of the salivary glands

Mouth
The mouth is primarily designed for taking in food, but it is also used for speaking and breathing

Parotid (Stensen's) duct
Opens into the cheek lining, next to the upper second molar teeth

Tongue
A mass of muscles, the tongue moves food around in the mouth, and also carries the taste buds

Teeth
A range of different teeth bite, slice, and grind up the food that comes into the mouth

Sublingual gland

Submandibular duct

Submandibular salivary gland
One of three pairs of large salivary glands which empty their secretions into the mouth through their ducts

Epiglottis
Flap of cartilage at the base of the tongue that folds backward during swallowing to cover the larynx

Esophagus
Contractions in the muscular wall of the esophagus carry food down to the stomach

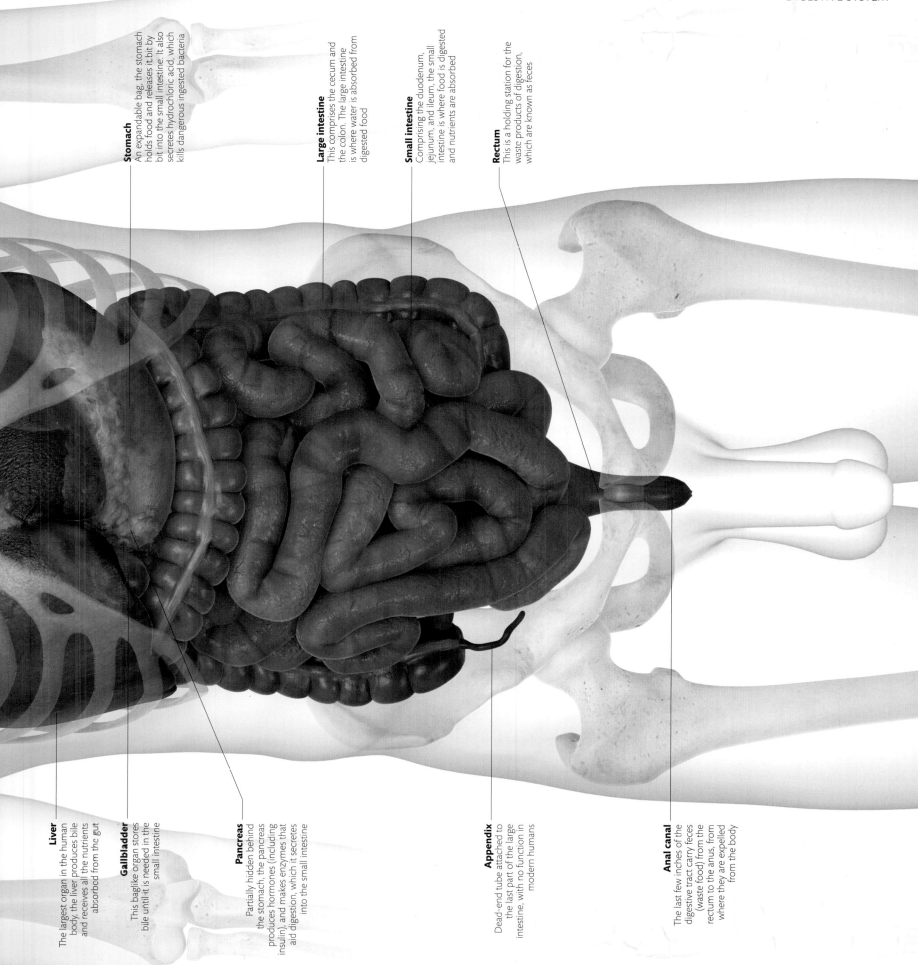

Stomach
An expandable bag, the stomach holds food and releases it bit by bit into the small intestine. It also secretes hydrochloric acid, which kills dangerous ingested bacteria

Large intestine
This comprises the cecum and the colon. The large intestine is where water is absorbed from digested food

Small intestine
Comprising the duodenum, jejunum, and ileum, the small intestine is where food is digested and nutrients are absorbed

Rectum
This is a holding station for the waste products of digestion, which are known as feces

Liver
The largest organ in the human body, the liver produces bile and receives all the nutrients absorbed from the gut

Gallbladder
This baglike organ stores bile until it is needed in the small intestine

Pancreas
Partially hidden behind the stomach, the pancreas produces hormones (including insulin), and makes enzymes that aid digestion, which it secretes into the small intestine

Appendix
Dead-end tube attached to the last part of the large intestine, with no function in modern humans

Anal canal
The last few inches of the digestive tract carry feces (waste food) from the rectum to the anus, from where they are expelled from the body

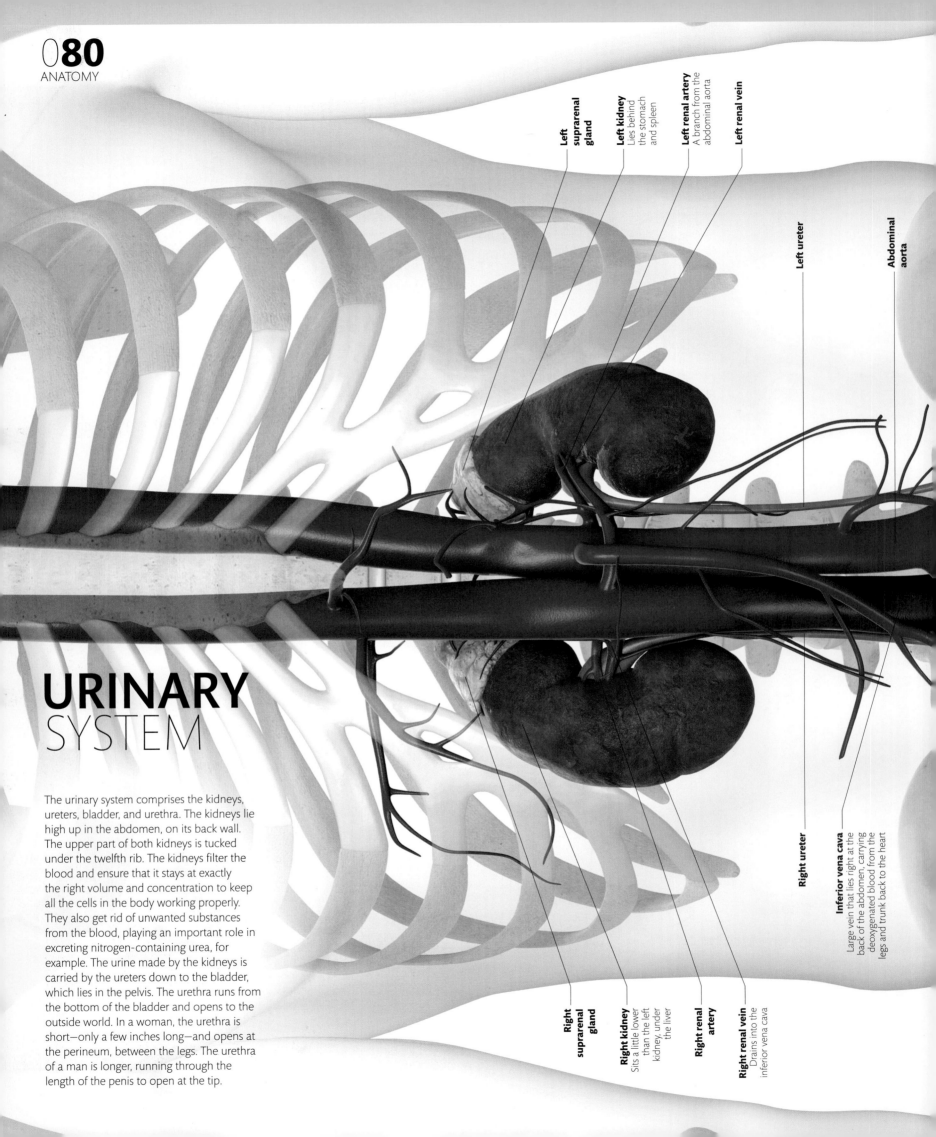

Left suprarenal gland

Left kidney
Lies behind the stomach and spleen

Left renal artery
A branch from the abdominal aorta

Left renal vein

Left ureter

Abdominal aorta

Right suprarenal gland

Right kidney
Sits a little lower than the left kidney, under the liver

Right renal artery

Right renal vein
Drains into the inferior vena cava

Right ureter

Inferior vena cava
Large vein that lies right at the back of the abdomen, carrying deoxygenated blood from the legs and trunk back to the heart

URINARY
SYSTEM

The urinary system comprises the kidneys, ureters, bladder, and urethra. The kidneys lie high up in the abdomen, on its back wall. The upper part of both kidneys is tucked under the twelfth rib. The kidneys filter the blood and ensure that it stays at exactly the right volume and concentration to keep all the cells in the body working properly. They also get rid of unwanted substances from the blood, playing an important role in excreting nitrogen-containing urea, for example. The urine made by the kidneys is carried by the ureters down to the bladder, which lies in the pelvis. The urethra runs from the bottom of the bladder and opens to the outside world. In a woman, the urethra is short—only a few inches long—and opens at the perineum, between the legs. The urethra of a man is longer, running through the length of the penis to open at the tip.

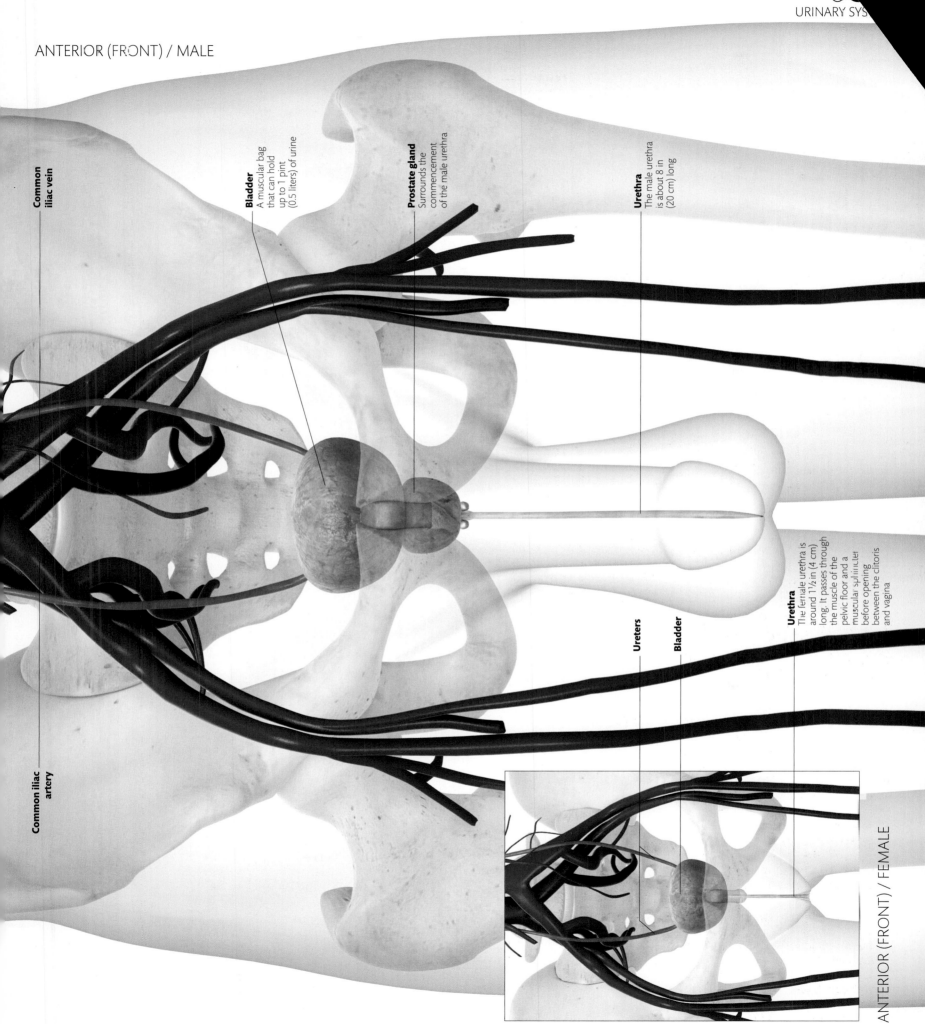

ANTERIOR (FRONT) / MALE

Common iliac vein

Bladder
A muscular bag that can hold up to 1 pint (0.5 liters) of urine

Prostate gland
Surrounds the commencement of the male urethra

Urethra
The male urethra is about 8 in (20 cm) long

Common iliac artery

Ureters

Bladder

Urethra
The female urethra is around 1½ in (4 cm) long. It passes through the muscle of the pelvic floor and a muscular sphincter before opening between the clitoris and vagina

ANTERIOR (FRONT) / FEMALE

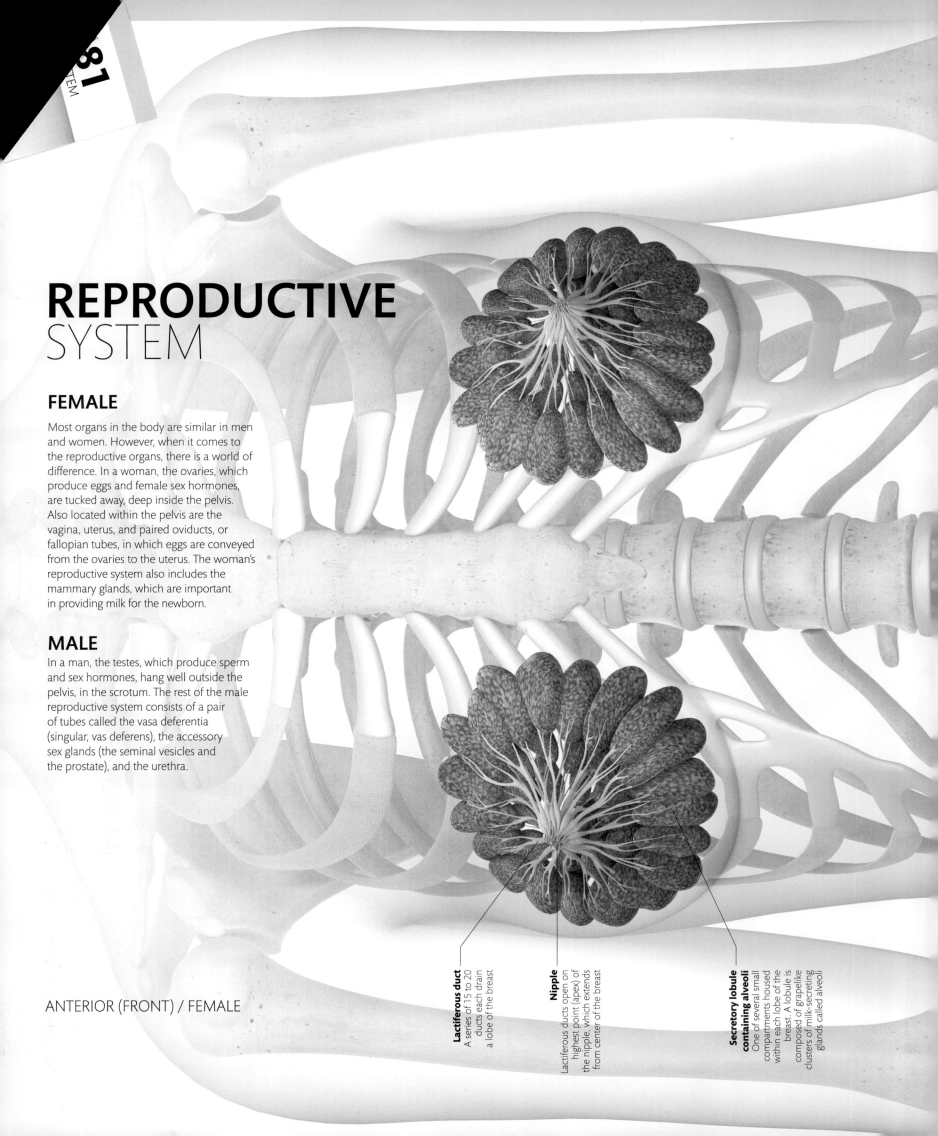

REPRODUCTIVE
SYSTEM

FEMALE

Most organs in the body are similar in men and women. However, when it comes to the reproductive organs, there is a world of difference. In a woman, the ovaries, which produce eggs and female sex hormones, are tucked away, deep inside the pelvis. Also located within the pelvis are the vagina, uterus, and paired oviducts, or fallopian tubes, in which eggs are conveyed from the ovaries to the uterus. The woman's reproductive system also includes the mammary glands, which are important in providing milk for the newborn.

MALE

In a man, the testes, which produce sperm and sex hormones, hang well outside the pelvis, in the scrotum. The rest of the male reproductive system consists of a pair of tubes called the vasa deferentia (singular, vas deferens), the accessory sex glands (the seminal vesicles and the prostate), and the urethra.

ANTERIOR (FRONT) / FEMALE

Lactiferous duct
A series of 15 to 20 ducts each drain a lobe of the breast

Nipple
Lactiferous ducts open on highest point (apex) of the nipple, which extends from center of the breast

Secretory lobule containing alveoli
One of several small compartments housed within each lobe of the breast. A lobule is composed of grapelike clusters of milk-secreting glands called alveoli

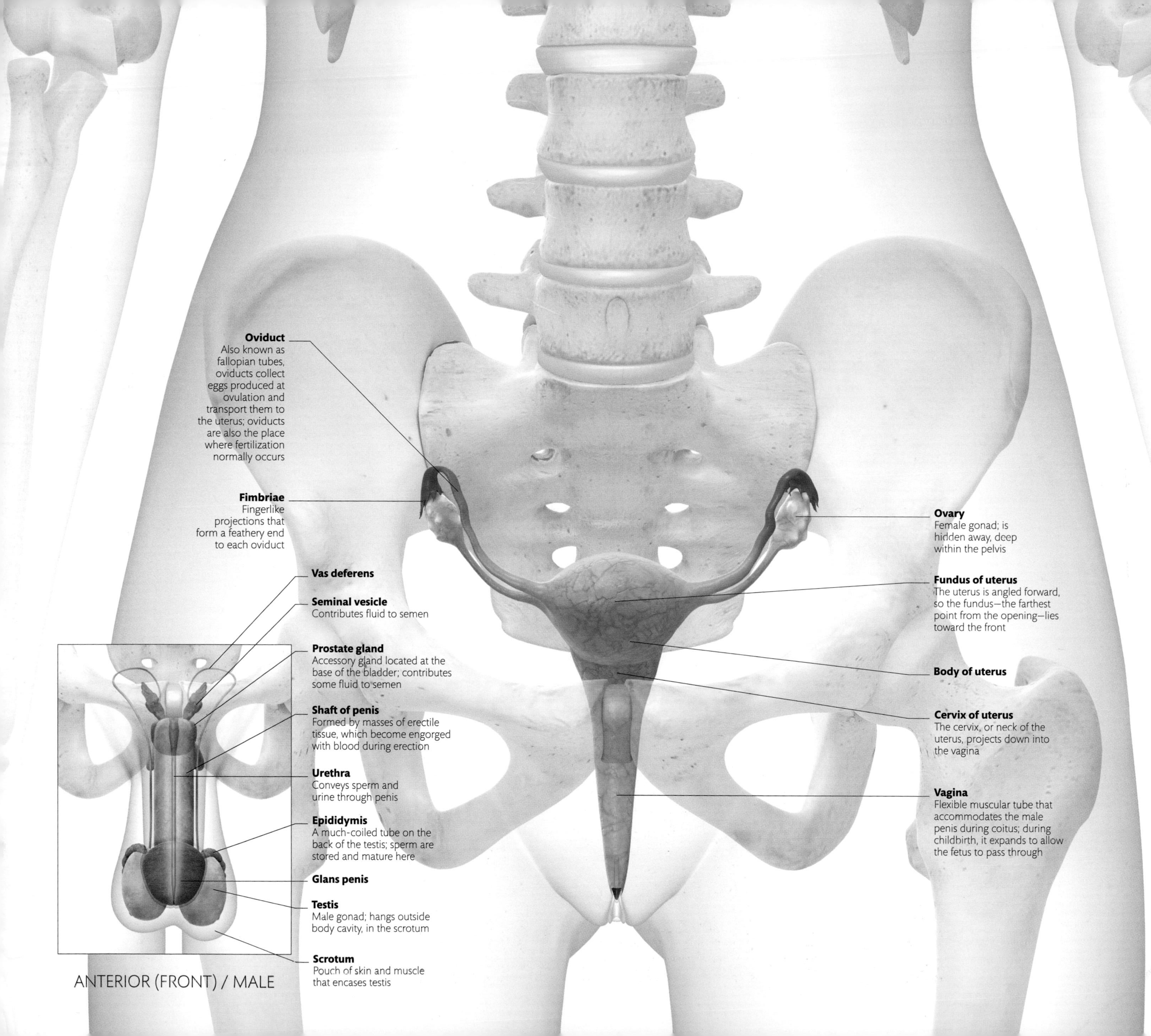

Ovary
Female gonad; is hidden away, deep within the pelvis

Fundus of uterus
The uterus is angled forward, so the fundus—the farthest point from the opening—lies toward the front

Body of uterus

Cervix of uterus
The cervix, or neck of the uterus, projects down into the vagina

Vagina
Flexible muscular tube that accommodates the male penis during coitus; during childbirth, it expands to allow the fetus to pass through

Oviduct
Also known as fallopian tubes, oviducts collect eggs produced at ovulation and transport them to the uterus; oviducts are also the place where fertilization normally occurs

Fimbriae
Fingerlike projections that form a feathery end to each oviduct

Vas deferens

Seminal vesicle
Contributes fluid to semen

Prostate gland
Accessory gland located at the base of the bladder; contributes some fluid to semen

Shaft of penis
Formed by masses of erectile tissue, which become engorged with blood during erection

Urethra
Conveys sperm and urine through penis

Epididymis
A much-coiled tube on the back of the testis; sperm are stored and mature here

Glans penis

Testis
Male gonad; hangs outside body cavity, in the scrotum

Scrotum
Pouch of skin and muscle that encases testis

ANTERIOR (FRONT) / MALE

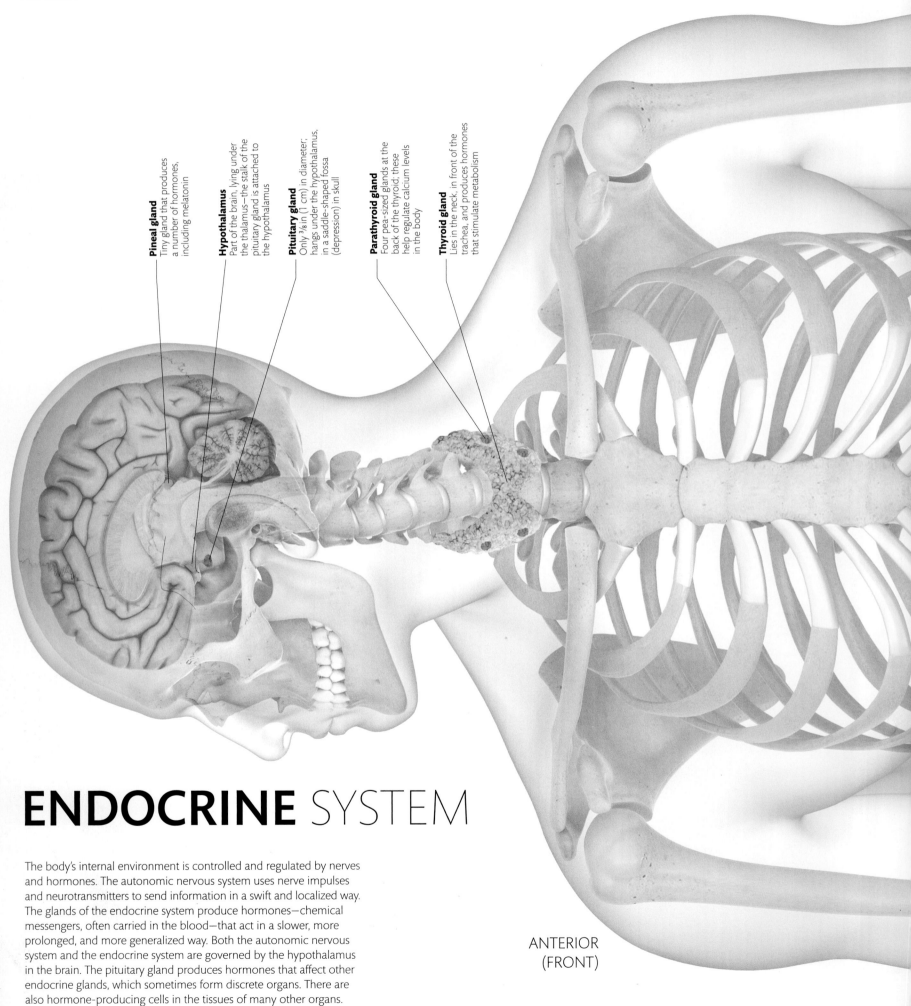

Pineal gland
Tiny gland that produces a number of hormones, including melatonin

Hypothalamus
Part of the brain, lying under the thalamus—the stalk of the pituitary gland is attached to the hypothalamus

Pituitary gland
Only ³⁄₈ in (1 cm) in diameter; hangs under the hypothalamus, in a saddle-shaped fossa (depression) in skull

Parathyroid gland
Four pea-sized glands at the back of the thyroid; these help regulate calcium levels in the body

Thyroid gland
Lies in the neck, in front of the trachea, and produces hormones that stimulate metabolism

ENDOCRINE SYSTEM

The body's internal environment is controlled and regulated by nerves and hormones. The autonomic nervous system uses nerve impulses and neurotransmitters to send information in a swift and localized way. The glands of the endocrine system produce hormones—chemical messengers, often carried in the blood—that act in a slower, more prolonged, and more generalized way. Both the autonomic nervous system and the endocrine system are governed by the hypothalamus in the brain. The pituitary gland produces hormones that affect other endocrine glands, which sometimes form discrete organs. There are also hormone-producing cells in the tissues of many other organs.

ANTERIOR
(FRONT)

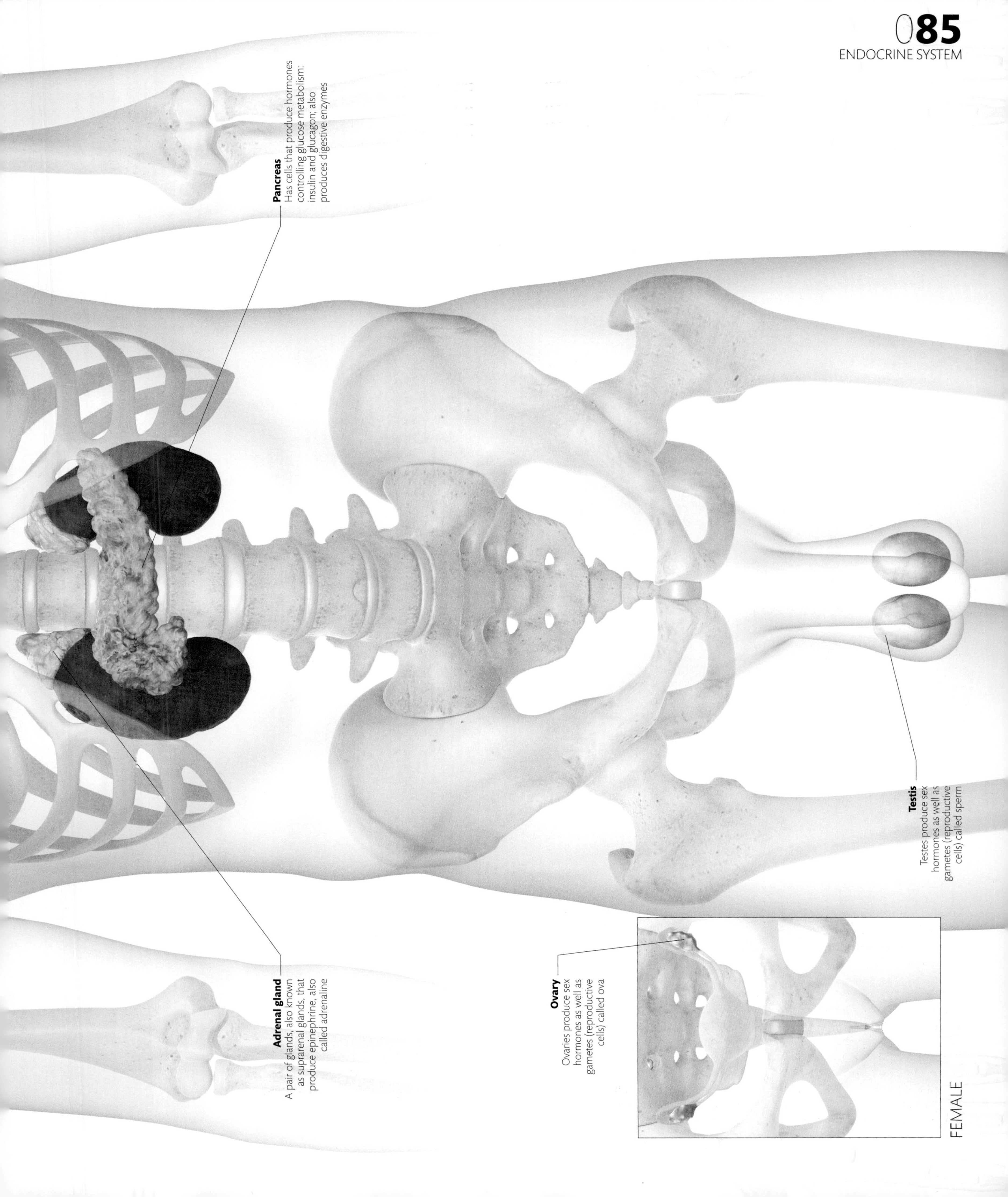

Pancreas
Has cells that produce hormones controlling glucose metabolism: insulin and glucagon; also produces digestive enzymes

Adrenal gland
A pair of glands, also known as suprarenal glands, that produce epinephrine, also called adrenaline

Testis
Testes produce sex hormones as well as gametes (reproductive cells) called sperm

Ovary
Ovaries produce sex hormones as well as gametes (reproductive cells) called ova

FEMALE

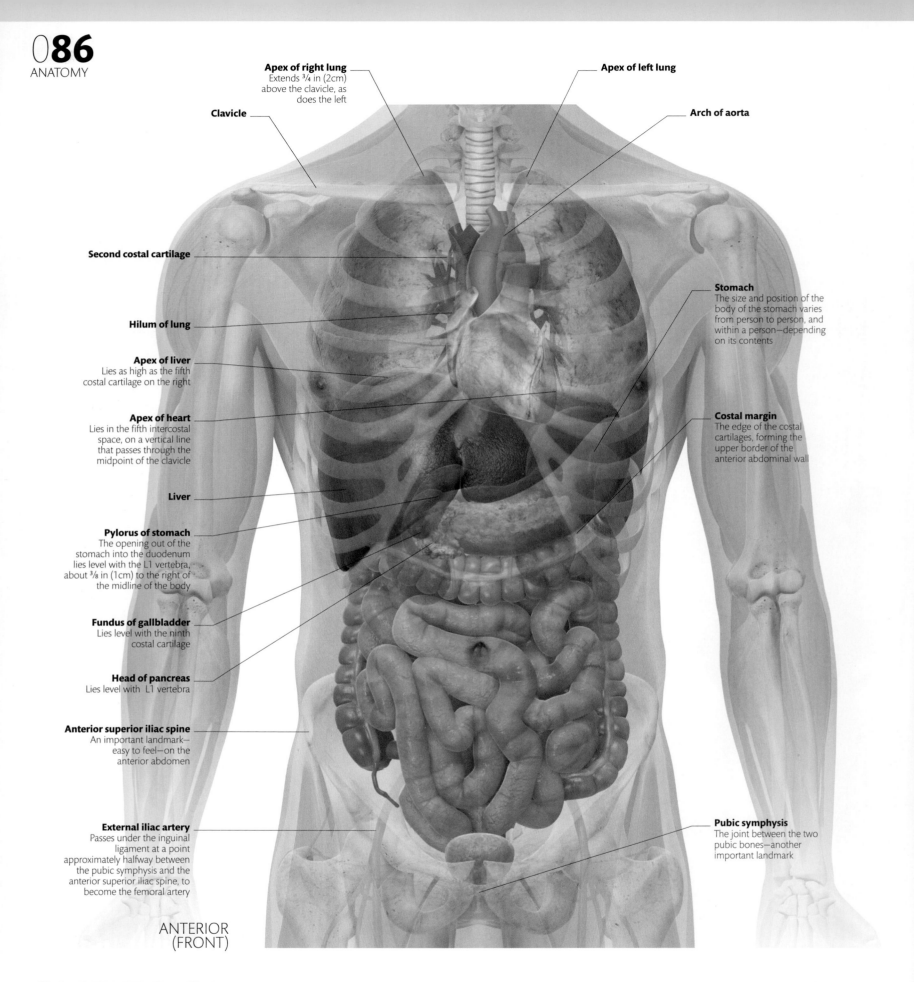

Apex of right lung
Extends ¾ in (2cm)
above the clavicle, as
does the left

Apex of left lung

Clavicle

Arch of aorta

Second costal cartilage

Stomach
The size and position of the
body of the stomach varies
from person to person, and
within a person—depending
on its contents

Hilum of lung

Apex of liver
Lies as high as the fifth
costal cartilage on the right

Costal margin
The edge of the costal
cartilages, forming the
upper border of the
anterior abdominal wall

Apex of heart
Lies in the fifth intercostal
space, on a vertical line
that passes through the
midpoint of the clavicle

Liver

Pylorus of stomach
The opening out of the
stomach into the duodenum
lies level with the L1 vertebra,
about ⅜ in (1cm) to the right of
the midline of the body

Fundus of gallbladder
Lies level with the ninth
costal cartilage

Head of pancreas
Lies level with L1 vertebra

Anterior superior iliac spine
An important landmark—
easy to feel—on the
anterior abdomen

External iliac artery
Passes under the inguinal
ligament at a point
approximately halfway between
the pubic symphysis and the
anterior superior iliac spine, to
become the femoral artery

Pubic symphysis
The joint between the two
pubic bones—another
important landmark

ANTERIOR
(FRONT)

SURFACE
ANATOMY

It's important for doctors and other clinicians to know exactly where particular organs and vessels lie in the body, in relation to bony landmarks such as the ribs and spinal vertebrae. A sound knowledge of surface anatomy provides the basis for clinical examination, allowing a doctor to know if an organ feels strange or enlarged, or where to feel for a pulse. Using only a stethoscope, a doctor can detect whether a particular lobe of a lung, or a specific valve in the heart, sounds normal or abnormal. Although medical imaging is now hugely helpful to diagnosis, surface anatomy and clinical examination still represent essential knowledge and an essential skill in medicine.

Vertebra

C5
C6
C7

T1
T2
T3
T4
T5
T6
T7
T8
T9
T10
T11
T12

L1
L2
L3
L4
L5

S1
S2
S3
S4
S5

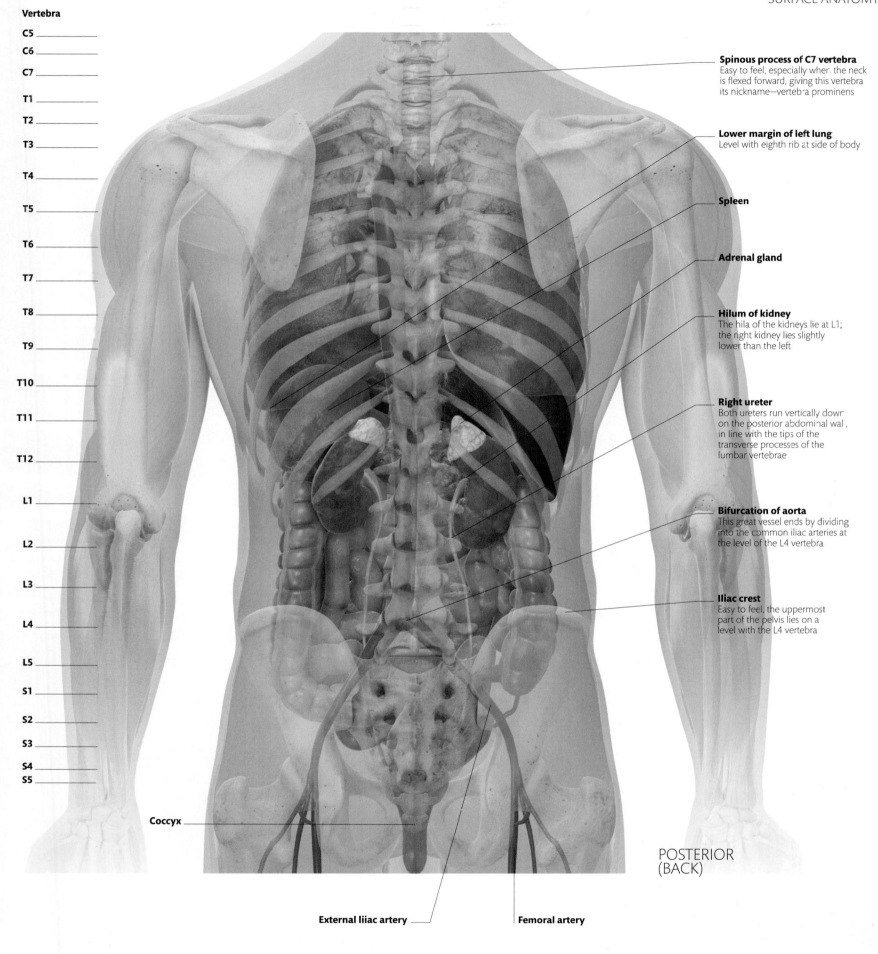

Spinous process of C7 vertebra
Easy to feel, especially when the neck is flexed forward, giving this vertebra its nickname—vertebra prominens

Lower margin of left lung
Level with eighth rib at side of body

Spleen

Adrenal gland

Hilum of kidney
The hila of the kidneys lie at L1; the right kidney lies slightly lower than the left

Right ureter
Both ureters run vertically down on the posterior abdominal wall, in line with the tips of the transverse processes of the lumbar vertebrae

Bifurcation of aorta
This great vessel ends by dividing into the common iliac arteries at the level of the L4 vertebra

Iliac crest
Easy to feel, the uppermost part of the pelvis lies on a level with the L4 vertebra

Coccyx

POSTERIOR
(BACK)

External iliac artery

Femoral artery

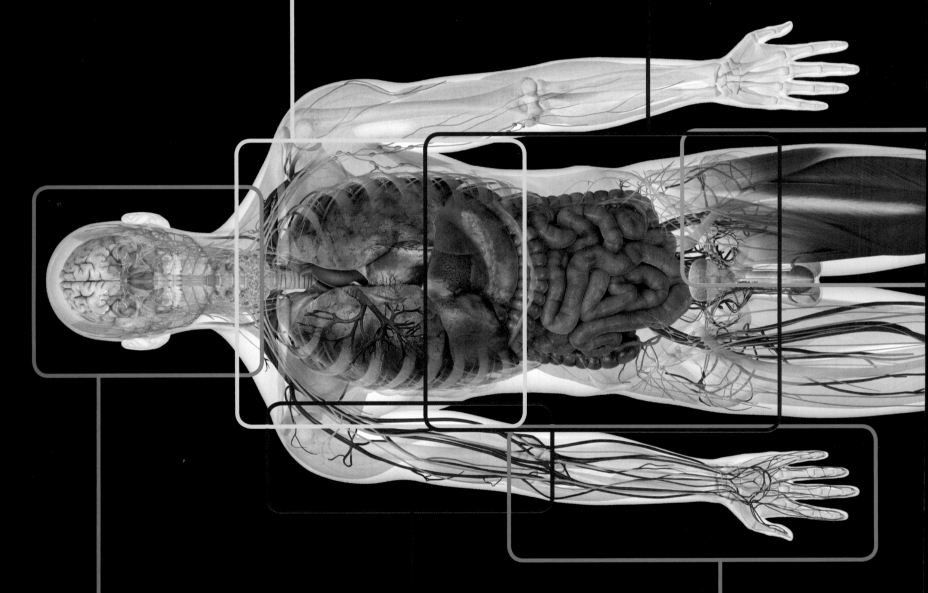

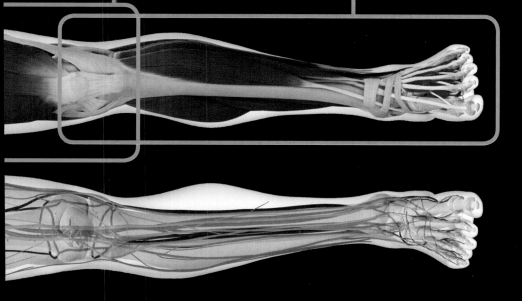

ANATOMY
ATLAS

The **Anatomy Atlas** splits the body into seven regions, starting with the head and neck and working down to the lower leg and foot. Each region is explored through the systems within it: skeletal, muscular, nervous, respiratory, cardiovascular, lymphatic and immune, endocrine, and reproductive. MRI scans at the end of each section show a series of real-life images through the body.

HEAD AND NECK
SKELETAL

The skull comprises the cranium and mandible. It houses and protects the brain and the eyes, ears, nose, and mouth. It encloses the first parts of the airway and of the alimentary canal, and provides attachment for the muscles of the head and neck. The cranium itself comprises more than 20 bones that meet each other at fibrous joints called sutures. In addition to the main bones labeled on these pages, there are sometimes extra bones along the sutures. In a young adult skull, the sutures are visible as tortuous lines between the cranial bones; they gradually fuse with age. The mandible of a newborn baby is in two halves, with a fibrous joint in the middle. The joint fuses during early infancy, so that the mandible becomes a single bone.

Frontal bone

Coronal suture
Where the frontal and parietal bones meet; crosses the skull's highest part (the crown)

Bregma
Where the sagittal and coronal sutures meet

Sagittal suture
Joint on the midline (sagittal plane) where parietal bones meet

Parietal bone
From the Latin for wall

Occipital bone

TOP

Parietal bones
Paired bones forming most of the roof and sides of skull

Sagittal suture

Lambdoid suture
Joint between occipital and parietal bones

Lambda
Point where the sagittal suture meets the lambdoid suture

Occipital bone
Forms lower part of back of skull, and back of cranial base

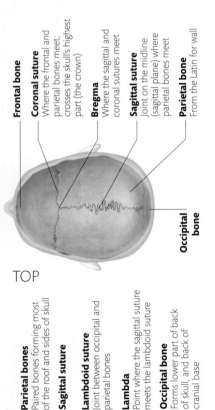

BACK

Superciliary arch
Also called the supraorbital ridge, or brow ridge; from the Latin for eyebrow

Nasal bone
Two small bones form the bony bridge of the nose

Orbit
Technical term for the eye-socket, from the Latin for wheel-track

Frontal process of maxilla
Rises up on the medial (inner) side of the orbit

Piriform aperture
Pear-shaped (piriform) opening; also called the anterior nasal aperture

Inferior nasal concha
Lowest of the three curled protrusions on the lateral wall of the nasal cavity

Zygomatic process of maxilla
Part of the maxilla that projects laterally (to the side)

Frontal bone

Glabella
Area between the two superciliary arches; glabella comes from the Latin for smooth, and refers to the bare area between the eyebrows

Supraorbital foramen
The supraorbital nerve passes through this hole to supply sensation to the forehead

Zygomatic process of frontal bone
Runs down to join the frontal process of the zygomatic bone

Superior orbital fissure
Gap between the sphenoid bone's greater and lesser wings, opening into the orbit

Inferior orbital fissure
Gap between the maxilla and the greater wing of the sphenoid bone, opening into the back of the orbit

Infraorbital foramen
Hole for infraorbital branch of maxillary nerve to supply sensation to the cheek

Nasal crest
Where the two maxillae meet; the vomer (part of the septum) sits on the crest

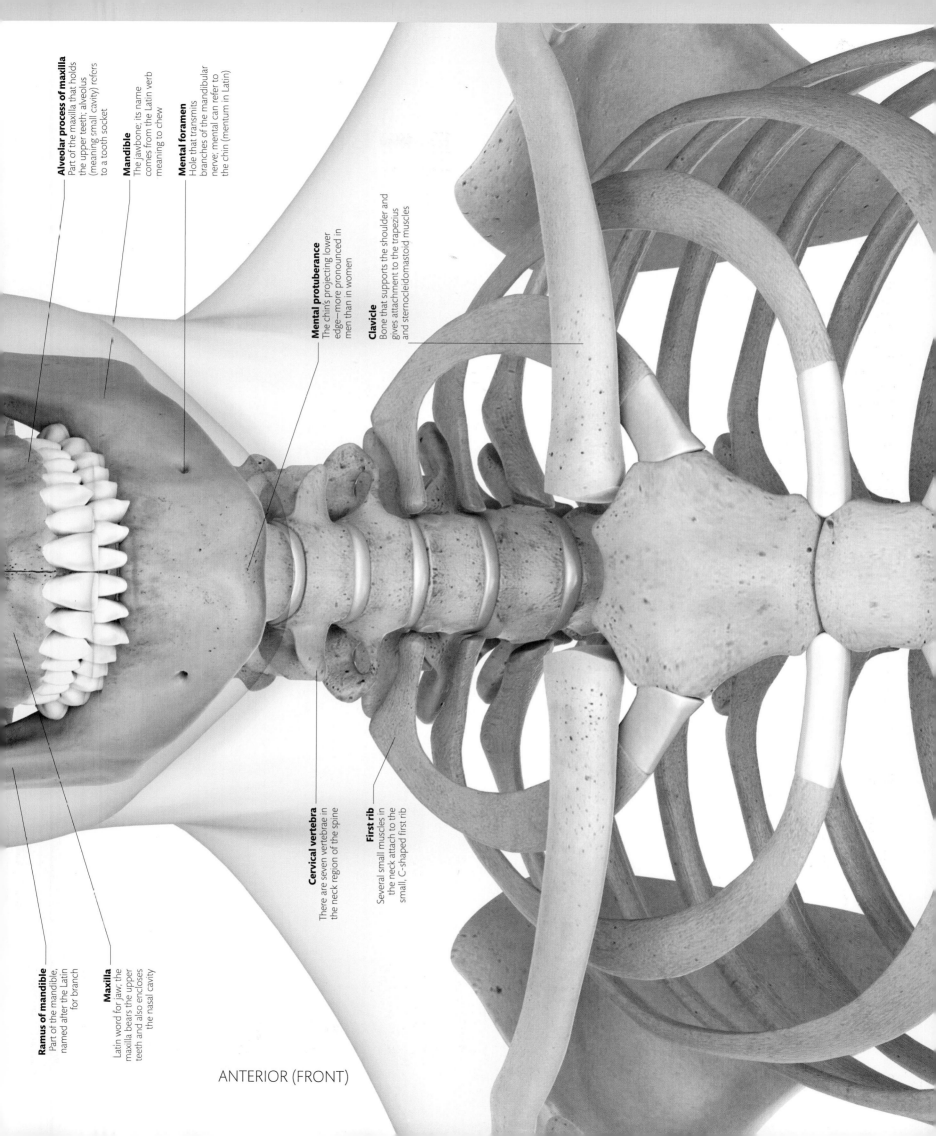

Alveolar process of maxilla
Part of the maxilla that holds the upper teeth; alveolus (meaning small cavity) refers to a tooth socket

Mandible
The jawbone; its name comes from the Latin verb meaning to chew

Mental foramen
Hole that transmits branches of the mandibular nerve; mental can refer to the chin (mentum in Latin)

Mental protuberance
The chin's projecting lower edge—more pronounced in men than in women

Clavicle
Bone that supports the shoulder and gives attachment to the trapezius and sternocleidomastoid muscles

Ramus of mandible
Part of the mandible, named after the Latin for branch

Maxilla
Latin word for jaw; the maxilla bears the upper teeth and also encloses the nasal cavity

Cervical vertebra
There are seven vertebrae in the neck region of the spine

First rib
Several small muscles in the neck attach to the small, C-shaped first rib

ANTERIOR (FRONT)

HEAD AND NECK
SKELETAL

The cervical spine includes seven vertebrae, the top two of which have specific names. The first vertebra, which supports the skull, is called the atlas, after the Greek god who carried the sky on his shoulders. Nodding movements of the head occur at the joint between the atlas and the skull. The second cervical vertebra is the axis, from the Greek word for axle, so-called because when you shake your head from side to side, the atlas rotates on the axis. In this side view, we can also see more of the bones that make up the cranium, as well as the temporomandibular (jaw) joint between the mandible and the skull. The hyoid bone is also visible. This small bone is a very important anchor for the muscles that form the tongue and the floor of the mouth, as well as muscles that attach to the larynx and pharynx.

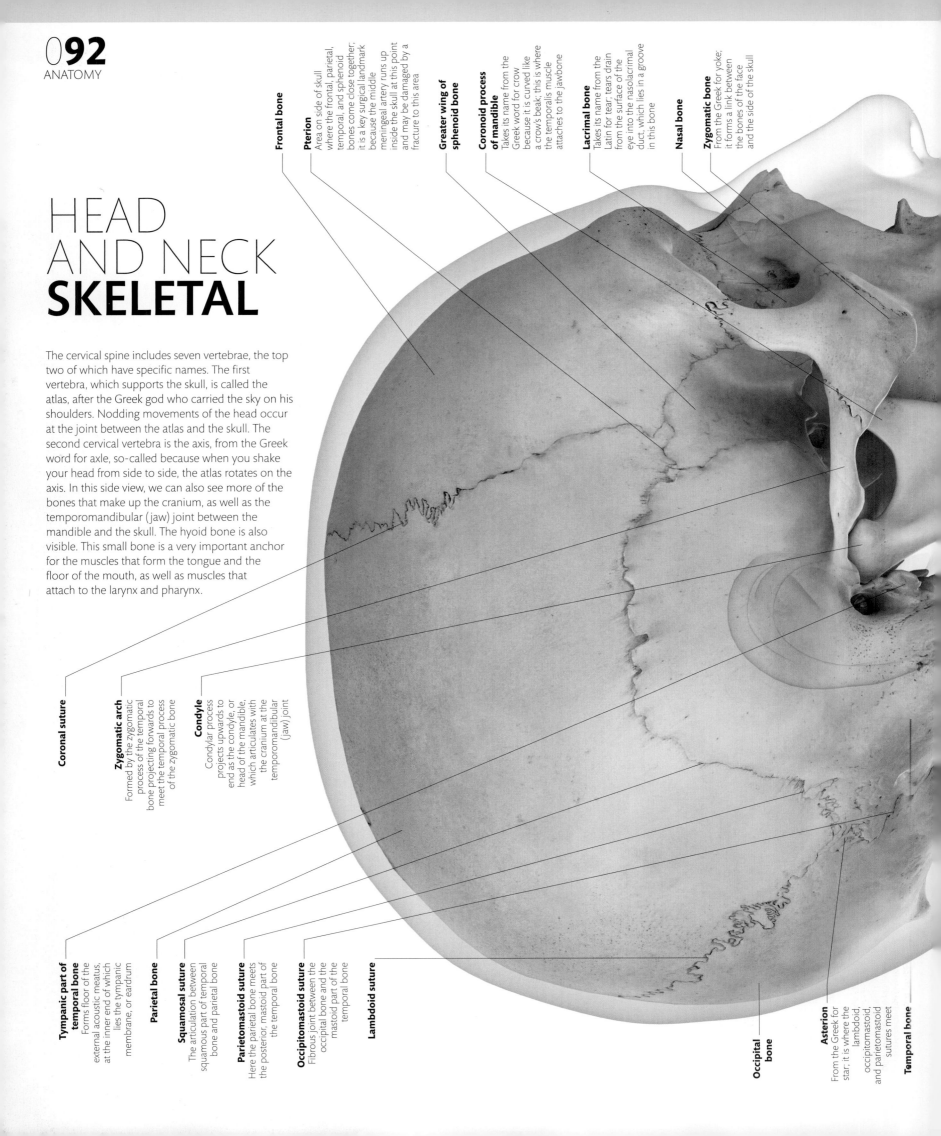

Frontal bone

Pterion
Area on side of skull where the frontal, parietal, temporal, and sphenoid bones come close together; it is a key surgical landmark because the middle meningeal artery runs up inside the skull at this point and may be damaged by a fracture to this area

Greater wing of sphenoid bone

Coronoid process of mandible
Takes its name from the Greek word for crow because it is curved like a crow's beak; this is where the temporalis muscle attaches to the jawbone

Lacrimal bone
Takes its name from the Latin for tear; tears drain from the surface of the eye into the nasolacrimal duct, which lies in a groove in this bone

Nasal bone

Zygomatic bone
From the Greek for yoke; it forms a link between the bones of the face and the side of the skull

Coronal suture

Zygomatic arch
Formed by the zygomatic process of the temporal bone projecting forwards to meet the temporal process of the zygomatic bone

Condyle
Condylar process projects upwards to end as the condyle, or head of the mandible, which articulates with the cranium at the temporomandibular (jaw) joint

Tympanic part of temporal bone
Forms floor of the external acoustic meatus, at the inner end of which lies the tympanic membrane, or eardrum

Parietal bone

Squamosal suture
The articulation between squamous part of temporal bone and parietal bone

Parietomastoid suture
Here the parietal bone meets the posterior, mastoid part of the temporal bone

Occipitomastoid suture
Fibrous joint between the occipital bone and the mastoid part of the temporal bone

Lambdoid suture

Occipital bone

Asterion
From the Greek for star; it is where the lambdoid, occipitomastoid, and parietomastoid sutures meet

Temporal bone

SIDE

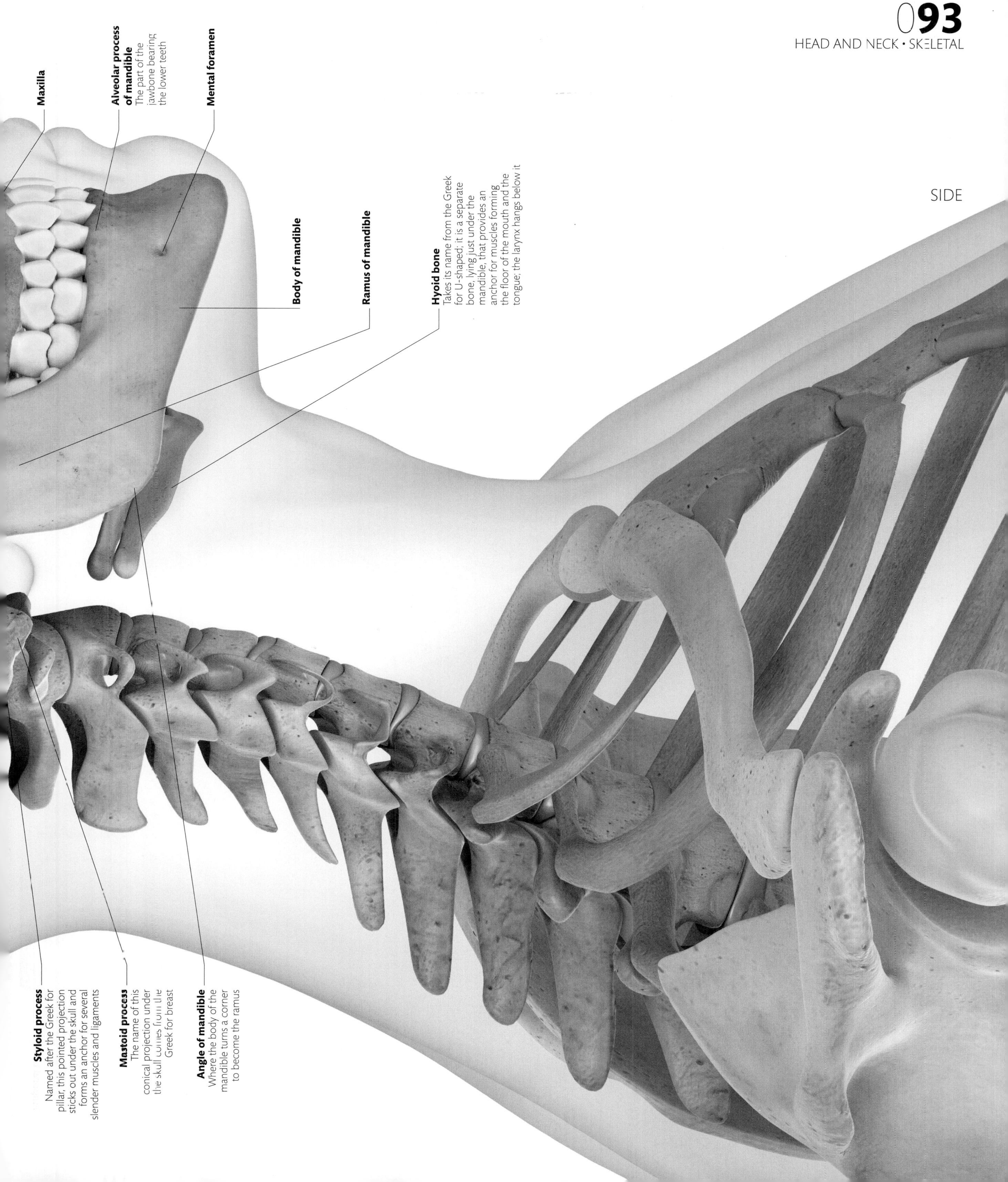

Maxilla

Alveolar process of mandible
The part of the jawbone bearing the lower teeth

Mental foramen

Body of mandible

Ramus of mandible

Hyoid bone
Takes its name from the Greek for U-shaped; it is a separate bone, lying just under the mandible, that provides an anchor for muscles forming the floor of the mouth and the tongue; the larynx hangs below it

Styloid process
Named after the Greek for pillar, this pointed projection sticks out under the skull and forms an anchor for several slender muscles and ligaments

Mastoid process
The name of this conical projection under the skull comes from the Greek for breast

Angle of mandible
Where the body of the mandible turns a corner to become the ramus

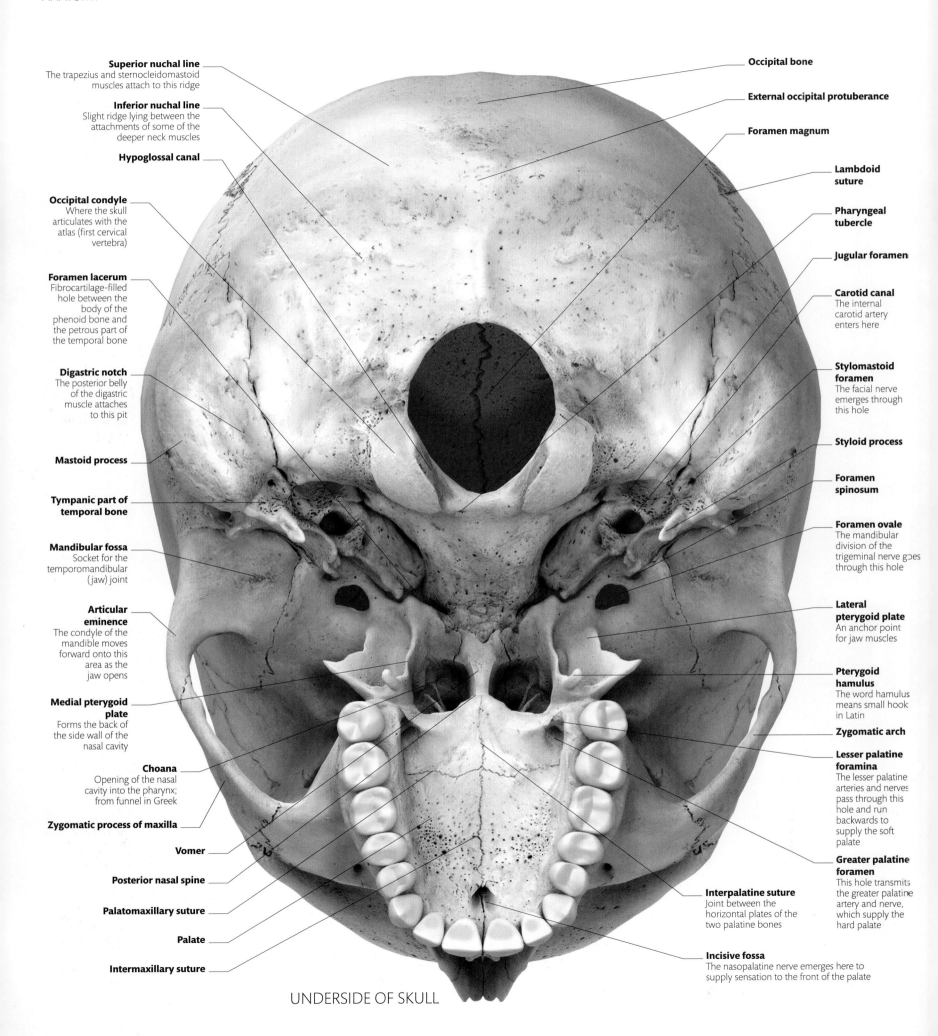

Superior nuchal line
The trapezius and sternocleidomastoid muscles attach to this ridge

Inferior nuchal line
Slight ridge lying between the attachments of some of the deeper neck muscles

Hypoglossal canal

Occipital condyle
Where the skull articulates with the atlas (first cervical vertebra)

Foramen lacerum
Fibrocartilage-filled hole between the body of the phenoid bone and the petrous part of the temporal bone

Digastric notch
The posterior belly of the digastric muscle attaches to this pit

Mastoid process

Tympanic part of temporal bone

Mandibular fossa
Socket for the temporomandibular (jaw) joint

Articular eminence
The condyle of the mandible moves forward onto this area as the jaw opens

Medial pterygoid plate
Forms the back of the side wall of the nasal cavity

Choana
Opening of the nasal cavity into the pharynx; from funnel in Greek

Zygomatic process of maxilla

Vomer

Posterior nasal spine

Palatomaxillary suture

Palate

Intermaxillary suture

Occipital bone

External occipital protuberance

Foramen magnum

Lambdoid suture

Pharyngeal tubercle

Jugular foramen

Carotid canal
The internal carotid artery enters here

Stylomastoid foramen
The facial nerve emerges through this hole

Styloid process

Foramen spinosum

Foramen ovale
The mandibular division of the trigeminal nerve goes through this hole

Lateral pterygoid plate
An anchor point for jaw muscles

Pterygoid hamulus
The word hamulus means small hook in Latin

Zygomatic arch

Lesser palatine foramina
The lesser palatine arteries and nerves pass through this hole and run backwards to supply the soft palate

Greater palatine foramen
This hole transmits the greater palatine artery and nerve, which supply the hard palate

Interpalatine suture
Joint between the horizontal plates of the two palatine bones

Incisive fossa
The nasopalatine nerve emerges here to supply sensation to the front of the palate

UNDERSIDE OF SKULL

HEAD AND NECK
SKELETAL

The most striking features of the skull viewed from these angles are the holes in it. In the middle, there is one large hole—the foramen magnum—through which the brain stem emerges to become the spinal cord. But there are also many smaller holes, most of them paired. Through these holes, the cranial nerves from the brain escape to supply the muscles, skin, and mucosa, and the glands of the head and neck. Blood vessels also pass through some holes, on their way to and from the brain. At the front, we can also see the upper teeth sitting in their sockets in the maxillae, and the bony, hard palate.

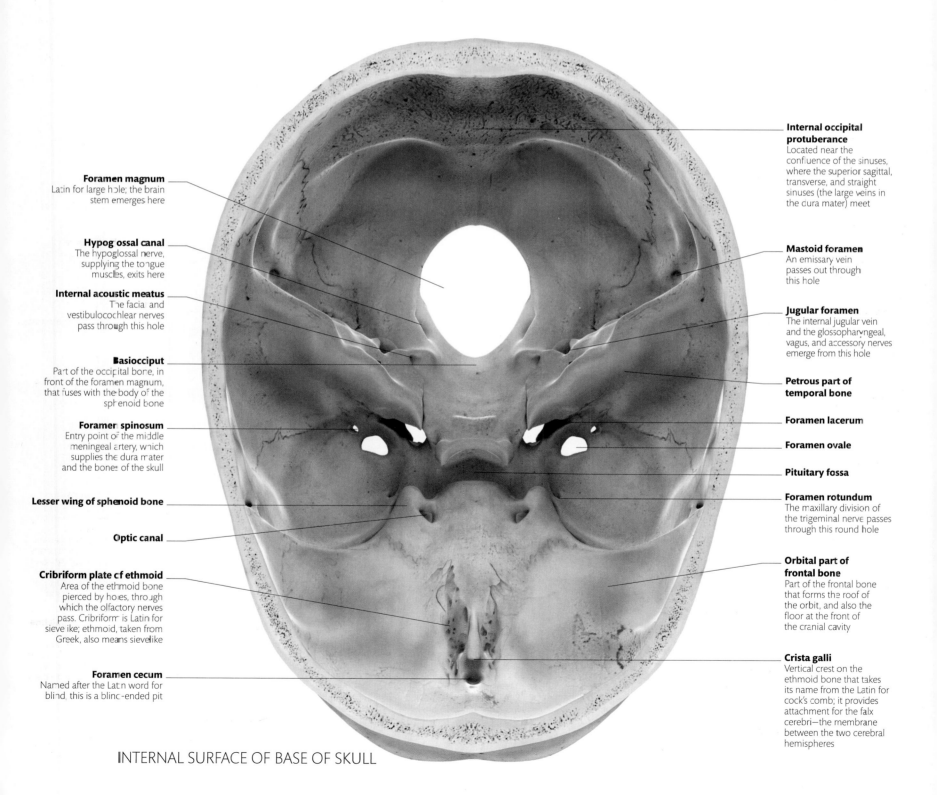

Foramen magnum
Latin for large hole; the brain stem emerges here

Hypoglossal canal
The hypoglossal nerve, supplying the tongue muscles, exits here

Internal acoustic meatus
The facial and vestibulocochlear nerves pass through this hole

Basiocciput
Part of the occipital bone, in front of the foramen magnum, that fuses with the body of the sphenoid bone

Foramen spinosum
Entry point of the middle meningeal artery, which supplies the dura mater and the bones of the skull

Lesser wing of sphenoid bone

Optic canal

Cribriform plate of ethmoid
Area of the ethmoid bone pierced by holes, through which the olfactory nerves pass. Cribriform is Latin for sievelike; ethmoid, taken from Greek, also means sievelike

Foramen cecum
Named after the Latin word for blind, this is a blind-ended pit

Internal occipital protuberance
Located near the confluence of the sinuses, where the superior sagittal, transverse, and straight sinuses (the large veins in the dura mater) meet

Mastoid foramen
An emissary vein passes out through this hole

Jugular foramen
The internal jugular vein and the glossopharyngeal, vagus, and accessory nerves emerge from this hole

Petrous part of temporal bone

Foramen lacerum

Foramen ovale

Pituitary fossa

Foramen rotundum
The maxillary division of the trigeminal nerve passes through this round hole

Orbital part of frontal bone
Part of the frontal bone that forms the roof of the orbit, and also the floor at the front of the cranial cavity

Crista galli
Vertical crest on the ethmoid bone that takes its name from the Latin for cock's comb; it provides attachment for the falx cerebri—the membrane between the two cerebral hemispheres

INTERNAL SURFACE OF BASE OF SKULL

HEAD AND NECK
SKELETAL

This section—right through the middle of the skull—lets us in on some secrets. We can clearly appreciate the size of the cranial cavity, which is almost completely filled by the brain, with just a small gap for membranes, fluid, and blood vessels. Some of those blood vessels leave deep grooves on the inner surface of the skull: we can trace the course of the large venous sinuses and the branches of the middle meningeal artery. We can also see that the skull bones are not solid, but contain trabecular bone (or diploe), which itself contains red marrow. Some skull bones also contain air spaces, like the sphenoidal sinus visible here. We can also appreciate the large size of the nasal cavity, hidden away inside the skull.

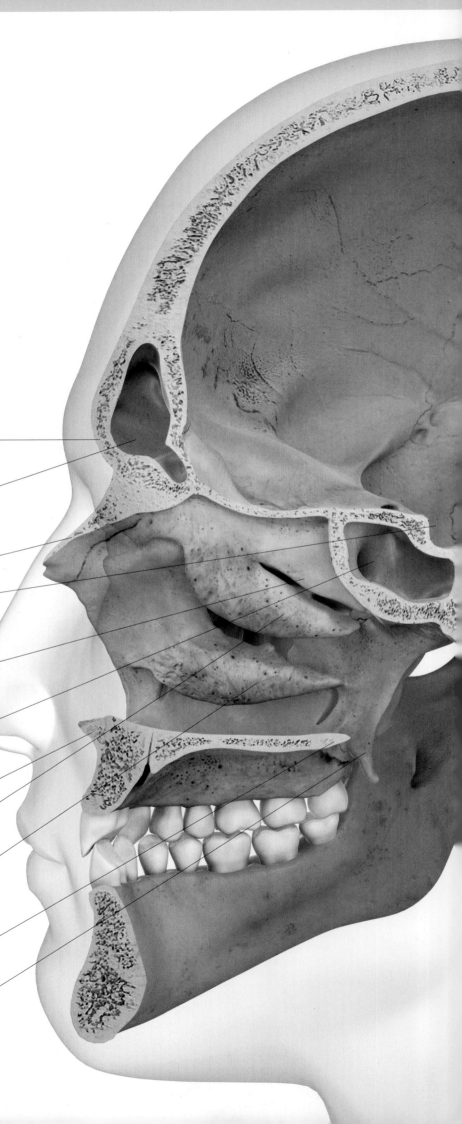

Frontal bone
Forms the anterior cranial fossa, where the frontal lobes of the brain lie, inside the skull

Frontal sinus
One of the paranasal air sinuses that drain into the nasal cavity, this is an air space within the frontal bone

Nasal bone

Pituitary fossa
Fossa is the Latin word for ditch; the pituitary gland occupies this small cavity on the upper surface of the sphenoid bone

Superior nasal concha
Part of the ethmoid bone, which forms the roof and upper sides of the nasal cavity

Sphenoidal sinus
Another paranasal air sinus; it lies within the body of the sphenoid bone

Anterior nasal crest

Middle nasal concha
Like the superior nasal concha, this is also part of the ethmoid bone

Inferior nasal concha
A separate bone, attached to the inner surface of the maxilla; the conchae increase the surface area of the nasal cavity

Palatine bone
Joins to the maxillae and forms the back of the hard palate

Pterygoid process
Sticking down from the greater wing of the sphenoid bone, theis process flanks the back of the nasal cavity and provides attachment for muscles of the palate and jaw

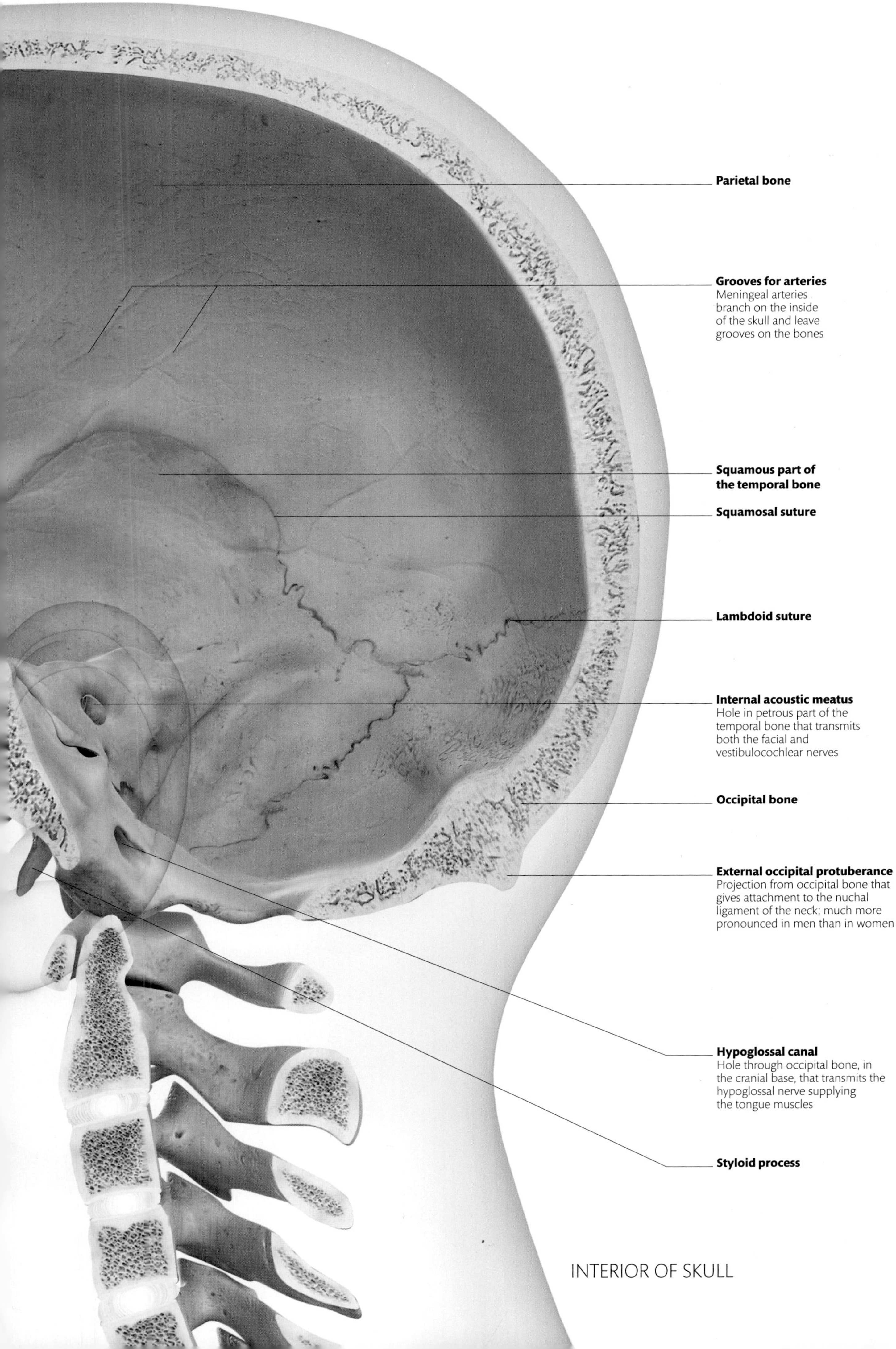

Parietal bone

Grooves for arteries
Meningeal arteries branch on the inside of the skull and leave grooves on the bones

Squamous part of the temporal bone

Squamosal suture

Lambdoid suture

Internal acoustic meatus
Hole in petrous part of the temporal bone that transmits both the facial and vestibulocochlear nerves

Occipital bone

External occipital protuberance
Projection from occipital bone that gives attachment to the nuchal ligament of the neck; much more pronounced in men than in women

Hypoglossal canal
Hole through occipital bone, in the cranial base, that transmits the hypoglossal nerve supplying the tongue muscles

Styloid process

INTERIOR OF SKULL

HEAD AND NECK
SKELETAL

In this view of the skull, we can clearly see that it is not one single bone, and we can also see how the various cranial bones fit together to produce the shape we are more familiar with. The butterfly-shaped sphenoid bone is right in the middle of the action—it forms part of the skull base, the orbits, and the side-walls of the skull, and it articulates with many of the other bones of the skull. The temporal bones also form part of the skull's base and side walls. The extremely dense petrous parts of the temporal bones contain and protect the delicate workings of the ear, including the tiny ossicles (malleus, incus, and stapes) that transmit vibrations from the eardrum to the inner ear.

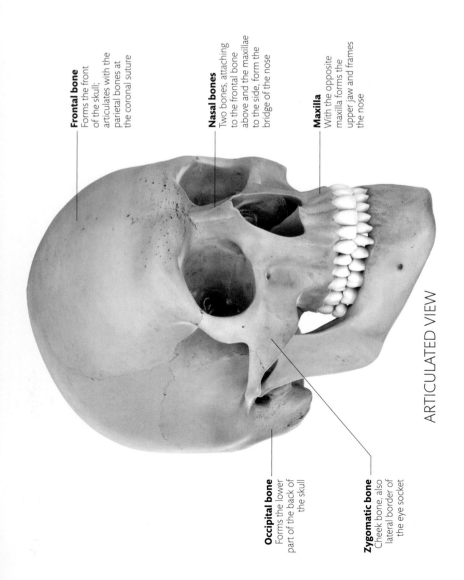

Frontal bone
Forms the front of the skull; articulates with the parietal bones at the coronal suture

Nasal bones
Two bones, attaching to the frontal bone above and the maxillae to the side, form the bridge of the nose

Maxilla
With the opposite maxilla forms the upper jaw and frames the nose

Occipital bone
Forms the lower part of the back of the skull

Zygomatic bone
Cheek bone, also lateral border of the eye socket

ARTICULATED VIEW

Parietal bone

Frontal bone
Forms joints with the parietal and sphenoid bones on the top and sides of the skull, and with the maxilla, nasal, lacrimal, and ethmoid bones below

Parietal bone
Forms the roof and side of the skull

Occipital bone

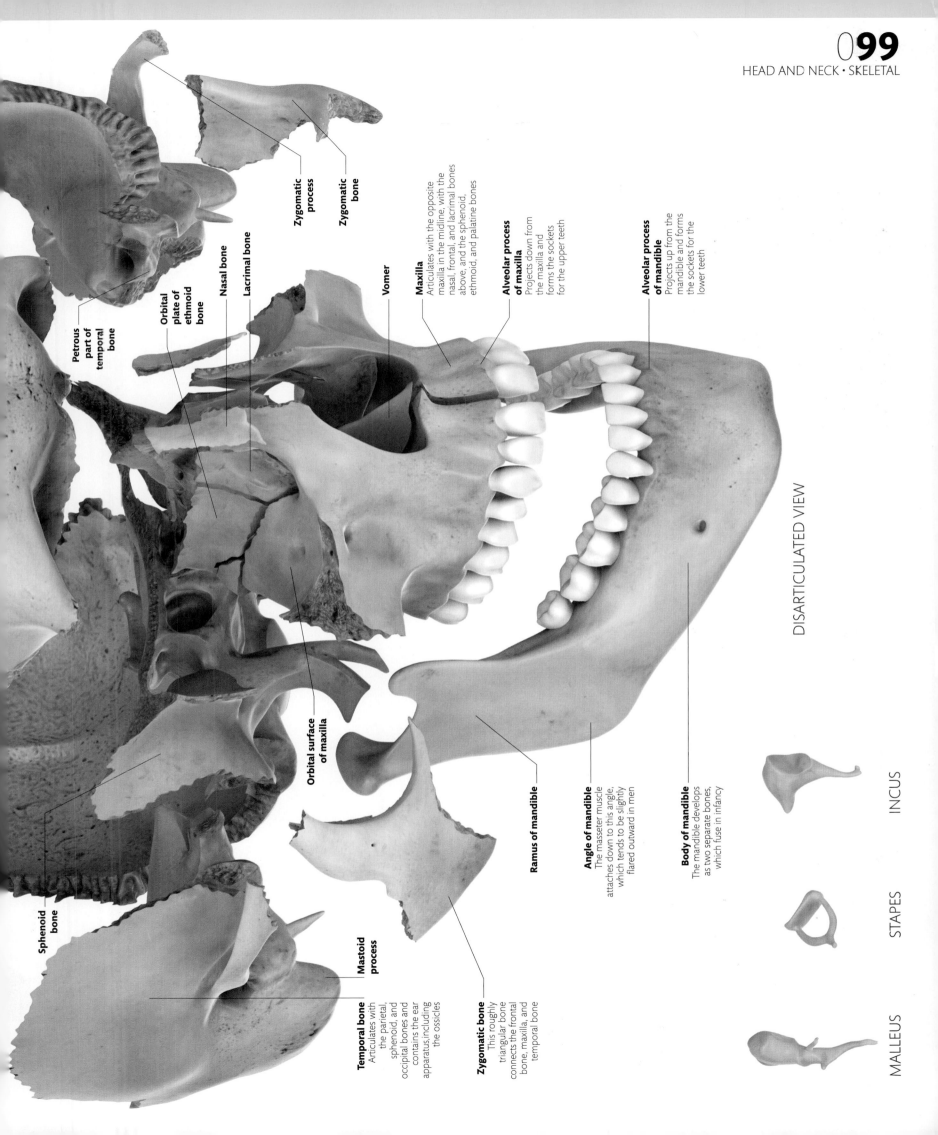

Petrous part of temporal bone

Orbital plate of ethmoid bone

Nasal bone

Lacrimal bone

Zygomatic process

Zygomatic bone

Vomer

Maxilla
Articulates with the opposite maxilla in the midline, with the nasal, frontal, and lacrimal bones above, and the sphenoid, ethmoid, and palatine bones

Alveolar process of maxilla
Projects down from the maxilla and forms the sockets for the upper teeth

Alveolar process of mandible
Projects up from the mandible and forms the sockets for the lower teeth

Orbital surface of maxilla

Ramus of mandible

Angle of mandible
The masseter muscle attaches down to this angle, which tends to be slightly flared outward in men

Body of mandible
The mandible develops as two separate bones, which fuse in infancy

Sphenoid bone

Mastoid process

Temporal bone
Articulates with the parietal, sphenoid, and occipital bones and contains the ear apparatus, including the ossicles

Zygomatic bone
This roughly triangular bone connects the frontal bone, maxilla, and temporal bone

DISARTICULATED VIEW

INCUS

STAPES

MALLEUS

HEAD AND NECK
MUSCULAR

The muscles of the face have very important functions. They open and close the apertures in our faces—our eyes, noses, and mouths. But they also play an extremely important role in communication, and this is why these muscles are often known, collectively, as "the muscles of facial expression." These muscles are attached to bone at one end and skin at the other. It is these muscles that allow us to raise our eyebrows in surprise, frown, or knit our brows in concentration, to scrunch up our noses in distaste, to smile gently or grin widely, and to pout. As we age, and our skin forms creases and wrinkles, these reflect the expressions we have used throughout our lives. The wrinkles and creases lie perpendicular to the direction of the underlying muscle fibers.

ANTERIOR (FRONT)

Nasalis
The upper part of this nasal muscle compresses the nose, while the lower part flares the nostrils

Levator labii superioris alaeque nasi
This small muscle with a very long name lifts the upper lip and the side of the nostril to produce an unpleasant sneer

Levator labii superioris
Raises the upper lip

Zygomaticus minor

Zygomaticus major
Both the zygomaticus major and minor attach from the zygomatic arch (cheek bone) to the side of the upper lip, and are used in smiling

Frontal belly of occipitofrontalis
Epicranial aponeurosis
Temporalis
Occipital belly of occipitofrontalis

TOP

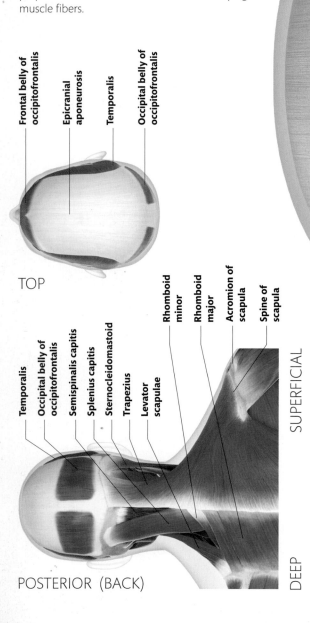

Temporalis
Occipital belly of occipitofrontalis
Semispinalis capitis
Splenius capitis
Sternocleidomastoid
Trapezius
Levator scapulae
Rhomboid minor
Rhomboid major
Acromion of scapula
Spine of scapula

SUPERFICIAL

DEEP

POSTERIOR (BACK)

Epicranial aponeurosis
This connects the frontal and occipital bellies of the occipitofrontalis muscle

Frontal belly of occipitofrontalis
Occipitofrontalis extends from the eyebrows to the superior nuchal line on the back of the skull, and can raise the eyebrows and move the scalp

Temporalis
One of the four paired muscles of mastication, or chewing; acts to close the mouth and bring the teeth together

Orbicularis oculi
These muscle fibers encircle the eye and act to close the eye

Cartilage of the external nose

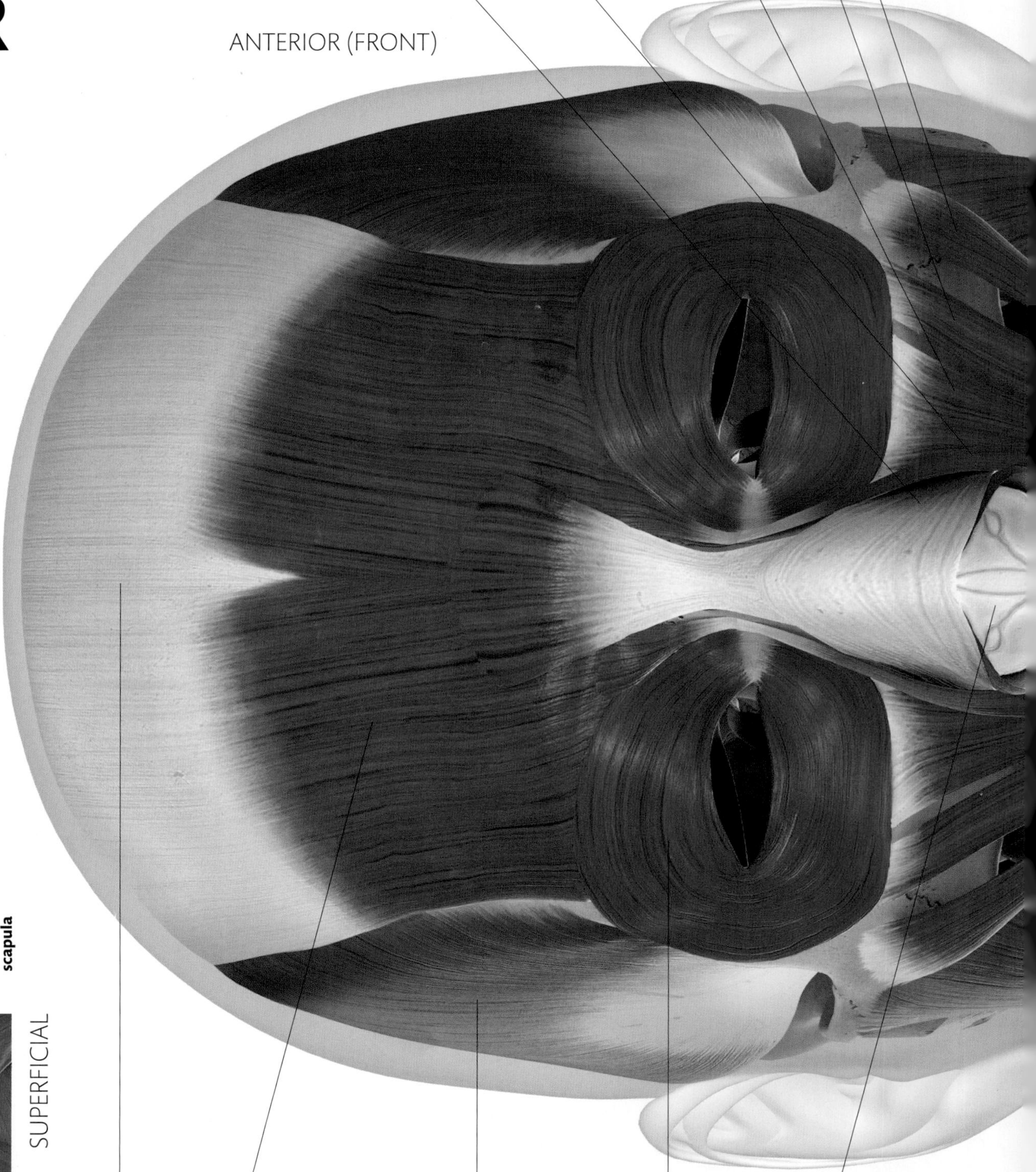

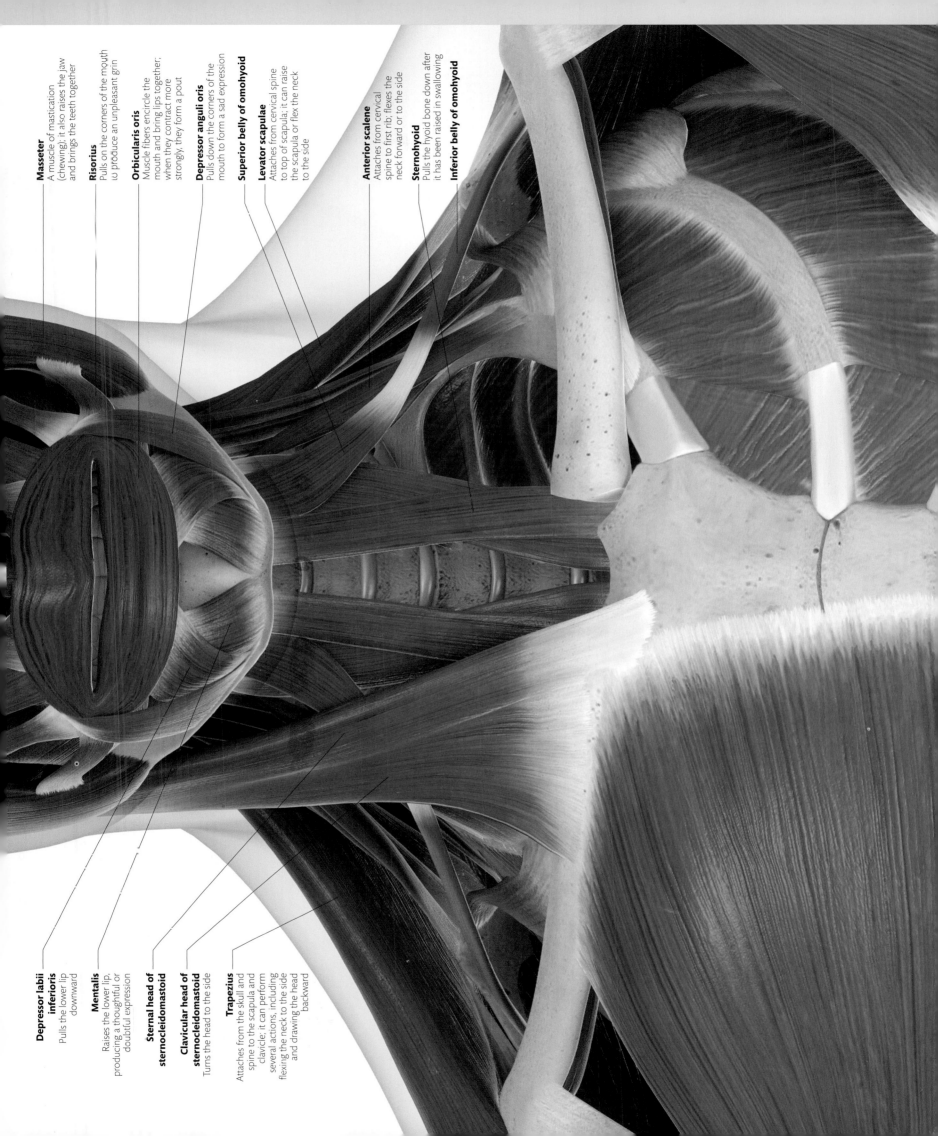

Masseter
A muscle of mastication (chewing); it also raises the jaw and brings the teeth together

Risorius
Pulls on the corners of the mouth to produce an unpleasant grin

Orbicularis oris
Muscle fibers encircle the mouth and bring lips together; when they contract more strongly, they form a pout

Depressor anguli oris
Pulls down the corners of the mouth to form a sad expression

Superior belly of omohyoid

Levator scapulae
Attaches from cervical spine to top of scapula; it can raise the scapula or flex the neck to the side

Anterior scalene
Attaches from cervical spine to first rib; flexes the neck forward or to the side

Sternohyoid
Pulls the hyoid bone down after it has been raised in swallowing

Inferior belly of omohyoid

Depressor labii inferioris
Pulls the lower lip downward

Mentalis
Raises the lower lip, producing a thoughtful or doubtful expression

Sternal head of sternocleidomastoid

Clavicular head of sternocleidomastoid
Turns the head to the side

Trapezius
Attaches from the skull and spine to the scapula and clavicle; it can perform several actions, including flexing the neck to the side and drawing the head backward

HEAD AND NECK
MUSCULAR

The muscles of mastication (chewing) attach from the skull to the mandible (jawbone), operating to open and shut the mouth, and to grind the teeth together to crush the food we eat. In this side view, we can see the two largest muscles of mastication, the temporalis and masseter muscles. Two smaller muscles attach to the inner surface of the mandible. Human jaws don't just open and close, they also move from side to side, and these four muscles act in concert to produce complex chewing movements. In this view, we can also see how the frontal bellies (fleshy central parts) of the occipitofrontalis muscle are connected to occipital bellies at the back of the head by a thin, flat tendon, or aponeurosis. This makes the entire scalp movable on the skull.

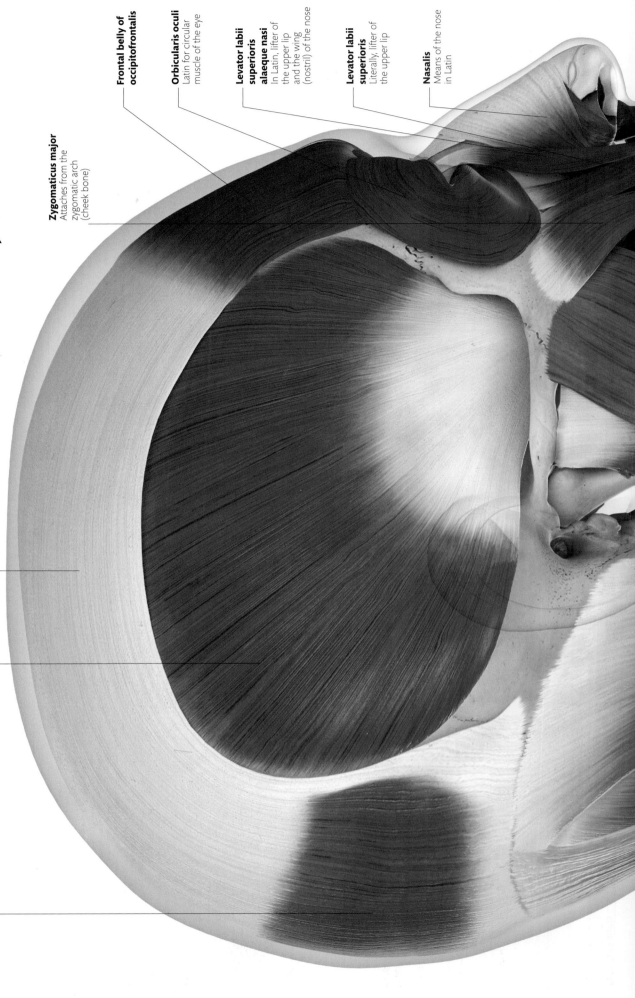

Frontal belly of occipitofrontalis

Orbicularis oculi
Latin for circular muscle of the eye

Levator labii superioris alaeque nasi
In Latin, lifter of the upper lip and the wing (nostril) of the nose

Levator labii superioris
Literally, lifter of the upper lip

Nasalis
Means of the nose in Latin

Zygomaticus major
Attaches from the zygomatic arch (cheek bone)

Epicranial aponeurosis

Temporalis
Attaches from the temporal bone of the skull to the coronoid process of the mandible (jawbone)

Occipital belly of occipitofrontalis

SIDE

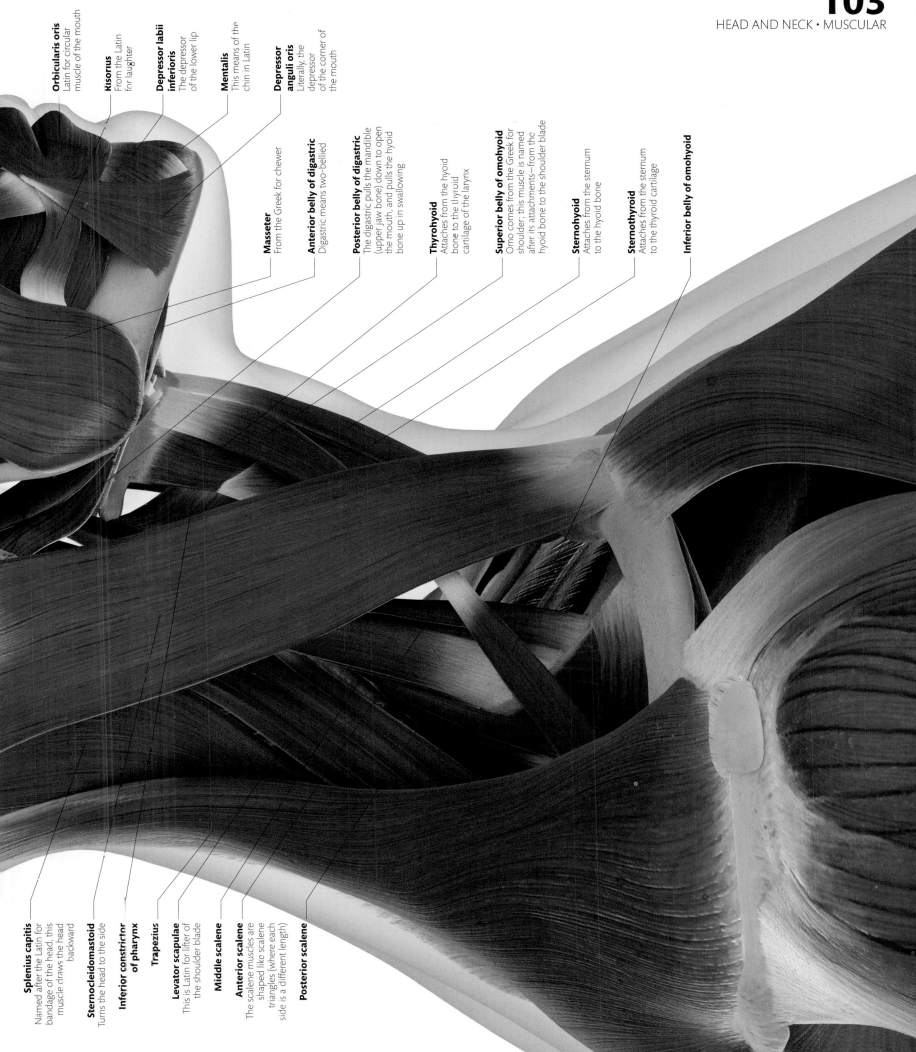

Orbicularis oris
Latin for circular muscle of the mouth

Risorius
From the Latin for laughter

Depressor labii inferioris
The depressor of the lower lip

Mentalis
This means of the chin in Latin

Depressor anguli oris
Literally, the depressor of the corner of the mouth

Masseter
From the Greek for chewer

Anterior belly of digastric
Digastric means two-bellied

Posterior belly of digastric
The digastric pulls the mandible (upper jaw bone) down to open the mouth, and pulls the hyoid bone up in swallowing

Thyrohyoid
Attaches from the hyoid bone to the thyroid cartilage of the larynx

Superior belly of omohyoid
Omo comes from the Greek for shoulder; this muscle is named after its attachments—from the hyoid bone to the shoulder blade

Sternohyoid
Attaches from the sternum to the hyoid bone

Sternothyroid
Attaches from the sternum to the thyroid cartilage

Inferior belly of omohyoid

Splenius capitis
Named after the Latin for bandage of the head, this muscle draws the head backward

Sternocleidomastoid
Turns the head to the side

Inferior constrictor of pharynx

Trapezius

Levator scapulae
This is Latin for lifter of the shoulder blade

Middle scalene

Anterior scalene
The scalene muscles are shaped like scalene triangles (where each side is a different length)

Posterior scalene

SAGITTAL SECTION

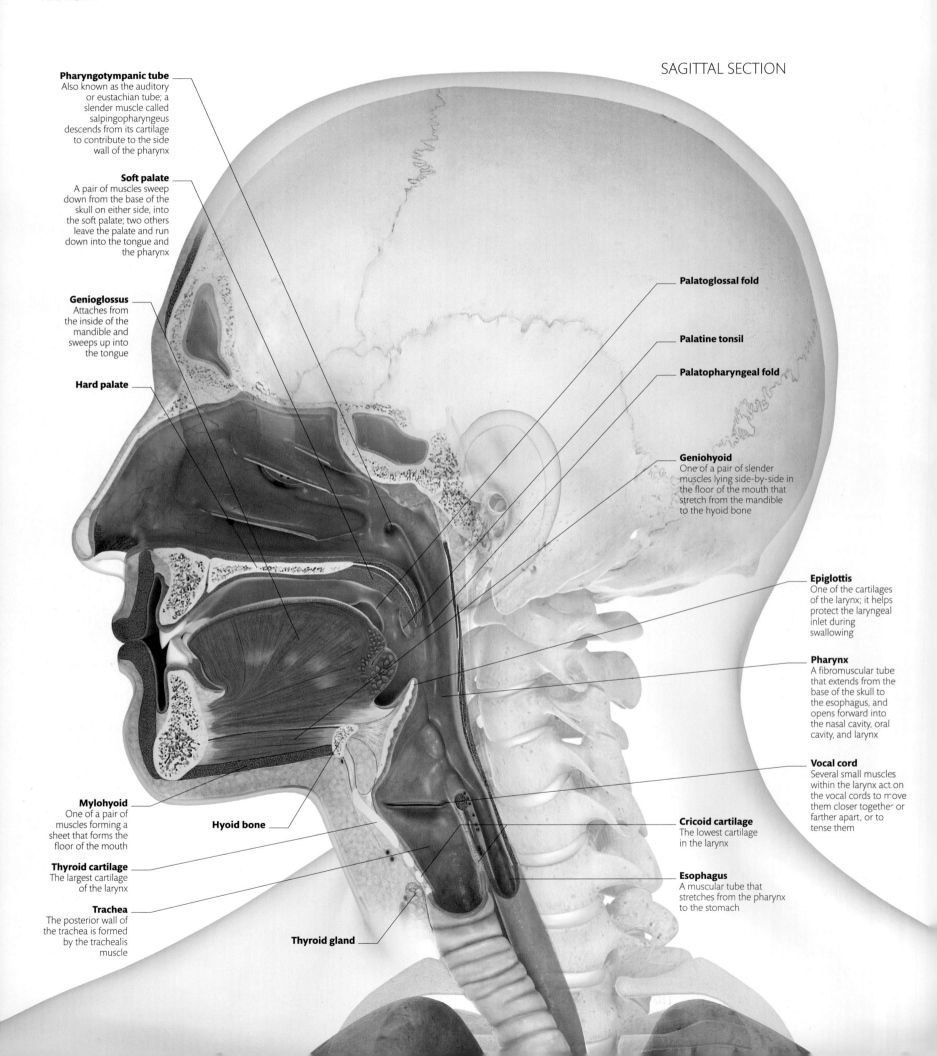

Pharyngotympanic tube
Also known as the auditory or eustachian tube; a slender muscle called salpingopharyngeus descends from its cartilage to contribute to the side wall of the pharynx

Soft palate
A pair of muscles sweep down from the base of the skull on either side, into the soft palate; two others leave the palate and run down into the tongue and the pharynx

Genioglossus
Attaches from the inside of the mandible and sweeps up into the tongue

Hard palate

Mylohyoid
One of a pair of muscles forming a sheet that forms the floor of the mouth

Thyroid cartilage
The largest cartilage of the larynx

Trachea
The posterior wall of the trachea is formed by the trachealis muscle

Hyoid bone

Thyroid gland

Palatoglossal fold

Palatine tonsil

Palatopharyngeal fold

Geniohyoid
One of a pair of slender muscles lying side-by-side in the floor of the mouth that stretch from the mandible to the hyoid bone

Epiglottis
One of the cartilages of the larynx; it helps protect the laryngeal inlet during swallowing

Pharynx
A fibromuscular tube that extends from the base of the skull to the esophagus, and opens forward into the nasal cavity, oral cavity, and larynx

Vocal cord
Several small muscles within the larynx act on the vocal cords to move them closer together or farther apart, or to tense them

Cricoid cartilage
The lowest cartilage in the larynx

Esophagus
A muscular tube that stretches from the pharynx to the stomach

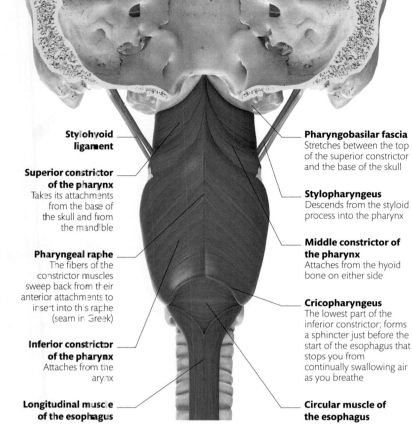

Stylohyoid ligament

Superior constrictor of the pharynx
Takes its attachments from the base of the skull and from the mandible

Pharyngeal raphe
The fibers of the constrictor muscles sweep back from their anterior attachments to insert into this raphe (seam in Greek)

Inferior constrictor of the pharynx
Attaches from the larynx

Longitudinal muscle of the esophagus

Pharyngobasilar fascia
Stretches between the top of the superior constrictor and the base of the skull

Stylopharyngeus
Descends from the styloid process into the pharynx

Middle constrictor of the pharynx
Attaches from the hyoid bone on either side

Cricopharyngeus
The lowest part of the inferior constrictor; forms a sphincter just before the start of the esophagus that stops you from continually swallowing air as you breathe

Circular muscle of the esophagus

PHARYNX POSTERIOR (BACK)

HEAD AND NECK
MUSCULAR

In the section through the head (opposite), we see the soft palate, tongue, pharynx, and larynx, all of which contain muscles. The soft palate comprises five pairs of muscles. When relaxed, it hangs down at the back of the mouth but, during swallowing, it thickens and is drawn upward to block off the airway. The tongue is a great mass of muscle, covered in mucosa. Some of its muscles arise from the hyoid bone and the mandible, and anchor it to these bones and move it around. Other muscle fibers are entirely within the tongue and change its shape. The pharyngeal muscles are important in swallowing, and the laryngeal muscles control the vocal cords. The muscles that move the eye can be seen on p.118.

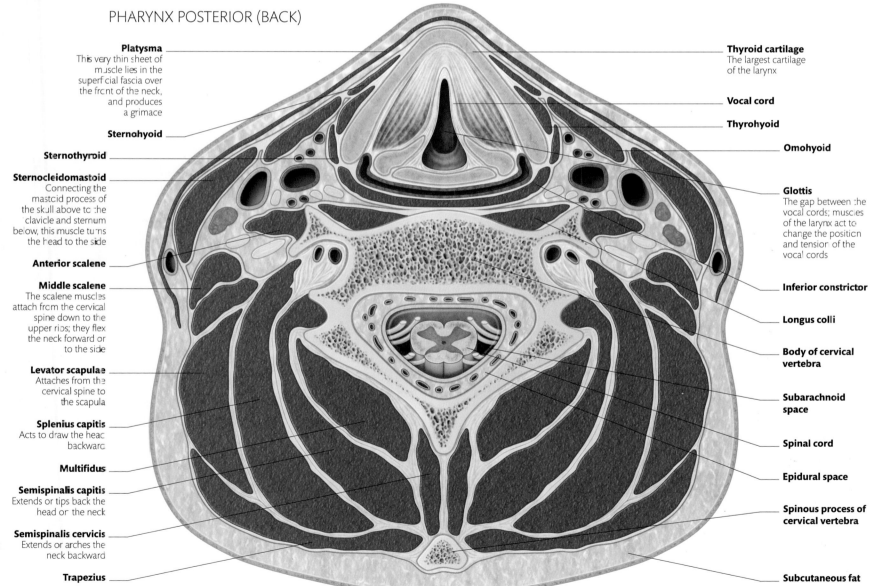

Platysma
This very thin sheet of muscle lies in the superficial fascia over the front of the neck, and produces a grimace

Sternohyoid

Sternothyroid

Sternocleidomastoid
Connecting the mastoid process of the skull above to the clavicle and sternum below, this muscle turns the head to the side

Anterior scalene

Middle scalene
The scalene muscles attach from the cervical spine down to the upper ribs; they flex the neck forward or to the side

Levator scapulae
Attaches from the cervical spine to the scapula

Splenius capitis
Acts to draw the head backward

Multifidus

Semispinalis capitis
Extends or tips back the head or the neck

Semispinalis cervicis
Extends or arches the neck backward

Trapezius

Thyroid cartilage
The largest cartilage of the larynx

Vocal cord

Thyrohyoid

Omohyoid

Glottis
The gap between the vocal cords; muscles of the larynx act to change the position and tension of the vocal cords

Inferior constrictor

Longus colli

Body of cervical vertebra

Subarachnoid space

Spinal cord

Epidural space

Spinous process of cervical vertebra

Subcutaneous fat

TRANSVERSE SECTION OF THE NECK AT THE VOCAL CORDS

HEAD AND NECK
NERVOUS

Compared to other animals, humans have massive brains for the size of our bodies. The human brain has grown larger and larger over the course of evolution, and it is now so overblown that the frontal lobes of the brain lie right over the top of the orbits that contain the eyes. Think about any other mammal, perhaps a dog or a cat for easy reference, and you will quickly realize what an odd shape the human head is—and most of that is a result of our huge brains. Looking at a side view of the brain, you can see all the lobes that make up each cerebral hemisphere: the frontal, parietal, temporal, and occipital lobes (individually colored, below). Tucked under the cerebral hemispheres at the back of the brain is the cerebellum (Latin for little brain). The brain stem leads down, through the foramen magnum of the skull, to the spinal cord.

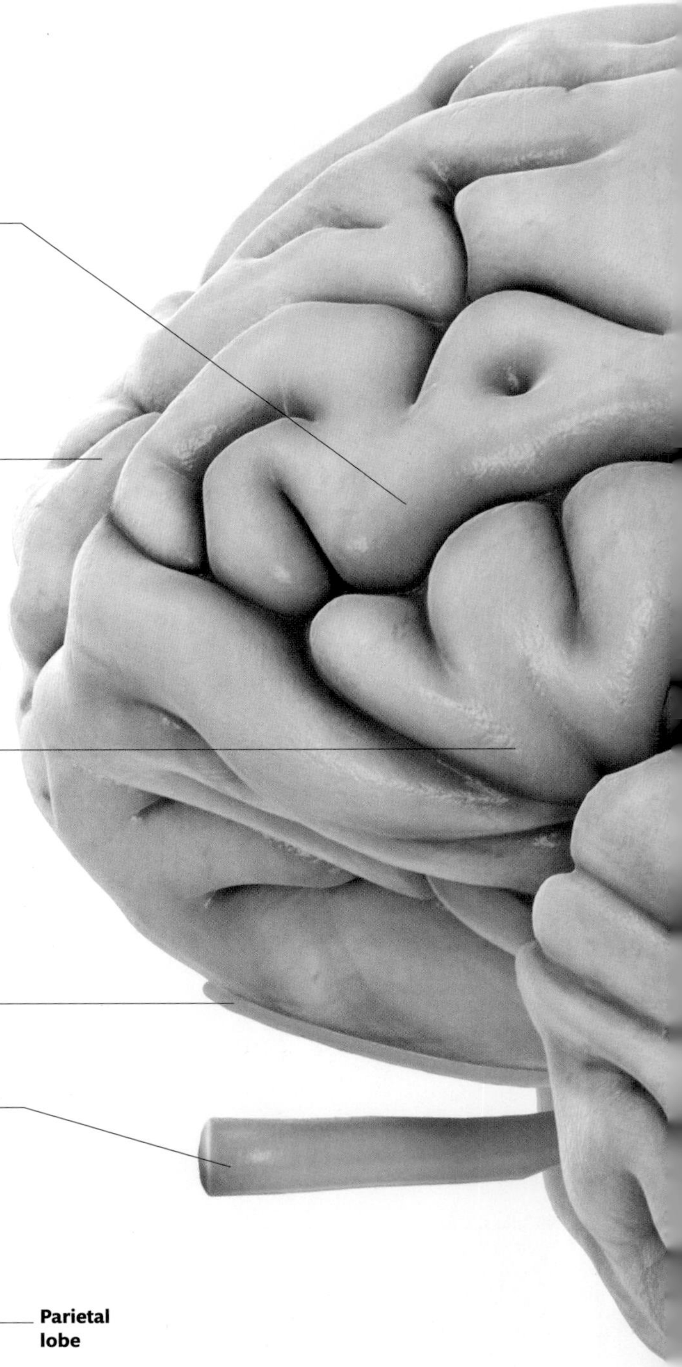

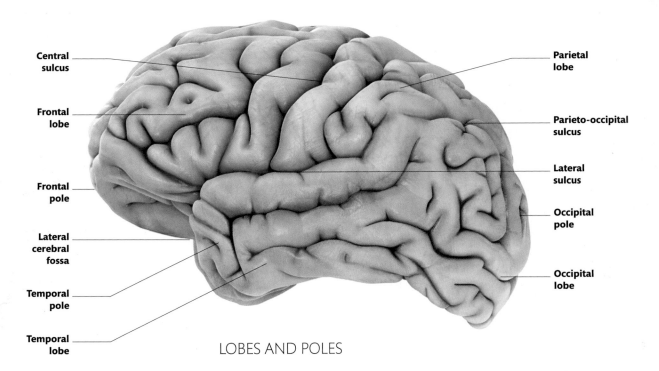

Middle frontal gyrus
The word gyrus comes from the Latin for ring or convolution, and is a term used for the scroll-like folds of the cerebral cortex

Superior frontal gyrus

Inferior frontal gyrus
Includes Broca's area, part of the cerebral cortex that is involved with generating speech

Olfactory bulb

Optic nerve
The second cranial nerve. It carries nerve fibres from the retina to the optic chiasma

Central sulcus

Frontal lobe

Frontal pole

Lateral cerebral fossa

Temporal pole

Temporal lobe

Parietal lobe

Parieto-occipital sulcus

Lateral sulcus

Occipital pole

Occipital lobe

LOBES AND POLES

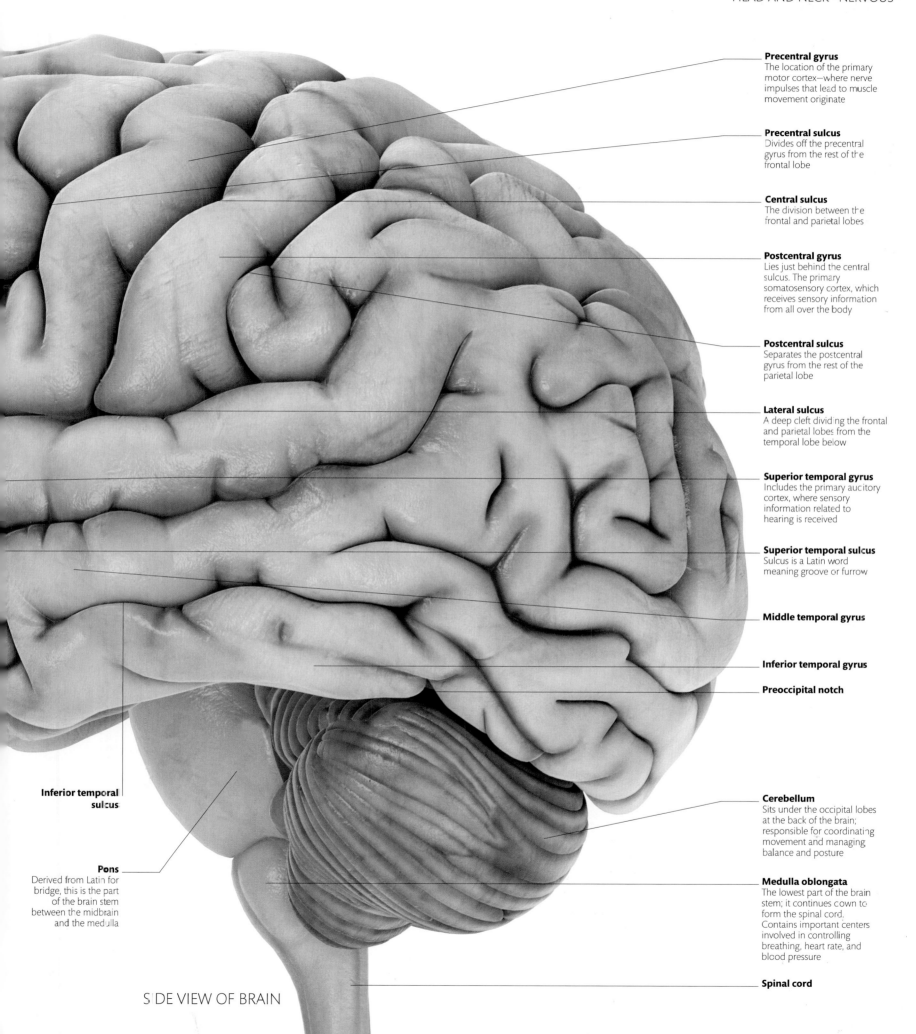

Precentral gyrus
The location of the primary motor cortex—where nerve impulses that lead to muscle movement originate

Precentral sulcus
Divides off the precentral gyrus from the rest of the frontal lobe

Central sulcus
The division between the frontal and parietal lobes

Postcentral gyrus
Lies just behind the central sulcus. The primary somatosensory cortex, which receives sensory information from all over the body

Postcentral sulcus
Separates the postcentral gyrus from the rest of the parietal lobe

Lateral sulcus
A deep cleft dividing the frontal and parietal lobes from the temporal lobe below

Superior temporal gyrus
Includes the primary auditory cortex, where sensory information related to hearing is received

Superior temporal sulcus
Sulcus is a Latin word meaning groove or furrow

Middle temporal gyrus

Inferior temporal gyrus

Preoccipital notch

Inferior temporal sulcus

Pons
Derived from Latin for bridge, this is the part of the brain stem between the midbrain and the medulla

Cerebellum
Sits under the occipital lobes at the back of the brain; responsible for coordinating movement and managing balance and posture

Medulla oblongata
The lowest part of the brain stem; it continues down to form the spinal cord. Contains important centers involved in controlling breathing, heart rate, and blood pressure

Spinal cord

SIDE VIEW OF BRAIN

HEAD AND NECK
NERVOUS

From an anatomist's point of view, the brain is quite an ugly and unprepossessing organ. It looks rather like a large, pinkish gray, wrinkled walnut—especially when viewed from above. The outer layer of gray matter, called the cortex, is highly folded. Underneath the brain we see some more detail, including some of the cranial nerves that emerge from the brain itself. To the naked eye, there is little to suggest that the brain is the most complicated organ in the human body. Its true complexity is only visible through a microscope, revealing billions of neurons that connect with each other to form the pathways that carry our senses, govern our actions, and create our minds.

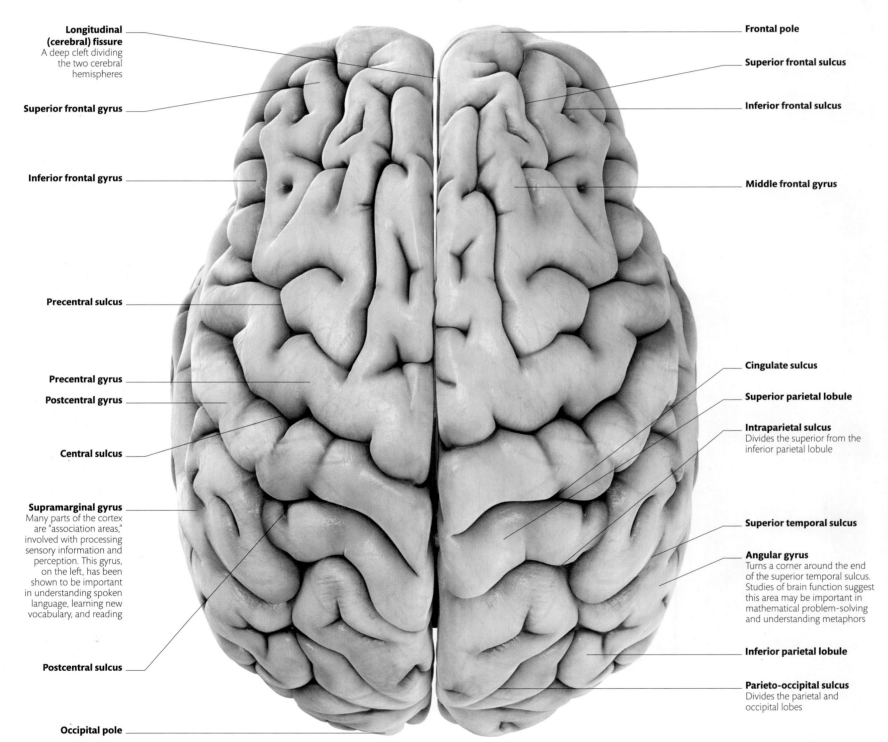

Longitudinal (cerebral) fissure
A deep cleft dividing the two cerebral hemispheres

Superior frontal gyrus

Inferior frontal gyrus

Precentral sulcus

Precentral gyrus

Postcentral gyrus

Central sulcus

Supramarginal gyrus
Many parts of the cortex are "association areas," involved with processing sensory information and perception. This gyrus, on the left, has been shown to be important in understanding spoken language, learning new vocabulary, and reading

Postcentral sulcus

Occipital pole

Frontal pole

Superior frontal sulcus

Inferior frontal sulcus

Middle frontal gyrus

Cingulate sulcus

Superior parietal lobule

Intraparietal sulcus
Divides the superior from the inferior parietal lobule

Superior temporal sulcus

Angular gyrus
Turns a corner around the end of the superior temporal sulcus. Studies of brain function suggest this area may be important in mathematical problem-solving and understanding metaphors

Inferior parietal lobule

Parieto-occipital sulcus
Divides the parietal and occipital lobes

TOP VIEW OF BRAIN

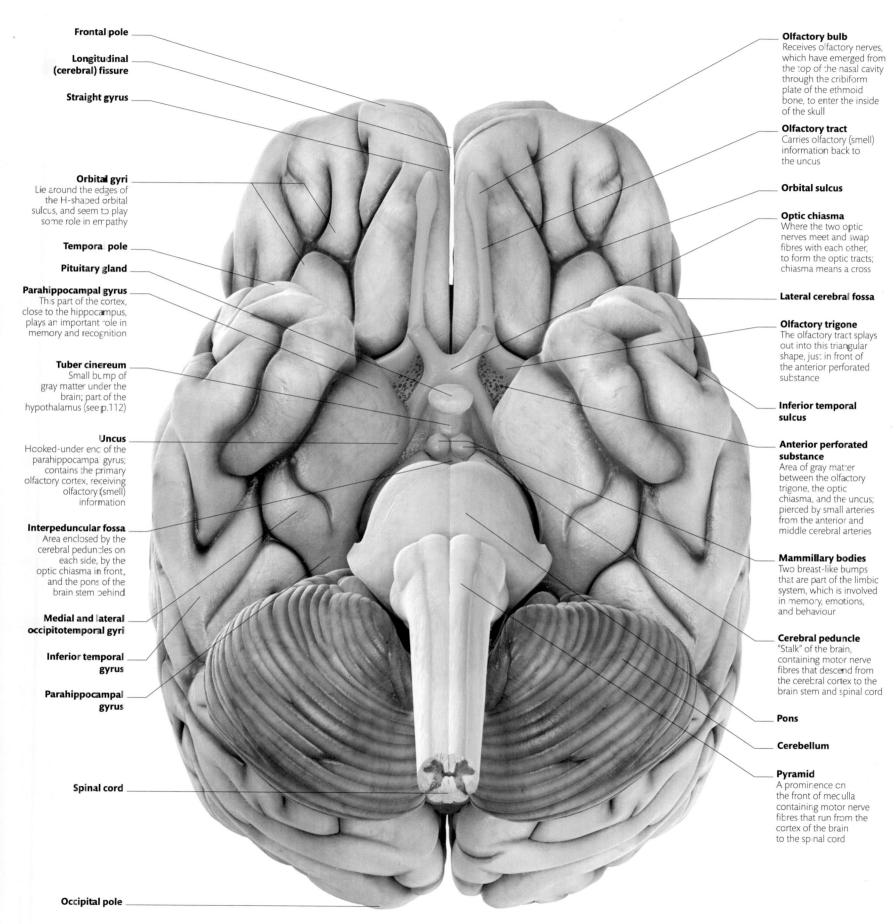

Frontal pole

Longitudinal (cerebral) fissure

Straight gyrus

Orbital gyri
Lie around the edges of the H-shaped orbital sulcus, and seem to play some role in empathy

Temporal pole

Pituitary gland

Parahippocampal gyrus
This part of the cortex, close to the hippocampus, plays an important role in memory and recognition

Tuber cinereum
Small bump of gray matter under the brain; part of the hypothalamus (see p.112)

Uncus
Hooked-under end of the parahippocampal gyrus; contains the primary olfactory cortex, receiving olfactory (smell) information

Interpeduncular fossa
Area enclosed by the cerebral peduncles on each side, by the optic chiasma in front, and the pons of the brain stem behind

Medial and lateral occipitotemporal gyri

Inferior temporal gyrus

Parahippocampal gyrus

Spinal cord

Occipital pole

Olfactory bulb
Receives olfactory nerves, which have emerged from the top of the nasal cavity through the cribiform plate of the ethmoid bone, to enter the inside of the skull

Olfactory tract
Carries olfactory (smell) information back to the uncus

Orbital sulcus

Optic chiasma
Where the two optic nerves meet and swap fibres with each other, to form the optic tracts; chiasma means a cross

Lateral cerebral fossa

Olfactory trigone
The olfactory tract splays out into this triangular shape, just in front of the anterior perforated substance

Inferior temporal sulcus

Anterior perforated substance
Area of gray matter between the olfactory trigone, the optic chiasma, and the uncus; pierced by small arteries from the anterior and middle cerebral arteries

Mammillary bodies
Two breast-like bumps that are part of the limbic system, which is involved in memory, emotions, and behaviour

Cerebral peduncle
"Stalk" of the brain, containing motor nerve fibres that descend from the cerebral cortex to the brain stem and spinal cord

Pons

Cerebellum

Pyramid
A prominence on the front of medulla containing motor nerve fibres that run from the cortex of the brain to the spinal cord

UNDERSIDE OF BRAIN

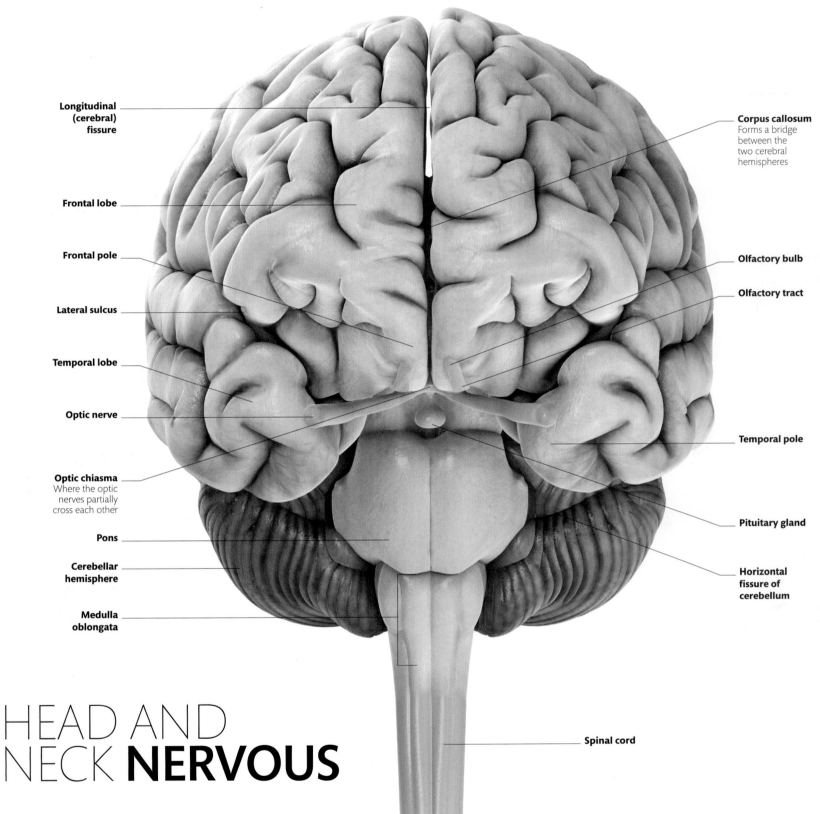

Longitudinal (cerebral) fissure

Frontal lobe

Frontal pole

Lateral sulcus

Temporal lobe

Optic nerve

Optic chiasma
Where the optic nerves partially cross each other

Pons

Cerebellar hemisphere

Medulla oblongata

Corpus callosum
Forms a bridge between the two cerebral hemispheres

Olfactory bulb

Olfactory tract

Temporal pole

Pituitary gland

Horizontal fissure of cerebellum

Spinal cord

HEAD AND NECK **NERVOUS**

The largest part of the brain, the cerebrum, is almost completely divided into two cerebral hemispheres. This division is clearly seen when viewing the brain from the front, back, or top. The fissure between the hemispheres runs deep, but at the bottom of it lies the corpus callosum, which forms a bridge between the two sides. Areas of the brain that receive and process certain types of information, or govern movements, can be very widely separated. The visual pathways from the eyes end in the cortex of the occipital lobe at the back of the brain, and visual information is also processed in this lobe. But the nerve impulses that eventually reach the muscles to move the eyes begin in the cortex of the brain's frontal lobe.

FRONT VIEW OF BRAIN

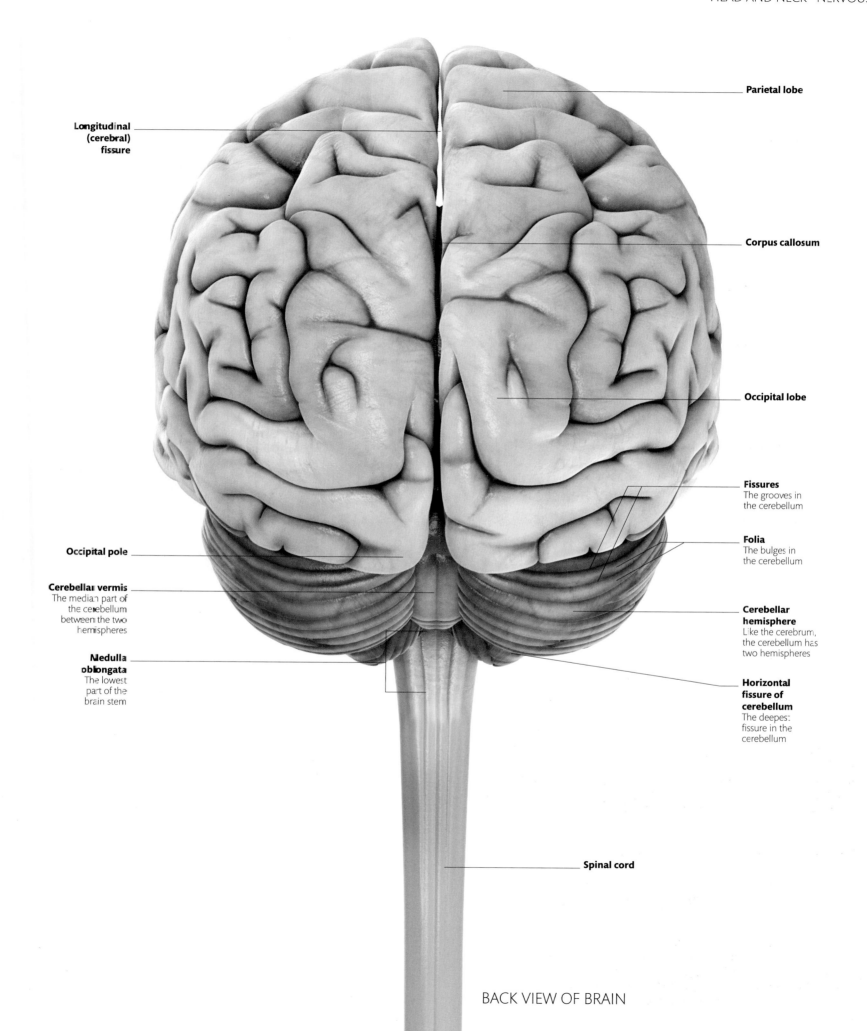

Longitudinal (cerebral) fissure

Parietal lobe

Corpus callosum

Occipital lobe

Fissures
The grooves in the cerebellum

Folia
The bulges in the cerebellum

Occipital pole

Cerebellar vermis
The median part of the cerebellum between the two hemispheres

Cerebellar hemisphere
Like the cerebrum, the cerebellum has two hemispheres

Medulla oblongata
The lowest part of the brain stem

Horizontal fissure of cerebellum
The deepest fissure in the cerebellum

Spinal cord

BACK VIEW OF BRAIN

Body of corpus callosum
The largest commissure (or bundle of connecting nerve fibers) between the two hemispheres, this forms the roofs of the lateral ventricles

Septum pellucidum
This "translucent partition" is a thin dividing wall between the two lateral ventricles

Superior frontal gyrus

Cingulate gyrus
"Cingulum" is the Latin for girdle and this gyrus wraps closely around the corpus callosum; it is part of the limbic system, which is involved with emotional responses and behaviors

Genu of corpus callosum
The anterior (front) end of the corpus callosum is bent over—"genu" means knee in Latin

Anterior commissure
A bundle of nerve fibers connecting parts of the two cerebral hemispheres

Optic chiasma
The crossover point where the two optic nerves meet and swap fibers, then part company as the optic tracts, which continue on each side of the brain toward the thalamus

Hypothalamus
Plays an important role in regulating the internal environment of the body, by keeping a check on body temperature, blood pressure, and blood sugar level, for instance

Pituitary gland
Produces many hormones and forms a link between the brain and endocrine system

Mammillary body
Part of the limbic system of the brain

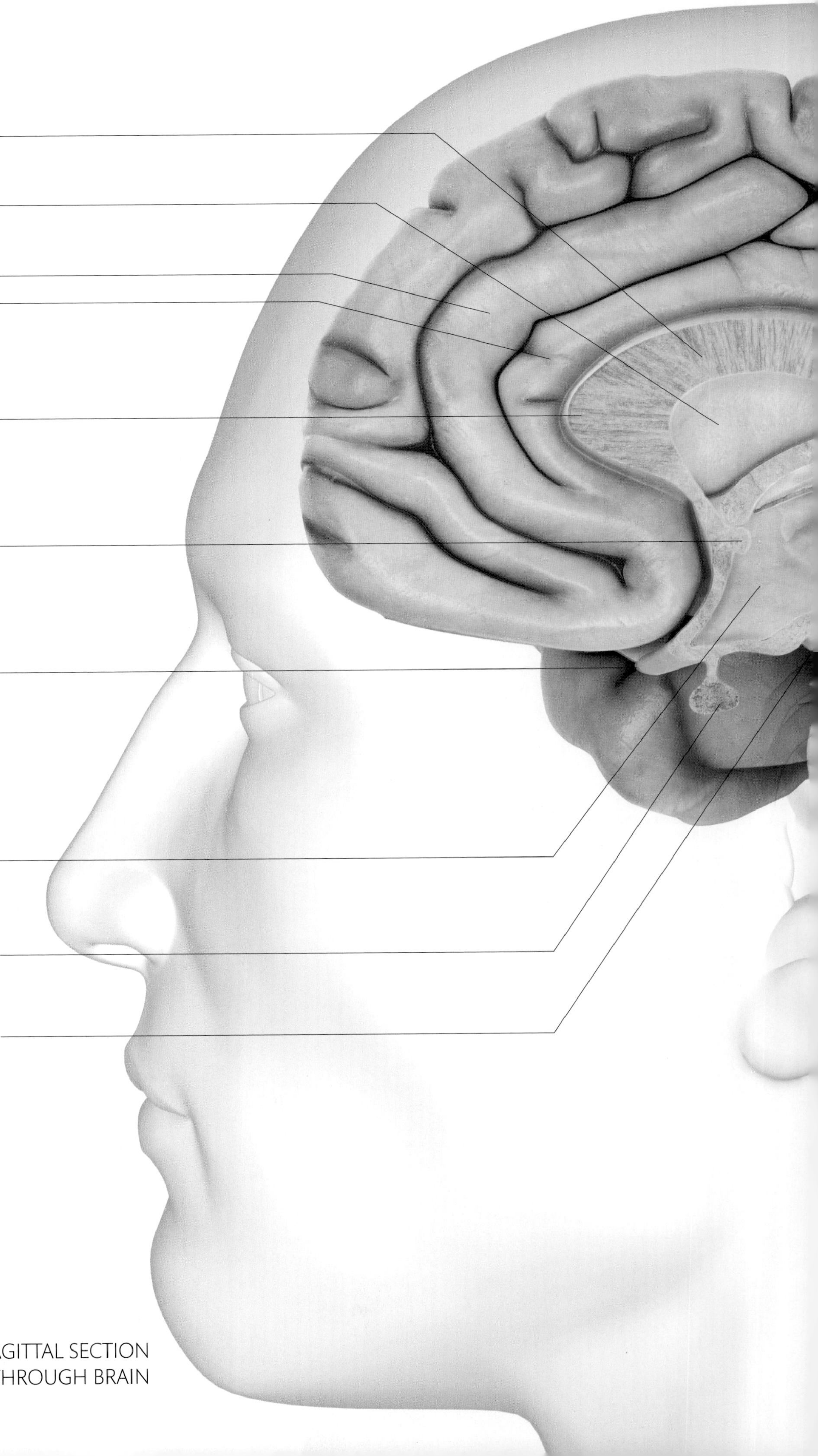

SAGITTAL SECTION
THROUGH BRAIN

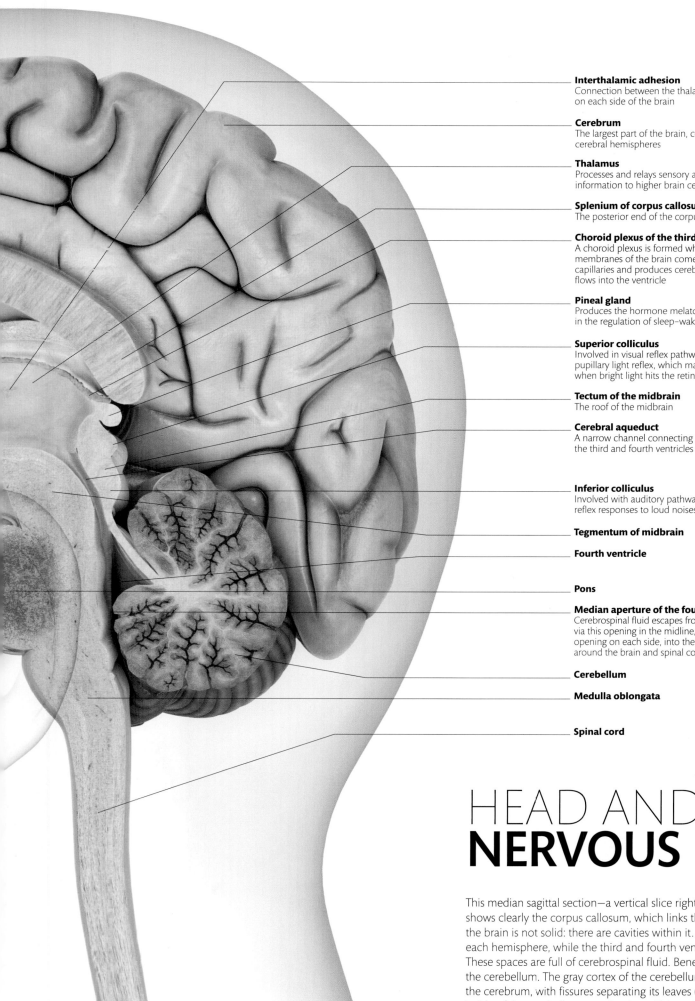

Interthalamic adhesion
Connection between the thalami
on each side of the brain

Cerebrum
The largest part of the brain, consisting of the two
cerebral hemispheres

Thalamus
Processes and relays sensory and motor
information to higher brain centers

Splenium of corpus callosum
The posterior end of the corpus callosum

Choroid plexus of the third ventricle
A choroid plexus is formed where the inner and outer
membranes of the brain come together; it is full of
capillaries and produces cerebrospinal fluid, which
flows into the ventricle

Pineal gland
Produces the hormone melatonin and is involved
in the regulation of sleep-wake cycles

Superior colliculus
Involved in visual reflex pathways, including the
pupillary light reflex, which makes the pupils constrict
when bright light hits the retina

Tectum of the midbrain
The roof of the midbrain

Cerebral aqueduct
A narrow channel connecting
the third and fourth ventricles

Inferior colliculus
Involved with auditory pathways, including
reflex responses to loud noises

Tegmentum of midbrain

Fourth ventricle

Pons

Median aperture of the fourth ventricle
Cerebrospinal fluid escapes from the fourth ventricle
via this opening in the midline, as well as through an
opening on each side, into the subarachnoid space
around the brain and spinal cord

Cerebellum

Medulla oblongata

Spinal cord

HEAD AND NECK
NERVOUS

This median sagittal section—a vertical slice right through the middle of the brain—
shows clearly the corpus callosum, which links the two hemispheres. We also see that
the brain is not solid: there are cavities within it. Two spaces (or ventricles) lie inside
each hemisphere, while the third and fourth ventricles are located on the midline.
These spaces are full of cerebrospinal fluid. Beneath and behind the cerebrum sits
the cerebellum. The gray cortex of the cerebellum is more finely folded than that of
the cerebrum, with fissures separating its leaves (or folia). Sliced through this way, the
inside of the cerebellum reveals a beautiful, treelike pattern. In this section, we can
also see clearly all the parts of the brainstem—the midbrain, pons, and medulla.

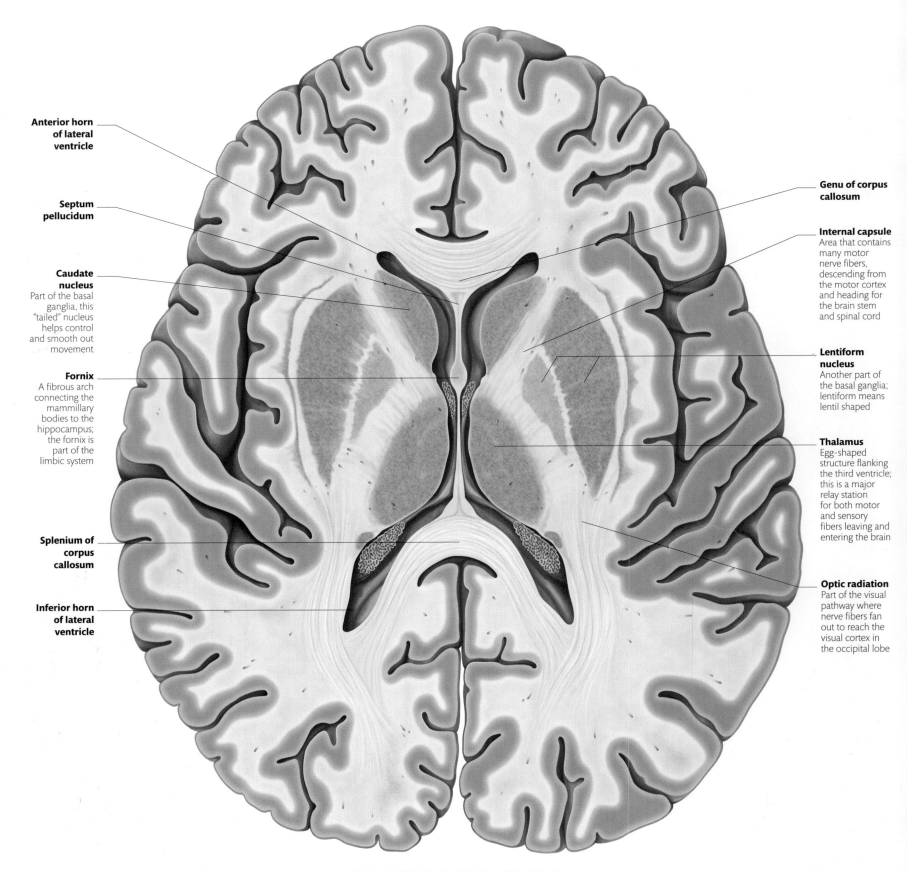

Anterior horn of lateral ventricle

Septum pellucidum

Caudate nucleus
Part of the basal ganglia, this "tailed" nucleus helps control and smooth out movement

Fornix
A fibrous arch connecting the mammillary bodies to the hippocampus; the fornix is part of the limbic system

Splenium of corpus callosum

Inferior horn of lateral ventricle

Genu of corpus callosum

Internal capsule
Area that contains many motor nerve fibers, descending from the motor cortex and heading for the brain stem and spinal cord

Lentiform nucleus
Another part of the basal ganglia; lentiform means lentil shaped

Thalamus
Egg-shaped structure flanking the third ventricle; this is a major relay station for both motor and sensory fibers leaving and entering the brain

Optic radiation
Part of the visual pathway where nerve fibers fan out to reach the visual cortex in the occipital lobe

TRANSVERSE SECTION OF BRAIN

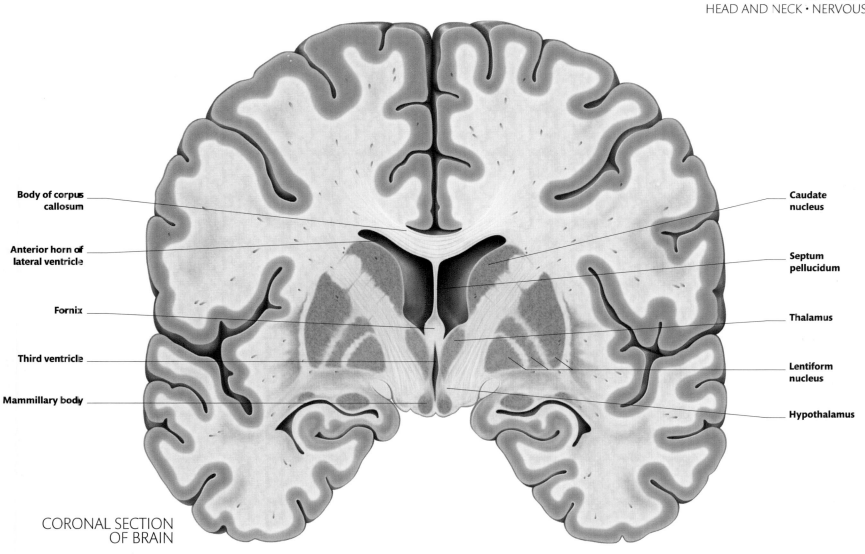

Body of corpus callosum

Anterior horn of lateral ventricle

Fornix

Third ventricle

Mammillary body

Caudate nucleus

Septum pellucidum

Thalamus

Lentiform nucleus

Hypothalamus

CORONAL SECTION OF BRAIN

HEAD AND NECK
NERVOUS

The brain is protected by three membranes called the meninges (which become inflamed in meningitis). The tough dura mater layer is the outermost covering, which surrounds the brain and the spinal cord. Under the dura mater is the cobweblike arachnoid mater layer. The delicate pia mater is a thin membrane on the surface of the brain. Between the pia mater and the arachnoid mater there is a slim gap —the subarachnoid space—which contains cerebrospinal fluid (CSF). Mainly produced by the choroid plexus in the brain's lateral ventricles, CSF flows through the third ventricle into the fourth where it can escape via small apertures into the subarachnoid space.

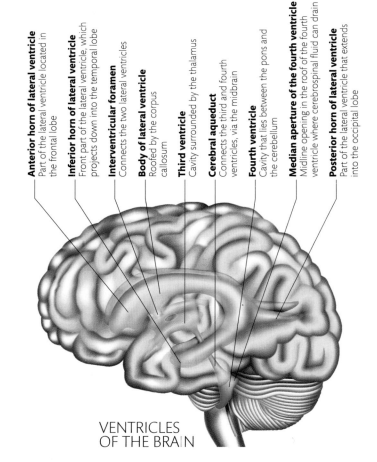

Anterior horn of lateral ventricle
Part of the lateral ventricle located in the frontal lobe

Inferior horn of lateral ventricle
Front part of the lateral ventricle, which projects down into the temporal lobe

Interventricular foramen
Connects the two lateral ventricles

Body of lateral ventricle
Roofed by the corpus callosum

Third ventricle
Cavity surrounded by the thalamus

Cerebral aqueduct
Connects the third and fourth ventricles, via the midbrain

Fourth ventricle
Cavity that lies between the pons and the cerebellum

Median aperture of the fourth ventricle
Midline opening in the roof of the fourth ventricle where cerebrospinal fluid can drain

Posterior horn of lateral ventricle
Part of the lateral ventricle that extends into the occipital lobe

VENTRICLES OF THE BRAIN

Falx cerebri

Pia mater
A thin membrane that is the innermost of the meninges, lining the brain itself

Arachnoid mater
Middle layer of the meninges

Arachnoid granulation
Pocket of the subarachnoid space, where cerebrospinal fluid flows back into the blood

Dura mater
Outer layer of the meninges; dura mater is Latin for hard mother

Superior sagittal sinus

Skull

MENINGES SECTION

HEAD AND NECK
NERVOUS

The 12 pairs of cranial nerves (the standard abbreviation for which is CN) emerge from the brain and brain stem, leaving through holes, or "foramina," in the base of the skull. Some nerves are purely sensory, some just have motor functions, but most contain a mixture of motor and sensory fibers. A few also contain autonomic nerve fibers. The olfactory nerve and the optic nerve attach to the brain itself. The other 10 pairs of cranial nerves emerge from the brain stem. All the cranial nerves supply parts of the head and neck, except the vagus nerve. This has branches in the neck, but then continues on to supply organs in the thorax and right down in the abdomen. Careful testing of cranial nerves, including tests of sight, eye and head movement, taste, and so on, can help doctors pinpoint neurological problems in the head and neck.

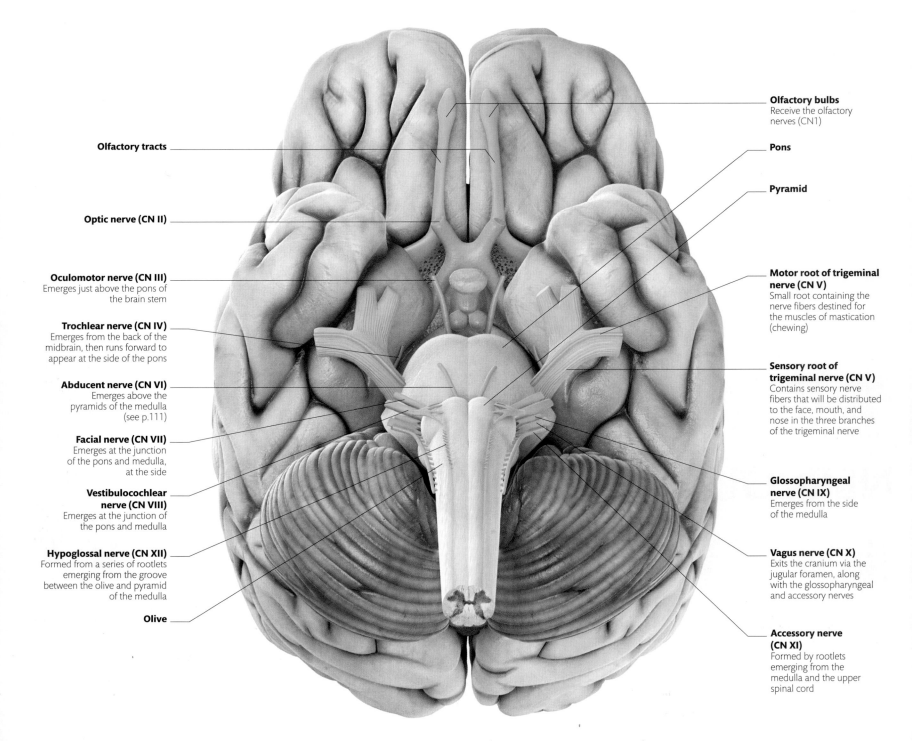

Olfactory tracts

Optic nerve (CN II)

Oculomotor nerve (CN III)
Emerges just above the pons of the brain stem

Trochlear nerve (CN IV)
Emerges from the back of the midbrain, then runs forward to appear at the side of the pons

Abducent nerve (CN VI)
Emerges above the pyramids of the medulla (see p.111)

Facial nerve (CN VII)
Emerges at the junction of the pons and medulla, at the side

Vestibulocochlear nerve (CN VIII)
Emerges at the junction of the pons and medulla

Hypoglossal nerve (CN XII)
Formed from a series of rootlets emerging from the groove between the olive and pyramid of the medulla

Olive

Olfactory bulbs
Receive the olfactory nerves (CN1)

Pons

Pyramid

Motor root of trigeminal nerve (CN V)
Small root containing the nerve fibers destined for the muscles of mastication (chewing)

Sensory root of trigeminal nerve (CN V)
Contains sensory nerve fibers that will be distributed to the face, mouth, and nose in the three branches of the trigeminal nerve

Glossopharyngeal nerve (CN IX)
Emerges from the side of the medulla

Vagus nerve (CN X)
Exits the cranium via the jugular foramen, along with the glossopharyngeal and accessory nerves

Accessory nerve (CN XI)
Formed by rootlets emerging from the medulla and the upper spinal cord

ORIGIN OF CRANIAL NERVES (UNDERSIDE OF BRAIN)

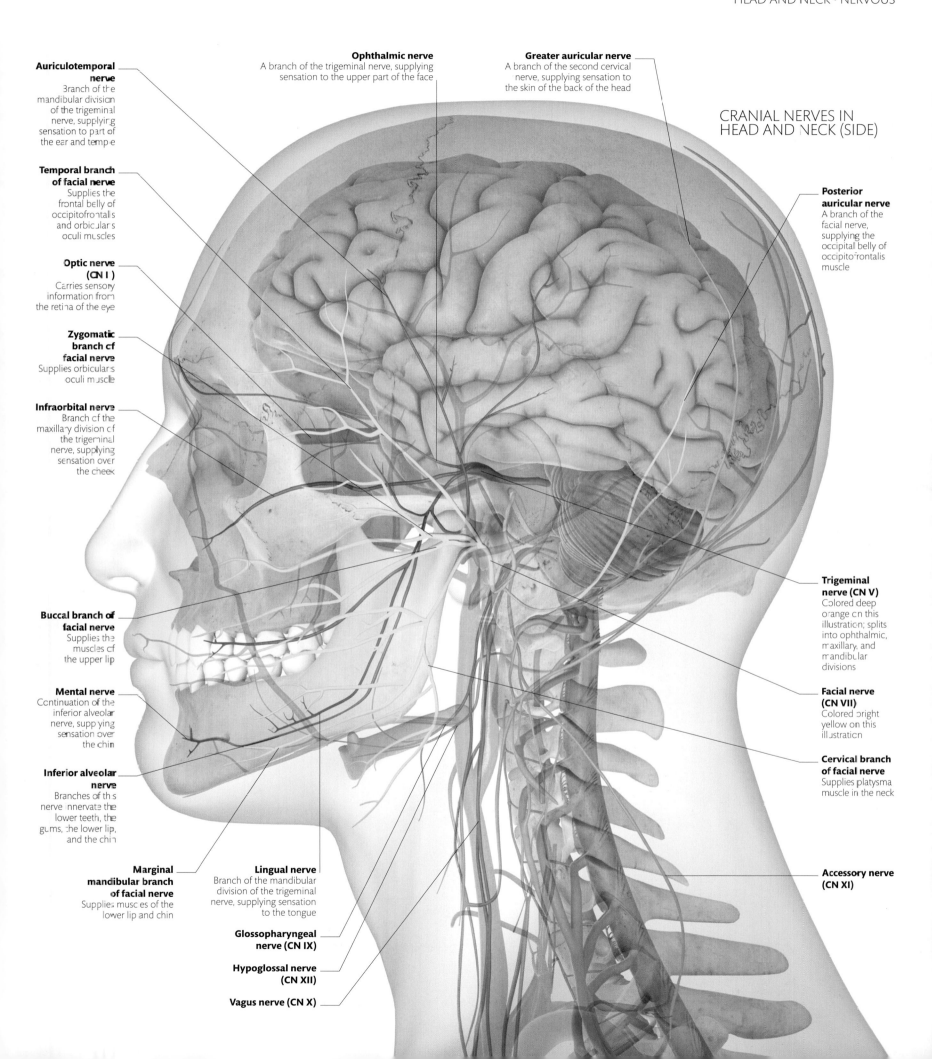

CRANIAL NERVES IN
HEAD AND NECK (SIDE)

Auriculotemporal nerve
Branch of the mandibular division of the trigeminal nerve, supplying sensation to part of the ear and temple

Temporal branch of facial nerve
Supplies the frontal belly of occipitofrontalis and orbicularis oculi muscles

Optic nerve (CN II)
Carries sensory information from the retina of the eye

Zygomatic branch of facial nerve
Supplies orbicularis oculi muscle

Infraorbital nerve
Branch of the maxillary division of the trigeminal nerve, supplying sensation over the cheek

Buccal branch of facial nerve
Supplies the muscles of the upper lip

Mental nerve
Continuation of the inferior alveolar nerve, supplying sensation over the chin

Inferior alveolar nerve
Branches of this nerve innervate the lower teeth, the gums, the lower lip, and the chin

Marginal mandibular branch of facial nerve
Supplies muscles of the lower lip and chin

Lingual nerve
Branch of the mandibular division of the trigeminal nerve, supplying sensation to the tongue

Glossopharyngeal nerve (CN IX)

Hypoglossal nerve (CN XII)

Vagus nerve (CN X)

Ophthalmic nerve
A branch of the trigeminal nerve, supplying sensation to the upper part of the face

Greater auricular nerve
A branch of the second cervical nerve, supplying sensation to the skin of the back of the head

Posterior auricular nerve
A branch of the facial nerve, supplying the occipital belly of occipitofrontalis muscle

Trigeminal nerve (CN V)
Colored deep orange on this illustration; splits into ophthalmic, maxillary, and mandibular divisions

Facial nerve (CN VII)
Colored bright yellow on this illustration

Cervical branch of facial nerve
Supplies platysma muscle in the neck

Accessory nerve (CN XI)

HEAD AND NECK
NERVOUS

EYE

The eyes are precious organs. They are well protected inside the eye sockets, or bony orbits, of the skull. They are also protected by the eyelids, and bathed in tears produced by the lacrimal glands. Each eyeball is only 1 in (2.5 cm) in diameter. The orbit provides an anchor for the muscles that move the eye, and the rest of the space inside the orbit is largely filled up with fat. Holes and fissures at the back of this bony cavern transmit nerves and blood vessels, including the optic nerve, which carries sensory information from the retina to the brain. Other nerves supply the eye muscles and the lacrimal glands, and even continue on to the face to supply sensation to the skin of the eyelids and forehead.

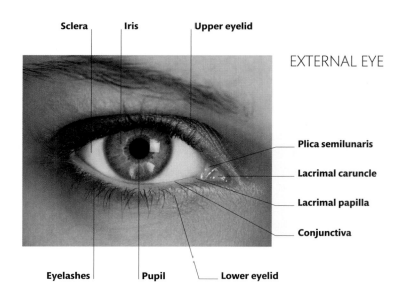

EXTERNAL EYE

Sclera — Iris — Upper eyelid

Plica semilunaris

Lacrimal caruncle

Lacrimal papilla

Conjunctiva

Eyelashes — Pupil — Lower eyelid

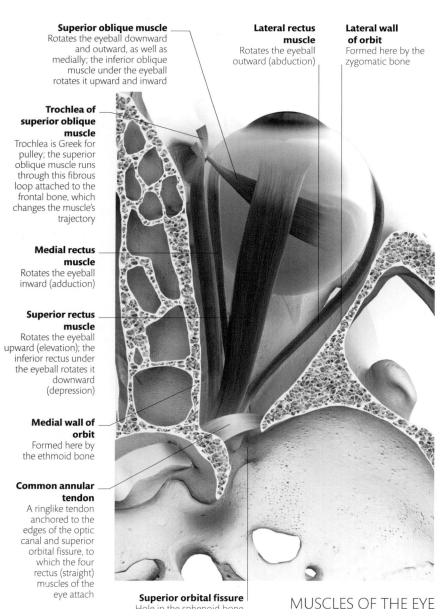

Superior oblique muscle
Rotates the eyeball downward and outward, as well as medially; the inferior oblique muscle under the eyeball rotates it upward and inward

Lateral rectus muscle
Rotates the eyeball outward (abduction)

Lateral wall of orbit
Formed here by the zygomatic bone

Trochlea of superior oblique muscle
Trochlea is Greek for pulley; the superior oblique muscle runs through this fibrous loop attached to the frontal bone, which changes the muscle's trajectory

Medial rectus muscle
Rotates the eyeball inward (adduction)

Superior rectus muscle
Rotates the eyeball upward (elevation); the inferior rectus under the eyeball rotates it downward (depression)

Medial wall of orbit
Formed here by the ethmoid bone

Common annular tendon
A ringlike tendon anchored to the edges of the optic canal and superior orbital fissure, to which the four rectus (straight) muscles of the eye attach

Superior orbital fissure
Hole in the sphenoid bone at the back of the orbit

MUSCLES OF THE EYE
(FROM ABOVE)

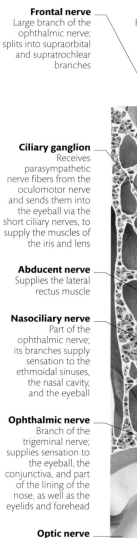

Frontal nerve
Large branch of the ophthalmic nerve; splits into supraorbital and supratrochlear branches

Supratrochlear nerve
Runs over the eyeball and up, out of the orbit, to supply sensation to the middle of the forehead

Supraorbital nerve
Runs forward, out of the orbit, and turns upward on the frontal bone to supply the upper eyelid

Lacrimal nerve
Supplies skin over the upper eyelid and lateral forehead

Ciliary ganglion
Receives parasympathetic nerve fibers from the oculomotor nerve and sends them into the eyeball via the short ciliary nerves, to supply the muscles of the iris and lens

Lacrimal gland

Abducent nerve
Supplies the lateral rectus muscle

Nasociliary nerve
Part of the ophthalmic nerve; its branches supply sensation to the ethmoidal sinuses, the nasal cavity, and the eyeball

Ophthalmic nerve
Branch of the trigeminal nerve; supplies sensation to the eyeball, the conjunctiva, and part of the lining of the nose, as well as the eyelids and forehead

Optic nerve
Carries sensory nerve fibers from the retina

Oculomotor nerve
Supplies all muscles that move the eye, except for the superior oblique and lateral rectus muscles

Trochlear nerve
Supplies the superior oblique muscle

NERVES OF THE ORBIT
(FROM ABOVE)

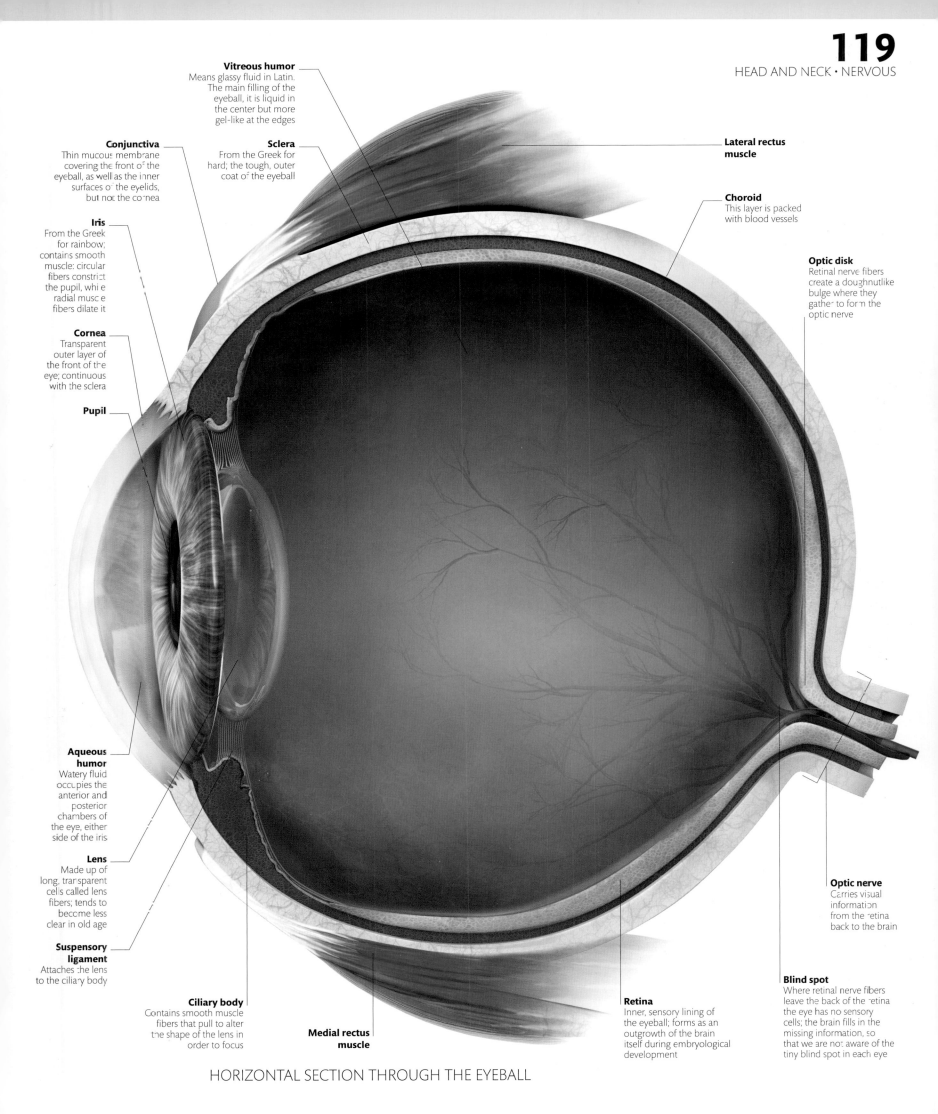

Vitreous humor
Means glassy fluid in Latin. The main filling of the eyeball, it is liquid in the center but more gel-like at the edges

Conjunctiva
Thin mucous membrane covering the front of the eyeball, as well as the inner surfaces of the eyelids, but not the cornea

Sclera
From the Greek for hard; the tough, outer coat of the eyeball

Lateral rectus muscle

Choroid
This layer is packed with blood vessels

Iris
From the Greek for rainbow; contains smooth muscle: circular fibers constrict the pupil, while radial muscle fibers dilate it

Optic disk
Retinal nerve fibers create a doughnutlike bulge where they gather to form the optic nerve

Cornea
Transparent outer layer of the front of the eye; continuous with the sclera

Pupil

Aqueous humor
Watery fluid occupies the anterior and posterior chambers of the eye, either side of the iris

Optic nerve
Carries visual information from the retina back to the brain

Lens
Made up of long, transparent cells called lens fibers; tends to become less clear in old age

Suspensory ligament
Attaches the lens to the ciliary body

Ciliary body
Contains smooth muscle fibers that pull to alter the shape of the lens in order to focus

Medial rectus muscle

Retina
Inner, sensory lining of the eyeball; forms as an outgrowth of the brain itself during embryological development

Blind spot
Where retinal nerve fibers leave the back of the retina the eye has no sensory cells; the brain fills in the missing information, so that we are not aware of the tiny blind spot in each eye

HORIZONTAL SECTION THROUGH THE EYEBALL

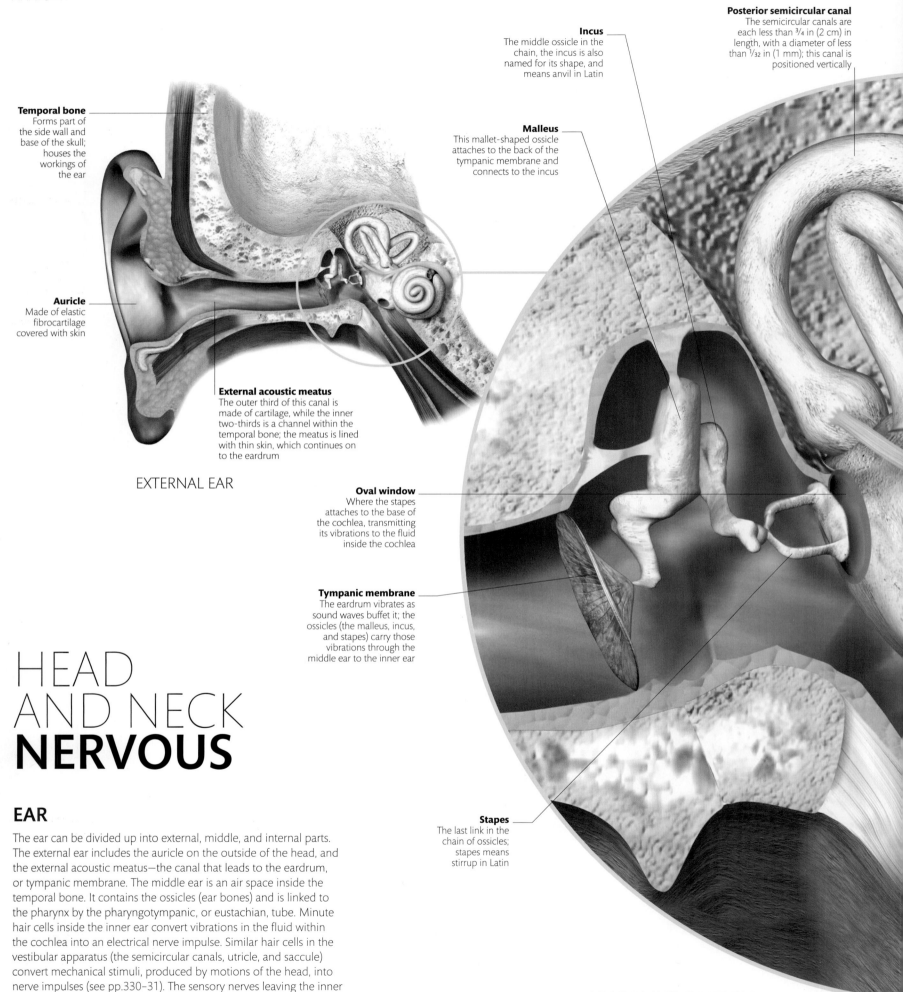

Temporal bone
Forms part of the side wall and base of the skull; houses the workings of the ear

Auricle
Made of elastic fibrocartilage covered with skin

External acoustic meatus
The outer third of this canal is made of cartilage, while the inner two-thirds is a channel within the temporal bone; the meatus is lined with thin skin, which continues on to the eardrum

EXTERNAL EAR

Incus
The middle ossicle in the chain, the incus is also named for its shape, and means anvil in Latin

Malleus
This mallet-shaped ossicle attaches to the back of the tympanic membrane and connects to the incus

Posterior semicircular canal
The semicircular canals are each less than ¾ in (2 cm) in length, with a diameter of less than ¹⁄₃₂ in (1 mm); this canal is positioned vertically

Oval window
Where the stapes attaches to the base of the cochlea, transmitting its vibrations to the fluid inside the cochlea

Tympanic membrane
The eardrum vibrates as sound waves buffet it; the ossicles (the malleus, incus, and stapes) carry those vibrations through the middle ear to the inner ear

Stapes
The last link in the chain of ossicles; stapes means stirrup in Latin

MIDDLE AND INNER EAR

HEAD
AND NECK
NERVOUS

EAR

The ear can be divided up into external, middle, and internal parts. The external ear includes the auricle on the outside of the head, and the external acoustic meatus—the canal that leads to the eardrum, or tympanic membrane. The middle ear is an air space inside the temporal bone. It contains the ossicles (ear bones) and is linked to the pharynx by the pharyngotympanic, or eustachian, tube. Minute hair cells inside the inner ear convert vibrations in the fluid within the cochlea into an electrical nerve impulse. Similar hair cells in the vestibular apparatus (the semicircular canals, utricle, and saccule) convert mechanical stimuli, produced by motions of the head, into nerve impulses (see pp.330–31). The sensory nerves leaving the inner ear join to form the vestibulocochlear nerve.

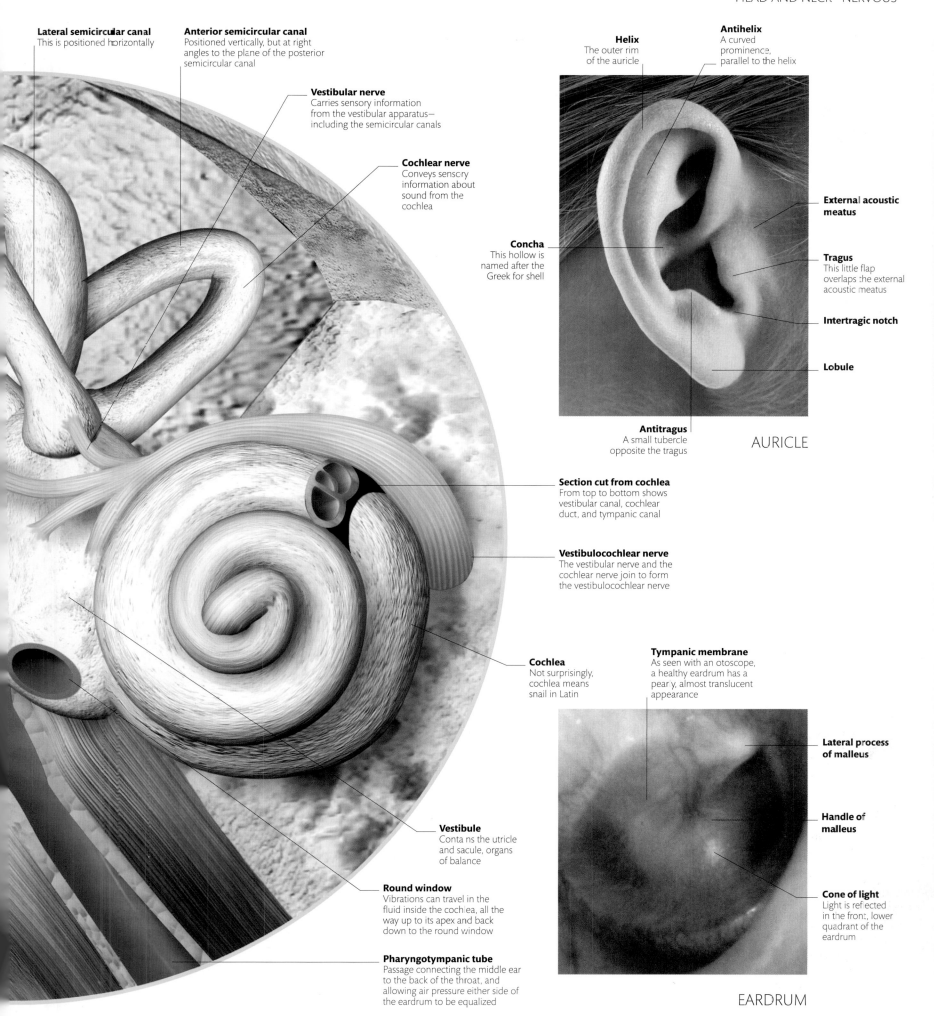

Lateral semicircular canal
This is positioned horizontally

Anterior semicircular canal
Positioned vertically, but at right angles to the plane of the posterior semicircular canal

Vestibular nerve
Carries sensory information from the vestibular apparatus— including the semicircular canals

Cochlear nerve
Conveys sensory information about sound from the cochlea

Helix
The outer rim of the auricle

Antihelix
A curved prominence, parallel to the helix

External acoustic meatus

Concha
This hollow is named after the Greek for shell

Tragus
This little flap overlaps the external acoustic meatus

Intertragic notch

Lobule

Antitragus
A small tubercle opposite the tragus

AURICLE

Section cut from cochlea
From top to bottom shows vestibular canal, cochlear duct, and tympanic canal

Vestibulocochlear nerve
The vestibular nerve and the cochlear nerve join to form the vestibulocochlear nerve

Cochlea
Not surprisingly, cochlea means snail in Latin

Tympanic membrane
As seen with an otoscope, a healthy eardrum has a pearly, almost translucent appearance

Lateral process of malleus

Handle of malleus

Vestibule
Contains the utricle and sacule, organs of balance

Round window
Vibrations can travel in the fluid inside the cochlea, all the way up to its apex and back down to the round window

Cone of light
Light is reflected in the front, lower quadrant of the eardrum

Pharyngotympanic tube
Passage connecting the middle ear to the back of the throat, and allowing air pressure either side of the eardrum to be equalized

EARDRUM

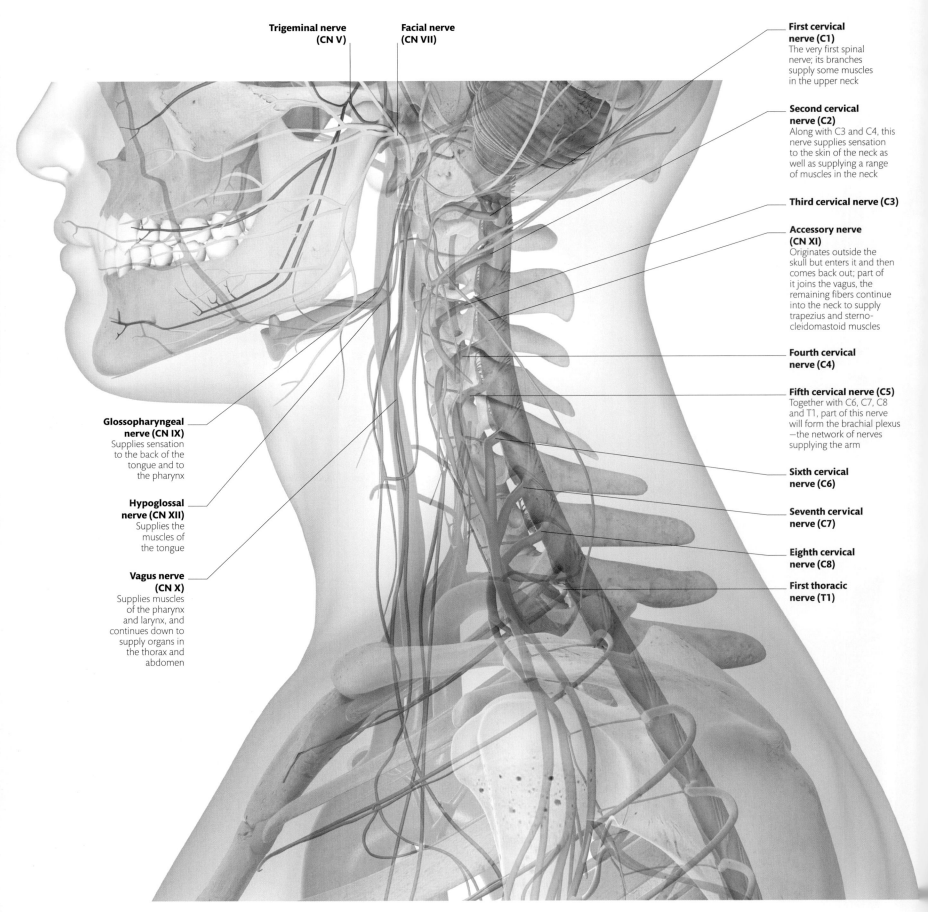

Trigeminal nerve (CN V)

Facial nerve (CN VII)

First cervical nerve (C1)
The very first spinal nerve; its branches supply some muscles in the upper neck

Second cervical nerve (C2)
Along with C3 and C4, this nerve supplies sensation to the skin of the neck as well as supplying a range of muscles in the neck

Third cervical nerve (C3)

Accessory nerve (CN XI)
Originates outside the skull but enters it and then comes back out; part of it joins the vagus, the remaining fibers continue into the neck to supply trapezius and sterno-cleidomastoid muscles

Fourth cervical nerve (C4)

Fifth cervical nerve (C5)
Together with C6, C7, C8 and T1, part of this nerve will form the brachial plexus —the network of nerves supplying the arm

Sixth cervical nerve (C6)

Seventh cervical nerve (C7)

Eighth cervical nerve (C8)

First thoracic nerve (T1)

Glossopharyngeal nerve (CN IX)
Supplies sensation to the back of the tongue and to the pharynx

Hypoglossal nerve (CN XII)
Supplies the muscles of the tongue

Vagus nerve (CN X)
Supplies muscles of the pharynx and larynx, and continues down to supply organs in the thorax and abdomen

NERVES OF THE NECK (SIDE)

HEAD AND NECK
NERVOUS

The last four cranial nerves all appear in the neck. The glossopharyngeal nerve supplies the parotid gland and the back of the tongue, then runs down to the pharynx. The vagus nerve is sandwiched between the common carotid artery and the internal jugular vein, and it gives branches to the pharynx and larynx before continuing down into the thorax. The accessory nerve supplies the sternocleidomastoid and trapezius muscles in the neck, while the last cranial nerve, the hypoglossal, dips down below the mandible, then curves back up to supply the muscles of the tongue. We can also see spinal nerves in the neck. The upper four cervical nerves supply neck muscles and skin, while the lower four contribute to the brachial plexus and are destined for the arm.

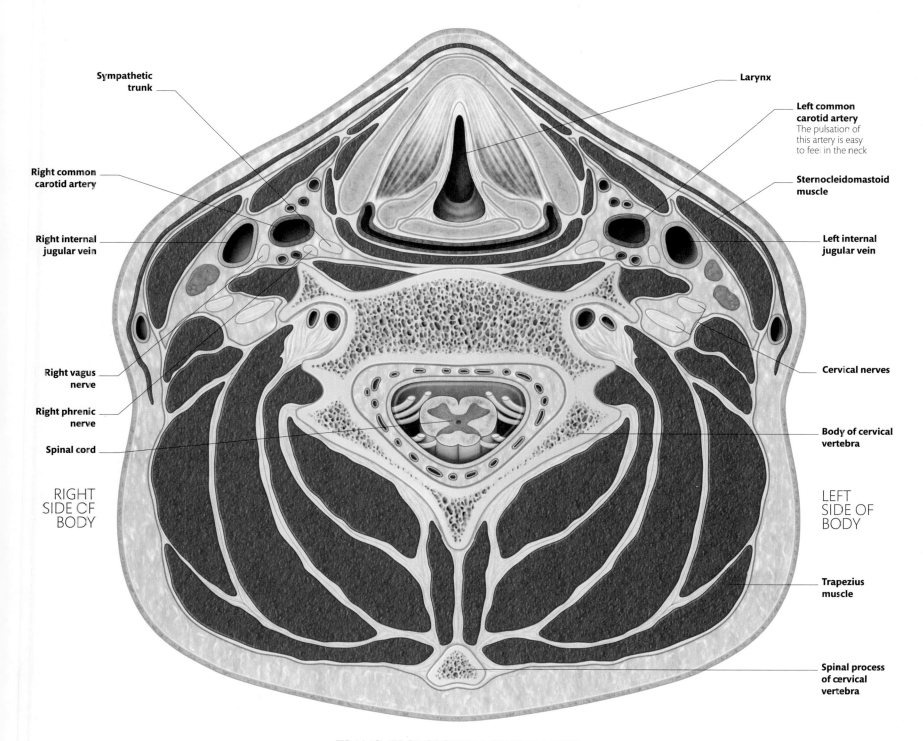

Sympathetic trunk

Larynx

Left common carotid artery
The pulsation of this artery is easy to feel in the neck

Right common carotid artery

Sternocleidomastoid muscle

Right internal jugular vein

Left internal jugular vein

Right vagus nerve

Cervical nerves

Right phrenic nerve

Spinal cord

Body of cervical vertebra

RIGHT SIDE OF BODY

LEFT SIDE OF BODY

Trapezius muscle

Spinal process of cervical vertebra

TRANSVERSE SECTION OF THE NECK

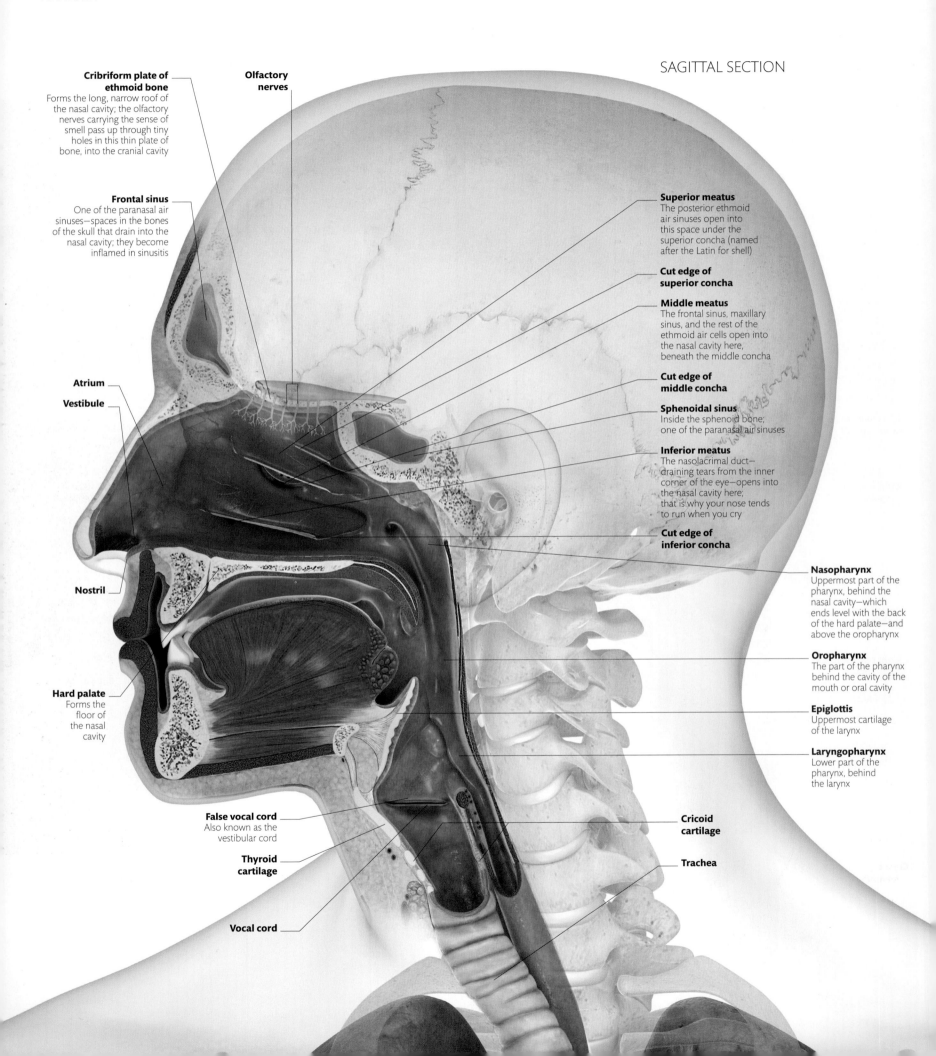

Cribriform plate of ethmoid bone
Forms the long, narrow roof of the nasal cavity; the olfactory nerves carrying the sense of smell pass up through tiny holes in this thin plate of bone, into the cranial cavity

Olfactory nerves

Frontal sinus
One of the paranasal air sinuses—spaces in the bones of the skull that drain into the nasal cavity; they become inflamed in sinusitis

Atrium

Vestibule

Nostril

Hard palate
Forms the floor of the nasal cavity

False vocal cord
Also known as the vestibular cord

Thyroid cartilage

Vocal cord

Superior meatus
The posterior ethmoid air sinuses open into this space under the superior concha (named after the Latin for shell)

Cut edge of superior concha

Middle meatus
The frontal sinus, maxillary sinus, and the rest of the ethmoid air cells open into the nasal cavity here, beneath the middle concha

Cut edge of middle concha

Sphenoidal sinus
Inside the sphenoid bone; one of the paranasal air sinuses

Inferior meatus
The nasolacrimal duct—draining tears from the inner corner of the eye—opens into the nasal cavity here; that is why your nose tends to run when you cry

Cut edge of inferior concha

Nasopharynx
Uppermost part of the pharynx, behind the nasal cavity—which ends level with the back of the hard palate—and above the oropharynx

Oropharynx
The part of the pharynx behind the cavity of the mouth or oral cavity

Epiglottis
Uppermost cartilage of the larynx

Laryngopharynx
Lower part of the pharynx, behind the larynx

Cricoid cartilage

Trachea

HEAD AND NECK
RESPIRATORY

When we take a breath, air is pulled in through our nostrils, into the nasal cavities. Here the air is cleaned, warmed, and moistened before its onward journey. The nasal cavities are divided by the thin partition of the nasal septum, which is composed of plates of cartilage and bone. The lateral walls of the nasal cavity are more elaborate, with bony curls (conchae) that increase the surface area over which the air flows. The nasal cavity is lined with mucosa, which produces mucus. This often undervalued substance does an important job of trapping particles and moistening the air. The nasal sinuses, also lined with mucosa, open via tiny orifices into the nasal cavity. Below and in front of the pharynx is the larynx—the organ of speech. The way that air passes through this can be modulated to produce sound.

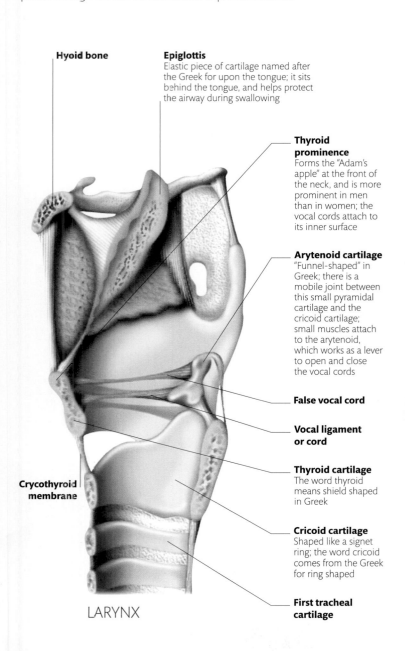

Hyoid bone

Epiglottis
Elastic piece of cartilage named after the Greek for upon the tongue; it sits behind the tongue, and helps protect the airway during swallowing

Thyroid prominence
Forms the "Adam's apple" at the front of the neck, and is more prominent in men than in women; the vocal cords attach to its inner surface

Arytenoid cartilage
"Funnel-shaped" in Greek; there is a mobile joint between this small pyramidal cartilage and the cricoid cartilage; small muscles attach to the arytenoid, which works as a lever to open and close the vocal cords

False vocal cord

Vocal ligament or cord

Thyroid cartilage
The word thyroid means shield shaped in Greek

Crycothyroid membrane

Cricoid cartilage
Shaped like a signet ring; the word cricoid comes from the Greek for ring shaped

First tracheal cartilage

LARYNX

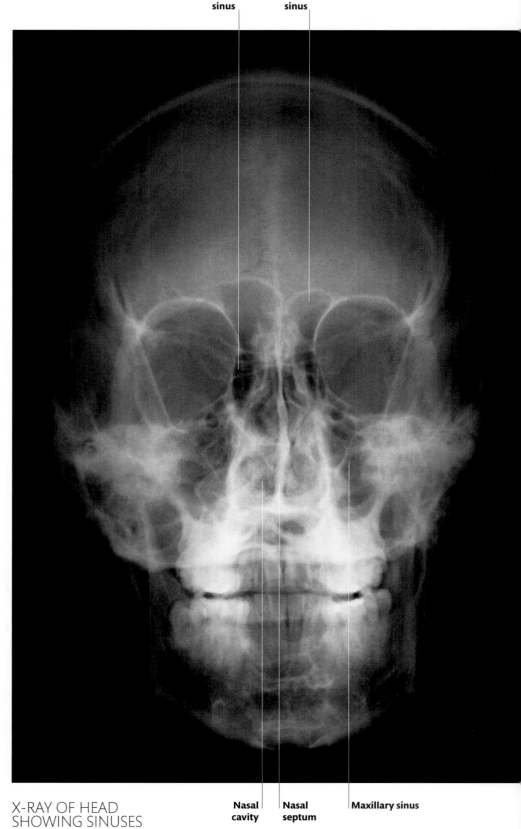

Ethmoid sinus

Frontal sinus

Nasal cavity

Nasal septum

Maxillary sinus

X-RAY OF HEAD
SHOWING SINUSES

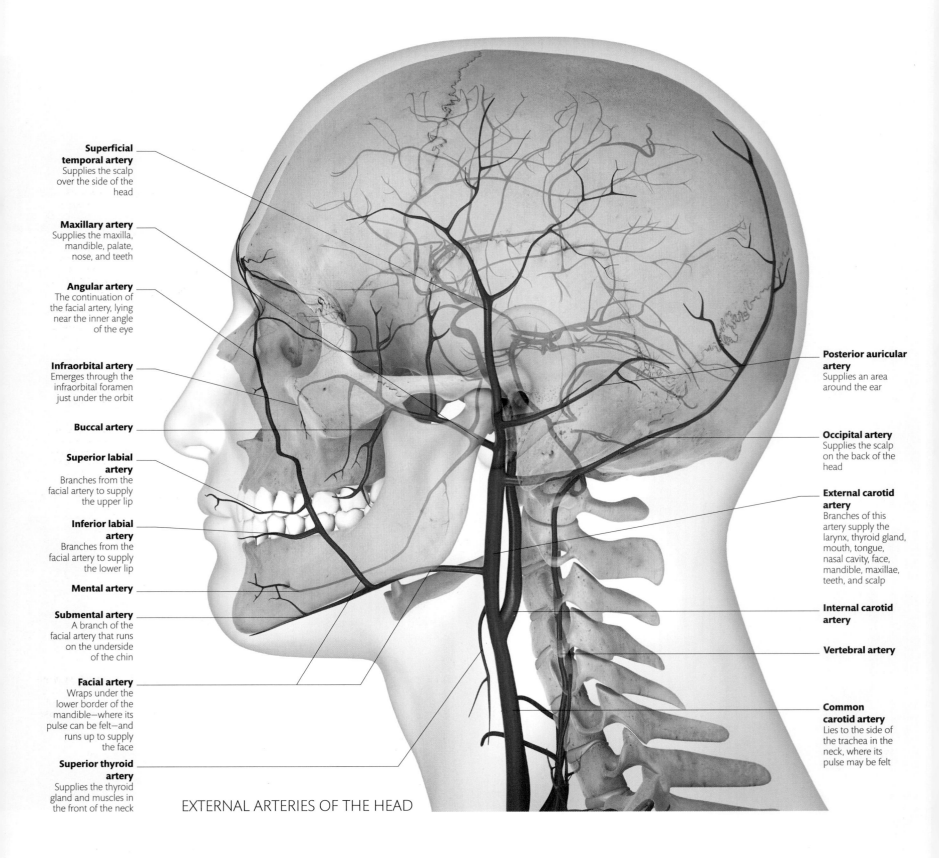

Superficial temporal artery
Supplies the scalp over the side of the head

Maxillary artery
Supplies the maxilla, mandible, palate, nose, and teeth

Angular artery
The continuation of the facial artery, lying near the inner angle of the eye

Infraorbital artery
Emerges through the infraorbital foramen just under the orbit

Buccal artery

Superior labial artery
Branches from the facial artery to supply the upper lip

Inferior labial artery
Branches from the facial artery to supply the lower lip

Mental artery

Submental artery
A branch of the facial artery that runs on the underside of the chin

Facial artery
Wraps under the lower border of the mandible—where its pulse can be felt—and runs up to supply the face

Superior thyroid artery
Supplies the thyroid gland and muscles in the front of the neck

Posterior auricular artery
Supplies an area around the ear

Occipital artery
Supplies the scalp on the back of the head

External carotid artery
Branches of this artery supply the larynx, thyroid gland, mouth, tongue, nasal cavity, face, mandible, maxillae, teeth, and scalp

Internal carotid artery

Vertebral artery

Common carotid artery
Lies to the side of the trachea in the neck, where its pulse may be felt

EXTERNAL ARTERIES OF THE HEAD

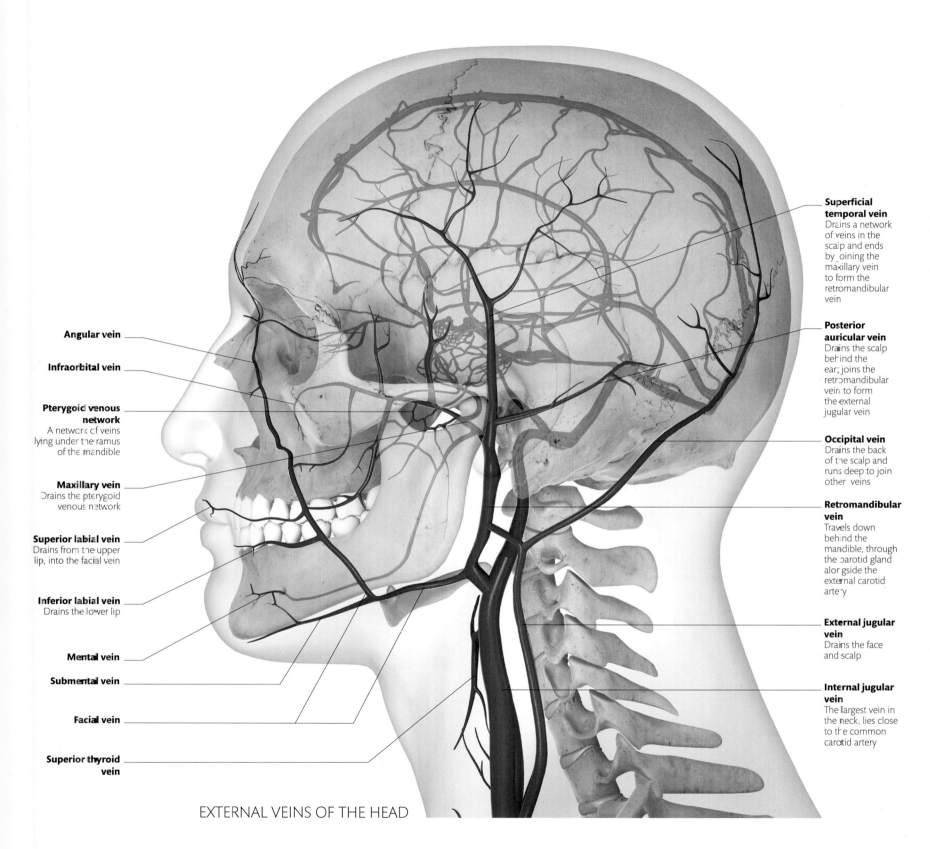

Superficial temporal vein
Drains a network of veins in the scalp and ends by joining the maxillary vein to form the retromandibular vein

Posterior auricular vein
Drains the scalp behind the ear; joins the retromandibular vein to form the external jugular vein

Occipital vein
Drains the back of the scalp and runs deep to join other veins

Retromandibular vein
Travels down behind the mandible, through the parotid gland alongside the external carotid artery

External jugular vein
Drains the face and scalp

Internal jugular vein
The largest vein in the neck, lies close to the common carotid artery

Angular vein

Infraorbital vein

Pterygoid venous network
A network of veins lying under the ramus of the mandible

Maxillary vein
Drains the pterygoid venous network

Superior labial vein
Drains from the upper lip, into the facial vein

Inferior labial vein
Drains the lower lip

Mental vein

Submental vein

Facial vein

Superior thyroid vein

EXTERNAL VEINS OF THE HEAD

HEAD AND NECK
CARDIOVASCULAR

The main vessels supplying oxygenated blood to the head and neck are the common carotid and vertebral arteries. The vertebral artery runs up through holes in the cervical vertebrae and eventually enters the skull through the foramen magnum. The common carotid artery runs up the neck and divides into two—the internal carotid artery supplies the brain, and the external carotid artery gives rise to a profusion of branches, some of which supply the thyroid gland, the mouth, tongue, and nasal cavity. Veins of the head and neck come together like river tributaries, draining into the large internal jugular vein, behind the sternocleidomastoid muscle, and into the subclavian vein, low in the neck.

The brain has a rich blood supply, which arrives via the internal carotid and vertebral arteries. The vertebral arteries join together to form the basilar artery. The internal carotid arteries and basilar artery join on the undersurface of the brain to form the Circle of Willis. From there, three pairs of cerebral arteries make their way into the brain. The veins of the brain and the skull drain into venous sinuses, which are enclosed within the dura mater (the outermost layer of the meninges) and form grooves on the inner surface of the skull. The sinuses join up and eventually drain out of the base of the skull, into the internal jugular vein.

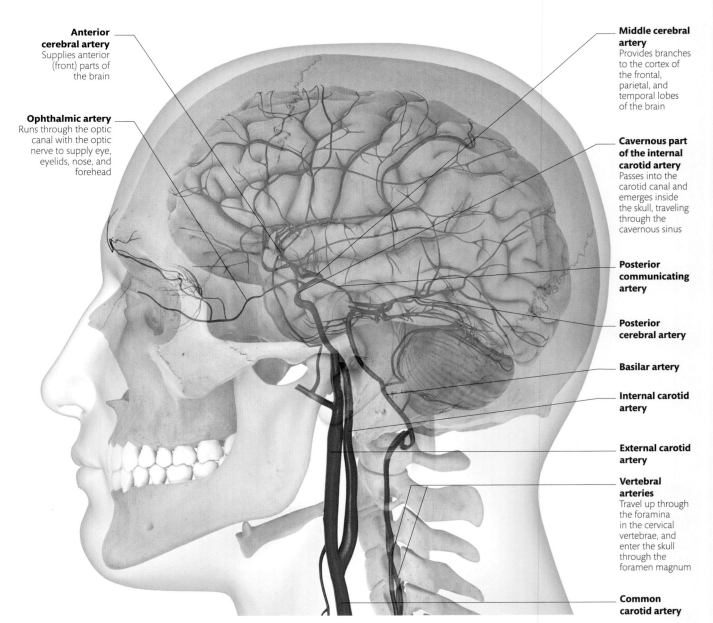

Anterior cerebral artery
Supplies anterior (front) parts of the brain

Ophthalmic artery
Runs through the optic canal with the optic nerve to supply eye, eyelids, nose, and forehead

Middle cerebral artery
Provides branches to the cortex of the frontal, parietal, and temporal lobes of the brain

Cavernous part of the internal carotid artery
Passes into the carotid canal and emerges inside the skull, traveling through the cavernous sinus

Posterior communicating artery

Posterior cerebral artery

Basilar artery

Internal carotid artery

External carotid artery

Vertebral arteries
Travel up through the foramina in the cervical vertebrae, and enter the skull through the foramen magnum

Common carotid artery

ARTERIES AROUND THE BRAIN

HEAD AND NECK
CARDIOVASCULAR

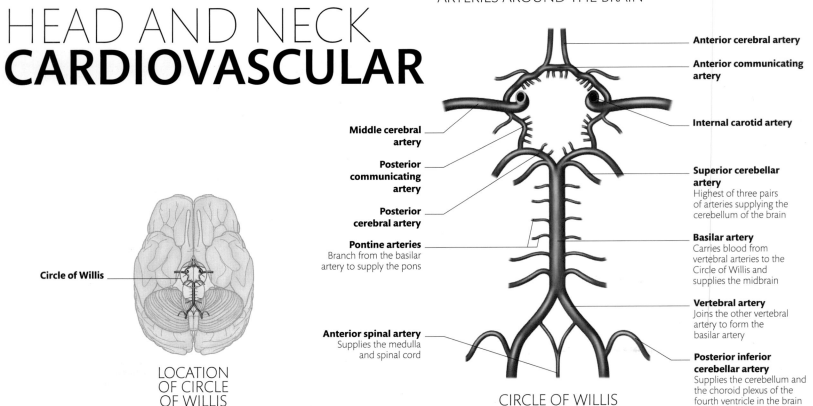

Circle of Willis

LOCATION OF CIRCLE OF WILLIS

Middle cerebral artery

Posterior communicating artery

Posterior cerebral artery

Pontine arteries
Branch from the basilar artery to supply the pons

Anterior spinal artery
Supplies the medulla and spinal cord

Anterior cerebral artery

Anterior communicating artery

Internal carotid artery

Superior cerebellar artery
Highest of three pairs of arteries supplying the cerebellum of the brain

Basilar artery
Carries blood from vertebral arteries to the Circle of Willis and supplies the midbrain

Vertebral artery
Joins the other vertebral artery to form the basilar artery

Posterior inferior cerebellar artery
Supplies the cerebellum and the choroid plexus of the fourth ventricle in the brain

CIRCLE OF WILLIS

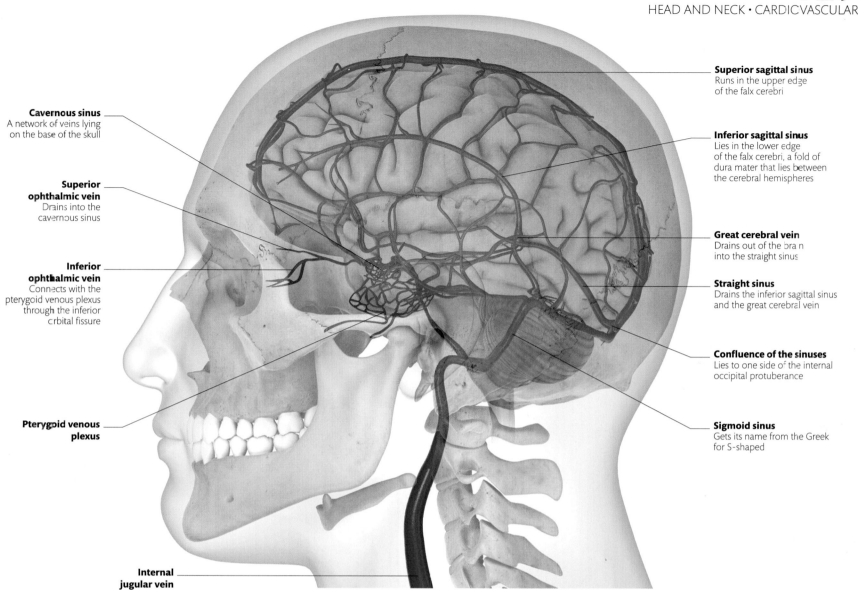

Superior sagittal sinus
Runs in the upper edge
of the falx cerebri

Inferior sagittal sinus
Lies in the lower edge
of the falx cerebri, a fold of
dura mater that lies between
the cerebral hemispheres

Cavernous sinus
A network of veins lying
on the base of the skull

**Superior
ophthalmic vein**
Drains into the
cavernous sinus

**Inferior
ophthalmic vein**
Connects with the
pterygoid venous plexus
through the inferior
orbital fissure

Great cerebral vein
Drains out of the brain
into the straight sinus

Straight sinus
Drains the inferior sagittal sinus
and the great cerebral vein

Confluence of the sinuses
Lies to one side of the internal
occipital protuberance

**Pterygoid venous
plexus**

Sigmoid sinus
Gets its name from the Greek
for S-shaped

**Internal
jugular vein**

VEINS AROUND THE BRAIN

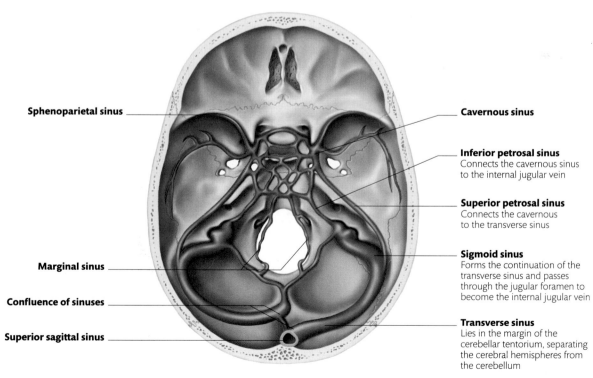

Sphenoparietal sinus

Cavernous sinus

Inferior petrosal sinus
Connects the cavernous sinus
to the internal jugular vein

Superior petrosal sinus
Connects the cavernous
to the transverse sinus

Marginal sinus

Sigmoid sinus
Forms the continuation of the
transverse sinus and passes
through the jugular foramen to
become the internal jugular vein

Confluence of sinuses

Superior sagittal sinus

Transverse sinus
Lies in the margin of the
cerebellar tentorium, separating
the cerebral hemispheres from
the cerebellum

DURAL VENOUS SINUSES

LYMPH NODES OF HEAD

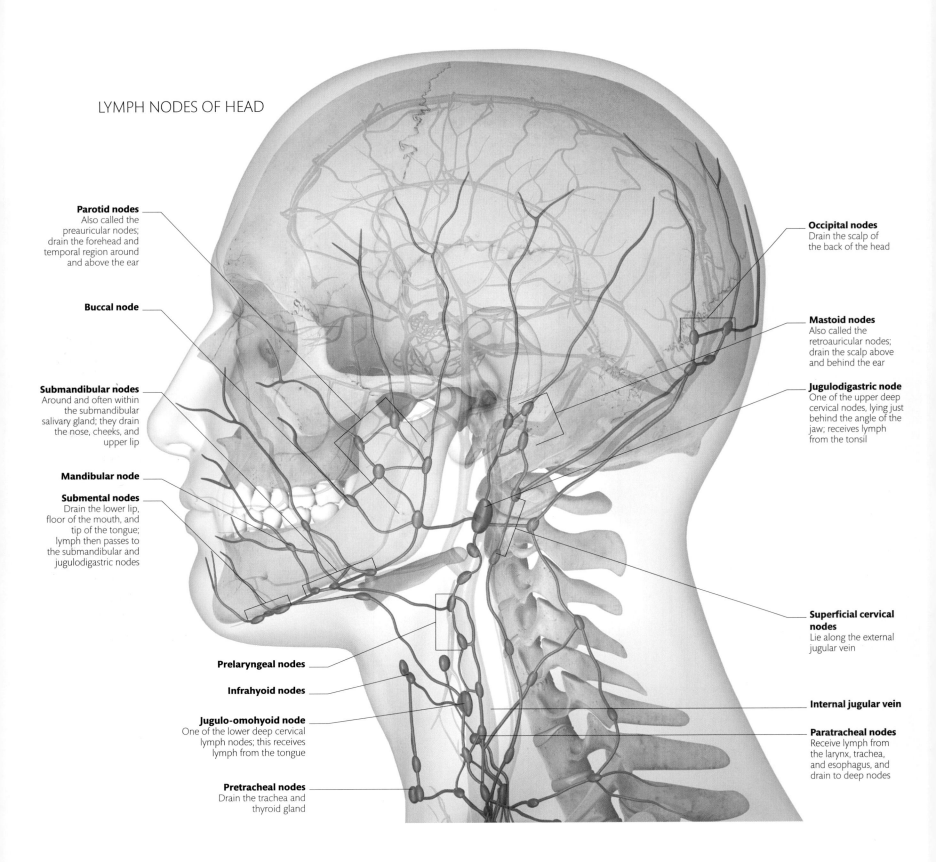

Parotid nodes
Also called the preauricular nodes; drain the forehead and temporal region around and above the ear

Buccal node

Submandibular nodes
Around and often within the submandibular salivary gland; they drain the nose, cheeks, and upper lip

Mandibular node

Submental nodes
Drain the lower lip, floor of the mouth, and tip of the tongue; lymph then passes to the submandibular and jugulodigastric nodes

Prelaryngeal nodes

Infrahyoid nodes

Jugulo-omohyoid node
One of the lower deep cervical lymph nodes; this receives lymph from the tongue

Pretracheal nodes
Drain the trachea and thyroid gland

Occipital nodes
Drain the scalp of the back of the head

Mastoid nodes
Also called the retroauricular nodes; drain the scalp above and behind the ear

Jugulodigastric node
One of the upper deep cervical nodes, lying just behind the angle of the jaw; receives lymph from the tonsil

Superficial cervical nodes
Lie along the external jugular vein

Internal jugular vein

Paratracheal nodes
Receive lymph from the larynx, trachea, and esophagus, and drain to deep nodes

HEAD AND NECK
LYMPHATIC AND IMMUNE

LOCATION OF TONSILS

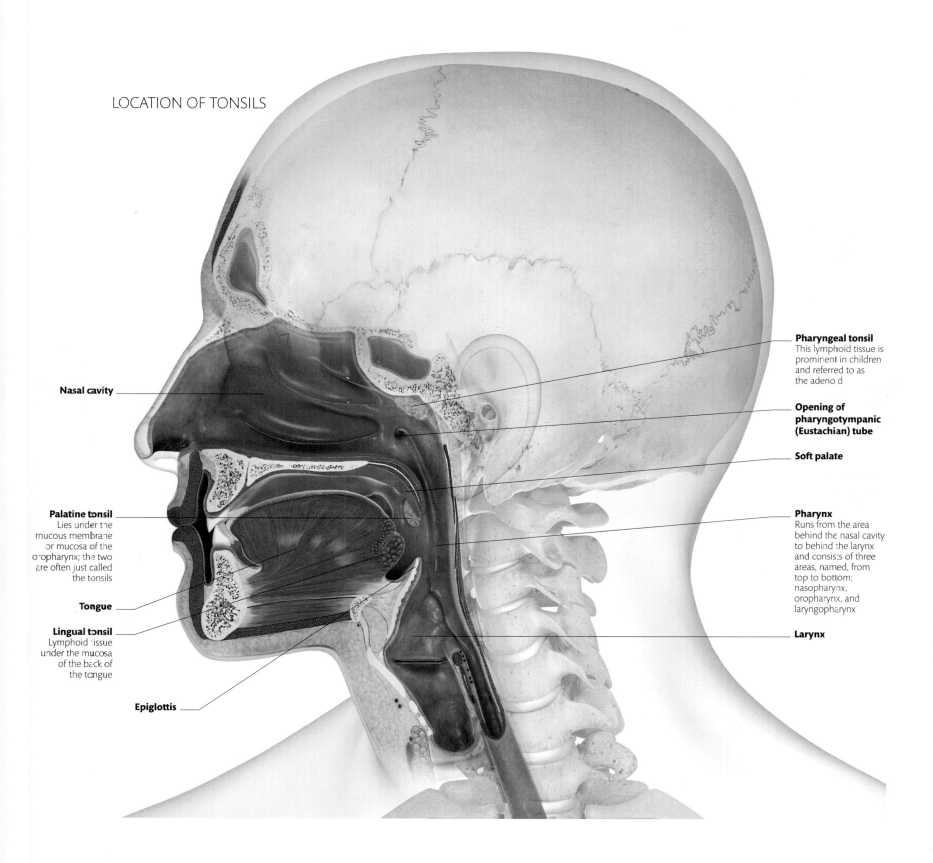

Nasal cavity

Palatine tonsil
Lies under the mucous membrane or mucosa of the oropharynx; the two are often just called the tonsils

Tongue

Lingual tonsil
Lymphoid tissue under the mucosa of the back of the tongue

Epiglottis

Pharyngeal tonsil
This lymphoid tissue is prominent in children and referred to as the adenoid

Opening of pharyngotympanic (Eustachian) tube

Soft palate

Pharynx
Runs from the area behind the nasal cavity to behind the larynx and consists of three areas, named, from top to bottom: nasopharynx, oropharynx, and laryngopharynx

Larynx

A ring of lymph nodes lies close to the skin where the head meets the neck, from the occipital nodes (against the skull at the back) to the submandibular and submental nodes (which are tucked under the jaw). Superficial nodes lie along the sides and front of the neck, and deep nodes are clustered around the internal jugular vein, under cover of the sternocleidomastoid muscle. Lymph from all other nodes passes to these deep ones, then into the jugular lymphatic trunk before draining back into veins in the base of the neck. Lymphoid tissue, in the form of the palatine, pharyngeal, and lingual tonsils, forms a protective ring around the upper parts of the respiratory and digestive tracts.

SAGITTAL SECTION

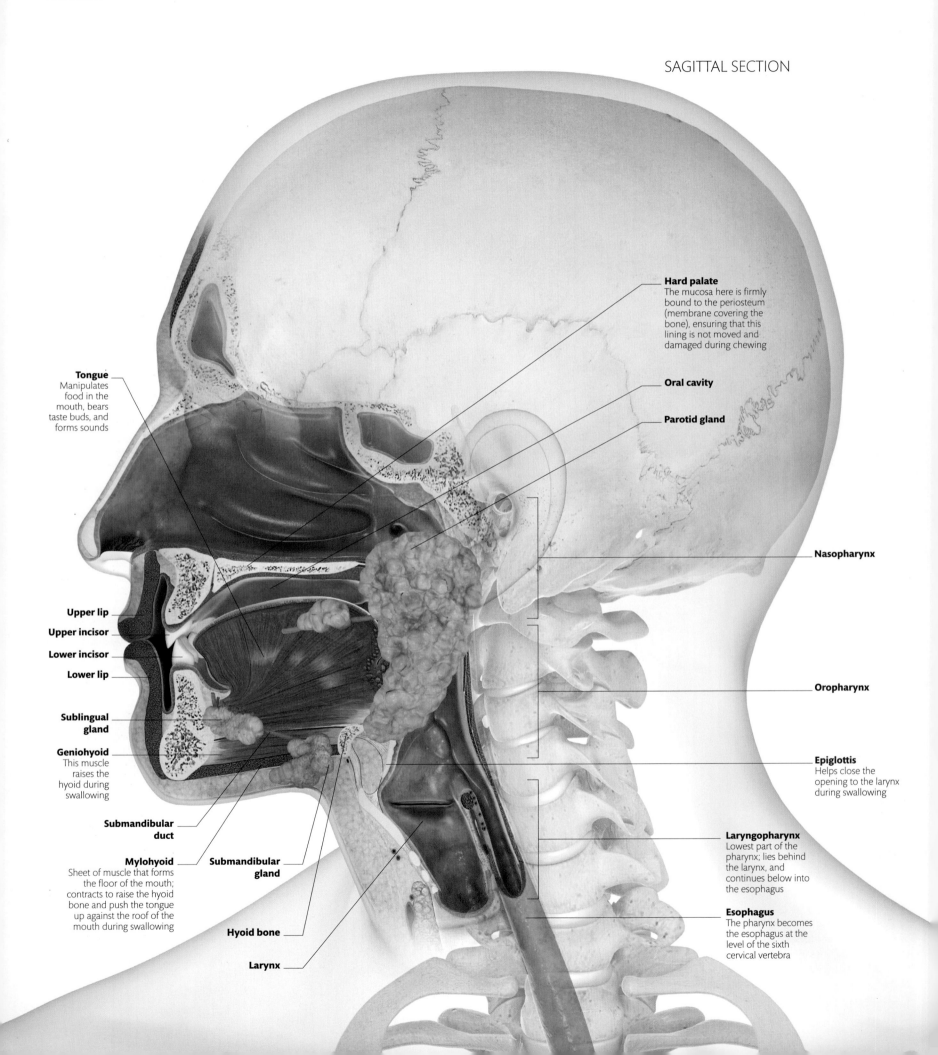

Hard palate
The mucosa here is firmly bound to the periosteum (membrane covering the bone), ensuring that this lining is not moved and damaged during chewing

Oral cavity

Parotid gland

Nasopharynx

Oropharynx

Epiglottis
Helps close the opening to the larynx during swallowing

Laryngopharynx
Lowest part of the pharynx; lies behind the larynx, and continues below into the esophagus

Esophagus
The pharynx becomes the esophagus at the level of the sixth cervical vertebra

Tongue
Manipulates food in the mouth, bears taste buds, and forms sounds

Upper lip

Upper incisor

Lower incisor

Lower lip

Sublingual gland

Geniohyoid
This muscle raises the hyoid during swallowing

Submandibular duct

Mylohyoid
Sheet of muscle that forms the floor of the mouth; contracts to raise the hyoid bone and push the tongue up against the roof of the mouth during swallowing

Submandibular gland

Hyoid bone

Larynx

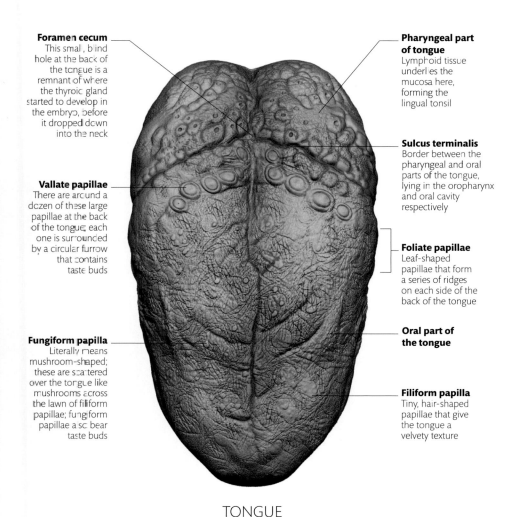

Foramen cecum
This small, blind hole at the back of the tongue is a remnant of where the thyroid gland started to develop in the embryo, before it dropped down into the neck

Pharyngeal part of tongue
Lymphoid tissue underlies the mucosa here, forming the lingual tonsil

Sulcus terminalis
Border between the pharyngeal and oral parts of the tongue, lying in the oropharynx and oral cavity respectively

Vallate papillae
There are around a dozen of these large papillae at the back of the tongue, each one is surrounded by a circular furrow that contains taste buds

Foliate papillae
Leaf-shaped papillae that form a series of ridges on each side of the back of the tongue

Oral part of the tongue

Fungiform papilla
Literally means mushroom-shaped; these are scattered over the tongue like mushrooms across the lawn of filiform papillae; fungiform papillae also bear taste buds

Filiform papilla
Tiny, hair-shaped papillae that give the tongue a velvety texture

TONGUE

HEAD AND NECK
DIGESTIVE

The mouth is the first part of the digestive tract, and it is here that the processes of mechanical and chemical digestion get underway. Your teeth grind each mouthful, and you have three pairs of major salivary glands—parotid, submandibular, and sublingual—that secrete saliva through ducts into the mouth. Saliva contains digestive enzymes that begin to chemically break down the food in your mouth. The tongue manipulates the food, and also has taste buds that allow you to quickly make the important distinction between delicious food and potentially harmful toxins. As you swallow, the tongue pushes up against the hard palate, the soft palate seals off the airway, and the muscular tube of the pharynx contracts in a wave to push the ball of food down into the esophagus, ready for the next stage of its journey.

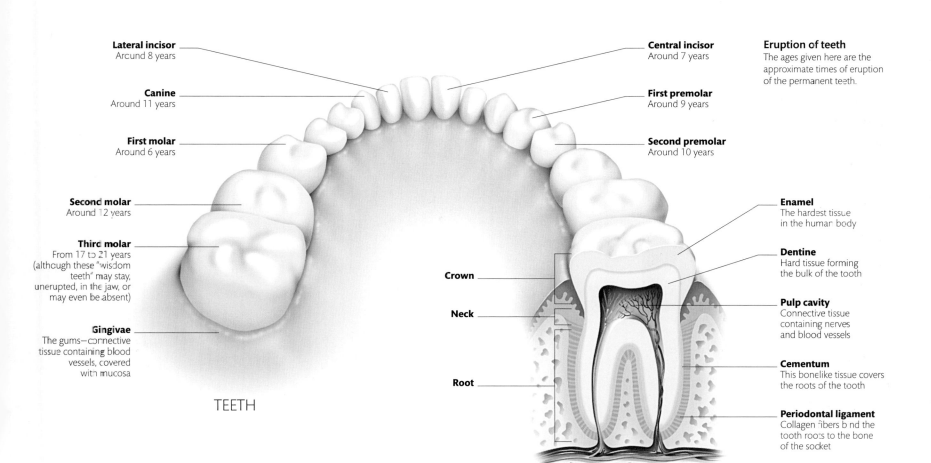

Lateral incisor
Around 8 years

Central incisor
Around 7 years

Eruption of teeth
The ages given here are the approximate times of eruption of the permanent teeth.

Canine
Around 11 years

First premolar
Around 9 years

First molar
Around 6 years

Second premolar
Around 10 years

Second molar
Around 12 years

Enamel
The hardest tissue in the human body

Third molar
From 17 to 21 years (although these "wisdom teeth" may stay, unerupted, in the jaw, or may even be absent)

Dentine
Hard tissue forming the bulk of the tooth

Crown

Pulp cavity
Connective tissue containing nerves and blood vessels

Neck

Cementum
This bonelike tissue covers the roots of the tooth

Gingivae
The gums—connective tissue containing blood vessels, covered with mucosa

Root

Periodontal ligament
Collagen fibers bind the tooth roots to the bone of the socket

TEETH

ENDOCRINE
SYSTEM

The insides of our bodies are regulated by the
autonomic nervous and endocrine systems. There
is overlap between these two systems, and their
functions are integrated and controlled within the
hypothalamus of the brain. The pituitary gland has
two lobes; its posterior lobe develops as a direct
extension of the hypothalamus (see pp.400–01).
Both lobes of the pituitary gland secrete hormones
into the bloodstream, in response to nerve signals
or blood-borne releasing factors from the
hypothalamus. Many of the pituitary hormones
act on other endocrine glands, including the
thyroid gland in the neck, the suprarenal glands
on top of the kidneys, and the ovaries or testes.

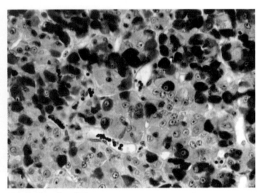

Pituitary gland tissue
Some hormone-secreting cells in the
anterior pituitary appear stained red in
this image, including those that produce
growth hormone, others are stained blue.

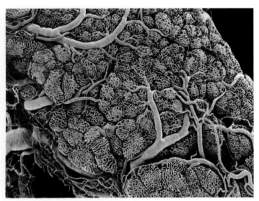

Thyroid blood supply
This resin cast of the thyroid gland
shows capillaries wrapped around
secretory cells (rounded), which release
hormones into the bloodstream.

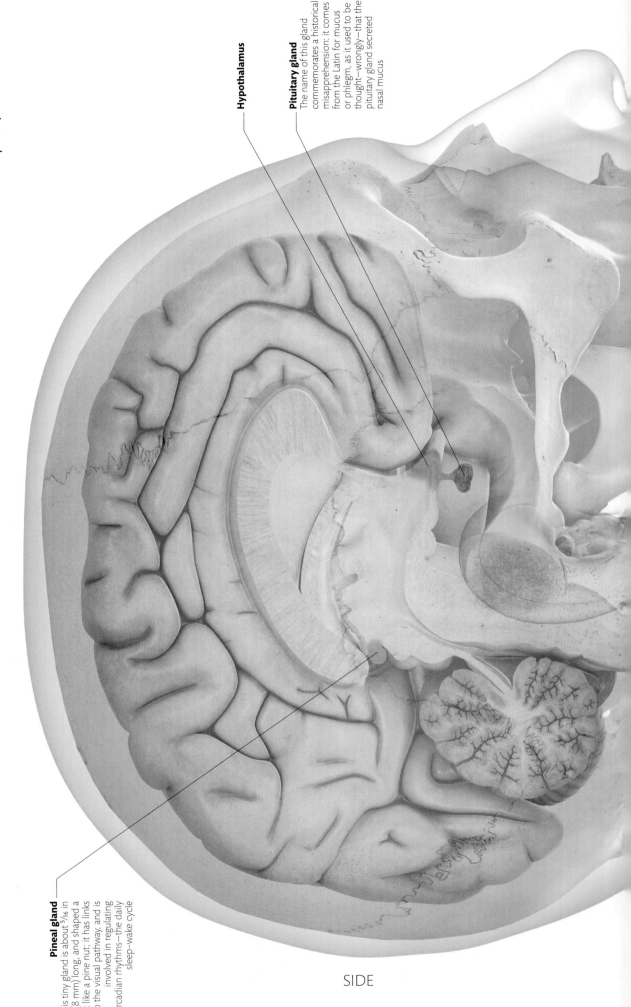

Hypothalamus

Pituitary gland
The name of this gland
commemorates a historical
misapprehension: it comes
from the Latin for mucus
or phlegm, as it used to be
thought–wrongly–that the
pituitary gland secreted
nasal mucus

Pineal gland
This tiny gland is about 5/16 in
(8 mm) long, and shaped a
bit like a pine nut; it has links
to the visual pathway, and is
involved in regulating
circadian rhythms–the daily
sleep-wake cycle

SIDE

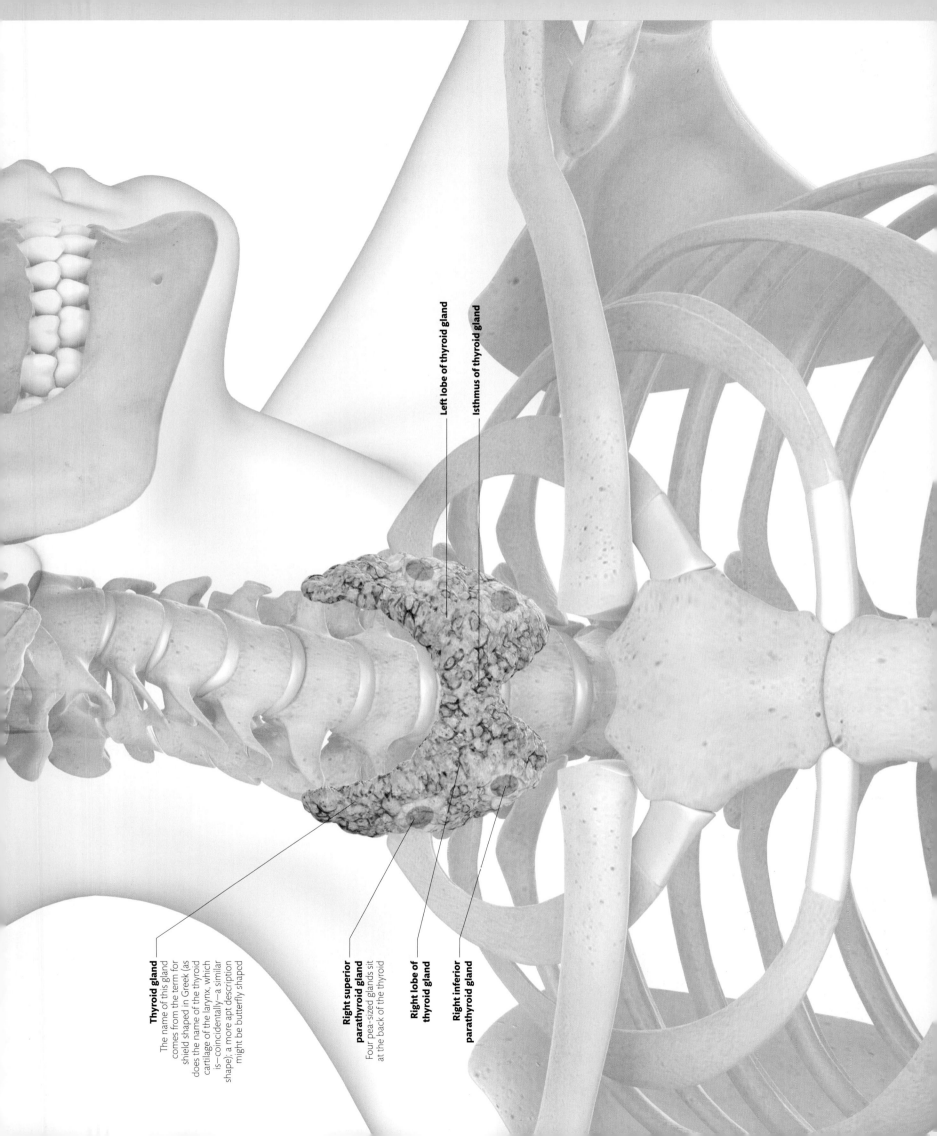

Left lobe of thyroid gland

Isthmus of thyroid gland

Thyroid gland
The name of this gland comes from the term for shield shaped in Greek (as does the name of the thyroid cartilage of the larynx, which is—coincidentally—a similar shape); a more apt description might be butterfly shaped

Right superior parathyroid gland
Four pea-sized glands sit at the back of the thyroid

Right lobe of thyroid gland

Right inferior parathyroid gland

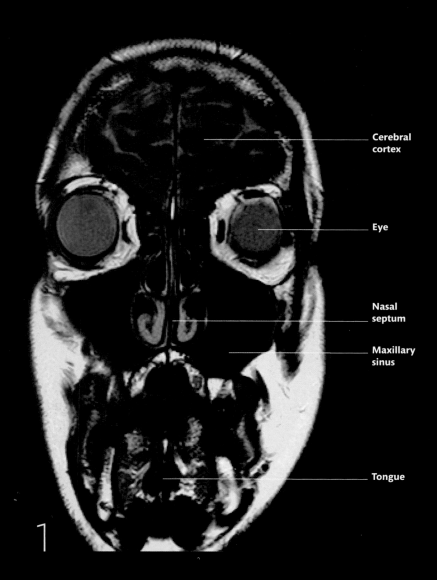

Cerebral cortex

Eye

Nasal septum

Maxillary sinus

Tongue

1

Cingulate gyrus

Frontal sinus

Meninges

Nasal cavity

Teeth

Soft palate

Tongue

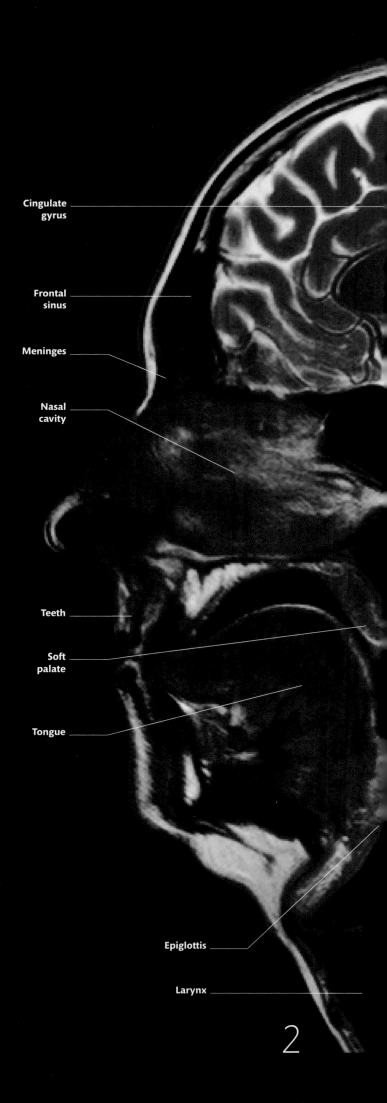

Epiglottis

Larynx

2

LEVELS OF SCANS

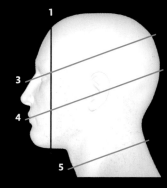

1
3
4
5

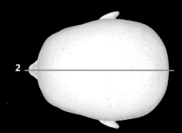

2

HEAD AND NECK **MRI**

The discovery of X-rays at the end of the 19th century suddenly created the possibility of looking inside the human body—without having to physically cut it open. Medical imaging is now an important diagnostic tool, as well as being used for the study of normal anatomy and physiology. In computed tomography (CT), X-rays are used to produce virtual sections or slices through the body. Another form of sectional imaging, using magnetic fields rather than X-rays to create images, is magnetic resonance imaging (MRI), as shown here. MRI is very useful for looking in detail at soft tissue, for instance, muscle, tendons, and the brain. Also seen clearly in these sections are the eyes (1 and 3), the tongue (1 and 2), the larynx, vertebrae, and spinal cord (2 and 5).

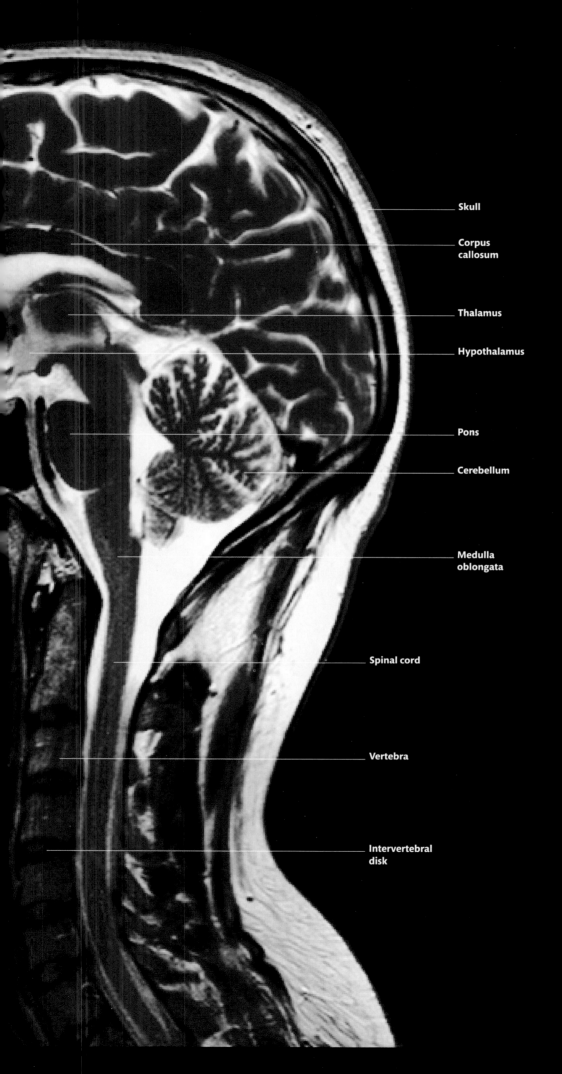

Skull

Corpus callosum

Thalamus

Hypothalamus

Pons

Cerebellum

Medulla oblongata

Spinal cord

Vertebra

Intervertebral disk

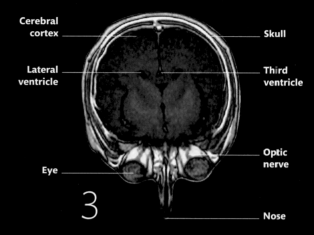

Cerebral cortex

Lateral ventricle

Eye

Skull

Third ventricle

Optic nerve

Nose

3

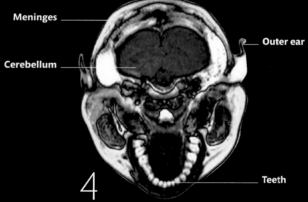

Meninges

Cerebellum

Outer ear

Teeth

4

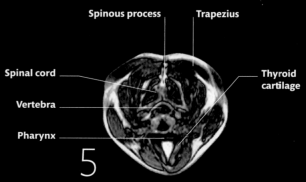

Spinous process

Trapezius

Spinal cord

Thyroid cartilage

Vertebra

Pharynx

5

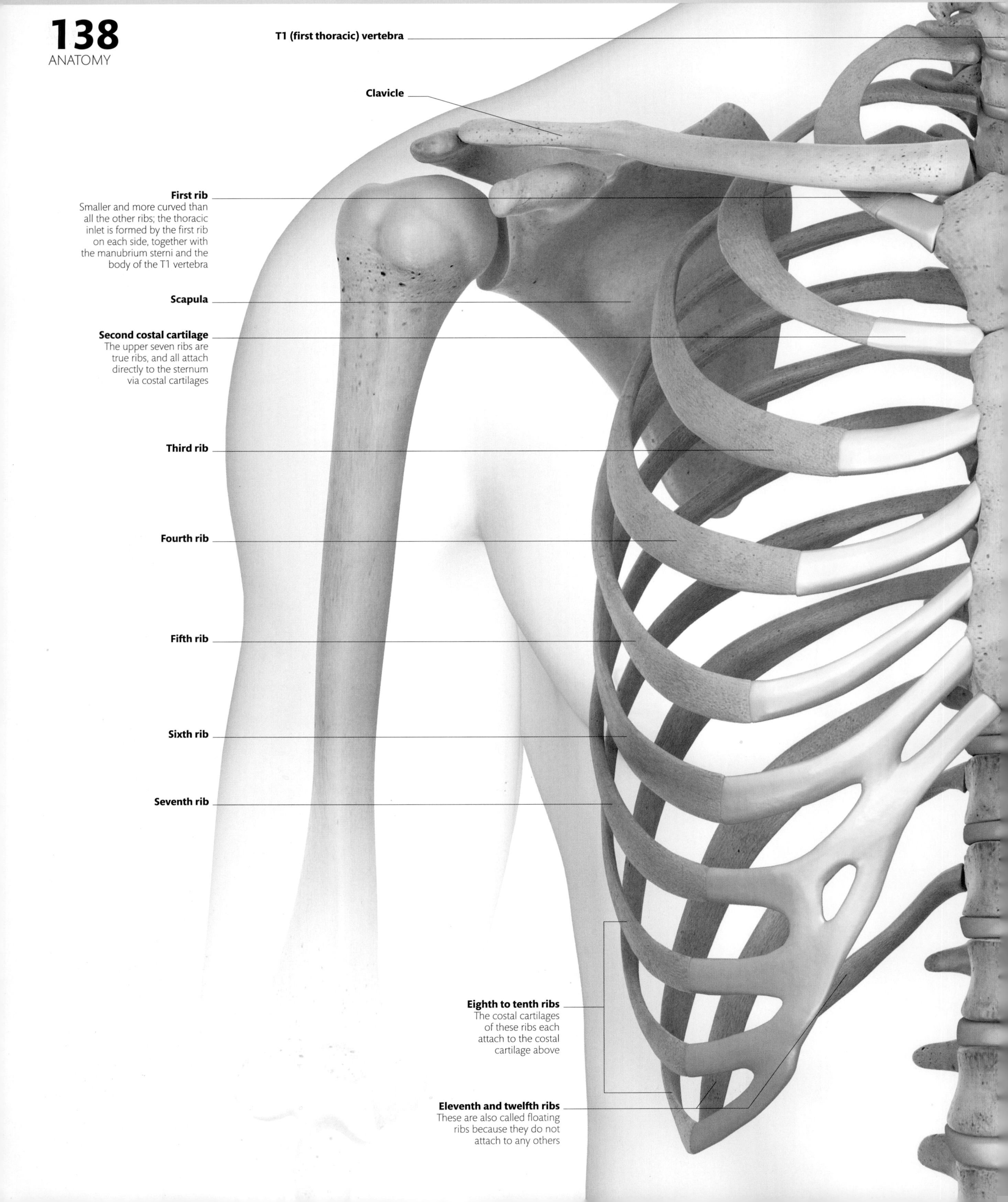

T1 (first thoracic) vertebra

Clavicle

First rib
Smaller and more curved than
all the other ribs; the thoracic
inlet is formed by the first rib
on each side, together with
the manubrium sterni and the
body of the T1 vertebra

Scapula

Second costal cartilage
The upper seven ribs are
true ribs, and all attach
directly to the sternum
via costal cartilages

Third rib

Fourth rib

Fifth rib

Sixth rib

Seventh rib

Eighth to tenth ribs
The costal cartilages
of these ribs each
attach to the costal
cartilage above

Eleventh and twelfth ribs
These are also called floating
ribs because they do not
attach to any others

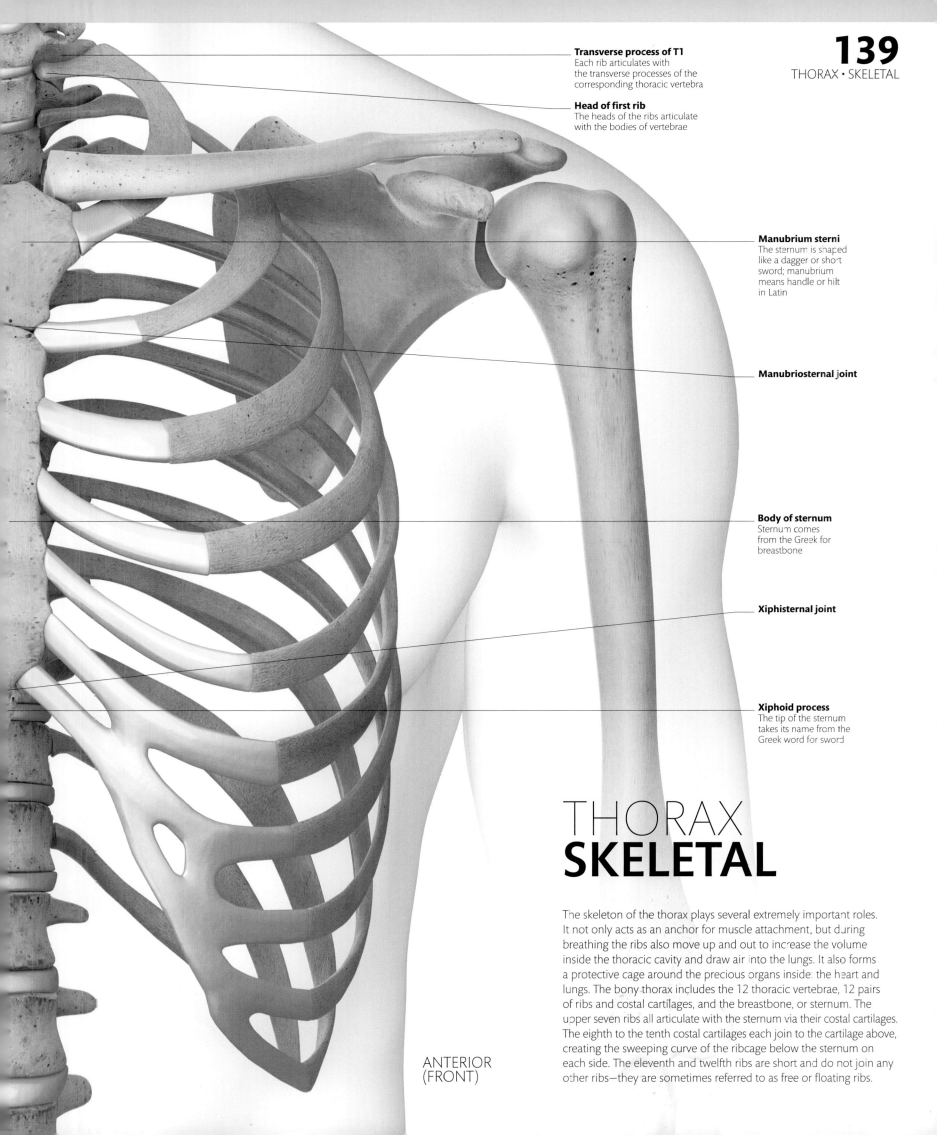

Transverse process of T1
Each rib articulates with
the transverse processes of the
corresponding thoracic vertebra

Head of first rib
The heads of the ribs articulate
with the bodies of vertebrae

Manubrium sterni
The sternum is shaped
like a dagger or short
sword; manubrium
means handle or hilt
in Latin

Manubriosternal joint

Body of sternum
Sternum comes
from the Greek for
breastbone

Xiphisternal joint

Xiphoid process
The tip of the sternum
takes its name from the
Greek word for sword

THORAX
SKELETAL

The skeleton of the thorax plays several extremely important roles.
It not only acts as an anchor for muscle attachment, but during
breathing the ribs also move up and out to increase the volume
inside the thoracic cavity and draw air into the lungs. It also forms
a protective cage around the precious organs inside: the heart and
lungs. The bony thorax includes the 12 thoracic vertebrae, 12 pairs
of ribs and costal cartilages, and the breastbone, or sternum. The
upper seven ribs all articulate with the sternum via their costal cartilages.
The eighth to the tenth costal cartilages each join to the cartilage above,
creating the sweeping curve of the ribcage below the sternum on
each side. The eleventh and twelfth ribs are short and do not join any
other ribs—they are sometimes referred to as free or floating ribs.

ANTERIOR
(FRONT)

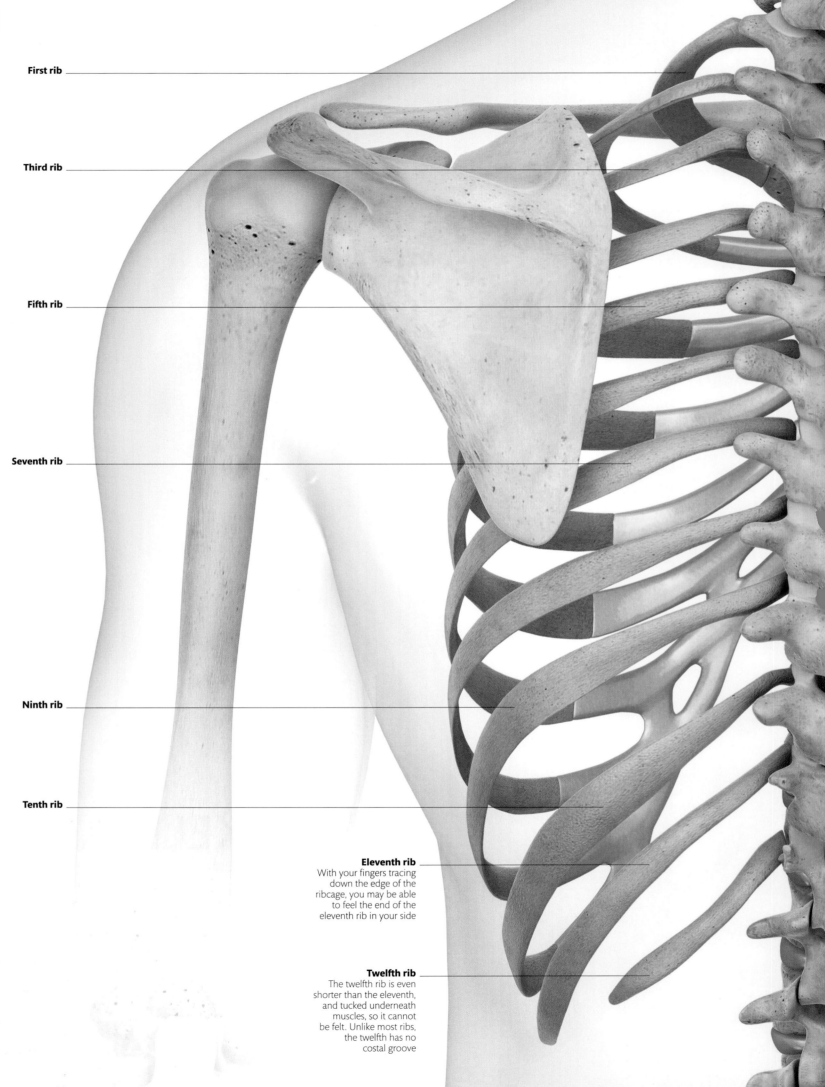

First rib

Third rib

Fifth rib

Seventh rib

Ninth rib

Tenth rib

Eleventh rib
With your fingers tracing
down the edge of the
ribcage, you may be able
to feel the end of the
eleventh rib in your side

Twelfth rib
The twelfth rib is even
shorter than the eleventh,
and tucked underneath
muscles, so it cannot
be felt. Unlike most ribs,
the twelfth has no
costal groove

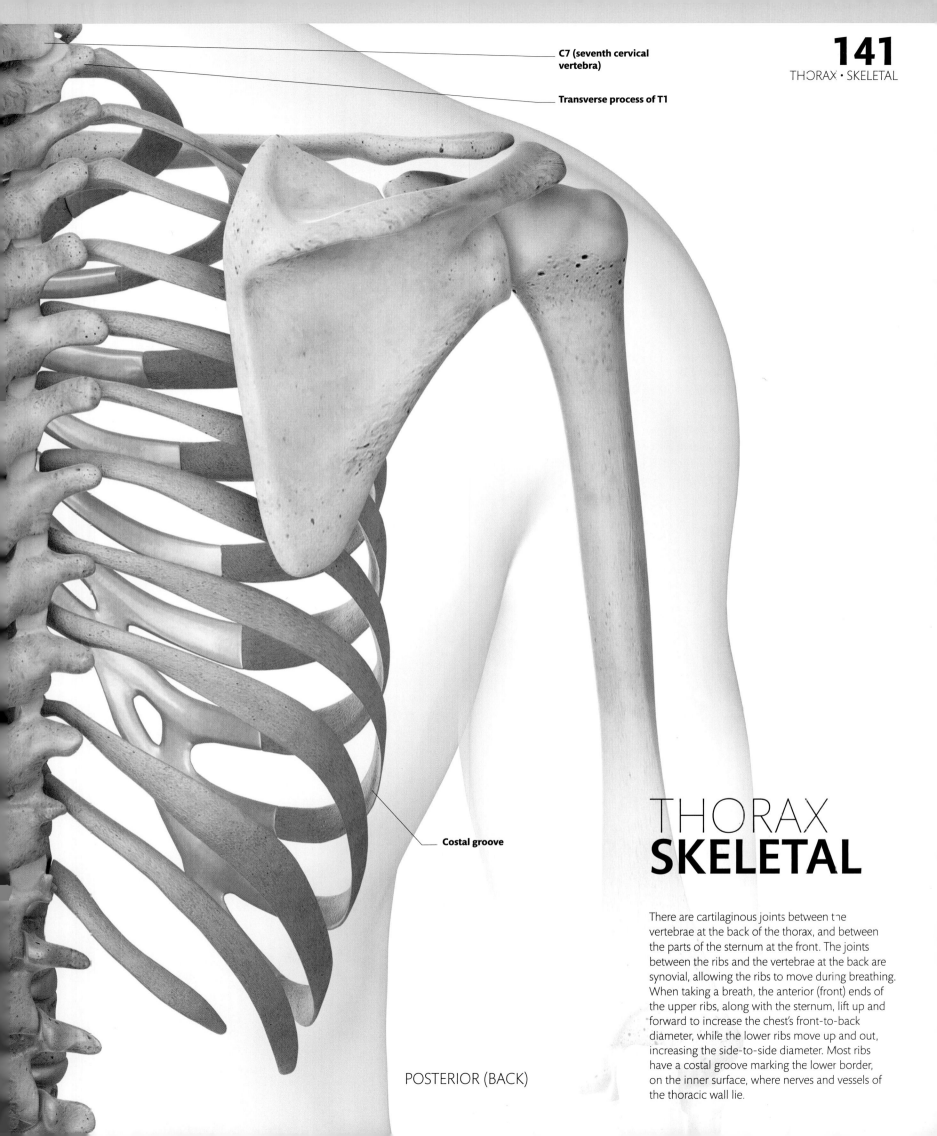

C7 (seventh cervical vertebra)

Transverse process of T1

Costal groove

THORAX
SKELETAL

There are cartilaginous joints between the vertebrae at the back of the thorax, and between the parts of the sternum at the front. The joints between the ribs and the vertebrae at the back are synovial, allowing the ribs to move during breathing. When taking a breath, the anterior (front) ends of the upper ribs, along with the sternum, lift up and forward to increase the chest's front-to-back diameter, while the lower ribs move up and out, increasing the side-to-side diameter. Most ribs have a costal groove marking the lower border, on the inner surface, where nerves and vessels of the thoracic wall lie.

POSTERIOR (BACK)

THORAX
SKELETAL

SPINE

The spine, or vertebral column, occupies a central position in the skeleton, and plays several extremely important roles: it supports the trunk, encloses and protects the spinal cord, provides sites for muscle attachment, and contains blood-forming bone marrow. The entire vertebral column is about 28 in (70 cm) long in men, and 24 in (60 cm) long in women. About a quarter of this length is made up by the cartilaginous intervertebral disks between the vertebrae. The number of vertebrae varies from 32 to 35, mostly due to variation in the number of small vertebrae that make up the coccyx. Although there is a general pattern for a vertebra—most possess a body, a neural arch, and spinous and transverse processes—there are recognizable features that mark out the vertebrae of each section of the spine.

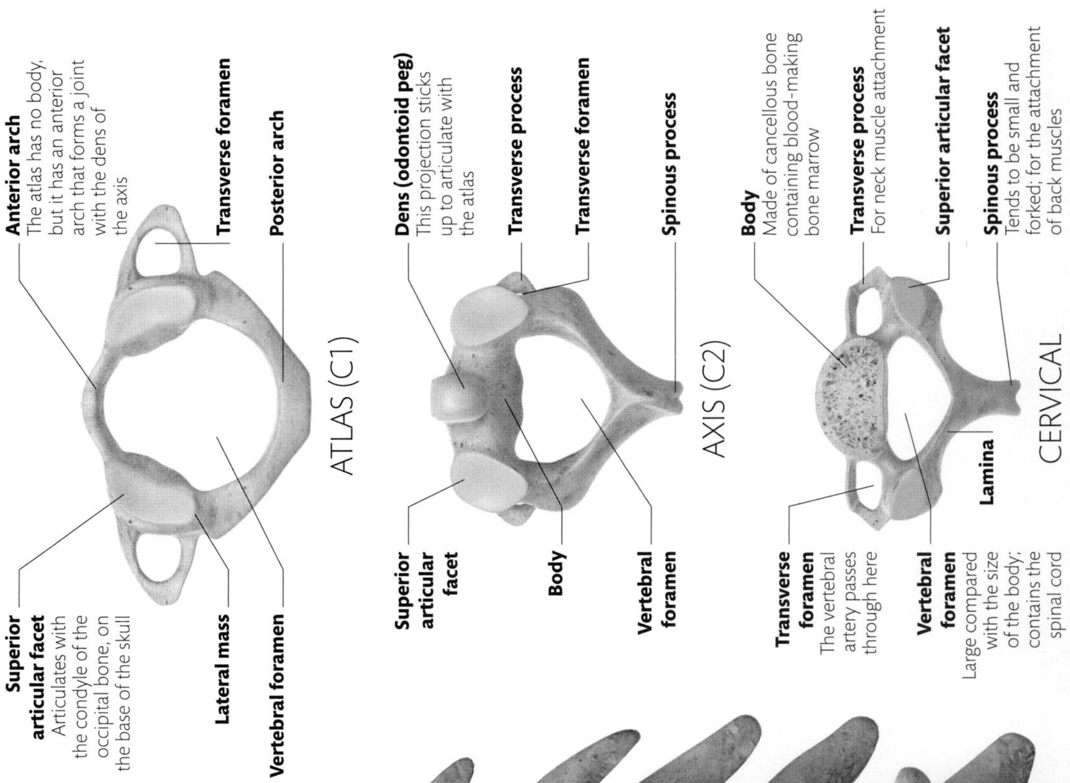

Anterior arch
The atlas has no body, but it has an anterior arch that forms a joint with the dens of the axis

Transverse foramen

Posterior arch

Superior articular facet
Articulates with the condyle of the occipital bone, on the base of the skull

Lateral mass

Vertebral foramen

ATLAS (C1)

Dens (odontoid peg)
This projection sticks up to articulate with the atlas

Transverse process

Transverse foramen

Spinous process

Superior articular facet

Body

Vertebral foramen

AXIS (C2)

Body
Made of cancellous bone containing blood-making bone marrow

Transverse process
For neck muscle attachment

Superior articular facet

Spinous process
Tends to be small and forked; for the attachment of back muscles

Transverse foramen
The vertebral artery passes through here

Vertebral foramen
Large compared with the size of the body; contains the spinal cord

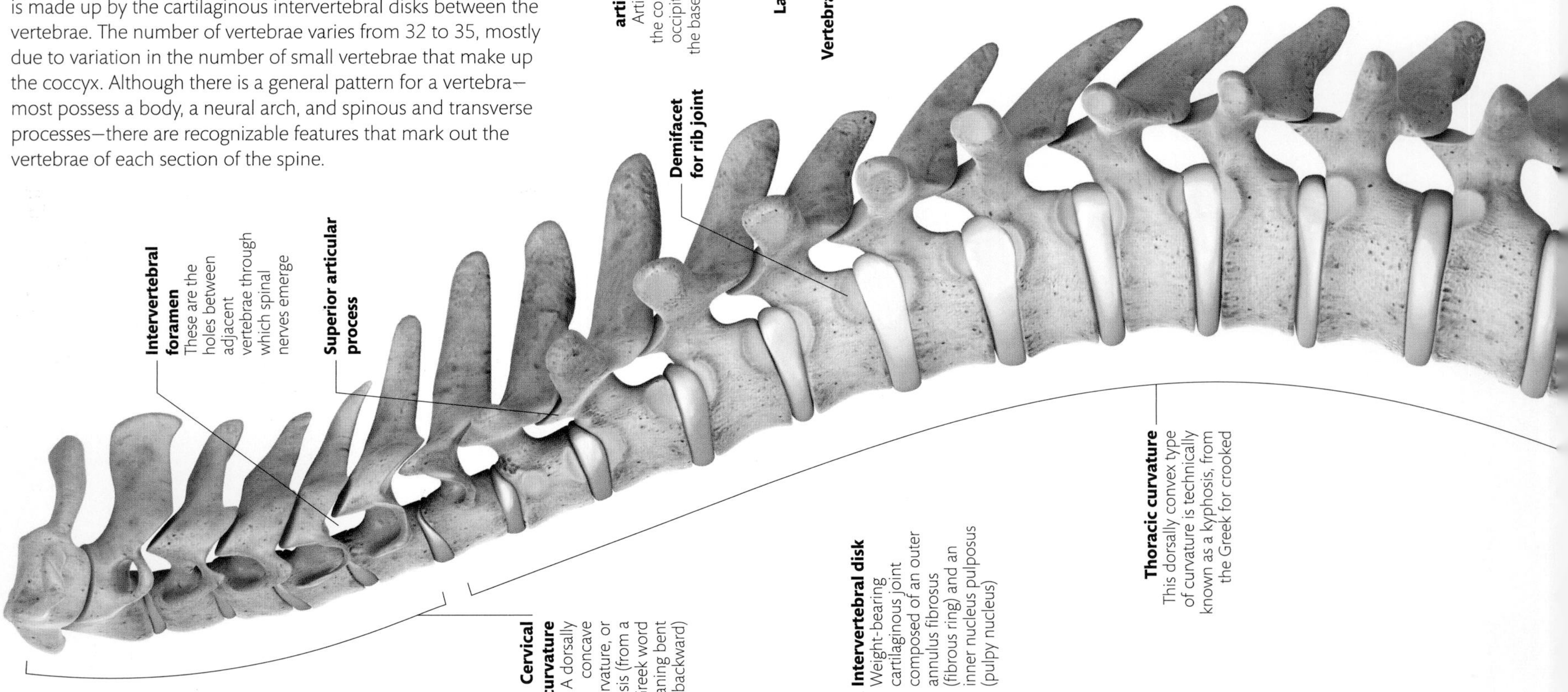

Lamina

CERVICAL

Intervertebral foramen
These are the holes between adjacent vertebrae through which spinal nerves emerge

Superior articular process

Demifacet for rib joint

Cervical curvature
A dorsally concave curvature, or lordosis (from a Greek word meaning bent backward)

Intervertebral disk
Weight-bearing cartilaginous joint composed of an outer annulus fibrosus (fibrous ring) and an inner nucleus pulposus (pulpy nucleus)

Thoracic curvature
This dorsally convex type of curvature is technically known as a kyphosis, from the Greek for crooked

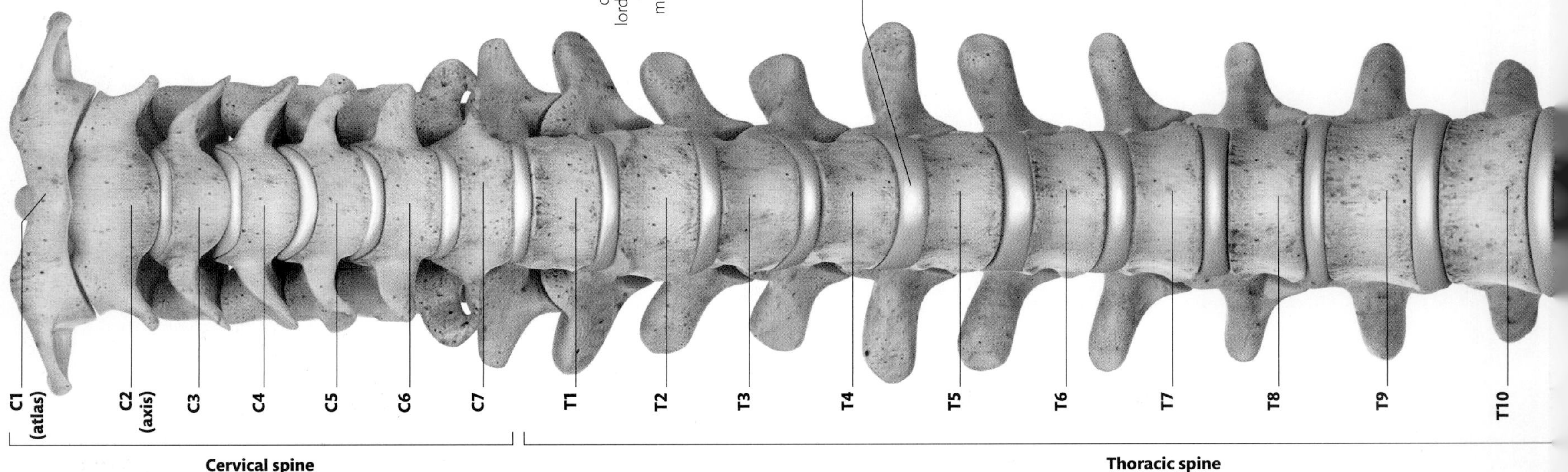

C1 (atlas)
C2 (axis)
C3
C4
C5
C6
C7
T1
T2
T3
T4
T5
T6
T7
T8
T9
T10

Cervical spine
(Seven vertebrae make up the spine in the neck)

Thoracic spine
(Twelve vertebrae, providing attachment for twelve pairs of ribs)

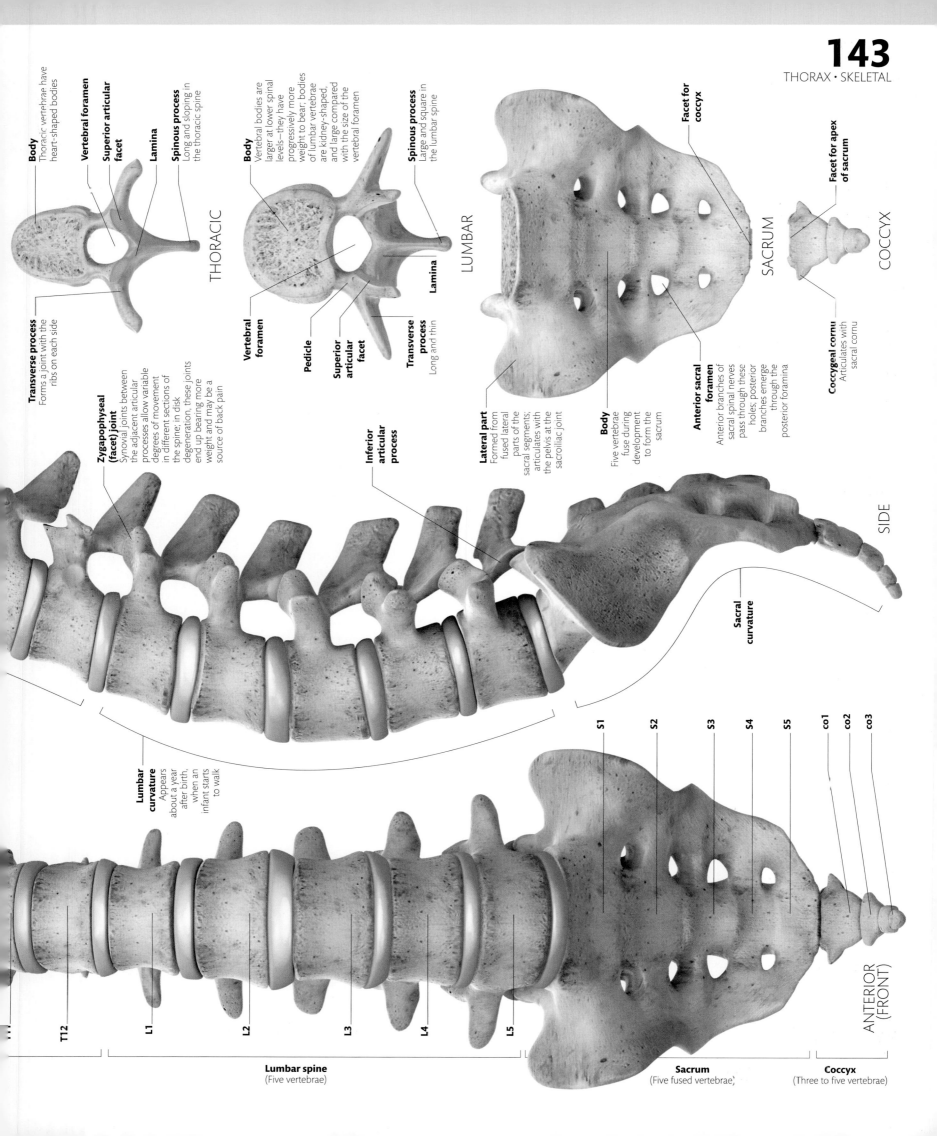

THORACIC

Body
Thoracic vertebrae have heart-shaped bodies

Vertebral foramen

Superior articular facet

Lamina

Spinous process
Long and sloping in the thoracic spine

Transverse process
Forms a joint with the ribs on each side

Zygapophyseal (facet) joint
Synovial joints between the adjacent articular processes allow variable degrees of movement in different sections of the spine; in disk degeneration, these joints end up bearing more weight and may be a source of back pain

LUMBAR

Body
Vertebral bodies are larger at lower spinal levels—they have progressively more weight to bear; bodies of lumbar vertebrae are kidney-shaped, and large compared with the size of the vertebral foramen

Spinous process
Large and square in the lumbar spine

Lamina

Transverse process
Long and thin

Superior articular facet

Pedicle

Vertebral foramen

Inferior articular process

SACRUM

Facet for coccyx

Lateral part
Formed from fused lateral parts of the sacral segments; articulates with the pelvis at the sacroiliac joint

Body
Five vertebrae fuse during development to form the sacrum

Anterior sacral foramen
Anterior branches of sacral spinal nerves pass through these holes; posterior branches emerge through the posterior foramina

COCCYX

Facet for apex of sacrum

Coccygeal cornu
Articulates with sacral cornu

SIDE

Sacral curvature

Lumbar curvature
Appears about a year after birth, when an infant starts to walk

S1
S2
S3
S4
S5
co1
co2
co3

ANTERIOR (FRONT)

T12
L1
L2
L3
L4
L5

Lumbar spine
(Five vertebrae)

Sacrum
(Five fused vertebrae)

Coccyx
(Three to five vertebrae)

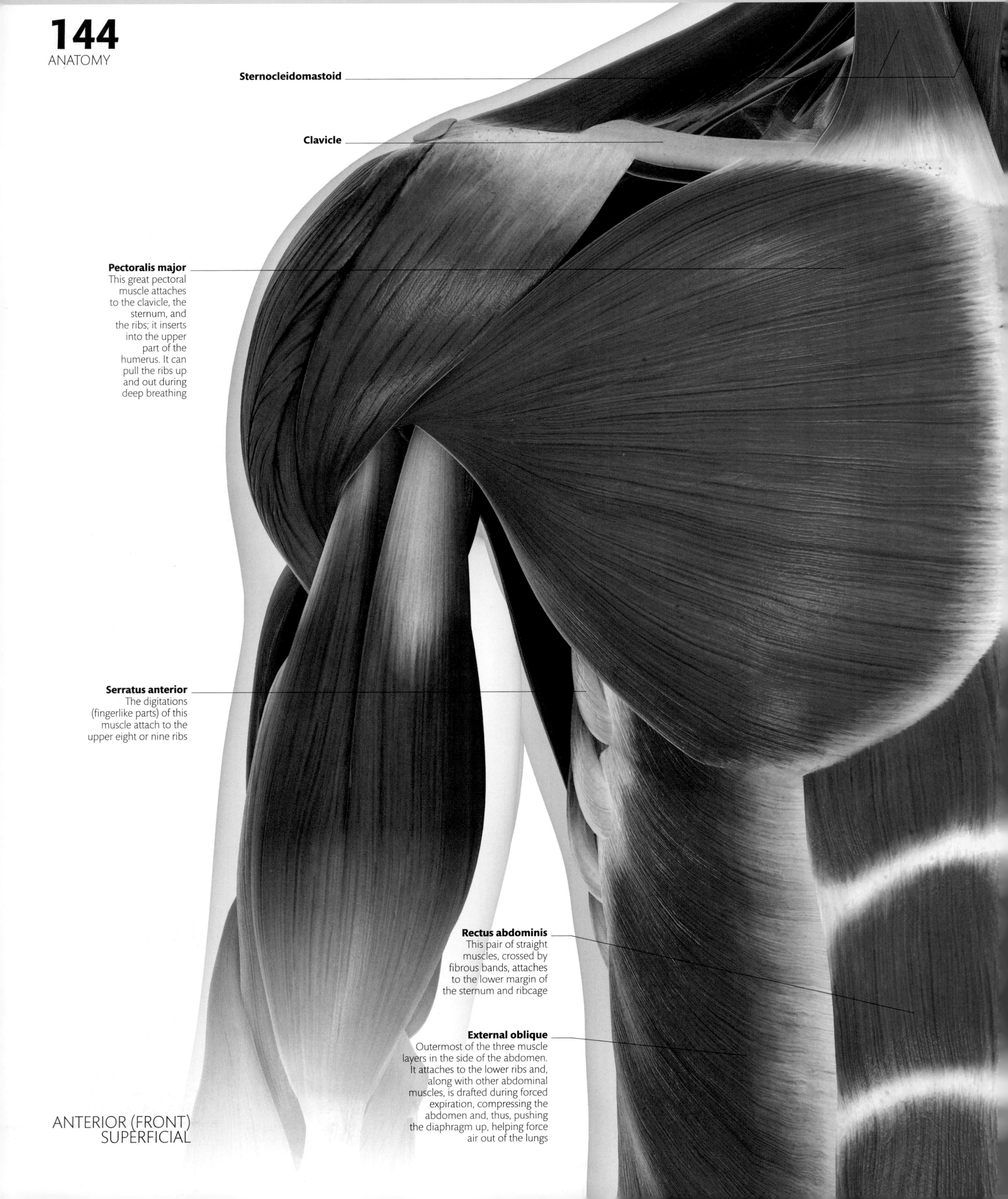

Sternocleidomastoid

Clavicle

Pectoralis major
This great pectoral
muscle attaches
to the clavicle, the
sternum, and
the ribs; it inserts
into the upper
part of the
humerus. It can
pull the ribs up
and out during
deep breathing

Serratus anterior
The digitations
(fingerlike parts) of this
muscle attach to the
upper eight or nine ribs

Rectus abdominis
This pair of straight
muscles, crossed by
fibrous bands, attaches
to the lower margin of
the sternum and ribcage

External oblique
Outermost of the three muscle
layers in the side of the abdomen.
It attaches to the lower ribs and,
along with other abdominal
muscles, is drafted during forced
expiration, compressing the
abdomen and, thus, pushing
the diaphragm up, helping force
air out of the lungs

ANTERIOR (FRONT)
SUPERFICIAL

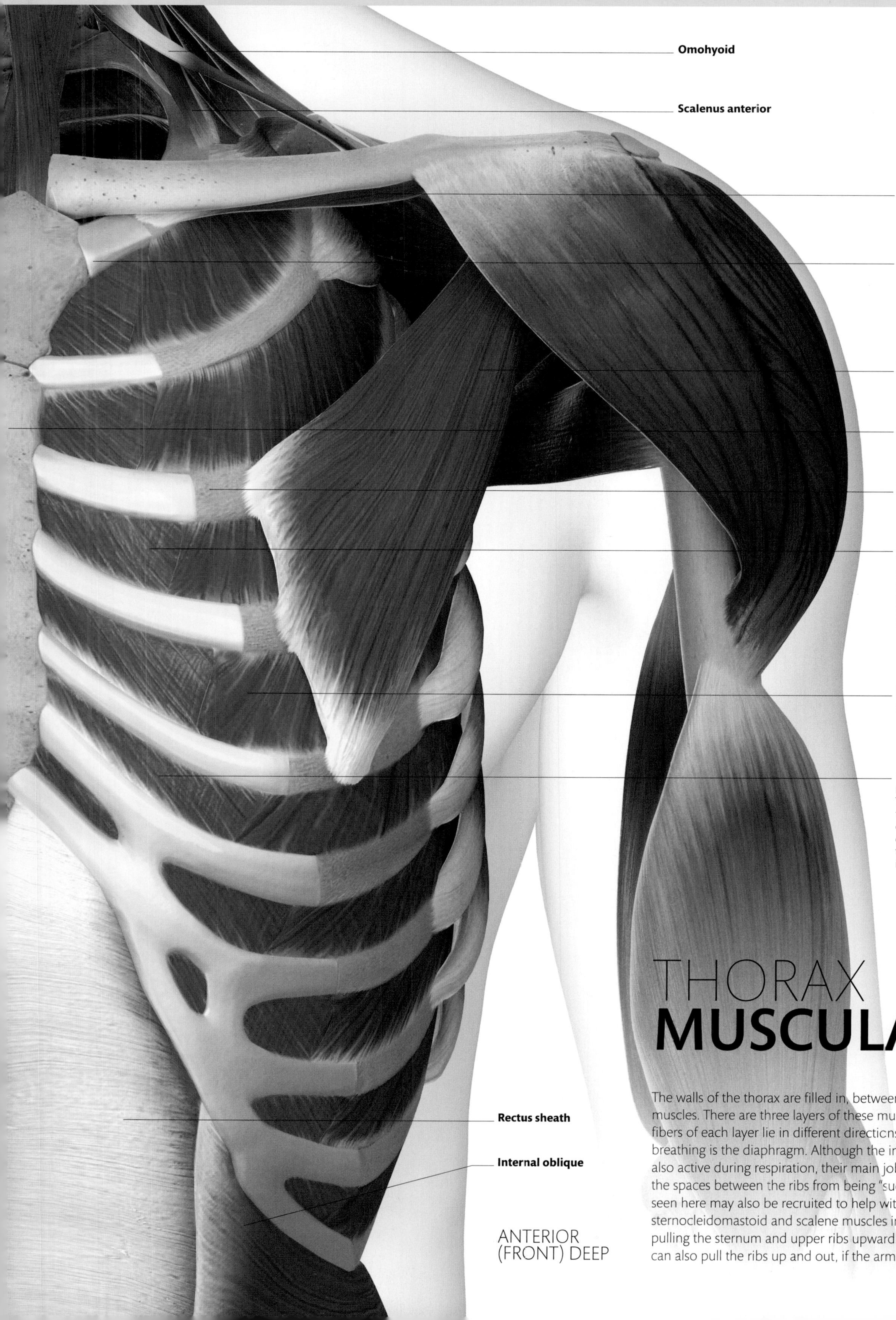

Omohyoid

Scalenus anterior

Subclavius

Costal cartilage

Pectoralis minor

Sternum

Rib

Intercostal muscles
Three layers of muscle occupy the intercostal spaces between the ribs: external, internal, and innermost intercostal muscles

External intercostal muscle

Internal intercostal muscle
The muscle fibers of this middle layer run diagonally in the opposite direction to those of the external intercostal muscle

Rectus sheath

Internal oblique

THORAX
MUSCULAR

The walls of the thorax are filled in, between the ribs, by the intercostal muscles. There are three layers of these muscles, and the muscle fibers of each layer lie in different directions. The main muscle for breathing is the diaphragm. Although the intercostal muscles are also active during respiration, their main job seems to be to prevent the spaces between the ribs from being "sucked in." Other muscles seen here may also be recruited to help with deep breathing. The sternocleidomastoid and scalene muscles in the neck can help by pulling the sternum and upper ribs upward. The pectoral muscles can also pull the ribs up and out, if the arm is held in a fixed position.

ANTERIOR
(FRONT) DEEP

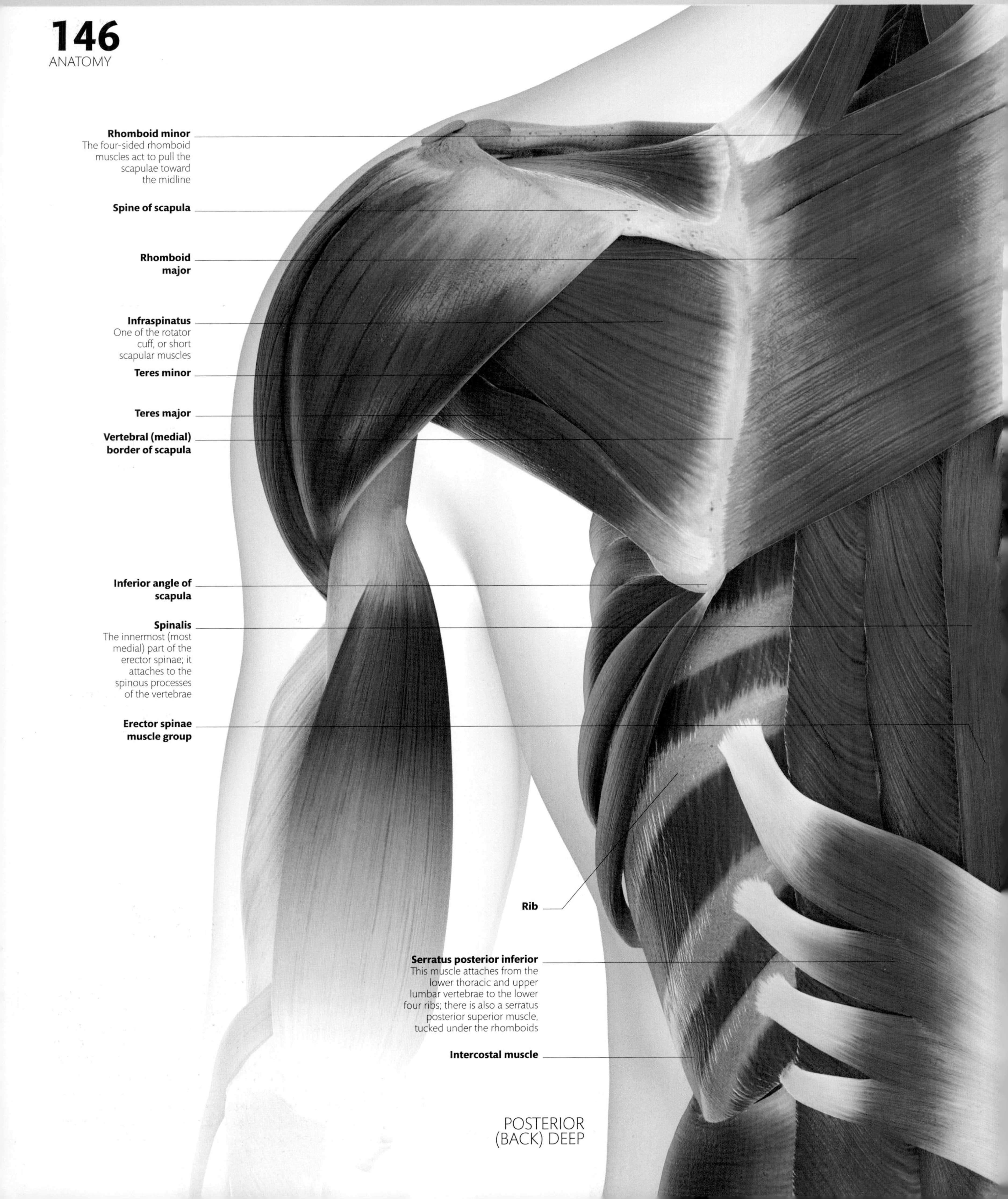

Rhomboid minor
The four-sided rhomboid
muscles act to pull the
scapulae toward
the midline

Spine of scapula

**Rhomboid
major**

Infraspinatus
One of the rotator
cuff, or short
scapular muscles

Teres minor

Teres major

**Vertebral (medial)
border of scapula**

**Inferior angle of
scapula**

Spinalis
The innermost (most
medial) part of the
erector spinae; it
attaches to the
spinous processes
of the vertebrae

**Erector spinae
muscle group**

Rib

Serratus posterior inferior
This muscle attaches from the
lower thoracic and upper
lumbar vertebrae to the lower
four ribs; there is also a serratus
posterior superior muscle,
tucked under the rhomboids

Intercostal muscle

POSTERIOR
(BACK) DEEP

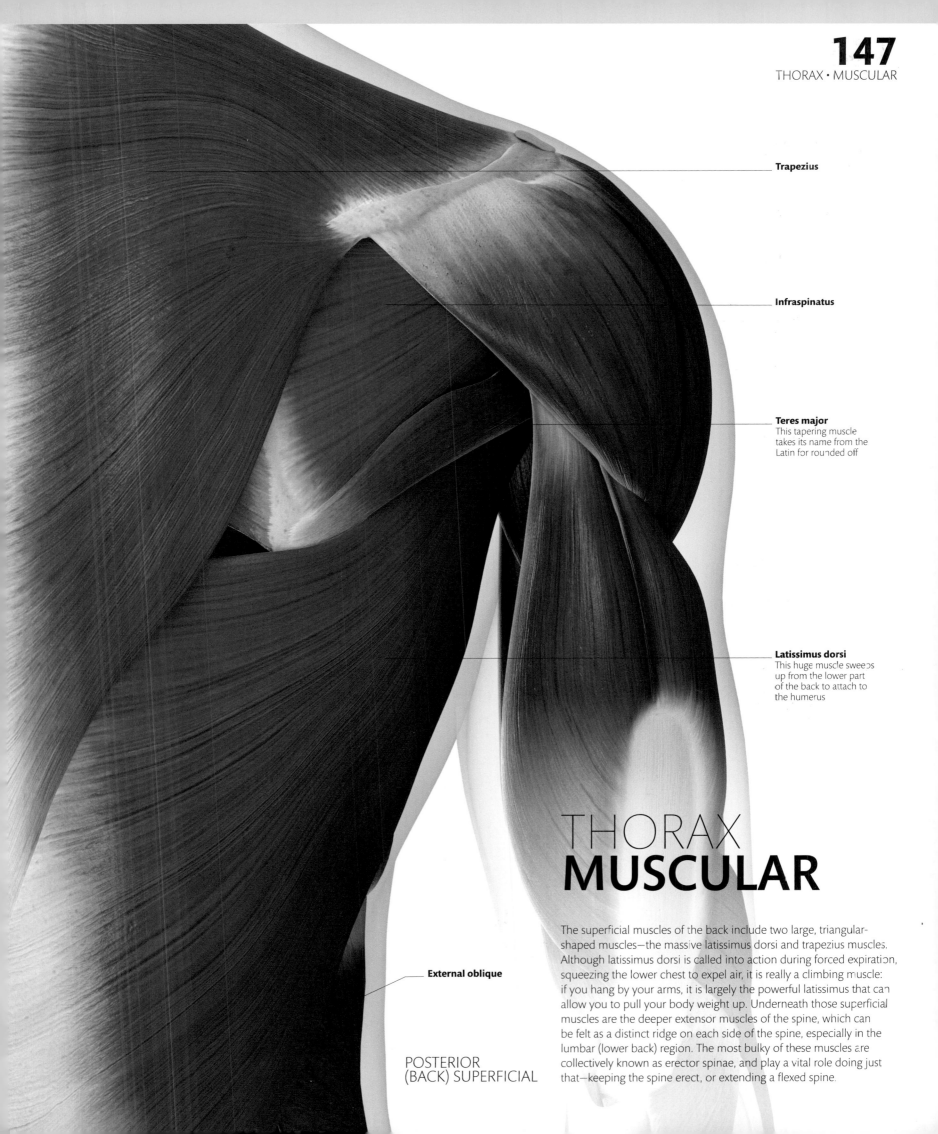

Trapezius

Infraspinatus

Teres major
This tapering muscle
takes its name from the
Latin for rounded off

Latissimus dorsi
This huge muscle sweeps
up from the lower part
of the back to attach to
the humerus

THORAX
MUSCULAR

The superficial muscles of the back include two large, triangular-
shaped muscles—the massive latissimus dorsi and trapezius muscles.
Although latissimus dorsi is called into action during forced expiration,
squeezing the lower chest to expel air, it is really a climbing muscle:
if you hang by your arms, it is largely the powerful latissimus that can
allow you to pull your body weight up. Underneath those superficial
muscles are the deeper extensor muscles of the spine, which can
be felt as a distinct ridge on each side of the spine, especially in the
lumbar (lower back) region. The most bulky of these muscles are
collectively known as erector spinae, and play a vital role doing just
that—keeping the spine erect, or extending a flexed spine.

External oblique

POSTERIOR
(BACK) SUPERFICIAL

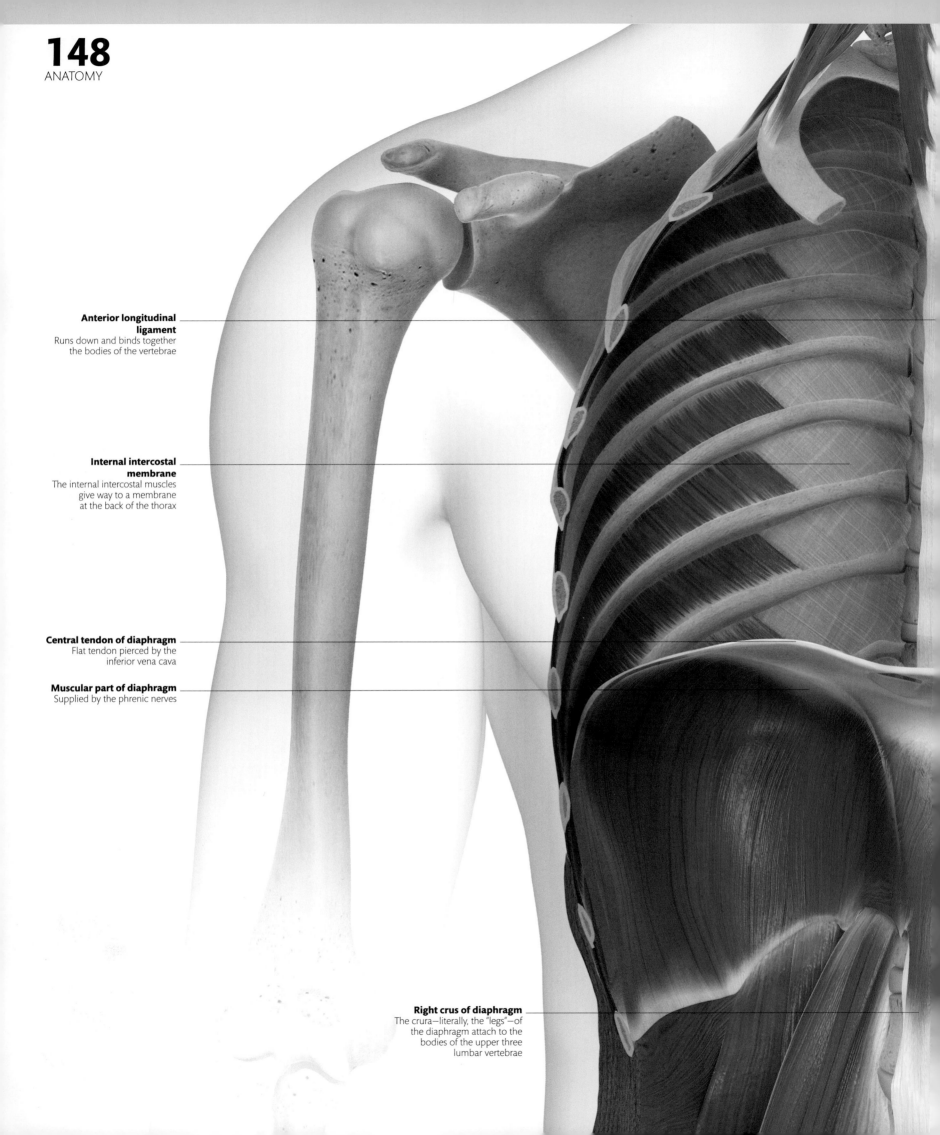

**Anterior longitudinal
ligament**
Runs down and binds together
the bodies of the vertebrae

**Internal intercostal
membrane**
The internal intercostal muscles
give way to a membrane
at the back of the thorax

Central tendon of diaphragm
Flat tendon pierced by the
inferior vena cava

Muscular part of diaphragm
Supplied by the phrenic nerves

Right crus of diaphragm
The crura—literally, the "legs"—of
the diaphragm attach to the
bodies of the upper three
lumbar vertebrae

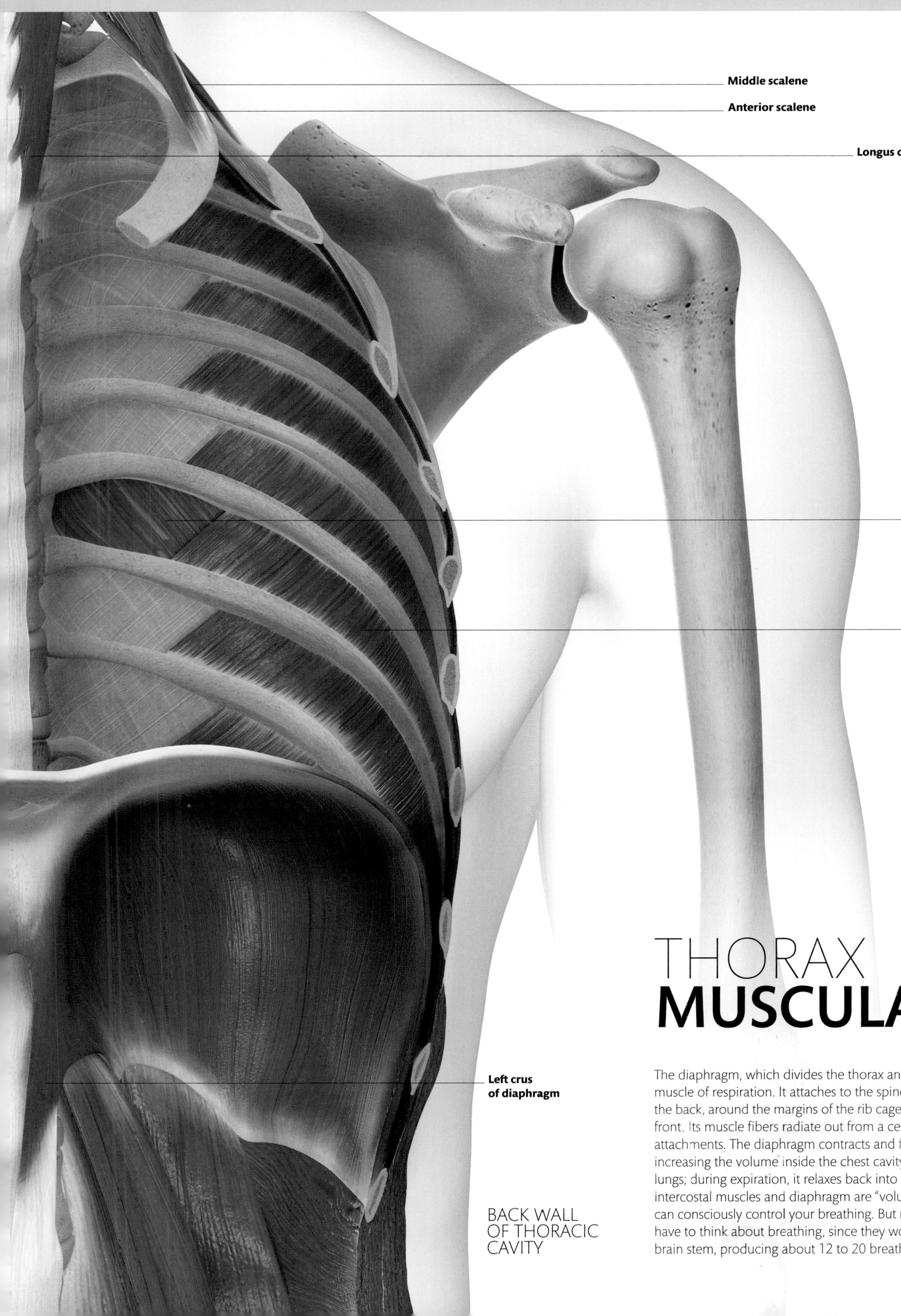

Middle scalene

Anterior scalene

Longus colli

External intercostal muscle
These muscles are replaced by a membrane around the front of the thorax. (Seen here after removal of internal intercostal membrane)

Internal intercostal muscle
The intercostal muscles are supplied by intercostal nerves

Left crus of diaphragm

BACK WALL OF THORACIC CAVITY

THORAX
MUSCULAR

The diaphragm, which divides the thorax and abdomen, is the main muscle of respiration. It attaches to the spine and to deep muscles in the back, around the margins of the rib cage, and to the sternum at the front. Its muscle fibers radiate out from a central, flat tendon to these attachments. The diaphragm contracts and flattens during inspiration, increasing the volume inside the chest cavity, and pulling air into the lungs; during expiration, it relaxes back into a domed shape. The intercostal muscles and diaphragm are "voluntary" muscle, and you can consciously control your breathing. But most of the time you don't have to think about breathing, since they work to a rhythm set by the brain stem, producing about 12 to 20 breaths per minute in an adult.

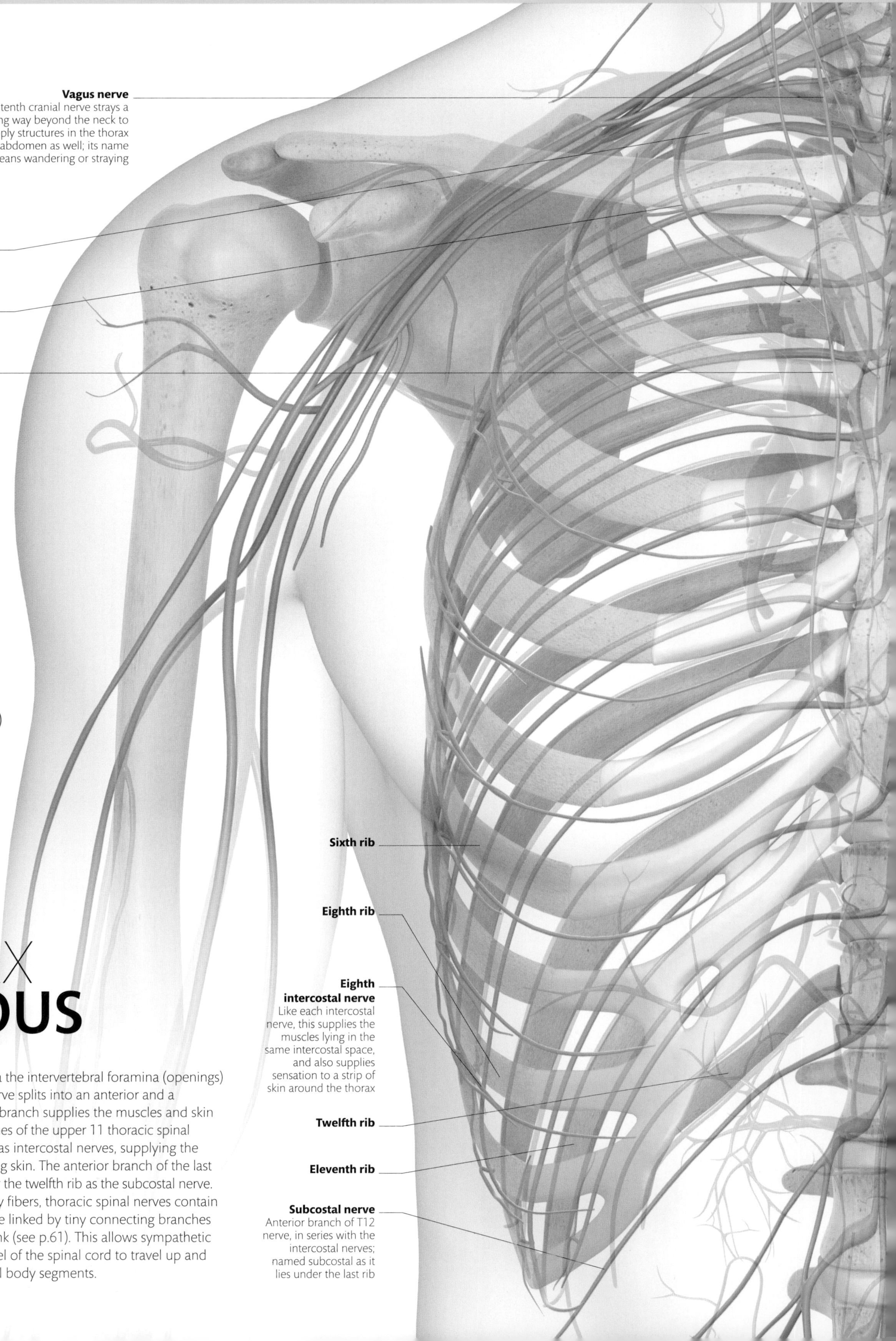

Vagus nerve
The tenth cranial nerve strays a long way beyond the neck to supply structures in the thorax and abdomen as well; its name means wandering or straying

First rib

First intercostal nerve
Anterior branch of T1 (first thoracic) spinal nerve

Phrenic nerve
Comes from the third, fourth, and fifth cervical nerves; supplies the muscle of the diaphragm and the membranes lining either side of it—the pleura on the thoracic side and peritoneum on the abdominal side

ANTERIOR (FRONT)

Sixth rib

Eighth rib

Eighth intercostal nerve
Like each intercostal nerve, this supplies the muscles lying in the same intercostal space, and also supplies sensation to a strip of skin around the thorax

Twelfth rib

Eleventh rib

Subcostal nerve
Anterior branch of T12 nerve, in series with the intercostal nerves; named subcostal as it lies under the last rib

THORAX
NERVOUS

Pairs of spinal nerves emerge via the intervertebral foramina (openings) between the vertebrae. Each nerve splits into an anterior and a posterior branch. The posterior branch supplies the muscles and skin of the back. The anterior branches of the upper 11 thoracic spinal nerves run, one under each rib, as intercostal nerves, supplying the intercostal muscles and overlying skin. The anterior branch of the last thoracic spinal nerve runs under the twelfth rib as the subcostal nerve. In addition to motor and sensory fibers, thoracic spinal nerves contain sympathetic nerve fibers that are linked by tiny connecting branches to the sympathetic chain or trunk (see p.61). This allows sympathetic nerves originating from one level of the spinal cord to travel up and down, and spread out to several body segments.

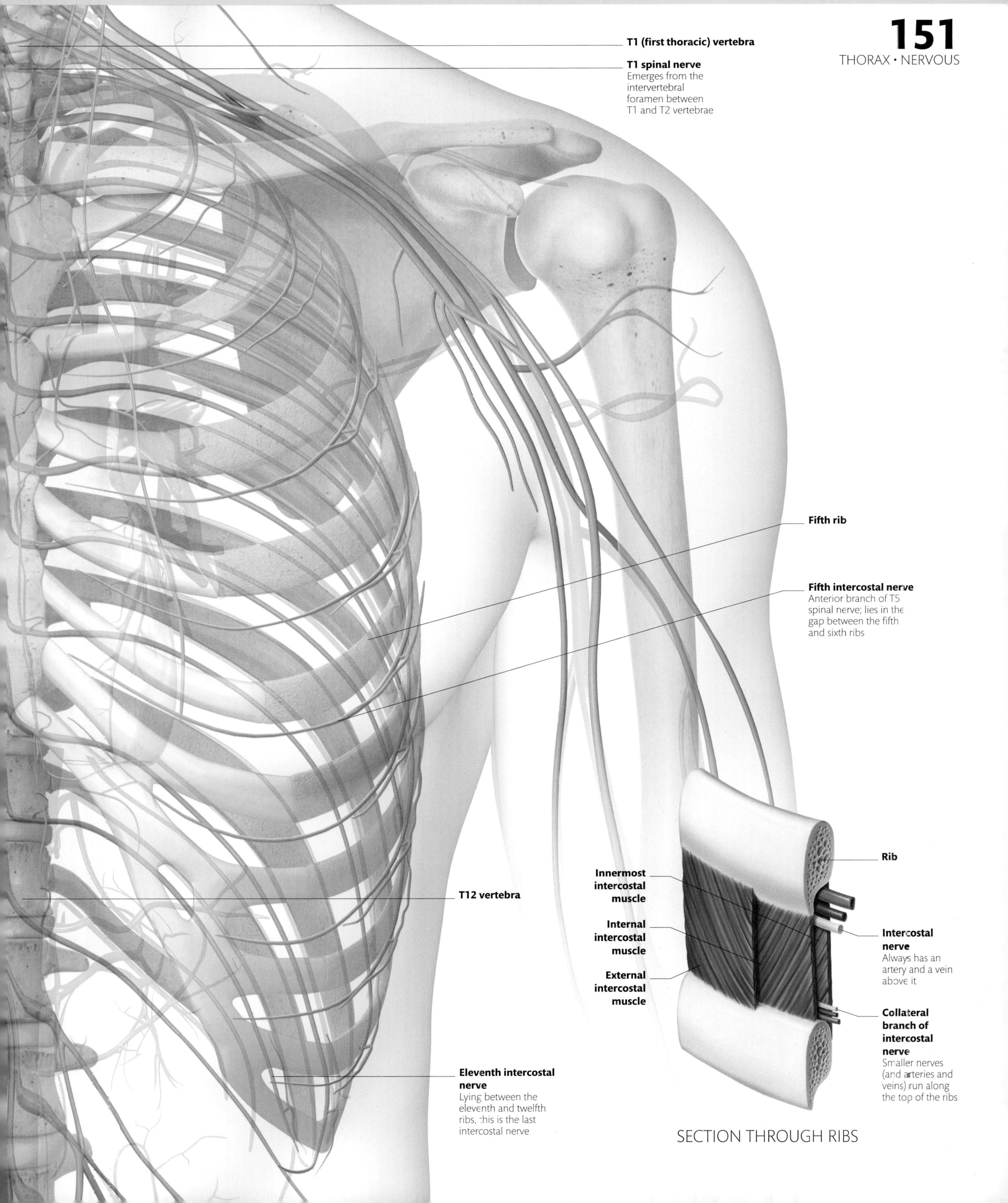

T1 (first thoracic) vertebra

T1 spinal nerve
Emerges from the
intervertebral
foramen between
T1 and T2 vertebrae

Fifth rib

Fifth intercostal nerve
Anterior branch of T5
spinal nerve; lies in the
gap between the fifth
and sixth ribs

Rib

**Innermost
intercostal
muscle**

**Internal
intercostal
muscle**

**External
intercostal
muscle**

**Intercostal
nerve**
Always has an
artery and a vein
above it

**Collateral
branch of
intercostal
nerve**
Smaller nerves
(and arteries and
veins) run along
the top of the ribs

T12 vertebra

**Eleventh intercostal
nerve**
Lying between the
eleventh and twelfth
ribs, this is the last
intercostal nerve

SECTION THROUGH RIBS

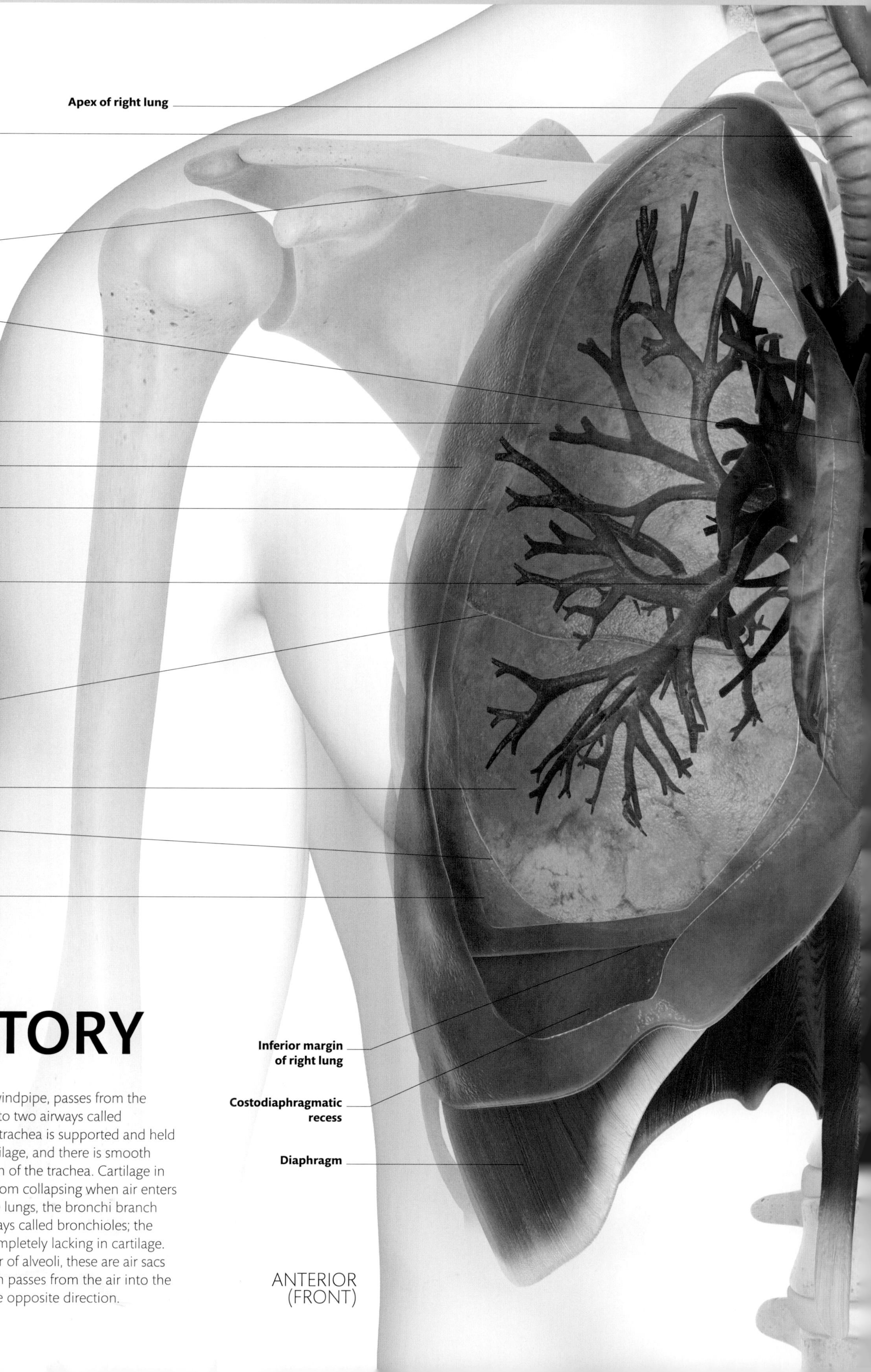

Apex of right lung

Trachea
Named after the Greek for
rough vessel, the trachea
is about 4¾ in (12 cm) long
and ½–¾ in (1.5–2 cm) wide
in an adult

Right clavicle
(cut away to show
lung behind)

Anterior margin
of right lung

Superior lobe of right lung

Parietal pleura

Visceral pleura

Bronchus of right lung
Several smaller bronchi branch
off the two main bronchi that
enter the lung by bifurcating
from the trachea; confusingly, the
word bronchus comes from the
Greek for windpipe

Horizontal fissure
Deep cleft that separates the
superior (upper) and middle
lobes of the right lung

Middle lobe of right lung

Oblique fissure of right lung
Separates the middle and inferior
(lower) lobes of the right lung

Inferior lobe of right lung

THORAX
RESPIRATORY

The trachea, commonly known as the windpipe, passes from the
neck into the thorax, where it divides into two airways called
bronchi—each supplying one lung. The trachea is supported and held
open by 15–20 C-shaped pieces of cartilage, and there is smooth
muscle in its wall that can alter the width of the trachea. Cartilage in
the walls of the bronchi prevent them from collapsing when air enters
the lungs under low pressure. Inside the lungs, the bronchi branch
and branch again, forming smaller airways called bronchioles; the
bronchioles are just muscular tubes, completely lacking in cartilage.
The smallest bronchioles end in a cluster of alveoli, these are air sacs
surrounded by capillaries, where oxygen passes from the air into the
blood, and carbon dioxide passes in the opposite direction.

Inferior margin
of right lung

Costodiaphragmatic
recess

Diaphragm

ANTERIOR
(FRONT)

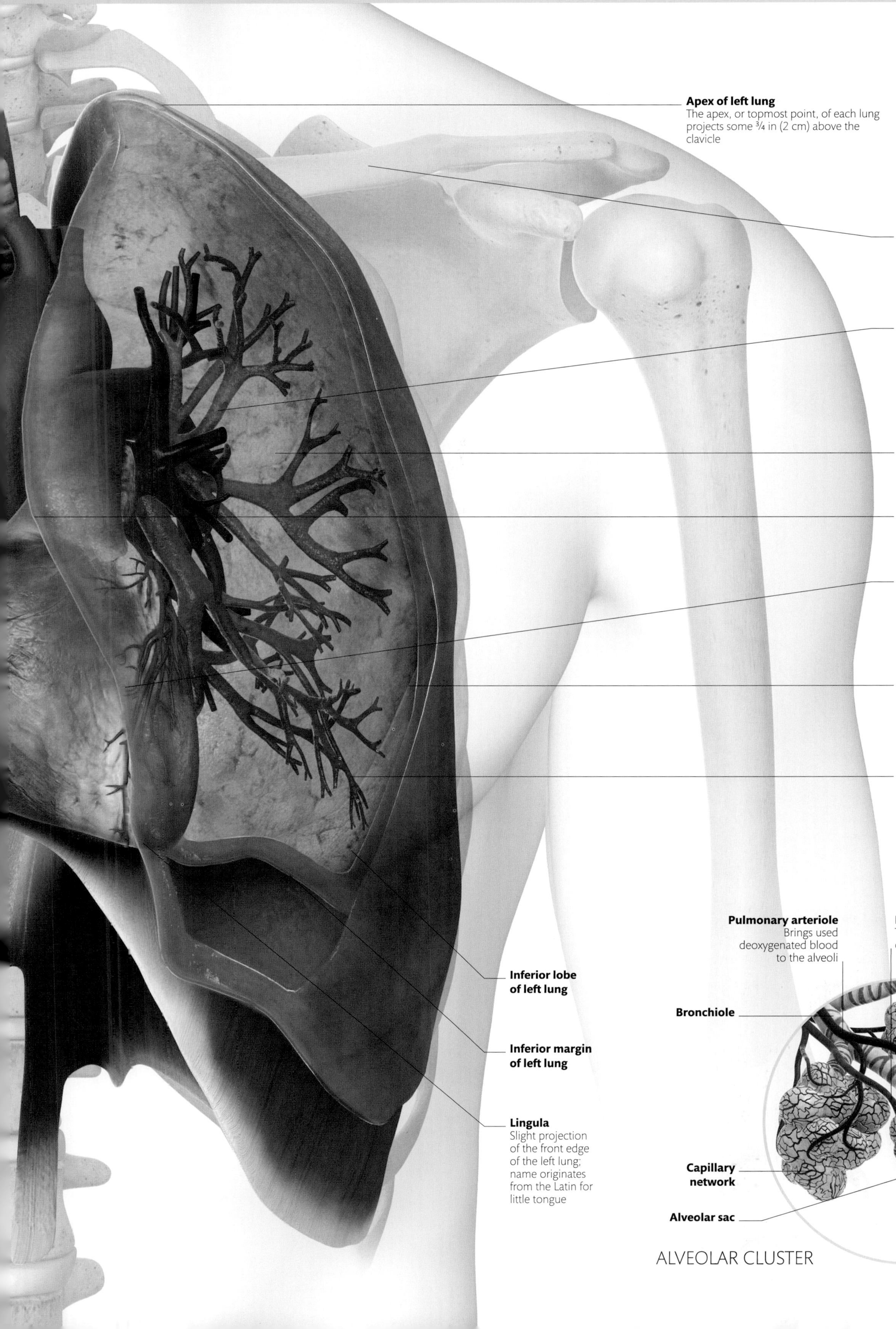

Apex of left lung
The apex, or topmost point, of each lung projects some ¾ in (2 cm) above the clavicle

**Left clavicle
(cut away to show
lung behind)**

Bronchus of left lung
Bronchi are lined with epithelium, which produces mucus to trap particles, and carpeted with tiny hairlike projections called cilia that waft mucus up and out of the lungs

Superior lobe of left lung

Anterior margin of left lung

Cardiac notch of left lung
Anterior edge of the left lung that curves inward slightly to accommodate the heart

Oblique fissure of left lung
Divides the superior and inferior lobes of the left lung

Bronchiole

Pulmonary arteriole
Brings used deoxygenated blood to the alveoli

Pulmonary venule
Takes away fresh, oxygenated blood

Bronchiole

**Inferior lobe
of left lung**

**Inferior margin
of left lung**

Lingula
Slight projection of the front edge of the left lung; name originates from the Latin for little tongue

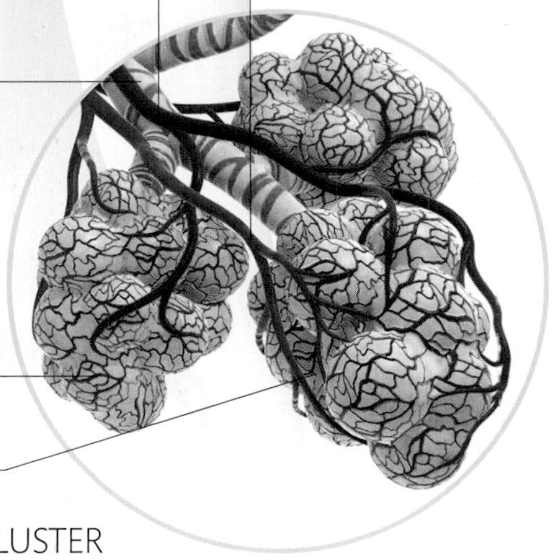

**Capillary
network**

Alveolar sac

ALVEOLAR CLUSTER

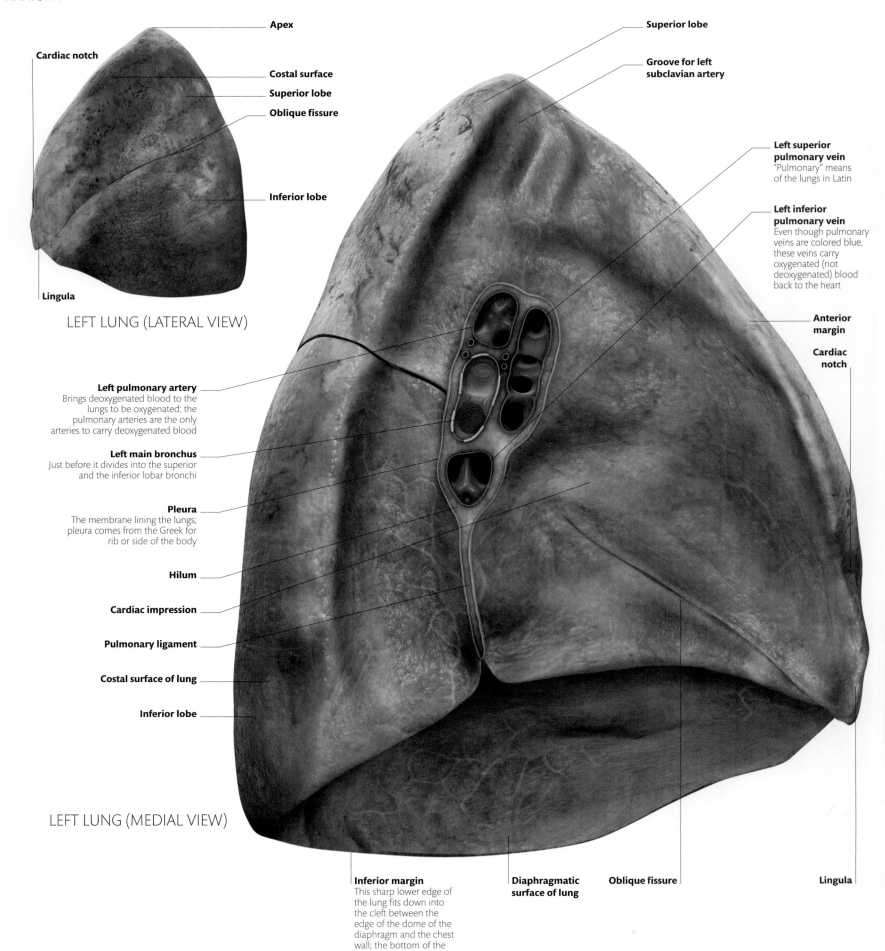

Apex

Cardiac notch

Costal surface

Superior lobe

Oblique fissure

Inferior lobe

Lingula

LEFT LUNG (LATERAL VIEW)

Superior lobe

Groove for left subclavian artery

Left superior pulmonary vein
"Pulmonary" means of the lungs in Latin

Left inferior pulmonary vein
Even though pulmonary veins are colored blue, these veins carry oxygenated (not deoxygenated) blood back to the heart

Anterior margin

Cardiac notch

Left pulmonary artery
Brings deoxygenated blood to the lungs to be oxygenated; the pulmonary arteries are the only arteries to carry deoxygenated blood

Left main bronchus
Just before it divides into the superior and the inferior lobar bronchi

Pleura
The membrane lining the lungs; pleura comes from the Greek for rib or side of the body

Hilum

Cardiac impression

Pulmonary ligament

Costal surface of lung

Inferior lobe

LEFT LUNG (MEDIAL VIEW)

Inferior margin
This sharp lower edge of the lung fits down into the cleft between the edge of the dome of the diaphragm and the chest wall; the bottom of the pleural cavity extends a couple more inches below the edge of the lung

Diaphragmatic surface of lung

Oblique fissure

Lingula

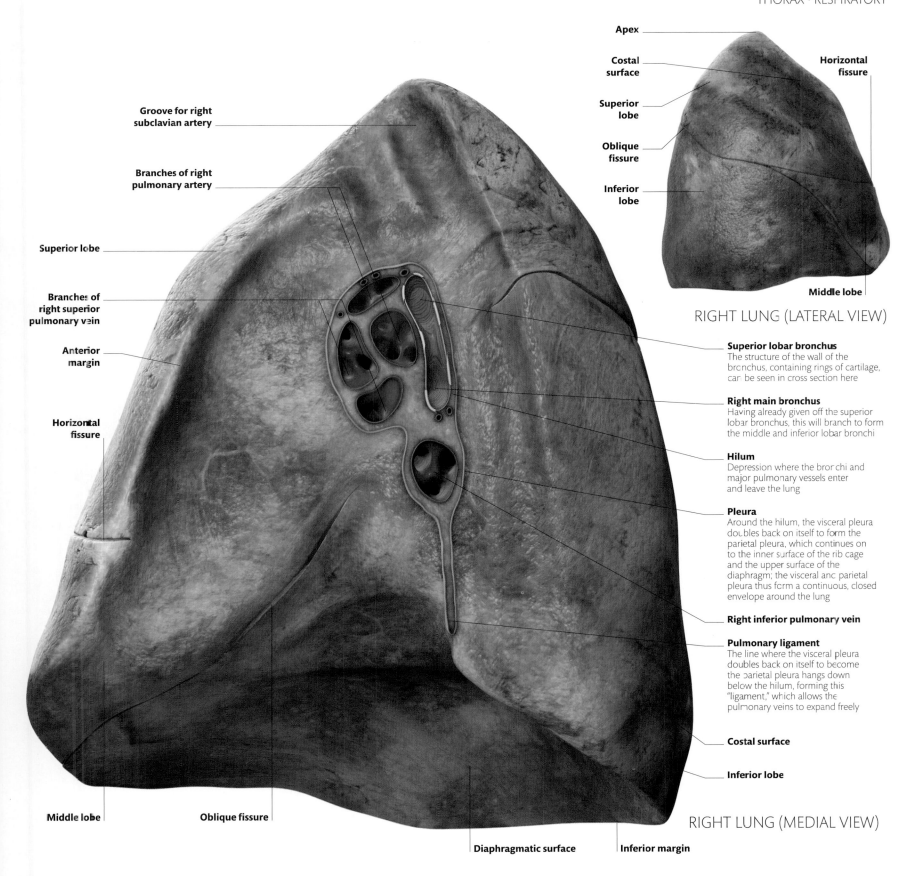

Apex

Costal surface

Superior lobe

Oblique fissure

Inferior lobe

Horizontal fissure

Middle lobe

RIGHT LUNG (LATERAL VIEW)

Groove for right subclavian artery

Branches of right pulmonary artery

Superior lobe

Branches of right superior pulmonary vein

Anterior margin

Horizontal fissure

Middle lobe

Oblique fissure

Diaphragmatic surface

Inferior margin

Superior lobar bronchus
The structure of the wall of the bronchus, containing rings of cartilage, can be seen in cross section here

Right main bronchus
Having already given off the superior lobar bronchus, this will branch to form the middle and inferior lobar bronchi

Hilum
Depression where the bronchi and major pulmonary vessels enter and leave the lung

Pleura
Around the hilum, the visceral pleura doubles back on itself to form the parietal pleura, which continues on to the inner surface of the rib cage and the upper surface of the diaphragm; the visceral and parietal pleura thus form a continuous, closed envelope around the lung

Right inferior pulmonary vein

Pulmonary ligament
The line where the visceral pleura doubles back on itself to become the parietal pleura hangs down below the hilum, forming this "ligament," which allows the pulmonary veins to expand freely

Costal surface

Inferior lobe

RIGHT LUNG (MEDIAL VIEW)

THORAX
RESPIRATORY

Each lung fits snugly inside its half of the thoracic cavity. The surface of each lung is covered with a thin pleural membrane (visceral pleura), and the inside of the chest wall is also lined with pleura (parietal pleura). Between the two pleural layers lies a thin film of lubricating fluid that allows the lungs to slide against the chest wall during breathing movements, but it also creates a fluid seal, effectively sticking the lungs to the ribs and the diaphragm.

Because of this seal, when you inhale, the lungs are pulled outward in all directions, and air rushes into them. The bronchi and blood vessels enter each lung at the hilum on its inner or medial surface. Although the two lungs may appear to be similar at first glance, there is some asymmetry. The left lung is concave to fit around the heart and has only two lobes, whereas the right lung has three lobes, delineated by two deep fissures.

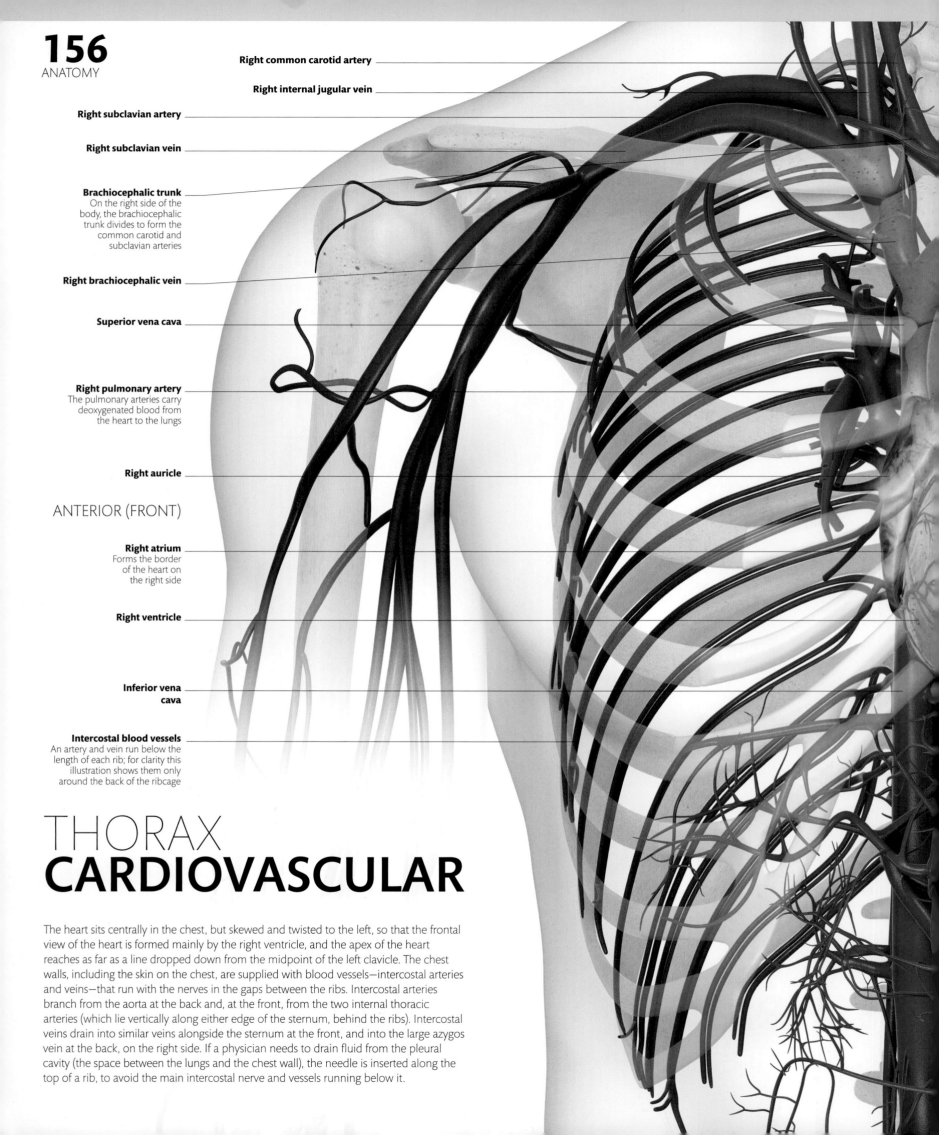

Right common carotid artery

Right internal jugular vein

Right subclavian artery

Right subclavian vein

Brachiocephalic trunk
On the right side of the body, the brachiocephalic trunk divides to form the common carotid and subclavian arteries

Right brachiocephalic vein

Superior vena cava

Right pulmonary artery
The pulmonary arteries carry deoxygenated blood from the heart to the lungs

Right auricle

ANTERIOR (FRONT)

Right atrium
Forms the border of the heart on the right side

Right ventricle

Inferior vena cava

Intercostal blood vessels
An artery and vein run below the length of each rib; for clarity this illustration shows them only around the back of the ribcage

THORAX
CARDIOVASCULAR

The heart sits centrally in the chest, but skewed and twisted to the left, so that the frontal view of the heart is formed mainly by the right ventricle, and the apex of the heart reaches as far as a line dropped down from the midpoint of the left clavicle. The chest walls, including the skin on the chest, are supplied with blood vessels—intercostal arteries and veins—that run with the nerves in the gaps between the ribs. Intercostal arteries branch from the aorta at the back and, at the front, from the two internal thoracic arteries (which lie vertically along either edge of the sternum, behind the ribs). Intercostal veins drain into similar veins alongside the sternum at the front, and into the large azygos vein at the back, on the right side. If a physician needs to drain fluid from the pleural cavity (the space between the lungs and the chest wall), the needle is inserted along the top of a rib, to avoid the main intercostal nerve and vessels running below it.

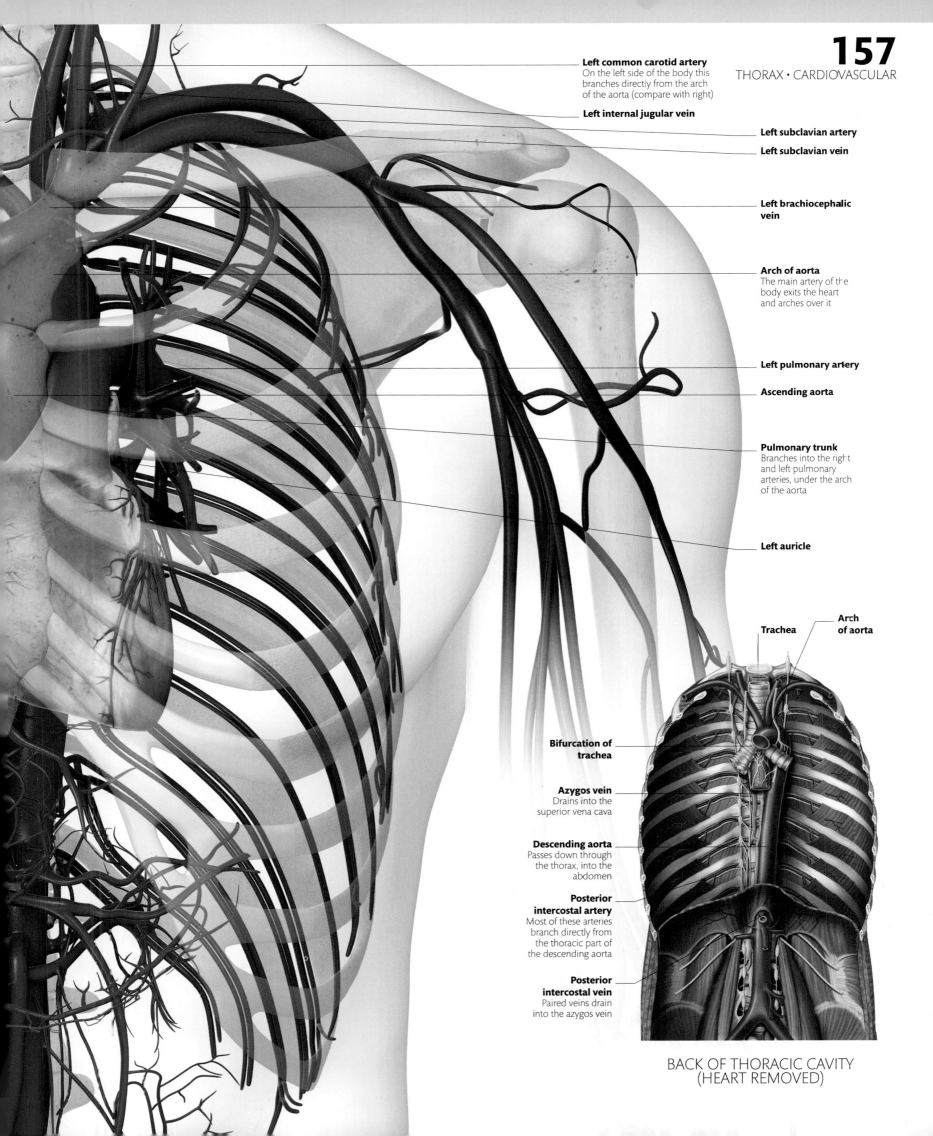

Left common carotid artery
On the left side of the body this
branches directly from the arch
of the aorta (compare with right)

Left internal jugular vein

Left subclavian artery

Left subclavian vein

**Left brachiocephalic
vein**

Arch of aorta
The main artery of the
body exits the heart
and arches over it

Left pulmonary artery

Ascending aorta

Pulmonary trunk
Branches into the right
and left pulmonary
arteries, under the arch
of the aorta

Left auricle

Trachea

**Arch
of aorta**

**Bifurcation of
trachea**

Azygos vein
Drains into the
superior vena cava

Descending aorta
Passes down through
the thorax, into the
abdomen

**Posterior
intercostal artery**
Most of these arteries
branch directly from
the thoracic part of
the descending aorta

**Posterior
intercostal vein**
Paired veins drain
into the azygos vein

BACK OF THORACIC CAVITY
(HEART REMOVED)

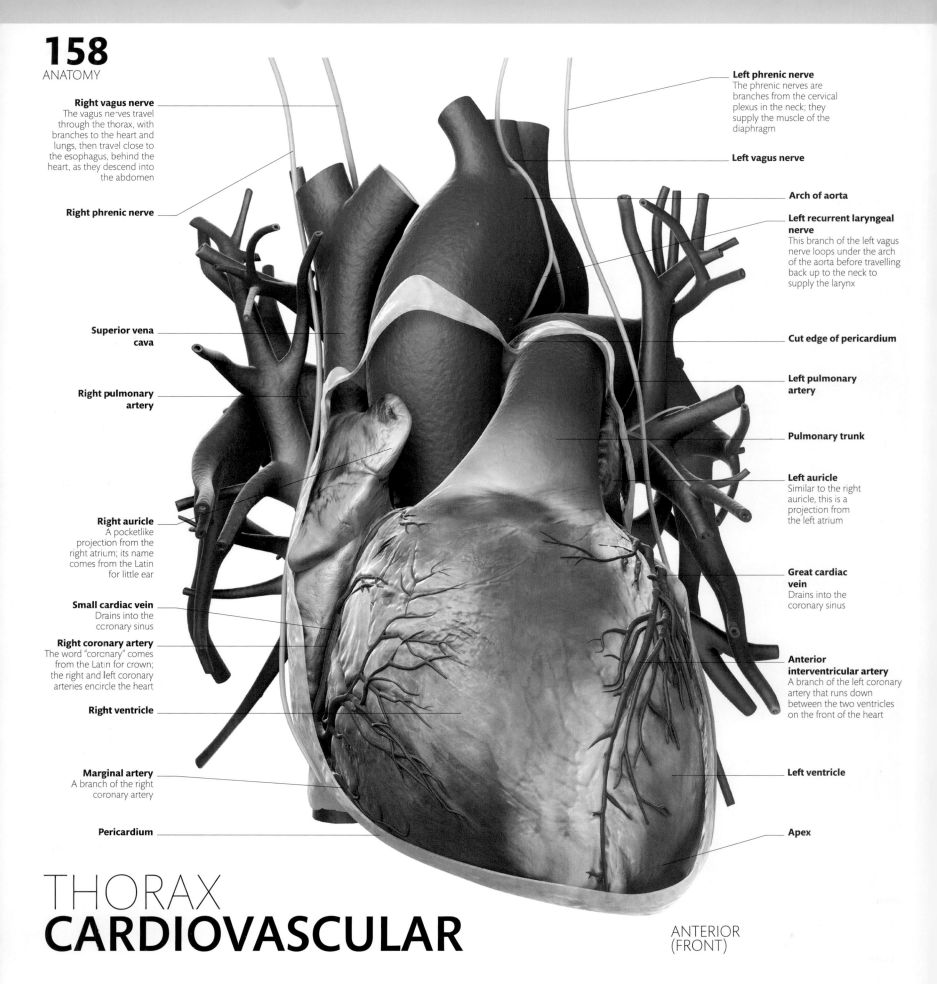

Right vagus nerve
The vagus nerves travel
through the thorax, with
branches to the heart and
lungs, then travel close to
the esophagus, behind the
heart, as they descend into
the abdomen

Right phrenic nerve

**Superior vena
cava**

**Right pulmonary
artery**

Right auricle
A pocketlike
projection from the
right atrium; its name
comes from the Latin
for little ear

Small cardiac vein
Drains into the
coronary sinus

Right coronary artery
The word "coronary" comes
from the Latin for crown;
the right and left coronary
arteries encircle the heart

Right ventricle

Marginal artery
A branch of the right
coronary artery

Pericardium

Left phrenic nerve
The phrenic nerves are
branches from the cervical
plexus in the neck; they
supply the muscle of the
diaphragm

Left vagus nerve

Arch of aorta

**Left recurrent laryngeal
nerve**
This branch of the left vagus
nerve loops under the arch
of the aorta before travelling
back up to the neck to
supply the larynx

Cut edge of pericardium

**Left pulmonary
artery**

Pulmonary trunk

Left auricle
Similar to the right
auricle, this is a
projection from
the left atrium

**Great cardiac
vein**
Drains into the
coronary sinus

**Anterior
interventricular artery**
A branch of the left coronary
artery that runs down
between the two ventricles
on the front of the heart

Left ventricle

Apex

THORAX
CARDIOVASCULAR

ANTERIOR
(FRONT)

The heart is encased in the pericardium. This has
a tough outer layer that is fused to the diaphragm
below and to the connective tissue around the
large blood vessels above the heart. Lining the
inside of this cylinder (and the outer surface of
the heart), is a thin membrane called the serous
pericardium. Between these two layers is a thin
film of fluid that lubricates the movement of the
heart as it beats. Inflammation of this membrane,
known as pericarditis, can be extremely painful.
Branches of the right and left coronary arteries,
which spring from the ascending aorta, supply the
heart muscle itself. The heart is drained by cardiac
veins, most of which drain into the coronary sinus.

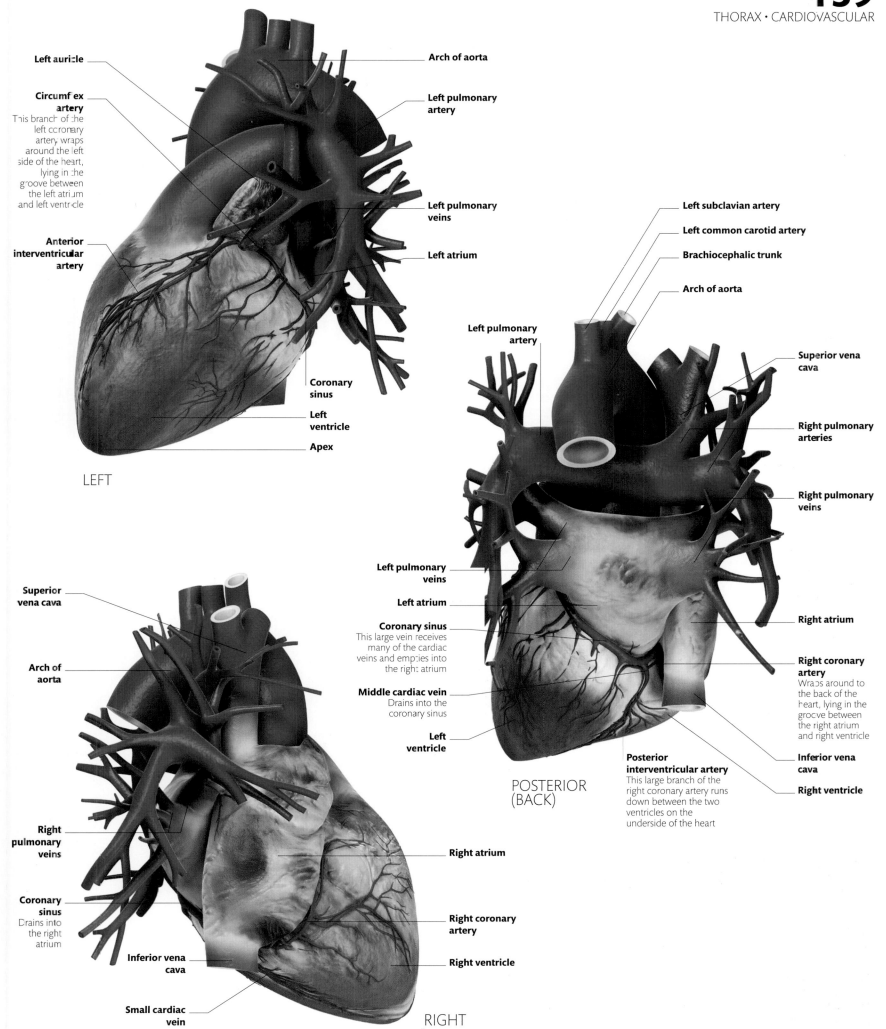

Left auricle

Arch of aorta

Left pulmonary artery

Circumflex artery
This branch of the left coronary artery wraps around the left side of the heart, lying in the groove between the left atrium and left ventricle

Left pulmonary veins

Left atrium

Anterior interventricular artery

Coronary sinus

Left ventricle

Apex

LEFT

Left subclavian artery

Left common carotid artery

Brachiocephalic trunk

Arch of aorta

Left pulmonary artery

Superior vena cava

Right pulmonary arteries

Right pulmonary veins

Left pulmonary veins

Left atrium

Right atrium

Coronary sinus
This large vein receives many of the cardiac veins and empties into the right atrium

Right coronary artery
Wraps around to the back of the heart, lying in the groove between the right atrium and right ventricle

Middle cardiac vein
Drains into the coronary sinus

Left ventricle

Posterior interventricular artery
This large branch of the right coronary artery runs down between the two ventricles on the underside of the heart

Inferior vena cava

Right ventricle

POSTERIOR (BACK)

Superior vena cava

Arch of aorta

Right pulmonary veins

Right atrium

Coronary sinus
Drains into the right atrium

Right coronary artery

Inferior vena cava

Right ventricle

Small cardiac vein

RIGHT

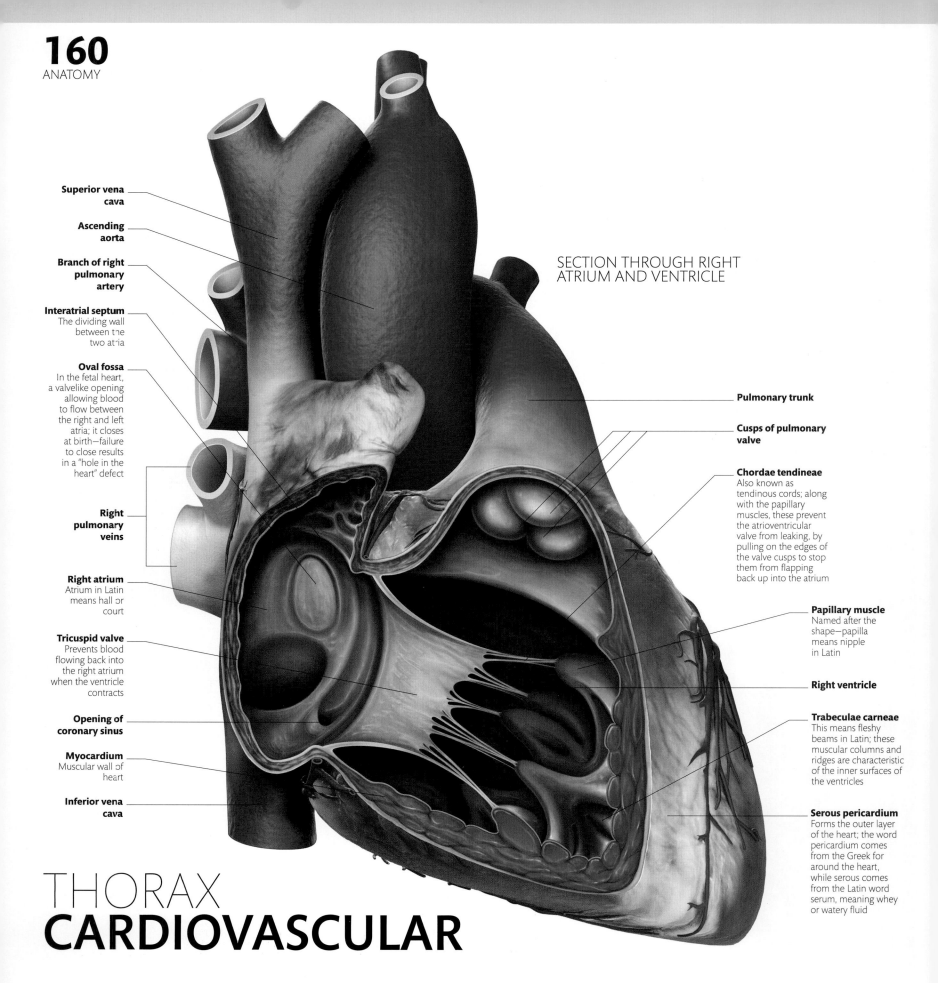

Superior vena cava

Ascending aorta

Branch of right pulmonary artery

Interatrial septum
The dividing wall between the two atria

Oval fossa
In the fetal heart, a valvelike opening allowing blood to flow between the right and left atria; it closes at birth—failure to close results in a "hole in the heart" defect

Right pulmonary veins

Right atrium
Atrium in Latin means hall or court

Tricuspid valve
Prevents blood flowing back into the right atrium when the ventricle contracts

Opening of coronary sinus

Myocardium
Muscular wall of heart

Inferior vena cava

SECTION THROUGH RIGHT ATRIUM AND VENTRICLE

Pulmonary trunk

Cusps of pulmonary valve

Chordae tendineae
Also known as tendinous cords; along with the papillary muscles, these prevent the atrioventricular valve from leaking, by pulling on the edges of the valve cusps to stop them from flapping back up into the atrium

Papillary muscle
Named after the shape—papilla means nipple in Latin

Right ventricle

Trabeculae carneae
This means fleshy beams in Latin; these muscular columns and ridges are characteristic of the inner surfaces of the ventricles

Serous pericardium
Forms the outer layer of the heart; the word pericardium comes from the Greek for around the heart, while serous comes from the Latin word serum, meaning whey or watery fluid

THORAX
CARDIOVASCULAR

The heart receives blood from veins and pumps it out through arteries. It has four chambers: two atria and two ventricles. The heart's left and right sides are separate. The right side receives deoxygenated blood from the body via the superior and inferior venae cavae, and pumps it to the lungs through the pulmonary trunk. The left gets oxygenated blood from the lungs via the pulmonary veins, and pumps it into the aorta for distribution. Each atrium opens into its corresponding ventricle via a valve (on the right, the tricuspid valve, and the bicuspid valve on the left), which shuts when the ventricle contracts, to stop blood flowing back into the atrium. The aorta and pulmonary trunk also have valves.

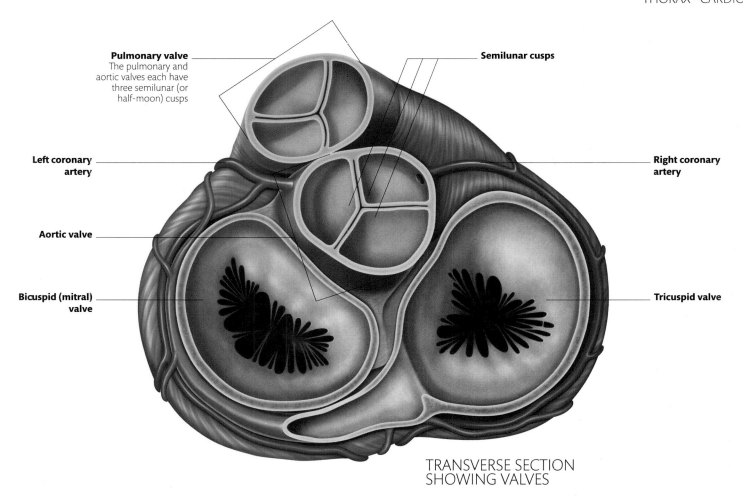

Pulmonary valve
The pulmonary and
aortic valves each have
three semilunar (or
half-moon) cusps

Semilunar cusps

**Left coronary
artery**

**Right coronary
artery**

Aortic valve

**Bicuspid (mitral)
valve**

Tricuspid valve

TRANSVERSE SECTION
SHOWING VALVES

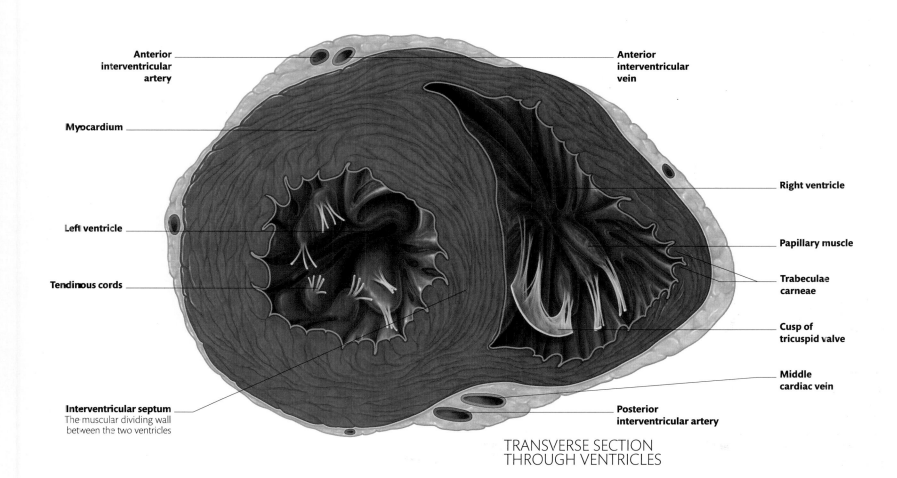

**Anterior
interventricular
artery**

**Anterior
interventricular
vein**

Myocardium

Right ventricle

Left ventricle

Papillary muscle

Tendinous cords

**Trabeculae
carneae**

**Cusp of
tricuspid valve**

**Middle
cardiac vein**

Interventricular septum
The muscular dividing wall
between the two ventricles

**Posterior
interventricular artery**

TRANSVERSE SECTION
THROUGH VENTRICLES

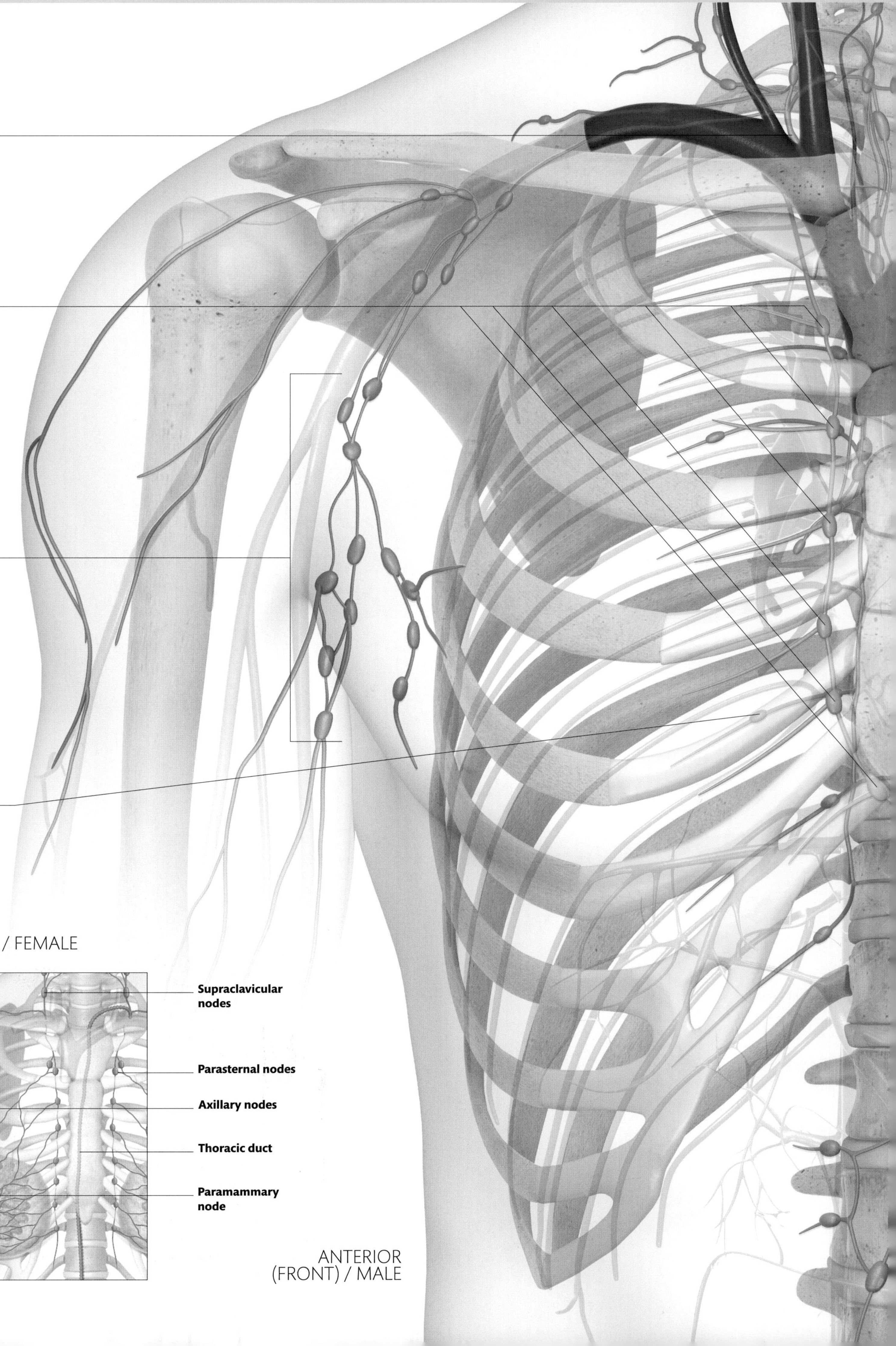

Right lymphatic duct
Lymph from the right arm and the right side of the neck and thorax drains into the junction of the right internal jugular and subclavian veins

Parasternal nodes
Also called internal thoracic nodes; these lie in the gaps between the ribs, either side of the sternum on the inside of the rib cage; they drain some of the lymph from the front of the thorax—including from the breast in a woman

Axillary nodes
Receive lymph from superficial tissues of the thorax, upper limb, and breast

Intercostal nodes
Sitting in the intercostal spaces between the ribs at the back of the rib cage, these drain lymph from the deeper tissues at the sides and back of the thorax

ANTERIOR (FRONT) / FEMALE

Supraclavicular nodes

Parasternal nodes

Axillary nodes

Thoracic duct

Paramammary node

ANTERIOR
(FRONT) / MALE

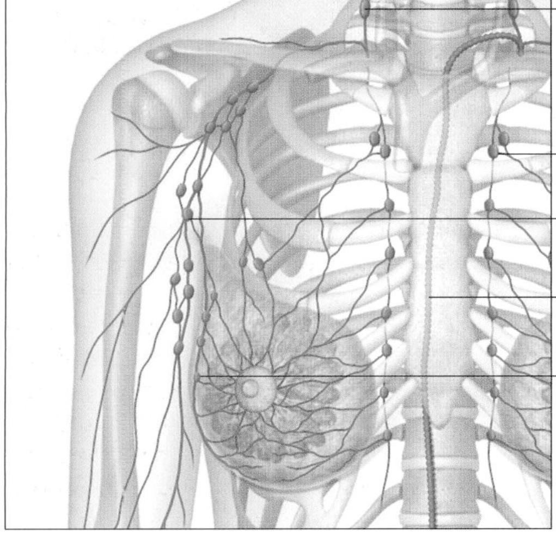

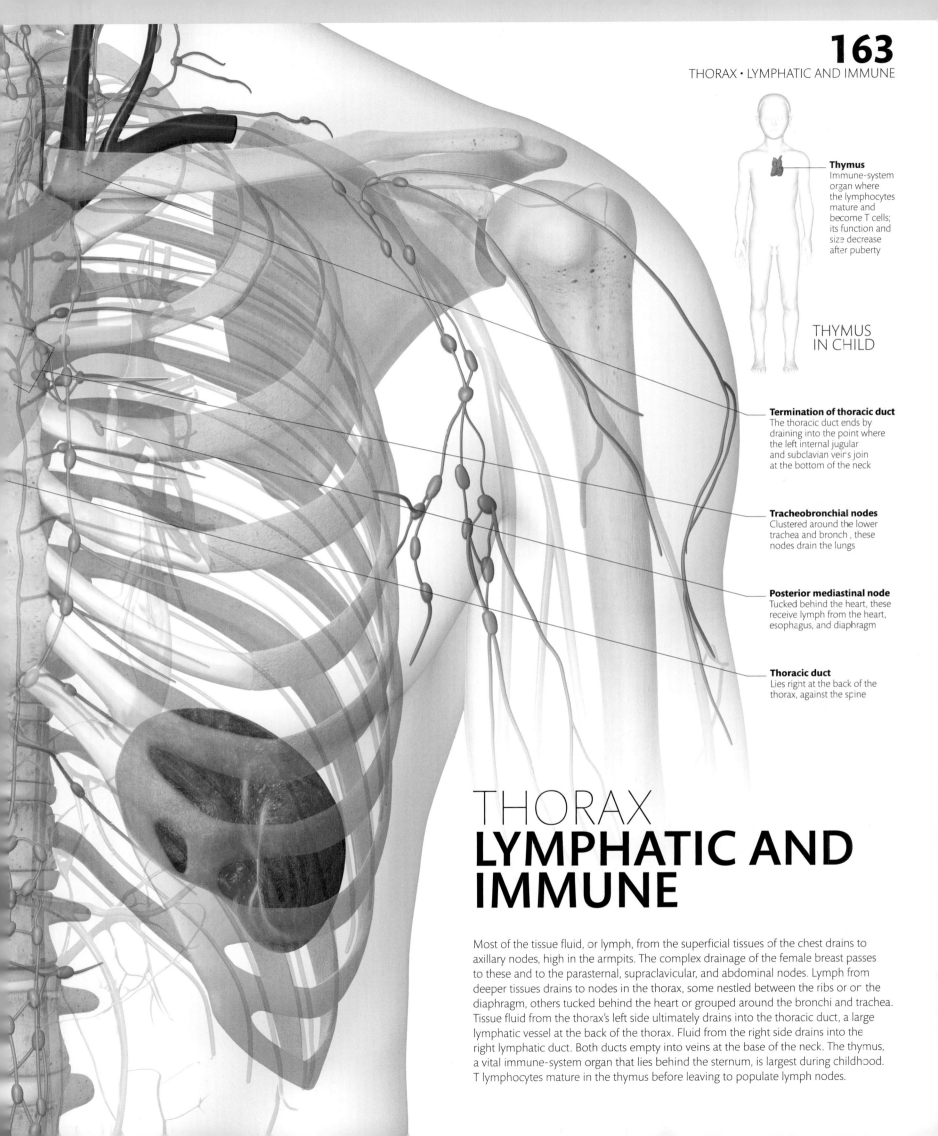

Thymus
Immune-system organ where the lymphocytes mature and become T cells; its function and size decrease after puberty

THYMUS IN CHILD

Termination of thoracic duct
The thoracic duct ends by draining into the point where the left internal jugular and subclavian veins join at the bottom of the neck

Tracheobronchial nodes
Clustered around the lower trachea and bronch , these nodes drain the lungs

Posterior mediastinal node
Tucked behind the heart, these receive lymph from the heart, esophagus, and diaphragm

Thoracic duct
Lies right at the back of the thorax, against the spine

THORAX
LYMPHATIC AND IMMUNE

Most of the tissue fluid, or lymph, from the superficial tissues of the chest drains to axillary nodes, high in the armpits. The complex drainage of the female breast passes to these and to the parasternal, supraclavicular, and abdominal nodes. Lymph from deeper tissues drains to nodes in the thorax, some nestled between the ribs or on the diaphragm, others tucked behind the heart or grouped around the bronchi and trachea. Tissue fluid from the thorax's left side ultimately drains into the thoracic duct, a large lymphatic vessel at the back of the thorax. Fluid from the right side drains into the right lymphatic duct. Both ducts empty into veins at the base of the neck. The thymus, a vital immune-system organ that lies behind the sternum, is largest during childhood. T lymphocytes mature in the thymus before leaving to populate lymph nodes.

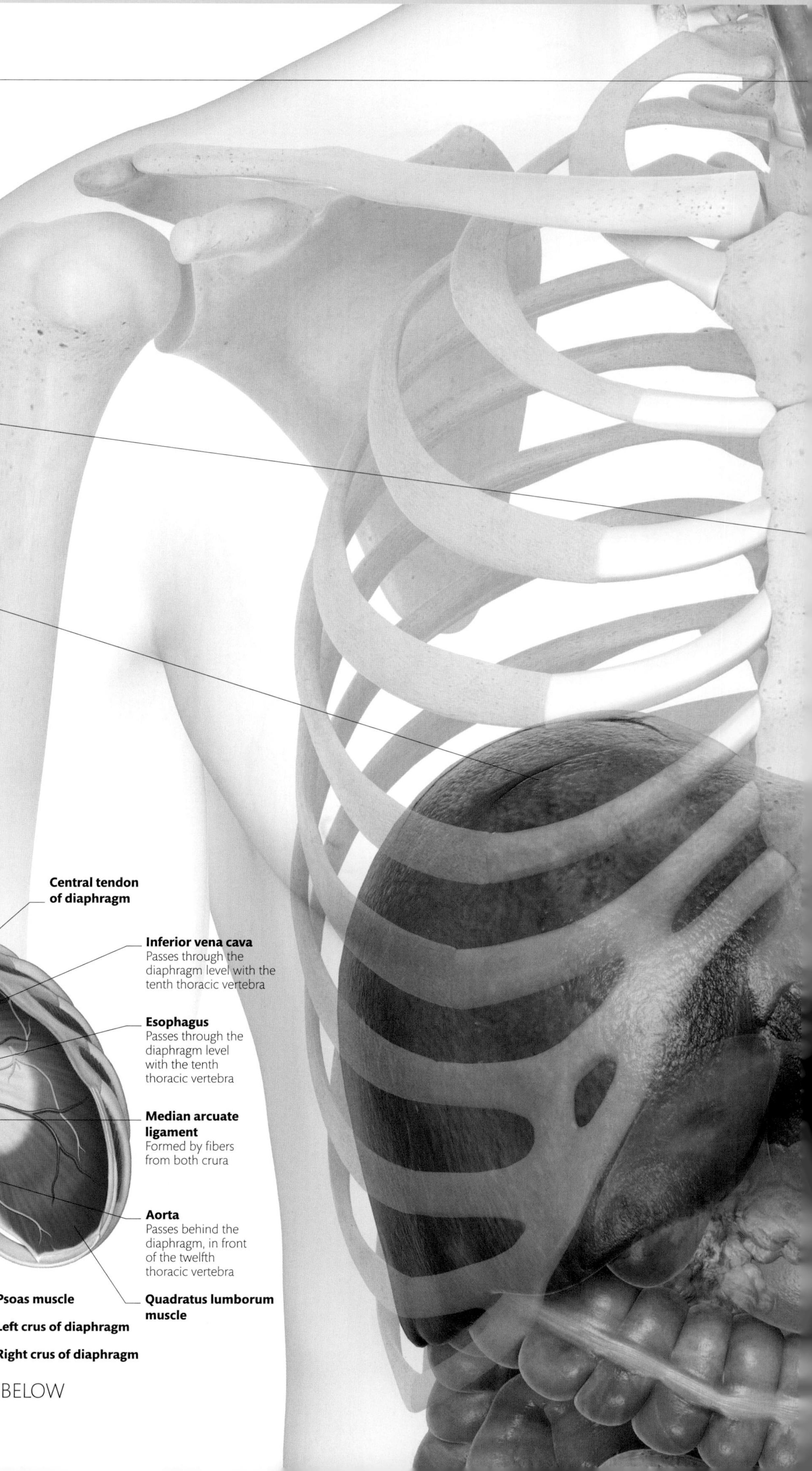

Esophagus
In the neck, the esophagus
lies behind the trachea

**Thoracic part of the
esophagus**
The esophagus is slightly
constricted here by the left
main bronchus, which
crosses in front of it

Liver
Lies under the right dome of
the diaphragm, and largely
under cover of the ribs

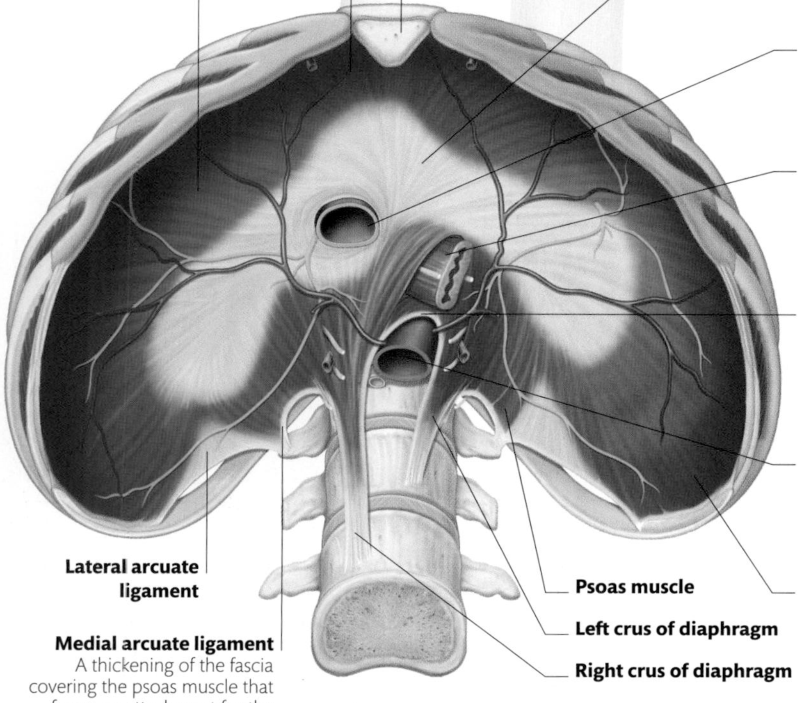

**Muscular part of
diaphragm**

Sternal part

Xiphoid process

**Central tendon
of diaphragm**

Inferior vena cava
Passes through the
diaphragm level with the
tenth thoracic vertebra

Esophagus
Passes through the
diaphragm level
with the tenth
thoracic vertebra

**Median arcuate
ligament**
Formed by fibers
from both crura

Aorta
Passes behind the
diaphragm, in front
of the twelfth
thoracic vertebra

**Lateral arcuate
ligament**

Psoas muscle

**Quadratus lumborum
muscle**

Medial arcuate ligament
A thickening of the fascia
covering the psoas muscle that
forms an attachment for the
muscle fibers of the diaphragm

Left crus of diaphragm

Right crus of diaphragm

DIAPHRAGM FROM BELOW

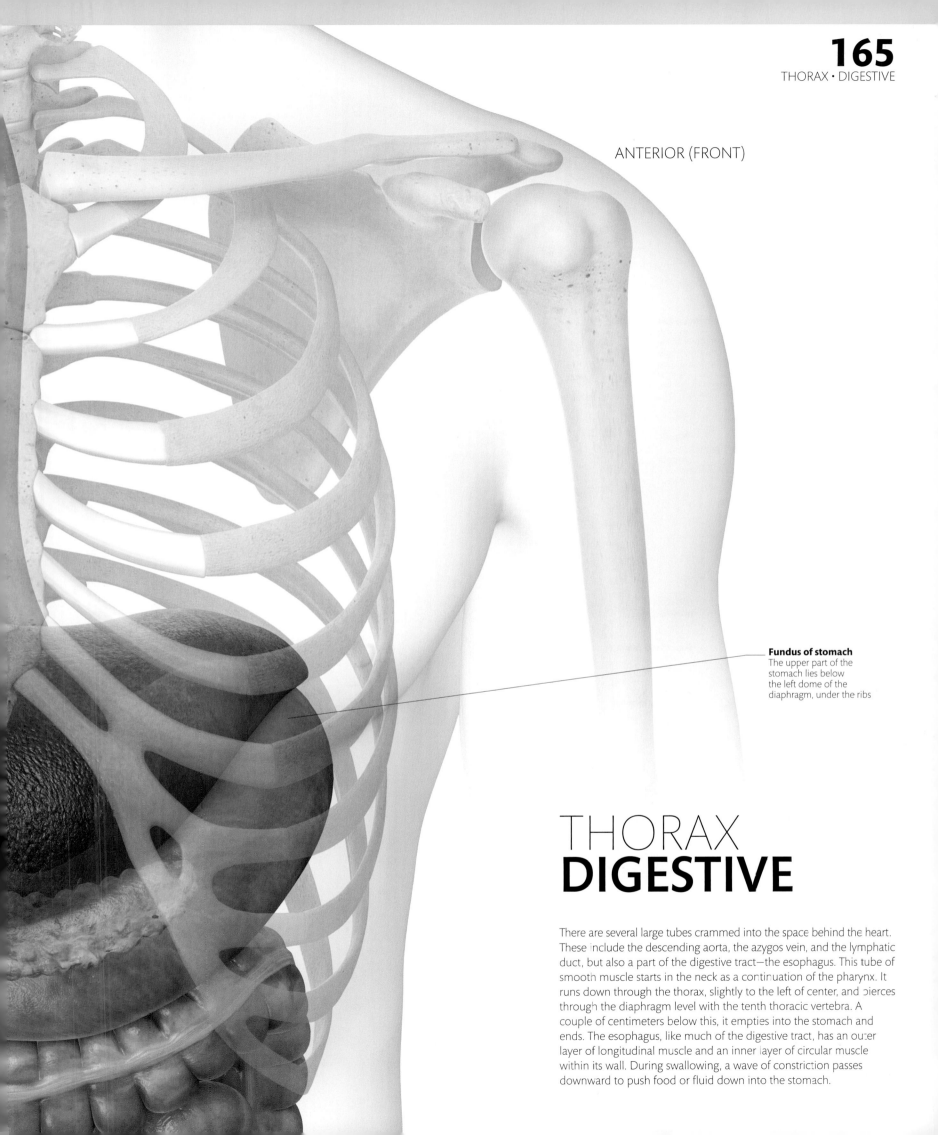

ANTERIOR (FRONT)

Fundus of stomach
The upper part of the
stomach lies below
the left dome of the
diaphragm, under the ribs

THORAX
DIGESTIVE

There are several large tubes crammed into the space behind the heart.
These include the descending aorta, the azygos vein, and the lymphatic
duct, but also a part of the digestive tract—the esophagus. This tube of
smooth muscle starts in the neck as a continuation of the pharynx. It
runs down through the thorax, slightly to the left of center, and pierces
through the diaphragm level with the tenth thoracic vertebra. A
couple of centimeters below this, it empties into the stomach and
ends. The esophagus, like much of the digestive tract, has an outer
layer of longitudinal muscle and an inner layer of circular muscle
within its wall. During swallowing, a wave of constriction passes
downward to push food or fluid down into the stomach.

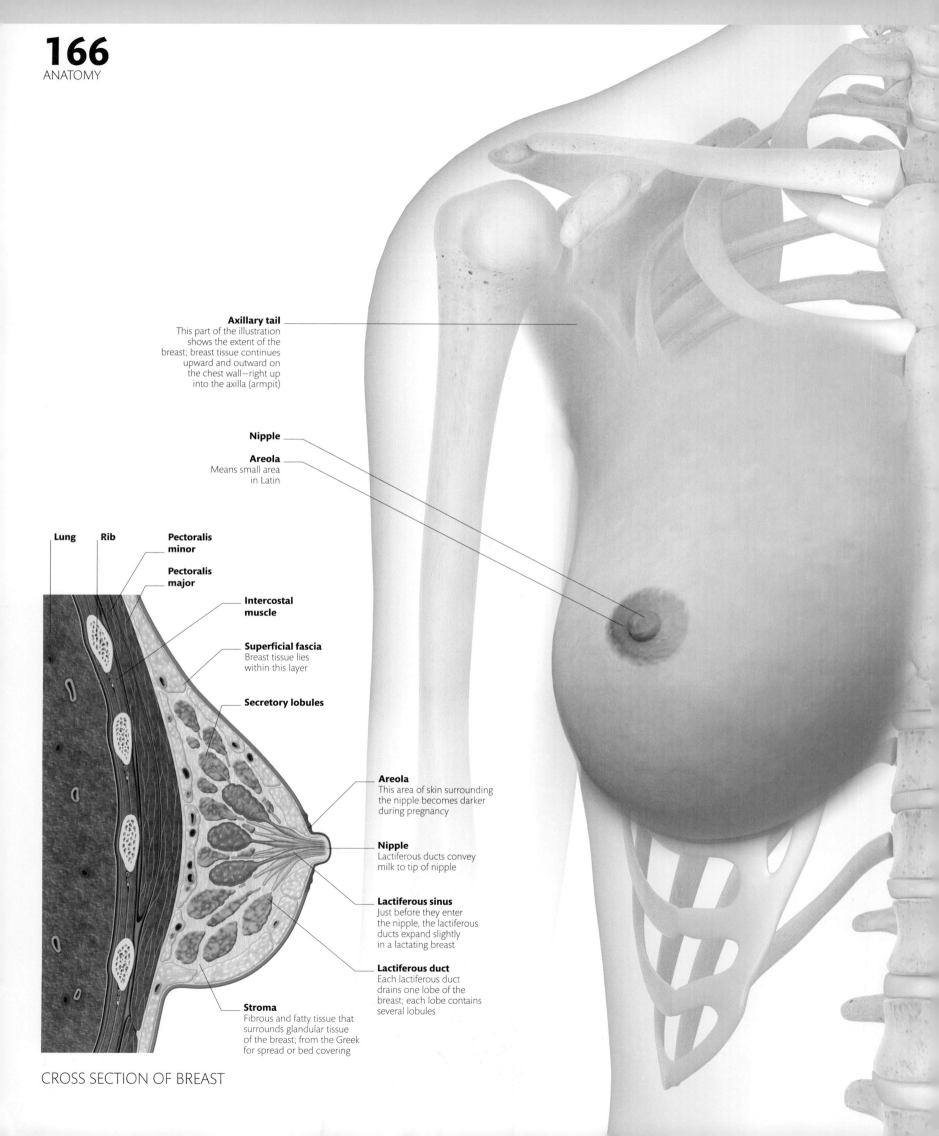

Axillary tail
This part of the illustration
shows the extent of the
breast; breast tissue continues
upward and outward on
the chest wall—right up
into the axilla (armpit)

Nipple

Areola
Means small area
in Latin

Lung   **Rib**

**Pectoralis
minor**

**Pectoralis
major**

**Intercostal
muscle**

Superficial fascia
Breast tissue lies
within this layer

Secretory lobules

Areola
This area of skin surrounding
the nipple becomes darker
during pregnancy

Nipple
Lactiferous ducts convey
milk to tip of nipple

Lactiferous sinus
Just before they enter
the nipple, the lactiferous
ducts expand slightly
in a lactating breast

Lactiferous duct
Each lactiferous duct
drains one lobe of the
breast; each lobe contains
several lobules

Stroma
Fibrous and fatty tissue that
surrounds glandular tissue
of the breast; from the Greek
for spread or bed covering

CROSS SECTION OF BREAST

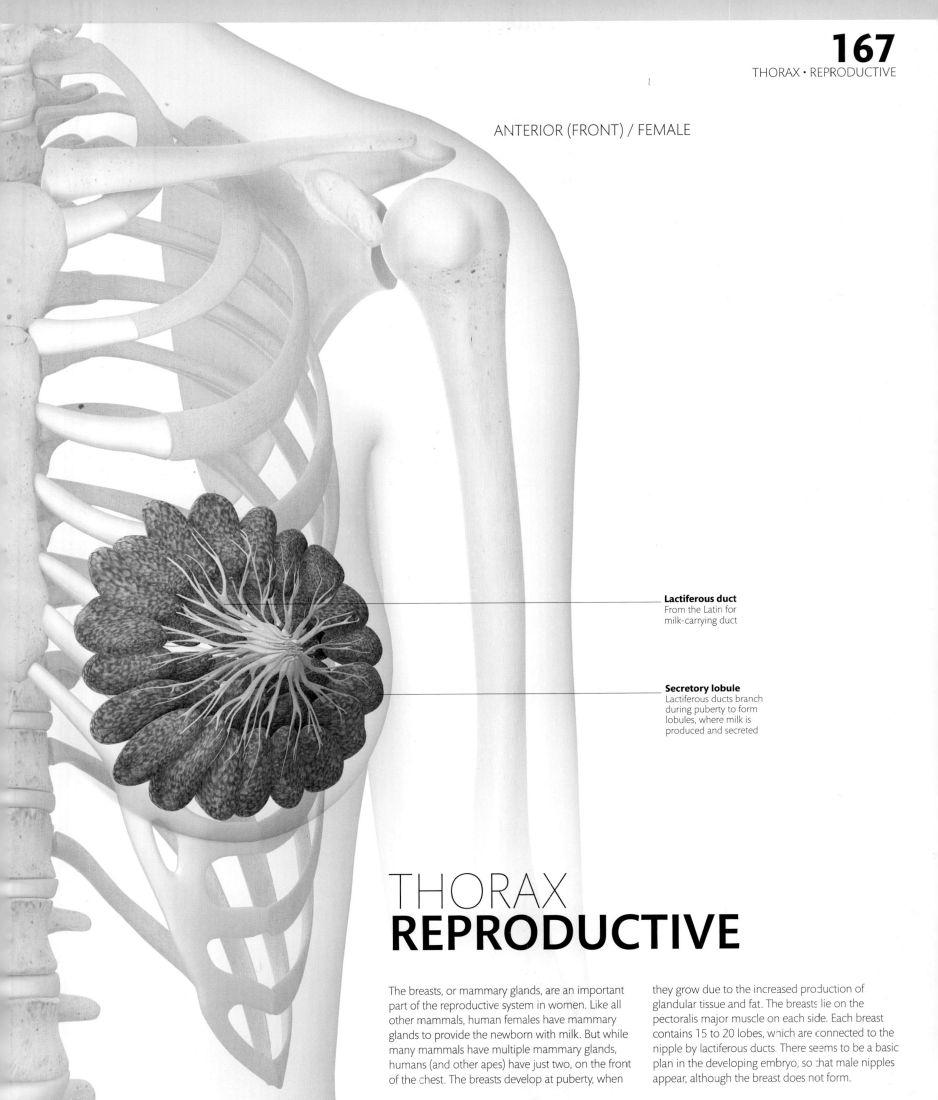

ANTERIOR (FRONT) / FEMALE

Lactiferous duct
From the Latin for
milk-carrying duct

Secretory lobule
Lactiferous ducts branch
during puberty to form
lobules, where milk is
produced and secreted

THORAX
REPRODUCTIVE

The breasts, or mammary glands, are an important part of the reproductive system in women. Like all other mammals, human females have mammary glands to provide the newborn with milk. But while many mammals have multiple mammary glands, humans (and other apes) have just two, on the front of the chest. The breasts develop at puberty, when they grow due to the increased production of glandular tissue and fat. The breasts lie on the pectoralis major muscle on each side. Each breast contains 15 to 20 lobes, which are connected to the nipple by lactiferous ducts. There seems to be a basic plan in the developing embryo, so that male nipples appear, although the breast does not form.

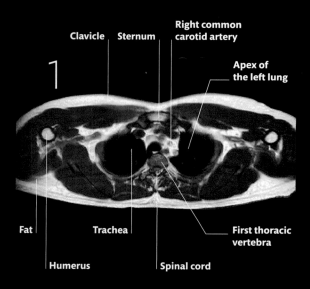

1

Clavicle   Sternum   Right common
                     carotid artery

Apex of
the left lung

Fat   Trachea   First thoracic
                vertebra

Humerus   Spinal cord

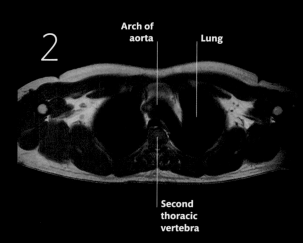

2

Arch of
aorta   Lung

Second
thoracic
vertebra

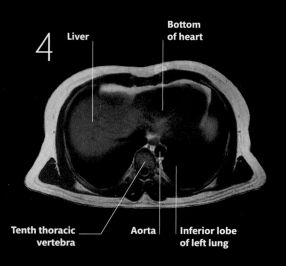

4

Liver   Bottom
        of heart

Tenth thoracic   Aorta   Inferior lobe
vertebra   of left lung

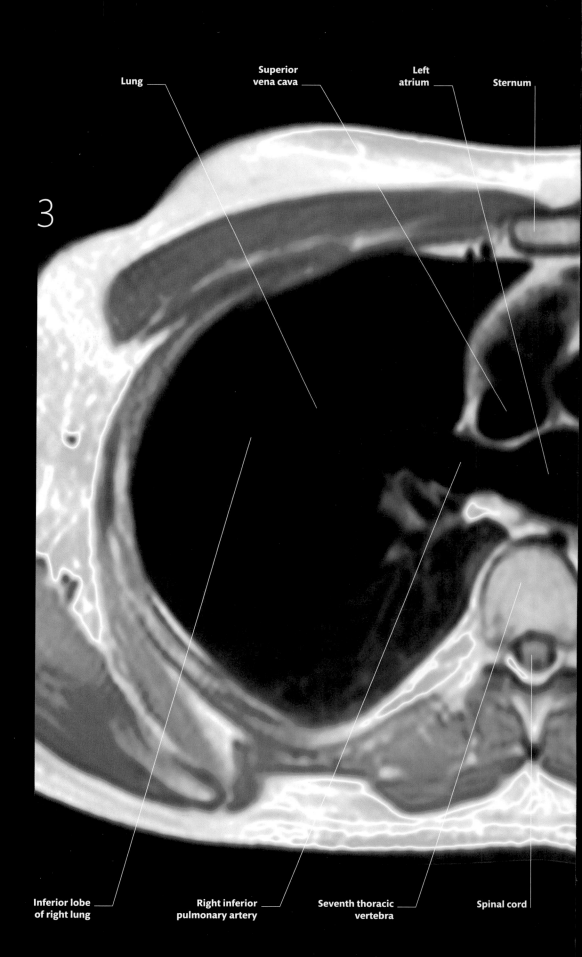

3

Lung   Superior
       vena cava   Left
                   atrium   Sternum

Inferior lobe
of right lung   Right inferior
                pulmonary artery   Seventh thoracic
                                   vertebra   Spinal cord

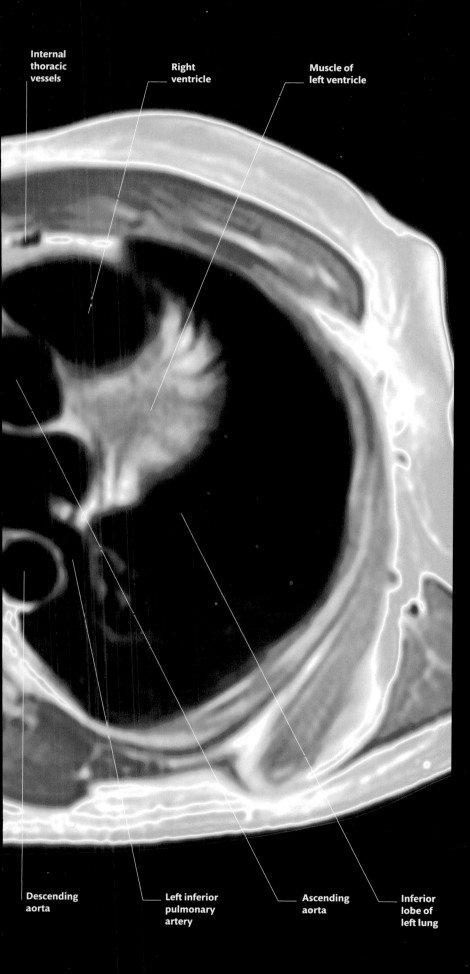

Internal thoracic vessels

Right ventricle

Muscle of left ventricle

Descending aorta

Left inferior pulmonary artery

Ascending aorta

Inferior lobe of left lung

LEVELS OF SCANS

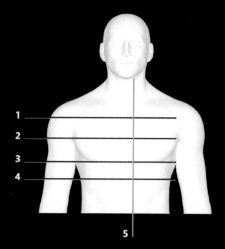

1
2
3
4

5

THORAX **MRI**

The axial, or transverse, sections through the chest (sections 1–4) show the heart and large blood vessels lying centrally within the thorax, flanked by the lungs, and all set within the protective, bony casing of the rib cage. Section 1 shows the clavicles, or collarbones, joining the sternum at the front, the apex (top) of the lungs, and the great vessels passing between the neck and the thorax. Section 2 is lower down in the chest, just above the heart, while section 3 shows the heart with detail of its different chambers. The aorta appears to be to the right of the spine in this image, rather than to the left, but this is the usual way in which scans are viewed. You need to imagine yourself standing at the foot of the bed, looking down at the patient. This means that the left side of the body appears on the right side of the image as you view it. Section 4 shows the very bottom of the heart, and the inferior lobes of the lungs.

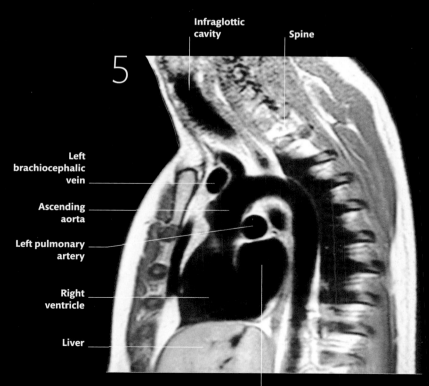

Infraglottic cavity

Spine

5

Left brachiocephalic vein

Ascending aorta

Left pulmonary artery

Right ventricle

Liver

Left atrium

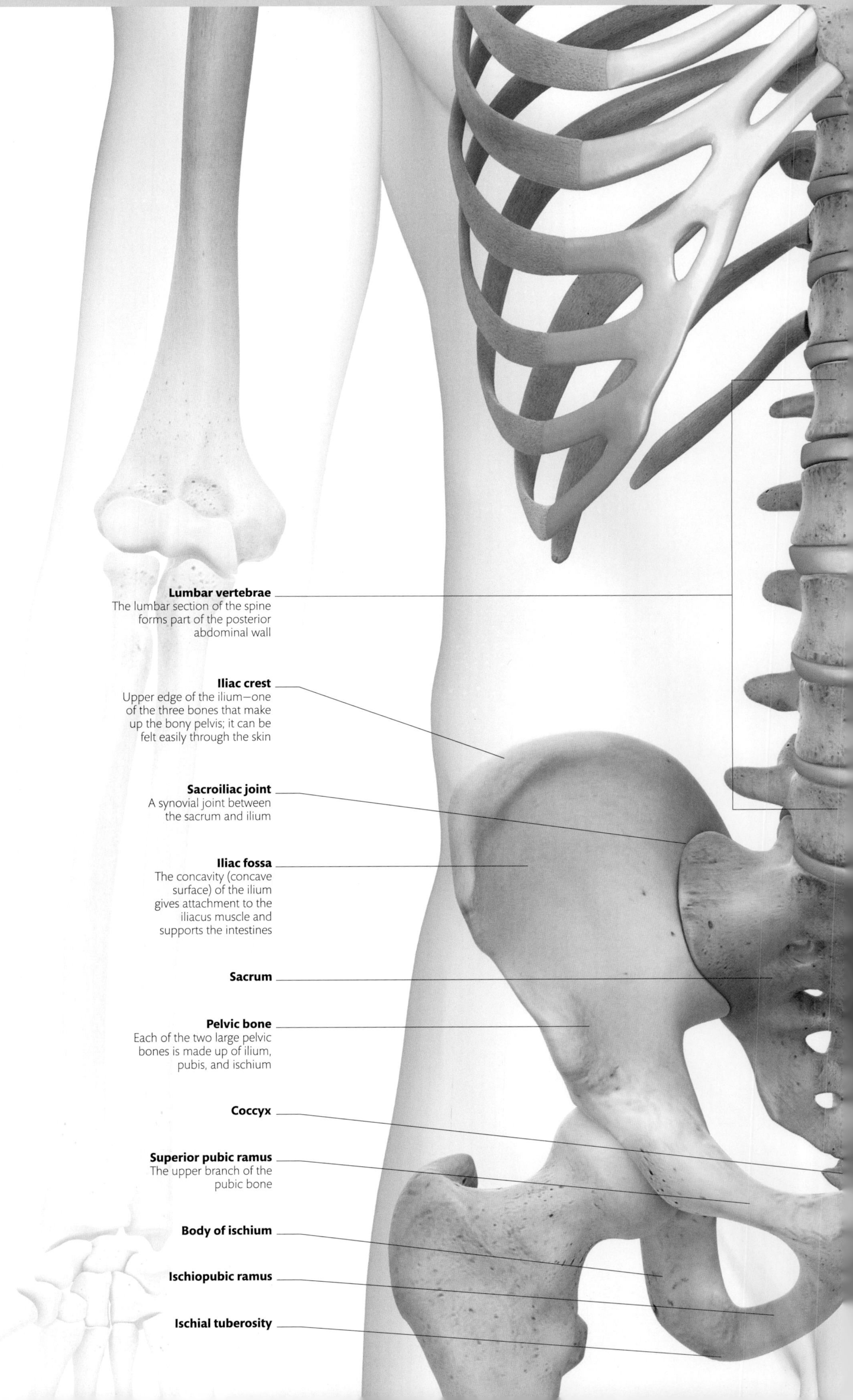

Lumbar vertebrae
The lumbar section of the spine forms part of the posterior abdominal wall

Iliac crest
Upper edge of the ilium—one of the three bones that make up the bony pelvis; it can be felt easily through the skin

Sacroiliac joint
A synovial joint between the sacrum and ilium

Iliac fossa
The concavity (concave surface) of the ilium gives attachment to the iliacus muscle and supports the intestines

Sacrum

Pelvic bone
Each of the two large pelvic bones is made up of ilium, pubis, and ischium

Coccyx

Superior pubic ramus
The upper branch of the pubic bone

Body of ischium

Ischiopubic ramus

Ischial tuberosity

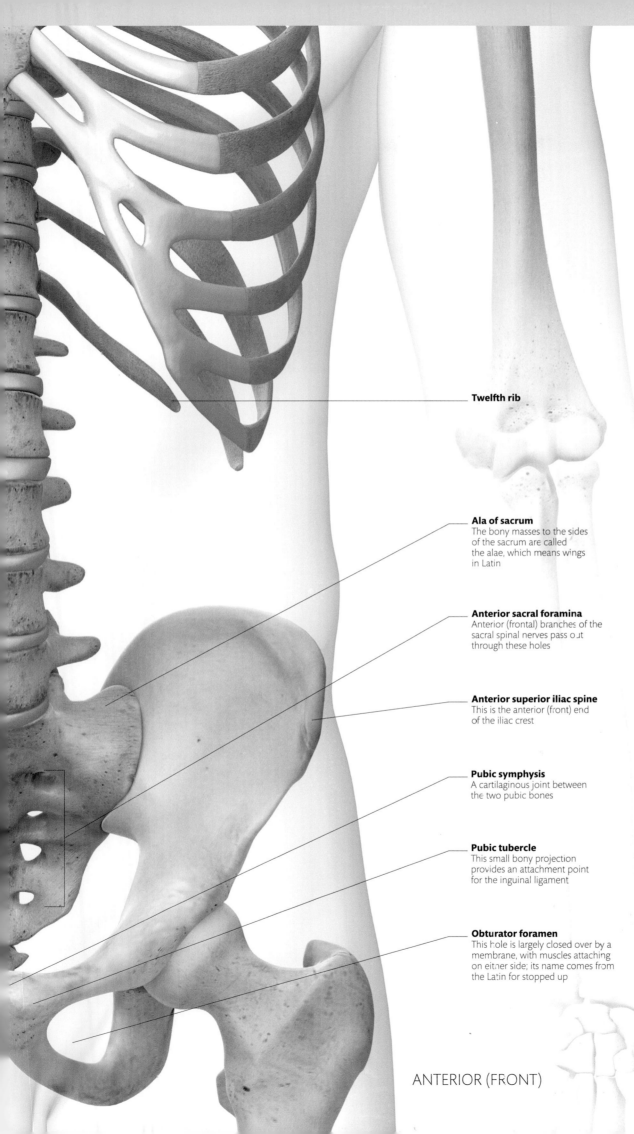

ABDOMEN AND PELVIS
SKELETAL

The bony boundaries of the abdomen include the five lumbar vertebrae at the back, the lower margin of the ribs above, and the pubic bones and iliac crest of the pelvic bones below. The abdominal cavity itself extends up under the rib cage, as high as the gap between the fifth and sixth ribs, due to the domed shape of the diaphragm. This means that some abdominal organs, such as the liver, stomach, and spleen, are, in fact, largely tucked up under the ribs. The pelvis is a basin shape, and is enclosed by the two pelvic (or innominate) bones, at the front and sides, and by the sacrum at the back. Each pelvic bone is made of three fused bones: the ilium at the rear, the ischium at the lower front, and the pubis above it.

Twelfth rib

Ala of sacrum
The bony masses to the sides of the sacrum are called the alae, which means wings in Latin

Anterior sacral foramina
Anterior (frontal) branches of the sacral spinal nerves pass out through these holes

Anterior superior iliac spine
This is the anterior (front) end of the iliac crest

Pubic symphysis
A cartilaginous joint between the two pubic bones

Pubic tubercle
This small bony projection provides an attachment point for the inguinal ligament

Obturator foramen
This hole is largely closed over by a membrane, with muscles attaching on either side; its name comes from the Latin for stopped up

ANTERIOR (FRONT)

ABDOMEN AND PELVIS
SKELETAL

The orientation of the facet joints (the joints between the vertebrae) of the lumbar spine restrict rotation of the vertebrae, but flexion and extension can occur freely. There is, however, rotation at the lumbosacral joint, which allows the pelvis to swing during walking. The sacroiliac joints are unusual in that they are synovial joints (which are usually very movable), yet they are particularly limited in their movement. This is because strong sacroiliac ligaments around the joints bind the ilium (part of the pelvic bone) tightly to the sacrum on each side. Lower down, the sacrospinous and sacrotuberous ligaments, stretching from the sacrum and coccyx to the ilium, provide additional support and stability.

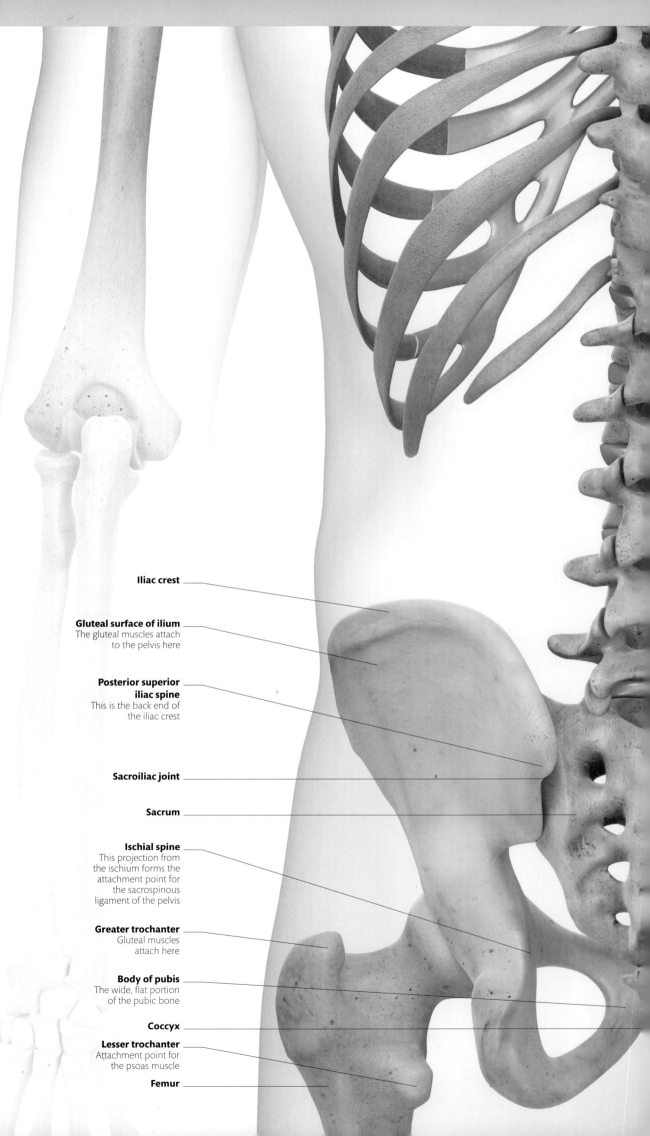

Iliac crest

Gluteal surface of ilium
The gluteal muscles attach to the pelvis here

Posterior superior iliac spine
This is the back end of the iliac crest

Sacroiliac joint

Sacrum

Ischial spine
This projection from the ischium forms the attachment point for the sacrospinous ligament of the pelvis

Greater trochanter
Gluteal muscles attach here

Body of pubis
The wide, flat portion of the pubic bone

Coccyx

Lesser trochanter
Attachment point for the psoas muscle

Femur

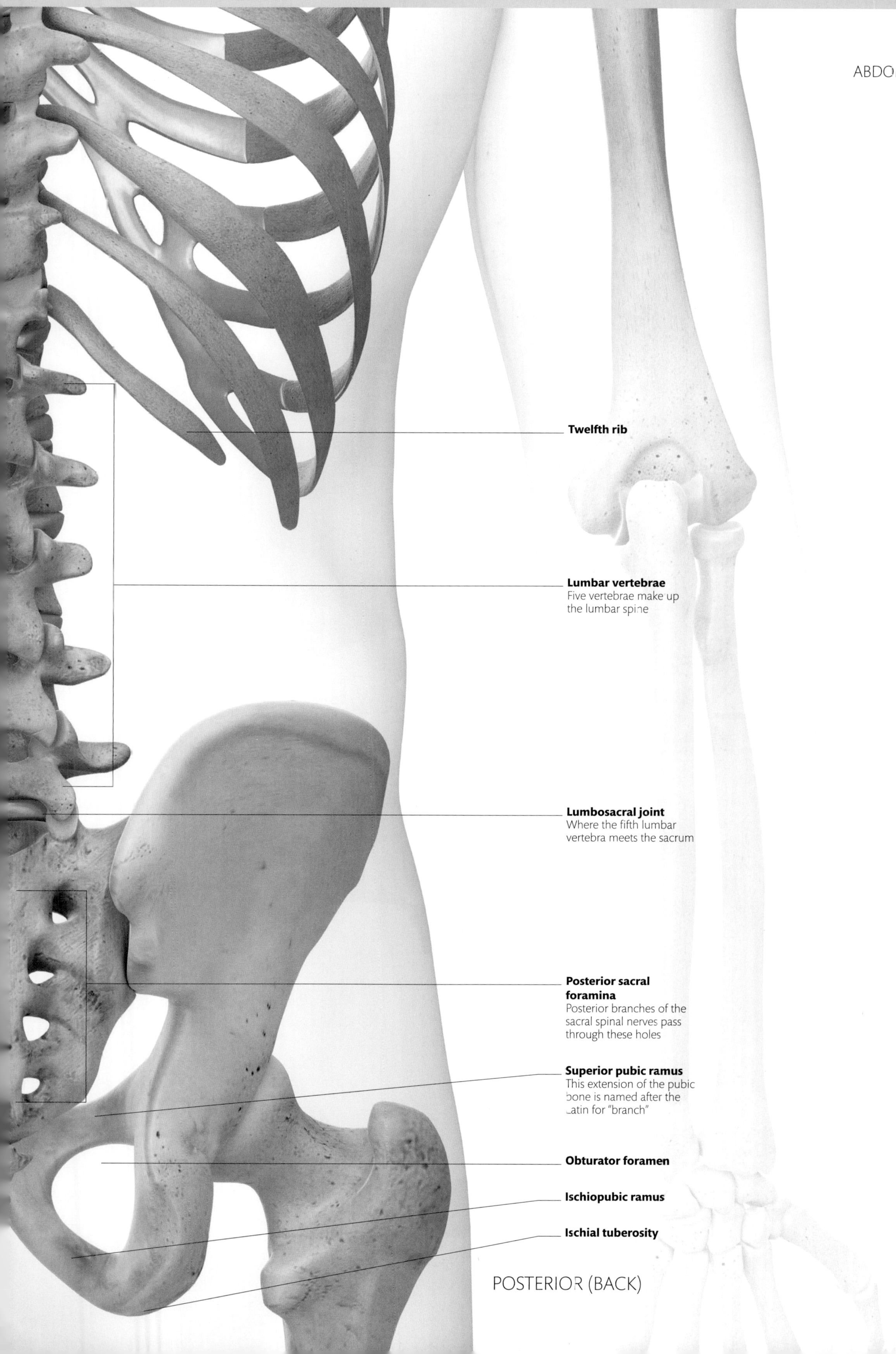

Twelfth rib

Lumbar vertebrae
Five vertebrae make up
the lumbar spine

Lumbosacral joint
Where the fifth lumbar
vertebra meets the sacrum

**Posterior sacral
foramina**
Posterior branches of the
sacral spinal nerves pass
through these holes

Superior pubic ramus
This extension of the pubic
bone is named after the
Latin for "branch"

Obturator foramen

Ischiopubic ramus

Ischial tuberosity

POSTERIOR (BACK)

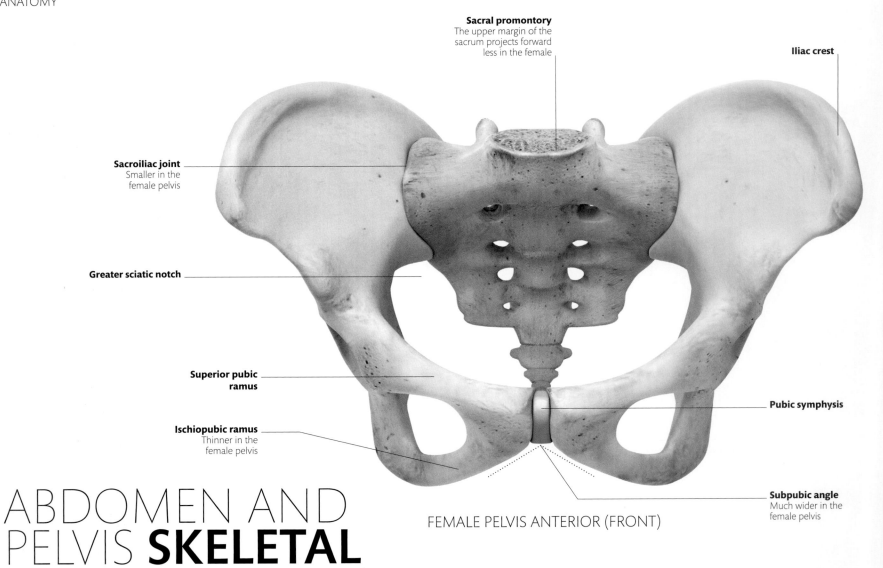

Sacral promontory
The upper margin of the sacrum projects forward less in the female

Iliac crest

Sacroiliac joint
Smaller in the female pelvis

Greater sciatic notch

Superior pubic ramus

Ischiopubic ramus
Thinner in the female pelvis

Pubic symphysis

Subpubic angle
Much wider in the female pelvis

FEMALE PELVIS ANTERIOR (FRONT)

ABDOMEN AND
PELVIS **SKELETAL**

The bony pelvis is the part of the skeleton that is most different between the sexes, because the pelvis in the female has to accommodate the birth canal, unlike the male pelvis. Comparing the pelvic bones of a man and a woman, there are obvious differences between the two. The shape of the ring formed by the sacrum and the two pelvic bones—the pelvic brim—tends to be a wide oval in the woman and much narrower and heart-shaped in a man. The subpubic angle, underneath the joint between the two pubic bones, is much narrower in a man than it is in with a woman. As with the rest of the skeleton, the pelvic bone also tends to be more chunky or robust in a man, with more obvious ridges where muscles attach.

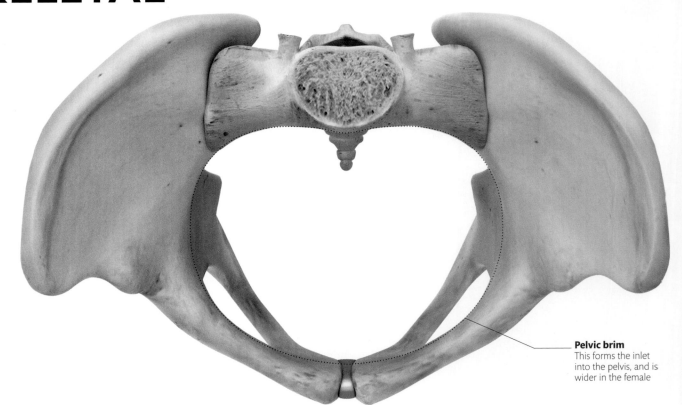

Pelvic brim
This forms the inlet into the pelvis, and is wider in the female

FEMALE PELVIS VIEWED FROM ABOVE

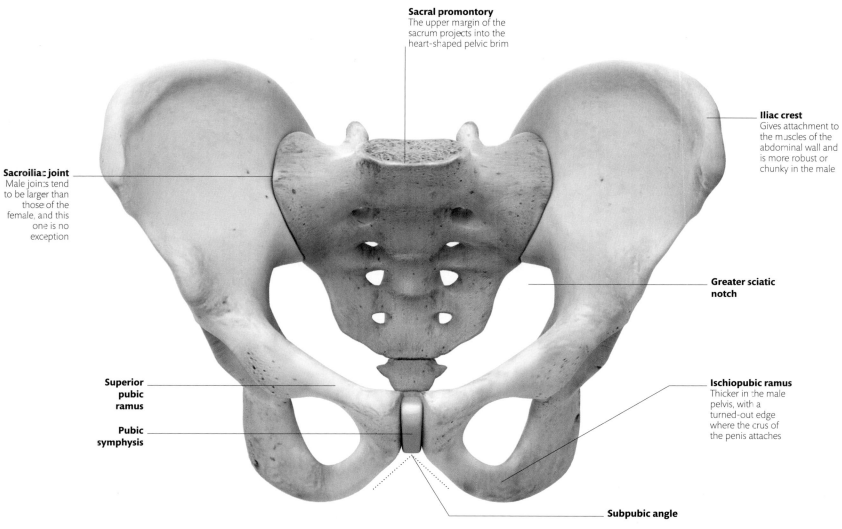

Sacral promontory
The upper margin of the sacrum projects into the heart-shaped pelvic brim

Iliac crest
Gives attachment to the muscles of the abdominal wall and is more robust or chunky in the male

Sacroiliac joint
Male joints tend to be larger than those of the female, and this one is no exception

Greater sciatic notch

Superior pubic ramus

Pubic symphysis

Ischiopubic ramus
Thicker in the male pelvis, with a turned-out edge where the crus of the penis attaches

Subpubic angle

MALE PELVIS ANTERIOR (FRONT)

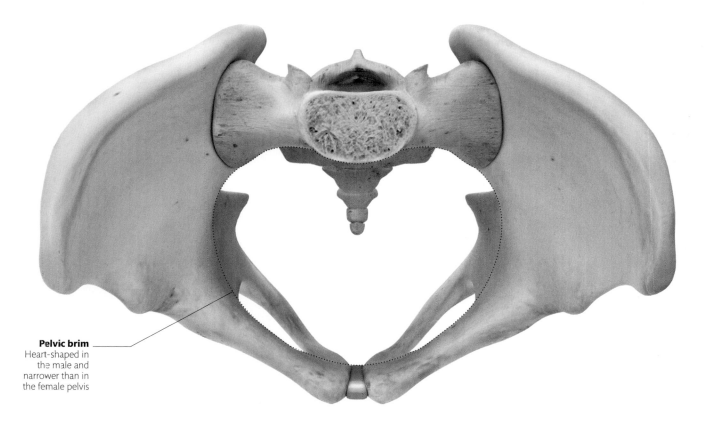

Pelvic brim
Heart-shaped in the male and narrower than in the female pelvis

MALE PELVIS VIEWED FROM ABOVE

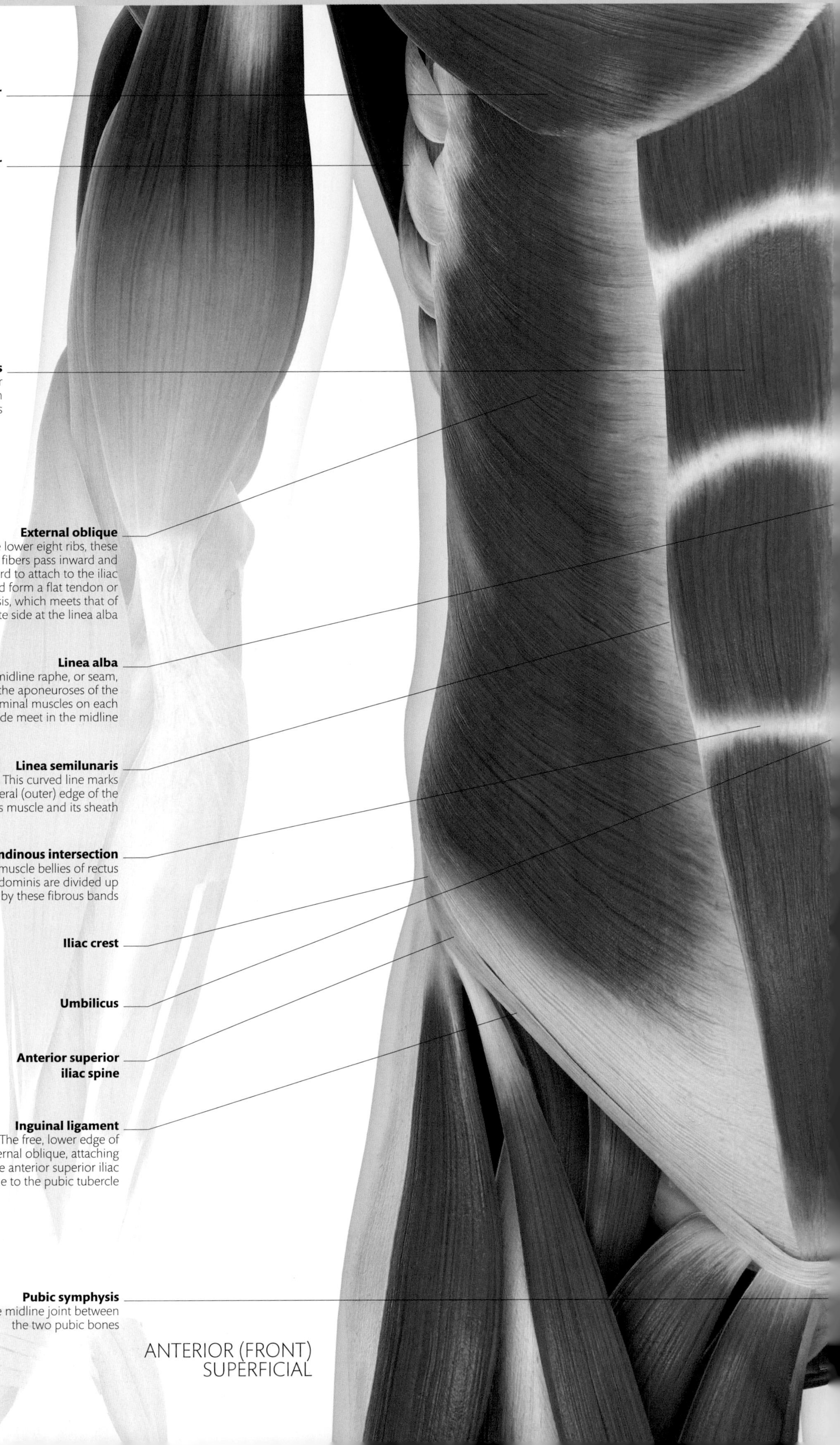

Pectoralis major

Serratus anterior

Rectus abdominis
Attaches from the lower
costal cartilages, down
to the pubic bones

External oblique
From the lower eight ribs, these
muscle fibers pass inward and
downward to attach to the iliac
crest, and form a flat tendon or
aponeurosis, which meets that of
the opposite side at the linea alba

Linea alba
The midline raphe, or seam,
where the aponeuroses of the
abdominal muscles on each
side meet in the midline

Linea semilunaris
This curved line marks
the lateral (outer) edge of the
rectus muscle and its sheath

Tendinous intersection
The muscle bellies of rectus
abdominis are divided up
by these fibrous bands

Iliac crest

Umbilicus

**Anterior superior
iliac spine**

Inguinal ligament
The free, lower edge of
the external oblique, attaching
from the anterior superior iliac
spine to the pubic tubercle

Pubic symphysis
The midline joint between
the two pubic bones

ANTERIOR (FRONT)
SUPERFICIAL

ABDOMEN AND PELVIS
MUSCULAR

The abdominal muscles can move the trunk—flexing the spine to the front or to the side, or twisting the abdomen from side to side. They are very important muscles in posture, helping support the upright spine when we are standing or sitting, and are also called into action when we lift heavy objects. Because they compress the abdomen and raise the pressure internally, they are involved during defecation, micturition (emptying the bladder), and in forced expiration of air from the lungs. Right at the front, lying either side of the midline, there are two straight, straplike rectus abdominis muscles. These muscles are each broken up by horizontal tendons: in a well-toned, slim person, this creates the much-sought-after "six-pack" appearance. Flanking the recti muscles on each side are three layers of broad, flat muscles.

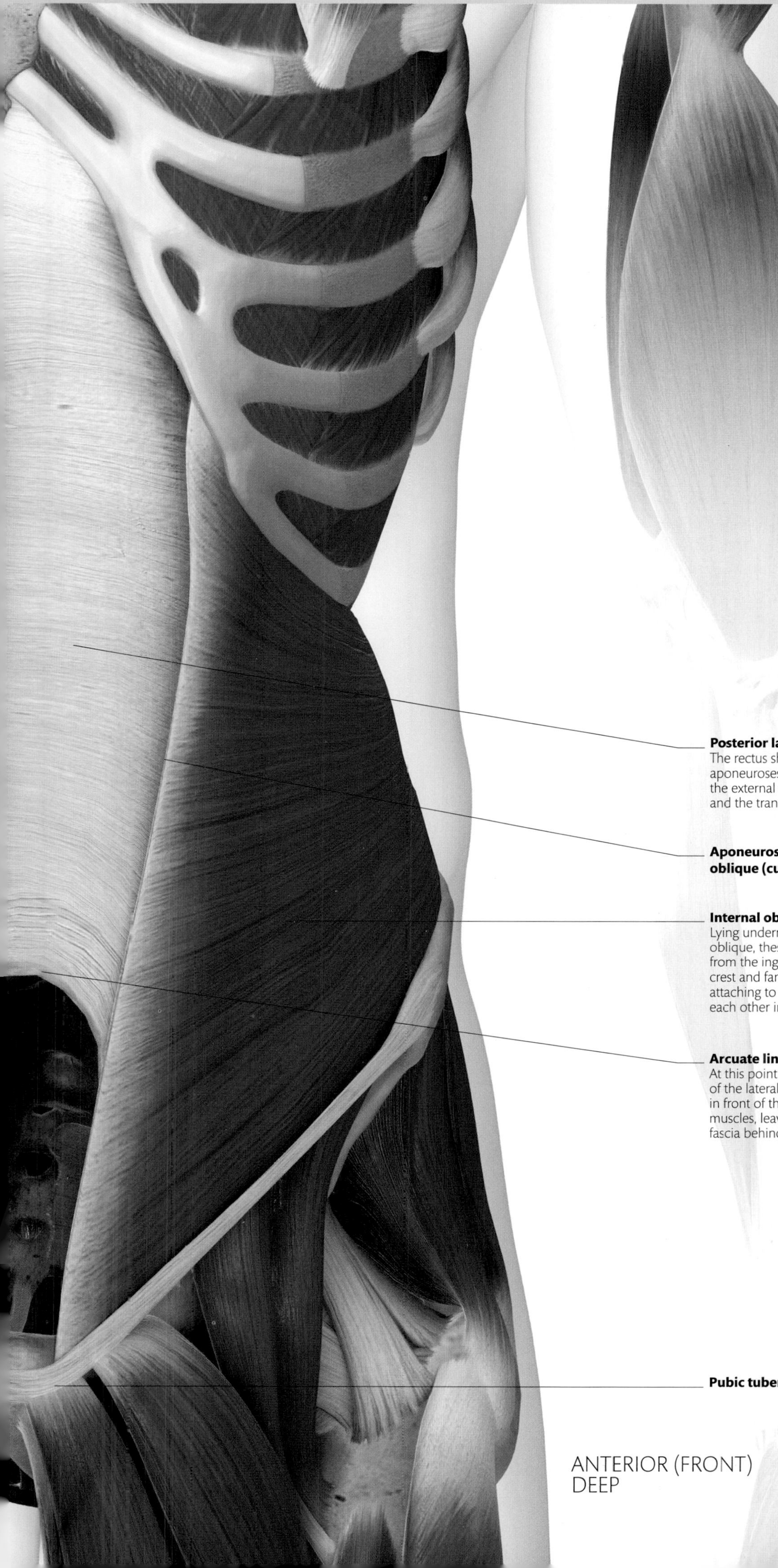

Posterior layer of rectus sheath
The rectus sheath is formed by the aponeuroses of the muscles to the sides: the external oblique, the internal oblique, and the transversus abdominis

Aponeurosis of internal oblique (cut edge)

Internal oblique
Lying underneath the external oblique, these muscle fibers spring from the inguinal ligament and iliac crest and fan inward and upward, attaching to the lower ribs and to each other in the midline

Arcuate line
At this point, all the aponeuroses of the lateral muscles swap to lie in front of the rectus abdominis muscles, leaving only a layer of fascia behind that muscle

Pubic tubercle

ANTERIOR (FRONT)
DEEP

ABDOMEN AND PELVIS
MUSCULAR

The most superficial muscle of the lower back is the incredibly broad latissimus dorsi. Underneath this, lying along the spine on each side, there is a large bulk of muscle that forms two ridges in the lumbar region in a well-toned person. This muscle mass is collectively known as the erector spinae, and its name suggests its importance in keeping the spine upright. When the spine is flexed forward, the erector spinae can pull it back into an upright position, and even take it further, into extension. The muscle can be divided up into three main strips on each side: iliocostalis, longissimus, and spinalis. Most of the muscle bulk of the buttock comes down to just one muscle: the fleshy gluteus maximus, which extends the hip joint. Hidden beneath the gluteus maximus are a range of smaller muscles that also move the hip.

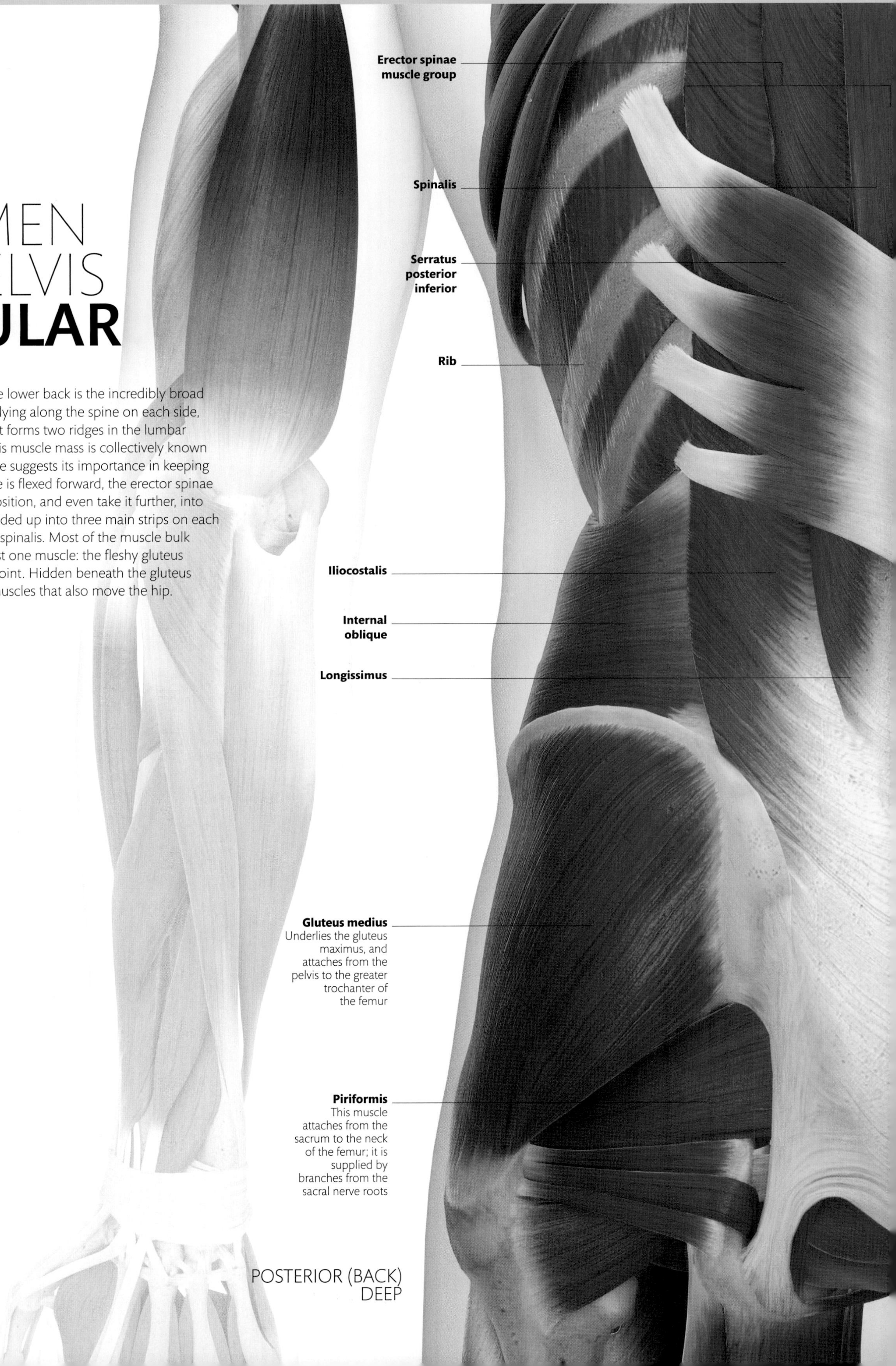

Erector spinae muscle group

Spinalis

Serratus posterior inferior

Rib

Iliocostalis

Internal oblique

Longissimus

Gluteus medius
Underlies the gluteus maximus, and attaches from the pelvis to the greater trochanter of the femur

Piriformis
This muscle attaches from the sacrum to the neck of the femur; it is supplied by branches from the sacral nerve roots

POSTERIOR (BACK)
DEEP

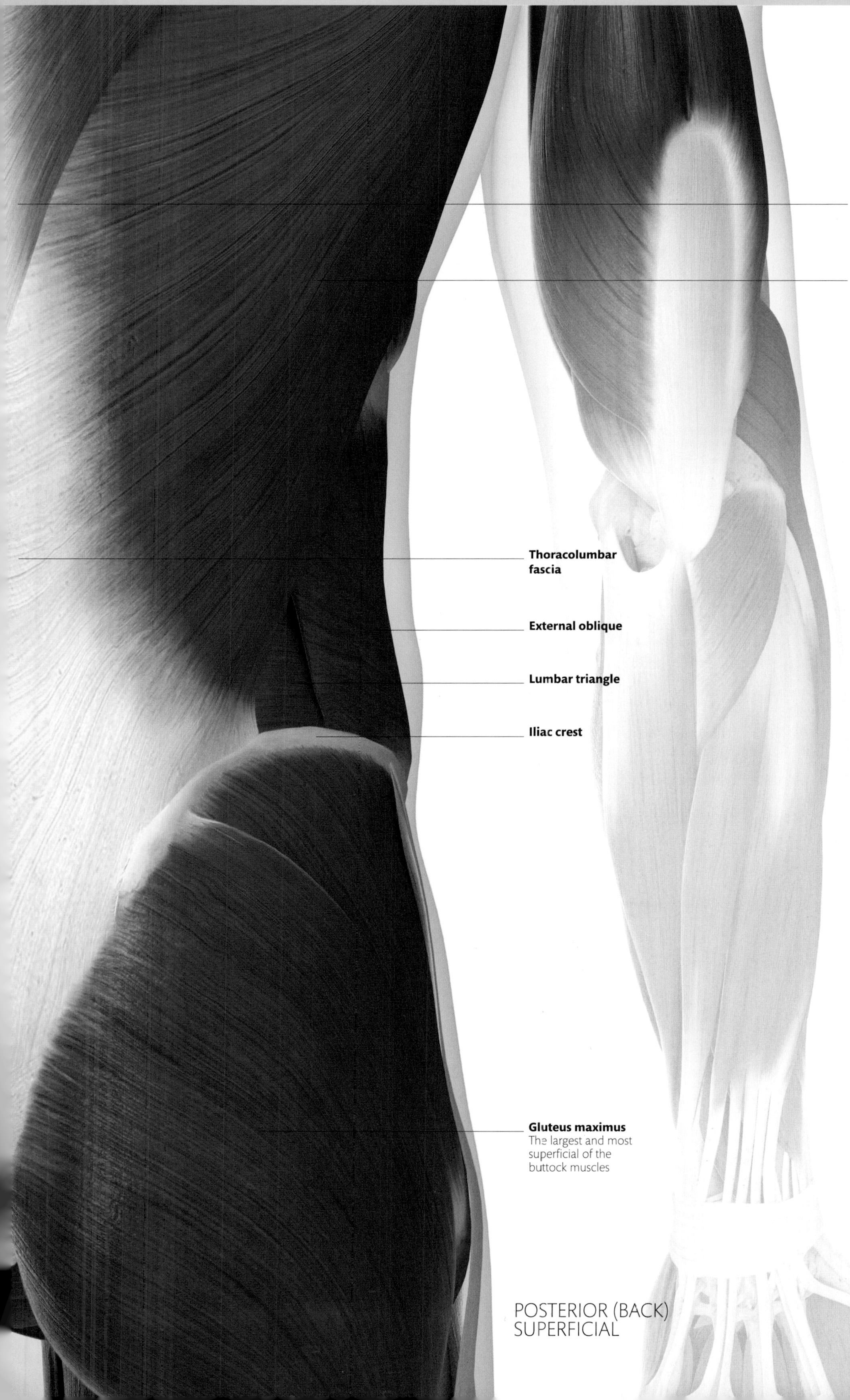

Trapezius

Latissimus dorsi
This massive muscle takes its attachment from a wide area: from the lower thoracic vertebrae, and from the lumbar vertebrae, sacrum, and iliac crest via the thoracolumbar fascia; its fibers converge on a narrow tendon that attaches to the humerus

Thoracolumbar fascia

External oblique

Lumbar triangle

Iliac crest

Gluteus maximus
The largest and most superficial of the buttock muscles

POSTERIOR (BACK)
SUPERFICIAL

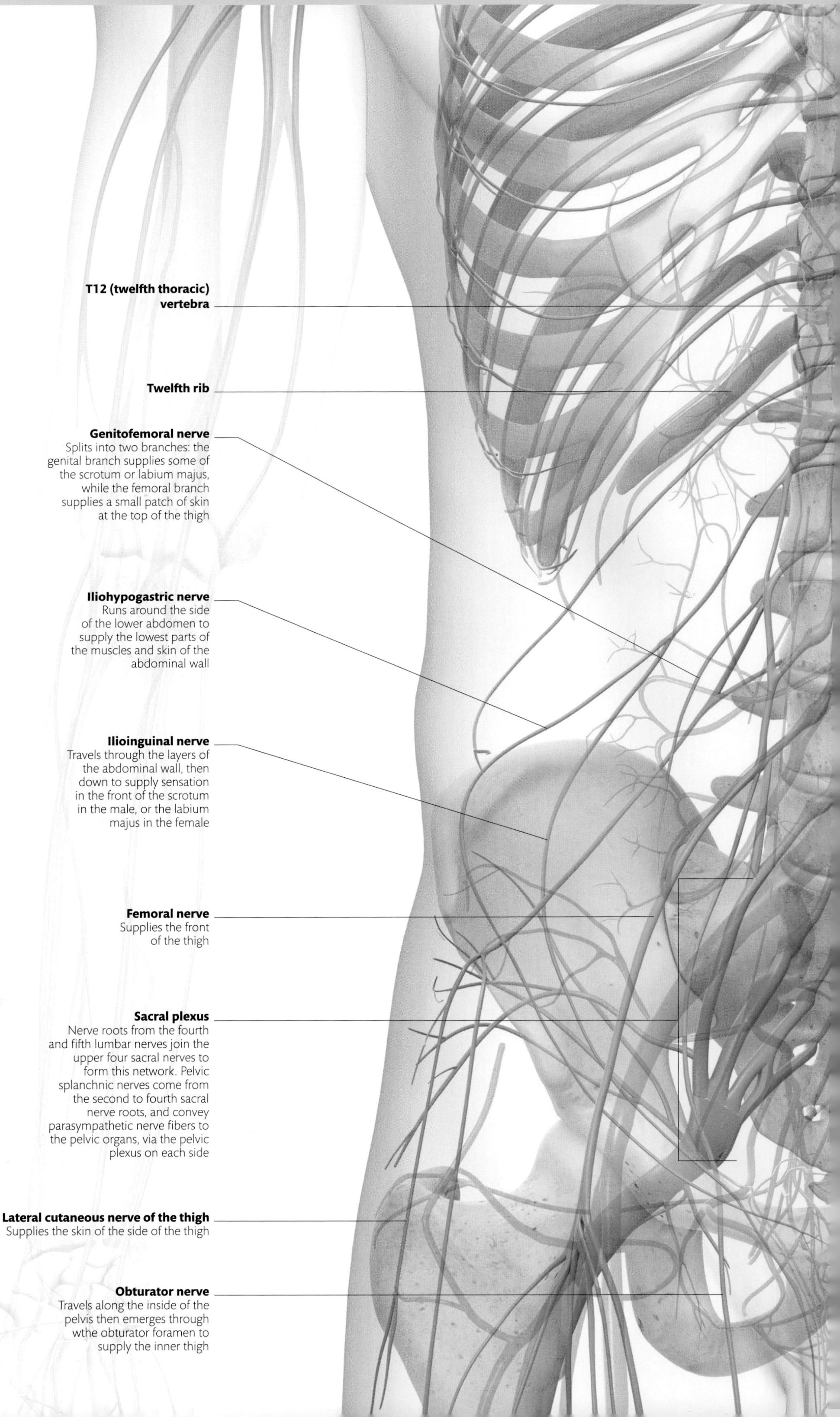

T12 (twelfth thoracic) vertebra

Twelfth rib

Genitofemoral nerve
Splits into two branches: the genital branch supplies some of the scrotum or labium majus, while the femoral branch supplies a small patch of skin at the top of the thigh

Iliohypogastric nerve
Runs around the side of the lower abdomen to supply the lowest parts of the muscles and skin of the abdominal wall

Ilioinguinal nerve
Travels through the layers of the abdominal wall, then down to supply sensation in the front of the scrotum in the male, or the labium majus in the female

Femoral nerve
Supplies the front of the thigh

Sacral plexus
Nerve roots from the fourth and fifth lumbar nerves join the upper four sacral nerves to form this network. Pelvic splanchnic nerves come from the second to fourth sacral nerve roots, and convey parasympathetic nerve fibers to the pelvic organs, via the pelvic plexus on each side

Lateral cutaneous nerve of the thigh
Supplies the skin of the side of the thigh

Obturator nerve
Travels along the inside of the pelvis then emerges through wthe obturator foramen to supply the inner thigh

ABDOMEN AND PELVIS **NERVOUS**

The lower intercostal nerves continue past the lower edges of the rib cage at the front to supply the muscles and skin of the abdominal wall. The lower parts of the abdomen are supplied by the subcostal and iliohypogastric nerves. The abdominal portion of the sympathetic trunk receives nerves from the thoracic and first two lumbar spinal nerves, and sends nerves back to all the spinal nerves. The lumbar spinal nerves emerge from the spine and run into the psoas major muscle at the back of the abdomen. Inside the muscle, the nerves join up and swap fibers to form a network or plexus. Branches of this lumbar plexus emerge around and through the psoas muscle and make their way into the thigh. Lower down, branches of the sacral plexus supply pelvic organs and enter the buttock. One of these branches, the sciatic nerve, is the largest nerve in the entire body. It supplies the back of the thigh, as well as the rest of the leg and foot.

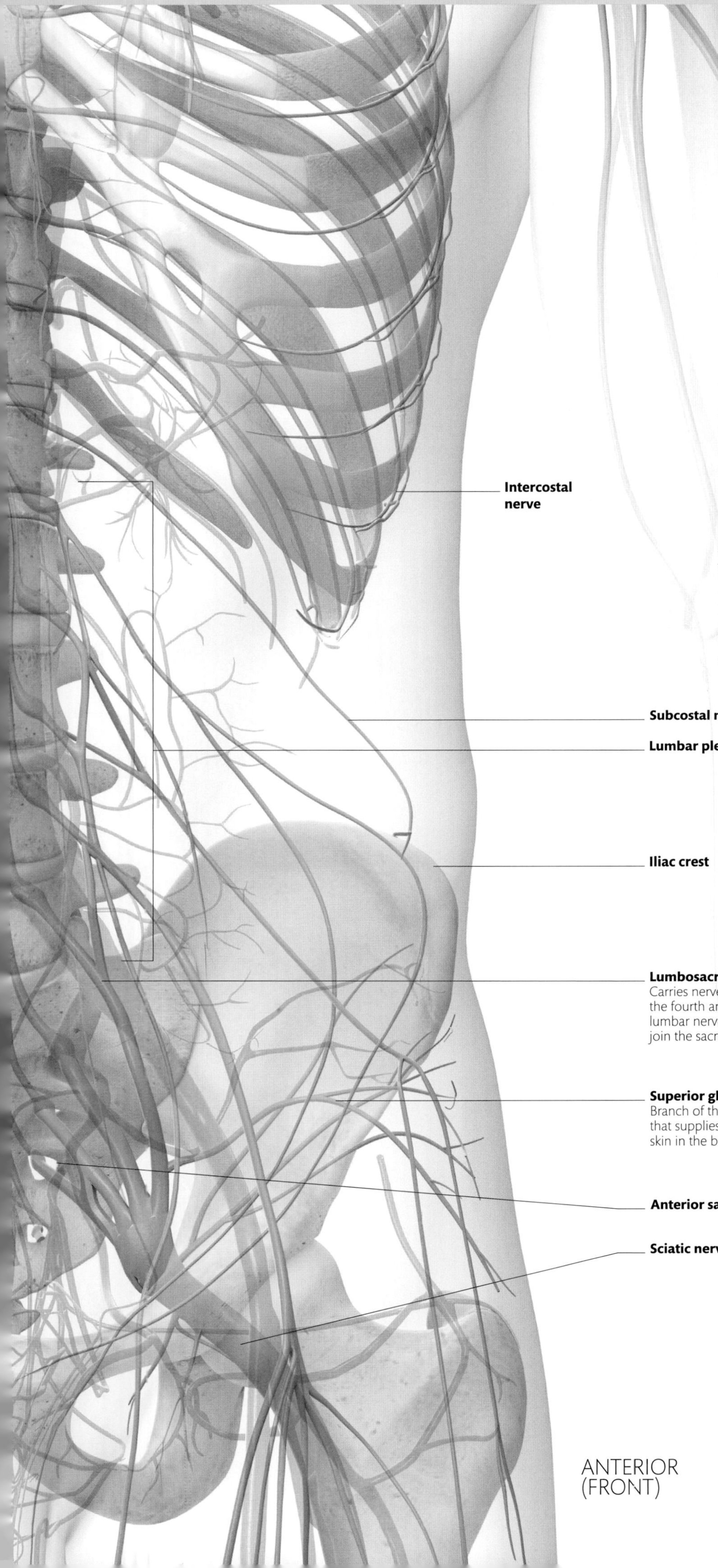

Intercostal nerve

Subcostal nerve

Lumbar plexus

Iliac crest

Lumbosacral trunk
Carries nerve fibers from the fourth and fifth lumbar nerves down to join the sacral plexus

Superior gluteal nerve
Branch of the sacral plexus that supplies muscles and skin in the buttock

Anterior sacral foramen

Sciatic nerve

ANTERIOR
(FRONT)

Spinal ganglion

Rami communicantes

Sympathetic ganglion

Sympathetic trunk

Spinal nerves

Spinal cord

SECTION OF SYMPATHETIC
TRUNK AND SPINAL CORD

ABDOMEN AND PELVIS
CARDIOVASCULAR

The aorta passes behind the diaphragm, level with the twelfth thoracic vertebra, and enters the abdomen. Pairs of arteries branch from the sides of the aorta to supply the walls of the abdomen, the kidneys, adrenal glands, and the testes or ovaries with oxygenated blood. A series of branches emerge from the front of the abdominal aorta to supply the abdominal organs: the celiac trunk gives branches to the liver, stomach, pancreas, and spleen, and the mesenteric arteries provide blood to the gut. The abdominal aorta ends by splitting into two, forming the common iliac arteries. Each of these then divides, in turn, forming an internal iliac artery (which supplies the pelvic organs) and an external iliac artery (which continues into the thigh, becoming the femoral artery). Lying to the right of the aorta is the major vein of the abdomen: the inferior vena cava.

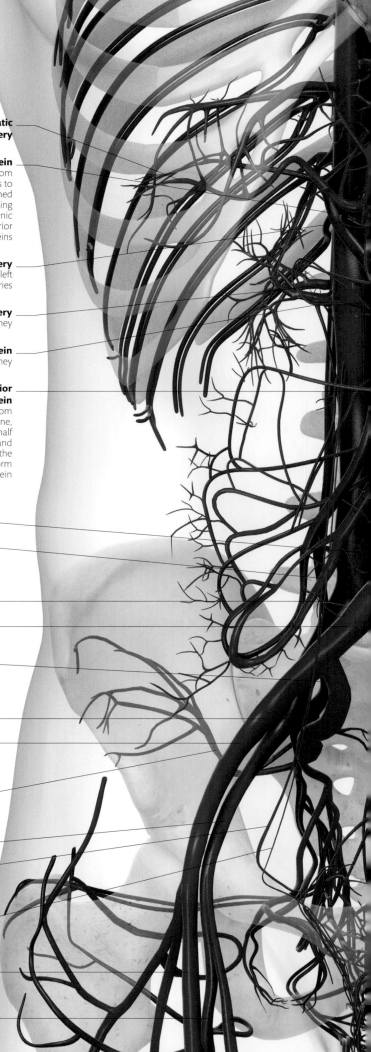

Right hepatic artery

Portal vein
Carries blood from the intestines to the liver; formed from the joining of the splenic and superior mesenteric veins

Common hepatic artery
Branches into right and left hepatic arteries

Right renal artery
Supplies the right kidney

Right renal vein
Drains the right kidney

Superior mesenteric vein
Drains blood from the small intestine, cecum, and half of the colon, and ends by joining the splenic vein to form the portal vein

Inferior vena cava

Ileocolic artery
Branch of the superior mesenteric artery supplying the end of the ileum, the cecum, the start of the ascending colon, and the appendix

Right common iliac vein

Right common iliac artery
Divides into the right external and internal iliac arteries

Right internal iliac artery
Provides branches to the bladder, rectum, perineum, and external genitals, muscles of the inner thigh, bone of the ilium and sacrum, and the buttock, as well as the uterus and vagina in a woman

Right internal iliac vein

Right external iliac artery
Gives a branch to the lower part of the anterior abdominal wall before passing over the pubic bone and under the inguinal ligament to become the femoral artery

Right superior gluteal artery
The largest branch of the internal iliac artery; passes out through the back of the pelvis to supply the upper buttock

Right external iliac vein

Right gonadal artery
In a woman, supplies the ovary on each side; in a man, extends to the scrotum to supply the testis

Right gonadal vein
Drains the ovary or testis and ends by joining the inferior vena cava

Right femoral artery
The main artery of the leg; the continuation of the external iliac artery in the thigh

ANTERIOR (FRONT)

Right femoral vein

Celiac trunk
Only just over ³⁄₈ in (1 cm) long, it quickly branches into the left gastric, splenic, and common hepatic arteries

Splenic artery
Supplies the spleen, as well as most of the pancreas and the upper part of the stomach

Splenic vein
Drains the spleen and receives other veins from the stomach and pancreas, as well as the inferior mesenteric vein

Left renal artery
Shorter than the right renal artery, this supplies the left kidney

Left renal vein
Longer than its counterpart on the right, this drains the left kidney and receives the left gonadal vein

Inferior mesenteric vein
Drains blood from the colon and rectum and ends by emptying into the splenic vein

Superior mesenteric artery
Branches within the mesentery to supply a great length of intestine, including all of the jejunum and ileum and half of the colon

Abdominal aorta
The thoracic aorta becomes the abdominal aorta as it passes behind the diaphragm, level with the twelfth thoracic vertebra

Inferior mesenteric artery
Supplies the last third of the transverse colon, the descending and sigmoid colon, and the rectum

Bifurcation of aorta
The abdominal aorta divides in front of the fourth lumbar vertebra

Superior rectal artery
The last branch of the inferior mesenteric artery passes down into the pelvis to supply the rectum

Left common iliac artery

Left common iliac vein
Formed from the union of the external and internal iliac veins

Left external iliac vein
The continuation of the femoral vein, after it has passed into the pelvis

Left internal iliac artery

Left external iliac artery

Left internal iliac vein
Drains the pelvic organs, perineum, and buttock

Left gonadal artery
Gonadal arteries branch from the aorta just below the renal arteries

Left gonadal vein
Drains the ovary or testis, and empties into the left renal vein

Left femoral artery

Left femoral vein
The main vein from the leg; becomes the external iliac vein

ABDOMEN AND PELVIS
LYMPHATIC AND IMMUNE

The deep lymph nodes of the abdomen are clustered around arteries. Nodes lying along each side of the aorta receive lymph from paired structures, such as the muscles of the abdominal wall, the kidneys and adrenal glands, and the testes or ovaries. Iliac nodes collect lymph returning from the legs and pelvis. Nodes clustered around the branches on the front of the aorta collect lymph from the gut and abdominal organs. Eventually, all this lymph from the legs, pelvis, and abdomen passes into a swollen lymphatic vessel called the cisterna chyli; this narrows down to become the thoracic duct, which runs up into the chest. Most lymph nodes are small, bean-sized structures, but the abdomen also contains a large and important organ of the immune system—the spleen.

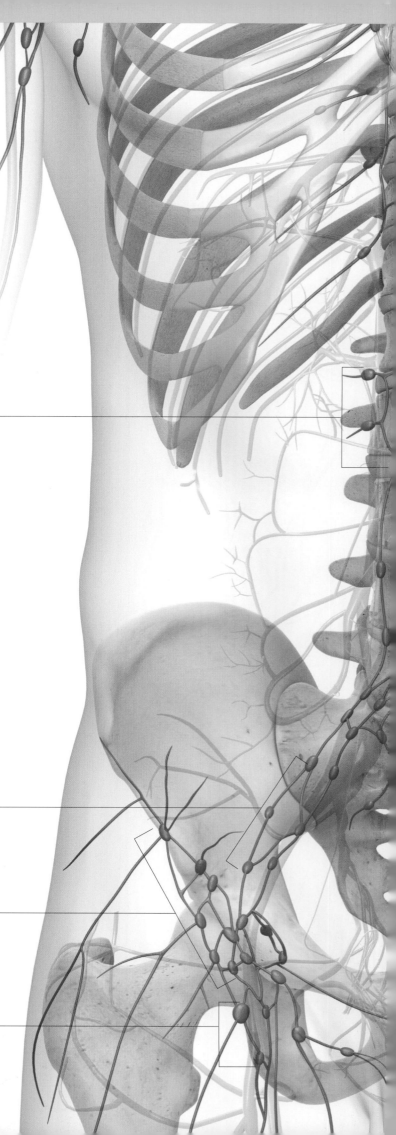

Lateral aortic nodes
Lying along each side of the aorta, these collect lymph from the kidneys, posterior abdominal wall, and pelvic viscera; they drain into the right and left intestinal trunks

External iliac nodes
Collect lymph from the inguinal nodes in the groin, from the perineum, and the inner thigh

Proximal superficial inguinal nodes
Lying just below the inguinal ligament, this upper group of superficial inguinal nodes receives lymph from the lower abdominal wall, below the umbilicus, as well as from the external genitalia

Distal superficial inguinal nodes
The lower nodes in the groin drain most of the superficial lymphatics of the thigh and leg

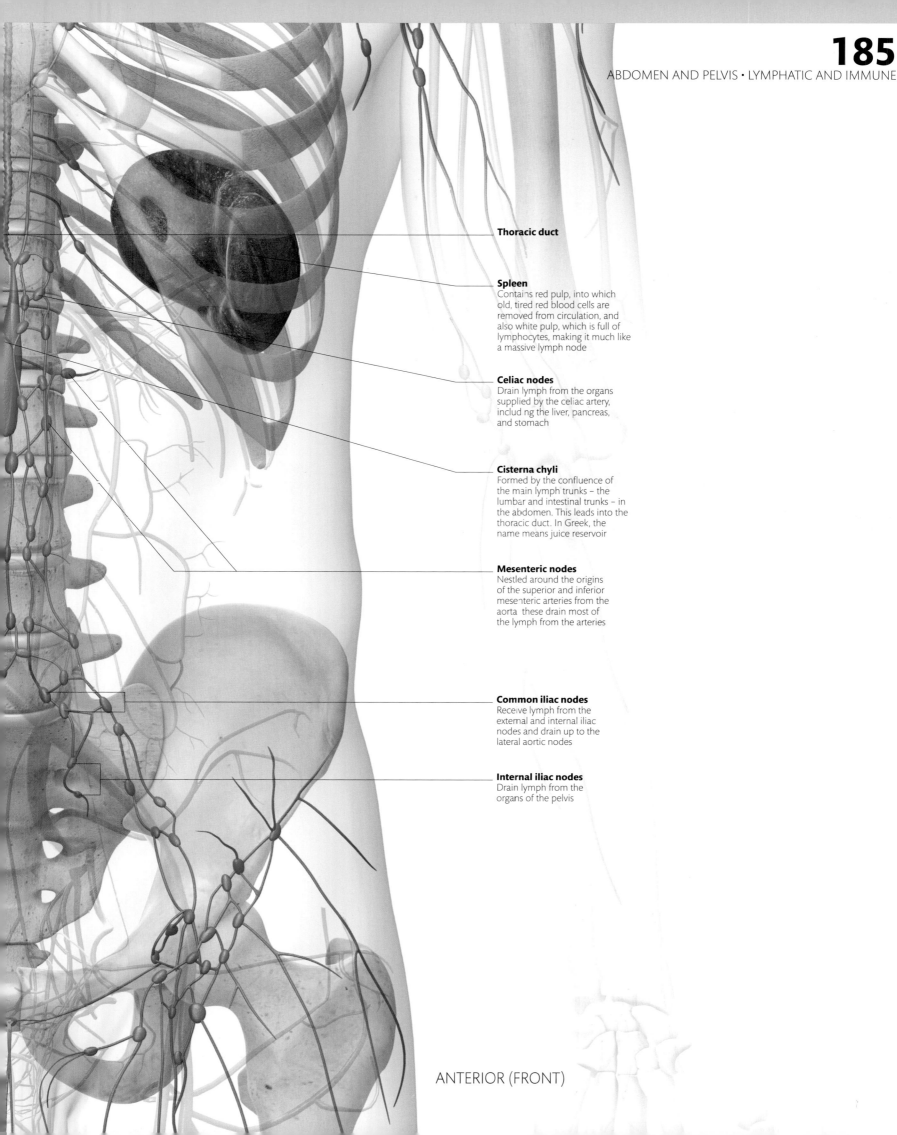

Thoracic duct

Spleen
Contains red pulp, into which
old, tired red blood cells are
removed from circulation, and
also white pulp, which is full of
lymphocytes, making it much like
a massive lymph node

Celiac nodes
Drain lymph from the organs
supplied by the celiac artery,
including the liver, pancreas,
and stomach

Cisterna chyli
Formed by the confluence of
the main lymph trunks – the
lumbar and intestinal trunks – in
the abdomen. This leads into the
thoracic duct. In Greek, the
name means juice reservoir

Mesenteric nodes
Nestled around the origins
of the superior and inferior
mesenteric arteries from the
aorta these drain most of
the lymph from the arteries

Common iliac nodes
Receive lymph from the
external and internal iliac
nodes and drain up to the
lateral aortic nodes

Internal iliac nodes
Drain lymph from the
organs of the pelvis

ANTERIOR (FRONT)

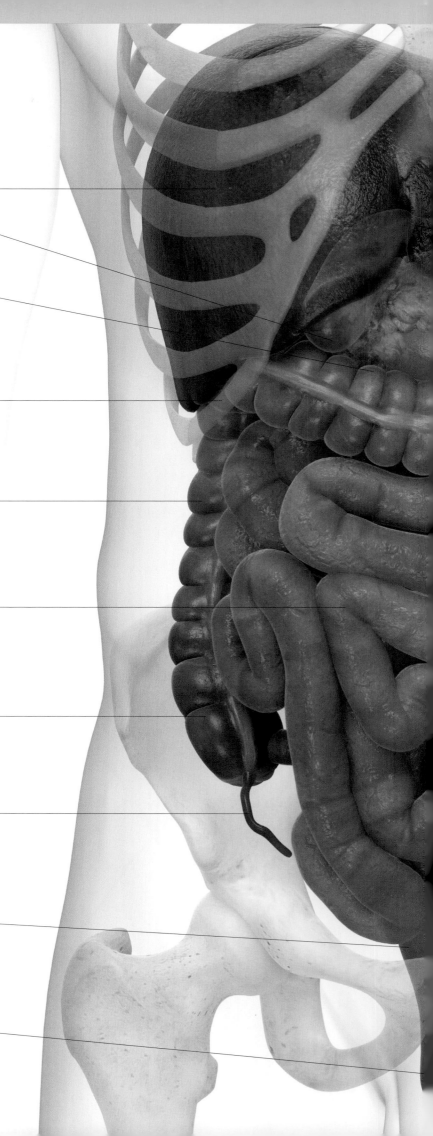

Right lobe of liver

Fundus of gallbladder
Bottom of the baglike
gallbladder, which just sticks
out under the liver

Transverse colon
Hanging down below the liver
and stomach, this part of the
colon has a mesentery (fold of the
peritoneum that connects the
intestines to the dorsal abdominal
wall) through which its blood
vessels and nerves travel

Hepatic flexure of colon
Junction between the ascending
and transverse colon, tucked up
under the liver

Ascending colon
This part of the large
intestine is firmly bound
down to the back wall
of the abdomen

Ileum
Lying mainly in the suprapubic region
of the abdomen, this part of the small
intestine is about 13 ft (4 m) long;
ileum simply means entrails in Latin

Cecum
First part of the large
intestine, lying in the right
iliac fossa of the abdomen

Appendix
Properly known as the vermiform
(wormlike) appendix; usually a few
centimeters long, it is full of
lymphoid tissue, and thus forms
part of the gut's immune system

Rectum
About 4¾ in (12 cm) long, this
penultimate part of the gut is
stretchy; it can expand to store
feces, until a convenient time for
emptying presents itself

Anal canal
Muscular sphincters in and around the anal
canal keep it closed; the sphincters relax during
defecation, as the diaphragm and abdominal
wall muscles contract to raise pressure in the
abdomen and force the feces out

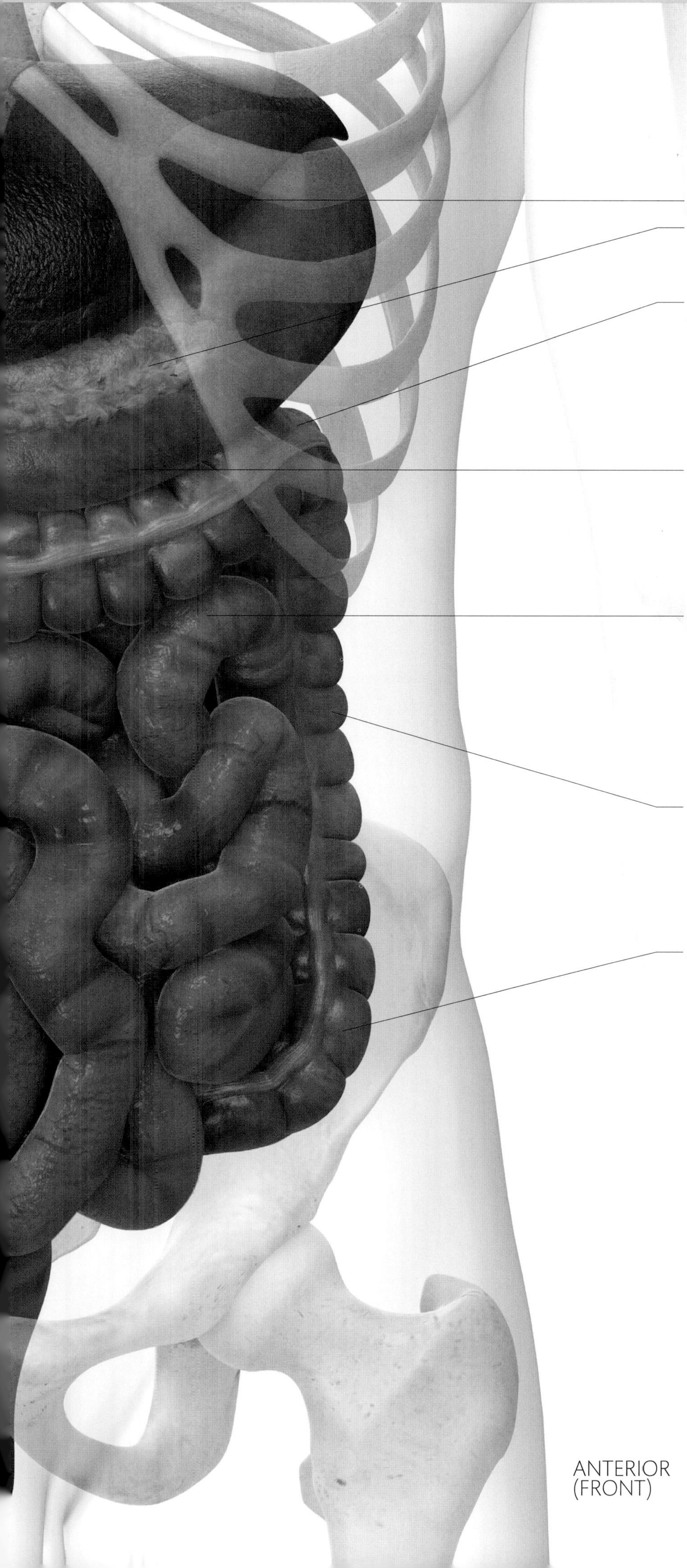

Left lobe of liver

Pancreas

Splenic flexure of colon
Junction between the
tranverse and descending
colon, close to the spleen
(spleen not shown here)

Stomach
The name comes originally
from the Greek for gullet, but
has come to mean this baglike
part of the digestive system, just
below the diaphragm

Jejunum
About 6½ ft (2 m) long, this part
of the small intestine is more
vascular (so slightly redder) than
the ileum, and lies mainly in the
umbilical region of the abdomen;
its name comes from the Latin for
empty—perhaps because food
passes through here quickly

Descending colon
Like the ascending colon,
this part of the large intestine
has no mesentery, and is
firmly bound to the back
wall of the abdomen

Sigmoid colon
This S-shaped part of
colon has a mesentery

ABDOMEN AND PELVIS
DIGESTIVE

With the organs in situ, it is clear how much the abdominal cavity
extends up under the ribs. The upper abdominal organs—the liver,
stomach, and spleen—are largely under cover of the rib cage. This
gives them some protection, but it also means that they are vulnerable
to injury if a lower rib is fractured. The large intestine forms an M
shape in the abdomen, starting with the cecum low down on the
right, and the ascending colon running up the right flank and tucking
under the liver. The transverse colon hangs down below the liver and
stomach, and the descending colon runs down the left side of the
abdomen. This becomes the S-shaped sigmoid colon, which runs
down into the pelvis to become the rectum. The coils of the small
intestine occupy the middle of the abdomen.

ANTERIOR
(FRONT)

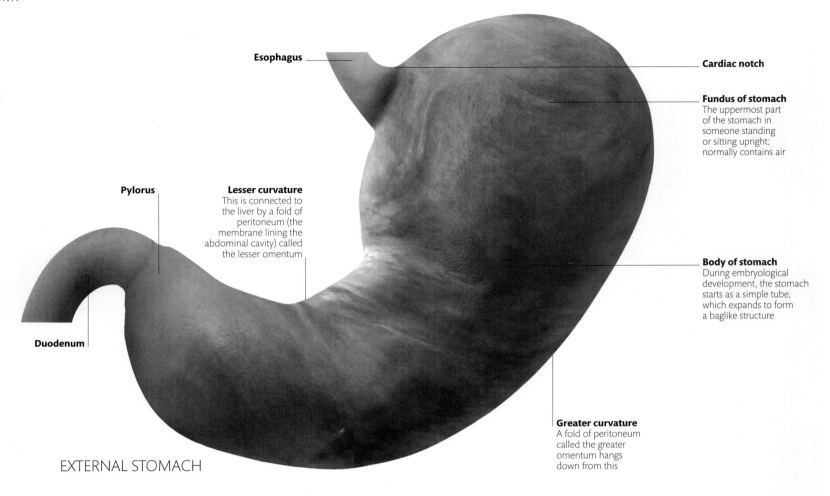

Esophagus

Cardiac notch

Fundus of stomach
The uppermost part of the stomach in someone standing or sitting upright; normally contains air

Pylorus

Lesser curvature
This is connected to the liver by a fold of peritoneum (the membrane lining the abdominal cavity) called the lesser omentum

Body of stomach
During embryological development, the stomach starts as a simple tube, which expands to form a baglike structure

Duodenum

Greater curvature
A fold of peritoneum called the greater omentum hangs down from this

EXTERNAL STOMACH

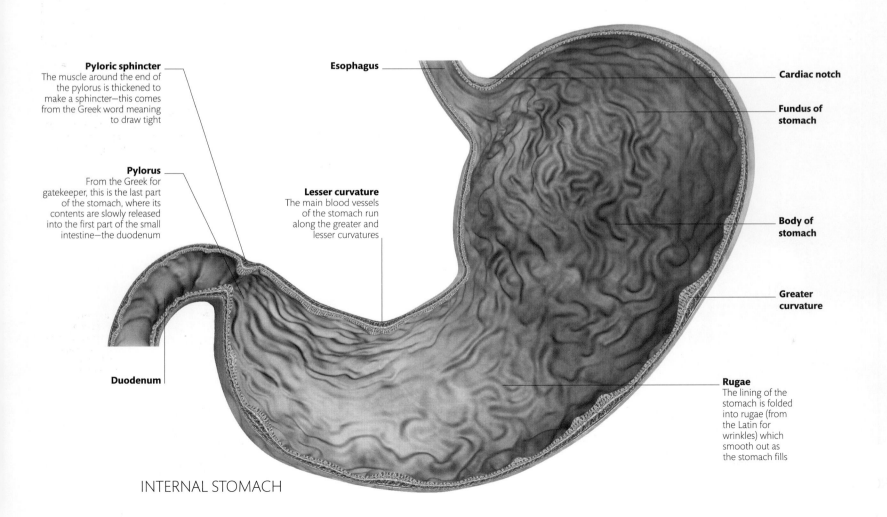

Pyloric sphincter
The muscle around the end of the pylorus is thickened to make a sphincter—this comes from the Greek word meaning to draw tight

Esophagus

Cardiac notch

Fundus of stomach

Pylorus
From the Greek for gatekeeper, this is the last part of the stomach, where its contents are slowly released into the first part of the small intestine—the duodenum

Lesser curvature
The main blood vessels of the stomach run along the greater and lesser curvatures

Body of stomach

Greater curvature

Duodenum

Rugae
The lining of the stomach is folded into rugae (from the Latin for wrinkles) which smooth out as the stomach fills

INTERNAL STOMACH

ABDOMEN AND PELVIS **DIGESTIVE**

The stomach is a muscular bag, where food is held before moving on to the intestines. Inside the stomach, food is exposed to a cocktail of hydrochloric acid, which kills off bacteria, and protein-digesting enzymes. The layered muscle of the stomach wall contracts to churn up its contents. Semidigested food is released from the stomach into the first part of the small intestine, the duodenum, where bile and pancreatic juices are added. Contractions in the intestine wall then push the liquid food into the jejunum and ileum, where digestion continues. What is left passes into the cecum, the beginning of the large intestine. In the colon, the next part of the large intestine, water is absorbed so that the gut contents become more solid. The resulting feces pass into the rectum, where they are stored until excretion.

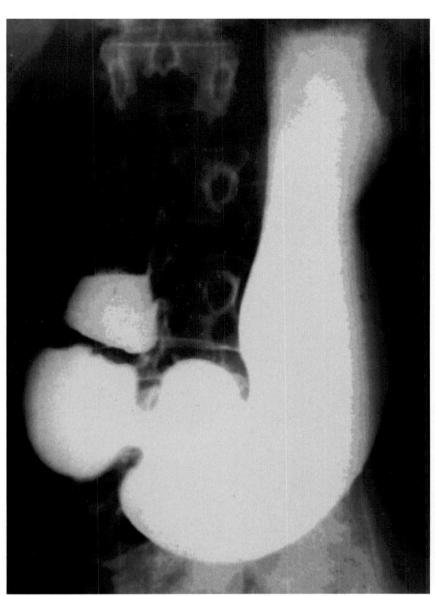

Barium meal
Colored X-ray showing the results of a barium meal, which is used to highlight the structure of the stomach and to reveal disorders of the digestive tract.

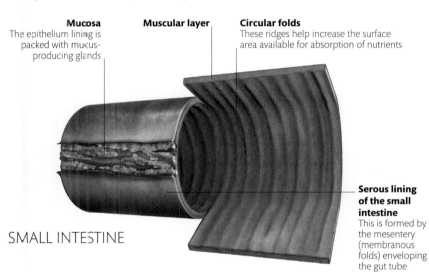

Mucosa
The epithelium lining is packed with mucus-producing glands

Muscular layer

Circular folds
These ridges help increase the surface area available for absorption of nutrients

Serous lining of the small intestine
This is formed by the mesentery (membranous folds) enveloping the gut tube

SMALL INTESTINE

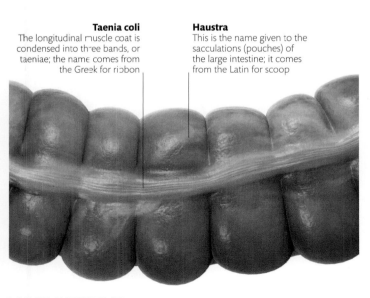

Taenia coli
The longitudinal muscle coat is condensed into three bands, or taeniae; the name comes from the Greek for ribbon

Haustra
This is the name given to the sacculations (pouches) of the large intestine; it comes from the Latin for scoop

LARGE INTESTINE

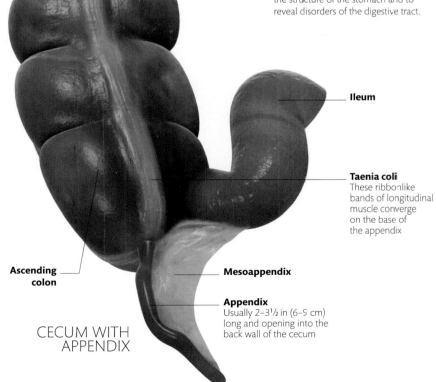

Ileum

Taenia coli
These ribbonlike bands of longitudinal muscle converge on the base of the appendix

Ascending colon

Mesoappendix

Appendix
Usually 2–3½ in (6–9 cm) long and opening into the back wall of the cecum

CECUM WITH APPENDIX

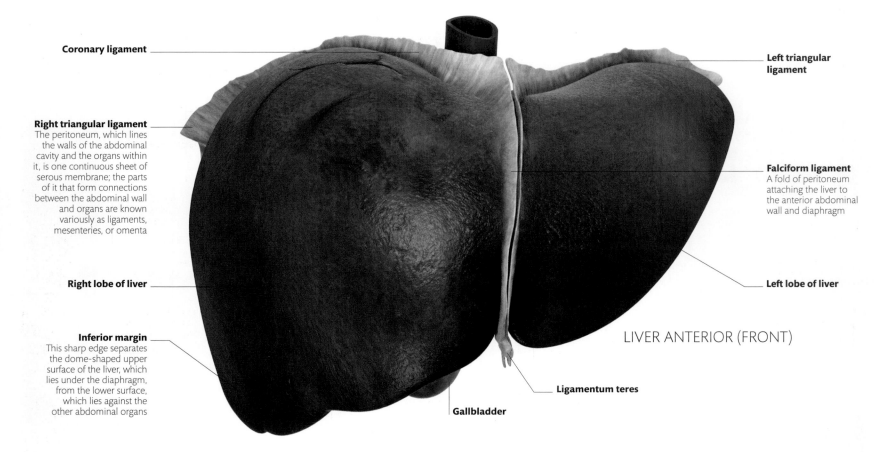

Coronary ligament

Left triangular ligament

Right triangular ligament
The peritoneum, which lines the walls of the abdominal cavity and the organs within it, is one continuous sheet of serous membrane; the parts of it that form connections between the abdominal wall and organs are known variously as ligaments, mesenteries, or omenta

Falciform ligament
A fold of peritoneum attaching the liver to the anterior abdominal wall and diaphragm

Right lobe of liver

Left lobe of liver

Inferior margin
This sharp edge separates the dome-shaped upper surface of the liver, which lies under the diaphragm, from the lower surface, which lies against the other abdominal organs

LIVER ANTERIOR (FRONT)

Ligamentum teres

Gallbladder

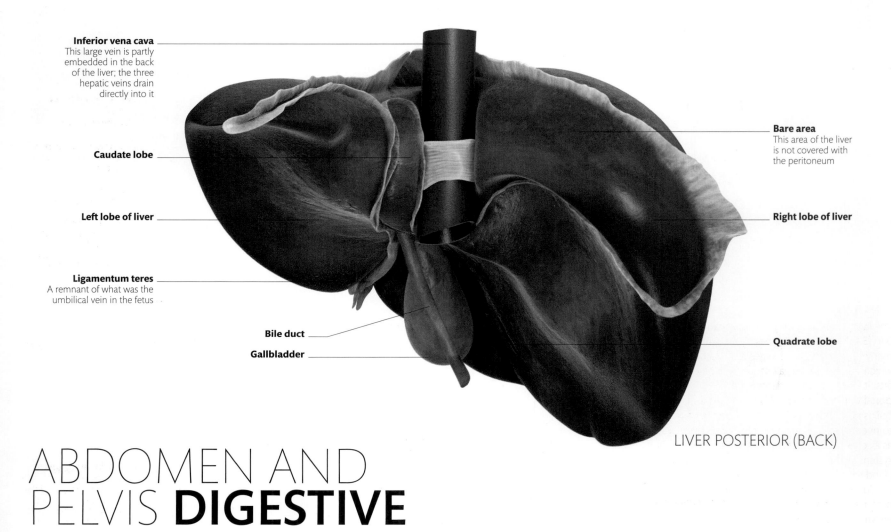

Inferior vena cava
This large vein is partly embedded in the back of the liver; the three hepatic veins drain directly into it

Bare area
This area of the liver is not covered with the peritoneum

Caudate lobe

Left lobe of liver

Right lobe of liver

Ligamentum teres
A remnant of what was the umbilical vein in the fetus

Bile duct

Gallbladder

Quadrate lobe

LIVER POSTERIOR (BACK)

ABDOMEN AND PELVIS **DIGESTIVE**

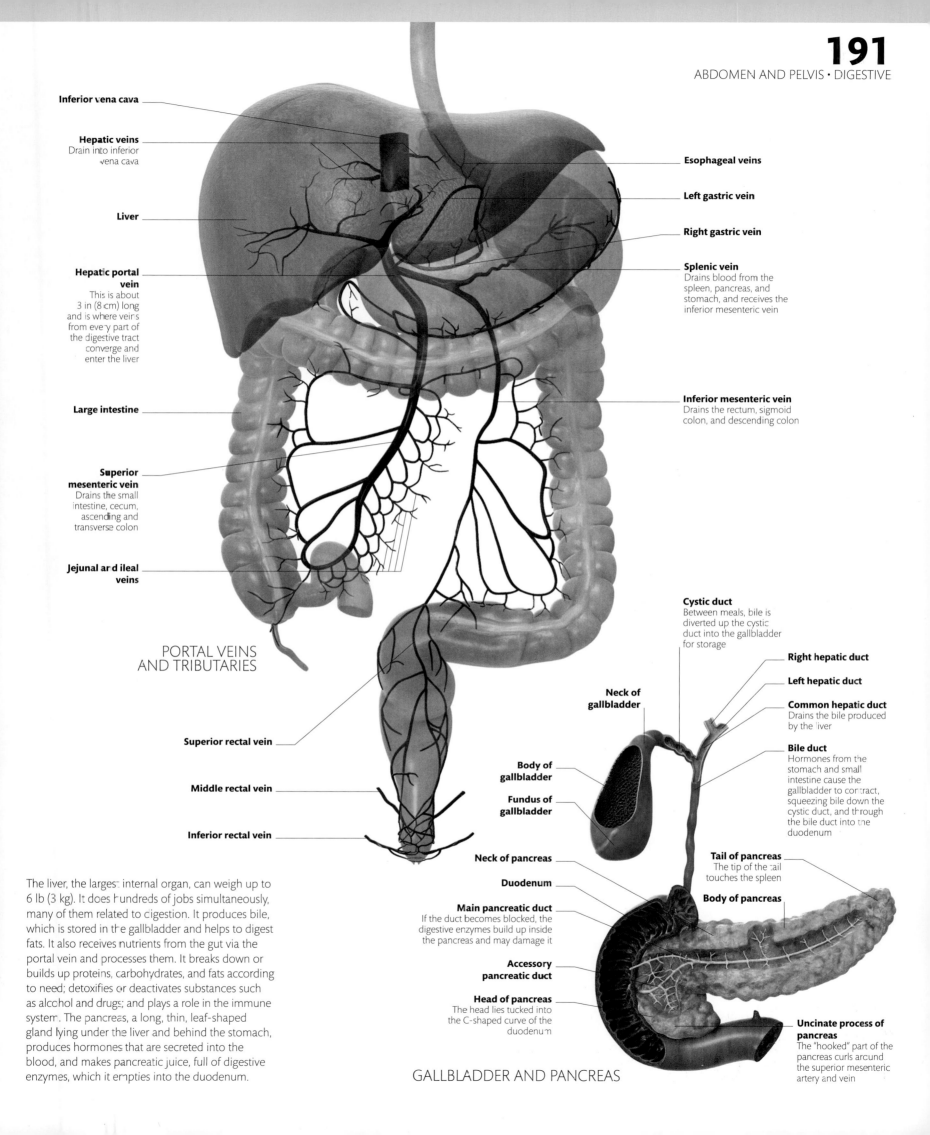

Inferior vena cava

Hepatic veins
Drain into inferior vena cava

Liver

Hepatic portal vein
This is about 3 in (8 cm) long and is where veins from every part of the digestive tract converge and enter the liver

Large intestine

Superior mesenteric vein
Drains the small intestine, cecum, ascending and transverse colon

Jejunal and ileal veins

Esophageal veins

Left gastric vein

Right gastric vein

Splenic vein
Drains blood from the spleen, pancreas, and stomach, and receives the inferior mesenteric vein

Inferior mesenteric vein
Drains the rectum, sigmoid colon, and descending colon

Superior rectal vein

Middle rectal vein

Inferior rectal vein

PORTAL VEINS AND TRIBUTARIES

Cystic duct
Between meals, bile is diverted up the cystic duct into the gallbladder for storage

Neck of gallbladder

Right hepatic duct

Left hepatic duct

Common hepatic duct
Drains the bile produced by the liver

Bile duct
Hormones from the stomach and small intestine cause the gallbladder to contract, squeezing bile down the cystic duct, and through the bile duct into the duodenum

Body of gallbladder

Fundus of gallbladder

Neck of pancreas

Duodenum

Main pancreatic duct
If the duct becomes blocked, the digestive enzymes build up inside the pancreas and may damage it

Accessory pancreatic duct

Head of pancreas
The head lies tucked into the C-shaped curve of the duodenum

Tail of pancreas
The tip of the tail touches the spleen

Body of pancreas

Uncinate process of pancreas
The "hooked" part of the pancreas curls around the superior mesenteric artery and vein

The liver, the largest internal organ, can weigh up to 6 lb (3 kg). It does hundreds of jobs simultaneously, many of them related to digestion. It produces bile, which is stored in the gallbladder and helps to digest fats. It also receives nutrients from the gut via the portal vein and processes them. It breaks down or builds up proteins, carbohydrates, and fats according to need; detoxifies or deactivates substances such as alcohol and drugs; and plays a role in the immune system. The pancreas, a long, thin, leaf-shaped gland lying under the liver and behind the stomach, produces hormones that are secreted into the blood, and makes pancreatic juice, full of digestive enzymes, which it empties into the duodenum.

GALLBLADDER AND PANCREAS

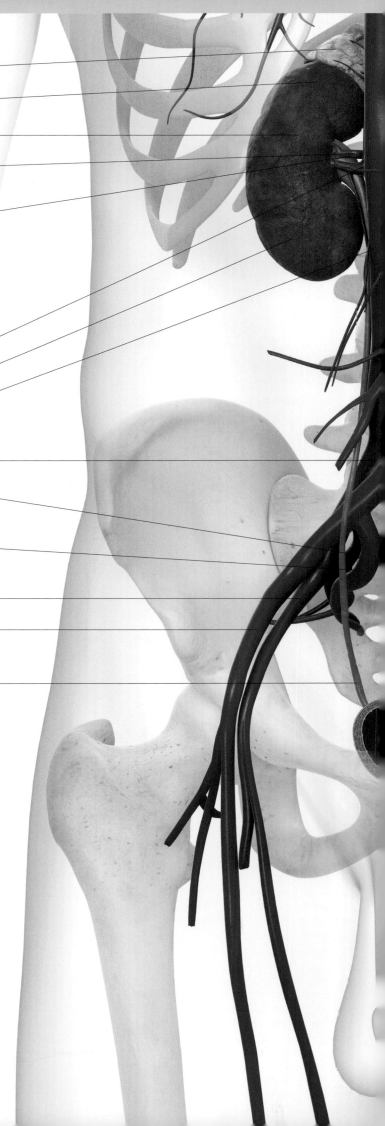

Adrenal gland

Upper pole

Right kidney

Right renal artery
Renal comes from the
Latin for kidney

Hilum
Where the artery enters
and the vein and ureter
exit the kidney; the word
just means small thing
in Latin, but is used in
botany to describe the
area on a seed where the
seed-vessel attaches,
such as the eye of a bean

Right renal vein

Lower pole

Inferior vena cava

Right common iliac vein

Right internal iliac vein
Veins from the bladder
eventually drain into the
internal iliac veins

Right internal iliac artery
Vesical branches of the
internal iliac artery supply
the bladder

Right external iliac vein

**Right external iliac
artery**

Right ureter
The two ureters are muscular tubes:
peristaltic (wavelike) contractions
pump urine down into the bladder,
even if you stand on your head; each
ureter is about 10 in (25 cm) long

ABDOMEN AND
PELVIS **URINARY**

The kidneys lie high up on the back wall of the abdomen, tucked up
under the twelfth ribs. A thick layer of perinephric fat surrounds and
protects each kidney. The kidneys filter the blood, which is carried to
them via the renal arteries. They remove waste from the blood, and
keep a tight check on blood volume and concentration. The urine
they produce collects first in cup-shaped calyces, which join to form
the renal pelvis. The urine then flows out of the kidneys and down
narrow, muscular tubes called ureters to the bladder in the pelvis.
The bladder is a muscular bag that can expand to hold up to about
1 pint (0.5 liters) of urine, and empties itself when the individual
decides it is convenient. The last part of the trip takes the urine
through the urethra to the outside world.

ANTERIOR
(FRONT)

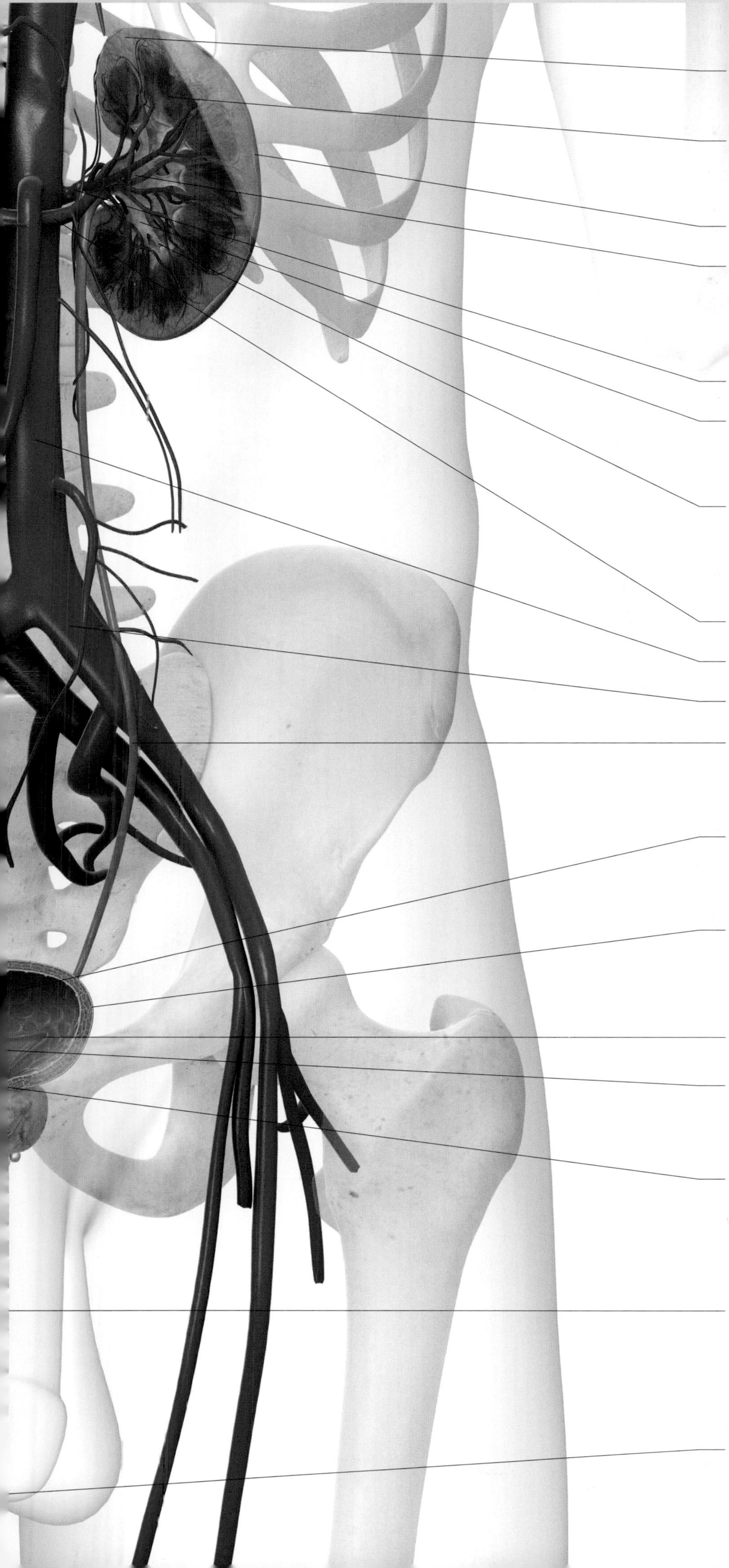

Renal cortex
Cortex means rind or bark; this
is the outer tissue of the kidney

Renal medullary pyramid
Medulla means marrow or pith; this core
tissue of the kidney is arranged as pyramids,
which look triangular in cross section

Left kidney

Renal pelvis
Collects all urine from the kidney, and
empties into the ureter; pelvis means
basin in Latin, and the renal pelvis should
not be confused with the bony pelvis –
also shaped like a large basin

Left renal artery

Major calyx
The major calyces collect urine from
the minor calyces, then themselves join
together to form the renal pelvis

Minor calyx
Calyx originally meant flower-covering in Greek, but
because it is similar to the Latin word for cup it is
used to describe cup-shaped structures in biology;
urine from the microscopic collecting tubules
of the kidney flows out into the minor calyces

Left renal vein

Abdominal aorta

Left common iliac artery

Left ureter
This name comes from the Greek for to make
water; the two ureters carry urine from the
kidneys to the bladder

Bladder
The empty bladder lies low down, in the true
pelvis, behind the pubic symphysis; as the
bladder fills, it expands up into the abdomen

Detrusor muscle
The criss-crossing smooth muscle
bundles of the bladder wall give
the inner surface of the bladder
a netlike appearance

Ureteric orifice

Trigone
The three-cornered region of the back
wall of the bladder, between the ureteric
orifices and the internal urethral orifice

Internal urethral orifice
Where the bladder opens into the urethra

Urethra
From the Greek for urinate; this
tube carries urine from the bladder
to the outside world, a distance
of around 1½ in (4 cm) in women,
and about 8 in (20 cm) in men (it
travels the length of the penis)

External urethral orifice
Where the urethra opens externally

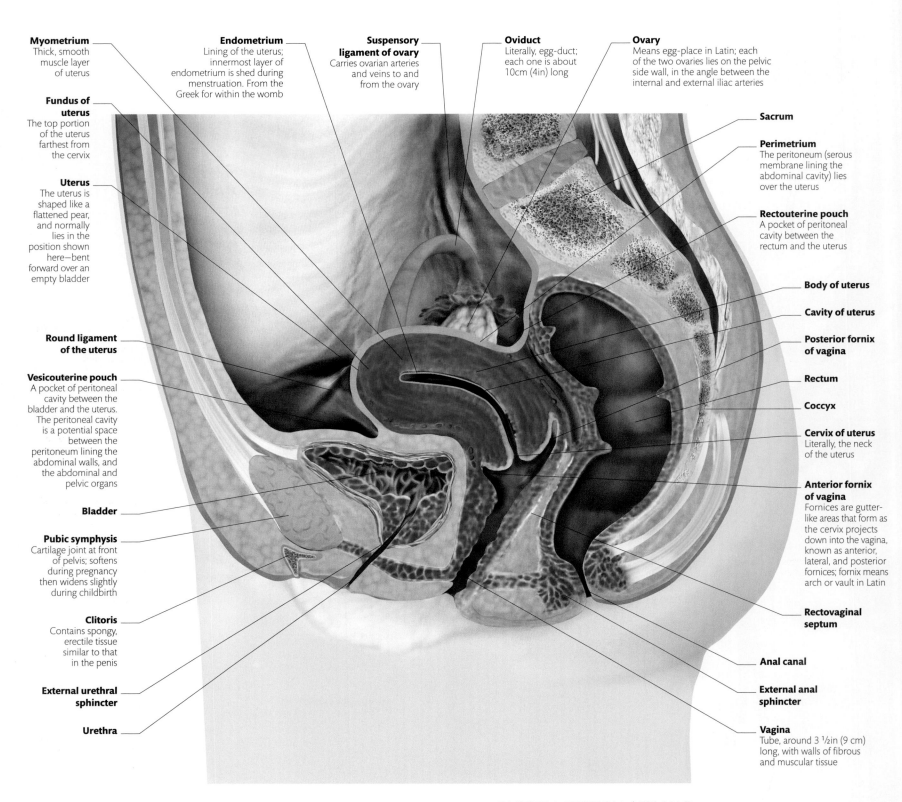

Myometrium
Thick, smooth muscle layer of uterus

Endometrium
Lining of the uterus; innermost layer of endometrium is shed during menstruation. From the Greek for within the womb

Suspensory ligament of ovary
Carries ovarian arteries and veins to and from the ovary

Oviduct
Literally, egg-duct; each one is about 10cm (4in) long

Ovary
Means egg-place in Latin; each of the two ovaries lies on the pelvic side wall, in the angle between the internal and external iliac arteries

Fundus of uterus
The top portion of the uterus farthest from the cervix

Uterus
The uterus is shaped like a flattened pear, and normally lies in the position shown here—bent forward over an empty bladder

Round ligament of the uterus

Vesicouterine pouch
A pocket of peritoneal cavity between the bladder and the uterus. The peritoneal cavity is a potential space between the peritoneum lining the abdominal walls, and the abdominal and pelvic organs

Bladder

Pubic symphysis
Cartilage joint at front of pelvis; softens during pregnancy then widens slightly during childbirth

Clitoris
Contains spongy, erectile tissue similar to that in the penis

External urethral sphincter

Urethra

Sacrum

Perimetrium
The peritoneum (serous membrane lining the abdominal cavity) lies over the uterus

Rectouterine pouch
A pocket of peritoneal cavity between the rectum and the uterus

Body of uterus

Cavity of uterus

Posterior fornix of vagina

Rectum

Coccyx

Cervix of uterus
Literally, the neck of the uterus

Anterior fornix of vagina
Fornices are gutter-like areas that form as the cervix projects down into the vagina, known as anterior, lateral, and posterior fornices; fornix means arch or vault in Latin

Rectovaginal septum

Anal canal

External anal sphincter

Vagina
Tube, around 3 ½in (9 cm) long, with walls of fibrous and muscular tissue

SAGITTAL SECTION / FEMALE

ABDOMEN AND PELVIS
REPRODUCTIVE

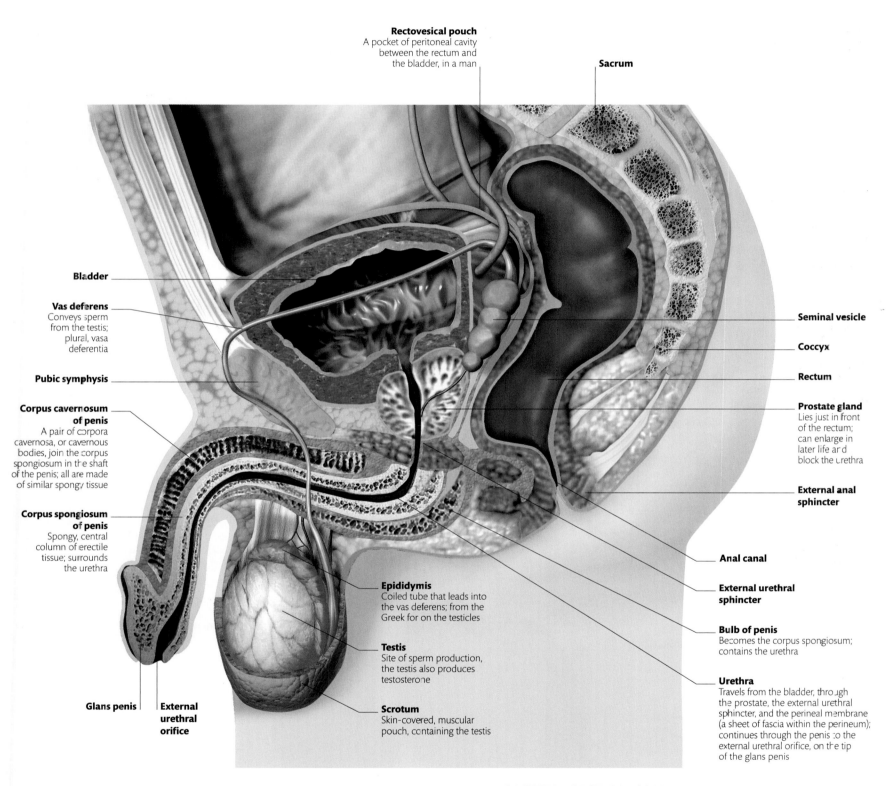

Rectovesical pouch
A pocket of peritoneal cavity between the rectum and the bladder, in a man

Sacrum

Bladder

Vas deferens
Conveys sperm from the testis; plural, vasa deferentia

Pubic symphysis

Corpus cavernosum of penis
A pair of corpora cavernosa, or cavernous bodies, join the corpus spongiosum in the shaft of the penis; all are made of similar spongy tissue

Corpus spongiosum of penis
Spongy, central column of erectile tissue; surrounds the urethra

Glans penis

External urethral orifice

Epididymis
Coiled tube that leads into the vas deferens; from the Greek for on the testicles

Testis
Site of sperm production, the testis also produces testosterone

Scrotum
Skin-covered, muscular pouch, containing the testis

Seminal vesicle

Coccyx

Rectum

Prostate gland
Lies just in front of the rectum; can enlarge in later life and block the urethra

External anal sphincter

Anal canal

External urethral sphincter

Bulb of penis
Becomes the corpus spongiosum; contains the urethra

Urethra
Travels from the bladder, through the prostate, the external urethral sphincter, and the perineal membrane (a sheet of fascia within the perineum); continues through the penis to the external urethral orifice, on the tip of the glans penis

SAGITTAL SECTION / MALE

The male and female reproductive systems are both comprised of a series of internal and external organs, although structurally these are very different. It is true that both sexes possess gonads (ovaries in women and testes in men) and a tract, or set of tubes, but the similarity ends there. When we look in detail at the anatomy of the pelvis in each sex, the differences are obvious. The pelvis of a man contains only part of the reproductive tract, as well as the lower parts of the digestive and urinary tracts, including the rectum and bladder. Beneath the bladder is the prostate gland; this is where the vasa deferentia, which bring sperm from the testis, empty into the urethra. A woman's pelvic cavity contains more of the reproductive tract than a man's. The vagina and uterus are situated between the bladder and rectum in the pelvis.

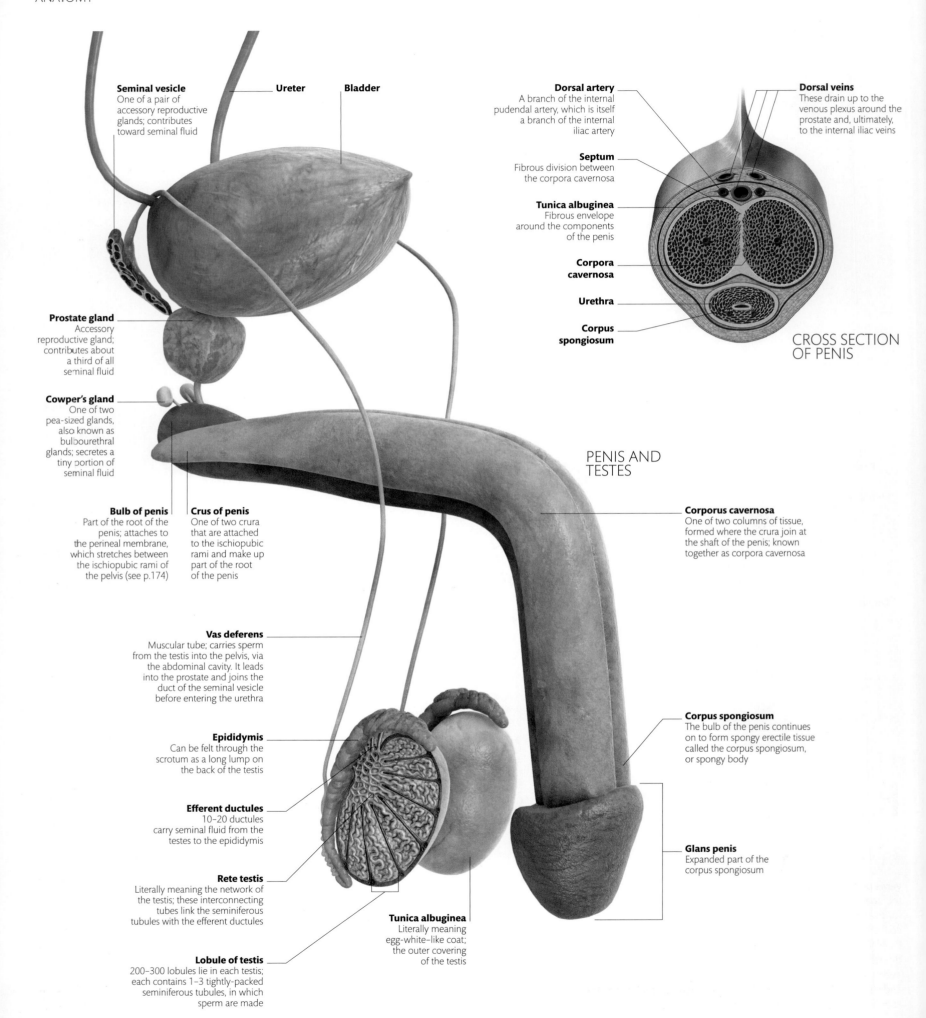

Seminal vesicle
One of a pair of accessory reproductive glands; contributes toward seminal fluid

Ureter

Bladder

Prostate gland
Accessory reproductive gland; contributes about a third of all seminal fluid

Cowper's gland
One of two pea-sized glands, also known as bulbourethral glands; secretes a tiny portion of seminal fluid

Bulb of penis
Part of the root of the penis; attaches to the perineal membrane, which stretches between the ischiopubic rami of the pelvis (see p.174)

Crus of penis
One of two crura that are attached to the ischiopubic rami and make up part of the root of the penis

Vas deferens
Muscular tube; carries sperm from the testis into the pelvis, via the abdominal cavity. It leads into the prostate and joins the duct of the seminal vesicle before entering the urethra

Epididymis
Can be felt through the scrotum as a long lump on the back of the testis

Efferent ductules
10–20 ductules carry seminal fluid from the testes to the epididymis

Rete testis
Literally meaning the network of the testis; these interconnecting tubes link the seminiferous tubules with the efferent ductules

Lobule of testis
200–300 lobules lie in each testis; each contains 1–3 tightly-packed seminiferous tubules, in which sperm are made

Tunica albuginea
Literally meaning egg-white–like coat; the outer covering of the testis

Dorsal artery
A branch of the internal pudendal artery, which is itself a branch of the internal iliac artery

Dorsal veins
These drain up to the venous plexus around the prostate and, ultimately, to the internal iliac veins

Septum
Fibrous division between the corpora cavernosa

Tunica albuginea
Fibrous envelope around the components of the penis

Corpora cavernosa

Urethra

Corpus spongiosum

CROSS SECTION OF PENIS

PENIS AND TESTES

Corporus cavernosa
One of two columns of tissue, formed where the crura join at the shaft of the penis; known together as corpora cavernosa

Corpus spongiosum
The bulb of the penis continues on to form spongy erectile tissue called the corpus spongiosum, or spongy body

Glans penis
Expanded part of the corpus spongiosum

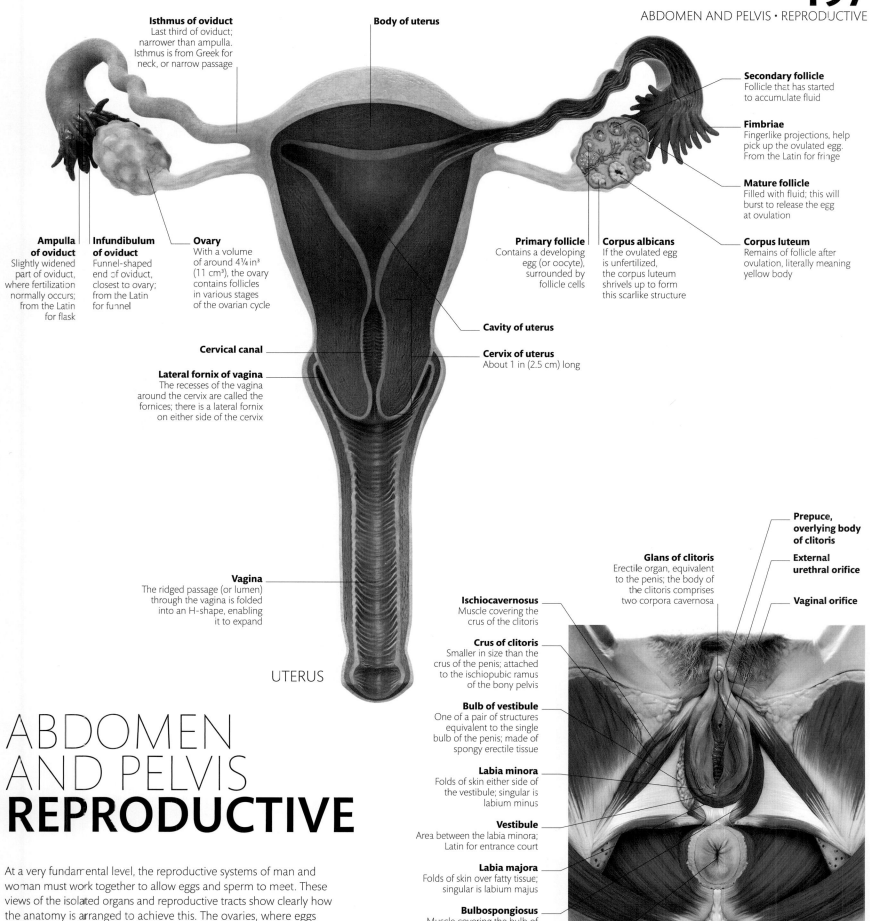

Isthmus of oviduct
Last third of oviduct;
narrower than ampulla.
Isthmus is from Greek for
neck, or narrow passage

Body of uterus

Secondary follicle
Follicle that has started
to accumulate fluid

Fimbriae
Fingerlike projections, help
pick up the ovulated egg.
From the Latin for fringe

Mature follicle
Filled with fluid; this will
burst to release the egg
at ovulation

Corpus luteum
Remains of follicle after
ovulation, literally meaning
yellow body

Ampulla of oviduct
Slightly widened
part of oviduct,
where fertilization
normally occurs;
from the Latin
for flask

Infundibulum of oviduct
Funnel-shaped
end of oviduct,
closest to ovary;
from the Latin
for funnel

Ovary
With a volume
of around 4¼ in³
(11 cm³), the ovary
contains follicles
in various stages
of the ovarian cycle

Primary follicle
Contains a developing
egg (or oocyte),
surrounded by
follicle cells

Corpus albicans
If the ovulated egg
is unfertilized,
the corpus luteum
shrivels up to form
this scarlike structure

Cavity of uterus

Cervical canal

Cervix of uterus
About 1 in (2.5 cm) long

Lateral fornix of vagina
The recesses of the vagina
around the cervix are called the
fornices; there is a lateral fornix
on either side of the cervix

Vagina
The ridged passage (or lumen)
through the vagina is folded
into an H-shape, enabling
it to expand

UTERUS

Prepuce, overlying body of clitoris

External urethral orifice

Vaginal orifice

Glans of clitoris
Erectile organ, equivalent
to the penis; the body of
the clitoris comprises
two corpora cavernosa

Ischiocavernosus
Muscle covering the
crus of the clitoris

Crus of clitoris
Smaller in size than the
crus of the penis; attached
to the ischiopubic ramus
of the bony pelvis

Bulb of vestibule
One of a pair of structures
equivalent to the single
bulb of the penis; made of
spongy erectile tissue

Labia minora
Folds of skin either side of
the vestibule; singular is
labium minus

Vestibule
Area between the labia minora;
Latin for entrance court

Labia majora
Folds of skin over fatty tissue;
singular is labium majus

Bulbospongiosus
Muscle covering the bulb of
vestibule; helps increase
pressure in the underlying
spongy tissue

Anus

EXTERNAL FEMALE
GENITALIA

ABDOMEN
AND PELVIS
REPRODUCTIVE

At a very fundamental level, the reproductive systems of man and
woman must work together to allow eggs and sperm to meet. These
views of the isolated organs and reproductive tracts show clearly how
the anatomy is arranged to achieve this. The ovaries, where eggs
(or ova) are produced, are deep inside the female pelvis. The eggs are
collected from the ovaries by a pair of tubes, the oviducts, and it is
usually here that fertilization takes place. The fertilized egg then
moves along the oviduct, dividing into a ball of cells. The embryo
eventually reaches the uterus, which is designed to accommodate
and support the growing fetus. The vagina provides both a way for
sperm to get in, and the route for the baby to get out at birth.

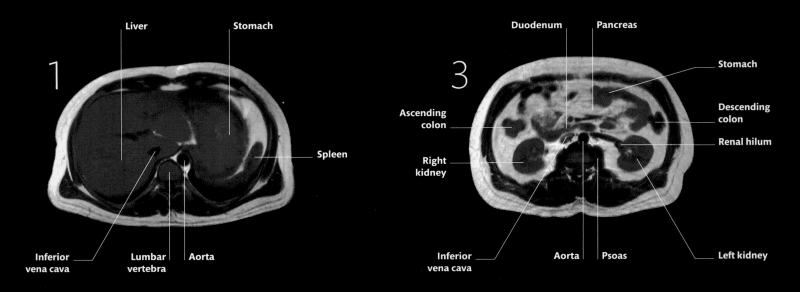

1

Liver

Stomach

Inferior
vena cava

Lumbar
vertebra

Aorta

Spleen

3

Duodenum

Pancreas

Stomach

Ascending
colon

Descending
colon

Right
kidney

Renal hilum

Inferior
vena cava

Aorta

Psoas

Left kidney

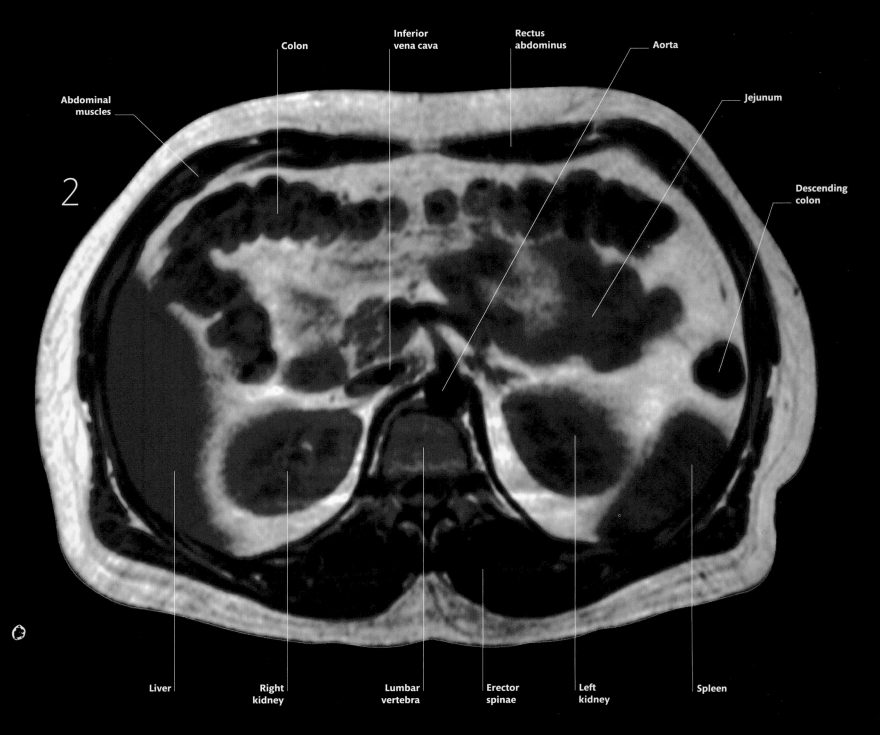

2

Abdominal
muscles

Colon

Inferior
vena cava

Rectus
abdominus

Aorta

Jejunum

Descending
colon

Liver

Right
kidney

Lumbar
vertebra

Erector
spinae

Left
kidney

Spleen

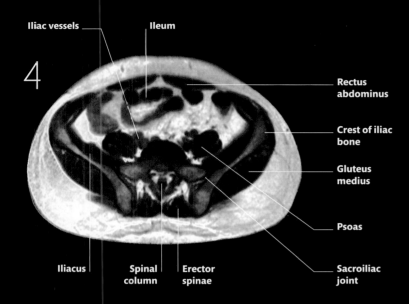

4

Iliac vessels | Ileum

Rectus abdominus

Crest of iliac bone

Gluteus medius

Psoas

Sacroiliac joint

Iliacus | Spinal column | Erector spinae

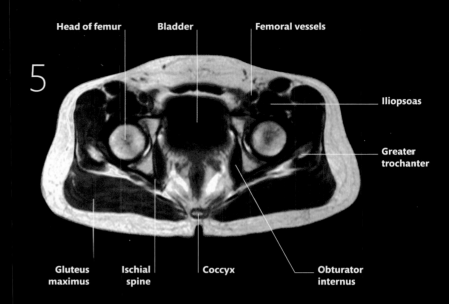

5

Head of femur | Bladder | Femoral vessels

Iliopsoas

Greater trochanter

Gluteus maximus | Ischial spine | Coccyx | Obturator internus

ABDOMEN AND PELVIS **MRI**

MRI is a useful way of looking at soft tissues—and for visualizing the organs of the abdomen and pelvis, which only appear as subtle shadows on a standard X-ray. In the series of axial or transverse sections through the abdomen and pelvis, we can clearly see the dense liver, and blood vessels branching within it (section 1); the right kidney lying close to the liver, and the left kidney close to the spleen (section 2); the kidneys at the level where the renal arteries enter them (section 3), with the stomach and pancreas lying in front; coils of small intestine, the ileum, resting in the lower part of the abdomen, cradled by the iliac bones (section 4); and the organs of the pelvis at the level of the hip joints (section 5). The sagittal view (section 6) shows how surprisingly shallow the abdominal cavity is, in front of the lumbar spine. In a slim person, it is possible to press down on the lower abdomen and feel the pulsations of the descending aorta—right at the back of the abdomen.

LEVELS OF SCANS

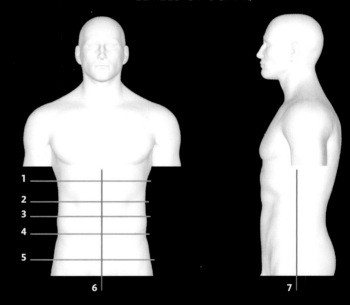

1
2
3
4
5

6 | 7

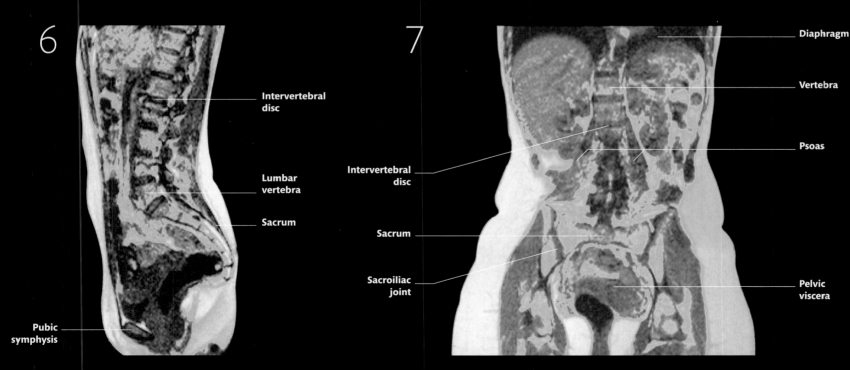

6

Intervertebral disc

Lumbar vertebra

Sacrum

Pubic symphysis

7

Diaphragm

Vertebra

Psoas

Intervertebral disc

Sacrum

Sacroiliac joint

Pelvic viscera

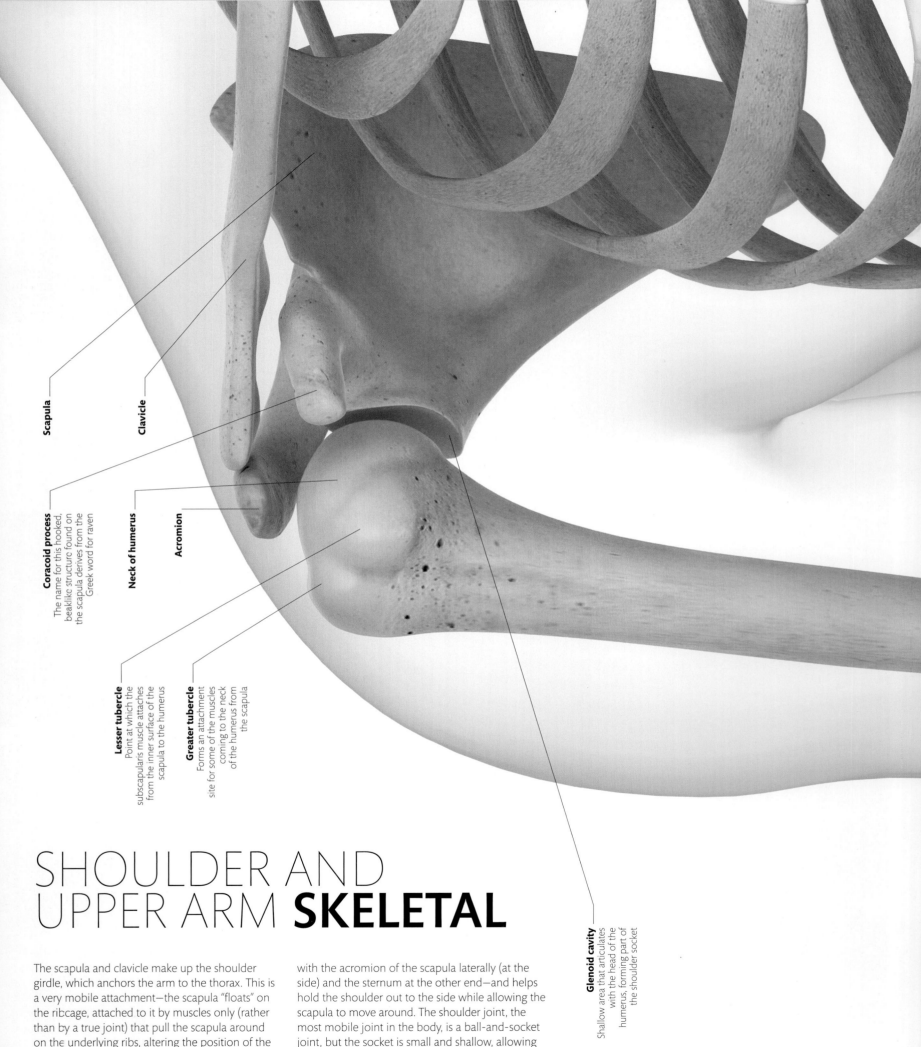

Scapula

Clavicle

Coracoid process
The name for this hooked, beaklike structure found on the scapula derives from the Greek word for raven

Neck of humerus

Acromion

Lesser tubercle
Point at which the subscapularis muscle attaches from the inner surface of the scapula to the humerus

Greater tubercle
Forms an attachment site for some of the muscles coming to the neck of the humerus from the scapula

Glenoid cavity
Shallow area that articulates with the head of the humerus, forming part of the shoulder socket

SHOULDER AND
UPPER ARM **SKELETAL**

The scapula and clavicle make up the shoulder girdle, which anchors the arm to the thorax. This is a very mobile attachment—the scapula "floats" on the ribcage, attached to it by muscles only (rather than by a true joint) that pull the scapula around on the underlying ribs, altering the position of the shoulder joint. The clavicle has joints—it articulates with the acromion of the scapula laterally (at the side) and the sternum at the other end—and helps hold the shoulder out to the side while allowing the scapula to move around. The shoulder joint, the most mobile joint in the body, is a ball-and-socket joint, but the socket is small and shallow, allowing the ball-shaped head of the humerus to move freely.

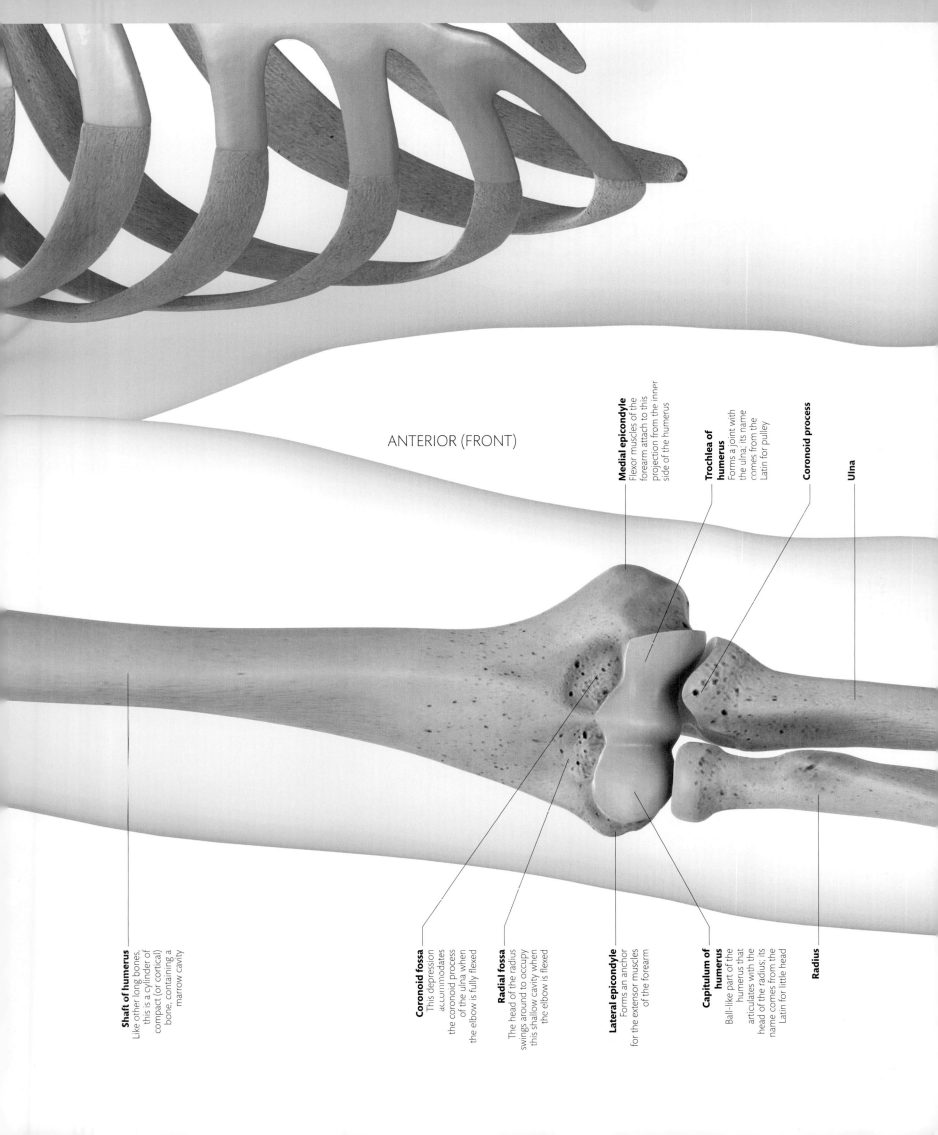

ANTERIOR (FRONT)

Medial epicondyle
Flexor muscles of the
forearm attach to this
projection from the inner
side of the humerus

**Trochlea of
humerus**
Forms a joint with
the ulna; its name
comes from the
Latin for pulley

Coronoid process

Ulna

Shaft of humerus
Like other long bones,
this is a cylinder of
compact (or cortical)
bone, containing a
marrow cavity

Coronoid fossa
This depression
accommodates
the coronoid process
of the ulna when
the elbow is fully flexed

Radial fossa
The head of the radius
swings around to occupy
this shallow cavity when
the elbow is flexed

Lateral epicondyle
Forms an anchor
for the extensor muscles
of the forearm

**Capitulum of
humerus**
Ball-like part of the
humerus that
articulates with the
head of the radius; its
name comes from the
Latin for little head

Radius

SHOULDER AND UPPER ARM **SKELETAL**

The back of the scapula is divided into two sections by its spine. The muscles that attach above this spine are called supraspinatus; those that attach below are called infraspinatus. They are part of the rotator cuff muscle group, which enables shoulder movements and stabilizes the shoulder joint. The spine of the scapula runs to the side and projects out above the shoulder joint to form the acromion, which can be easily felt on the top of the shoulder. The scapula rests in the position shown here when the arm is hanging at the side of the body. If the arm is abducted (raised to the side), the entire scapula rotates so that the glenoid cavity points upward and the inferior angle moves outward.

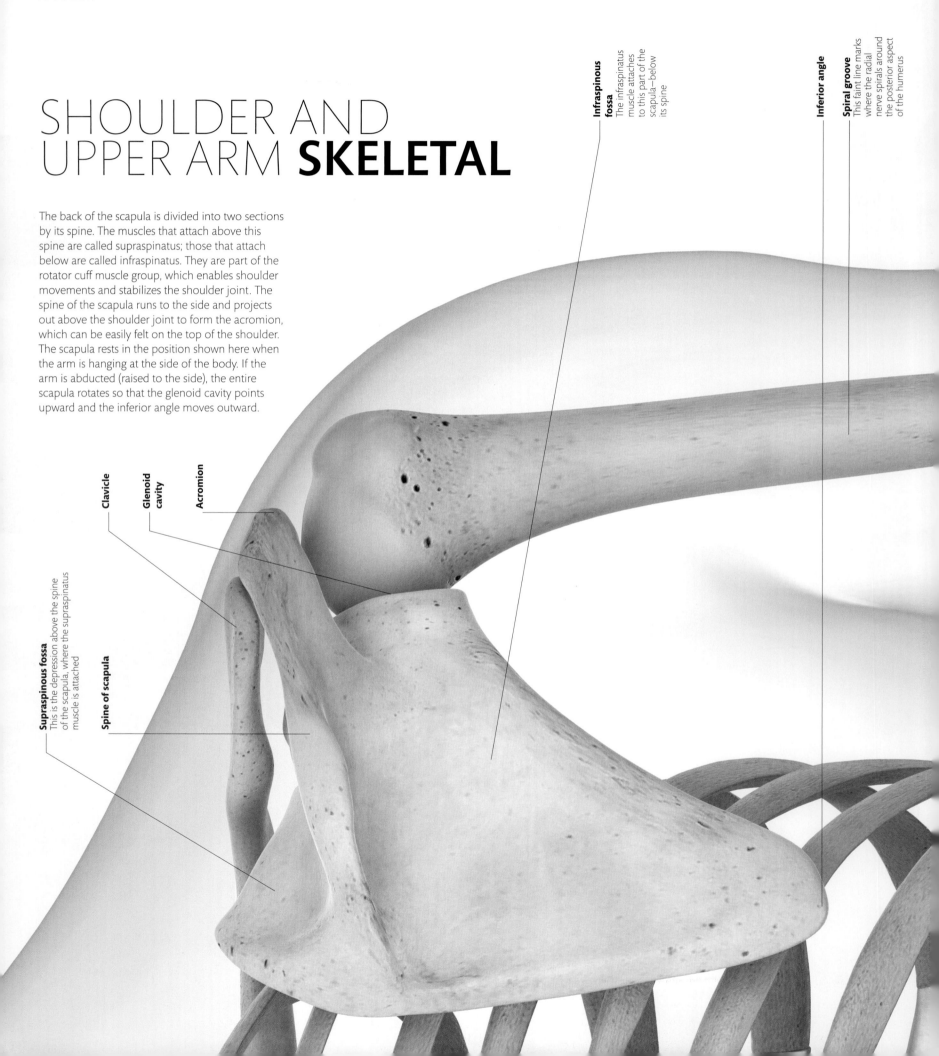

Infraspinous fossa
The infraspinatus muscle attaches to this part of the scapula—below its spine

Inferior angle

Spiral groove
This faint line marks where the radial nerve spirals around the posterior aspect of the humerus

Clavicle

Glenoid cavity

Acromion

Supraspinous fossa
This is the depression above the spine of the scapula, where the supraspinatus muscle is attached

Spine of scapula

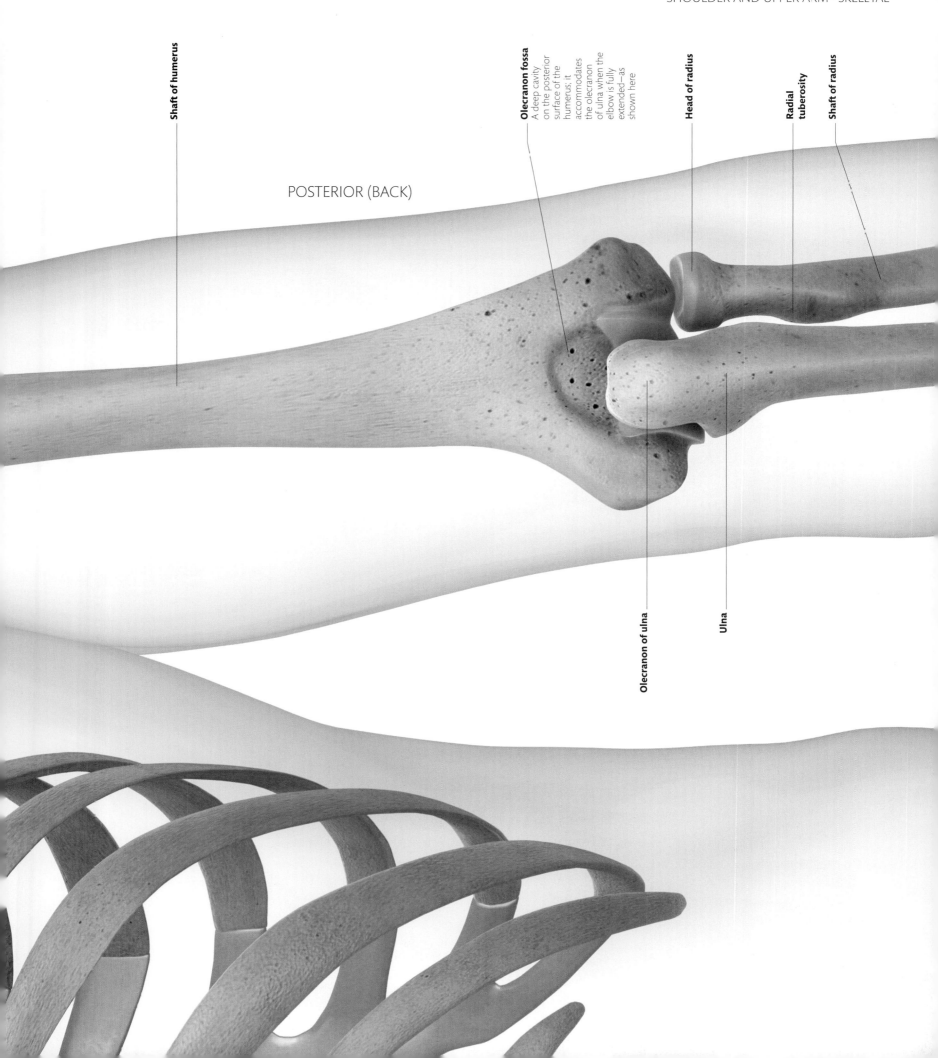

Shaft of humerus

Olecranon fossa
A deep cavity on the posterior surface of the humerus; it accommodates the olecranon of ulna when the elbow is fully extended—as shown here

Head of radius

Radial tuberosity

Shaft of radius

POSTERIOR (BACK)

Olecranon of ulna

Ulna

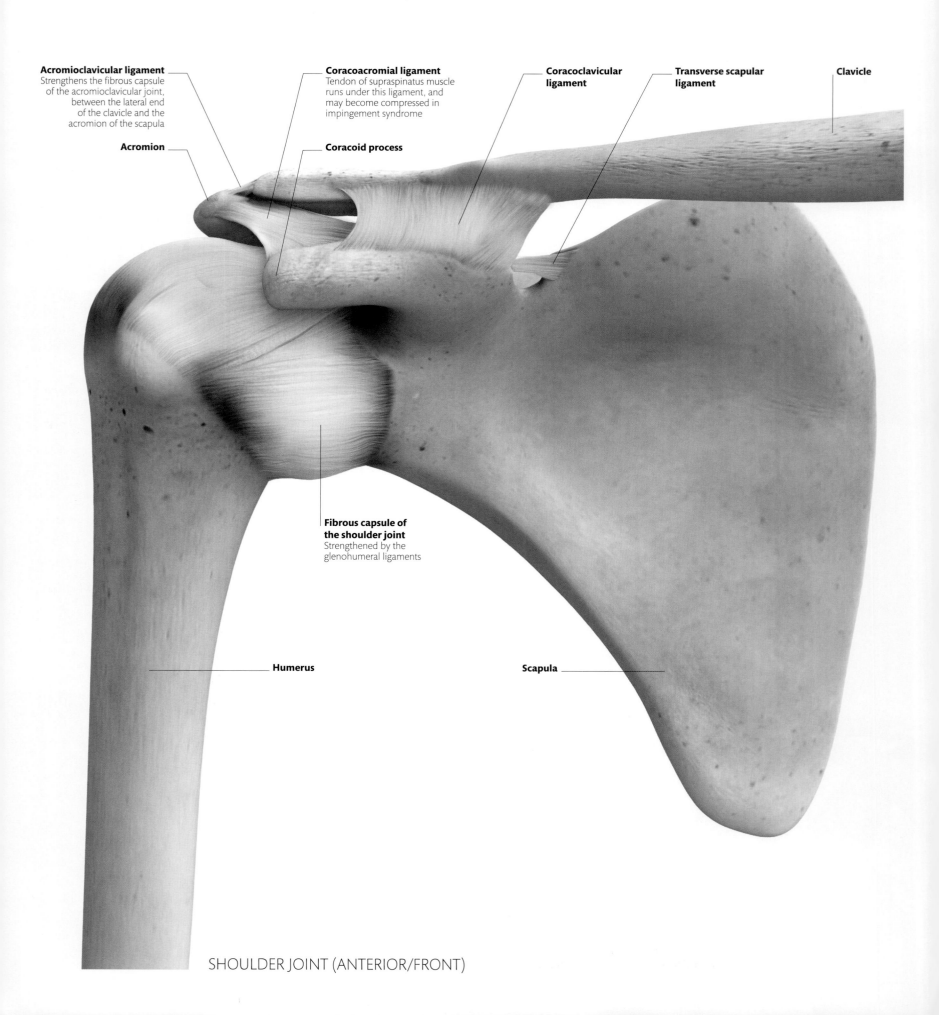

Acromioclavicular ligament
Strengthens the fibrous capsule
of the acromioclavicular joint,
between the lateral end
of the clavicle and the
acromion of the scapula

Coracoacromial ligament
Tendon of supraspinatus muscle
runs under this ligament, and
may become compressed in
impingement syndrome

**Coracoclavicular
ligament**

**Transverse scapular
ligament**

Clavicle

Acromion

Coracoid process

**Fibrous capsule of
the shoulder joint**
Strengthened by the
glenohumeral ligaments

Humerus

Scapula

SHOULDER JOINT (ANTERIOR/FRONT)

SHOULDER AND UPPER ARM **SKELETAL**

In any joint, there is always a play off between mobility and stability. The extremely mobile shoulder joint is therefore naturally unstable, and so it is not surprising that this is the most commonly dislocated joint in the body. The coracoacromial arch, formed by the acromion and coracoid process of the scapula with the strong coracoacromial ligament stretching between them, prevents upward dislocation; when the head of the humerus dislocates, it usually does so in a downward direction. The elbow joint is formed by the articulation of the humerus with the forearm bones: the trochlea articulates with the ulna, and the capitulum with the head of the radius. The elbow is a hinge joint, stabilized by collateral ligaments on each side.

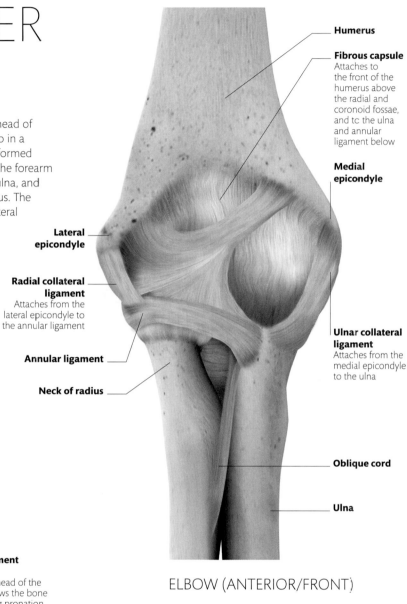

Humerus

Fibrous capsule
Attaches to the front of the humerus above the radial and coronoid fossae, and to the ulna and annular ligament below

Medial epicondyle

Lateral epicondyle

Radial collateral ligament
Attaches from the lateral epicondyle to the annular ligament

Annular ligament

Neck of radius

Ulnar collateral ligament
Attaches from the medial epicondyle to the ulna

Oblique cord

Ulna

ELBOW (ANTERIOR/FRONT)

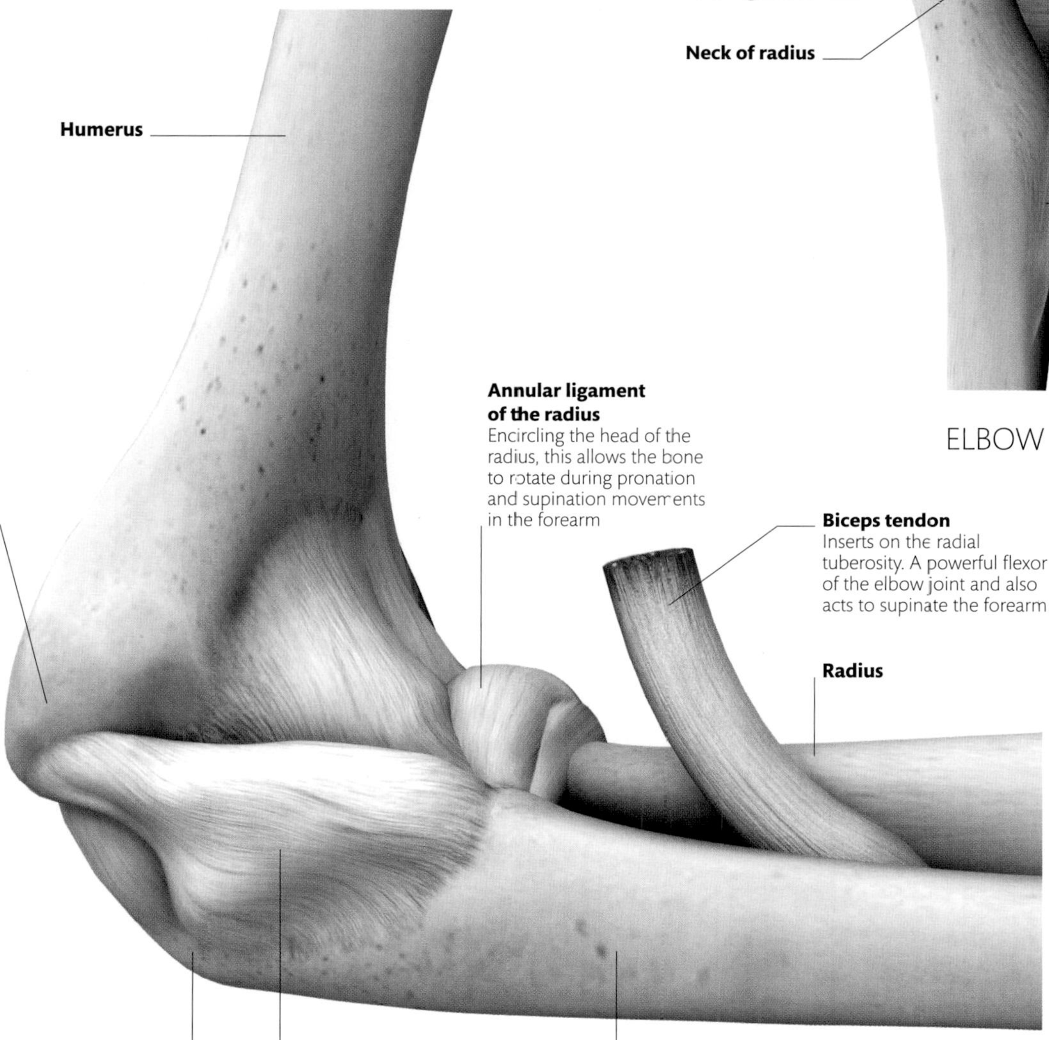

Humerus

Medial epicondyle
Also forms the common flexor origin—the attachment of many of the forearm flexor muscles

Annular ligament of the radius
Encircling the head of the radius, this allows the bone to rotate during pronation and supination movements in the forearm

Biceps tendon
Inserts on the radial tuberosity. A powerful flexor of the elbow joint and also acts to supinate the forearm

Radius

Olecranon of ulna

Ulnar collateral ligament

Ulna

ELBOW (LATERAL/SIDE)

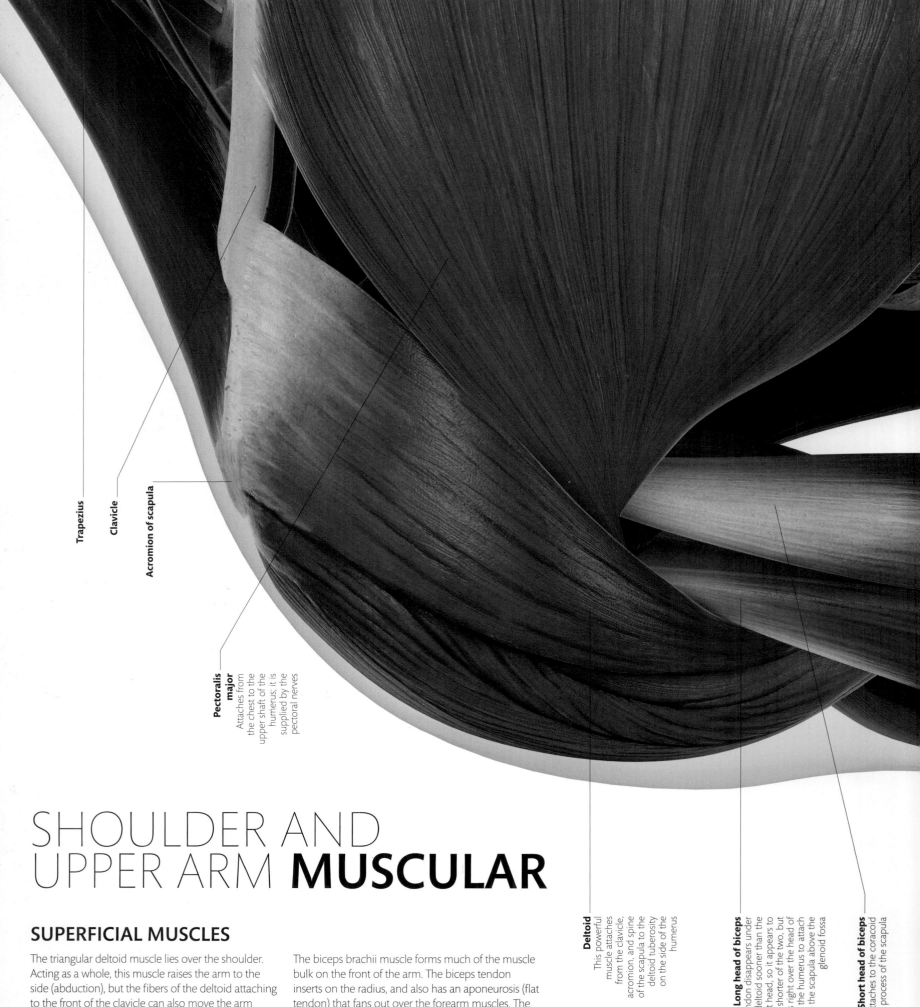

Trapezius

Clavicle

Acromion of scapula

Pectoralis major
Attaches from the chest to the upper shaft of the humerus; it is supplied by the pectoral nerves

Deltoid
This powerful muscle attaches from the clavicle, acromion, and spine of the scapula to the deltoid tuberosity on the side of the humerus

Long head of biceps
This tendon disappears under the deltoid sooner than the short head, so it appears to be the shorter of the two, but it runs right over the head of the humerus to attach to the scapula above the glenoid fossa

Short head of biceps
Attaches to the coracoid process of the scapula

SHOULDER AND UPPER ARM **MUSCULAR**

SUPERFICIAL MUSCLES

The triangular deltoid muscle lies over the shoulder. Acting as a whole, this muscle raises the arm to the side (abduction), but the fibers of the deltoid attaching to the front of the clavicle can also move the arm forward. The pectoralis major muscle can also act on the shoulder joint, flexing the arm forward or pulling it in to the side of the chest (adduction).

The biceps brachii muscle forms much of the muscle bulk on the front of the arm. The biceps tendon inserts on the radius, and also has an aponeurosis (flat tendon) that fans out over the forearm muscles. The biceps is a powerful flexor of the elbow, and can also rotate the radius to position the lower arm so the palm faces upward (supination).

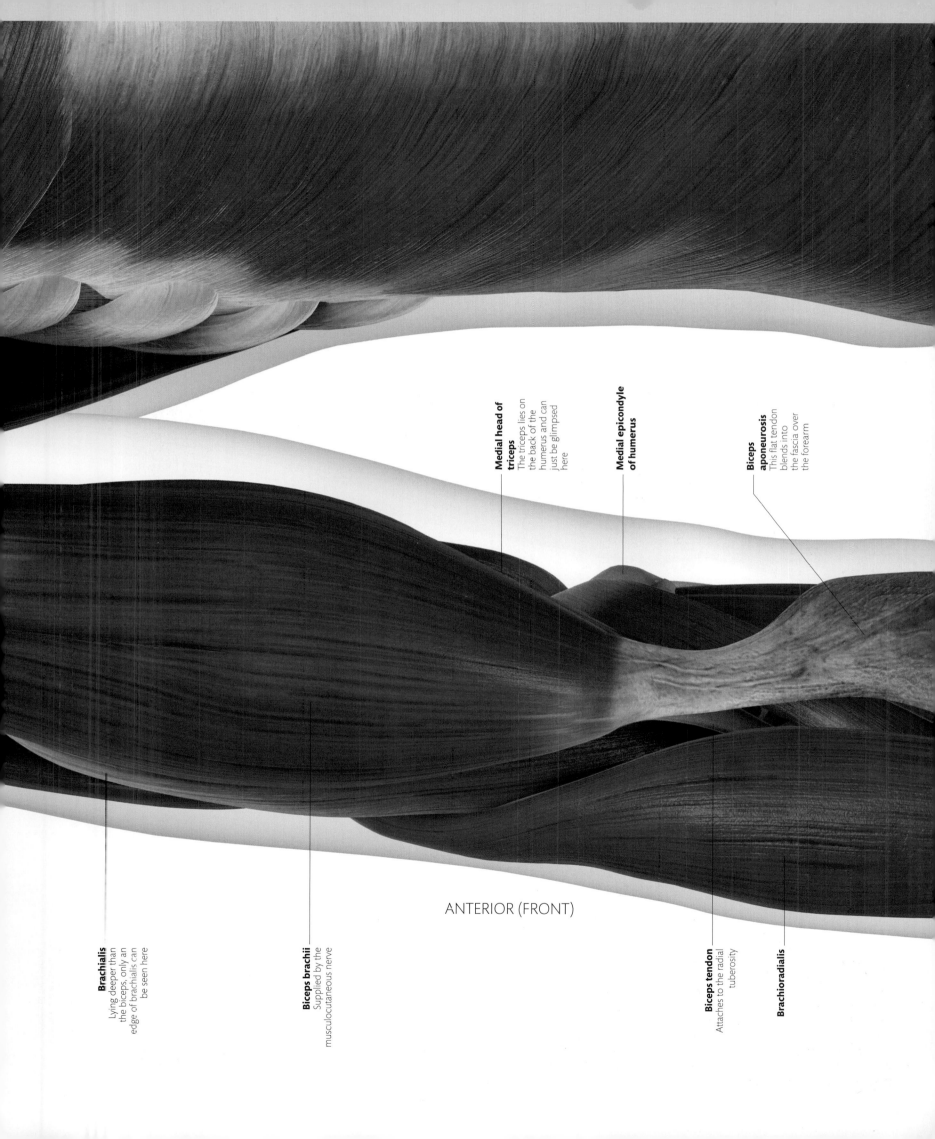

Medial head of triceps
The triceps lies on the back of the humerus and can just be glimpsed here

Medial epicondyle of humerus

Biceps aponeurosis
This flat tendon blends into the fascia over the forearm

ANTERIOR (FRONT)

Brachialis
Lying deeper than the biceps, only an edge of brachialis can be seen here

Biceps brachii
Supplied by the musculocutaneous nerve

Biceps tendon
Attaches to the radial tuberosity

Brachioradialis

SHOULDER AND UPPER ARM **MUSCULAR**

SUPERFICIAL MUSCLES

The posterior fibers of the deltoid attach from the spine of the scapula (shoulder blade) down to the humerus, and this part of the muscle can draw back the arm or extend it. Latissimus dorsi (a broad muscle attaching from the back of the trunk and ending in a narrow tendon that secures onto the humerus) can also extend the arm. The triceps brachii muscle is the sole extensor of the elbow. In a superficial dissection (represented in this view), only two of the three heads of the triceps can be seen—the long and lateral heads. The triceps tendon attaches to the leverlike olecranon of the ulna, which forms the bony knob at the back of the elbow.

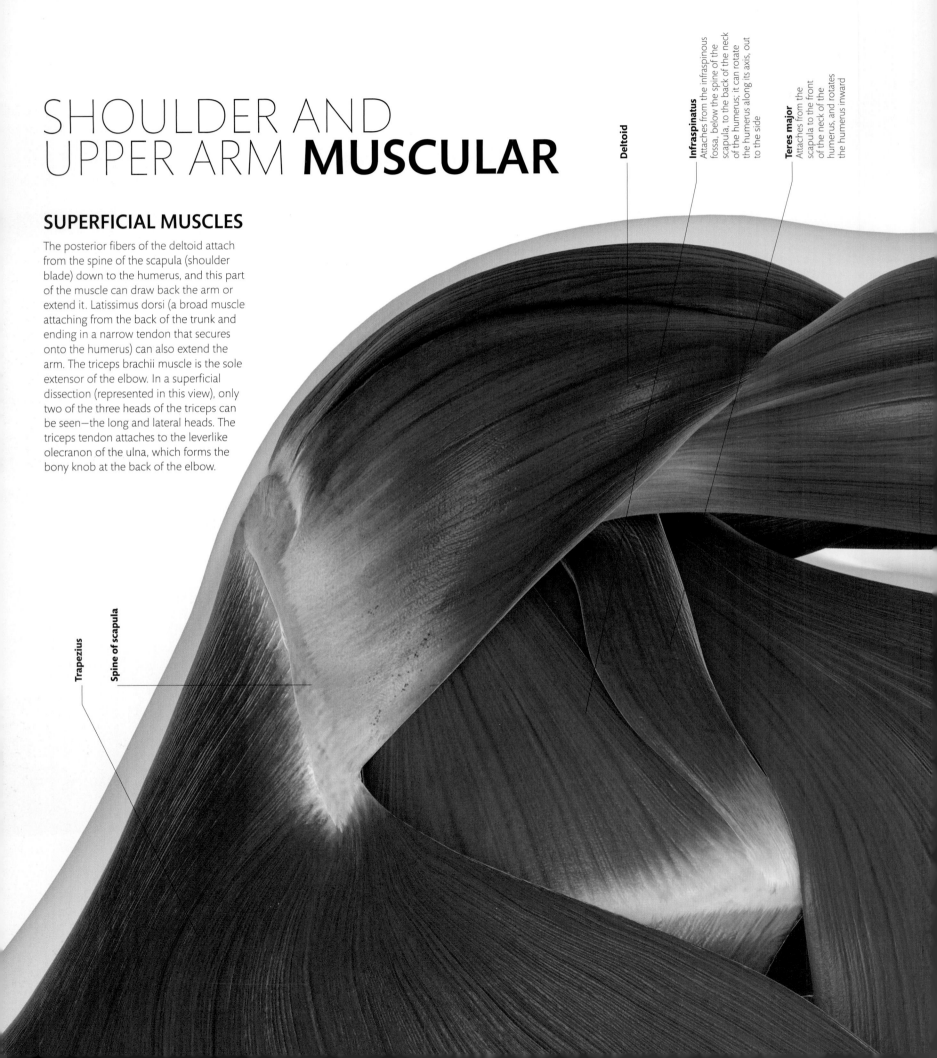

Deltoid

Infraspinatus
Attaches from the infraspinous fossa, below the spine of the scapula, to the back of the neck of the humerus; it can rotate the humerus along its axis, out to the side

Teres major
Attaches from the scapula to the front of the neck of the humerus, and rotates the humerus inward

Trapezius

Spine of scapula

POSTERIOR (BACK)

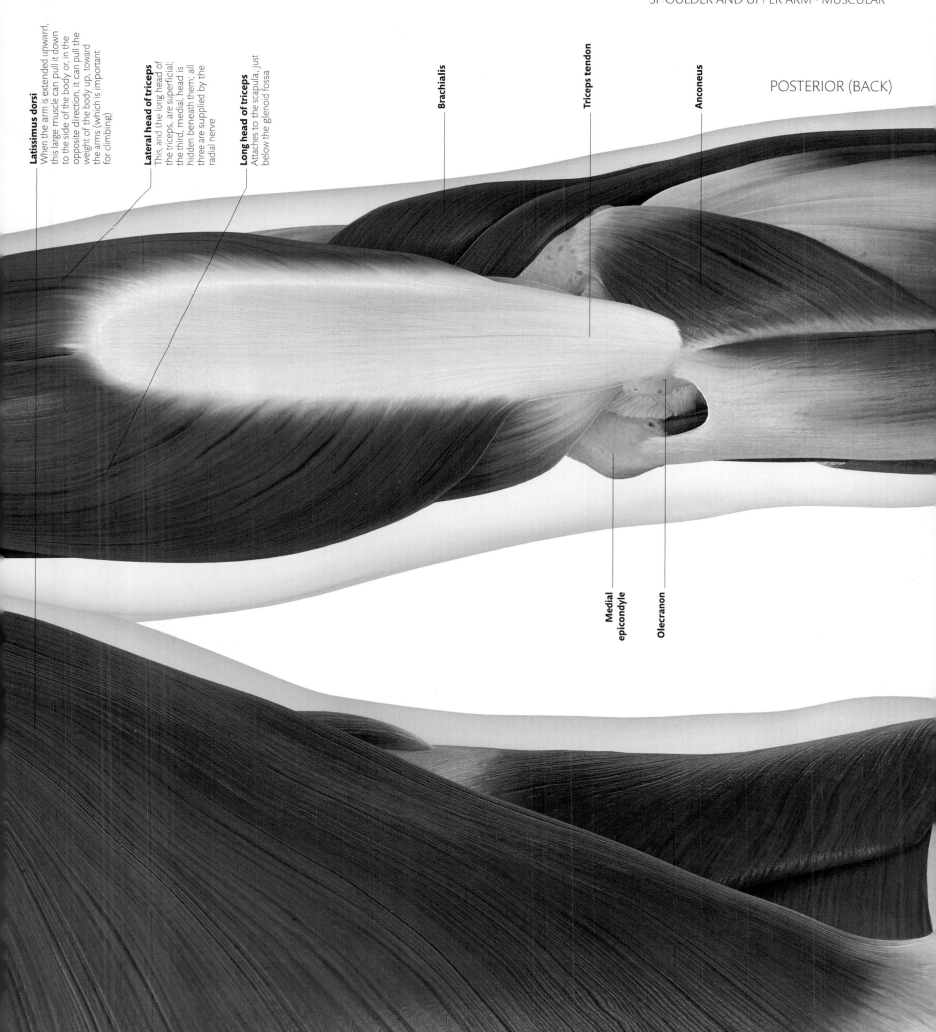

Latissimus dorsi
When the arm is extended upward, this large muscle can pull it down to the side of the body or, in the opposite direction, it can pull the weight of the body up, toward the arms (which is important for climbing)

Lateral head of triceps
This, and the long head of the triceps, are superficial; the third, medial, head is hidden beneath them; all three are supplied by the radial nerve

Long head of triceps
Attaches to the scapula, just below the glenoid fossa

Brachialis

Triceps tendon

Anconeus

Medial epicondyle

Olecranon

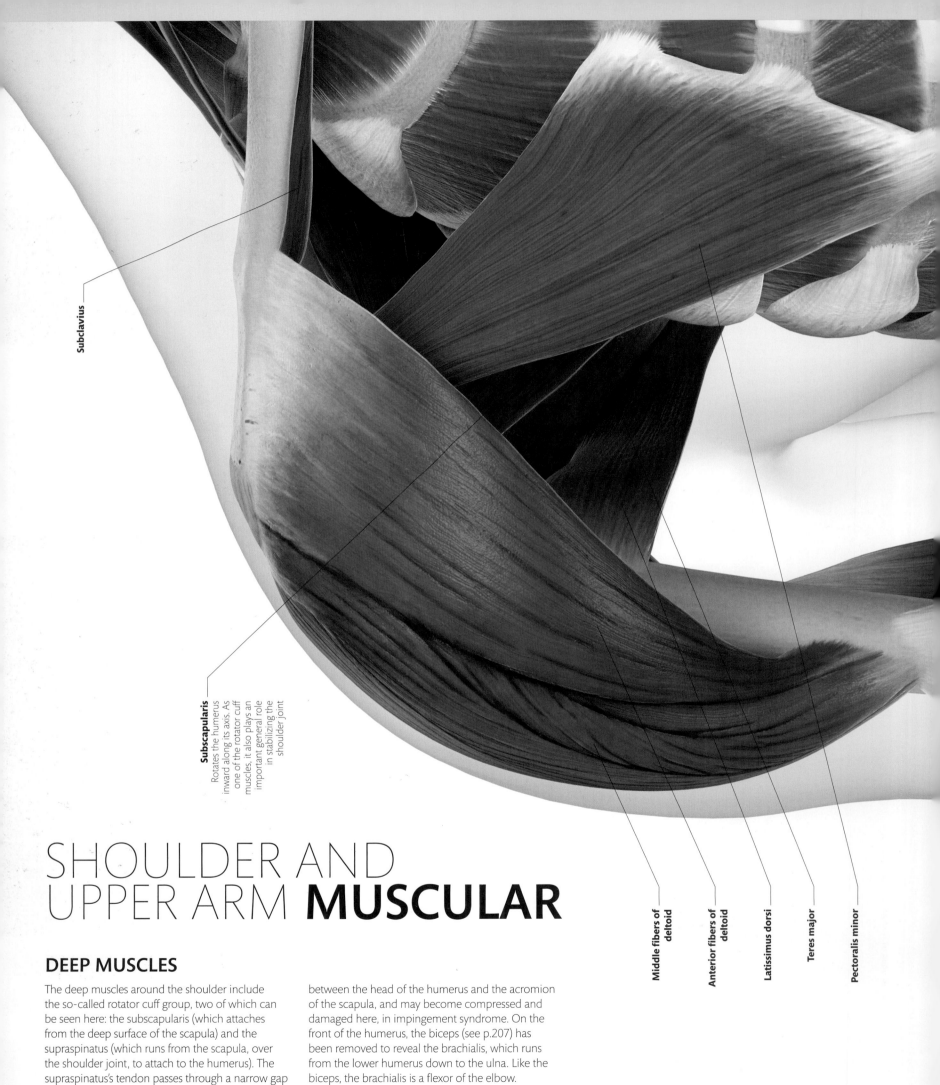

Subclavius

Subscapularis
Rotates the humerus inward along its axis. As one of the rotator cuff muscles, it also plays an important general role in stabilizing the shoulder joint

Middle fibers of deltoid

Anterior fibers of deltoid

Latissimus dorsi

Teres major

Pectoralis minor

SHOULDER AND UPPER ARM **MUSCULAR**

DEEP MUSCLES

The deep muscles around the shoulder include the so-called rotator cuff group, two of which can be seen here: the subscapularis (which attaches from the deep surface of the scapula) and the supraspinatus (which runs from the scapula, over the shoulder joint, to attach to the humerus). The supraspinatus's tendon passes through a narrow gap between the head of the humerus and the acromion of the scapula, and may become compressed and damaged here, in impingement syndrome. On the front of the humerus, the biceps (see p.207) has been removed to reveal the brachialis, which runs from the lower humerus down to the ulna. Like the biceps, the brachialis is a flexor of the elbow.

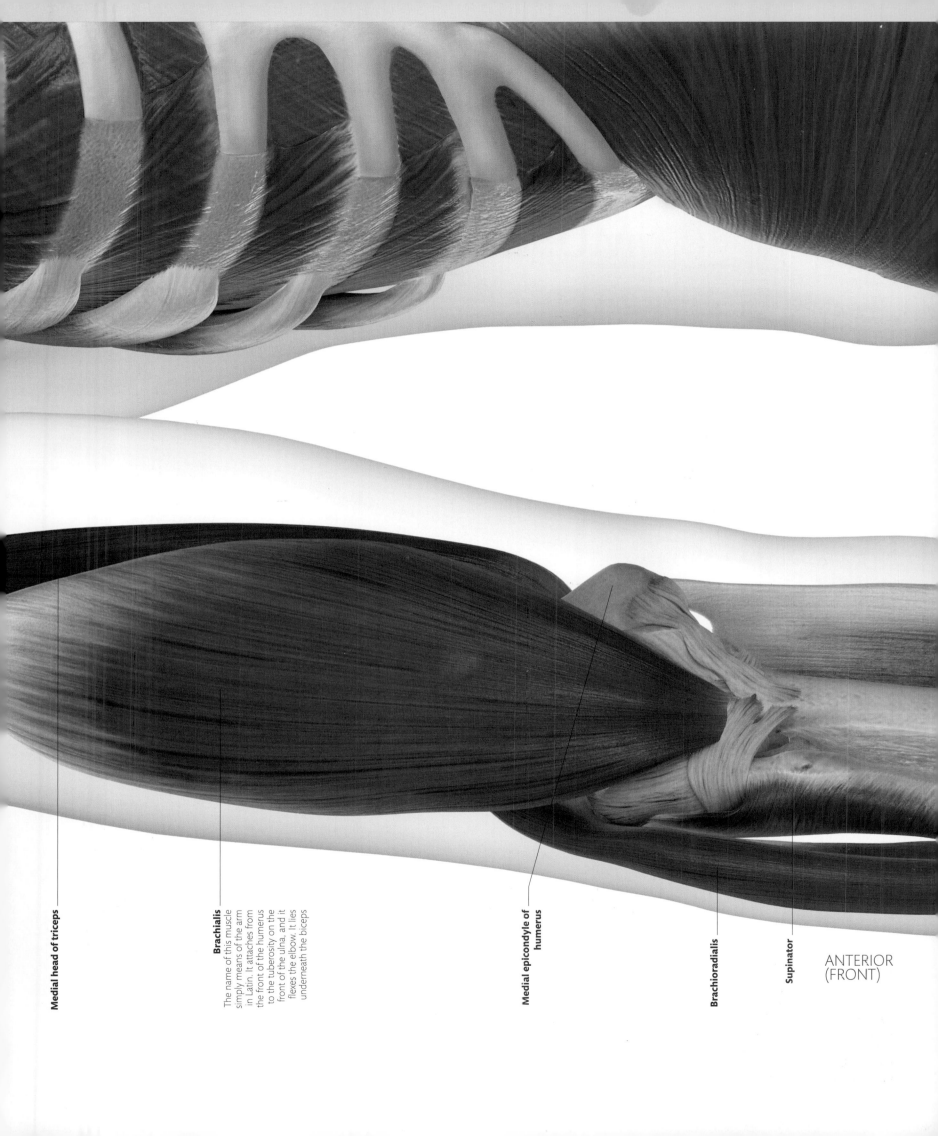

Medial head of triceps

Brachialis
The name of this muscle simply means of the arm in Latin. It attaches from the front of the humerus to the tuberosity on the front of the ulna, and it flexes the elbow. It lies underneath the biceps

Medial epicondyle of humerus

Brachioradialis

Supinator

ANTERIOR
(FRONT)

SHOULDER AND
UPPER ARM **MUSCULAR**

DEEP MUSCLES

More of the rotator cuff muscles—the supraspinatus, infraspinatus, and teres minor—can be seen from the back. In addition to moving the shoulder joint in various directions, including rotation, these muscles are important in helping stabilize the shoulder joint: they hug the head of the humerus into its socket during movements at the shoulder. On the back of the arm, a deeper view reveals the third, medial head of the triceps, which attaches from the back of the humerus. It joins with the lateral and long heads to form the triceps tendon, attaching to the olecranon. Most of the forearm muscles take their attachment from the epicondyles of the humerus, just above the elbow, but the brachioradialis and extensor carpi radialis longus have higher origins from the side of the humerus, as shown here.

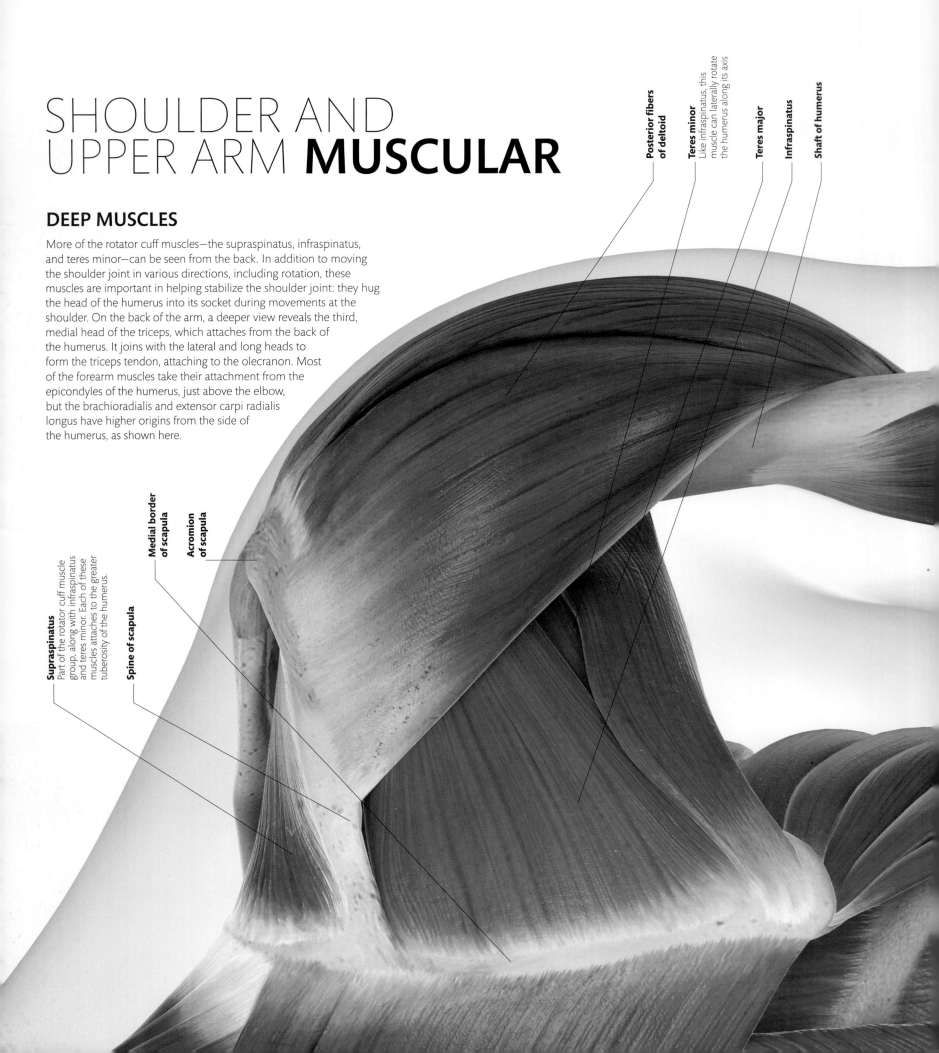

Posterior fibers of deltoid

Teres minor
Like infraspinatus, this muscle can laterally rotate the humerus along its axis

Teres major

Infraspinatus

Shaft of humerus

Medial border of scapula

Acromion of scapula

Supraspinatus
Part of the rotator cuff muscle group, along with infraspinatus and teres minor. Each of these muscles attaches to the greater tuberosity of the humerus.

Spine of scapula

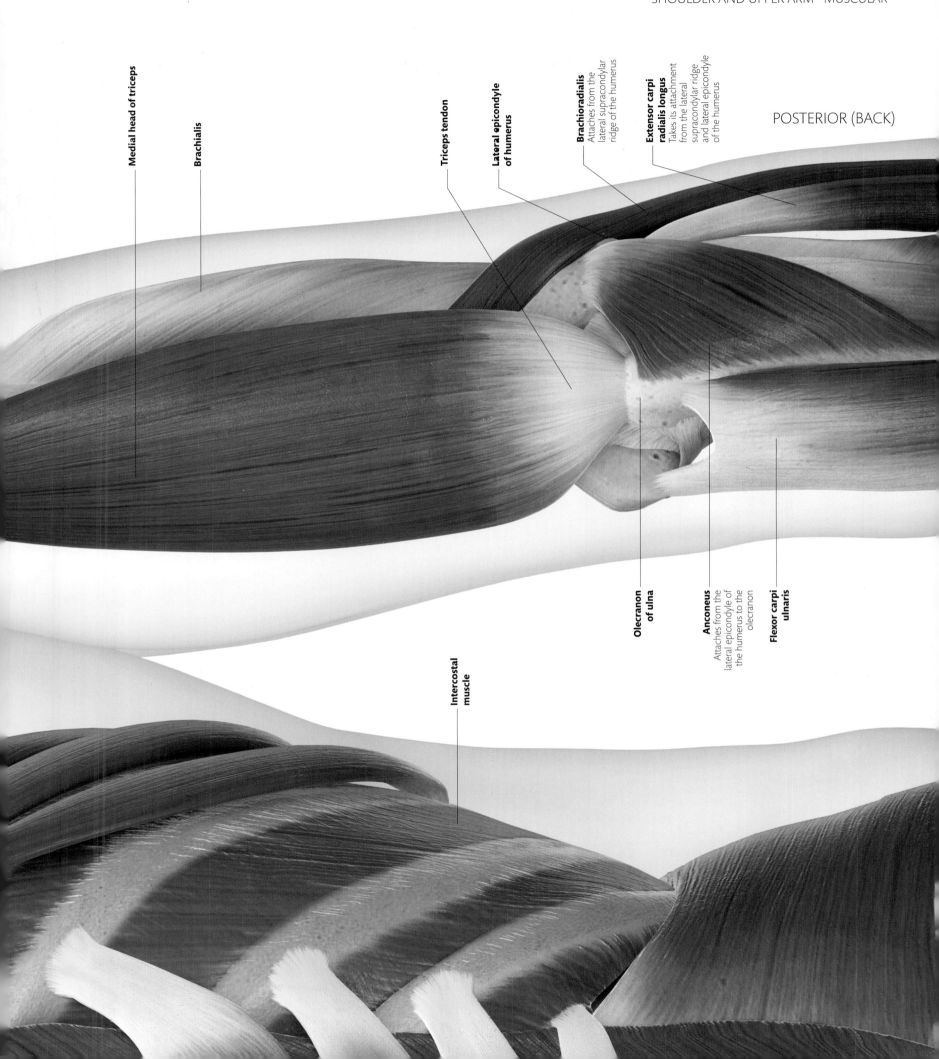

POSTERIOR (BACK)

Medial head of triceps

Brachialis

Triceps tendon

Lateral epicondyle of humerus

Brachioradialis
Attaches from the lateral supracondylar ridge of the humerus

Extensor carpi radialis longus
Takes its attachment from the lateral supracondylar ridge and lateral epicondyle of the humerus

Olecranon of ulna

Anconeus
Attaches from the lateral epicondyle of the humerus to the olecranon

Flexor carpi ulnaris

Intercostal muscle

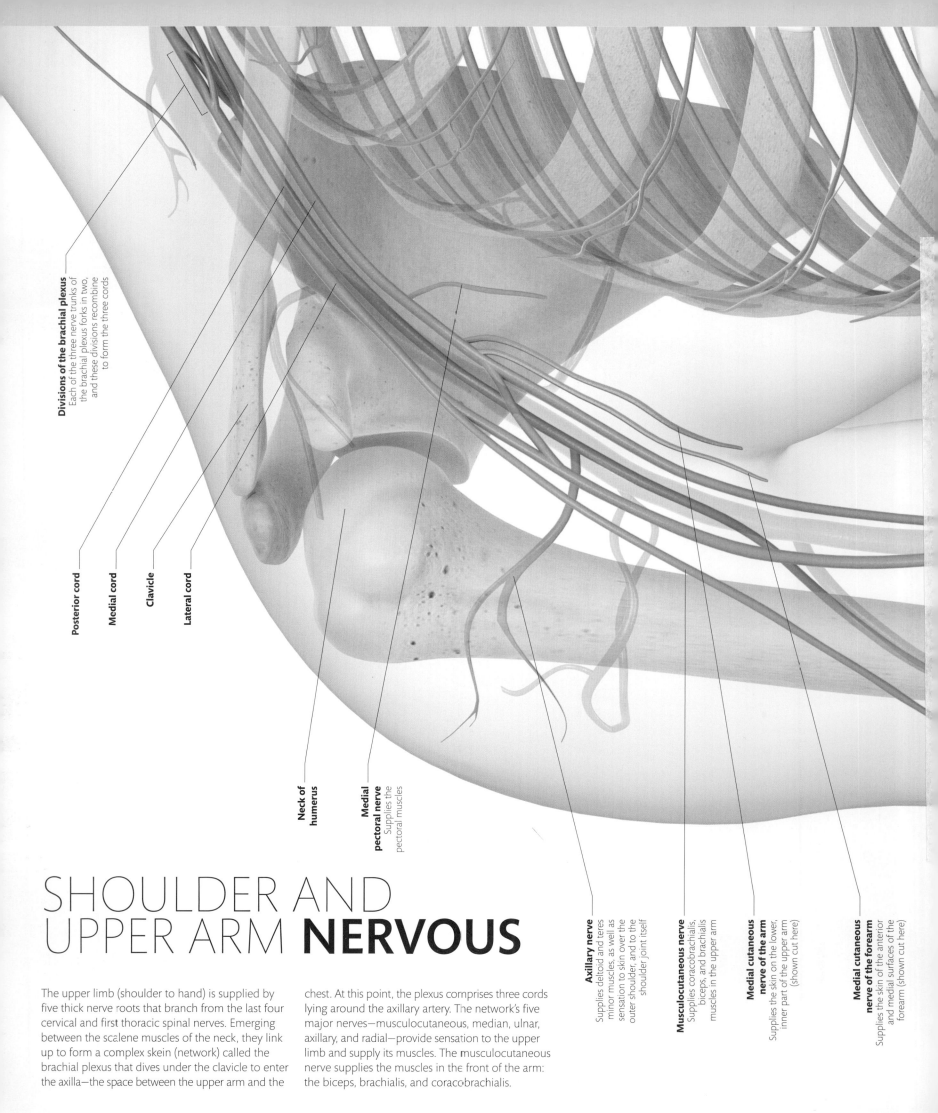

Divisions of the brachial plexus
Each of the three nerve trunks of the brachial plexus forks in two, and these divisions recombine to form the three cords

Posterior cord

Medial cord

Clavicle

Lateral cord

Neck of humerus

Medial pectoral nerve
Supplies the pectoral muscles

Axillary nerve
Supplies deltoid and teres minor muscles, as well as sensation to skin over the outer shoulder, and to the shoulder joint itself

Musculocutaneous nerve
Supplies coracobrachialis, biceps, and brachialis muscles in the upper arm

Medial cutaneous nerve of the arm
Supplies the skin on the lower, inner part of the upper arm (shown cut here)

Medial cutaneous nerve of the forearm
Supplies the skin of the anterior and medial surfaces of the forearm (shown cut here)

SHOULDER AND UPPER ARM **NERVOUS**

The upper limb (shoulder to hand) is supplied by five thick nerve roots that branch from the last four cervical and first thoracic spinal nerves. Emerging between the scalene muscles of the neck, they link up to form a complex skein (network) called the brachial plexus that dives under the clavicle to enter the axilla—the space between the upper arm and the chest. At this point, the plexus comprises three cords lying around the axillary artery. The network's five major nerves—musculocutaneous, median, ulnar, axillary, and radial—provide sensation to the upper limb and supply its muscles. The musculocutaneous nerve supplies the muscles in the front of the arm: the biceps, brachialis, and coracobrachialis.

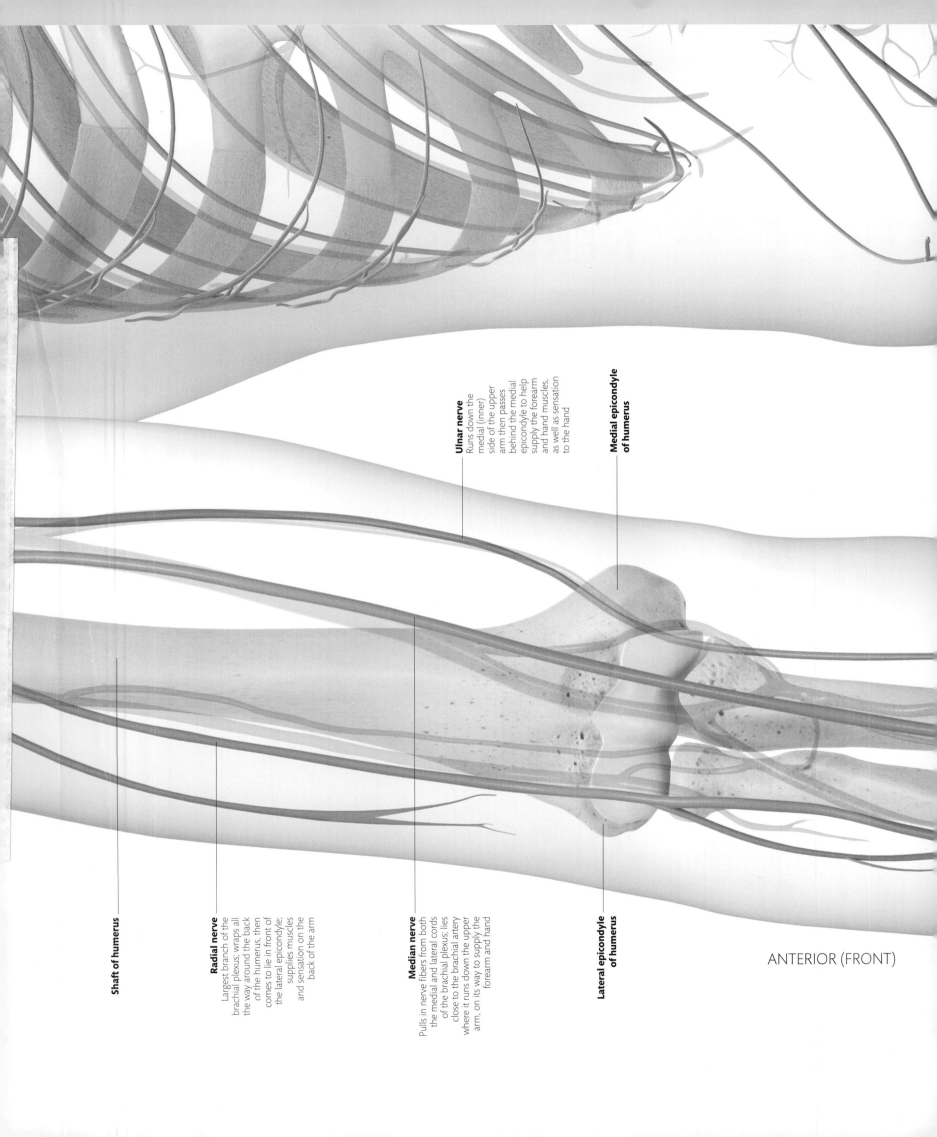

Ulnar nerve
Runs down the medial (inner) side of the upper arm then passes behind the medial epicondyle to help supply the forearm and hand muscles, as well as sensation to the hand

Medial epicondyle of humerus

Shaft of humerus

Radial nerve
Largest branch of the brachial plexus; wraps all the way around the back of the humerus, then comes to lie in front of the lateral epicondyle; supplies muscles and sensation on the back of the arm

Median nerve
Pulls in nerve fibers from both the medial and lateral cords of the brachial plexus; lies close to the brachial artery where it runs down the upper arm, on its way to supply the forearm and hand

Lateral epicondyle of humerus

ANTERIOR (FRONT)

SHOULDER AND
UPPER ARM **NERVOUS**

The axillary and radial nerves emerge from the back of the brachial plexus and run behind the humerus. The axillary nerve wraps around the neck of the humerus, just underneath the shoulder joint, and supplies the deltoid muscle. The radial nerve—the largest branch of the brachial plexus—supplies all the extensor muscles in the upper arm and in the forearm. It spirals around the back of the humerus, lying right against the bone, and sends branches to supply the heads of the triceps. The radial nerve then continues in its spiral, running forward to lie just in front of the medial epicondyle of the humerus at the elbow.

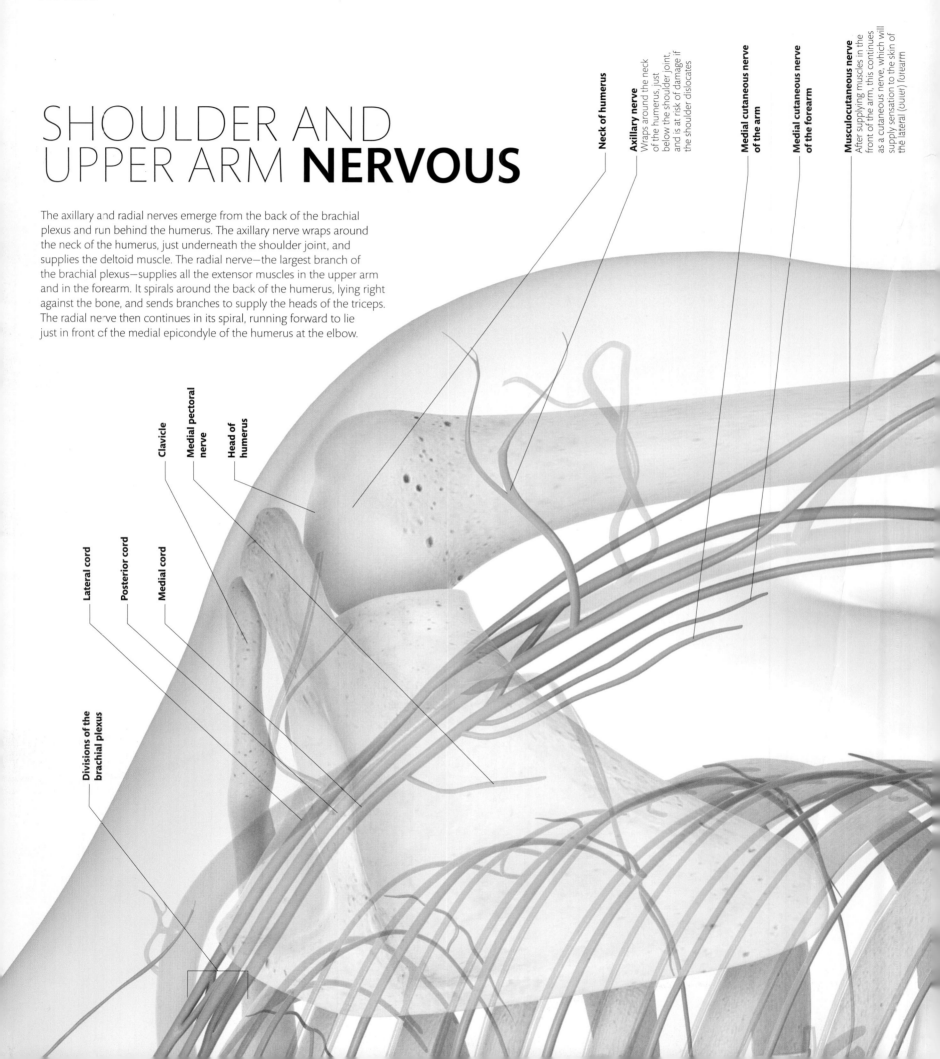

Neck of humerus

Axillary nerve
Wraps around the neck of the humerus, just below the shoulder joint, and is at risk of damage if the shoulder dislocates

Medial cutaneous nerve of the arm

Medial cutaneous nerve of the forearm

Musculocutaneous nerve
After supplying muscles in the front of the arm, this continues as a cutaneous nerve, which will supply sensation to the skin of the lateral (outer) forearm

Clavicle

Medial pectoral nerve

Head of humerus

Lateral cord

Posterior cord

Medial cord

Divisions of the brachial plexus

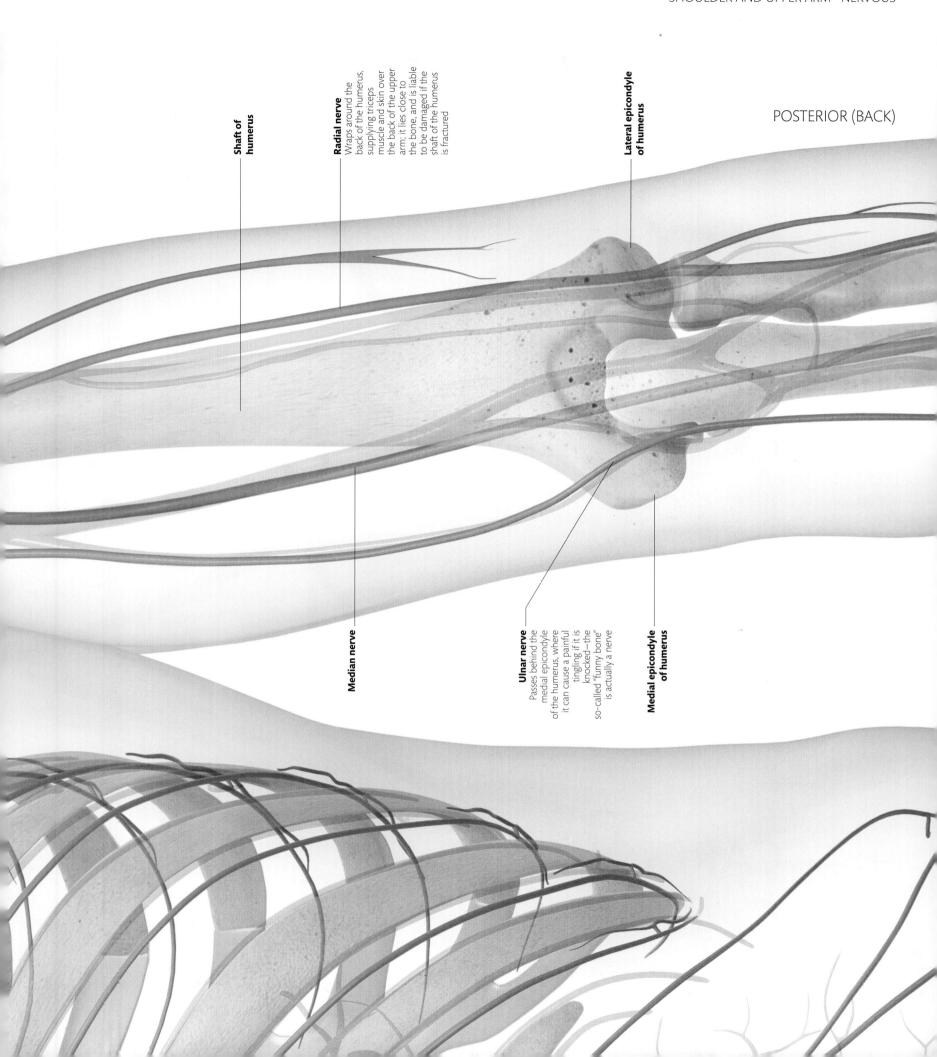

POSTERIOR (BACK)

Shaft of humerus

Radial nerve
Wraps around the back of the humerus, supplying triceps muscle and skin over the back of the upper arm; it lies close to the bone, and is liable to be damaged if the shaft of the humerus is fractured

Lateral epicondyle of humerus

Median nerve

Ulnar nerve
Passes behind the medial epicondyle of the humerus, where it can cause a painful tingling if it is knocked—the so-called "funny bone" is actually a nerve

Medial epicondyle of humerus

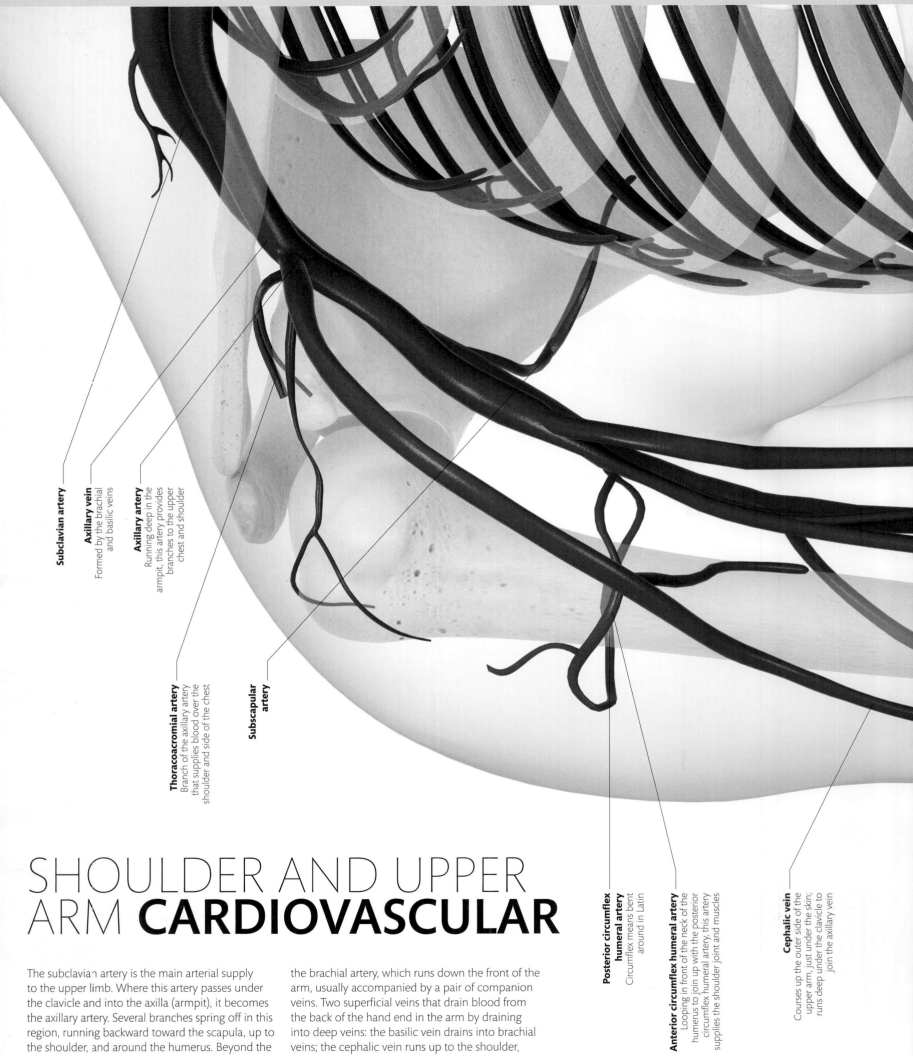

Subclavian artery

Axillary vein
Formed by the brachial
and basilic veins

Axillary artery
Running deep in the
armpit, this artery provides
branches to the upper
chest and shoulder

Thoracoacromial artery
Branch of the axillary artery
that supplies blood over the
shoulder and side of the chest

**Subscapular
artery**

**Posterior circumflex
humeral artery**
Circumflex means bent
around in Latin

Anterior circumflex humeral artery
Looping in front of the neck of the
humerus to join up with the posterior
circumflex humeral artery, this artery
supplies the shoulder joint and muscles

Cephalic vein
Courses up the outer side of the
upper arm, just under the skin;
runs deep under the clavicle to
join the axillary vein

SHOULDER AND UPPER
ARM **CARDIOVASCULAR**

The subclavian artery is the main arterial supply
to the upper limb. Where this artery passes under
the clavicle and into the axilla (armpit), it becomes
the axillary artery. Several branches spring off in this
region, running backward toward the scapula, up to
the shoulder, and around the humerus. Beyond the
armpit, the name of the axillary artery changes to

the brachial artery, which runs down the front of the
arm, usually accompanied by a pair of companion
veins. Two superficial veins that drain blood from
the back of the hand end in the arm by draining
into deep veins: the basilic vein drains into brachial
veins; the cephalic vein runs up to the shoulder,
then plunges deeper to join the axillary vein.

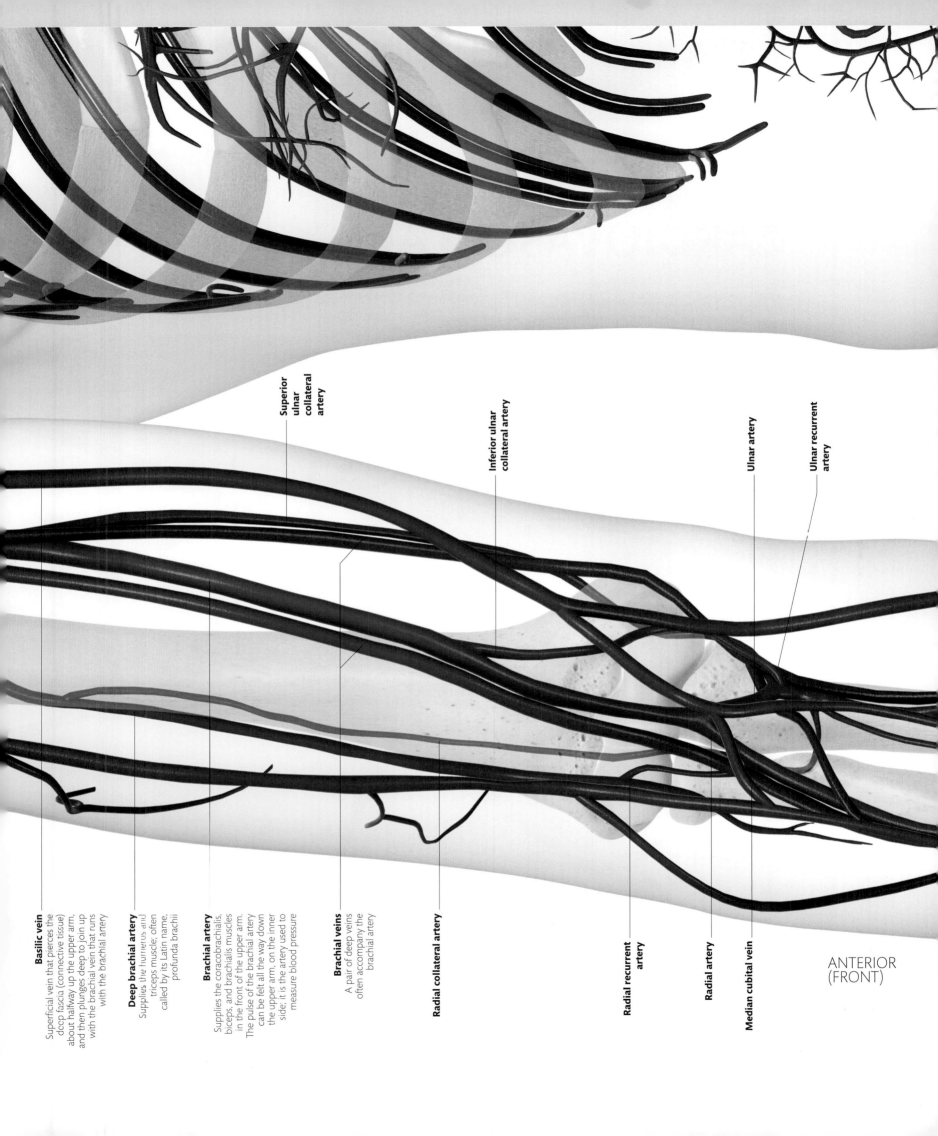

Basilic vein
Superficial vein that pierces the deep fascia (connective tissue) about halfway up the upper arm, and then plunges deep to join up with the brachial vein that runs with the brachial artery

Deep brachial artery
Supplies the humerus and triceps muscle, often called by its Latin name, profunda brachii

Brachial artery
Supplies the coracobrachialis, biceps, and brachialis muscles in the front of the upper arm. The pulse of the brachial artery can be felt all the way down the upper arm, on the inner side; it is the artery used to measure blood pressure

Brachial veins
A pair of deep veins often accompany the brachial artery

Radial collateral artery

Radial recurrent artery

Radial artery

Median cubital vein

Superior ulnar collateral artery

Inferior ulnar collateral artery

Ulnar artery

Ulnar recurrent artery

ANTERIOR
(FRONT)

SHOULDER AND UPPER ARM **CARDIOVASCULAR**

Various branches from the axillary and brachial arteries supply the back of the shoulder and upper arm. The posterior circumflex humeral artery, which runs with the axillary nerve, curls around the upper end of the humerus. The deep brachial artery runs with the radial nerve, spiraling around the back of the bone. From this artery, and from the brachial artery itself, collateral branches run down the arm and join up, or anastomose, with recurrent branches running back up from the ulnar and radial arteries of the forearm. There are also anastomoses (links) between branches of the subclavian and axillary arteries around the shoulder. Anastomoses like this, where branches from different regions join up, can provide alternative routes through which blood can flow if the main vessel becomes squashed or blocked.

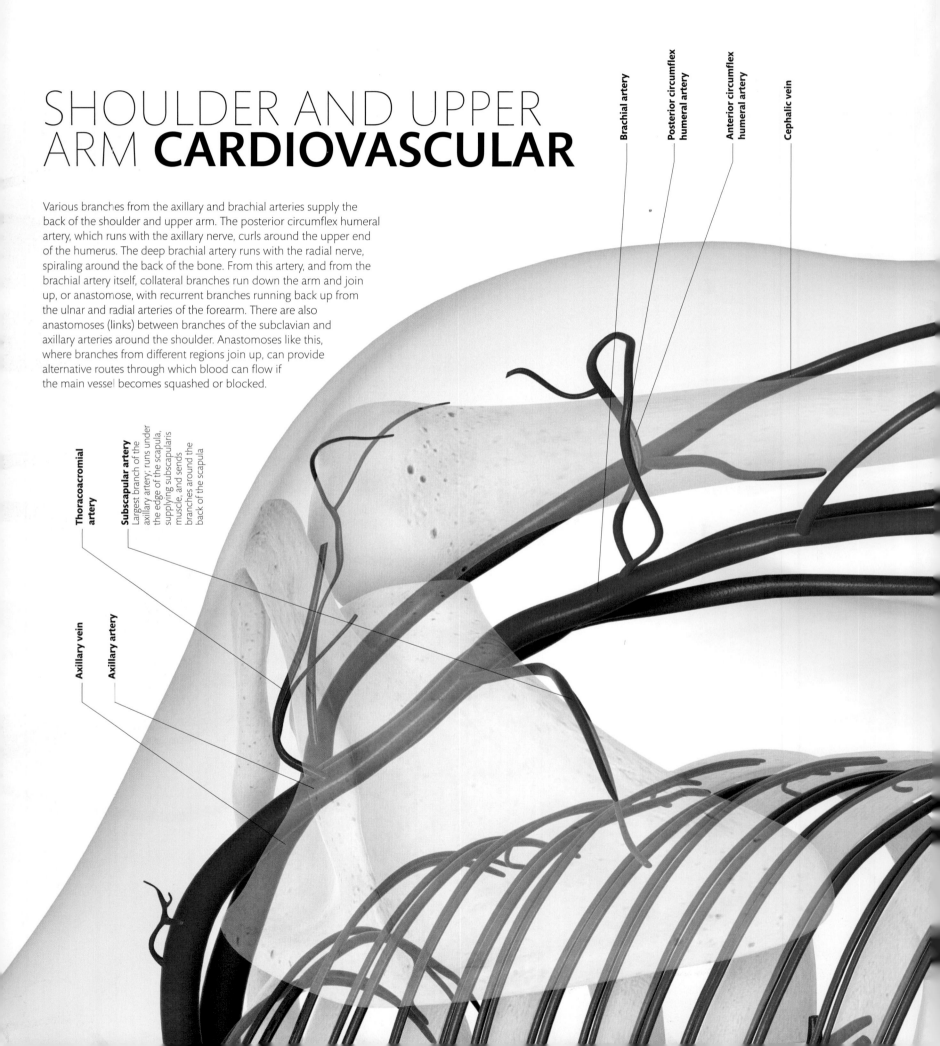

Brachial artery

Posterior circumflex humeral artery

Anterior circumflex humeral artery

Cephalic vein

Thoracoacromial artery

Subscapular artery
Largest branch of the axillary artery; runs under the edge of the scapula, supplying subscapularis muscle, and sends branches around the back of the scapula

Axillary vein

Axillary artery

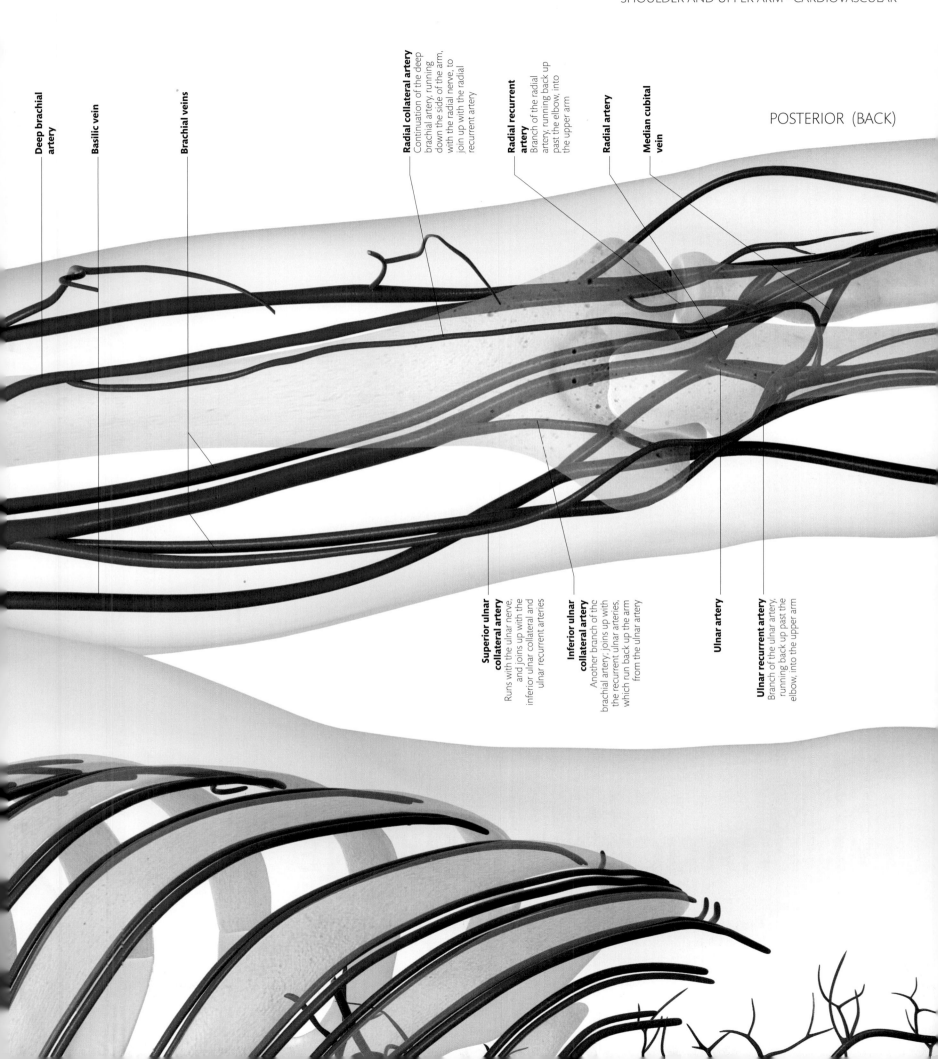

Deep brachial artery

Basilic vein

Brachial veins

Radial collateral artery
Continuation of the deep brachial artery, running down the side of the arm, with the radial nerve, to join up with the radial recurrent artery

Radial recurrent artery
Branch of the radial artery, running back up past the elbow, into the upper arm

Radial artery

Median cubital vein

POSTERIOR (BACK)

Superior ulnar collateral artery
Runs with the ulnar nerve, and joins up with the inferior ulnar collateral and ulnar recurrent arteries

Inferior ulnar collateral artery
Another branch of the brachial artery; joins up with the recurrent ulnar arteries, which run back up the arm from the ulnar artery

Ulnar artery

Ulnar recurrent artery
Branch of the ulnar artery, running back up past the elbow, into the upper arm

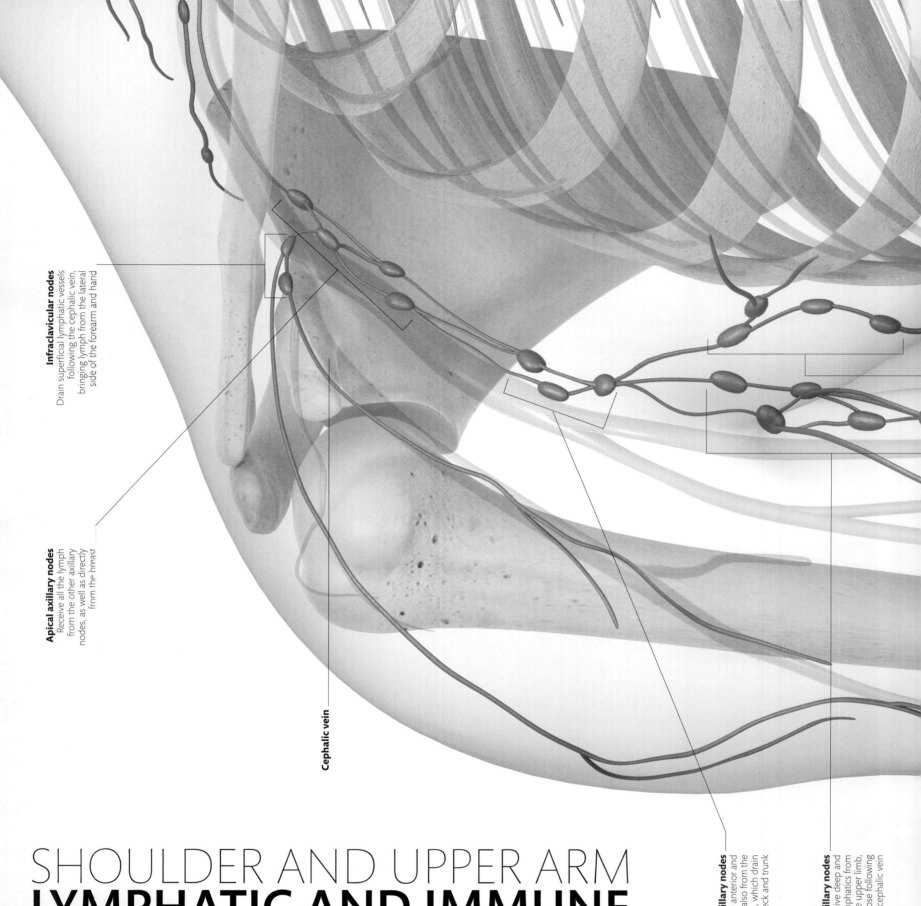

Infraclavicular nodes
Drain superficial lymphatic vessels following the cephalic vein, bringing lymph from the lateral side of the forearm and hand

Apical axillary nodes
Receive all the lymph from the other axillary nodes, as well as directly from the breast

Cephalic vein

Central axillary nodes
Receive lymph from the anterior and lateral axillary nodes; also from the posterior axillary nodes, which drain the back of the neck and trunk

Lateral axillary nodes
Receive deep and superficial lymphatics from most of the upper limb, apart from those following the cephalic vein

SHOULDER AND UPPER ARM
LYMPHATIC AND IMMUNE

Ultimately, all the lymph from the hand, forearm, and arm drains to the axillary nodes in the armpit. But there are a few nodes, lower in the arm, that lymph may pass through on its way to the axilla. The supratrochlear nodes lie in the subcutaneous fat on the inner arm, above the elbow. They collect lymph that has drained from the medial side of the hand and forearm. The infraclavicular nodes, lying along the cephalic vein, below the clavicle, receive lymphatics draining from the thumb and the lateral side of the forearm and arm. Axillary nodes drain lymph from the arm and receive it from the chest wall. They may become infiltrated with cancerous cells spreading from a tumor in the breast.

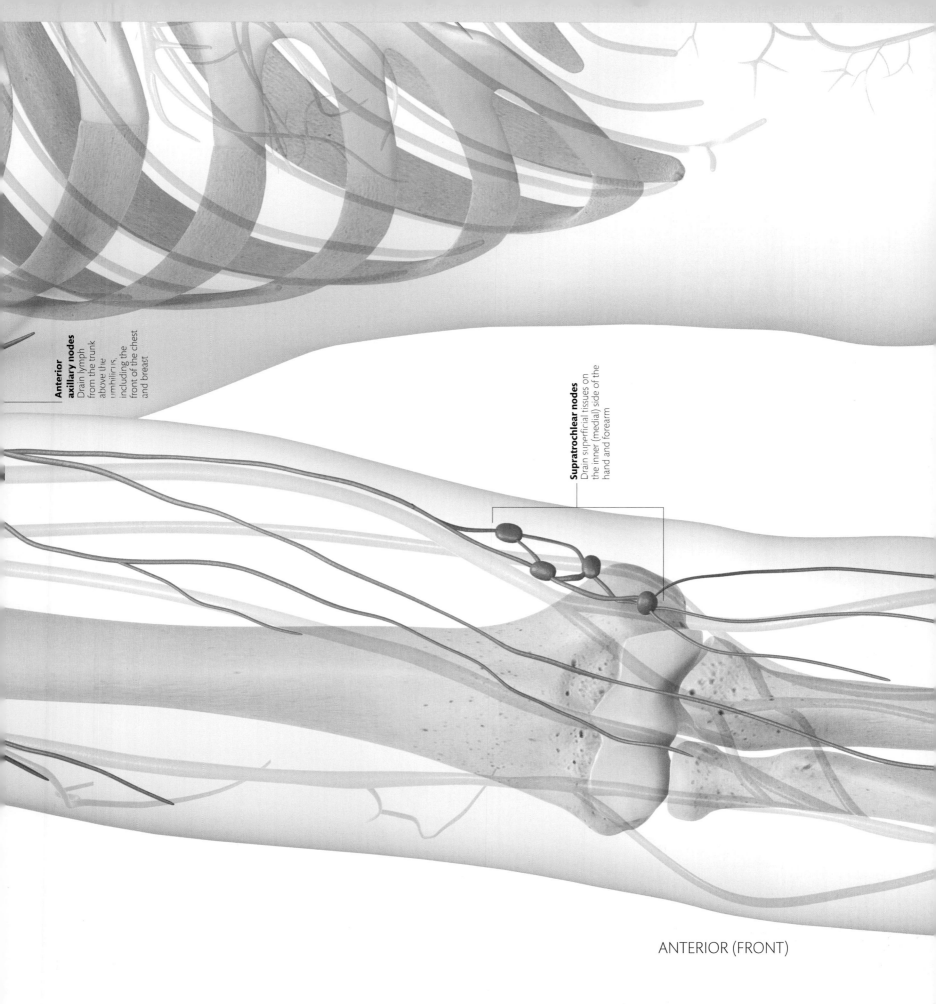

Anterior axillary nodes
Drain lymph from the trunk above the umbilicus, including the front of the chest and breast

Supratrochlear nodes
Drain superficial tissues on the inner (medial) side of the hand and forearm

ANTERIOR (FRONT)

SHOULDER AND UPPER ARM
SHOULDER

The shoulder joint is extremely mobile, allowing the arm to be moved into a wide range of positions. We've inherited this mobility from our ancient ancestors: arboreal apes that needed flexible arms to help them move around in trees. In fact, modern humans are still very good climbers, and the ability to raise the arm above the shoulder is also very useful in throwing and swimming. However, mobility comes at a price. In addition to being a common site for dislocation, the shoulder is frequently affected by degenerative changes.

The most common cause of pain in the shoulder is rotator cuff disease, where muscles and tendons around the joint may become trapped, frayed, and even ruptured. The fluid-filled bursa under the acromion may also be affected.

The axilla (from the Latin for "armpit") is a pyramidal space between the upper part of the humerus and the side of the chest. It contains important nerves and arteries, making their way from the neck, passing under the clavicle, and down into the arm.

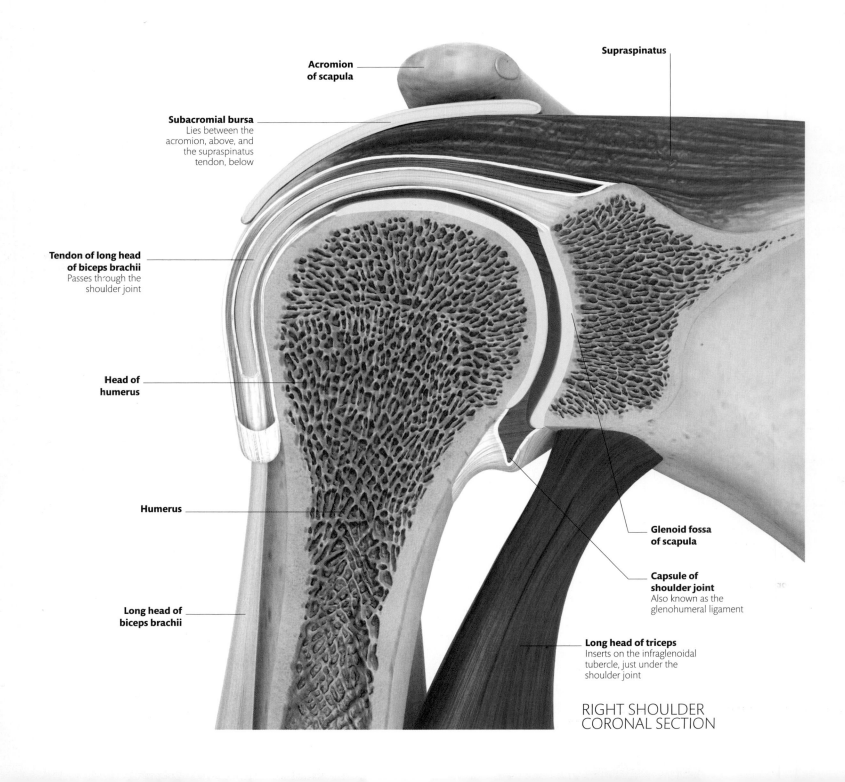

Acromion of scapula

Supraspinatus

Subacromial bursa
Lies between the acromion, above, and the supraspinatus tendon, below

Tendon of long head of biceps brachii
Passes through the shoulder joint

Head of humerus

Humerus

Long head of biceps brachii

Glenoid fossa of scapula

Capsule of shoulder joint
Also known as the glenohumeral ligament

Long head of triceps
Inserts on the infraglenoidal tubercle, just under the shoulder joint

RIGHT SHOULDER
CORONAL SECTION

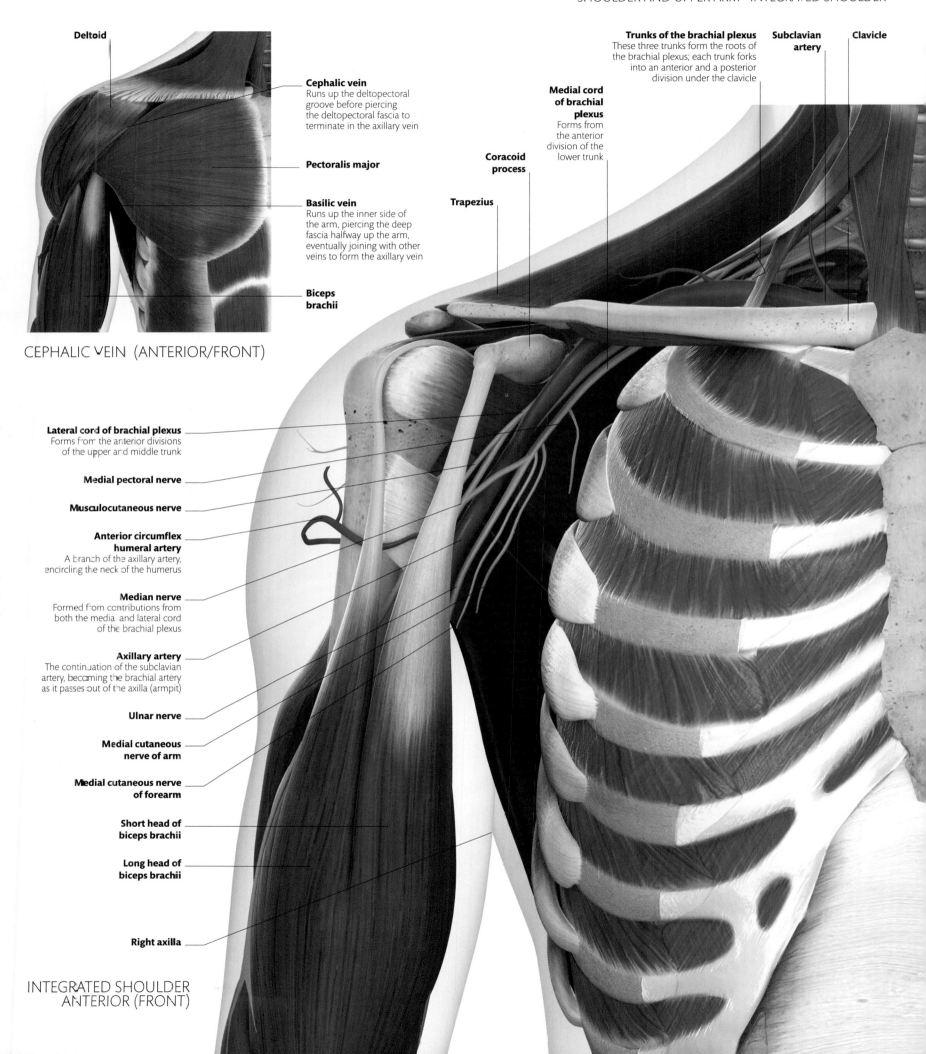

Deltoid

Cephalic vein
Runs up the deltopectoral groove before piercing the deltopectoral fascia to terminate in the axillary vein

Pectoralis major

Basilic vein
Runs up the inner side of the arm, piercing the deep fascia halfway up the arm, eventually joining with other veins to form the axillary vein

Biceps brachii

CEPHALIC VEIN (ANTERIOR/FRONT)

Trunks of the brachial plexus
These three trunks form the roots of the brachial plexus; each trunk forks into an anterior and a posterior division under the clavicle

Subclavian artery

Clavicle

Medial cord of brachial plexus
Forms from the anterior division of the lower trunk

Coracoid process

Trapezius

Lateral cord of brachial plexus
Forms from the anterior divisions of the upper and middle trunk

Medial pectoral nerve

Musculocutaneous nerve

Anterior circumflex humeral artery
A branch of the axillary artery, encircling the neck of the humerus

Median nerve
Formed from contributions from both the medial and lateral cord of the brachial plexus

Axillary artery
The continuation of the subclavian artery, becoming the brachial artery as it passes out of the axilla (armpit)

Ulnar nerve

Medial cutaneous nerve of arm

Medial cutaneous nerve of forearm

Short head of biceps brachii

Long head of biceps brachii

Right axilla

INTEGRATED SHOULDER
ANTERIOR (FRONT)

SHOULDER AND UPPER ARM
ELBOW

The anatomy of the elbow area is clinically very important. There are important nerves here that may become trapped, causing problems in the forearm and hand. The ulnar nerve may become trapped in the cubital tunnel—behind the medial epicondyle of the humerus, where the nerve passes between the humeral and ulnar heads of flexor carpi ulnaris muscle. Much more rarely, the median nerve can be trapped in front of the elbow, where it passes between the heads of pronator teres muscle.

The superficial veins in the front of the elbow, in the area known as the cubital fossa, are common sites for venepuncture—taking blood. The brachial artery, lying medial to the tendon of biceps at the elbow, can be felt as a pulse, and is also commonly used for measurement of blood pressure.

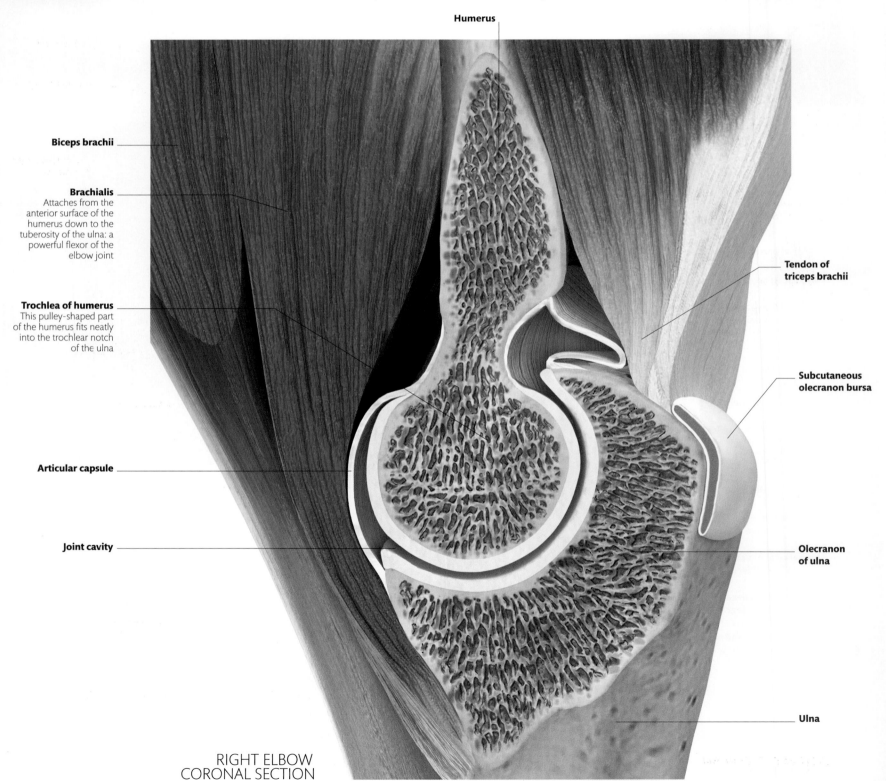

Humerus

Biceps brachii

Brachialis
Attaches from the anterior surface of the humerus down to the tuberosity of the ulna: a powerful flexor of the elbow joint

Trochlea of humerus
This pulley-shaped part of the humerus fits neatly into the trochlear notch of the ulna

Articular capsule

Joint cavity

Tendon of triceps brachii

Subcutaneous olecranon bursa

Olecranon of ulna

Ulna

RIGHT ELBOW
CORONAL SECTION

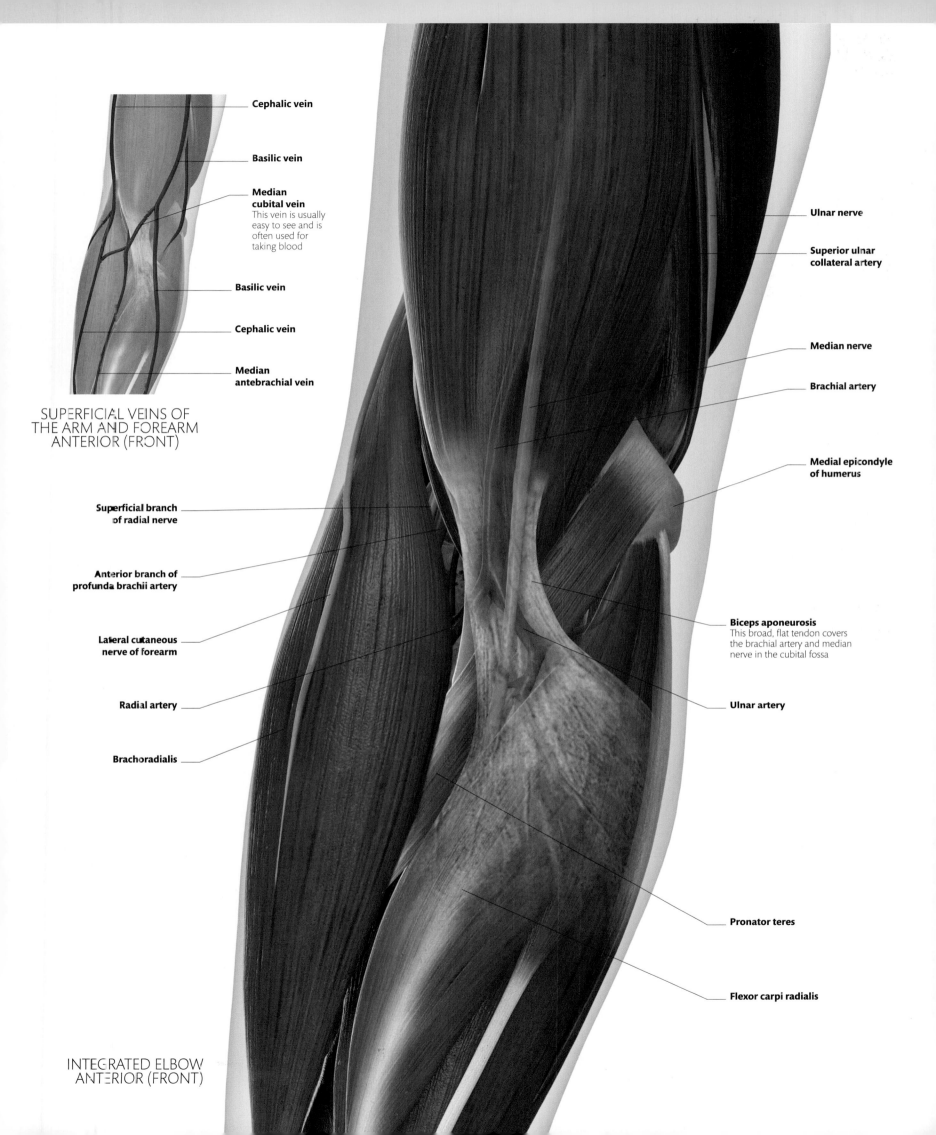

Cephalic vein

Basilic vein

Median cubital vein
This vein is usually easy to see and is often used for taking blood

Basilic vein

Cephalic vein

Median antebrachial vein

SUPERFICIAL VEINS OF
THE ARM AND FOREARM
ANTERIOR (FRONT)

Superficial branch of radial nerve

Anterior branch of profunda brachii artery

Lateral cutaneous nerve of forearm

Radial artery

Brachoradialis

Ulnar nerve

Superior ulnar collateral artery

Median nerve

Brachial artery

Medial epicondyle of humerus

Biceps aponeurosis
This broad, flat tendon covers the brachial artery and median nerve in the cubital fossa

Ulnar artery

Pronator teres

Flexor carpi radialis

INTEGRATED ELBOW
ANTERIOR (FRONT)

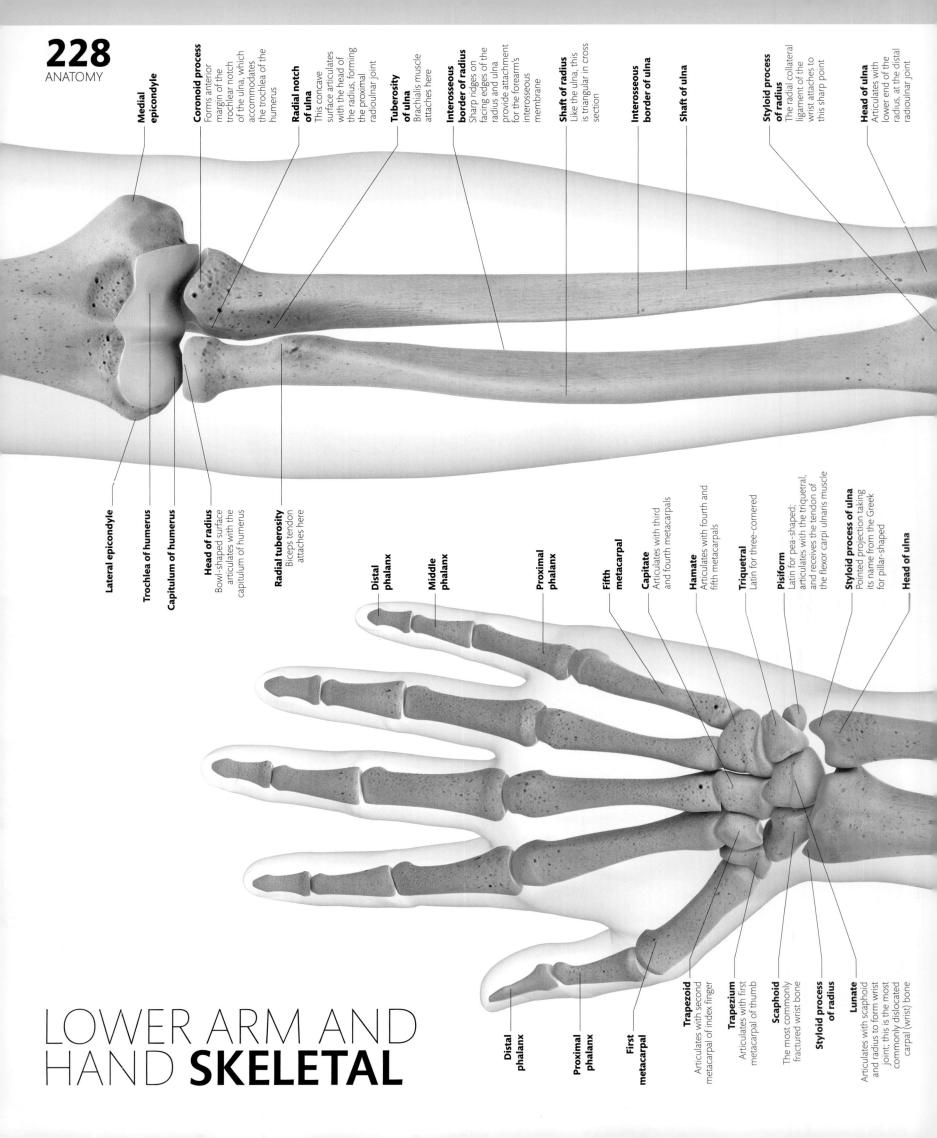

Medial epicondyle

Coronoid process
Forms anterior margin of the trochlear notch of the ulna, which accommodates the trochlea of the humerus

Radial notch of ulna
This concave surface articulates with the head of the radius, forming the proximal radioulnar joint

Tuberosity of ulna
Brachialis muscle attaches here

Interosseous border of radius
Sharp ridges on facing edges of the radius and ulna provide attachment for the forearm's interosseous membrane

Shaft of radius
Like the ulna, this is triangular in cross section

Interosseous border of ulna

Shaft of ulna

Styloid process of radius
The radial collateral ligament of the wrist attaches to this sharp point

Head of ulna
Articulates with lower end of the radius, at the distal radioulnar joint

Lateral epicondyle

Trochlea of humerus

Capitulum of humerus

Head of radius
Bowl-shaped surface articulates with the capitulum of humerus

Radial tuberosity
Biceps tendon attaches here

Distal phalanx

Middle phalanx

Proximal phalanx

Fifth metacarpal

Capitate
Articulates with third and fourth metacarpals

Hamate
Articulates with fourth and fifth metacarpals

Triquetral
Latin for three-cornered

Pisiform
Latin for pea-shaped; articulates with the triquetral, and receives the tendon of the flexor carpi ulnaris muscle

Styloid process of ulna
Pointed projection taking its name from the Greek for pillar-shaped

Head of ulna

Distal phalanx

Proximal phalanx

First metacarpal

Trapezoid
Articulates with second metacarpal of index finger

Trapezium
Articulates with first metacarpal of thumb

Scaphoid
The most commonly fractured wrist bone

Styloid process of radius

Lunate
Articulates with scaphoid and radius to form wrist joint; this is the most commonly dislocated carpal (wrist) bone

LOWER ARM AND HAND **SKELETAL**

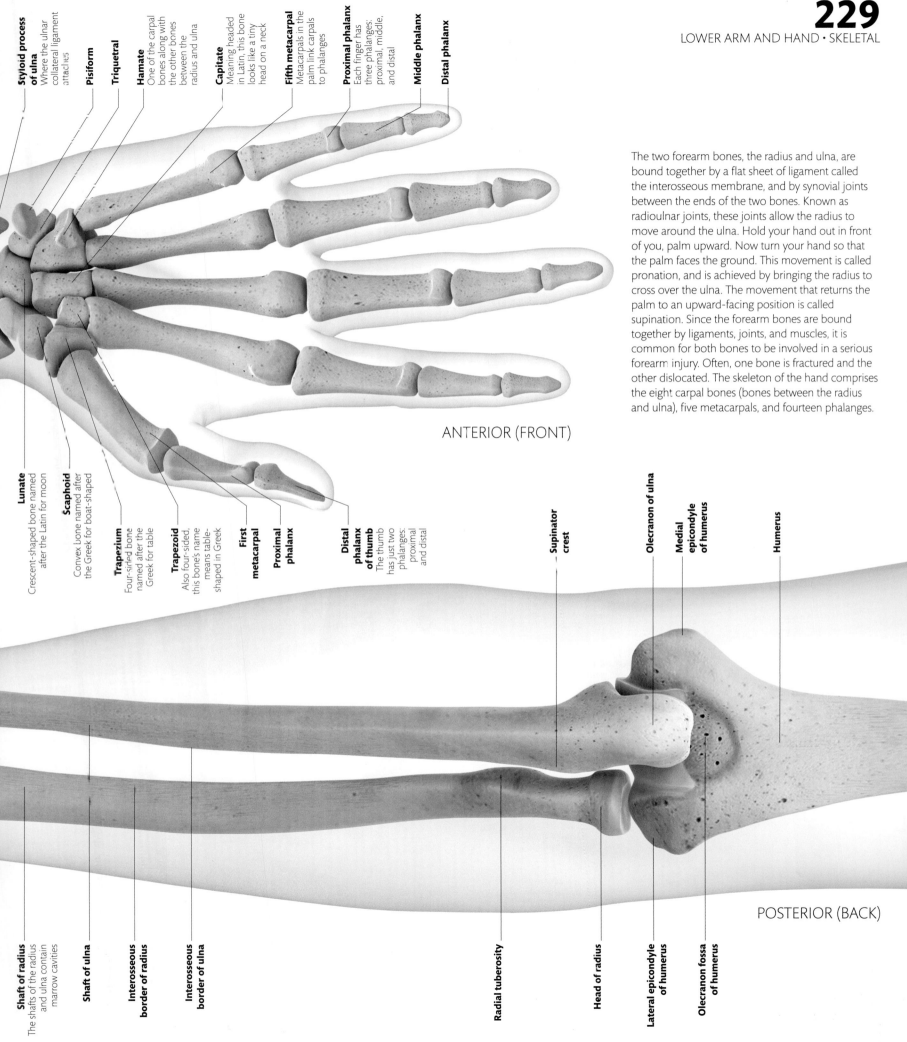

Styloid process of ulna
Where the ulnar collateral ligament attaches

Pisiform

Triquetral

Hamate
One of the carpal bones along with the other bones between the radius and ulna

Capitate
Meaning headed in Latin, this bone looks like a tiny head on a neck

Fifth metacarpal
Metacarpals in the palm link carpals to phalanges

Proximal phalanx
Each finger has three phalanges: proximal, middle, and distal

Middle phalanx

Distal phalanx

The two forearm bones, the radius and ulna, are bound together by a flat sheet of ligament called the interosseous membrane, and by synovial joints between the ends of the two bones. Known as radioulnar joints, these joints allow the radius to move around the ulna. Hold your hand out in front of you, palm upward. Now turn your hand so that the palm faces the ground. This movement is called pronation, and is achieved by bringing the radius to cross over the ulna. The movement that returns the palm to an upward-facing position is called supination. Since the forearm bones are bound together by ligaments, joints, and muscles, it is common for both bones to be involved in a serious forearm injury. Often, one bone is fractured and the other dislocated. The skeleton of the hand comprises the eight carpal bones (bones between the radius and ulna), five metacarpals, and fourteen phalanges.

ANTERIOR (FRONT)

Lunate
Crescent-shaped bone named after the Latin for moon

Scaphoid
Convex bone named after the Greek for boat-shaped

Trapezium
Four-sided bone named after the Greek for table

Trapezoid
Also four-sided, this bone's name means table-shaped in Greek

First metacarpal

Proximal phalanx

Distal phalanx of thumb
The thumb has just two phalanges: proximal and distal

Supinator crest

Olecranon of ulna

Medial epicondyle of humerus

Humerus

POSTERIOR (BACK)

Shaft of radius
The shafts of the radius and ulna contain marrow cavities

Shaft of ulna

Interosseous border of radius

Interosseous border of ulna

Radial tuberosity

Head of radius

Lateral epicondyle of humerus

Olecranon fossa of humerus

LOWER ARM AND HAND **SKELETAL**

HAND AND WRIST JOINTS

The radius widens out at its distal (lower) end to form the wrist joint with the closest two carpal bones, the lunate and scaphoid. This joint allows flexion, extension, adduction, and abduction (see p.34). There are also synovial joints (see p.49) between the carpal bones in the wrist, which increase the range of motion during wrist flexion and extension. Synovial joints between metacarpals and phalanges allow us to spread or close our fingers, as well as flexing or extending the whole finger. Joints between the individual finger bones or phalanges enable fingers to bend and straighten.

In common with many other primates, humans have opposable thumbs. The joints at the base of the thumb are shaped differently from those of the fingers. The joint between the metacarpal of the thumb and the wrist bones is especially mobile and allows the thumb to be brought across the palm of the hand so that the tip of the thumb can touch the other fingertips.

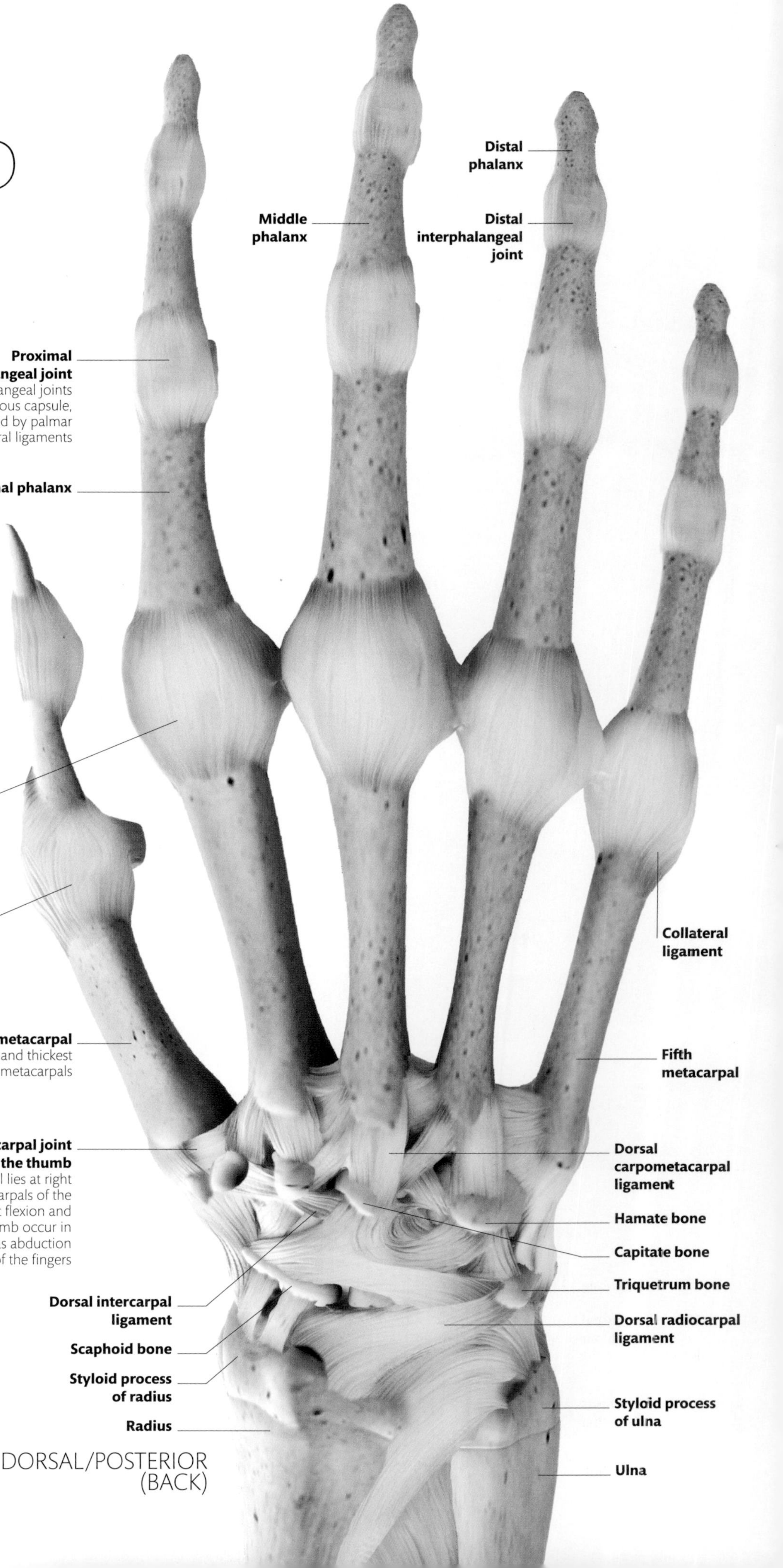

Distal phalanx

Middle phalanx

Distal interphalangeal joint

Proximal interphalangeal joint
The interphalangeal joints have a fibrous capsule, strengthened by palmar and collateral ligaments

Proximal phalanx

Metacarpophalangeal joint
These joints allow about 90 degrees of flexion, a very small amount of extension, and about 30 degrees of abduction and adduction of the metacarpals

Metacarpophalangeal joint of thumb
Allows about 60 degrees of flexion, a little extension, as well as abduction and adduction

Collateral ligament

Joint capsule

Metacarpophalangeal joint

First metacarpal
The shortest and thickest of the metacarpals

Fifth metacarpal

Proximal interphalangeal joint

Carpometacarpal joint of the thumb
The first metacarpal lies at right angles to the metacarpals of the fingers, so that flexion and extension of the thumb occur in the same plane as abduction and adduction of the fingers

Dorsal carpometacarpal ligament

Hamate bone

Capitate bone

Triquetrum bone

Dorsal intercarpal ligament

Dorsal radiocarpal ligament

Scaphoid bone

Distal interphalangeal joint

Styloid process of radius

Styloid process of ulna

Radius

Ulna

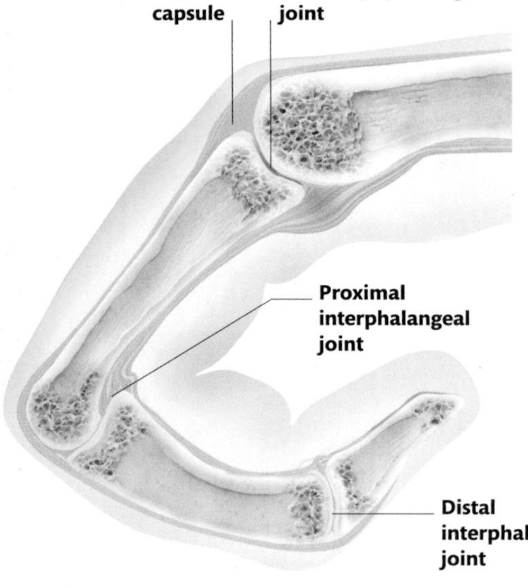

FINGER (SAGITTAL SECTION)

DORSAL/POSTERIOR (BACK)

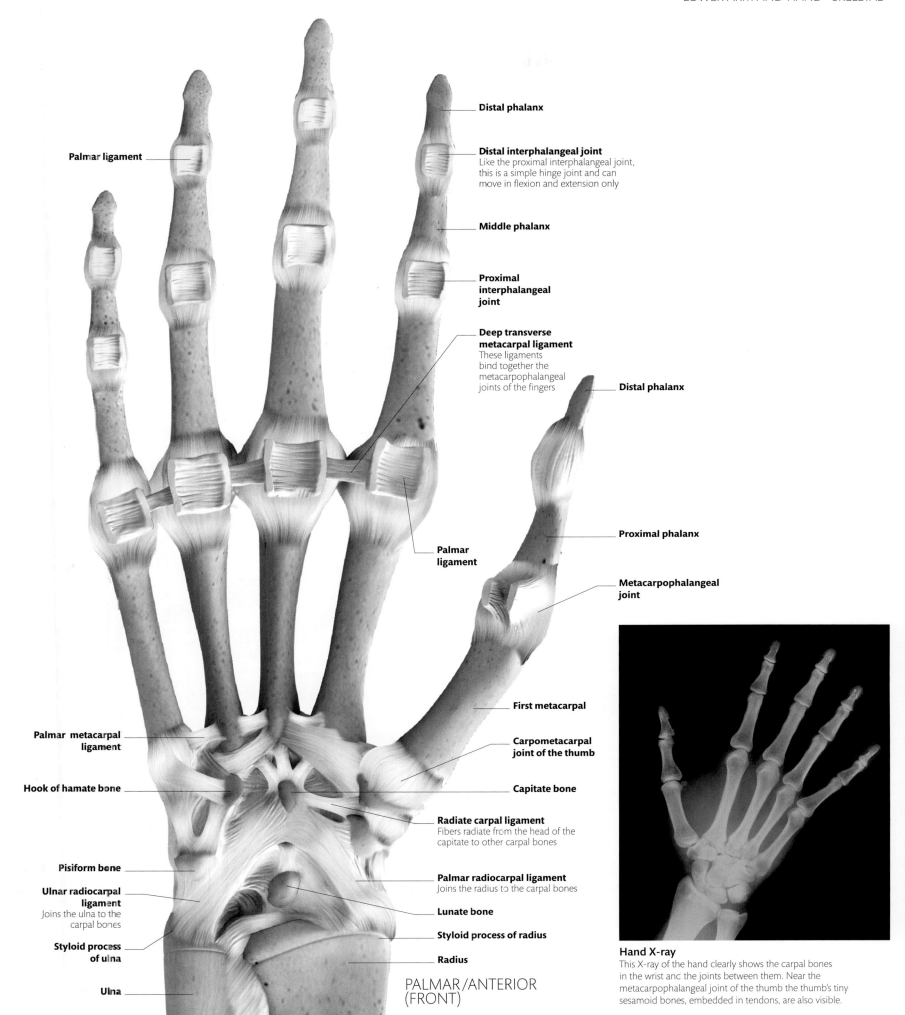

Distal phalanx

Distal interphalangeal joint
Like the proximal interphalangeal joint, this is a simple hinge joint and can move in flexion and extension only

Middle phalanx

Proximal interphalangeal joint

Deep transverse metacarpal ligament
These ligaments bind together the metacarpophalangeal joints of the fingers

Distal phalanx

Palmar ligament

Proximal phalanx

Palmar ligament

Metacarpophalangeal joint

First metacarpal

Palmar metacarpal ligament

Carpometacarpal joint of the thumb

Hook of hamate bone

Capitate bone

Radiate carpal ligament
Fibers radiate from the head of the capitate to other carpal bones

Pisiform bone

Ulnar radiocarpal ligament
Joins the ulna to the carpal bones

Palmar radiocarpal ligament
Joins the radius to the carpal bones

Lunate bone

Styloid process of radius

Styloid process of ulna

Radius

Ulna

PALMAR/ANTERIOR
(FRONT)

Hand X-ray
This X-ray of the hand clearly shows the carpal bones in the wrist and the joints between them. Near the metacarpophalangeal joint of the thumb the thumb's tiny sesamoid bones, embedded in tendons, are also visible.

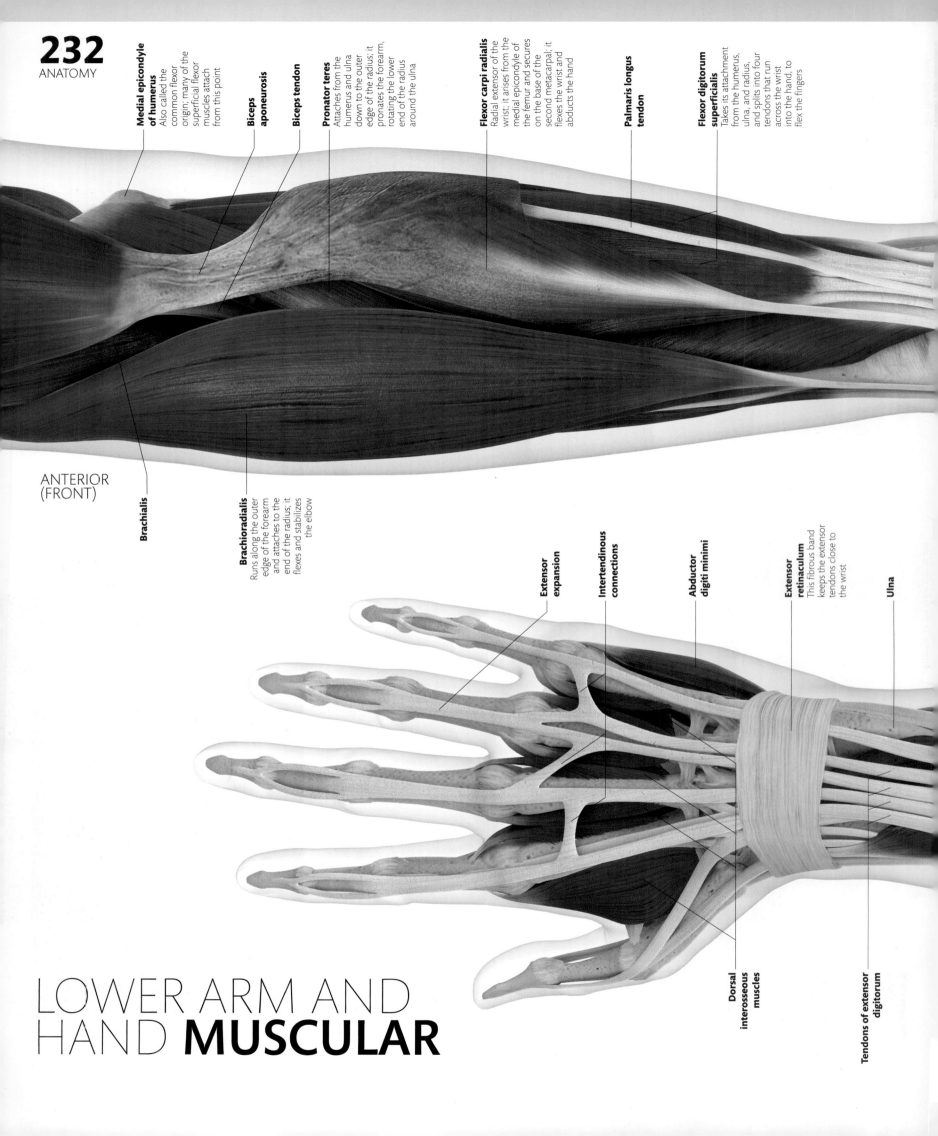

Medial epicondyle of humerus
Also called the common flexor origin; many of the superficial flexor muscles attach from this point

Biceps aponeurosis

Biceps tendon

Pronator teres
Attaches from the humerus and ulna down to the outer edge of the radius; it pronates the forearm, rotating the lower end of the radius around the ulna

Flexor carpi radialis
Radial extensor of the wrist; it arises from the medial epicondyle of the femur and secures on the base of the second metacarpal; it flexes the wrist and abducts the hand

Palmaris longus tendon

Flexor digitorum superficialis
Takes its attachment from the humerus, ulna, and radius, and splits into four tendons that run across the wrist into the hand, to flex the fingers

ANTERIOR
(FRONT)

Brachialis

Brachioradialis
Runs along the outer edge of the forearm and attaches to the end of the radius; it flexes and stabilizes the elbow

Extensor expansion

Intertendinous connections

Abductor digiti minimi

Extensor retinaculum
This fibrous band keeps the extensor tendons close to the wrist

Ulna

Dorsal interosseous muscles

Tendons of extensor digitorum

LOWER ARM AND HAND **MUSCULAR**

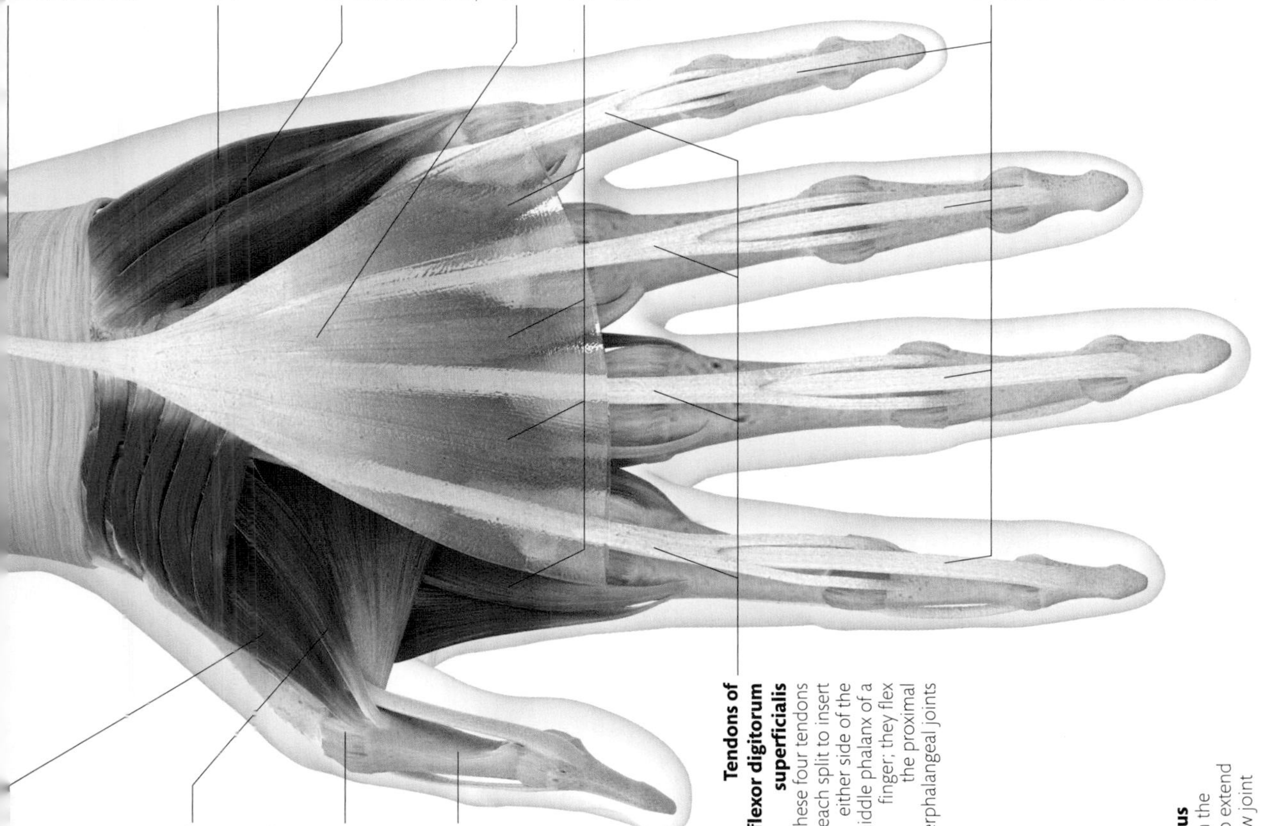

This fibrous band keeps the flexor tendons close to the wrist and stops them from bow stringing outward

Abductor digiti minimi

Flexor digiti minimi brevis
Short flexor of the little finger; it flexes the little finger's metacarpophalangeal joint

Palmar aponeurosis

Lumbricals
These small muscles are named after the Latin for worm

Tendons of flexor digitorum profundus
These tendons emerge through the superficial tendon and continue on, to attach to a distal phalanx; they flex the distal interphalangeal joints of the fingers

SUPERFICIAL MUSCLES

There are five superficial muscles on the front of the forearm, all taking their attachment from the medial epicondyle of the humerus. Pronator teres attaches across to the radius, and can pull this bone into pronation (held with the palm turned downward). The other muscles run farther down the forearm, becoming slender tendons that attach around the wrist, or continue into the hand. Flexor digitorum superficialis splits into four tendons, one for each finger. On the back of the forearm, seven superficial extensor muscles attach to the lateral epicondyle of the humerus. Most of these tendons run down to the wrist or into the hand.

Flexor pollicis brevis
Attaches to the base of the proximal phalanx of the thumb; it flexes the thumb's metacarpophalangeal joint

Metacarpophalangeal joint

First proximal phalanx

Tendons of flexor digitorum superficialis
These four tendons each split to insert either side of the middle phalanx of a finger; they flex the proximal interphalangeal joints

Attaches to the outer side of the base of the proximal phalanx of the thumb; with the palm facing up, it pulls the thumb away from the palm and fingers

Anconeus
Acts with the triceps to extend the elbow joint

Olecranon

Triceps

POSTERIOR
(BACK)

Extensor digiti minimi
The tendon of this extensor of the little finger joins the tendon of the extensor digitorum on the back of the little finger

Extensor digitorum
Extensor of the fingers; it takes its attachment from the lateral epicondyle and becomes four tendons that fan out over the back of the fingers, forming the "extensor expansion"

Extensor carpi ulnaris
Ulnar extensor of the wrist; it arises from the lateral epicondyle and attaches to the base of the fifth metacarpal; it extends the wrist and adducts the hand

Extensor carpi radialis brevis
Short extensor of the wrist; attaches from the lateral epicondyle to the third metacarpal in the hand

Extensor carpi radialis longus
Long extensor of the wrist; it attaches from the lateral supracondylar ridge all the way down to the base of the second metacarpal

Lateral epicondyle of humerus
Referred to as the common extensor origin—many forearm extensor muscles attach here

Brachioradialis

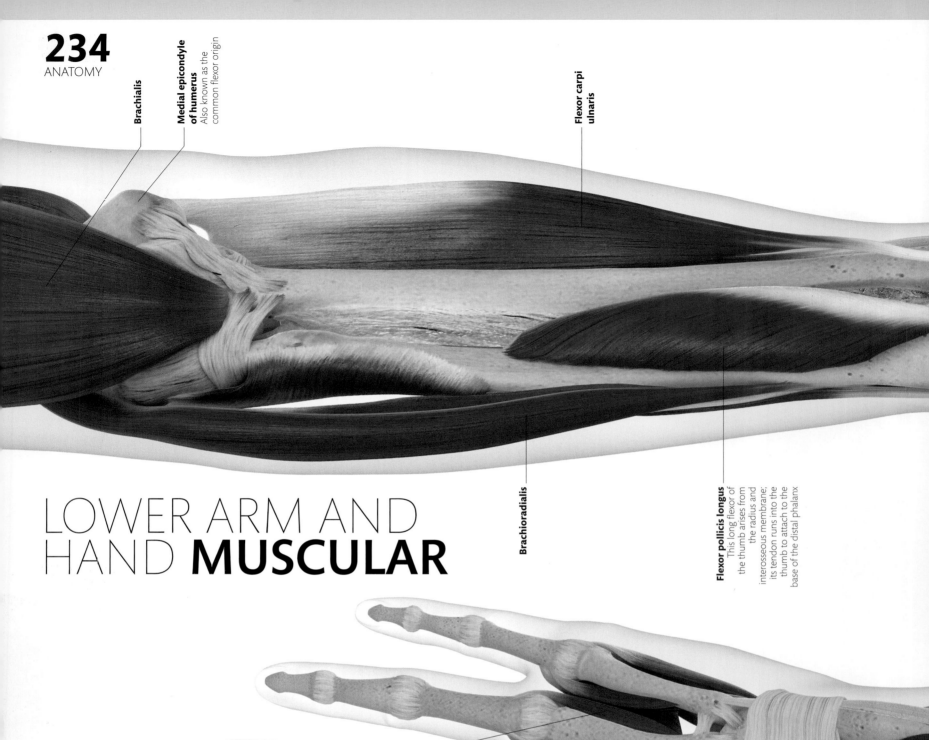

Brachialis

Medial epicondyle of humerus
Also known as the common flexor origin

Flexor carpi ulnaris

Brachioradialis

Flexor pollicis longus
This long flexor of the thumb arises from the radius and interosseous membrane; its tendon runs into the thumb to attach to the base of the distal phalanx

LOWER ARM AND HAND **MUSCULAR**

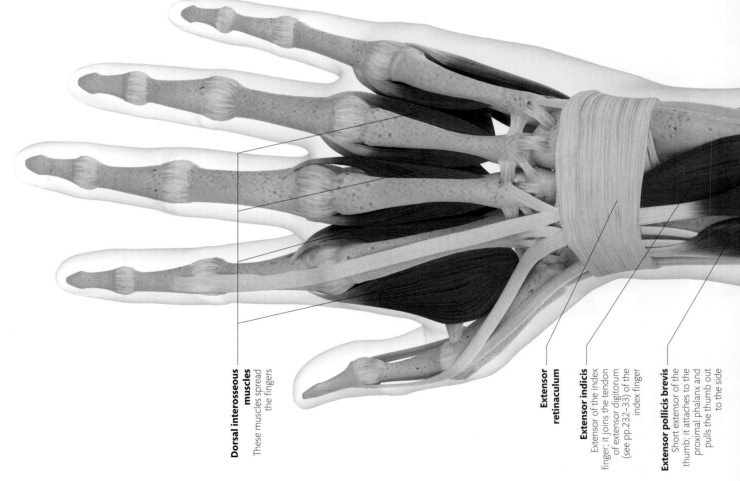

Dorsal interosseous muscles
These muscles spread the fingers

Extensor retinaculum

Extensor indicis
Extensor of the index finger; it joins the tendon of extensor digitorum (see pp.232–33) of the index finger

Extensor pollicis brevis
Short extensor of the thumb; it attaches to the proximal phalanx and pulls the thumb out to the side

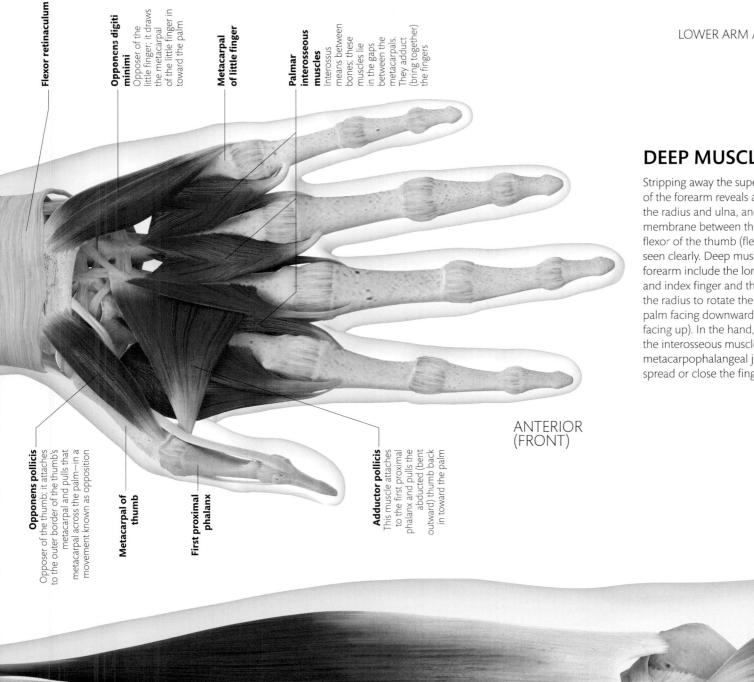

Flexor retinaculum

Opponens digiti minimi
Opposer of the little finger; it draws the metacarpal of the little finger in toward the palm

Metacarpal of little finger

Palmar interosseous muscles
Interossus means between bones; these muscles lie in the gaps between the metacarpals. They adduct (bring together) the fingers

Opponens pollicis
Opposer of the thumb; it attaches to the outer border of the thumb's metacarpal and pulls that metacarpal across the palm—in a movement known as opposition

Metacarpal of thumb

First proximal phalanx

Adductor pollicis
This muscle attaches to the first proximal phalanx and pulls the abducted (bent outward) thumb back in toward the palm

ANTERIOR
(FRONT)

DEEP MUSCLES

Stripping away the superficial muscles on the front of the forearm reveals a deeper layer attaching to the radius and ulna, and to the interosseous membrane between the bones. The long, quill-like flexor of the thumb (flexor pollicis longus) can be seen clearly. Deep muscles on the back of the forearm include the long extensors of the thumb and index finger and the supinator, which pulls on the radius to rotate the pronated arm (held with palm facing downward) into supination (with palm facing up). In the hand, a deep dissection reveals the interosseous muscles that act on the metacarpophalangeal joints in order to either spread or close the fingers.

POSTERIOR
(BACK)

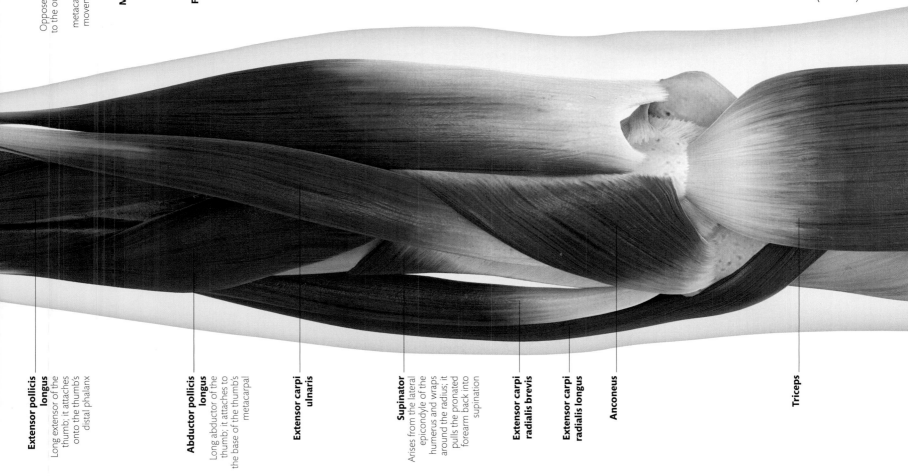

Extensor pollicis longus
Long extensor of the thumb; it attaches onto the thumb's distal phalanx

Abductor pollicis longus
Long abductor of the thumb; it attaches to the base of the thumb's metacarpal

Extensor carpi ulnaris

Supinator
Arises from the lateral epicondyle of the humerus and wraps around the radius; it pulls the pronated forearm back into supination

Extensor carpi radialis brevis

Extensor carpi radialis longus

Anconeus

Triceps

LOWER ARM AND HAND **NERVOUS**

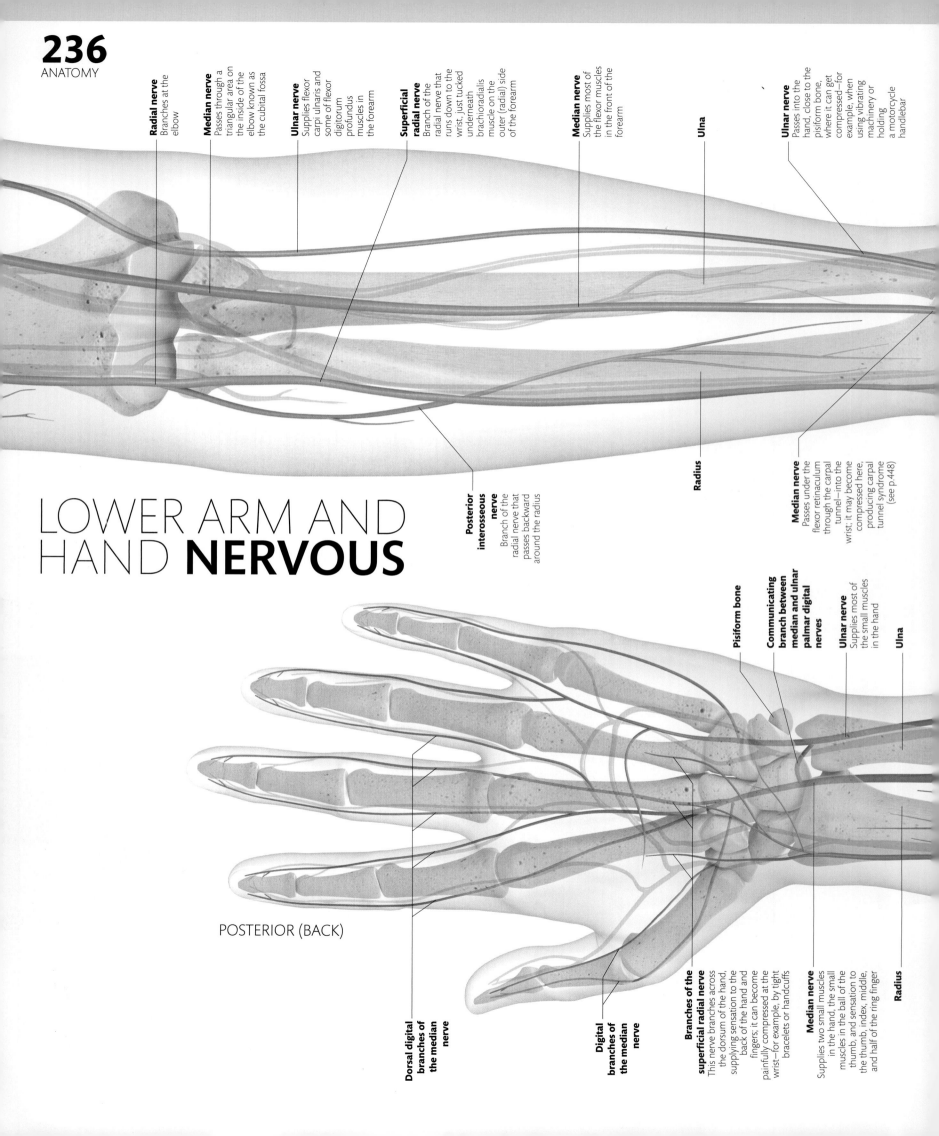

Radial nerve
Branches at the elbow

Median nerve
Passes through a triangular area on the inside of the elbow known as the cubital fossa

Ulnar nerve
Supplies flexor carpi ulnaris and some of flexor digitorum profundus muscles in the forearm

Superficial radial nerve
Branch of the radial nerve that runs down to the wrist, just tucked underneath brachioradialis muscle on the outer (radial) side of the forearm

Median nerve
Supplies most of the flexor muscles in the front of the forearm

Ulna

Ulnar nerve
Passes into the hand, close to the pisiform bone, where it can get compressed—for example, when using vibrating machinery or holding a motorcycle handlebar

Posterior interosseous nerve
Branch of the radial nerve that passes backward around the radius

Radius

Median nerve
Passes under the flexor retinaculum through the carpal tunnel—into the wrist; it may become compressed here, producing carpal tunnel syndrome (see p.448)

Pisiform bone

Communicating branch between median and ulnar palmar digital nerves

Ulnar nerve
Supplies most of the small muscles in the hand

Ulna

POSTERIOR (BACK)

Dorsal digital branches of the median nerve

Branches of the superficial radial nerve
This nerve branches across the dorsum of the hand, supplying sensation to the back of the hand and fingers; it can become painfully compressed at the wrist—for example, by tight bracelets or handcuffs

Digital branches of the median nerve

Median nerve
Supplies two small muscles in the hand, the small muscles in the ball of the thumb, and sensation to the thumb, index, middle, and half of the ring finger

Radius

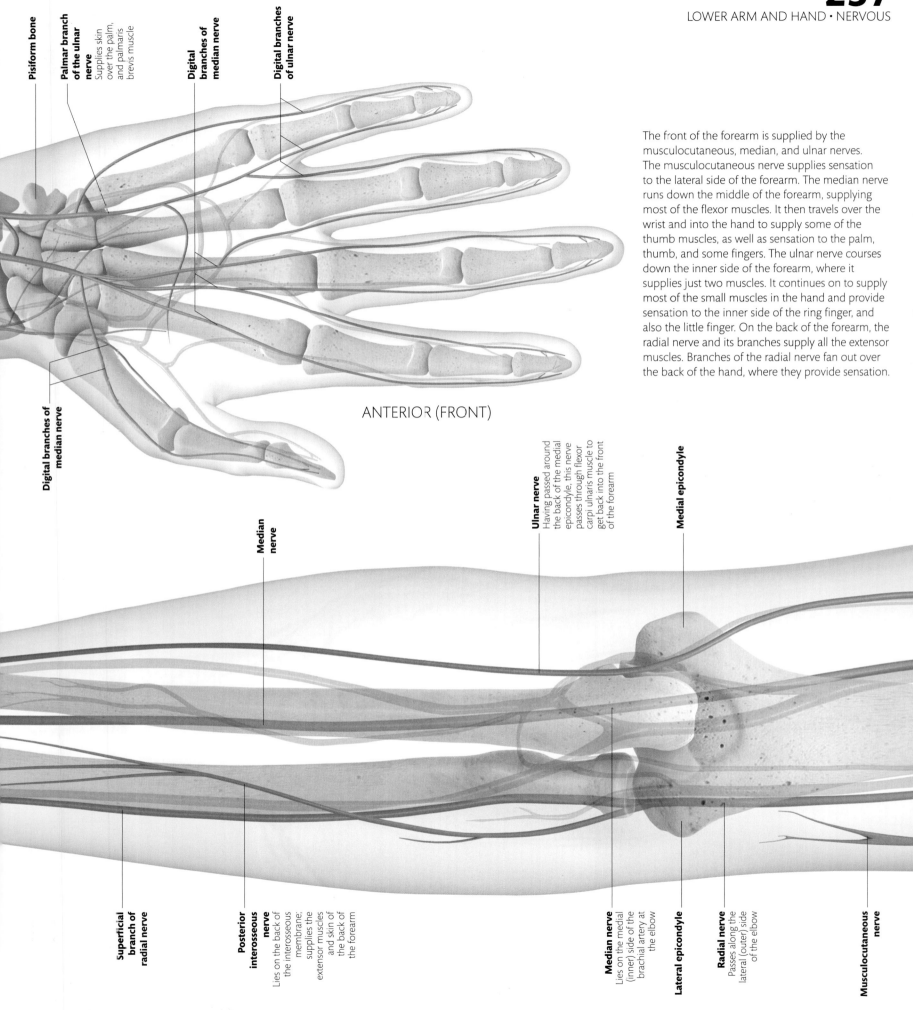

Pisiform bone

Palmar branch of the ulnar nerve
Supplies skin over the palm, and palmaris brevis muscle

Digital branches of median nerve

Digital branches of ulnar nerve

Digital branches of median nerve

ANTERIOR (FRONT)

The front of the forearm is supplied by the musculocutaneous, median, and ulnar nerves. The musculocutaneous nerve supplies sensation to the lateral side of the forearm. The median nerve runs down the middle of the forearm, supplying most of the flexor muscles. It then travels over the wrist and into the hand to supply some of the thumb muscles, as well as sensation to the palm, thumb, and some fingers. The ulnar nerve courses down the inner side of the forearm, where it supplies just two muscles. It continues on to supply most of the small muscles in the hand and provide sensation to the inner side of the ring finger, and also the little finger. On the back of the forearm, the radial nerve and its branches supply all the extensor muscles. Branches of the radial nerve fan out over the back of the hand, where they provide sensation.

Median nerve

Ulnar nerve
Having passed around the back of the medial epicondyle, this nerve passes through flexor carpi ulnaris muscle to get back into the front of the forearm

Medial epicondyle

Superficial branch of radial nerve

Posterior interosseous nerve
Lies on the back of the interosseous membrane; supplies the extensor muscles and skin of the back of the forearm

Median nerve
Lies on the medial (inner) side of the brachial artery at the elbow

Lateral epicondyle

Radial nerve
Passes along the lateral (outer) side of the elbow

Musculocutaneous nerve

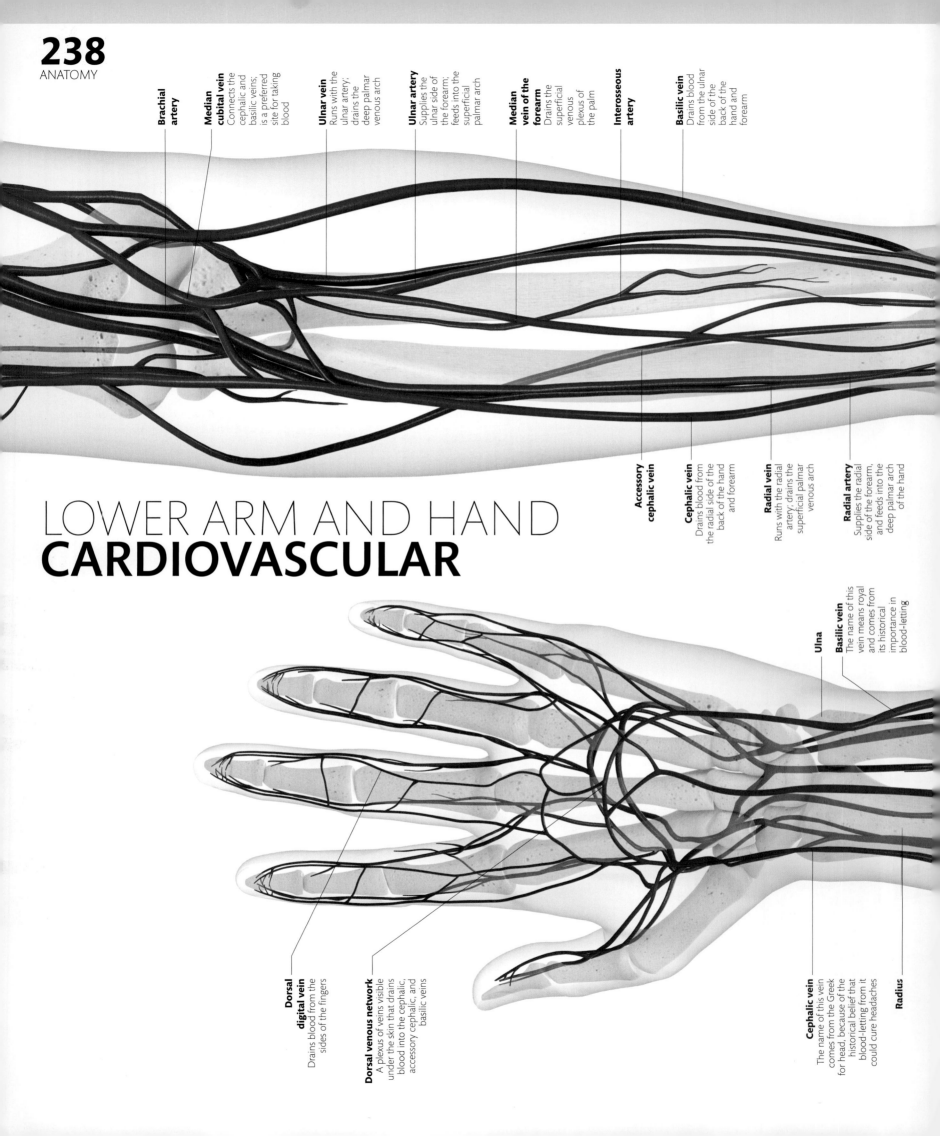

Brachial artery

Median cubital vein
Connects the cephalic and basilic veins; is a preferred site for taking blood

Ulnar vein
Runs with the ulnar artery; drains the deep palmar venous arch

Ulnar artery
Supplies the ulnar side of the forearm; feeds into the superficial palmar arch

Median vein of the forearm
Drains the superficial venous plexus of the palm

Interosseous artery

Basilic vein
Drains blood from the ulnar side of the back of the hand and forearm

Accessory cephalic vein

Cephalic vein
Drains blood from the radial side of the back of the hand and forearm

Radial vein
Runs with the radial artery; drains the superficial palmar venous arch

Radial artery
Supplies the radial side of the forearm, and feeds into the deep palmar arch of the hand

LOWER ARM AND HAND
CARDIOVASCULAR

Ulna

Basilic vein
The name of this vein means royal and comes from its historical importance in blood-letting

Dorsal digital vein
Drains blood from the sides of the fingers

Dorsal venous network
A plexus of veins visible under the skin that drains blood into the cephalic, accessory cephalic, and basilic veins

Cephalic vein
The name of this vein comes from the Greek for head, because of the historical belief that blood-letting from it could cure headaches

Radius

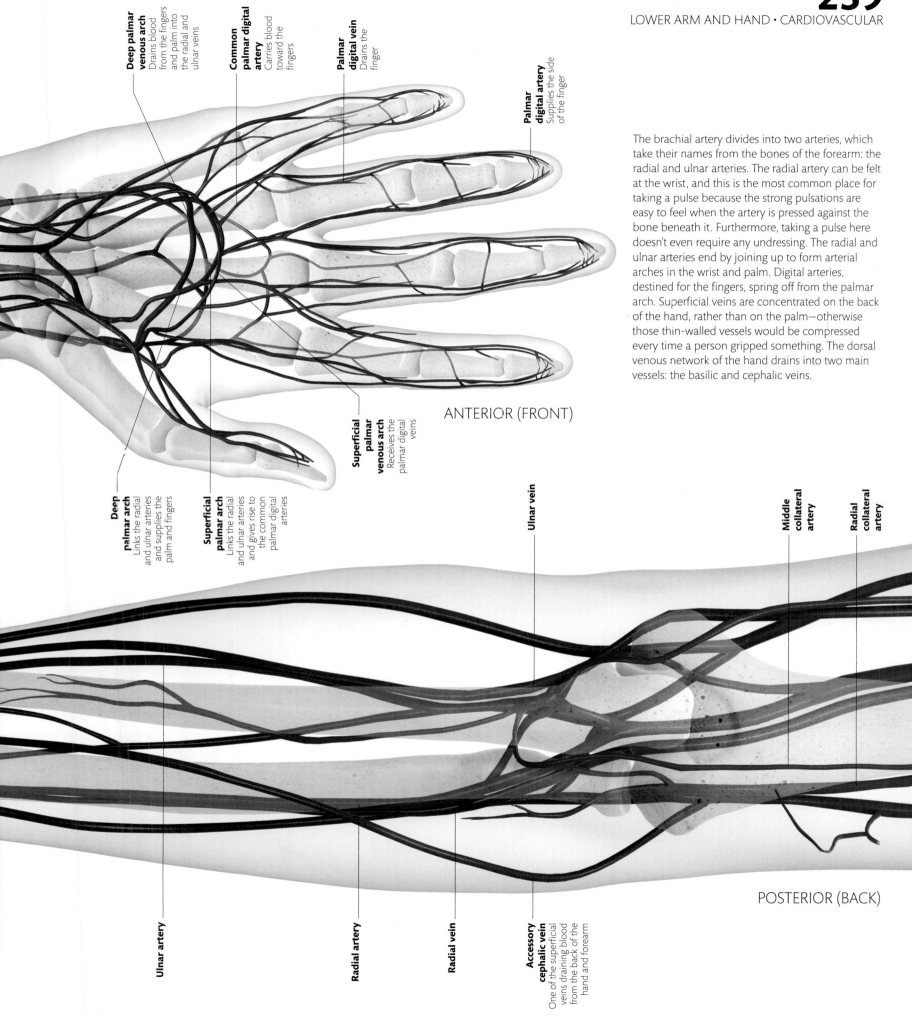

Deep palmar venous arch
Drains blood from the fingers and palm into the radial and ulnar veins

Common palmar digital artery
Carries blood toward the fingers

Palmar digital vein
Drains the finger

Palmar digital artery
Supplies the side of the finger

The brachial artery divides into two arteries, which take their names from the bones of the forearm: the radial and ulnar arteries. The radial artery can be felt at the wrist, and this is the most common place for taking a pulse because the strong pulsations are easy to feel when the artery is pressed against the bone beneath it. Furthermore, taking a pulse here doesn't even require any undressing. The radial and ulnar arteries end by joining up to form arterial arches in the wrist and palm. Digital arteries, destined for the fingers, spring off from the palmar arch. Superficial veins are concentrated on the back of the hand, rather than on the palm—otherwise those thin-walled vessels would be compressed every time a person gripped something. The dorsal venous network of the hand drains into two main vessels: the basilic and cephalic veins.

ANTERIOR (FRONT)

Superficial palmar venous arch
Receives the palmar digital veins

Deep palmar arch
Links the radial and ulnar arteries and supplies the palm and fingers

Superficial palmar arch
Links the radial and ulnar arteries and gives rise to the common palmar digital arteries

Ulnar vein

Middle collateral artery

Radial collateral artery

POSTERIOR (BACK)

Ulnar artery

Radial artery

Radial vein

Accessory cephalic vein
One of the superficial veins draining blood from the back of the hand and forearm

LOWER ARM AND HAND
HAND

The superficial veins that form the dorsal venous plexus, on the back of the hand, are common sites for cannulation—where a small plastic tube is inserted into a vein to deliver fluid directly into the circulation.

Under the skin of the palm of the hand there are long flexor tendons running down from muscles in the forearm to the fingers and thumb, as well as short muscles that arise near the wrist or deep in the palm and insert into the phalanges. The group of short muscles around the base of the thumb form a prominent bulge known as the thenar eminence. On the opposite side of the hand, the smaller muscles running up to the little finger form the hypothenar eminence. The ulnar and radial arteries form connections in the palm of the hand, which in turn lead to the digital arteries that supply the fingers and thumb.

Dorsal digital veins
Drain blood from the sides of the fingers and thumb

Dorsal venous plexus
These superficial veins are usually easy to see and a common site for cannulation (where a tube is inserted to administer fluids or drugs)

Median vein of the forearm

Basilic vein
Drains the medial (ulnar) side of the dorsal venous plexus of the hand

Cephalic vein
Drains the lateral (radial) side of the dorsal venous plexus of the hand

SUPERFICIAL VEINS OF THE RIGHT HAND
(DORSAL VIEW)

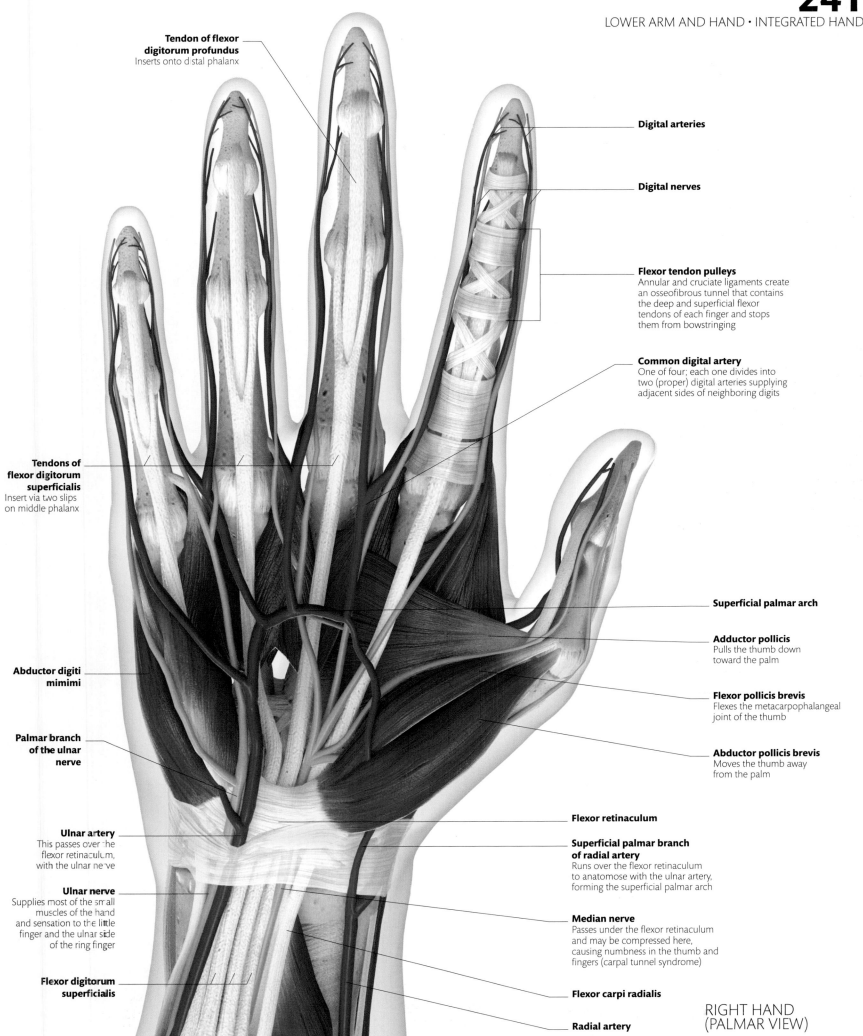

Tendon of flexor digitorum profundus
Inserts onto distal phalanx

Digital arteries

Digital nerves

Flexor tendon pulleys
Annular and cruciate ligaments create an osseofibrous tunnel that contains the deep and superficial flexor tendons of each finger and stops them from bowstringing

Common digital artery
One of four; each one divides into two (proper) digital arteries supplying adjacent sides of neighboring digits

Tendons of flexor digitorum superficialis
Insert via two slips on middle phalanx

Superficial palmar arch

Adductor pollicis
Pulls the thumb down toward the palm

Abductor digiti mimimi

Flexor pollicis brevis
Flexes the metacarpophalangeal joint of the thumb

Palmar branch of the ulnar nerve

Abductor pollicis brevis
Moves the thumb away from the palm

Flexor retinaculum

Ulnar artery
This passes over the flexor retinaculum, with the ulnar nerve

Superficial palmar branch of radial artery
Runs over the flexor retinaculum to anastomose with the ulnar artery, forming the superficial palmar arch

Ulnar nerve
Supplies most of the small muscles of the hand and sensation to the little finger and the ulnar side of the ring finger

Median nerve
Passes under the flexor retinaculum and may be compressed here, causing numbness in the thumb and fingers (carpal tunnel syndrome)

Flexor digitorum superficialis

Flexor carpi radialis

Radial artery

RIGHT HAND
(PALMAR VIEW)

LOWER ARM AND HAND
MRI

These scans of the arm, forearm, and hand show how tightly packed the structures are. Section 1 reveals the bones of the wrist—the carpals—interlocking like a jigsaw. The wrist joint itself is the articulation between the radius and the scaphoid and lunate bones. In section 2, part of the elbow joint is visible, with the bowl-shaped head of the radius cupping the rounded end of the humerus. Muscles in the forearm are grouped into two sets, flexors on the front and extensors behind the forearm bones and interosseous membrane. Compare sections 3–8 with sections through the leg (see pp.286–87)—both limbs have a single bone (humerus or femur) in the upper part, two bones in the lower part (radius and ulna in the forearm; tibia and fibula in the lower leg), a set of bones in the wrist and ankle (carpals and tarsals), fanning out to five digits at the end of the limb. Evolutionarily, these elements developed from the rays of a fish fin.

LEVELS OF SCANS

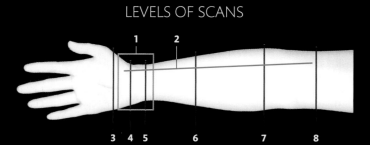

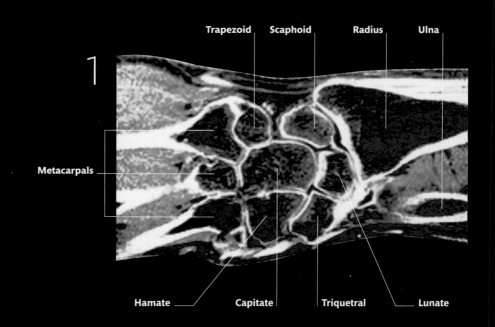

1 — Trapezoid · Scaphoid · Radius · Ulna · Metacarpals · Hamate · Capitate · Triquetral · Lunate

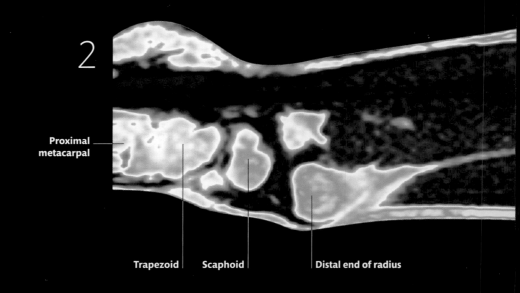

2 — Proximal metacarpal · Trapezoid · Scaphoid · Distal end of radius

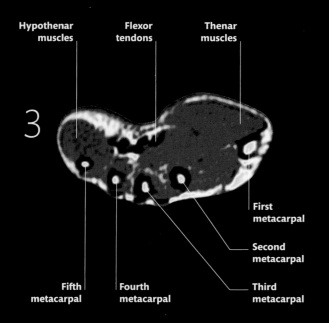

3 — Hypothenar muscles · Flexor tendons · Thenar muscles · First metacarpal · Second metacarpal · Third metacarpal · Fourth metacarpal · Fifth metacarpal

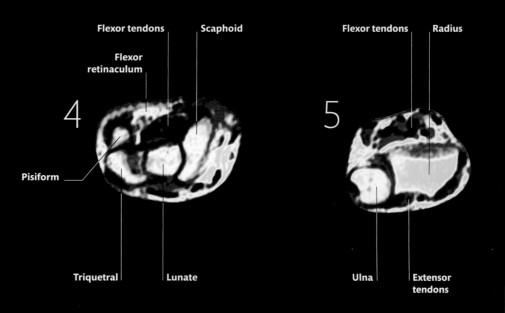

4 — Flexor tendons · Scaphoid · Flexor retinaculum · Pisiform · Triquetral · Lunate

5 — Flexor tendons · Radius · Ulna · Extensor tendons

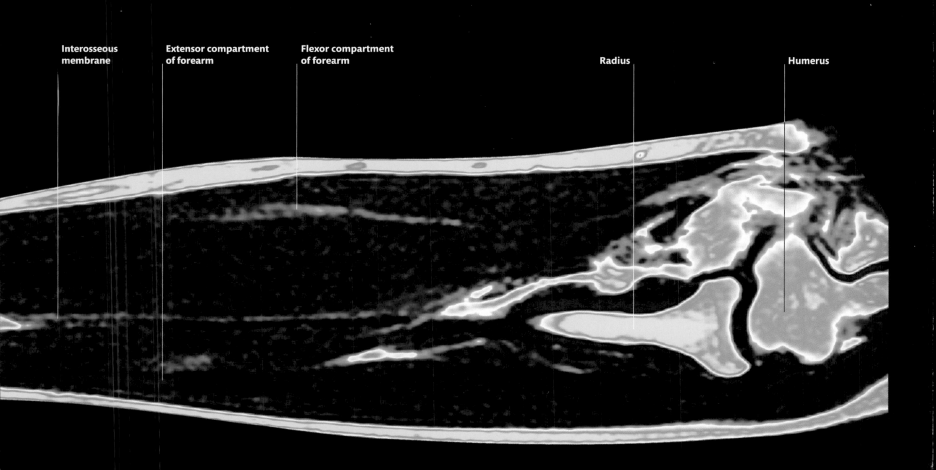

Interosseous membrane

Extensor compartment of forearm

Flexor compartment of forearm

Radius

Humerus

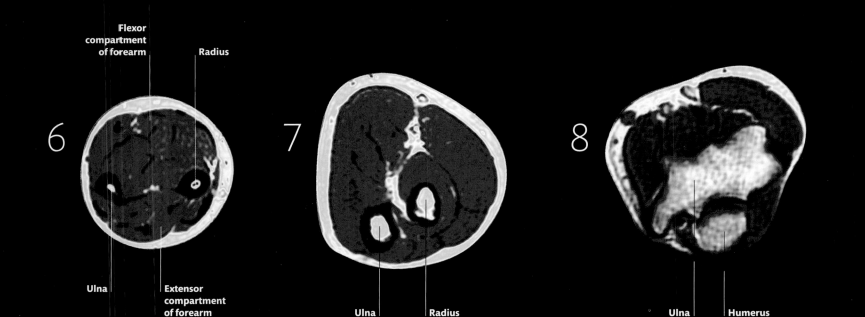

Flexor compartment of forearm

Radius

6

7

8

Ulna

Extensor compartment of forearm

Ulna

Radius

Ulna

Humerus

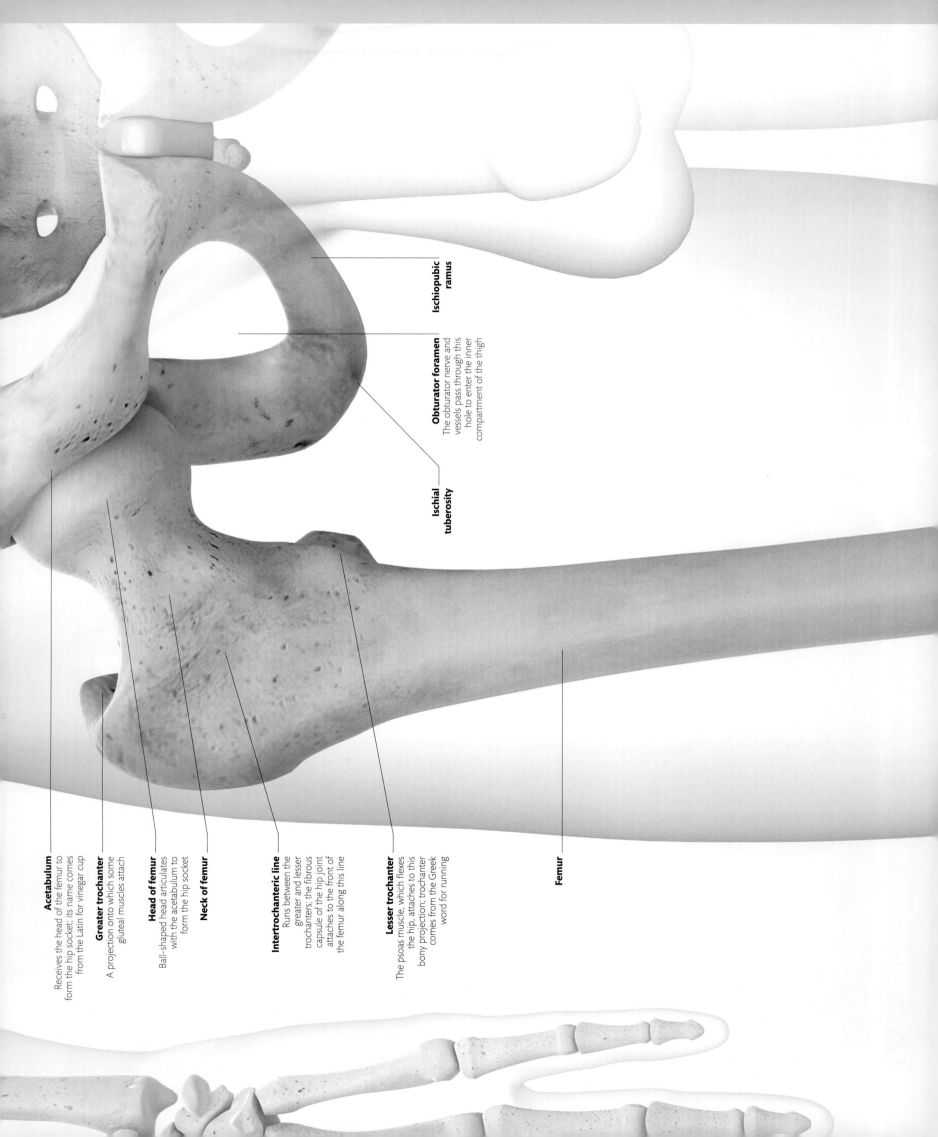

Ischiopubic ramus

Obturator foramen
The obturator nerve and vessels pass through this hole to enter the inner compartment of the thigh

Ischial tuberosity

Acetabulum
Receives the head of the femur to form the hip socket; its name comes from the Latin for vinegar cup

Greater trochanter
A projection onto which some gluteal muscles attach

Head of femur
Ball-shaped head articulates with the acetabulum to form the hip socket

Neck of femur

Intertrochanteric line
Runs between the greater and lesser trochanters; the fibrous capsule of the hip joint attaches to the front of the femur along this line

Lesser trochanter
The psoas muscle, which flexes the hip, attaches to this bony projection; trochanter comes from the Greek word for running

Femur

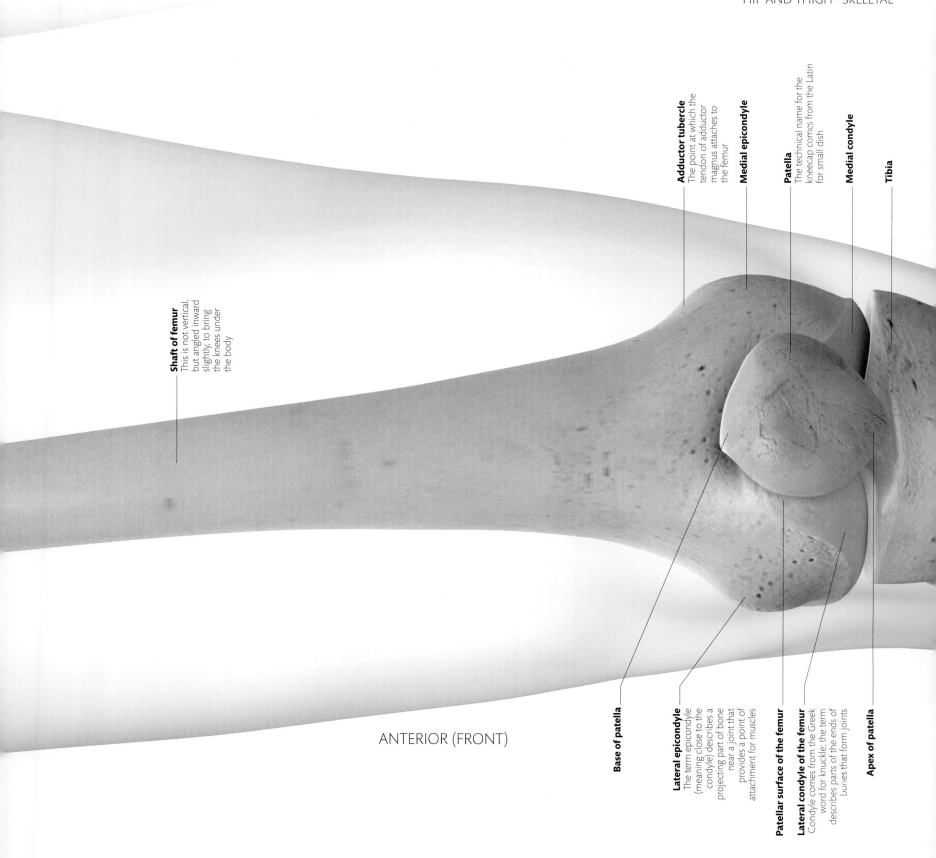

Adductor tubercle
The point at which the tendon of adductor magnus attaches to the femur

Medial epicondyle

Patella
The technical name for the kneecap comes from the Latin for small dish

Medial condyle

Tibia

Shaft of femur
This is not vertical, but angled inward slightly, to bring the knees under the body

Base of patella

Lateral epicondyle
The term epicondyle (meaning close to the condyle) describes a projecting part of bone near a joint that provides a point of attachment for muscles

Patellar surface of the femur

Lateral condyle of the femur
Condyle comes from the Greek word for knuckle; the term describes parts of the ends of bones that form joints

Apex of patella

ANTERIOR (FRONT)

HIP AND THIGH
SKELETAL

The leg or, to be anatomically precise, the lower limb, is attached to the spine by the pelvic bones. This is a much more stable arrangement than that of the shoulder girdle, which anchors the arm, because the legs and pelvis must bear our body weight as we stand or move around. The sacroiliac joint provides a strong attachment between the ilium of the pelvis and the sacrum, and the hip joint is a much deeper and more stable ball-and-socket joint than that in the shoulder. The neck of the femur joins the head at an obtuse angle. A slightly raised diagonal line on the front of the neck (the intertrochanteric line) shows where the fibrous capsule of the hip joint attaches to the bone.

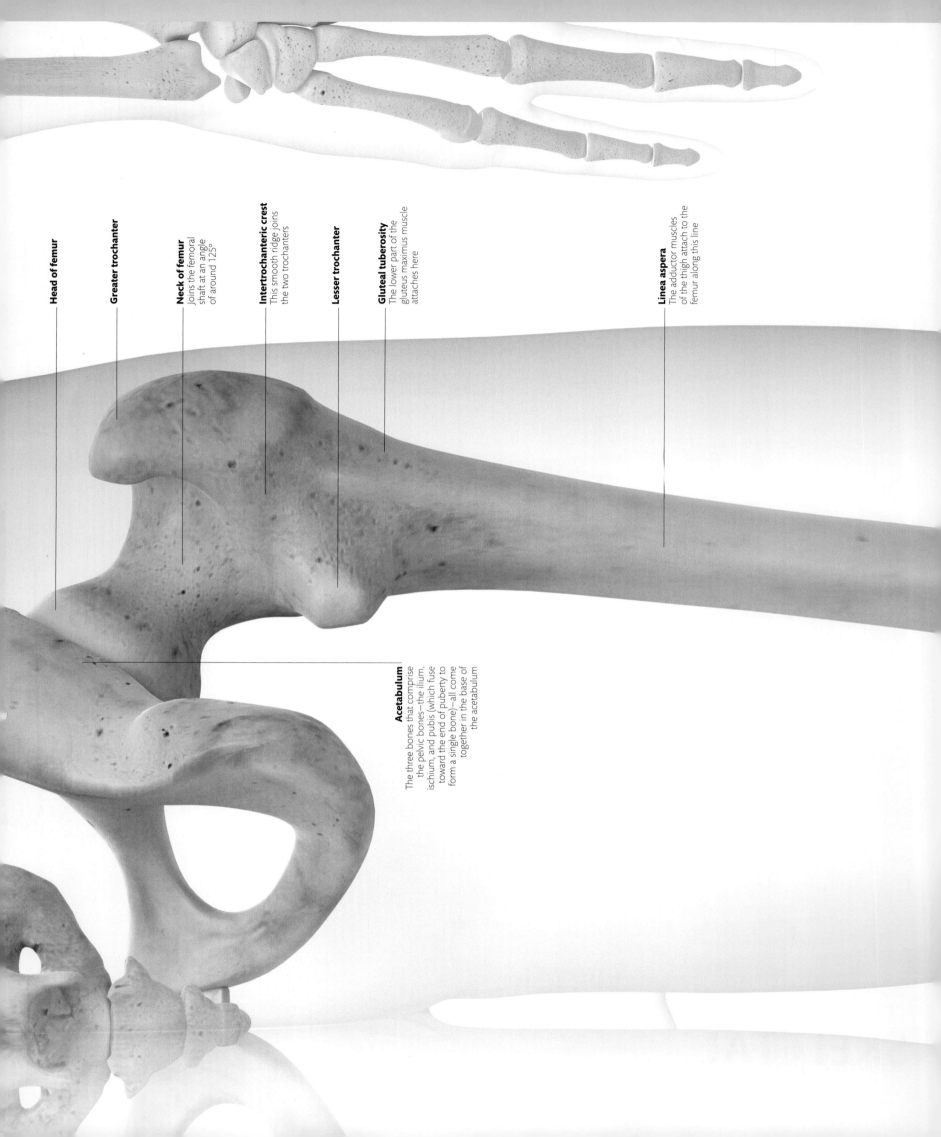

Head of femur

Greater trochanter

Neck of femur
Joins the femoral shaft at an angle of around 125°

Intertrochanteric crest
This smooth ridge joins the two trochanters

Lesser trochanter

Gluteal tuberosity
The lower part of the gluteus maximus muscle attaches here

Linea aspera
The adductor muscles of the thigh attach to the femur along this line

Acetabulum
The three bones that comprise the pelvic bones—the ilium, ischium, and pubis (which fuse toward the end of puberty to form a single bone)—all come together in the base of the acetabulum

HIP AND THIGH
SKELETAL

The shaft of the femur (thighbone) is cylindrical, with a marrow cavity. The linea aspera runs down along the back of the femoral shaft. This line is where the inner thigh's adductor muscles attach to the femur. Parts of the quadriceps muscle also wrap right around the back of the femur to attach to the linea aspera. At the bottom—or distal—end, toward the knee, the femur widens to form the knee joint with the tibia and the patella. From the back, the distal end of the femur has a distinct double-knuckle shape, with two condyles (rounded projections) that articulate with the tibia.

POSTERIOR (BACK)

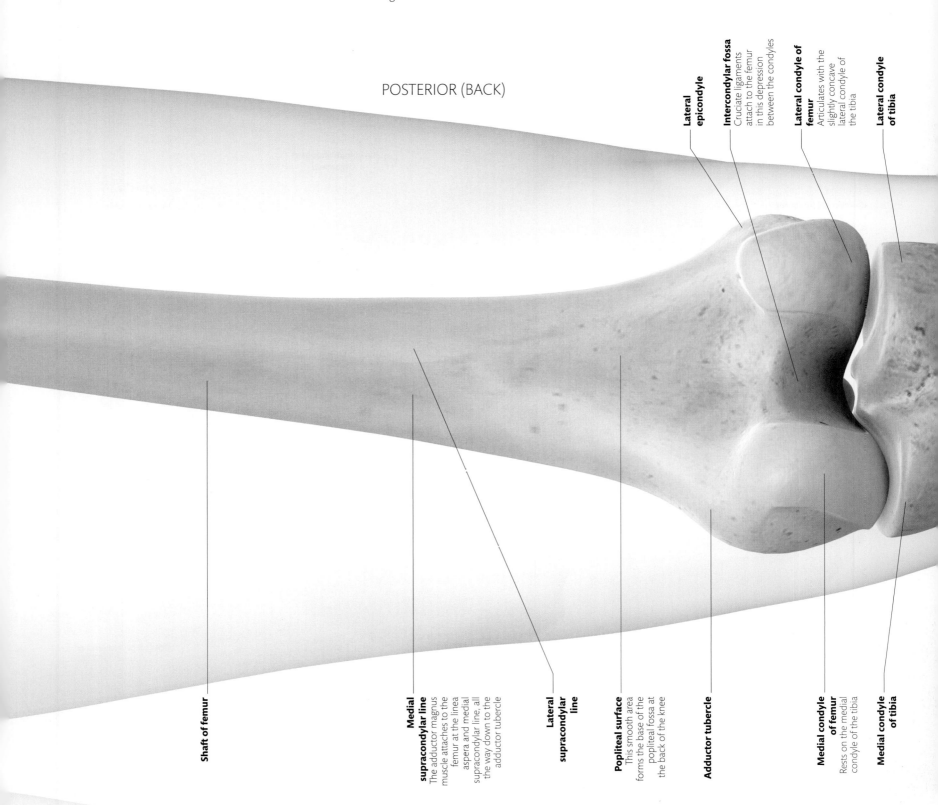

Lateral epicondyle

Intercondylar fossa
Cruciate ligaments attach to the femur in this depression between the condyles

Lateral condyle of femur
Articulates with the slightly concave lateral condyle of the tibia

Lateral condyle of tibia

Shaft of femur

Medial supracondylar line
The adductor magnus muscle attaches to the femur at the linea aspera and medial supracondylar line, all the way down to the adductor tubercle

Lateral supracondylar line

Popliteal surface
This smooth area forms the base of the popliteal fossa at the back of the knee

Adductor tubercle

Medial condyle of femur
Rests on the medial condyle of the tibia

Medial condyle of tibia

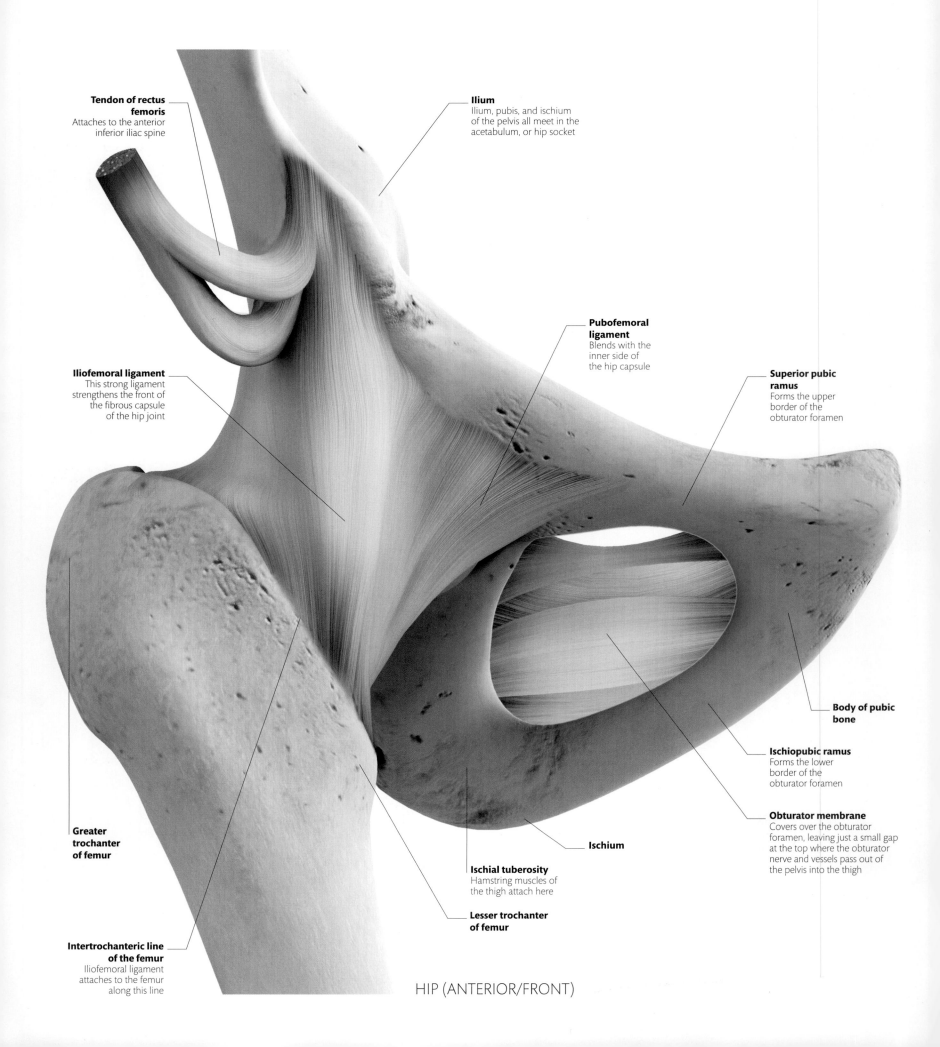

Tendon of rectus femoris
Attaches to the anterior inferior iliac spine

Ilium
Ilium, pubis, and ischium of the pelvis all meet in the acetabulum, or hip socket

Pubofemoral ligament
Blends with the inner side of the hip capsule

Iliofemoral ligament
This strong ligament strengthens the front of the fibrous capsule of the hip joint

Superior pubic ramus
Forms the upper border of the obturator foramen

Body of pubic bone

Ischiopubic ramus
Forms the lower border of the obturator foramen

Greater trochanter of femur

Obturator membrane
Covers over the obturator foramen, leaving just a small gap at the top where the obturator nerve and vessels pass out of the pelvis into the thigh

Ischium

Ischial tuberosity
Hamstring muscles of the thigh attach here

Lesser trochanter of femur

Intertrochanteric line of the femur
Iliofemoral ligament attaches to the femur along this line

HIP (ANTERIOR/FRONT)

HIP AND THIGH
SKELETAL

The hip joint is very stable. The socket of the hip joint is formed by the acetabulum, which is deepened by the acetabular labrum and transverse ligament. The fibrous capsule of the hip joint attaches from the labrum to the neck of the femur. It is strengthened by ligaments that attach from the neck of the femur to the pelvic bone. These are the iliofemoral and pubofemoral ligaments at the front, and the ischiofemoral ligament at the back. Inside the joint capsule, a small ligament attaches from the edge of the acetabulum (hip socket) to the head of the femur.

The hip is a large weight-bearing joint, and a common site for osteoarthritis. Since the hip is such a stable joint, it rarely becomes dislocated. Dislocation happens only with considerable force, and often with associated pelvic fractures.

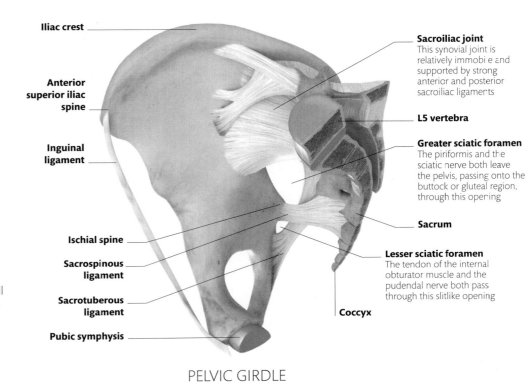

Iliac crest

Anterior superior iliac spine

Inguinal ligament

Ischial spine

Sacrospinous ligament

Sacrotuberous ligament

Pubic symphysis

Sacroiliac joint
This synovial joint is relatively immobile and supported by strong anterior and posterior sacroiliac ligaments

L5 vertebra

Greater sciatic foramen
The piriformis and the sciatic nerve both leave the pelvis, passing onto the buttock or gluteal region, through this opening

Sacrum

Lesser sciatic foramen
The tendon of the internal obturator muscle and the pudendal nerve both pass through this slitlike opening

Coccyx

PELVIC GIRDLE

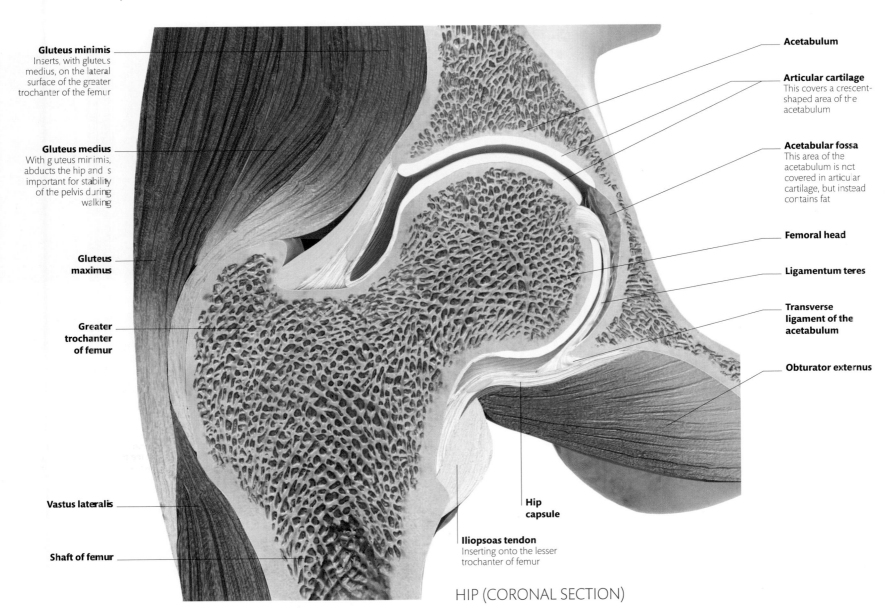

Gluteus minimis
Inserts, with gluteus medius, on the lateral surface of the greater trochanter of the femur

Gluteus medius
With gluteus minimis, abducts the hip and is important for stability of the pelvis during walking

Gluteus maximus

Greater trochanter of femur

Vastus lateralis

Shaft of femur

Acetabulum

Articular cartilage
This covers a crescent-shaped area of the acetabulum

Acetabular fossa
This area of the acetabulum is not covered in articular cartilage, but instead contains fat

Femoral head

Ligamentum teres

Transverse ligament of the acetabulum

Obturator externus

Hip capsule

Iliopsoas tendon
Inserting onto the lesser trochanter of femur

HIP (CORONAL SECTION)

Inguinal ligament

Adductor longus
Attaches from the pubis to the middle third of the linea aspera, a ridge on the back of the femur

Gracilis
This long, thin muscle attaches from the pubis down to the inner (medial) surface of the tibia, and adducts the thigh

Iliopsoas

Pubic symphysis

Pectineus
This muscle attaches from the pubic bone to the femur, and flexes and adducts the hip

Tensor fasciae latae
Tensor of the deep fascia; it attaches from the iliac crest on top of the pelvis and inserts into the iliotibial tract. It helps steady the thigh while standing upright

Sartorius
Named after the Latin for tailor, this muscle flexes, abducts, and laterally rotates the hip while flexing the knee—producing a cross-legged position, apparently the traditional posture of tailors

Iliotibial tract
A thickening of the deep fascia over the outer (lateral) thigh, reaching from the iliac crest to the tibia

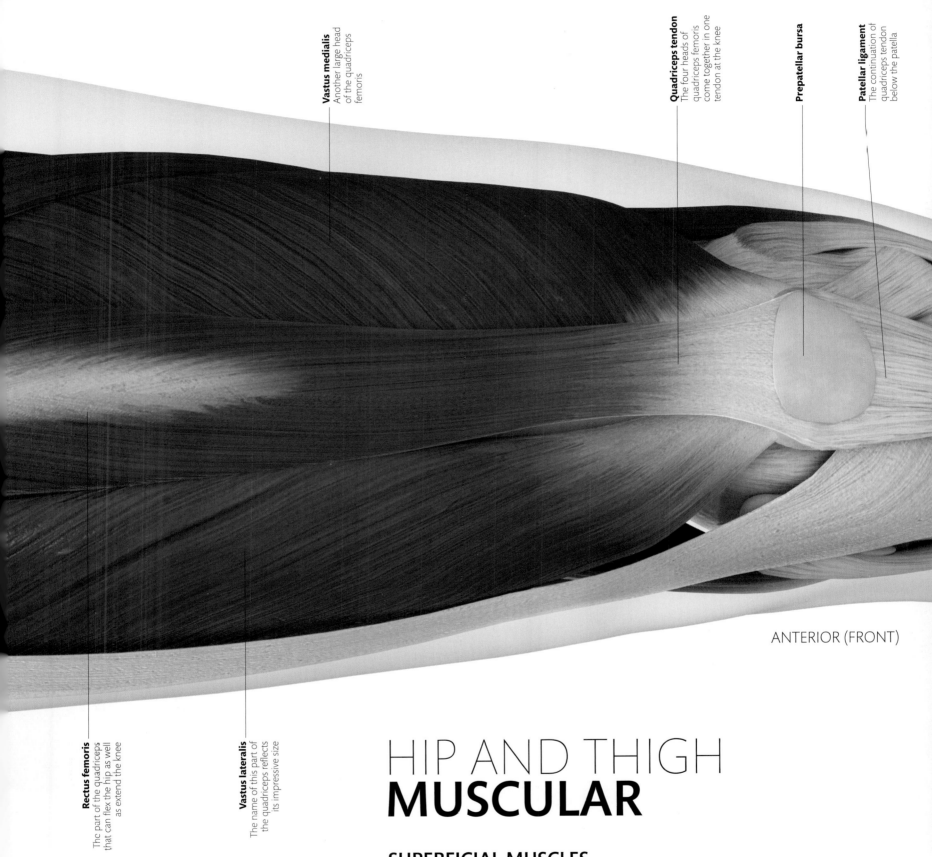

Vastus medialis
Another large head of the quadriceps femoris

Quadriceps tendon
The four heads of quadriceps femoris come together in one tendon at the knee

Prepatellar bursa

Patellar ligament
The continuation of quadriceps tendon below the patella

Rectus femoris
The part of the quadriceps that can flex the hip as well as extend the knee

Vastus lateralis
The name of this part of the quadriceps reflects its impressive size

ANTERIOR (FRONT)

HIP AND THIGH
MUSCULAR

SUPERFICIAL MUSCLES

Most of the muscle bulk on the front of the leg is the four-headed quadriceps femoris. Three of its heads can be seen in a superficial dissection of the thigh: the rectus femoris, vastus lateralis, and vastus medialis. The quadriceps extends the knee, but it can also flex the hip, since the rectus femoris part has an attachment from the pelvis, above the hip joint.

The patella is embedded in the quadriceps tendon; this may protect the tendon from wear and tear, but it also helps to give the quadriceps good leverage in extending the knee. The part of the tendon below the patella is usually called the patellar ligament. Tapping this with a tendon hammer produces a reflex contraction in the quadriceps—the "knee jerk."

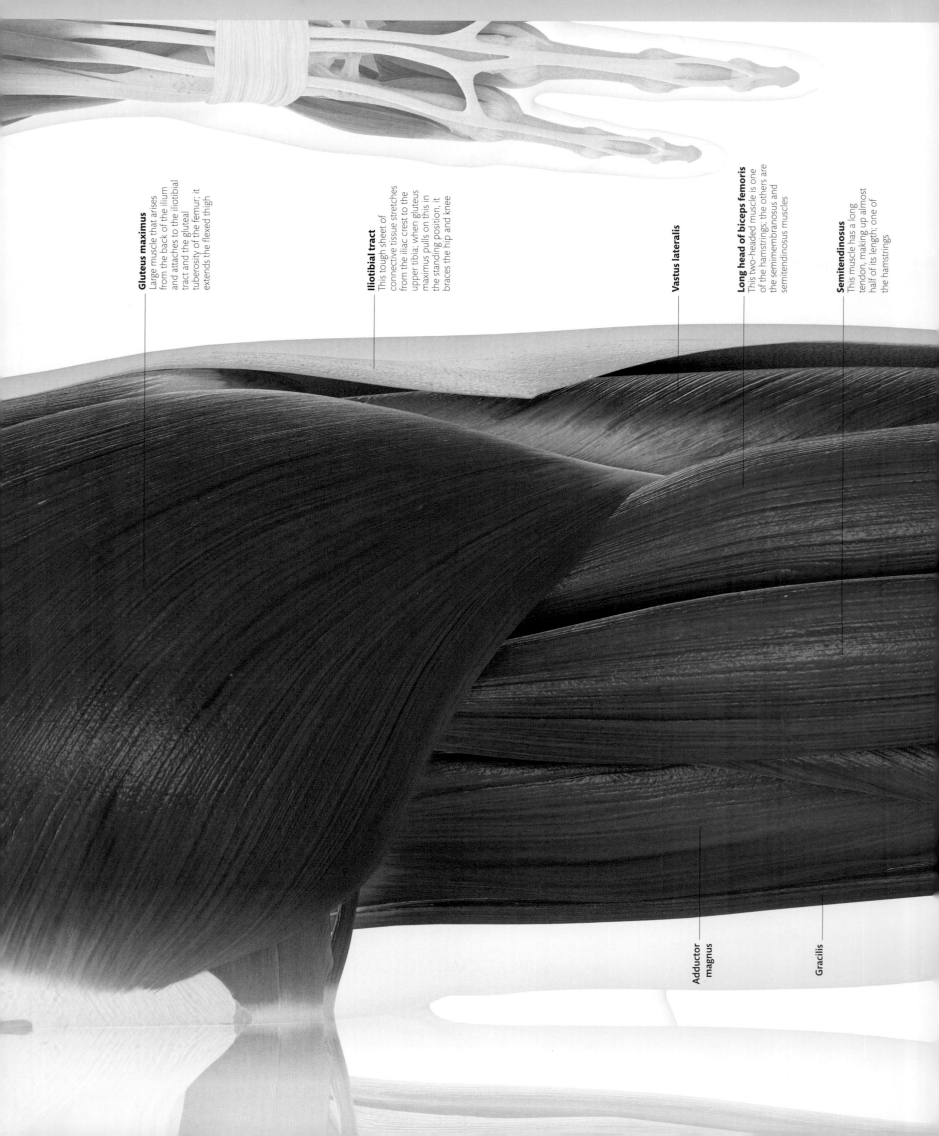

Gluteus maximus
Large muscle that arises from the back of the ilium and attaches to the iliotibial tract and the gluteal tuberosity of the femur; it extends the flexed thigh

Iliotibial tract
This tough sheet of connective tissue stretches from the iliac crest to the upper tibia; when gluteus maximus pulls on this in the standing position, it braces the hip and knee

Vastus lateralis

Long head of biceps femoris
This two-headed muscle is one of the hamstrings; the others are the semimembranosus and semitendinosus muscles

Semitendinosus
This muscle has a long tendon, making up almost half of its length; one of the hamstrings

Adductor magnus

Gracilis

HIP AND THIGH
MUSCULAR

SUPERFICIAL MUSCLES

On the back of the hip and thigh, a superficial dissection reveals the large gluteus maximus, an extensor of the hip joint, and the three hamstrings. The gluteus maximus acts to extend the hip joint, swinging the leg backward. While it doesn't really contribute to gentle walking, it is very important in running, and also when the hip is being extended from a flexed position, such as when getting up from sitting on the floor or when climbing the stairs. The hamstrings—the semimembranosus, semitendinosus, and biceps femoris muscles—attach from the ischial tuberosity of the pelvis and sweep down the back of the thigh to the tibia and fibula. They are the main flexors of the knee.

POSTERIOR (BACK)

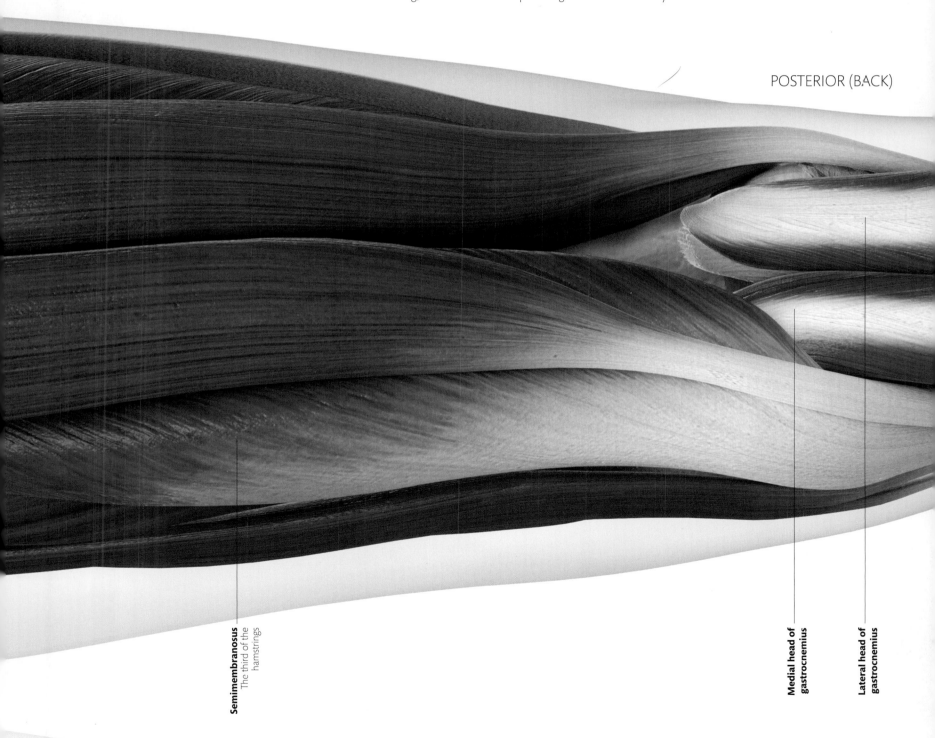

Semimembranosus
The third of the hamstrings

Medial head of gastrocnemius

Lateral head of gastrocnemius

Gluteus medius

Superior pubic ramus

Iliacus

Psoas major

Pectineus

Adductor longus

Adductor brevis
Tucked in behind adductor longus and pectineus, this "short adductor" attaches from the upper part of the pubis to the linea aspera, the ridge on the back of the femur

Gracilis

Adductor magnus
This muscle attaches, via a wide aponeurosis (band of fibrous tissue), to the entire length of the linea aspera, the ridge on the back of the femur

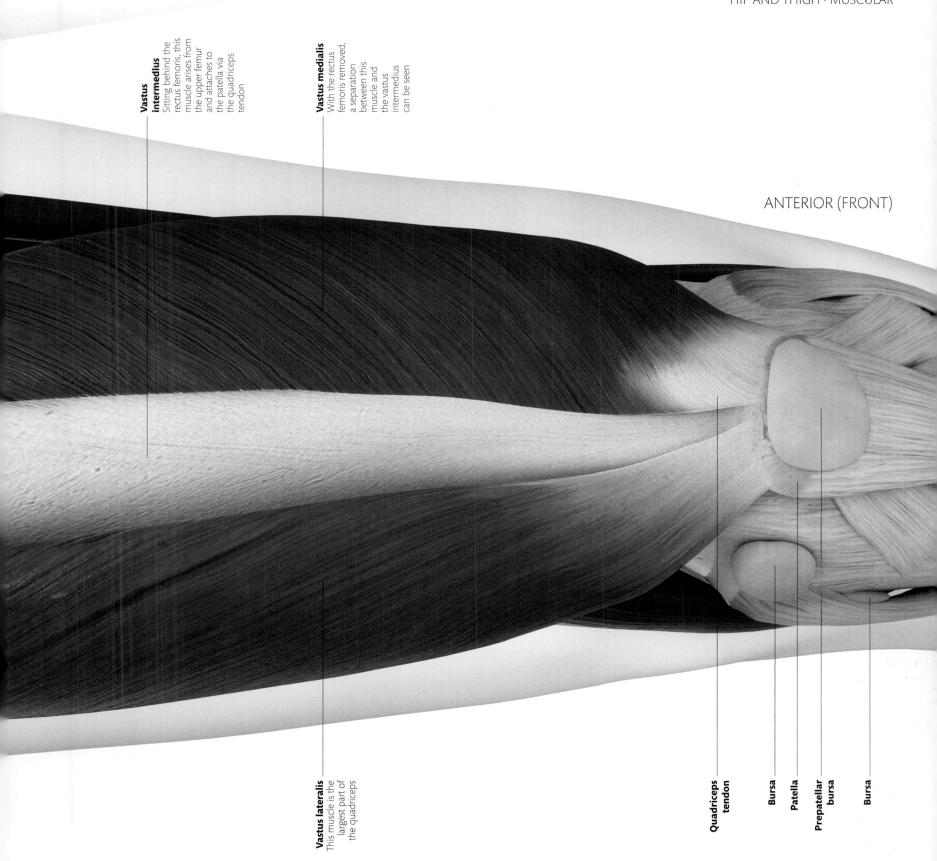

Vastus intermedius
Sitting behind the rectus femoris, this muscle arises from the upper femur and attaches to the patella via the quadriceps tendon

Vastus medialis
With the rectus femoris removed, a separation between this muscle and the vastus intermedius can be seen

ANTERIOR (FRONT)

Vastus lateralis
This muscle is the largest part of the quadriceps

Quadriceps tendon

Bursa

Patella

Prepatellar bursa

Bursa

HIP AND THIGH
MUSCULAR

DEEP MUSCLES

With the rectus femoris and sartorius muscles stripped away, we can see the deep, fourth head of the quadriceps, known as vastus intermedius. The adductor muscles that bring the thighs together can also be seen clearly, including the gracilis, which is long and slender, as its name suggests. The largest adductor muscle—the adductor magnus—has a hole in its tendon, through which the main artery of the leg (the femoral artery) passes. The adductor tendons attach from the pubis and ischium of the pelvis, and the sporting injuries referred to as "groin pulls" are often tears in these particular tendons.

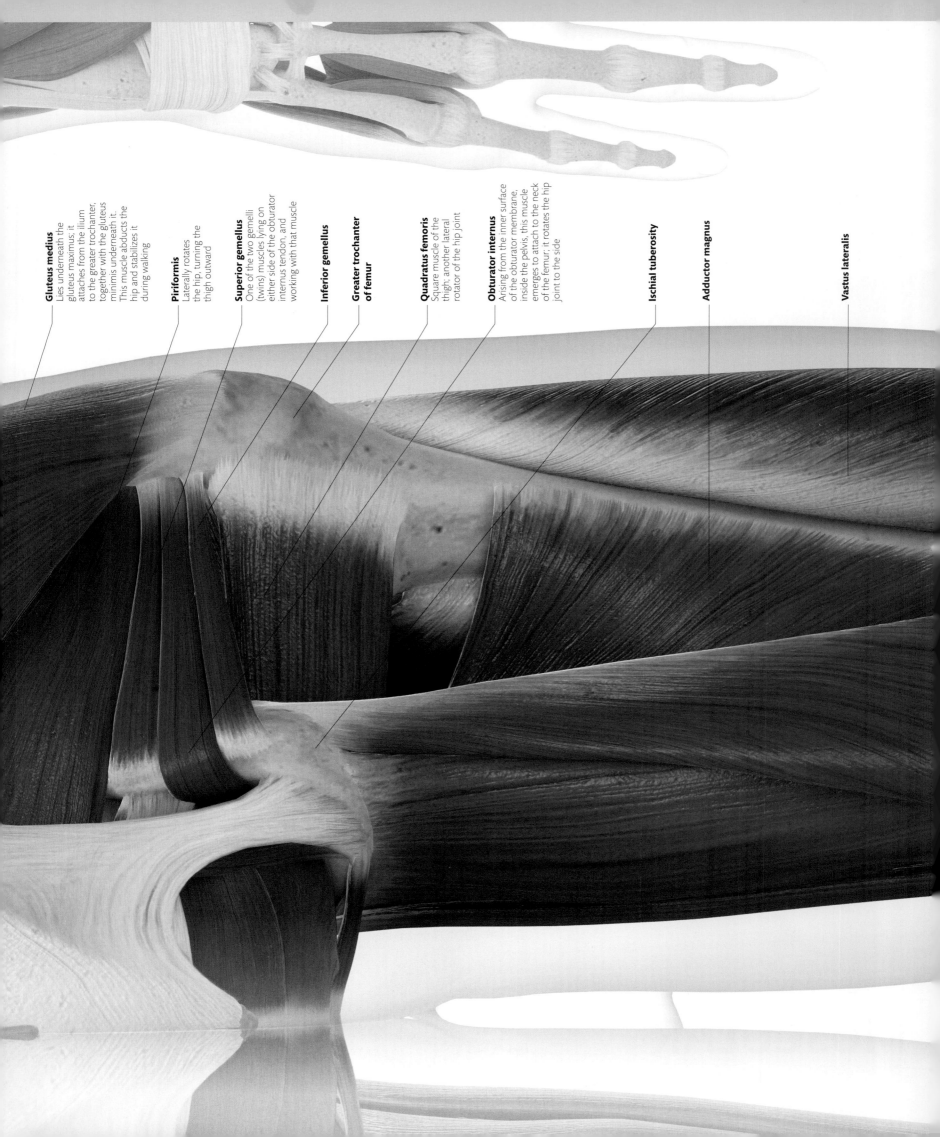

Gluteus medius
Lies underneath the gluteus maximus; it attaches from the ilium to the greater trochanter, together with the gluteus minimis underneath it. This muscle abducts the hip and stabilizes it during walking

Piriformis
Laterally rotates the hip, turning the thigh outward

Superior gemellus
One of the two gemelli (twins) muscles lying on either side of the obturator internus tendon, and working with that muscle

Inferior gemellus

Greater trochanter of femur

Quadratus femoris
Square muscle of the thigh; another lateral rotator of the hip joint

Obturator internus
Arising from the inner surface of the obturator membrane, inside the pelvis, this muscle emerges to attach to the neck of the femur; it rotates the hip joint to the side

Ischial tuberosity

Adductor magnus

Vastus lateralis

HIP AND THIGH
MUSCULAR

DEEP MUSCLES

On the back of the hip, with the gluteus maximus removed, the short muscles that rotate the hip out to the side are clearly revealed. These include the piriformis, obturator internus, and quadratus femoris muscles. With the long head of the biceps femoris removed, we can now see the deeper, short head attaching to the linea aspera on the back of the femur. The semitendinosus muscle has also been cut away to reveal the semimembranosus underneath it, with its flat, membranelike tendon at the top. Popliteus muscle is also visible at the back of the knee joint, as is one of the many fluid filled bursae around the knee.

POSTERIOR (BACK)

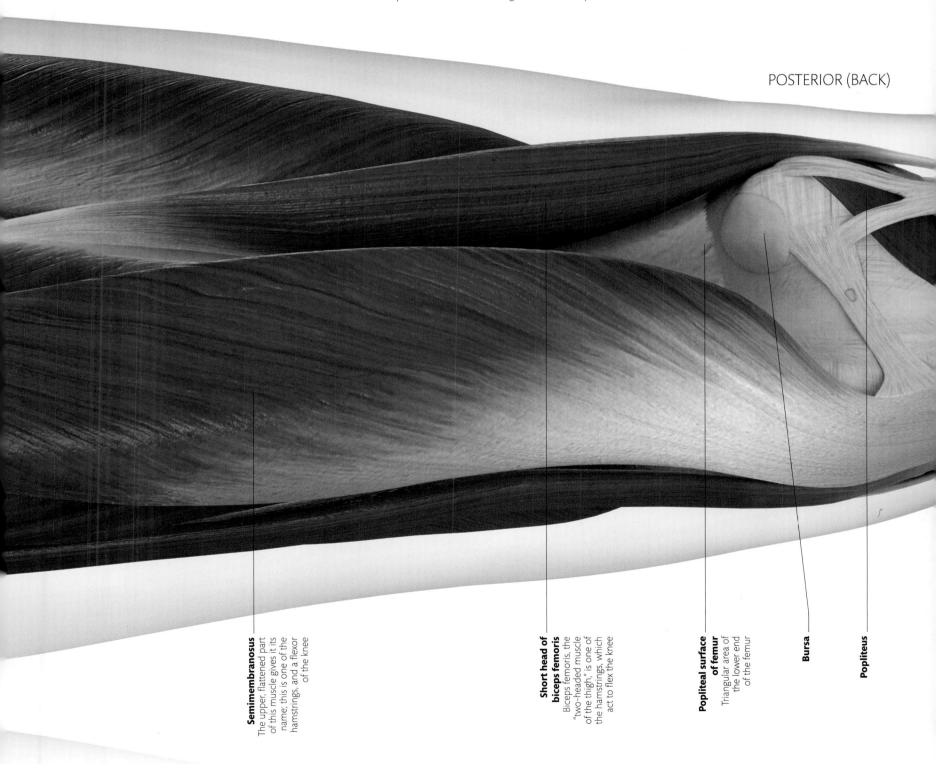

Semimembranosus
The upper, flattened part of this muscle gives it its name; this is one of the hamstrings, and a flexor of the knee

Short head of biceps femoris
Biceps femoris, the "two-headed muscle of the thigh," is one of the hamstrings, which act to flex the knee

Popliteal surface of femur
Triangular area of the lower end of the femur

Bursa

Popliteus

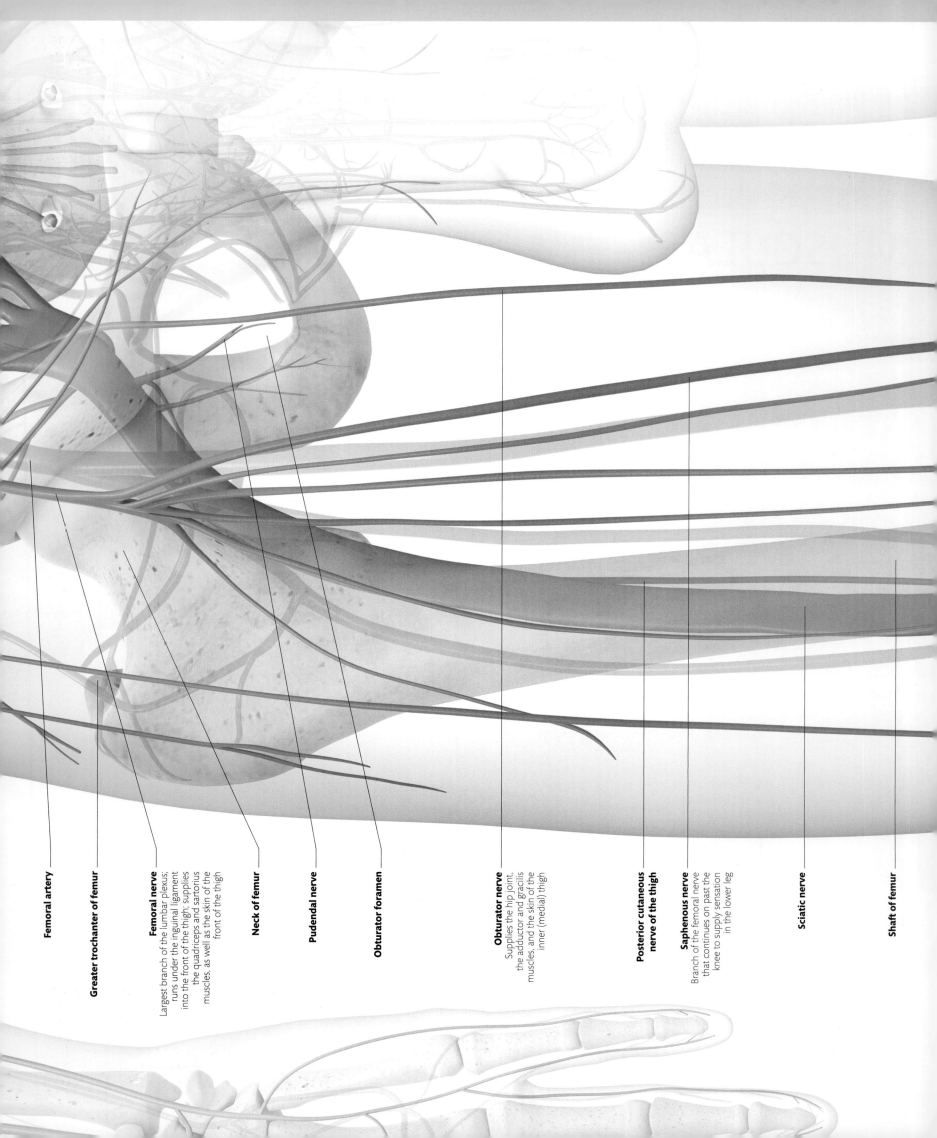

Femoral artery

Greater trochanter of femur

Femoral nerve
Largest branch of the lumbar plexus;
runs under the inguinal ligament
into the front of the thigh; supplies
the quadriceps and sartorius
muscles, as well as the skin of the
front of the thigh

Neck of femur

Pudendal nerve

Obturator foramen

Obturator nerve
Supplies the hip joint,
the adductor and gracilis
muscles, and the skin of the
inner (medial) thigh

**Posterior cutaneous
nerve of the thigh**

Saphenous nerve
Branch of the femoral nerve
that continues on past the
knee to supply sensation
in the lower leg

Sciatic nerve

Shaft of femur

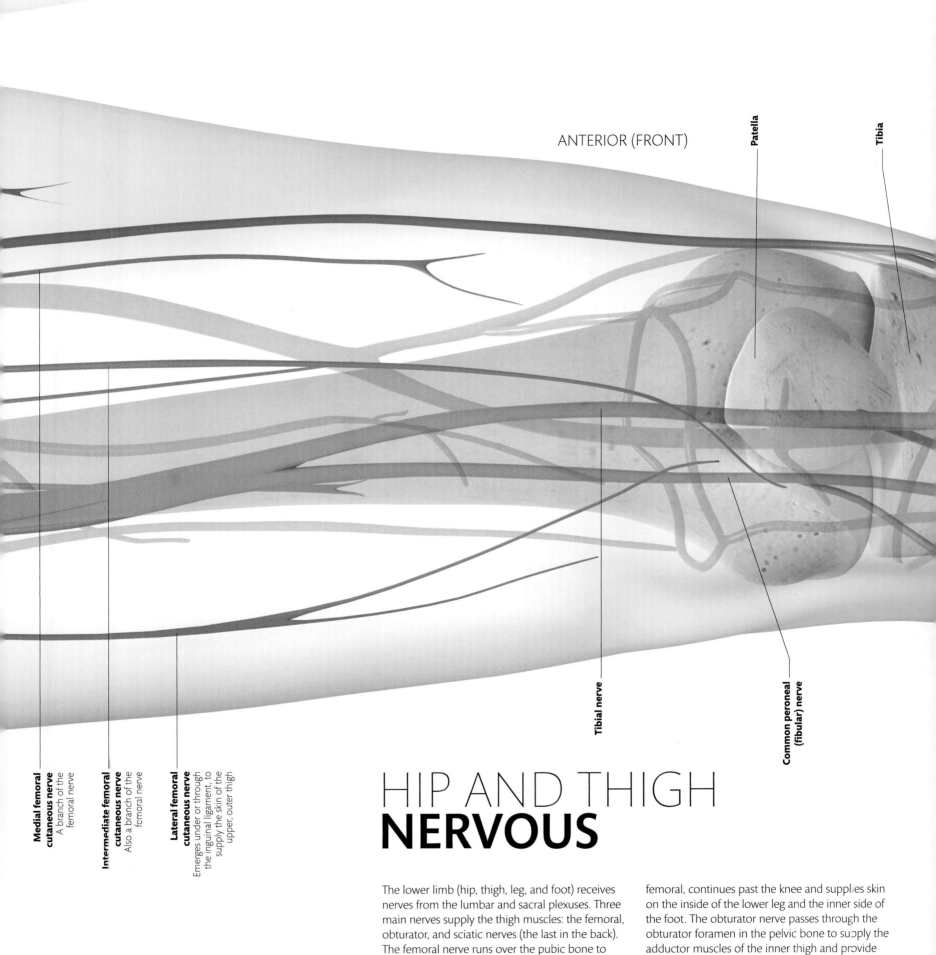

ANTERIOR (FRONT)

Patella

Tibia

Tibial nerve

Common peroneal (fibular) nerve

Medial femoral cutaneous nerve
A branch of the femoral nerve

Intermediate femoral cutaneous nerve
Also a branch of the femoral nerve

Lateral femoral cutaneous nerve
Emerges under or through the inguinal ligament, to supply the skin of the upper, outer thigh

HIP AND THIGH
NERVOUS

The lower limb (hip, thigh, leg, and foot) receives nerves from the lumbar and sacral plexuses. Three main nerves supply the thigh muscles: the femoral, obturator, and sciatic nerves (the last in the back). The femoral nerve runs over the pubic bone to supply the quadriceps and sartorius muscles in the front. The saphenous nerve, a slender branch of the femoral, continues past the knee and supplies skin on the inside of the lower leg and the inner side of the foot. The obturator nerve passes through the obturator foramen in the pelvic bone to supply the adductor muscles of the inner thigh and provide sensation to the skin there. Some smaller nerves just supply skin, such as the femoral cutaneous nerves.

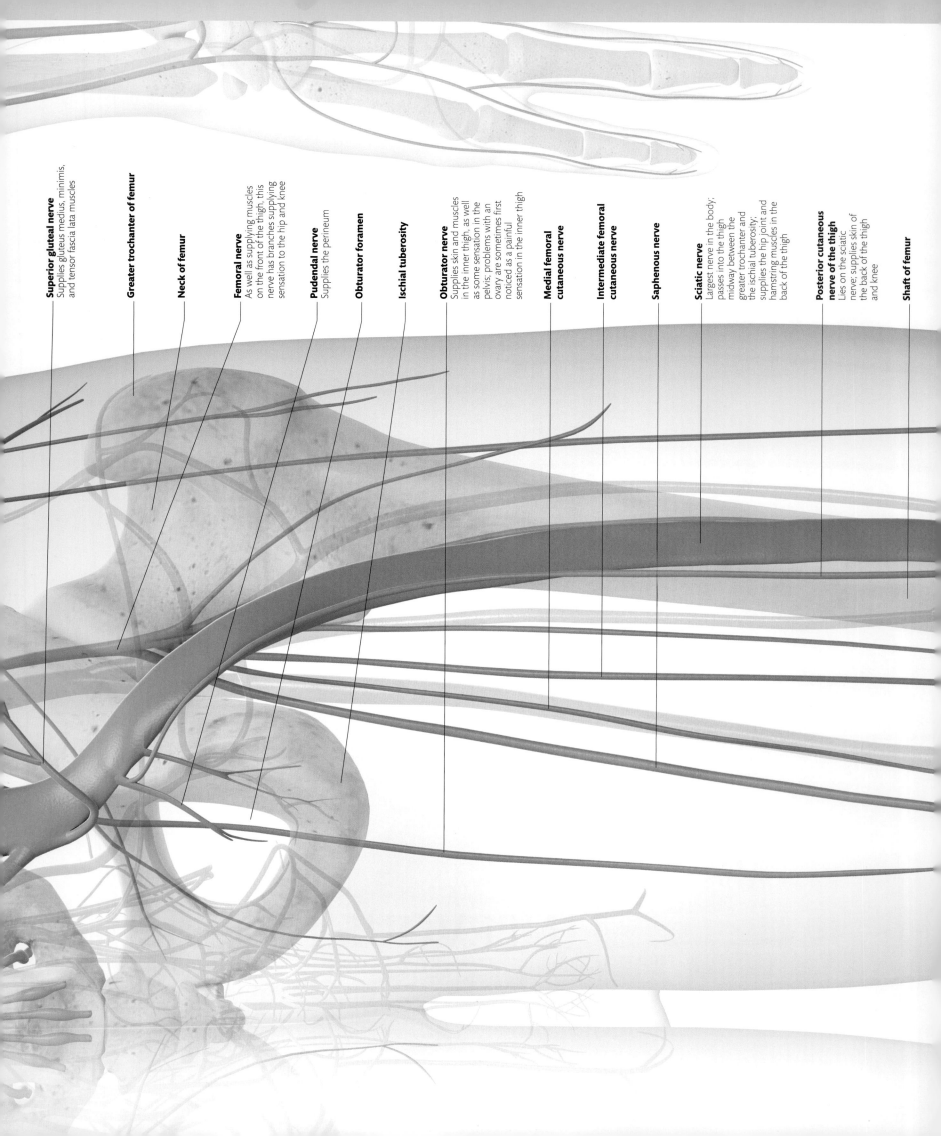

Superior gluteal nerve
Supplies gluteus medius, minimis, and tensor fascia lata muscles

Greater trochanter of femur

Neck of femur

Femoral nerve
As well as supplying muscles on the front of the thigh, this nerve has branches supplying sensation to the hip and knee

Pudendal nerve
Supplies the perineum

Obturator foramen

Ischial tuberosity

Obturator nerve
Supplies skin and muscles in the inner thigh, as well as some sensation in the pelvis; problems with an ovary are sometimes first noticed as a painful sensation in the inner thigh

Medial femoral cutaneous nerve

Intermediate femoral cutaneous nerve

Saphenous nerve

Sciatic nerve
Largest nerve in the body; passes into the thigh midway between the greater trochanter and the ischial tuberosity; supplies the hip joint and hamstring muscles in the back of the thigh

Posterior cutaneous nerve of the thigh
Lies on the sciatic nerve; supplies skin of the back of the thigh and knee

Shaft of femur

HIP AND THIGH
NERVOUS

Gluteal nerves from the sacral plexus emerge via the greater sciatic foramen, at the back of the pelvis, to supply the muscles and skin of the buttock. The sciatic nerve also emerges through the greater sciatic foramen into the buttock. The gluteus maximus is a good site for injections into a muscle, but these should be given in the upper, outer part of the buttock to make sure the needle is away from the sciatic nerve. The sciatic nerve runs down the back of the thigh, supplying the hamstrings. In most people, the sciatic nerve runs halfway down the thigh then splits into two branches, the tibial and common peroneal nerves. These continue into the popliteal fossa (back of the knee) and on into the lower leg.

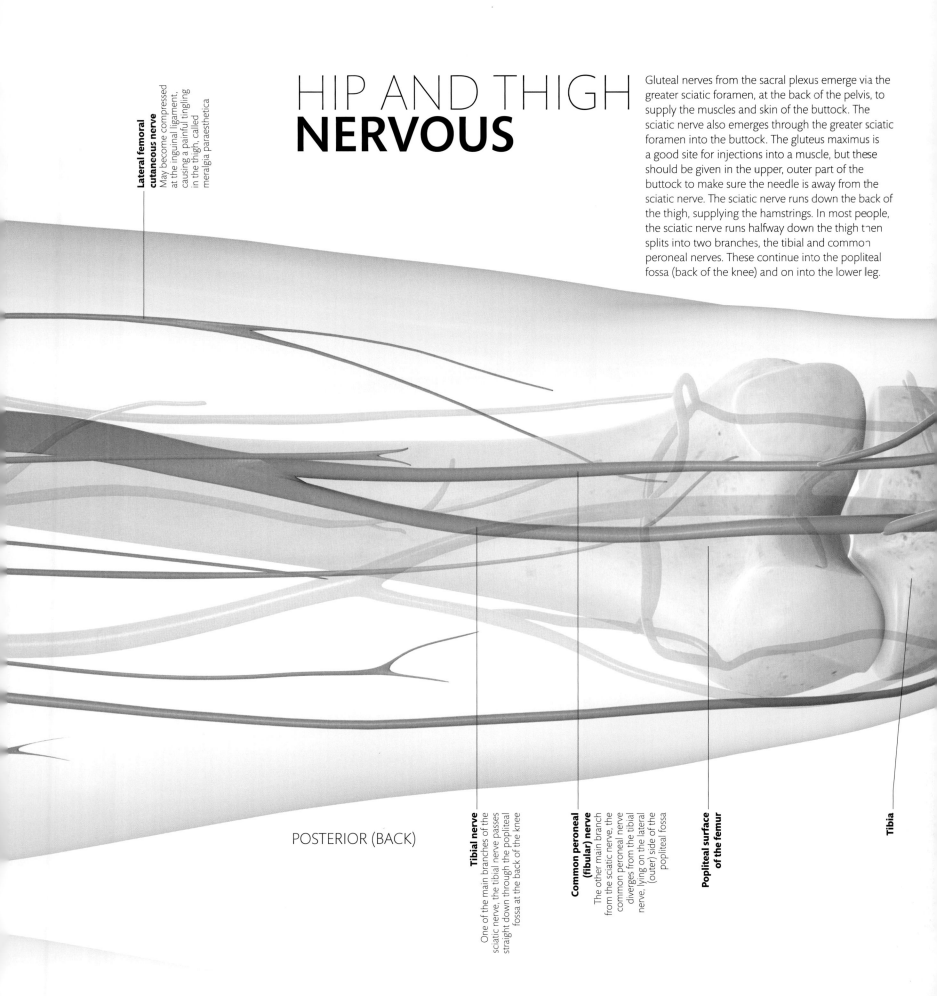

Lateral femoral cutaneous nerve
May become compressed at the inguinal ligament, causing a painful tingling in the thigh, called meralgia paraesthetica

POSTERIOR (BACK)

Tibial nerve
One of the main branches of the sciatic nerve, the tibial nerve passes straight down through the popliteal fossa at the back of the knee

Common peroneal (fibular) nerve
The other main branch from the sciatic nerve, the common peroneal nerve diverges from the tibial nerve, lying on the lateral (outer) side of the popliteal fossa

Popliteal surface of the femur

Tibia

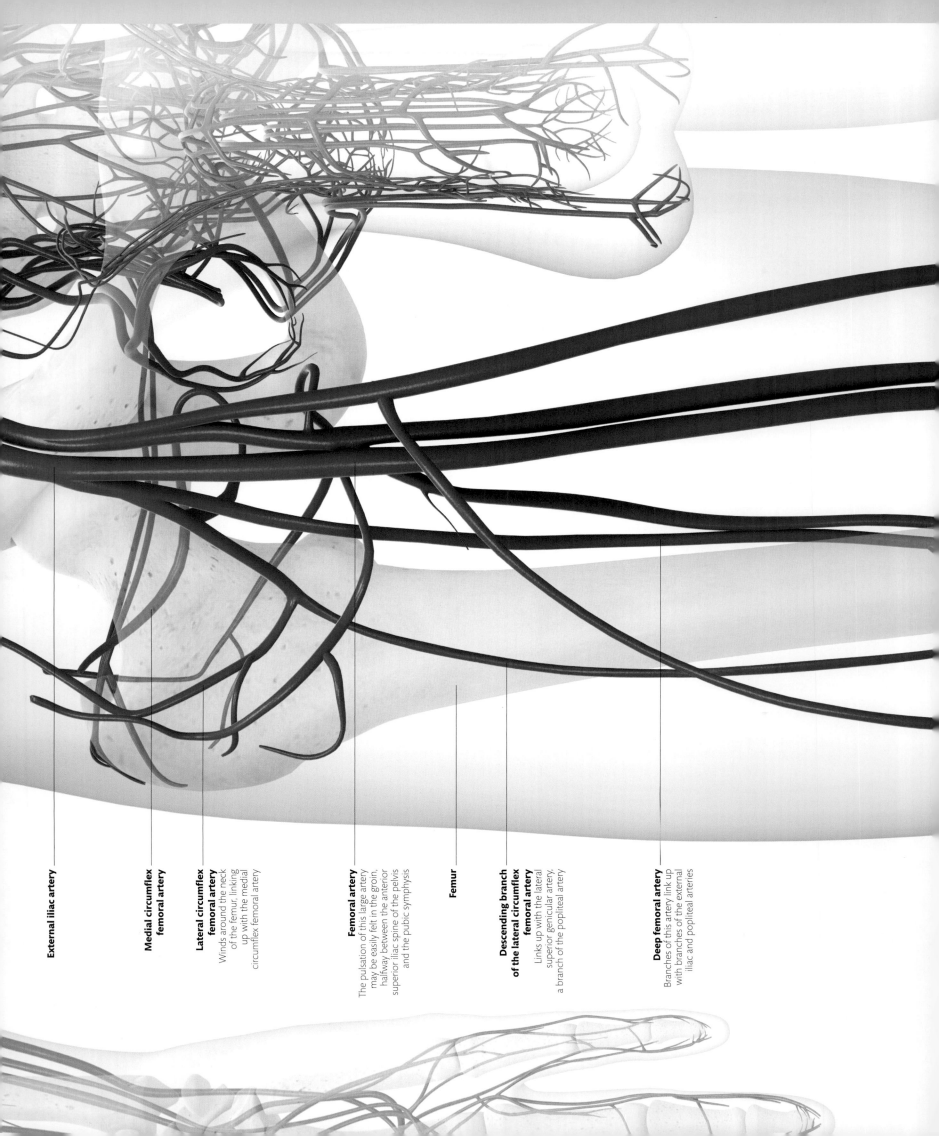

External iliac artery

Medial circumflex femoral artery

Lateral circumflex femoral artery
Winds around the neck of the femur, linking up with the medial circumflex femoral artery

Femoral artery
The pulsation of this large artery may be easily felt in the groin, halfway between the anterior superior iliac spine of the pelvis and the pubic symphysis

Femur

Descending branch of the lateral circumflex femoral artery
Links up with the lateral superior genicular artery, a branch of the popliteal artery

Deep femoral artery
Branches of this artery link up with branches of the external iliac and popliteal arteries

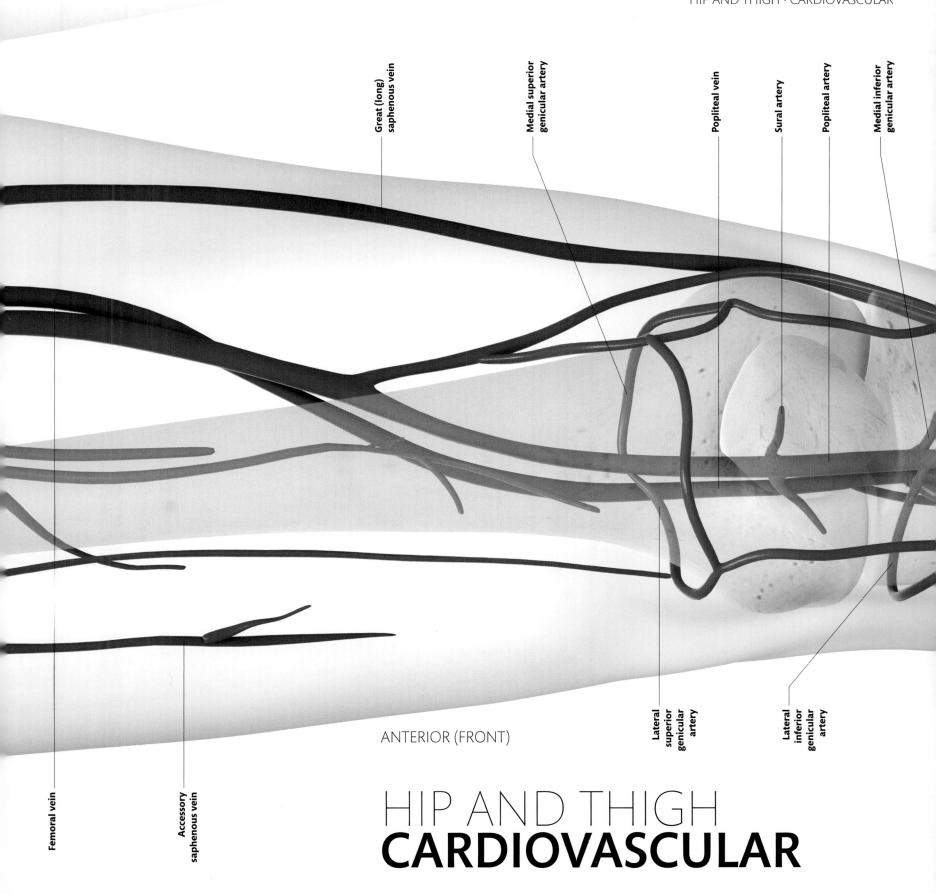

Great (long) saphenous vein

Medial superior genicular artery

Popliteal vein

Sural artery

Popliteal artery

Medial inferior genicular artery

Lateral superior genicular artery

Lateral inferior genicular artery

ANTERIOR (FRONT)

Femoral vein

Accessory saphenous vein

HIP AND THIGH
CARDIOVASCULAR

As the external iliac artery runs over the pubic bone and underneath the inguinal ligament, its name changes to the femoral artery—the main vessel carrying blood to the lower limb. The femoral artery lies exactly halfway along a line between the anterior superior iliac spine of the pelvis and the pubic symphysis. It has a large branch, the deep femoral artery, that supplies the muscles of the thigh. The femoral artery then runs toward the inner thigh, passing through the hole in the adductor magnus tendon, where its name changes to the popliteal artery. Deep veins run with the arteries, but—just as in the arm—there are also superficial veins. The great (or long) saphenous vein drains up the inner side of the leg and thigh, and ends by joining the femoral vein near the hip.

External iliac artery

Branch of internal iliac artery

Medial circumflex femoral artery

Lateral circumflex femoral artery

Perforating artery

Descending branch of the lateral circumflex femoral artery

Femur

Femoral artery

Femoral vein

Deep femoral artery

Accessory saphenous vein

HIP AND THIGH
CARDIOVASCULAR

In this back view, gluteal branches of the internal iliac artery can be clearly seen, emerging through the greater sciatic foramen to supply the buttock. The muscles and skin of the inner part and back of the thigh are supplied by branches of the deep femoral artery. These are known as the perforating arteries because they pierce through the adductor magnus muscle. Higher up, the circumflex femoral arteries encircle the femur. The popliteal artery, formed after the femoral artery passes through the hiatus (gap) in adductor magnus, lies on the back of the femur, deep to the popliteal vein.

POSTERIOR (BACK)

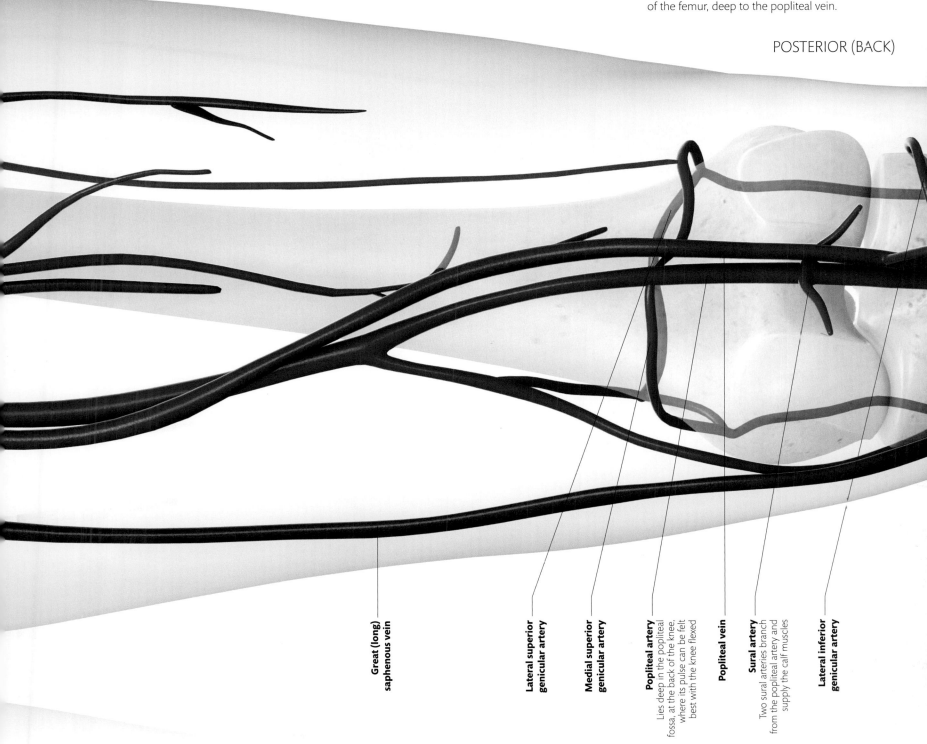

Great (long) saphenous vein

Lateral superior genicular artery

Medial superior genicular artery

Popliteal artery
Lies deep in the popliteal fossa, at the back of the knee, where its pulse can be felt best with the knee flexed

Popliteal vein

Sural artery
Two sural arteries branch from the popliteal artery and supply the calf muscles

Lateral inferior genicular artery

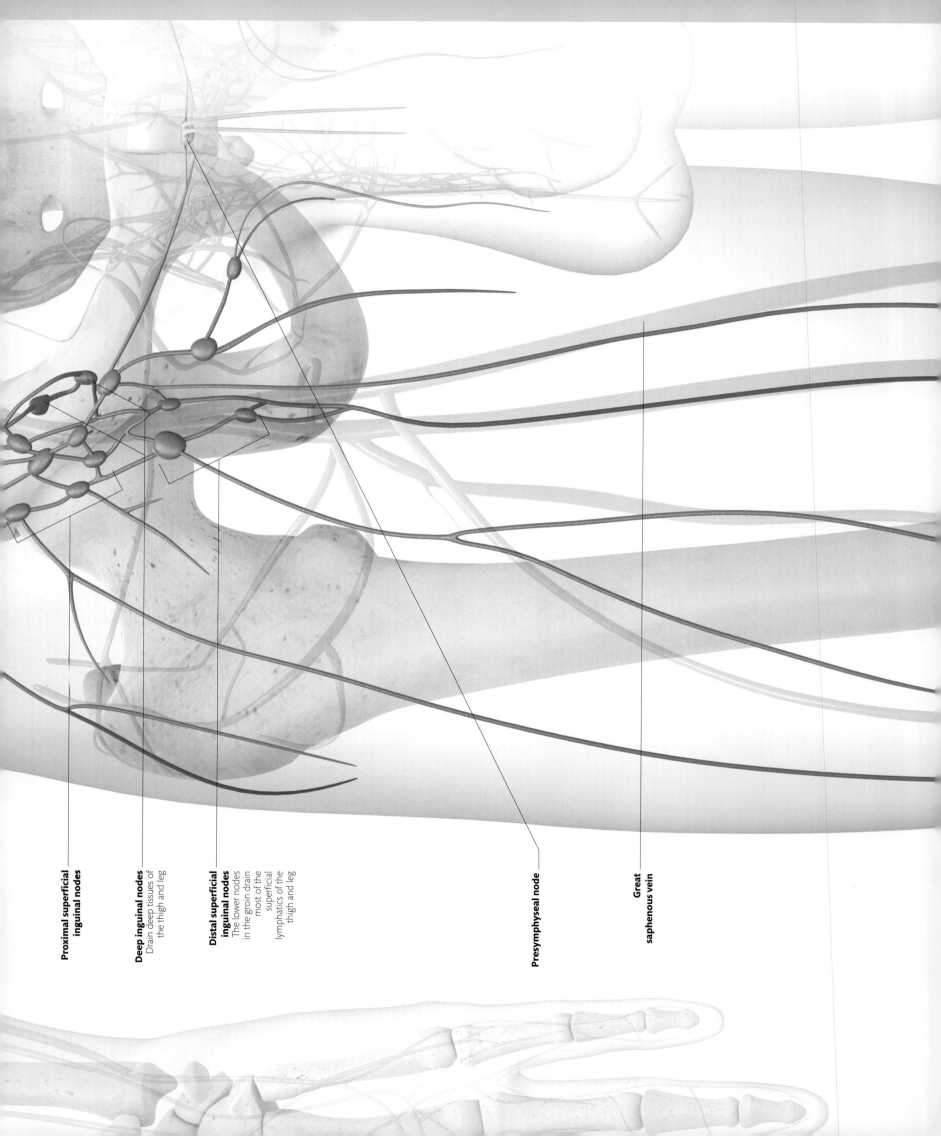

Proximal superficial inguinal nodes

Deep inguinal nodes
Drain deep tissues of the thigh and leg

Distal superficial inguinal nodes
The lower nodes in the groin drain most of the superficial lymphatics of the thigh and leg

Presymphyseal node

Great saphenous vein

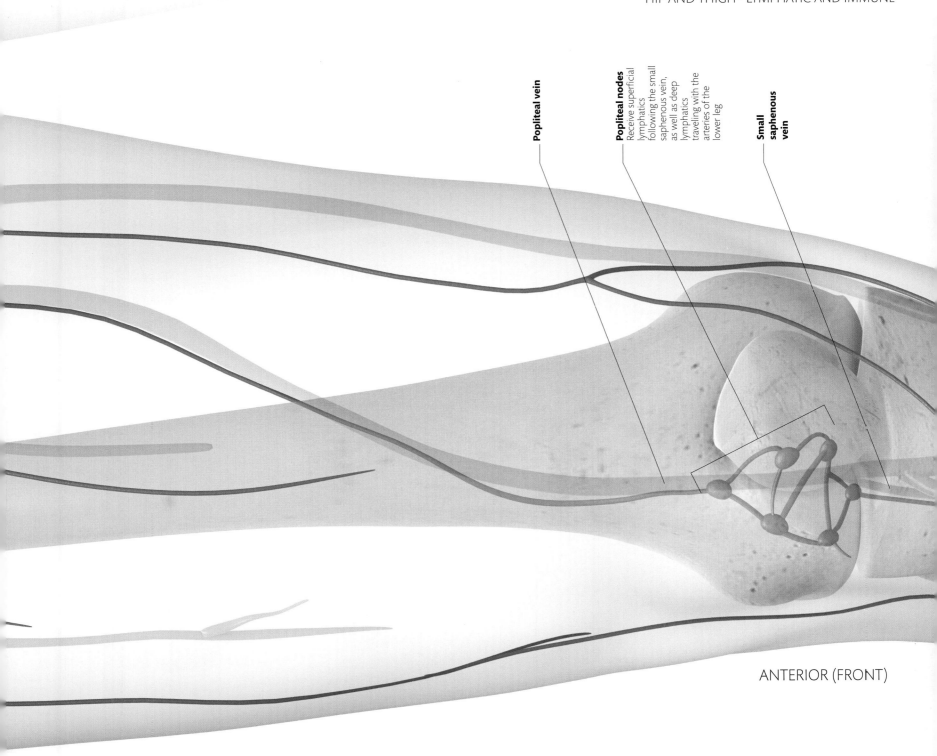

Popliteal vein

Popliteal nodes
Receive superficial lymphatics following the small saphenous vein, as well as deep lymphatics traveling with the arteries of the lower leg

Small saphenous vein

ANTERIOR (FRONT)

HIP AND THIGH
LYMPHATIC AND IMMUNE

Most lymph from the thigh, leg, and foot passes through the inguinal group of lymph nodes, which are in the groin. But lymph from the deep tissues of the buttock passes straight to nodes inside the pelvis (see p.184), along the internal and common iliac arteries. Eventually, all the lymph from the leg reaches the lateral aortic nodes, on the back wall of the abdomen. As in the arm, there are groups of nodes clustered around points at which superficial veins drain into deep veins. Popliteal nodes are close to the drainage of the small saphenous vein into the popliteal vein, while the superficial inguinal nodes lie close to the great saphenous vein, just before it empties into the femoral vein.

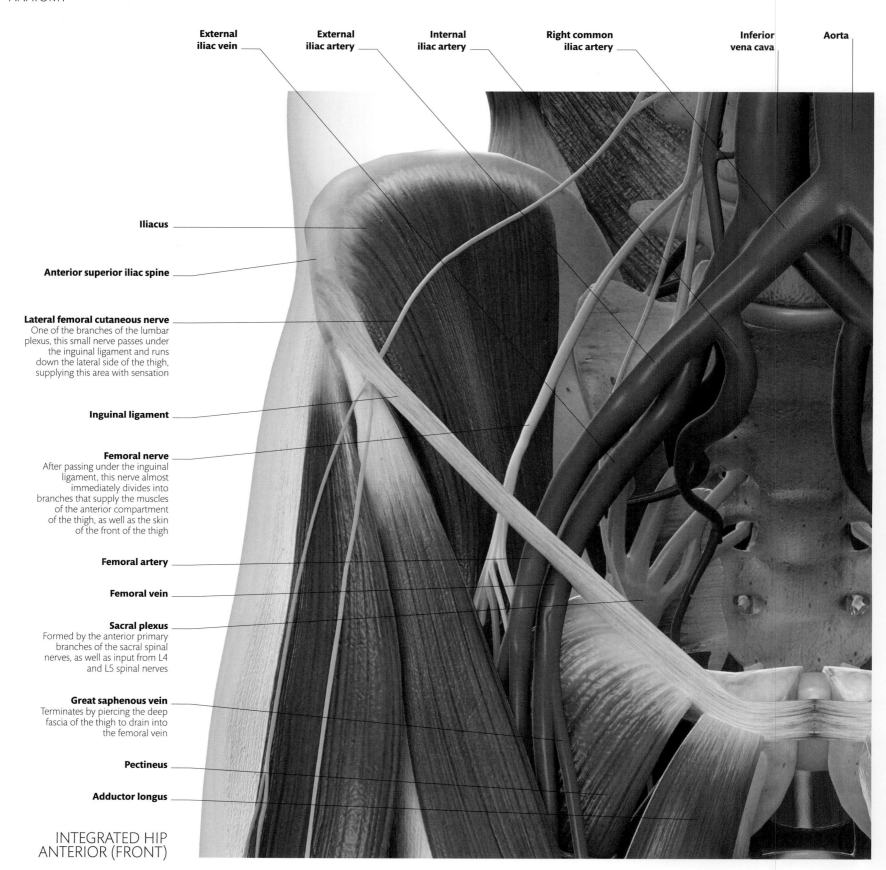

External iliac vein

External iliac artery

Internal iliac artery

Right common iliac artery

Inferior vena cava

Aorta

Iliacus

Anterior superior iliac spine

Lateral femoral cutaneous nerve
One of the branches of the lumbar plexus, this small nerve passes under the inguinal ligament and runs down the lateral side of the thigh, supplying this area with sensation

Inguinal ligament

Femoral nerve
After passing under the inguinal ligament, this nerve almost immediately divides into branches that supply the muscles of the anterior compartment of the thigh, as well as the skin of the front of the thigh

Femoral artery

Femoral vein

Sacral plexus
Formed by the anterior primary branches of the sacral spinal nerves, as well as input from L4 and L5 spinal nerves

Great saphenous vein
Terminates by piercing the deep fascia of the thigh to drain into the femoral vein

Pectineus

Adductor longus

INTEGRATED HIP
ANTERIOR (FRONT)

HIP AND THIGH **HIP**

There are many clinically important nerves and vessels around the hip. The femoral triangle in front of the hip, framed by sartorius muscle, adductor longus muscle, and the inguinal ligament, contains the femoral nerve, artery, and vein. The pulse of the femoral artery is easily felt in this area. The long saphenous vein, which runs up the medial or inner side of the leg and thigh terminates here, by draining into the femoral vein.

The gluteal region, or buttock area, stretches from the iliac crest above, to the gluteal fold below. Beneath the gluteus maximus, several nerves and arteries make their way out of the pelvis into the gluteal region. These include the sciatic nerve, which emerges under the piriformis muscle.

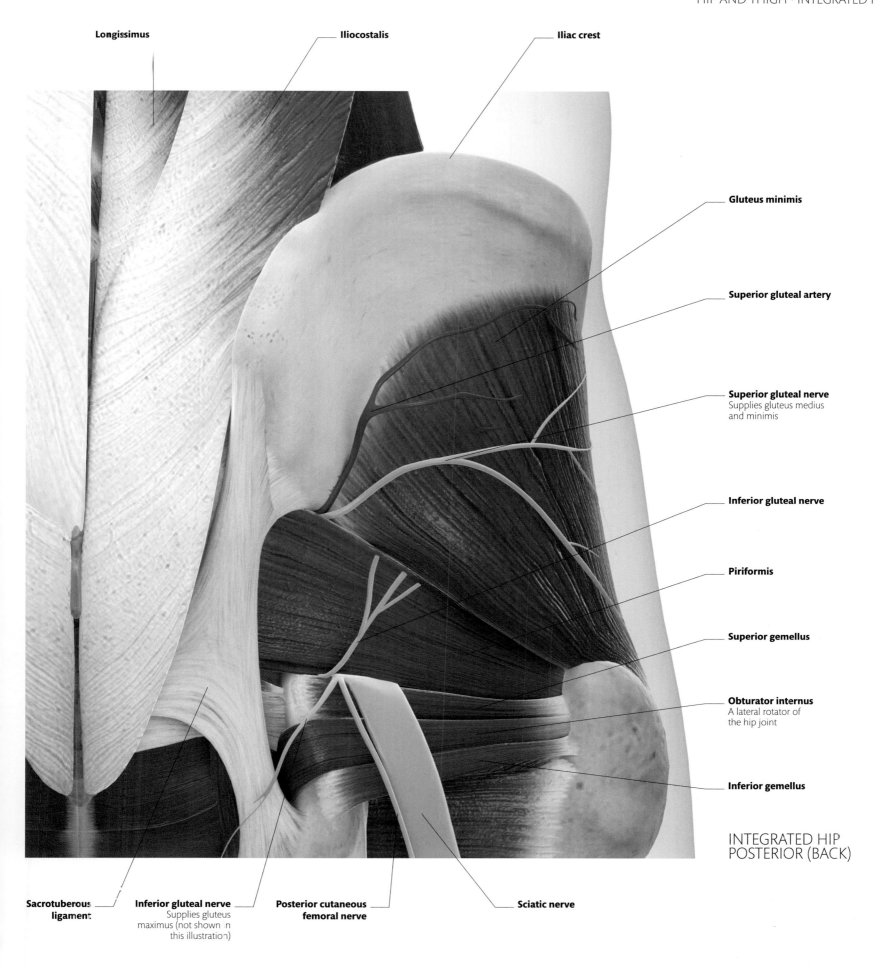

Longissimus

Iliocostalis

Iliac crest

Gluteus minimis

Superior gluteal artery

Superior gluteal nerve
Supplies gluteus medius
and minimis

Inferior gluteal nerve

Piriformis

Superior gemellus

Obturator internus
A lateral rotator of
the hip joint

Inferior gemellus

**Sacrotuberous
ligament**

Inferior gluteal nerve
Supplies gluteus
maximus (not shown in
this illustration)

**Posterior cutaneous
femoral nerve**

Sciatic nerve

INTEGRATED HIP
POSTERIOR (BACK)

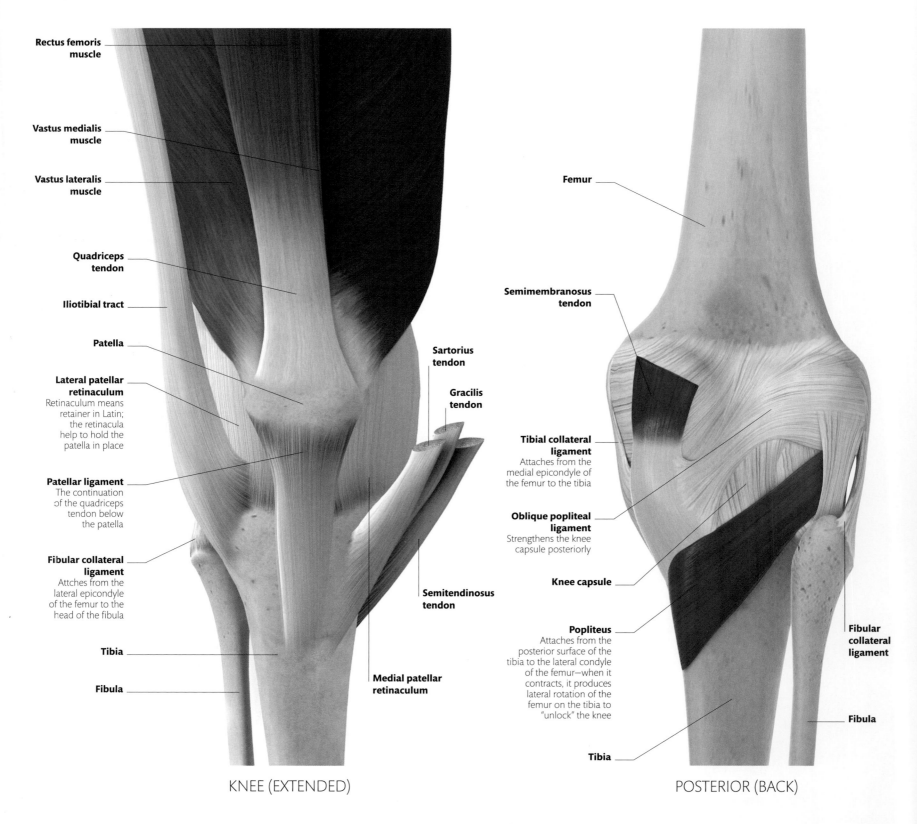

Rectus femoris muscle

Vastus medialis muscle

Vastus lateralis muscle

Quadriceps tendon

Iliotibial tract

Patella

Lateral patellar retinaculum
Retinaculum means retainer in Latin; the retinacula help to hold the patella in place

Patellar ligament
The continuation of the quadriceps tendon below the patella

Fibular collateral ligament
Attches from the lateral epicondyle of the femur to the head of the fibula

Tibia

Fibula

Sartorius tendon

Gracilis tendon

Semitendinosus tendon

Medial patellar retinaculum

KNEE (EXTENDED)

Femur

Semimembranosus tendon

Tibial collateral ligament
Attaches from the medial epicondyle of the femur to the tibia

Oblique popliteal ligament
Strengthens the knee capsule posteriorly

Knee capsule

Popliteus
Attaches from the posterior surface of the tibia to the lateral condyle of the femur—when it contracts, it produces lateral rotation of the femur on the tibia to "unlock" the knee

Fibular collateral ligament

Fibula

Tibia

POSTERIOR (BACK)

HIP AND THIGH **KNEE**

The knee joint is formed by the articulation of the femur with the tibia and patella. Although primarily a hinge joint, some sliding and axial rotation also occurs, alongside the main, hingelike motion of flexion and extension. These complex movements are reflected by the complexity of the joint. Inside the knee joint, the cruciate ligaments form attachments between the femur and the tibia, crossing over each other as their name suggests. There are two crescent-shaped pads of fibrocartilage lying on the articular facets of the tibia: the medial and lateral meniscus. Around the joint, tucked between bone, ligaments, and tendons, are many small bursae—bags of synovial fluid that help to keep everything moving smoothly.

The area behind the knee joint, known as the popliteal fossa, contains a large amount of fat, but also important nerves and vessels making their way between the thigh and the leg.

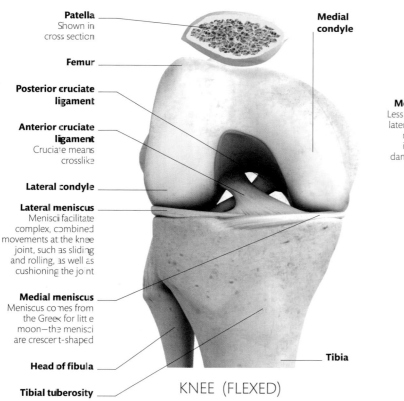

Patella
Shown in cross section

Femur

Posterior cruciate ligament

Anterior cruciate ligament
Cruciate means crosslike

Lateral condyle

Lateral meniscus
Menisci facilitate complex, combined movements at the knee joint, such as sliding and rolling, as well as cushioning the joint

Medial meniscus
Meniscus comes from the Greek for little moon—the menisci are crescent-shaped

Head of fibula

Tibial tuberosity

Medial condyle

Tibia

KNEE (FLEXED)

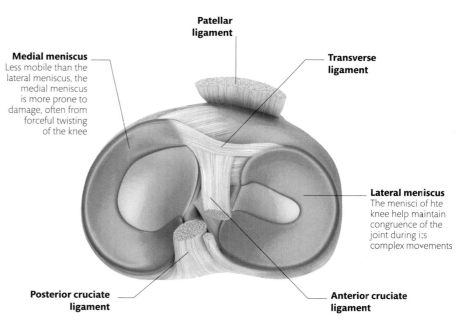

Medial meniscus
Less mobile than the lateral meniscus, the medial meniscus is more prone to damage, often from forceful twisting of the knee

Patellar ligament

Transverse ligament

Lateral meniscus
The menisci of hte knee help maintain congruence of the joint during its complex movements

Posterior cruciate ligament

Anterior cruciate ligament

TIBIAL PLATEAU (SUPERIOR VIEW)

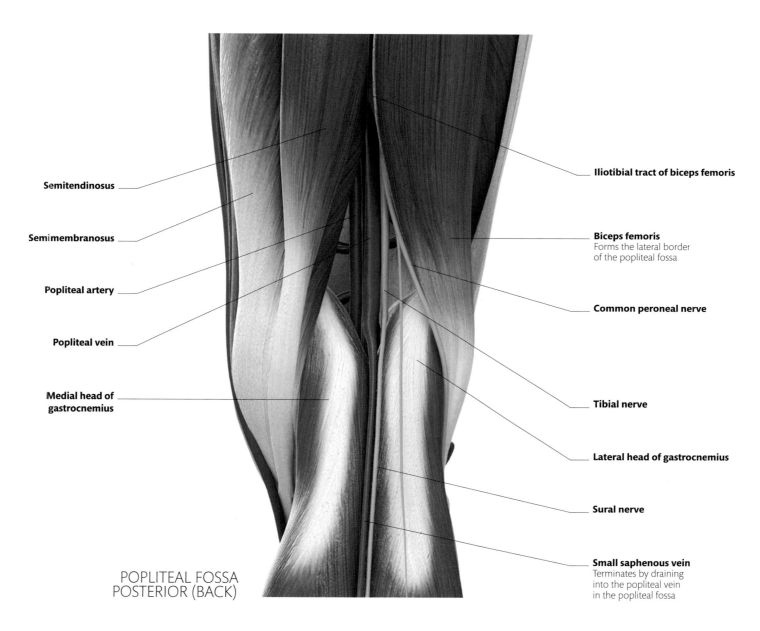

Semitendinosus

Semimembranosus

Popliteal artery

Popliteal vein

Medial head of gastrocnemius

Iliotibial tract of biceps femoris

Biceps femoris
Forms the lateral border of the popliteal fossa

Common peroneal nerve

Tibial nerve

Lateral head of gastrocnemius

Sural nerve

Small saphenous vein
Terminates by draining into the popliteal vein in the popliteal fossa

POPLITEAL FOSSA
POSTERIOR (BACK)

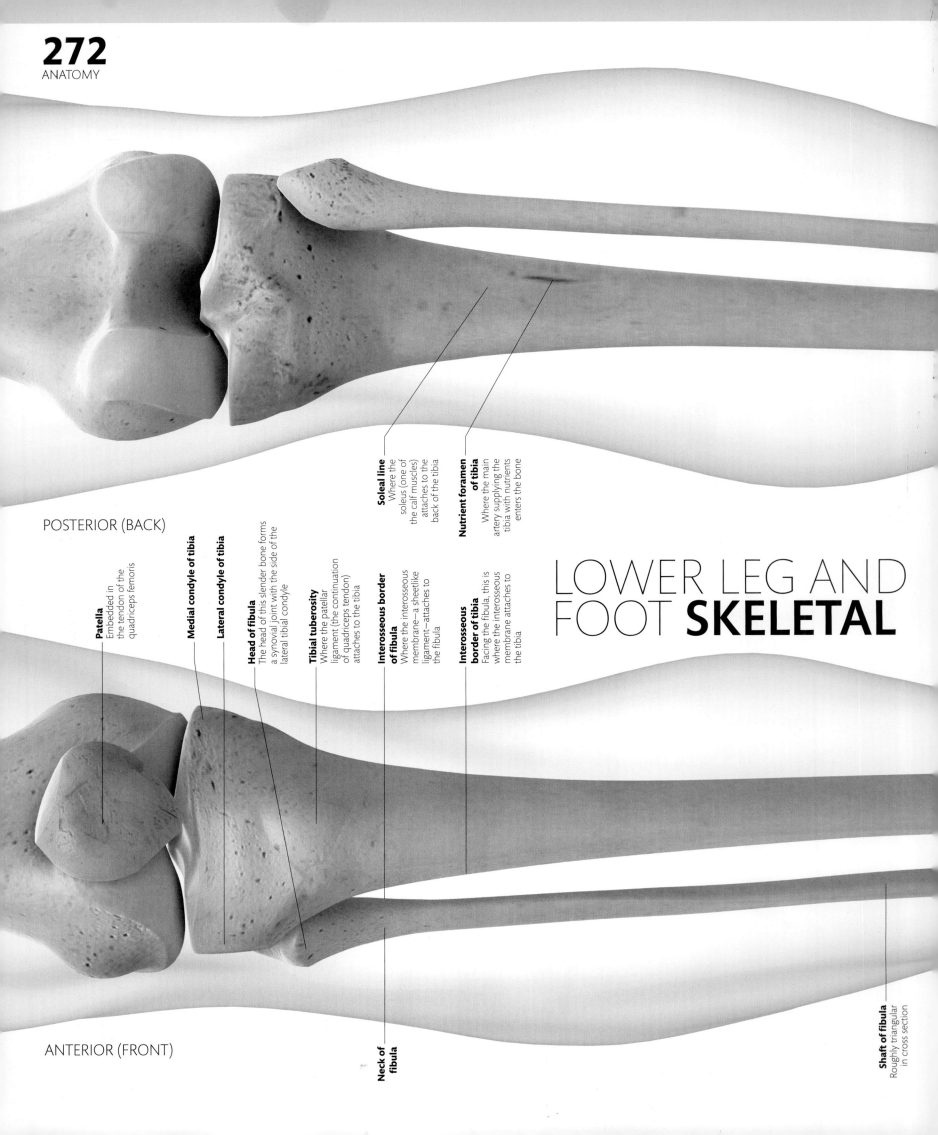

POSTERIOR (BACK)

Soleal line
Where the soleus (one of the calf muscles) attaches to the back of the tibia

Nutrient foramen of tibia
Where the main artery supplying the tibia with nutrients enters the bone

Patella
Embedded in the tendon of the quadriceps femoris

Medial condyle of tibia

Lateral condyle of tibia

Head of fibula
The head of this slender bone forms a synovial joint with the side of the lateral tibial condyle

Tibial tuberosity
Where the patellar ligament (the continuation of quadriceps tendon) attaches to the tibia

Interosseous border of fibula
Where the interosseous membrane—a sheetlike ligament—attaches to the fibula

Interosseous border of tibia
Facing the fibula, this is where the interosseous membrane attaches to the tibia

LOWER LEG AND FOOT **SKELETAL**

ANTERIOR (FRONT)

Neck of fibula

Shaft of fibula
Roughly triangular in cross section

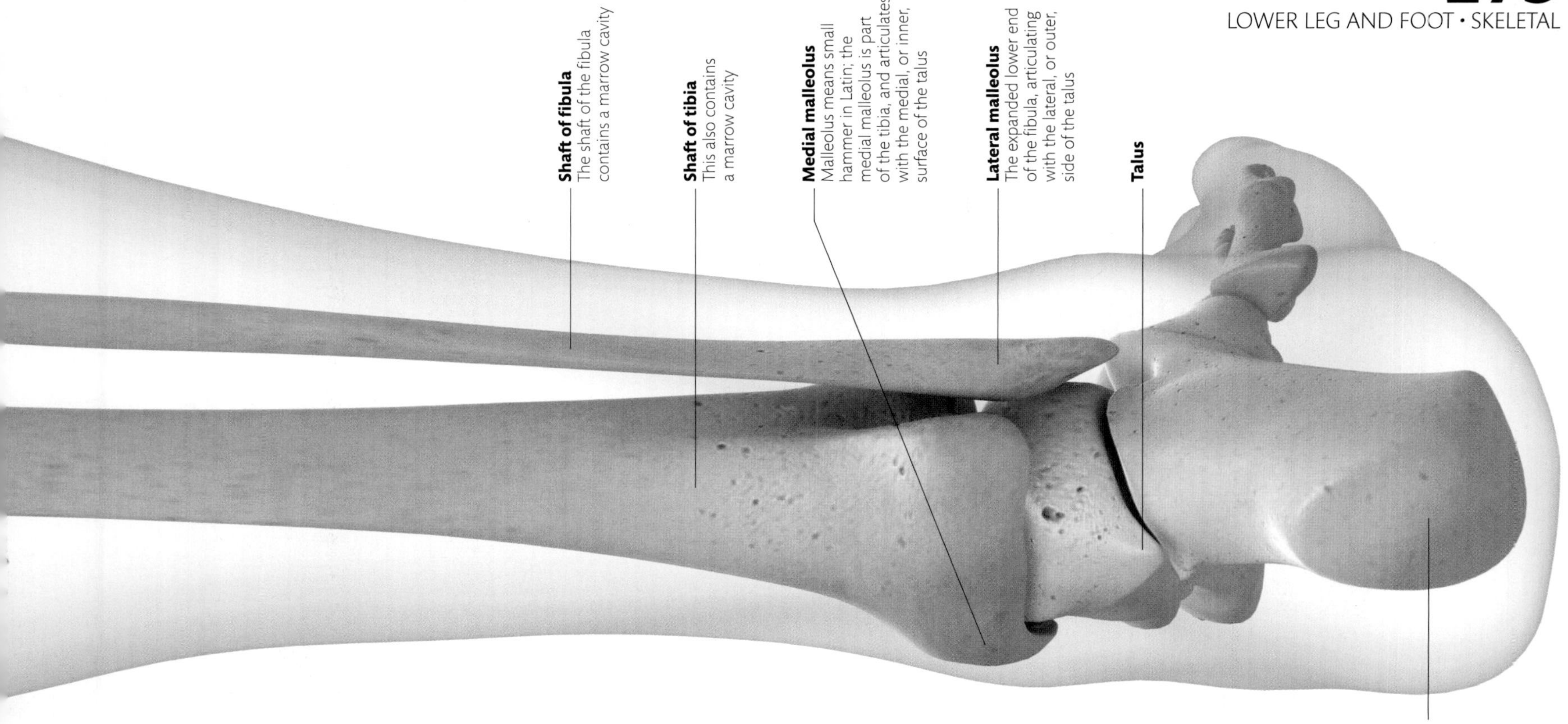

Shaft of fibula
The shaft of the fibula contains a marrow cavity

Shaft of tibia
This also contains a marrow cavity

Medial malleolus
Malleolus means small hammer in Latin; the medial malleolus is part of the tibia, and articulates with the medial, or inner, surface of the talus

Lateral malleolus
The expanded lower end of the fibula, articulating with the lateral, or outer, side of the talus

Talus

The tibia is the main weight-bearing bone of the lower leg. The fibula, which attaches to the tibia below the knee joint, provides extra areas for the attachment of muscles in the shin and calf and also forms part of the ankle joint. The foot comprises the tarsal bones, metatarsals, and phalanges. The arrangement of these bones is very similar to that of the carpals, metacarpals, and phalanges in the hand. In fact, each limb can be seen to be constructed to a common plan, with a limb girdle providing attachment to the thorax or spine, a single long bone in the first segment, two long bones in the second, a collection of small bones (at the wrist or ankle), and a fan of long, slender bones forming fingers or toes.

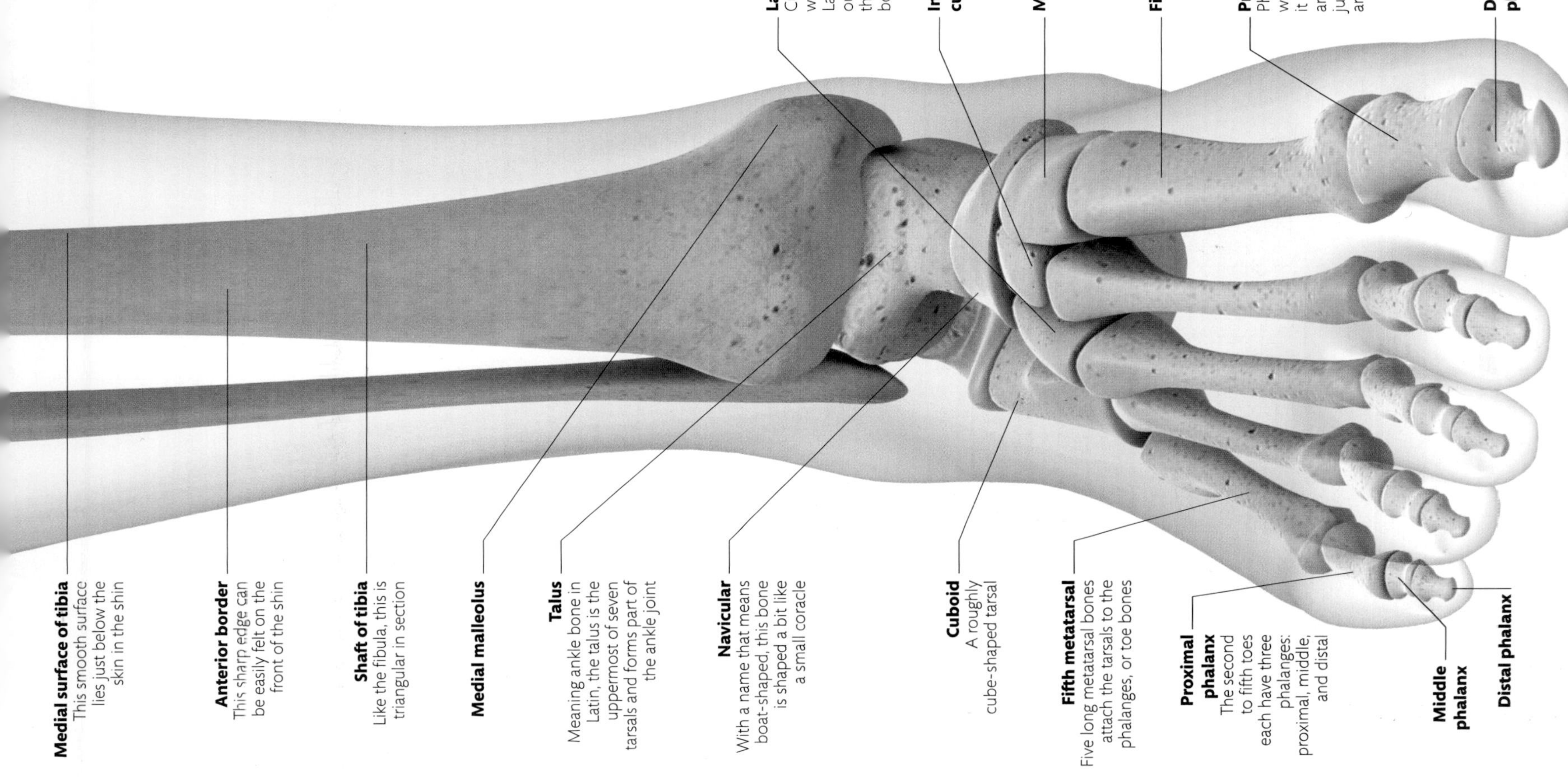

Lateral cuneiform
Cuneiform means wedge-shaped in Latin; this is the outermost of the three cuneiform bones in the foot

Intermediate cuneiform

Medial cuneiform

First metatarsal

Proximal phalanx
Phalanx comes from a Greek word for a line of infantry, and it refers to both the finger and toe bones; the big toe has just two phalanges: proximal and distal

Distal phalanx

Calcaneus
Meaning heel bone in Latin, this is the largest tarsal bone, projecting posteriorly to form a lever to which the Achilles tendon attaches

Medial surface of tibia
This smooth surface lies just below the skin in the shin

Anterior border
This sharp edge can be easily felt on the front of the shin

Shaft of tibia
Like the fibula, this is triangular in section

Medial malleolus

Talus
Meaning ankle bone in Latin, the talus is the uppermost of seven tarsals and forms part of the ankle joint

Navicular
With a name that means boat-shaped, this bone is shaped a bit like a small coracle

Cuboid
A roughly cube-shaped tarsal

Fifth metatarsal
Five long metatarsal bones attach the tarsals to the phalanges, or toe bones

Proximal phalanx
The second to fifth toes each have three phalanges: proximal, middle, and distal

Middle phalanx

Distal phalanx

LATERAL (OUTSIDE)

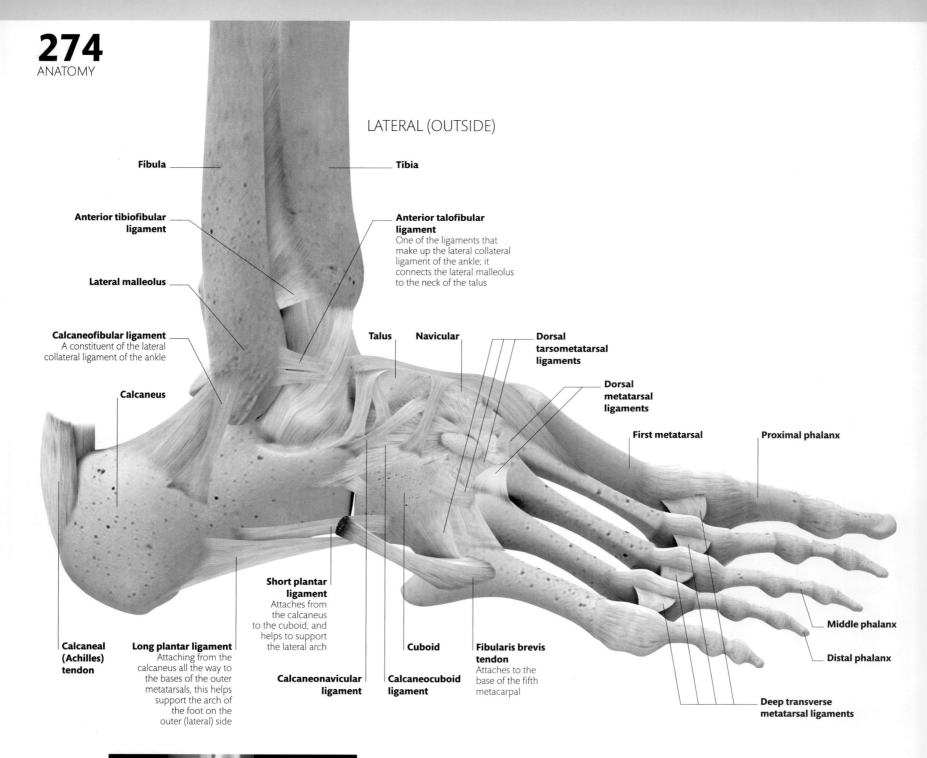

Fibula

Tibia

Anterior tibiofibular ligament

Anterior talofibular ligament
One of the ligaments that make up the lateral collateral ligament of the ankle; it connects the lateral malleolus to the neck of the talus

Lateral malleolus

Calcaneofibular ligament
A constituent of the lateral collateral ligament of the ankle

Talus

Navicular

Dorsal tarsometatarsal ligaments

Calcaneus

Dorsal metatarsal ligaments

First metatarsal

Proximal phalanx

Short plantar ligament
Attaches from the calcaneus to the cuboid, and helps to support the lateral arch

Middle phalanx

Distal phalanx

Calcaneal (Achilles) tendon

Long plantar ligament
Attaching from the calcaneus all the way to the bases of the outer metatarsals, this helps support the arch of the foot on the outer (lateral) side

Calcaneonavicular ligament

Cuboid

Calcaneocuboid ligament

Fibularis brevis tendon
Attaches to the base of the fifth metacarpal

Deep transverse metatarsal ligaments

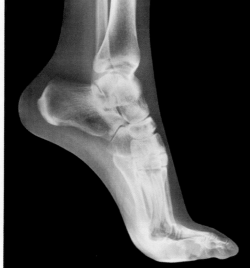

X-ray on tiptoe
This X-ray shows the foot in action. The calf muscles are pulling up on the lever of the calcaneus to flex the ankle down (plantarflex), while the metatarsophalangeal joints are extended.

LOWER LEG AND FOOT **SKELETAL**

The ankle joint is a simple hinge joint. The lower ends of the tibia and fibula are firmly bound together by ligaments, forming a strong fibrous joint, and making a spanner shape that neatly sits around the nut of the talus. The joint is stabilized by strong collateral ligaments on either side. The talus forms synovial joints (see p.49) with the calcaneus beneath it, and the navicular bone in front of it.

Level with the joint between the talus and the navicular is a joint between the calcaneus and the cuboid. These joints together allow the foot to be angled inward or outward—these movements are called inversion and eversion respectively. The skeleton of the foot is a sprung structure, with the bones forming arches, held together by ligaments and also supported by tendons.

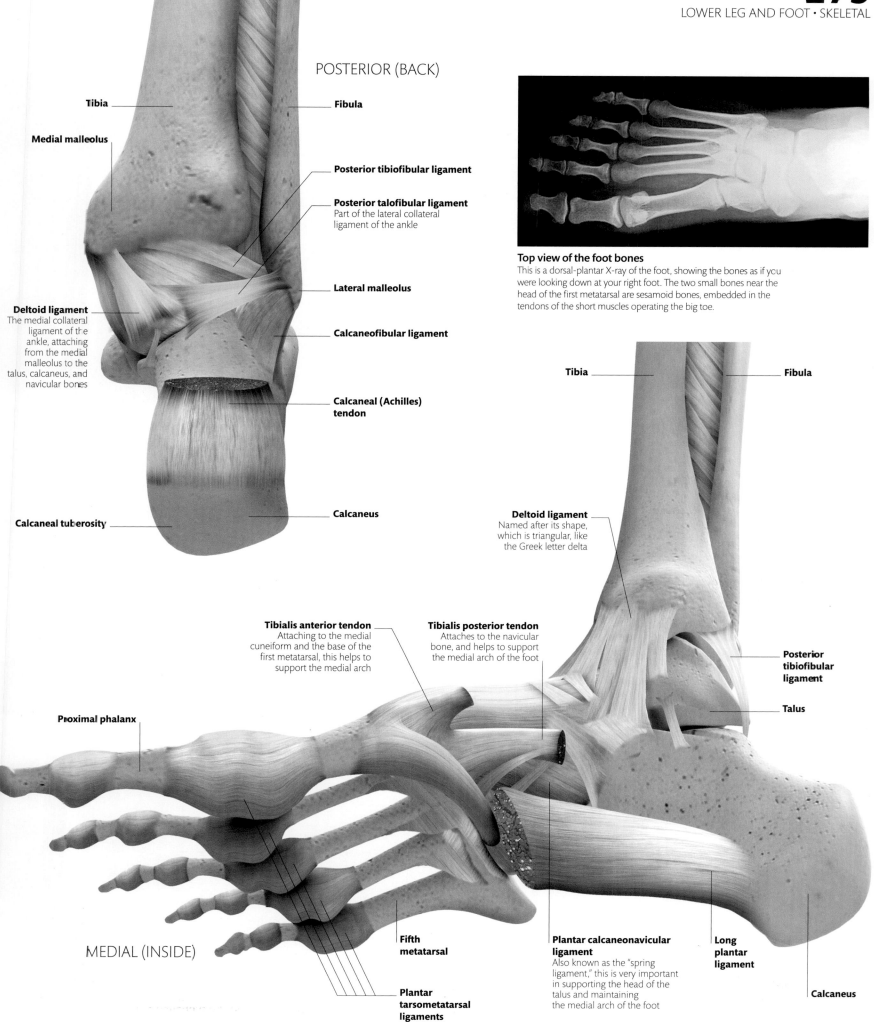

POSTERIOR (BACK)

Tibia

Fibula

Medial malleolus

Posterior tibiofibular ligament

Posterior talofibular ligament
Part of the lateral collateral
ligament of the ankle

Deltoid ligament
The medial collateral
ligament of the
ankle, attaching
from the medial
malleolus to the
talus, calcaneus, and
navicular bones

Lateral malleolus

Calcaneofibular ligament

**Calcaneal (Achilles)
tendon**

Calcaneal tuberosity

Calcaneus

Top view of the foot bones
This is a dorsal-plantar X-ray of the foot, showing the bones as if you
were looking down at your right foot. The two small bones near the
head of the first metatarsal are sesamoid bones, embedded in the
tendons of the short muscles operating the big toe.

Tibia

Fibula

Deltoid ligament
Named after its shape,
which is triangular, like
the Greek letter delta

**Posterior
tibiofibular
ligament**

Talus

Tibialis anterior tendon
Attaching to the medial
cuneiform and the base of the
first metatarsal, this helps to
support the medial arch

Tibialis posterior tendon
Attaches to the navicular
bone, and helps to support
the medial arch of the foot

Proximal phalanx

**Fifth
metatarsal**

MEDIAL (INSIDE)

**Plantar
tarsometatarsal
ligaments**

**Plantar calcaneonavicular
ligament**
Also known as the "spring
ligament," this is very important
in supporting the head of the
talus and maintaining
the medial arch of the foot

**Long
plantar
ligament**

Calcaneus

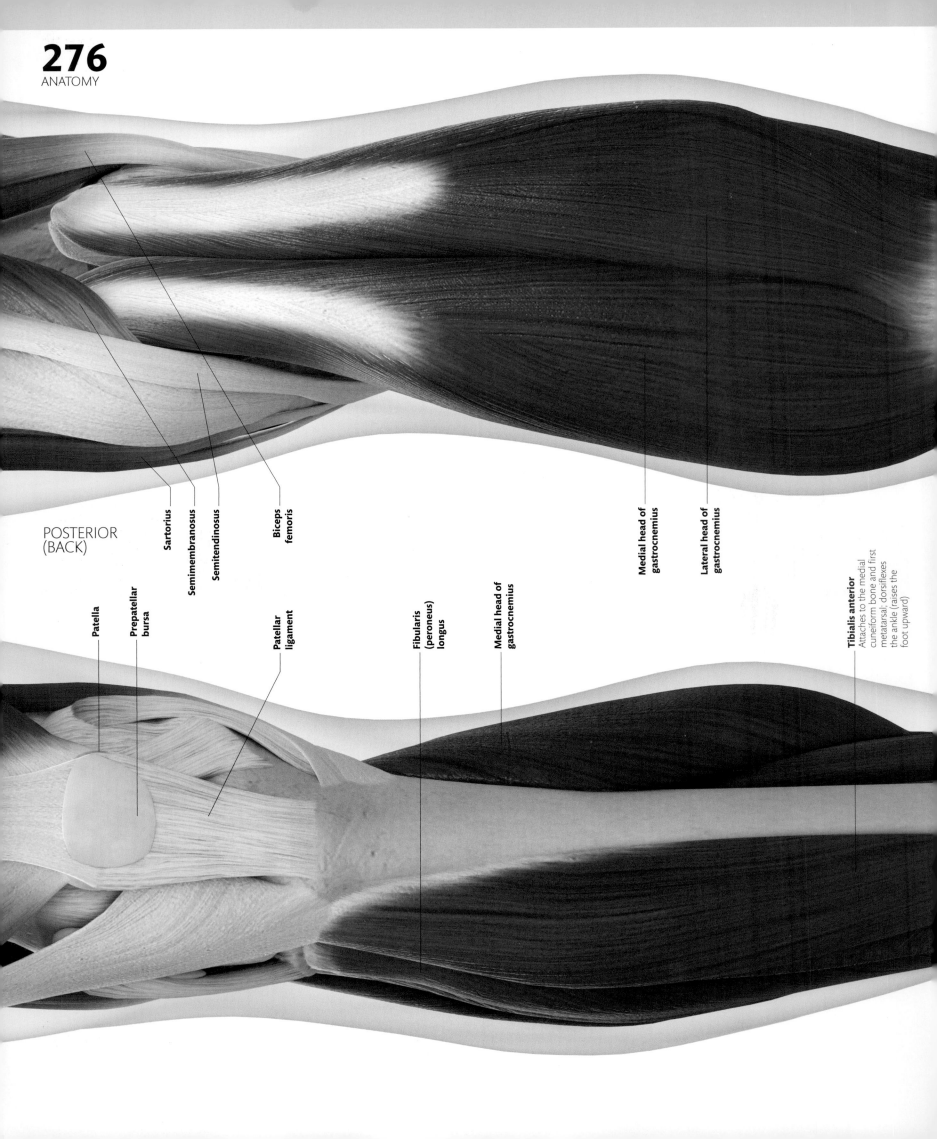

POSTERIOR
(BACK)

Sartorius

Semimembranosus

Semitendinosus

Biceps femoris

Medial head of gastrocnemius

Lateral head of gastrocnemius

Patella

Prepatellar bursa

Patellar ligament

Fibularis (peroneus) longus

Medial head of gastrocnemius

Tibialis anterior
Attaches to the medial cuneiform bone and first metatarsal; dorsiflexes the ankle (raises the foot upward)

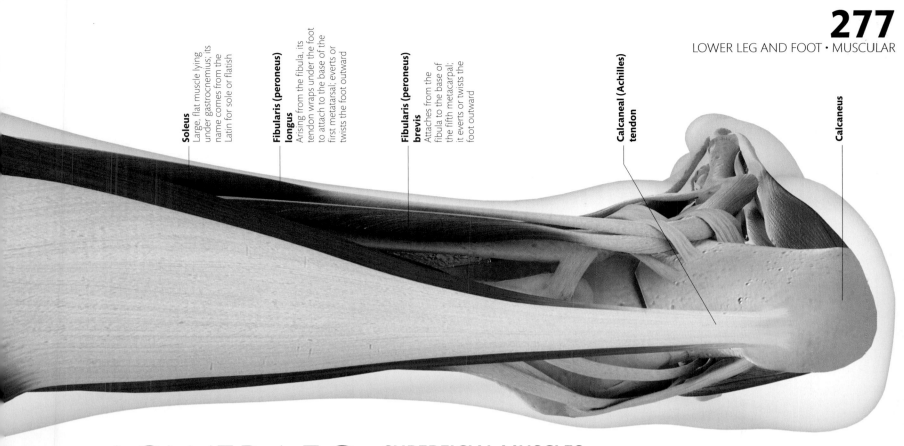

Soleus
Large, flat muscle lying under the gastrocnemius; its name comes from the Latin for sole or flatfish

Fibularis (peroneus) longus
Arising from the fibula, its tendon wraps under the foot to attach to the base of the first metatarsal; everts or twists the foot outward

Fibularis (peroneus) brevis
Attaches from the fibula to the base of the fifth metacarpal; it everts or twists the foot outward

Calcaneal (Achilles) tendon

Calcaneus

LOWER LEG AND FOOT
MUSCULAR

SUPERFICIAL MUSCLES

You can feel the medial surface of the tibia easily, just under the skin on the front of your lower leg, on the inner side. Move your fingers outward, and you feel the sharp border of the bone, and then a soft wedge of muscles alongside it. These muscles have tendons that run down to the foot. They can pull the foot upward at the ankle, in a movement called dorsiflexion. Some extensor tendons continue all the way to the toes. There are much bulkier muscles on the back of the leg, and these form the calf. The gastrocnemius, and the soleus underneath it, are large muscles that join together to form the Achilles tendon. They pull up on the lever of the calcaneus, pushing the ball of the foot down. They are involved as the foot pushes off from the ground during walking and running.

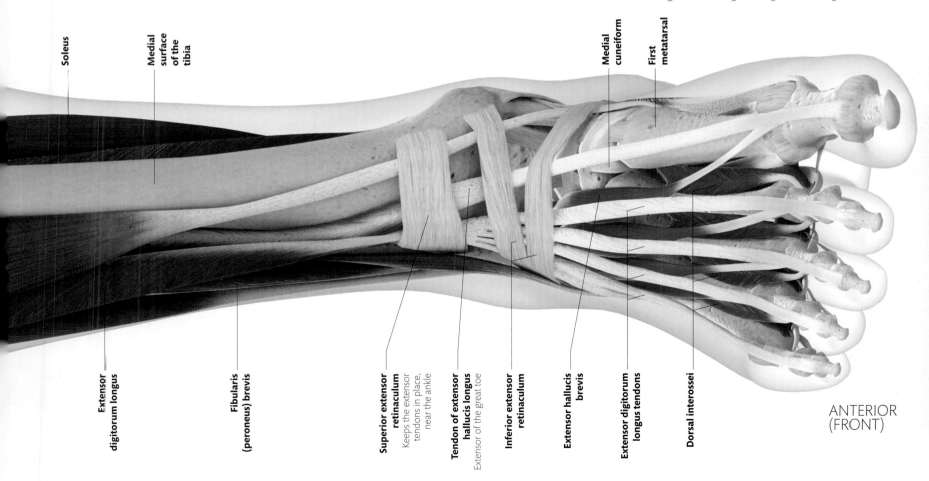

Soleus

Medial surface of the tibia

Medial cuneiform

First metatarsal

Extensor digitorum longus

Fibularis (peroneus) brevis

Superior extensor retinaculum
Keeps the extensor tendons in place, near the ankle

Tendon of extensor hallucis longus
Extensor of the great toe

Inferior extensor retinaculum

Extensor hallucis brevis

Extensor digitorum longus tendons

Dorsal interossei

ANTERIOR (FRONT)

LOWER LEG AND FOOT **MUSCULAR**

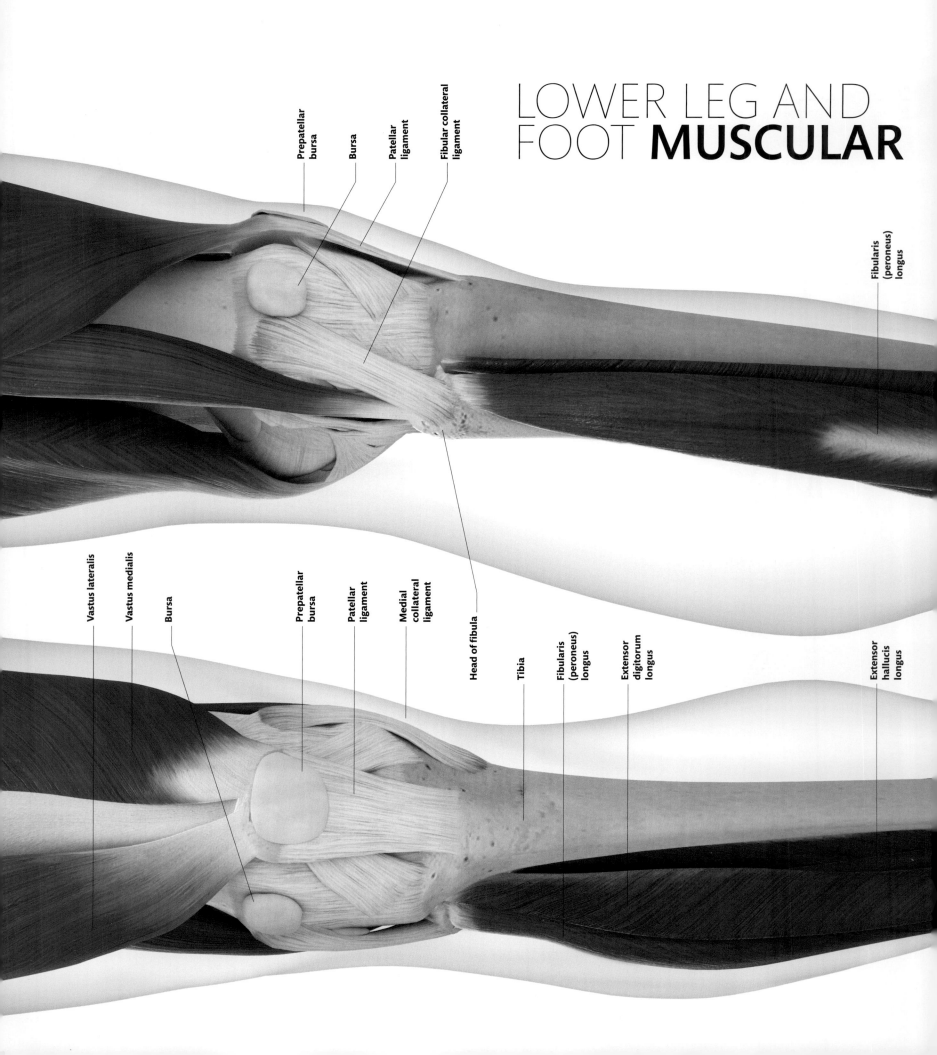

Prepatellar bursa

Bursa

Patellar ligament

Fibular collateral ligament

Fibularis (peroneus) longus

Vastus lateralis

Vastus medialis

Bursa

Prepatellar bursa

Patellar ligament

Medial collateral ligament

Head of fibula

Tibia

Fibularis (peroneus) longus

Extensor digitorum longus

Extensor hallucis longus

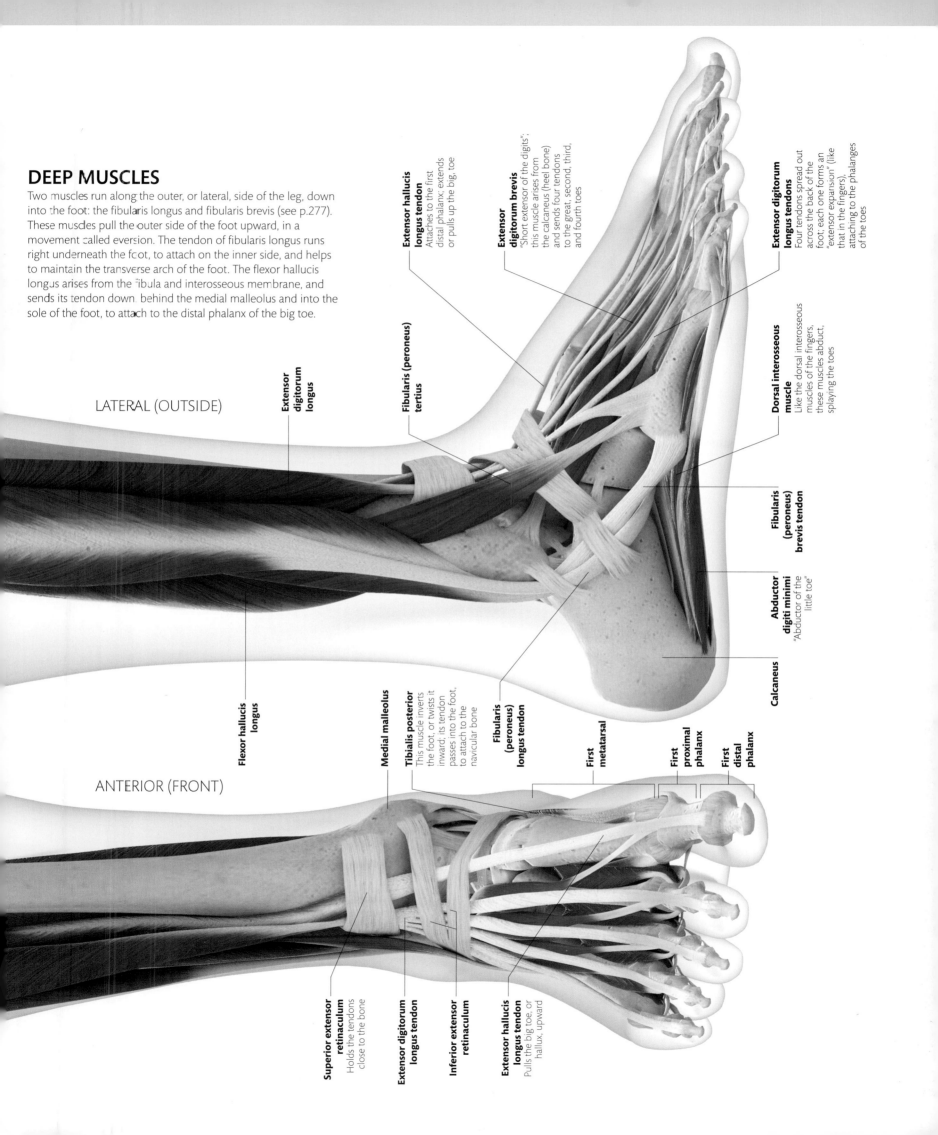

DEEP MUSCLES

Two muscles run along the outer, or lateral, side of the leg, down into the foot: the fibularis longus and fibularis brevis (see p.277). These muscles pull the outer side of the foot upward, in a movement called eversion. The tendon of fibularis longus runs right underneath the foot, to attach on the inner side, and helps to maintain the transverse arch of the foot. The flexor hallucis longus arises from the fibula and interosseous membrane, and sends its tendon down behind the medial malleolus and into the sole of the foot, to attach to the distal phalanx of the big toe.

LATERAL (OUTSIDE)

ANTERIOR (FRONT)

Extensor hallucis longus tendon
Attaches to the first distal phalanx; extends or pulls up the big, toe

Extensor digitorum brevis
"Short extensor of the digits"; this muscle arises from the calcaneus (heel bone) and sends four tendons to the great, second, third, and fourth toes

Extensor digitorum longus tendons
Four tendons spread out across the back of the foot; each one forms an "extensor expansion" (like that in the fingers), attaching to the phalanges of the toes

Extensor digitorum longus

Fibularis (peroneus) tertius

Dorsal interosseous muscle
Like the dorsal interosseous muscles of the fingers, these muscles abduct, splaying the toes

Fibularis (peroneus) brevis tendon

Abductor digiti minimi
"Abductor of the little toe"

Calcaneus

Flexor hallucis longus

Medial malleolus

Tibialis posterior
This muscle inverts the foot, or twists it inward; its tendon passes into the foot, to attach to the navicular bone

Fibularis (peroneus) longus tendon

First metatarsal

First proximal phalanx

First distal phalanx

Superior extensor retinaculum
Holds the tendons close to the bone

Extensor digitorum longus tendon

Inferior extensor retinaculum

Extensor hallucis longus tendon
Pulls the big toe, or hallux, upward

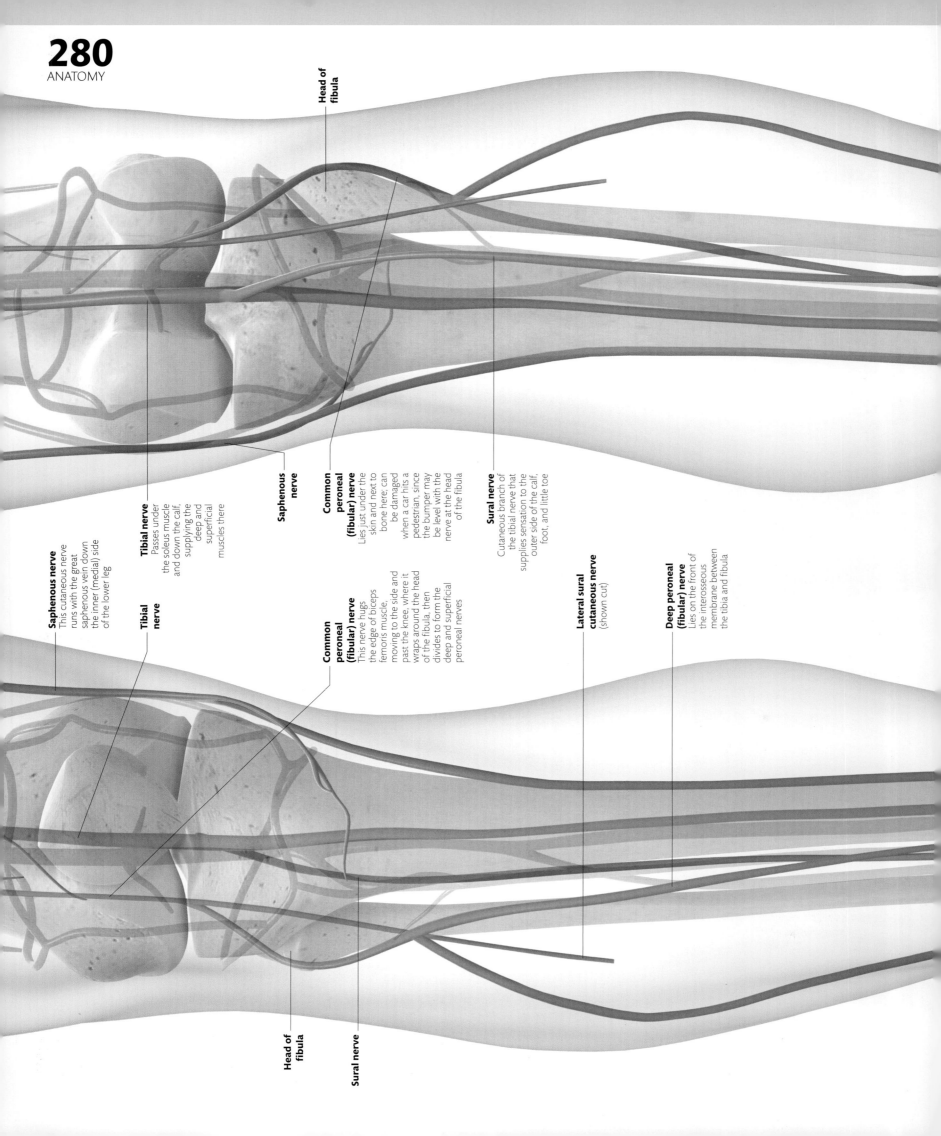

Head of fibula

Saphenous nerve
This cutaneous nerve runs with the great saphenous vein down the inner (medial) side of the lower leg

Tibial nerve
Passes under the soleus muscle and down the calf, supplying the deep and superficial muscles there

Saphenous nerve

Common peroneal (fibular) nerve
Lies just under the skin and next to bone here; can be damaged when a car hits a pedestrian, since the bumper may be level with the nerve at the head of the fibula

Sural nerve
Cutaneous branch of the tibial nerve that supplies sensation to the outer side of the calf, foot, and little toe

Tibial nerve

Common peroneal (fibular) nerve
This nerve hugs the edge of biceps femoris muscle, moving to the side and past the knee, where it wraps around the head of the fibula, then divides to form the deep and superficial peroneal nerves

Lateral sural cutaneous nerve
(shown cut)

Deep peroneal (fibular) nerve
Lies on the front of the interosseous membrane between the tibia and fibula

Head of fibula

Sural nerve

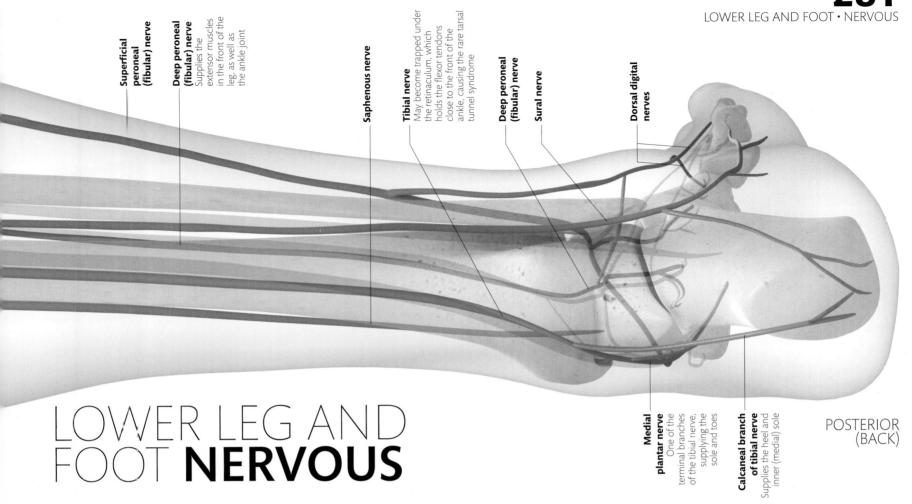

Superficial peroneal (fibular) nerve

Deep peroneal (fibular) nerve
Supplies the extensor muscles in the front of the leg, as well as the ankle joint

Saphenous nerve

Tibial nerve
May become trapped under the retinaculum, which holds the flexor tendons close to the front of the ankle, causing the rare tarsal tunnel syndrome

Deep peroneal (fibular) nerve

Sural nerve

Dorsal digital nerves

Medial plantar nerve
One of the terminal branches of the tibial nerve, supplying the sole and toes

Calcaneal branch of tibial nerve
Supplies the heel and inner (medial) sole

POSTERIOR (BACK)

LOWER LEG AND FOOT **NERVOUS**

The common peroneal nerve runs past the knee and wraps around the neck of the fibula. Then it splits into the deep and superficial peroneal nerves. The deep peroneal nerve supplies the extensor muscles of the shin, then fans out to provide sensation to the skin at the back of the foot. The superficial peroneal nerve stays on the side of the leg and supplies the peroneal muscles. The tibial nerve runs through the popliteal fossa (back of the knee), under the soleus muscle, and between the deep and superficial calf muscles, which it supplies. It continues behind the medial malleolus and under the foot, then splits into two plantar nerves that supply the small muscles of the foot and the skin of the sole.

Medial malleolus

Lateral plantar nerve
With the medial plantar nerve, supplies the muscles and skin of the sole and toes

ANTERIOR (FRONT)

Superficial peroneal (fibular) nerve
Supplies the peroneus longus and brevis muscles in the lower leg

Tibial nerve
Runs behind the medial malleolus

Saphenous nerve
Runs in front of the medial malleolus, to supply sensation to the inner (medial) side of the foot

Lateral branch of superficial peroneal nerve
With the medial branch, supplies skin over the top of the foot and toes

Medial branch of superficial peroneal nerve

Deep peroneal (fibular) nerve
Runs with the dorsal artery of the foot, and supplies the skin of the first web-space

Dorsal digital nerves
Branches of the superficial peroneal nerve

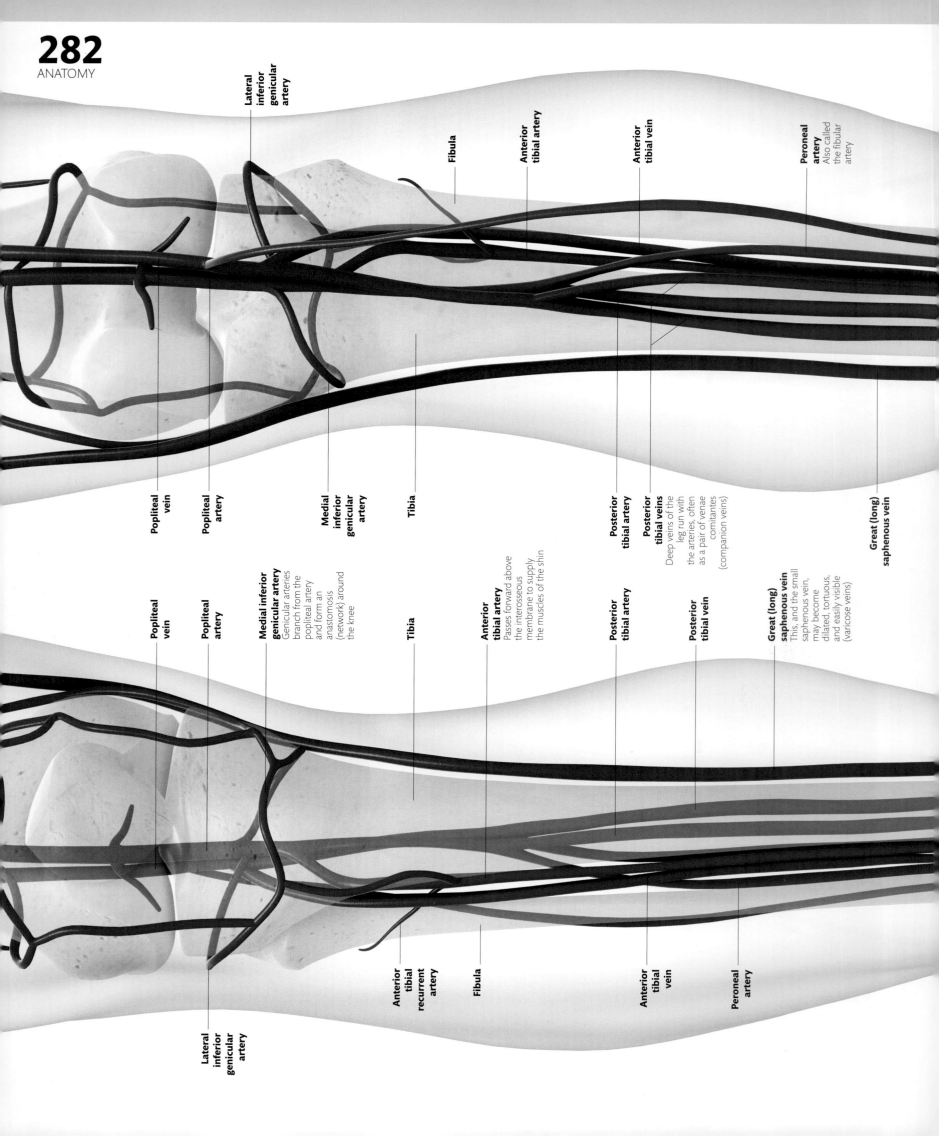

Lateral
inferior
genicular
artery

Fibula

Anterior
tibial artery

Anterior
tibial vein

Peroneal
artery
Also called
the fibular
artery

Popliteal
vein

Popliteal
artery

Medial
inferior
genicular
artery

Tibia

Posterior
tibial artery

Posterior
tibial veins
Deep veins of the
leg run with
the arteries, often
as a pair of venae
comitantes
(companion veins)

Great (long)
saphenous vein

Popliteal
vein

Popliteal
artery

Medial inferior
genicular artery
Genicular arteries
branch from the
popliteal artery
and form an
anastomosis
(network) around
the knee

Tibia

Anterior
tibial artery
Passes forward above
the interosseous
membrane to supply
the muscles of the shin

Posterior
tibial artery

Posterior
tibial vein

Great (long)
saphenous vein
This, and the small
saphenous vein,
may become
dilated, tortuous,
and easily visible
(varicose veins)

Anterior
tibial
recurrent
artery

Fibula

Anterior
tibial vein

Peroneal
artery

Lateral
inferior
genicular
artery

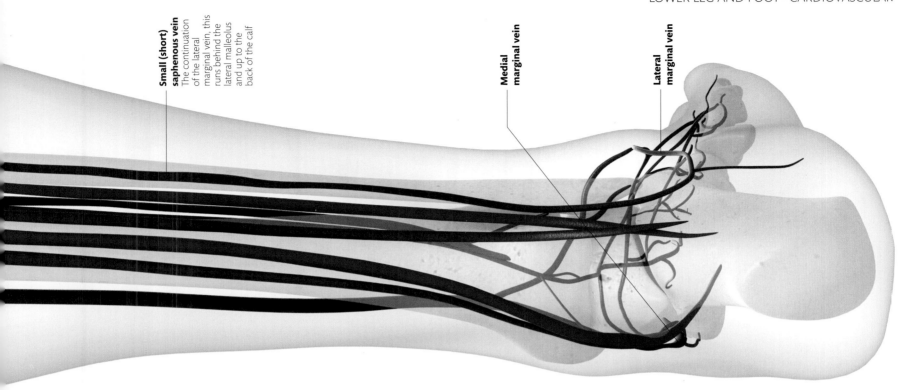

Small (short) saphenous vein
The continuation of the lateral marginal vein, this runs behind the lateral malleolus and up to the back of the calf

Medial marginal vein

Lateral marginal vein

POSTERIOR (BACK)

LOWER LEG AND FOOT
CARDIOVASCULAR

The popliteal artery runs deep across the back of the knee, dividing into two branches: the anterior and posterior tibial arteries. The former runs forward, piercing the interosseous membrane between the tibia and fibula, to supply the extensor muscles of the shin. It runs down past the ankle, onto the top of the foot, as the dorsalis pedis artery. The latter gives off a peroneal branch, supplying the muscles and skin on the leg's outer side. The posterior tibial artery itself continues in the calf, running with the tibial nerve and, like the nerve, divides into plantar branches to supply the sole of the foot. A network of superficial veins on the back of the foot is drained by the saphenous veins.

ANTERIOR (FRONT)

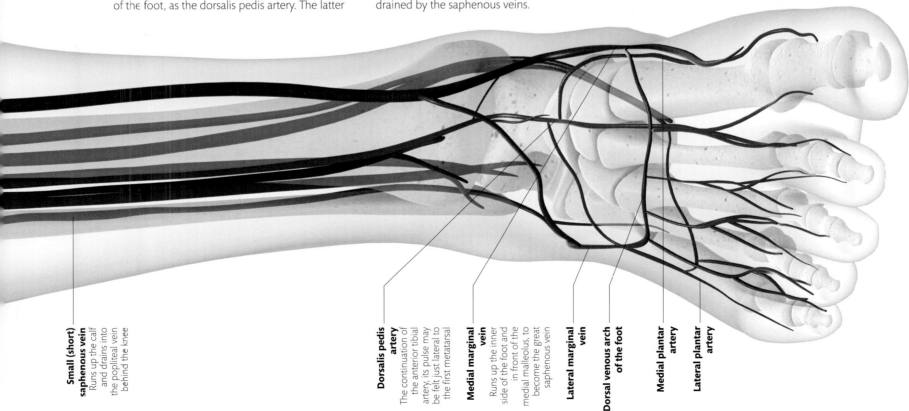

Small (short) saphenous vein
Runs up the calf and drains into the popliteal vein behind the knee

Dorsalis pedis artery
The continuation of the anterior tibial artery, its pulse may be felt just lateral to the first metatarsal

Medial marginal vein
Runs up the inner side of the foot and in front of the medial malleolus, to become the great saphenous vein

Lateral marginal vein

Dorsal venous arch of the foot

Medial plantar artery

Lateral plantar artery

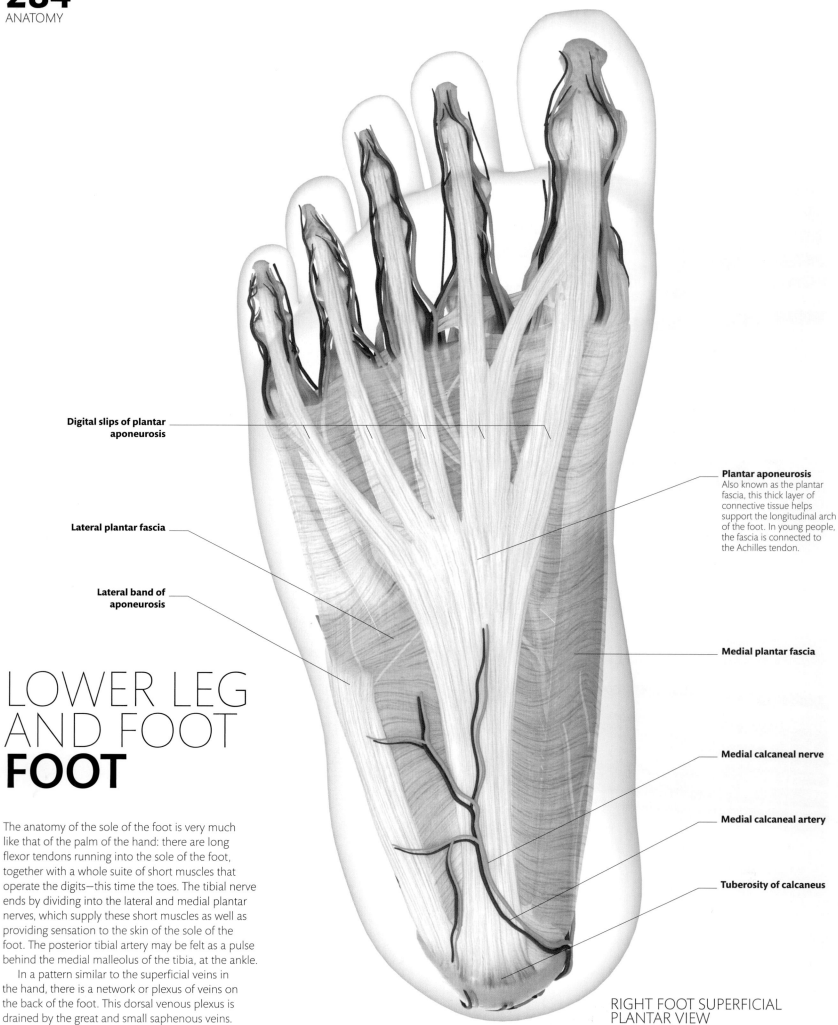

Digital slips of plantar
aponeurosis

Lateral plantar fascia

Lateral band of
aponeurosis

Plantar aponeurosis
Also known as the plantar
fascia, this thick layer of
connective tissue helps
support the longitudinal arch
of the foot. In young people,
the fascia is connected to
the Achilles tendon.

Medial plantar fascia

Medial calcaneal nerve

Medial calcaneal artery

Tuberosity of calcaneus

LOWER LEG AND FOOT
FOOT

The anatomy of the sole of the foot is very much
like that of the palm of the hand: there are long
flexor tendons running into the sole of the foot,
together with a whole suite of short muscles that
operate the digits—this time the toes. The tibial nerve
ends by dividing into the lateral and medial plantar
nerves, which supply these short muscles as well as
providing sensation to the skin of the sole of the
foot. The posterior tibial artery may be felt as a pulse
behind the medial malleolus of the tibia, at the ankle.

In a pattern similar to the superficial veins in
the hand, there is a network or plexus of veins on
the back of the foot. This dorsal venous plexus is
drained by the great and small saphenous veins.

RIGHT FOOT SUPERFICIAL
PLANTAR VIEW

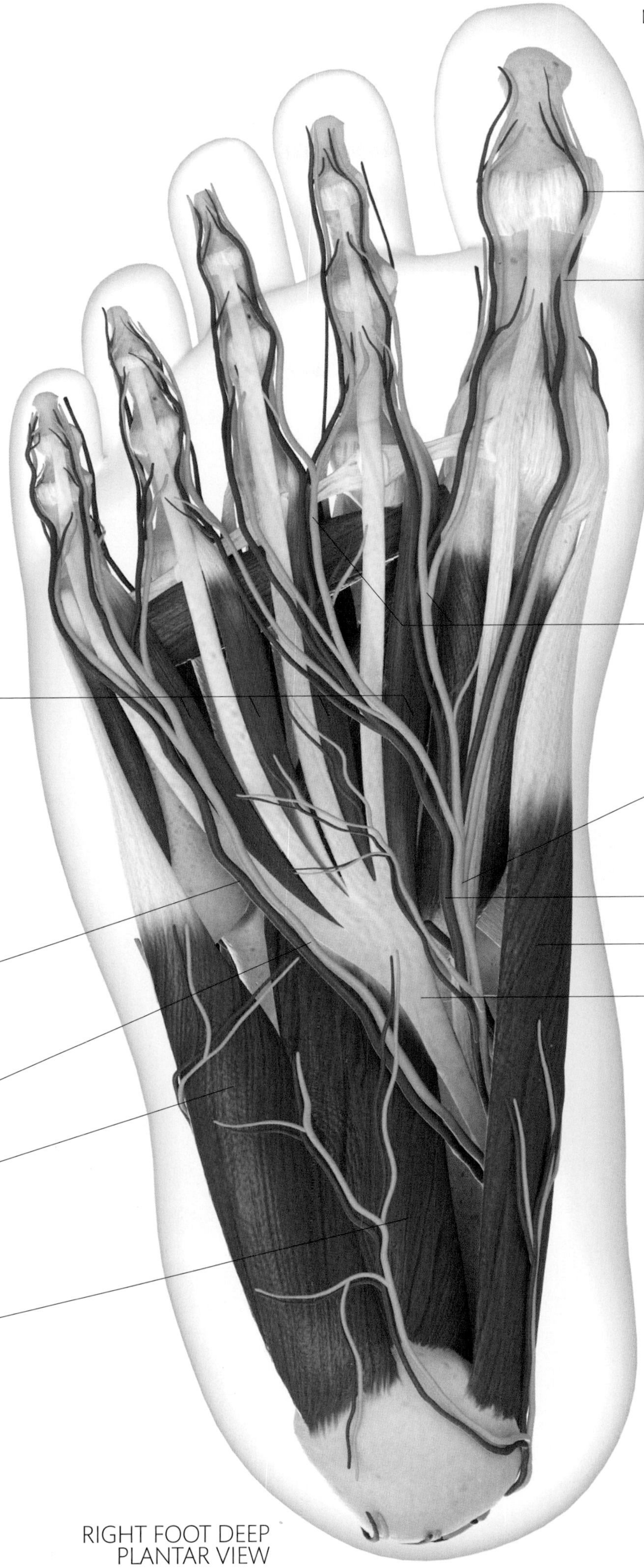

Proper plantar digital artery

Proper plantar digital nerve

Common plantar digital nerves

Lumbrical muscles

Medial plantar nerve
The plantar nerves are the
terminal branches of the tibial
nerve and supply the small
muscles of the sole of the foot
and the overlying skin

Medial plantar artery

Abductor hallucis

Lateral plantar artery
The plantar arteries are
branches of the posterial
tibial artery

Flexor digitorum longus

Lateral plantar nerve

Abductor digiti minimi

Quadratus plantae
Attaching from the calcaneus
to the tendon of flexor
digitorum longus, this muscle
helps redirect the line of pull
of the flexor tendons

RIGHT FOOT DEEP
PLANTAR VIEW

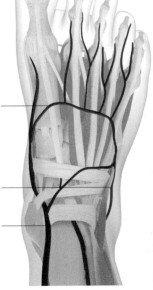

**Dorsal
venous arch**

**Small saphenous
vein**

**Great saphenous
vein**

VEINS (DORSAL)

1

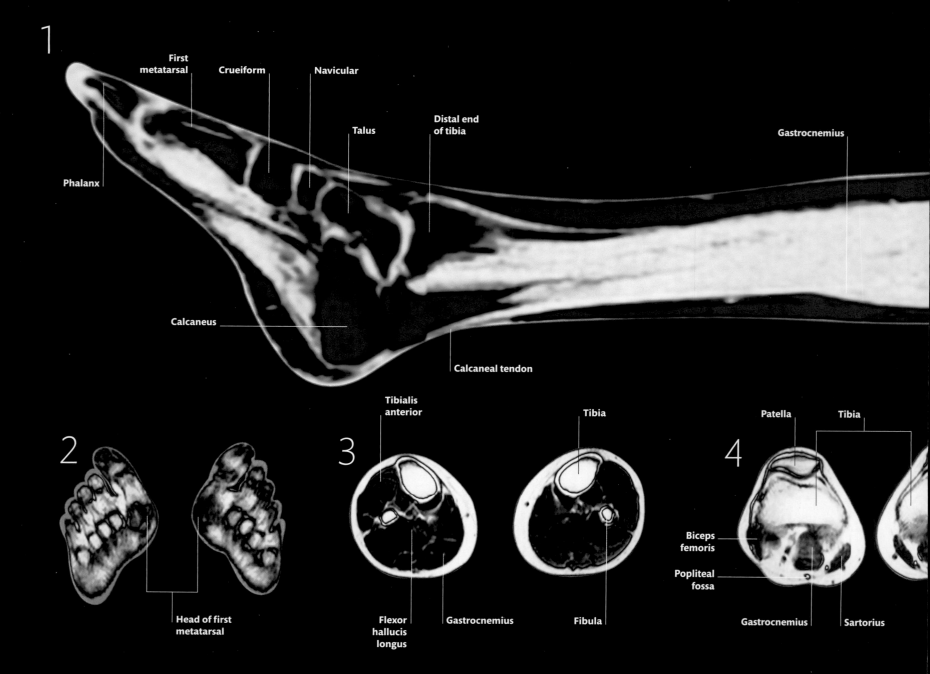

First metatarsal

Crueiform

Navicular

Distal end of tibia

Gastrocnemius

Talus

Phalanx

Calcaneus

Calcaneal tendon

2

Head of first metatarsal

3

Tibialis anterior

Tibia

Flexor hallucis longus

Gastrocnemius

Fibula

4

Patella

Tibia

Biceps femoris

Popliteal fossa

Gastrocnemius

Sartorius

LOWER LIMB AND FOOT **MRI**

The sequence of axial and transverse sections through the thigh and lower leg show how the muscles are arranged around the bones. Groups of muscles are bound together with fascia—fibrous packing tissue—forming three compartments in the thigh (the flexor, extensor, and adductor muscles), and three in the lower leg (flexor, extensor, and peroneal or fibular muscles). Nerves and deep blood vessels are also packaged together in sheaths of fascia, forming "neurovascular bundles." Section 2 shows the bones of the forefoot, while the tightly packed muscles surrounding the tibia and fibula in the lower leg are visible in section 3. At the knee joint, shown in section 4, the patella can be seen to fit neatly against the reciprocal shape of the femoral condyles. The neurovascular bundle is clearly visible here, at the back of the knee, in a space known as the popliteal fossa—with the hamstring muscles on either side. Sections 5 and 6, through the middle and upper thigh, show the powerful quadriceps and hamstring muscles surrounding the thigh bone, or femur.

LEVELS OF SCANS

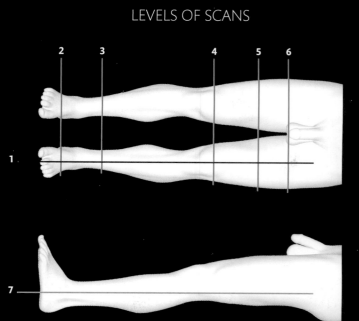

2   3   4   5   6

1

7

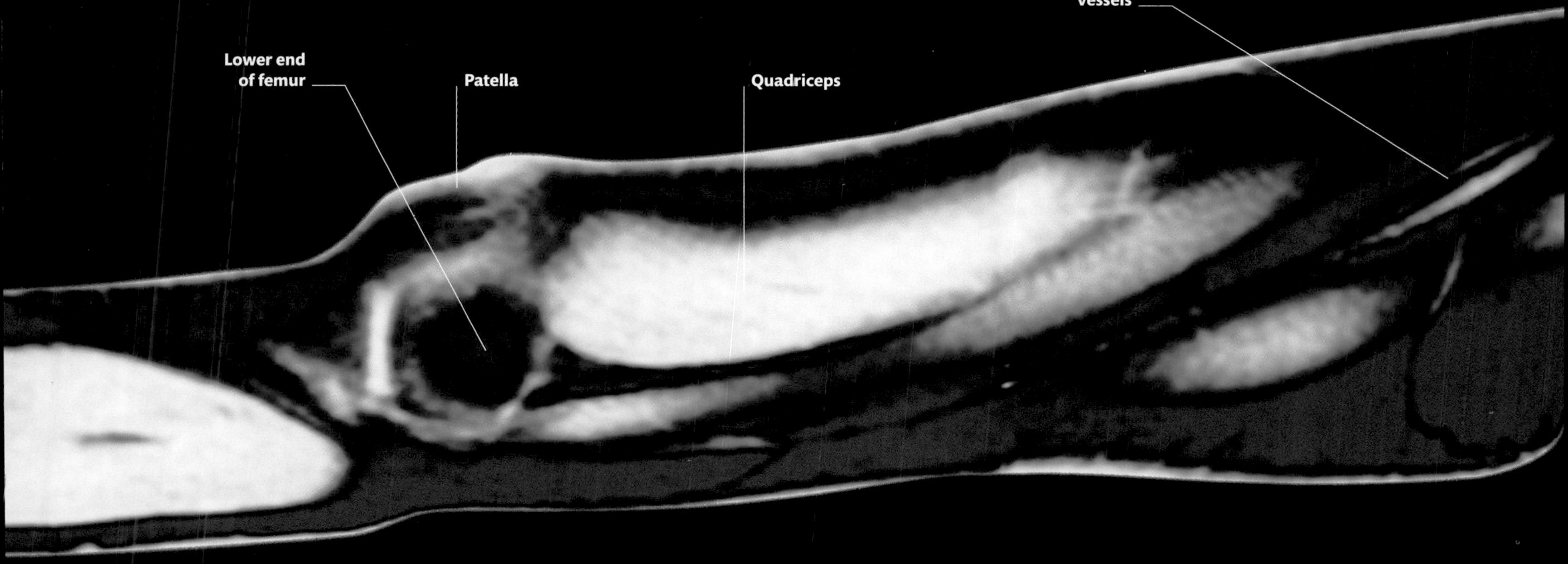

Femoral vessels

Lower end of femur

Patella

Quadriceps

5

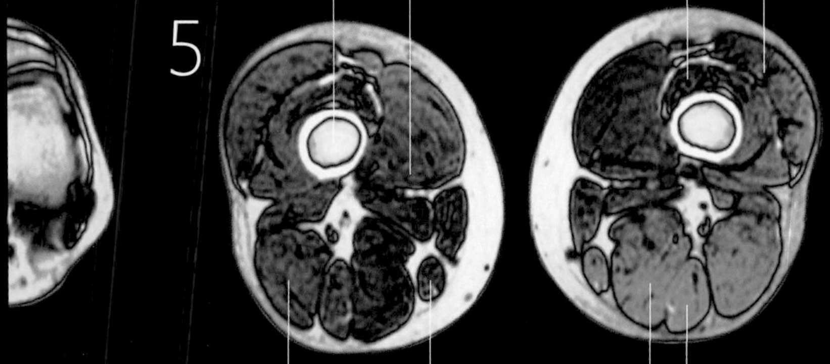

Femur

Vastus medialis

Vastus intermedius

Vastus lateralis

Biceps femoris

Gracilis

Semimembranosus

Semitendinosus

6

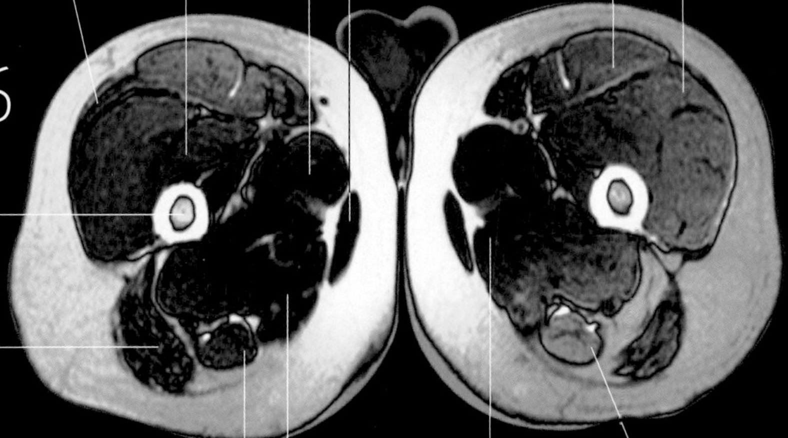

Tensor fasciae latae

Vastus medialis

Adductor longus

Gracilis

Rectus femoris

Vastus lateralis

Femur

Gluteus maximus

Semitendinosus

Adductor magnus

Adductor brevis

Long head of biceps femoris

7

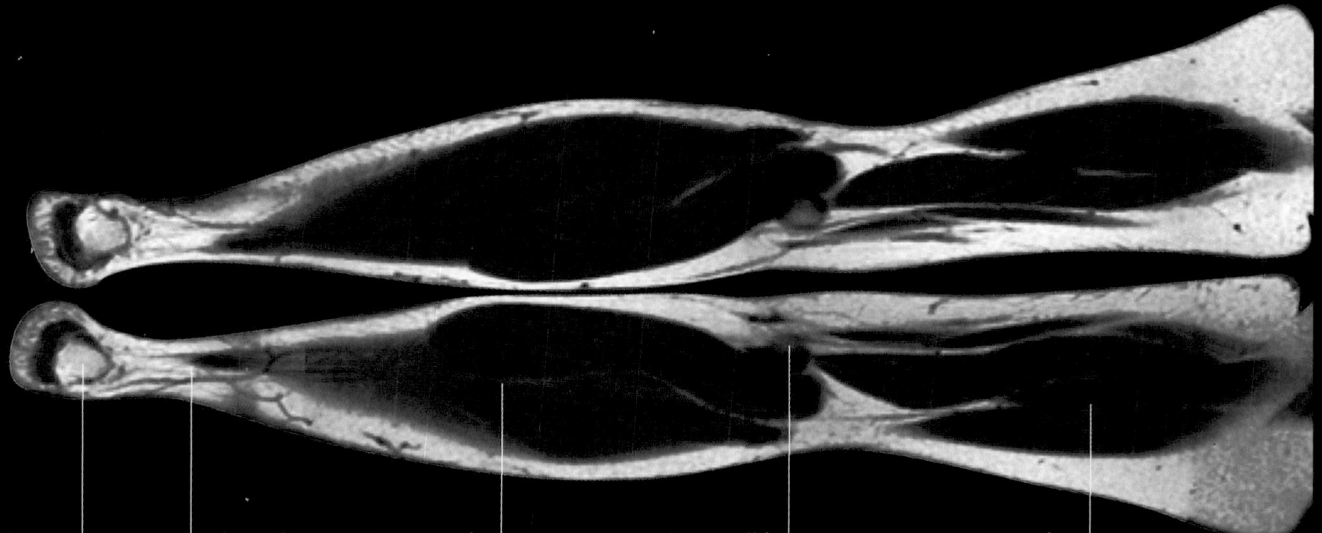

Calcaneus

Calcaneal tendon

Gastrocnemius

Tibia

Hamstrings

how the body works

The workings of the body begin at a molecular level—even a conscious perception can be traced to miniscule biochemical reactions at a cell wall. A myriad of processes are underway in the body at any given time, from the involuntary basics of staying alive to deliberate movement.

288
HOW THE BODY WORKS

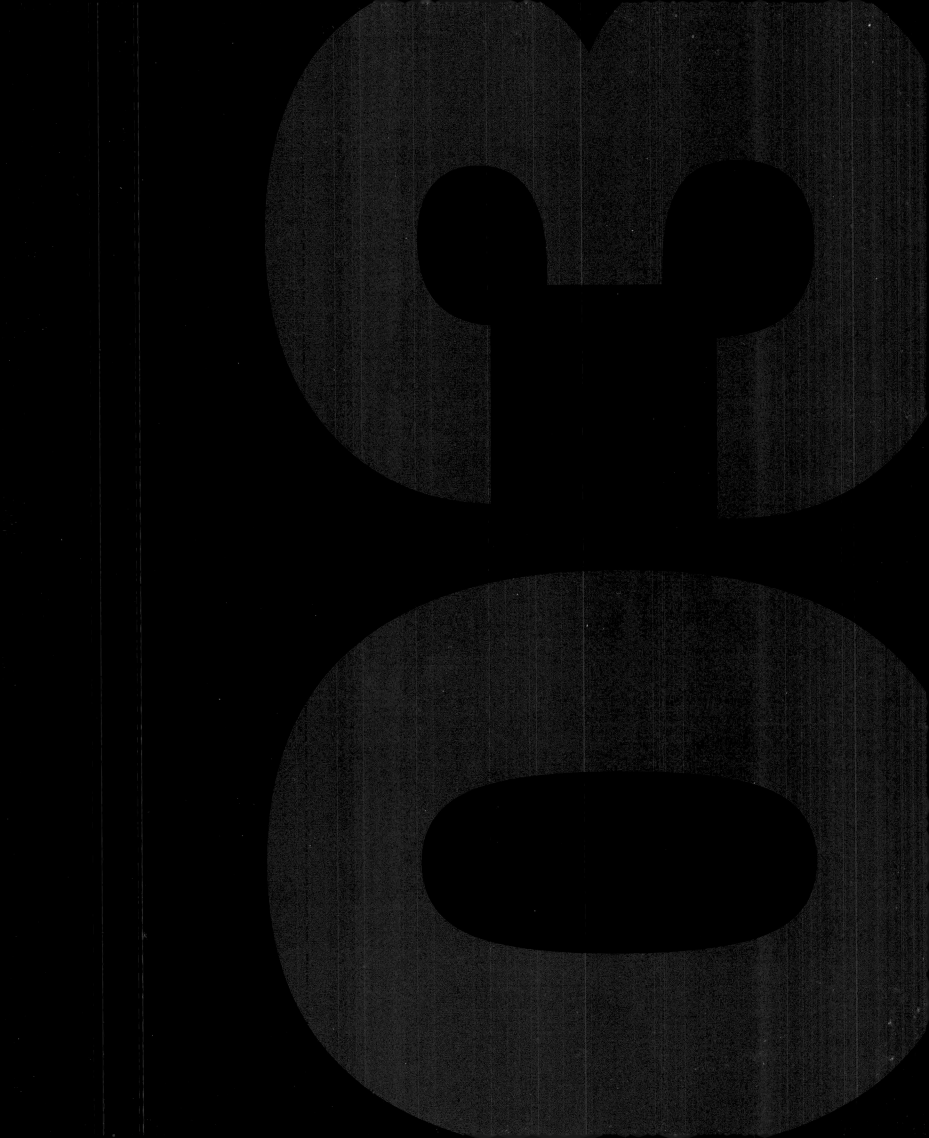

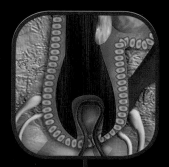

HAIR

Thick head hairs help keep the head warm; fine body hairs increase the skin's sensitivity. All visible hair is in fact dead; hairs are only alive at the root from which they grow. Hair doesn't grow continuously; it follows a cycle of growth and rest.

SKIN

Every month the skin renews its outer layer completely. Skin's texture is individual so each person's fingerprints are unique.

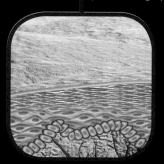

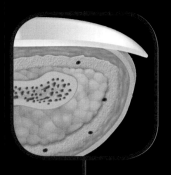

NAILS

Constantly growing and self-repairing, nails not only protect fingers and toes but also enhance their sensitivity.

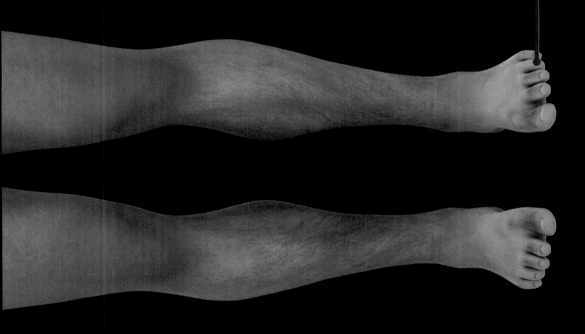

SKIN, HAIR, AND NAILS

The body is protected by an outer layer of skin, hair, and nails, all of which owe their toughness to the presence of a fibrous protein called keratin. The hair's luster and skin's radiance reveal aspects of health and lifestyle, such as diet.

SKIN, HAIR, AND NAILS

Also known as the integumentary system, the skin and its derivatives, hair and nails, form the body's outer covering. Skin in particular has a number of functions, including sensation, temperature regulation, making vitamin D, and protecting the body's internal tissues.

PROTECTION

As an organ that wraps around the body like a living overcoat, the skin is charged with a number of protective roles. These are carried out largely by the epidermis, the skin's upper layer. The uppermost part of the epidermis consists of dead, flattened cells that are packed with a tough, waterproof protein called keratin. The epidermis provides a physical barrier that is self-repairing, prevents damage from being caused to internal body tissues, and, by waterproofing, prevents water from leaking into or escaping from those tissues. It also filters out harmful sun rays.

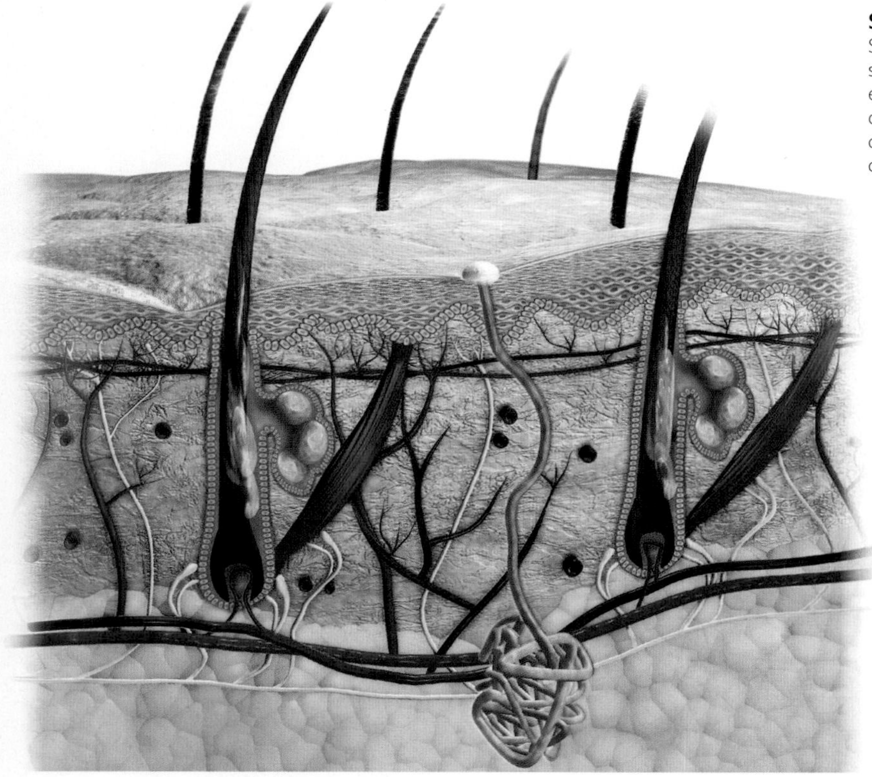

Skin structure
Shown here in cross section, the skin consists of two layers, a thinner epidermis made of epithelial cells overlying a thicker, connective tissue dermis. Beneath the dermis is a layer of heat-retaining fat.

Epidermis
Upper protective layer; consists largely of tough, flattened cells

Dermis
Contains blood vessels, sweat glands, and sensory receptors

Subcutaneous fat
Insulates and acts as a shock absorber and energy reserve

SKIN REPAIR

Because it covers the body's surface, skin is easily damaged. However, small nicks and cuts are rapidly sealed by the skin's self-repair system, thereby preventing entry by dirt and pathogens. When the skin is pierced, damaged cells release chemicals that attract platelets, which trigger clot formation; neutrophils, which engulf pathogens; and fibroblasts, which repair connective tissues.

Injury
A small cut in the skin causes bleeding. Damaged cells release chemicals that attract repair and defense cells.

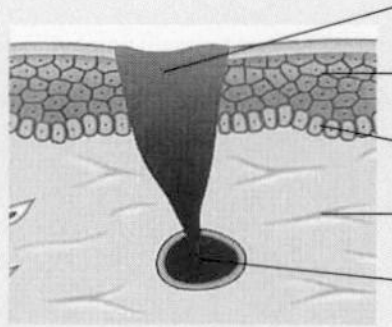

Injury site
Epidermis
Basal layer
Dermis
Severed vessel

Clotting
Platelets convert fibrinogen into fibers that trap blood cells to form a clot and stop bleeding.

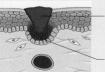

Blood clot
Fibroblast

Plugging
The blood clot shrinks and plugs the wound. Fibroblasts multiply and repair damaged tissues.

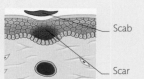

Blood clot contracts
New tissue

Scabbing
As tissues are repaired, they are protected by the dried clot or scab, which eventually falls off.

Scab
Scar

UV PROTECTION

The sun's rays contain a range of forms of radiation, including visible light and infrared and ultraviolet (UV) rays. One form of UV radiation called UVB can damage the DNA in basal epidermal cells and may trigger skin cancer. The skin protects itself from UV damage by producing a brown-black pigment called melanin that absorbs and filters out UVB radiation. It is produced by cells called melanocytes that are interspersed among "ordinary" cells, or keratinocytes, in the basal epidermis.

Melanin release
Melanin is made in membrane-bound bodies called melanosomes. These migrate along the dendrites of melanocytes to the upper parts of neighboring cells, where they release melanin granules.

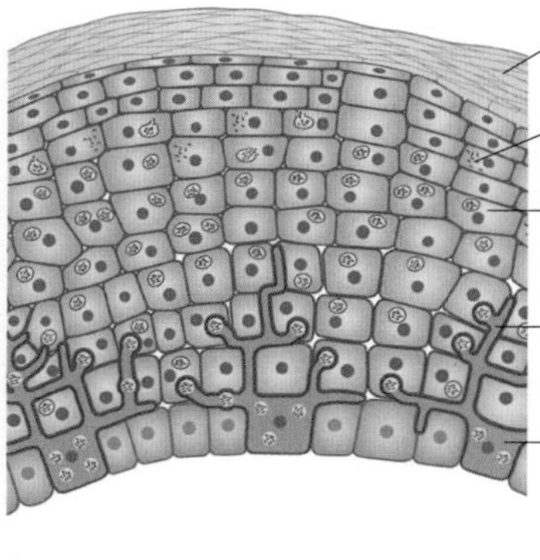

Surface
Dead, flat cells

Melanin granules
Disperse in keratinocyte

Keratinocyte
Epidermal cell

Dendrite
Distributes melanosomes to keratinocytes

Melanocyte
Cell that makes melanosomes

THICKNESS

Skin varies in thickness depending on its location on the body's surface. Thickness ranges from around $\frac{1}{64}$ in (0.5 mm) for the delicate skin of the eyelids and lips, to $\frac{3}{16}$ in (4 mm) on the underside of the feet (more in people who always walk barefoot), reflecting the considerable wear and tear experienced in that region. Although the dermis makes up most of the skin's thickness, it is the tough, keratinized epidermis that thickens more in skin exposed to most friction.

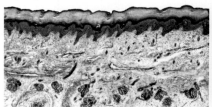

Thin skin
This section through eyelid skin shows how much thinner the epidermis—demarcated by the jagged line under the mauve zone—is than the dermis.

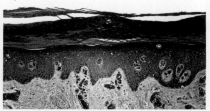

Thick skin
In this section through the skin covering the sole of the foot, the epidermal layer (purple) has become thickened as a protective measure.

SENSATION

The skin is a sense organ that detects the different aspects of "touch." It responds to external stimuli, sending signals to the sensory area of the brain (see p.335) that enables us to "describe" our surroundings. The skin is not a special sense organ, like the eye where sensory receptors are concentrated in one specific place, but a general sense organ that has receptors distributed throughout the skin. Some areas of skin, such as the fingertips and lips, have many more receptors than, say, the back of the leg, and are therefore much more sensitive. Most receptors are mechanoreceptors that send nerve impulses to the brain when they are physically pulled or squashed. Some are thermoreceptors that detect changes in temperature. Others are nociceptors, or pain receptors (see p.325), that detect chemicals released when skin is damaged.

Skin sensors

The position of each type of receptor in the dermis suits its particular role. Large receptors deep in the dermis detect pressure, while smaller receptors near the skin's surface pick up light touch. Receptors consist of the ends of neurons; these may be surrounded by a connective tissue capsule (encapsulated) or not (unencapsulated or free).

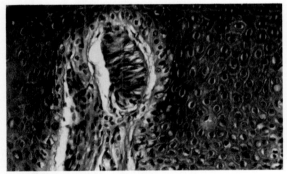

Fingertip receptor
This microscopic section through the skin of a fingertip shows a Meissner's corpuscle, one of its many sensory receptors, pushing into the epidermis and surrounded by densely packed epidermal cells.

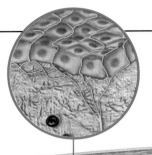

Free nerve endings
These branching, free endings may penetrate the epidermis. Some react to heat and cold, enabling a person to detect temperature changes; others are nociceptors that detect pain.

Merkel's disk
Free neuron endings associated with disklike epidermal cells, Merkel's disks are found at the dermis–epidermis border. They detect very faint touch and light pressure.

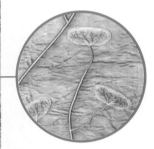

Ruffini's corpuscle
Consisting of branching neuron endings surrounded by a capsule, Ruffini's corpuscles detect stretching of the skin and deep, continuous pressure. In the fingertips they detect sliding movements, aiding grip.

Meissner's corpuscle
An encapsulated receptor that is more common in highly sensitive areas of hairless skin, such as on the fingertips, palms, soles, eyelids, nipples, and lips. It is sensitive to faint touch and light pressure.

Pacinian corpuscle
This big, egg-shaped receptor set deep in the dermis has a neuron ending surrounded by layers, resembling a cut onion. Squashed by outside forces, it detects stronger, sustained pressure as well as vibrations.

THERMOREGULATION

Controlled by the autonomic nervous system (see p.311), the skin plays an important part in regulating internal body temperature so that it is maintained at a constant 98.6° F (37° C) for optimal cell activity. It does this in two main ways: by constricting or dilating blood vessels in the dermis; and by sweating. The erection and flattening of hairs is a mammalian feature that no longer has a purpose in humans, apart from producing goosebumps.

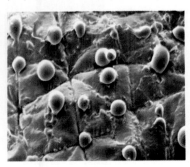

Sweat
Tiny droplets of sweat released onto the skin's surface from sweat glands evaporate, drawing heat from the body and cooling it down when hot.

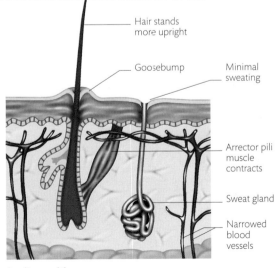

Hair stands more upright

Goosebump

Minimal sweating

Arrector pili muscle contracts

Sweat gland

Narrowed blood vessels

Feeling cold
Blood vessels constrict (narrow), reducing blood flow so that less heat escapes through the skin. Sweat glands produce little sweat when the body is cold, and heat is retained by the body.

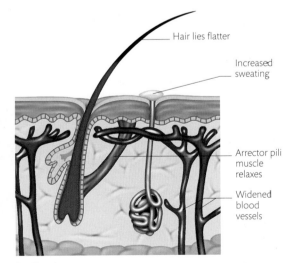

Hair lies flatter

Increased sweating

Arrector pili muscle relaxes

Widened blood vessels

Feeling hot
Blood vessels dilate (widen), increasing blood flow to the skin so more heat escapes through its surface. Copious sweating draws heat from the body to cool it down.

GRIP

The undersides of the hands and feet are the only areas of the skin that are covered by epidermal ridges separated by fine parallel grooves, which together form curved patterns on the skin that are unique to each individual. Epidermal ridges increase friction and greatly improve the ability of the hands and feet to grip surfaces. Well supplied with sweat glands, these ridges, notably on the fingers, leave behind sweat marks known as fingerprints that can be used to identify individuals.

Sweat pores
The crests of the epidermal ridges are covered with sweat pores

Epidermal ridges
This close-up view shows tightly packed epidermal ridges on the underside of the fingertips.

SKIN RENEWAL

The upper part of the epidermis, which consists of dead, flattened cells, is continually being worn away as skin flakes. Thousands of cells are shed every minute. Lost flakes are replaced by cells in the basal layer of the epidermis that divide actively by mitosis (see p.21) to create new cells. As these cells push upward toward the surface of the skin, they bind tightly together, fill with tough keratin, and eventually flatten and die, forming a scaly, interlocking barrier. The whole process takes about a month.

Layers of the epidermis
The cells that make up the different layers of the epidermis include the boxlike basal cells, spiky prickle cells, squashed granular cells, and dead surface layer cells.

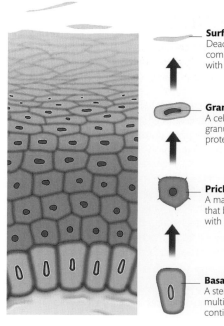

Surface layer cell
Dead, flattened cell completely filled with keratin

Granular cell
A cell containing granules of the protein keratin

Prickle cell
A many-sided cell that binds closely with its neighbors

Basal cell
A stem cell that multiplies continuously

SKIN COLOR

The color of a person's skin depends on the amount and distribution of melanin pigment in their skin. Melanin is made and packaged into melanosomes by melanocytes. Each melanocyte has branching dendrites that contact nearby keratinocytes, and through which melanosomes are released. Darker skin has larger (not more) melanocytes that produce more melanosomes, releasing melanin, which is distributed throughout the keratinocytes.

Lighter skin has smaller melanocytes and little distribution of melanin. UV rays in sunlight stimulate melanin production in all skin colors to produce a sun tan.

From dark to light
This comparison of dark-, intermediate-, and light-colored skin shows clearly the differences in melanocyte size and in melanosome and melanin distribution that produce a variety of skin colors.

8.8 pounds

The **weight** of the skin of an average adult, making it the body's **heaviest organ**.

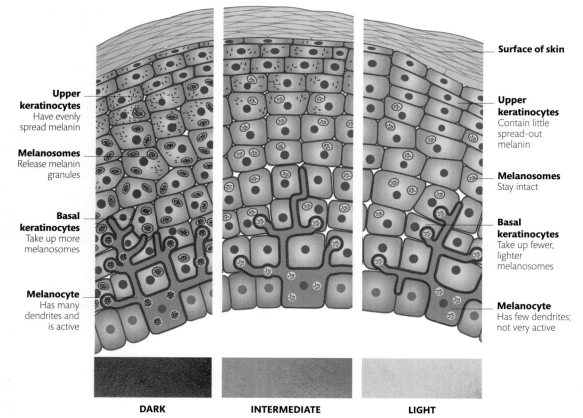

Upper keratinocytes
Have evenly spread melanin

Melanosomes
Release melanin granules

Basal keratinocytes
Take up more melanosomes

Melanocyte
Has many dendrites and is active

Surface of skin

Upper keratinocytes
Contain little spread-out melanin

Melanosomes
Stay intact

Basal keratinocytes
Take up fewer, lighter melanosomes

Melanocyte
Has few dendrites; not very active

DARK   **INTERMEDIATE**   **LIGHT**

VITAMIN D SYNTHESIS

In addition to being obtained from the diet, vitamin D is also made in the skin using sunlight. UVB rays passing through the epidermis convert 7-cholesterol into cholecalciferol, a relatively inactive form of vitamin D. This is carried by the blood to the kidneys, where it is converted into calcitriol, or active vitamin D_3. Since melanin filters UV light, people with darker skin need more UV radiation to make the same amount of vitamin D. UV radiation can be measured using an index.

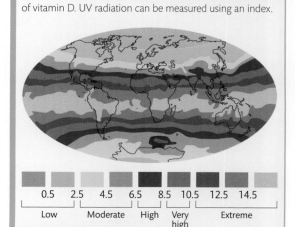

0.5	2.5	4.5	6.5	8.5	10.5	12.5	14.5
Low	Moderate		High	Very high	Extreme		

Radiation by UV index
This map indicates the different amounts of UV radiation from the sun around the globe each day. A dark-skinned person with a poor diet in a low UV area could suffer from vitamin D deficiency.

HAIR FUNCTIONS

The human body is covered with millions of hairs, with more than 100,000 on the scalp alone. The only hairless places are the lips, nipples, undersides of the hands and feet, and parts of the genitals. In our hairier ancestors, body hair gave insulation; that role is now provided by clothes. There are two main types of hair: thick, terminal hairs such as those on the head or in the nostrils of all ages, and in the armpits and pubic areas of adults; and short, very fine vellus hairs, found covering most of the body of children and in women. Hair has different functions according to where it is growing.

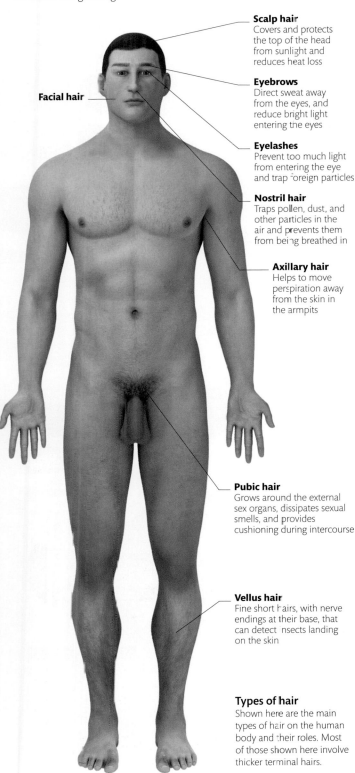

Facial hair

Scalp hair
Covers and protects the top of the head from sunlight and reduces heat loss

Eyebrows
Direct sweat away from the eyes, and reduce bright light entering the eyes

Eyelashes
Prevent too much light from entering the eye and trap foreign particles

Nostril hair
Traps pollen, dust, and other particles in the air and prevents them from being breathed in

Axillary hair
Helps to move perspiration away from the skin in the armpits

Pubic hair
Grows around the external sex organs, dissipates sexual smells, and provides cushioning during intercourse

Vellus hair
Fine short hairs, with nerve endings at their base, that can detect insects landing on the skin

Types of hair
Shown here are the main types of hair on the human body and their roles. Most of those shown here involve thicker terminal hairs.

HAIR GROWTH

Hairs are rods of keratinized, dead cells that grow from deep pits called follicles in the dermis. The hair shaft grows above the skin's surface, while its root is below the surface. At its base, the hair root expands into a hair bulb that contains actively dividing cells. As new cells are produced, they push upward, making the hair increase in length. Hair growth happens in a cycle that involves growth and resting phases. During the growth phase scalp hairs grow by about ³⁄₈ in (1 cm) each month, and last between 3 and 5 years, until they fall out. In the resting phase, growth halts and the hair eventually separates from its base. About 100 head hairs are lost daily and are replaced by new growth.

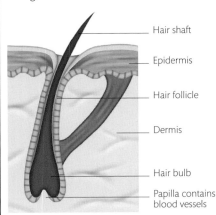

Hair shaft

Epidermis

Hair follicle

Dermis

Hair bulb

Papilla contains blood vessels

Resting phase
When the hair reaches maximum length, the resting phase, which lasts a few months, begins; cells in the hair root stop dividing, the root shrinks, and the hair shaft stops extending.

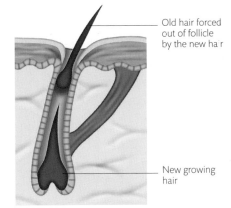

Old hair forced out of follicle by the new hair

New growing hair

Growth phase
Once the resting stage ends, cells in the base of the hair follicle start dividing and a new hair sprouts. Its rapidly extending shaft pushes the old hair out of the follicle.

NAILS

These hard plates cover and protect the sensitive tips of the fingers and toes. Fingernails also help the fingers grip small objects, and scratch itches. Each nail has a root, embedded in the skin, a body, and a free edge. Nail cells produced by the matrix push forward, becoming filled with keratin as the nail slides over the nail bed. Fingernails grow three times faster than toenails, and faster in summer than in winter.

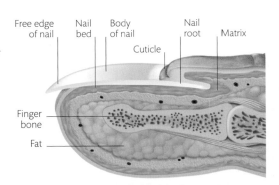

Free edge of nail

Nail bed

Body of nail

Cuticle

Nail root

Matrix

Finger bone

Fat

KERATIN

Nails are made of dead, flattened cells filled with the tough, structural protein keratin. This micrograph shows how those flattened cells form thin, interlocking plates that give nails their hardness but also make them translucent, so that the pinkness of the underlying dermis shows through. Keratin is also found in hair shafts and in epidermal cells, from which both nails and hairs are derived.

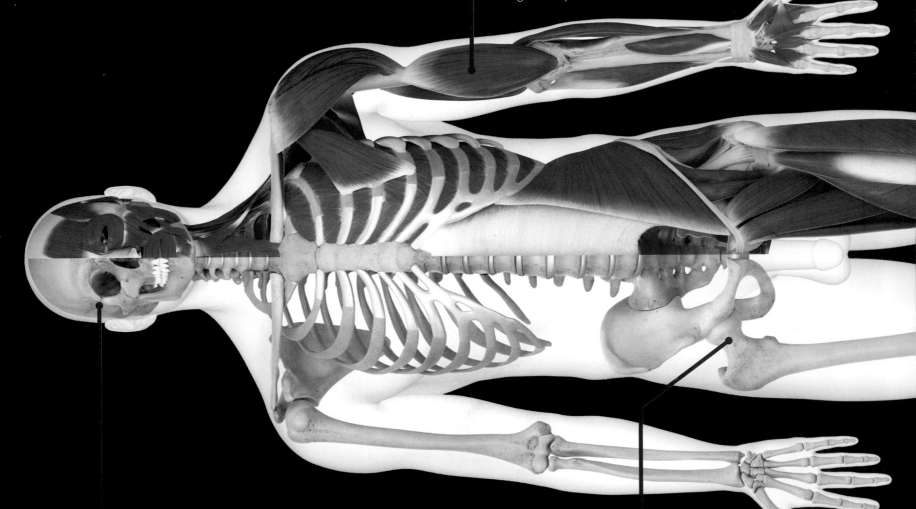

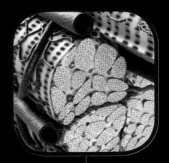

MUSCLE

Skeletal muscle contains thick
and thin myofilaments that
allow it to contract powerfully,
enabling the body to move.

BONE

The skeleton has about 206
bones. Bones are very strong,
and some contain marrow
which produces red blood cells.

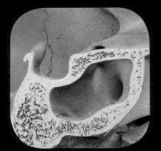

LIGAMENT

Joining one bone to another,
ligaments are elastic to allow
free movement but tough
enough to keep joints stable.

TENDON

Tough, elastic tendons connect muscle to bone. They are strong to withstand the pull of muscles and stay anchored to bone.

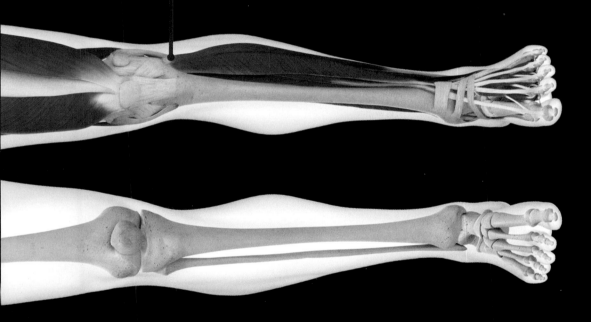

MUSCULOSKELETAL SYSTEM

An integrated system of bones, muscles, tendons, and ligaments allows the body to perform movements, from those that move the whole body, such as walking, to the more delicate finger strokes of typing on a keyboard.

THE WORKING SKELETON

Far from being an inert structure, the skeleton is a strong yet light, flexible living framework that supports the body, protects delicate internal organs, and makes movement possible. In addition, our bones store minerals, while red bone marrow produces blood cells.

SKELETAL DIVISIONS

To make its parts and functions easier to describe, the skeleton can be grouped into two divisions, the axial and appendicular skeletons. Containing 80 of the body's 206 bones, the axial skeleton makes up the long axis that runs down the center of the body, providing protection and support. It consists of the skull, backbone, ribs, and sternum. The appendicular skeleton, which contains 126 bones, allows us to move from place to place and to manipulate objects. It consists of the bones of the upper and lower limbs and the bony girdles that attach them to the axial skeleton. The pectoral or shoulder girdles, each made of a scapula and clavicle, attach the upper arm bones to the rest of the skeleton. The stronger pelvic girdle, made up of two hipbones joined to each other and the sacrum, anchors the thighbones.

Axis and attachments
This color-coded skeleton shows clearly the axial skeleton forming the central core of the skeleton to which the appendicular skeleton is appended, or attached.

KEY

    Appendicular skeleton

    Axial skeleton

SUPPORT

Denied the support of a skeleton, the body would collapse in a heap. The skeleton provides a substructure that shapes the body and holds it up, whether it is sitting, standing, or in another position (see right). Within the skeleton itself, different aspects of support can be identified. The backbone, as the body's main axis, supports the trunk, with its uppermost section, the neck, bearing the weight of the head. It provides attachment points for the rib cage, which supports the wall of the thorax, or chest. The backbone also positions the head and trunk above, and transmits their weight through the pelvis to the legs, the pillars that support our weight when we stand. The pelvis itself supports the organs of the lower abdomen such as the bladder and intestines.

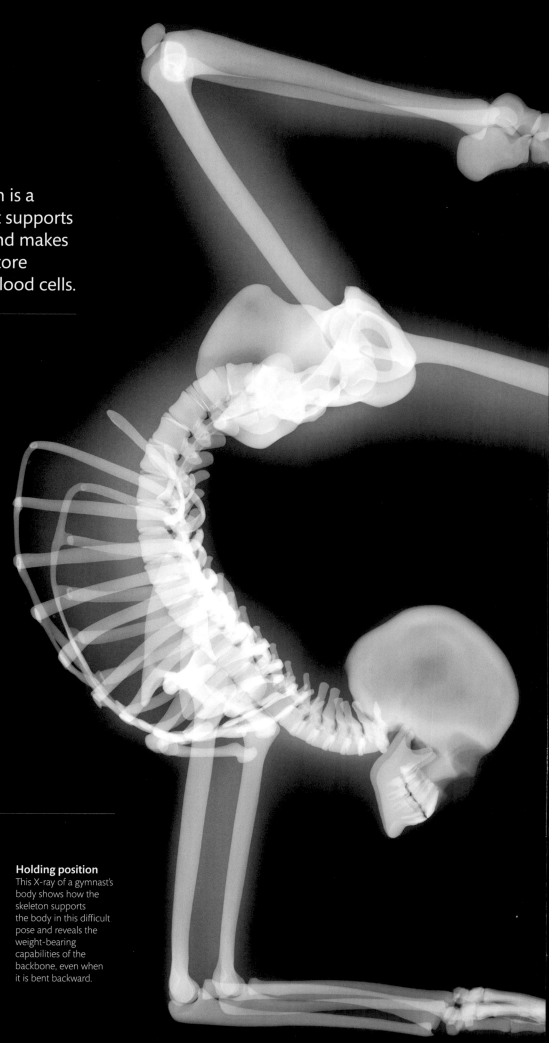

Holding position
This X-ray of a gymnast's body shows how the skeleton supports the body in this difficult pose and reveals the weight-bearing capabilities of the backbone, even when it is bent backward.

MOVEMENT

The human skeleton is not a rigid, inflexible structure. Where its bones meet, they form joints, most of which are flexible and allow movement. The range of movement any one joint permits depends on various factors including the conformation of the joint, and how tightly ligaments and skeletal muscles hold it together. Each bone has specific points to which skeletal muscles are attached by tendons. Muscles contract to pull bones in order to create an array of movements as diverse as running, grasping objects, and breathing.

Skillful moves
Dancers train for years to give their joints the flexibility, and their muscles the strength, to create graceful, carefully controlled, and well-balanced movements such as these.

PROTECTION

Body organs, such as the brain and heart, would be easily damaged were it not for the protection afforded by the skeleton, particularly by the skull and rib cage. The skull is constructed from interlocking bones, eight of which form the helmetlike cranium, a strong, self-bracing structure that surrounds the brain. The bones of the cranium also house the inner parts of the ears and, together with facial bones, create the protective orbits that accommodate the eyeballs. The rib cage is a cone-shaped protective cage that shapes the thorax, or chest, and protects the heart and lungs, as well as the major blood vessels—including the aorta and the superior and inferior venae cavae—within the thoracic cavity. It also lends a good degree of protection to the liver, stomach, and other upper abdominal organs.

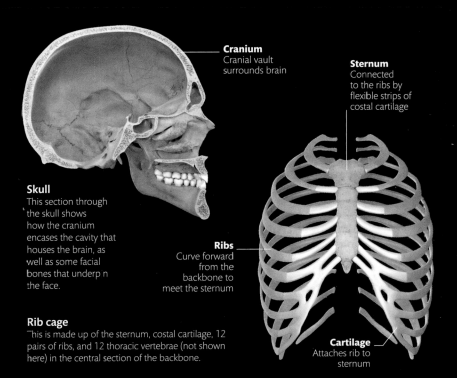

Cranium
Cranial vault surrounds brain

Sternum
Connected to the ribs by flexible strips of costal cartilage

Skull
This section through the skull shows how the cranium encases the cavity that houses the brain, as well as some facial bones that underpin the face.

Ribs
Curve forward from the backbone to meet the sternum

Rib cage
This is made up of the sternum, costal cartilage, 12 pairs of ribs, and 12 thoracic vertebrae (not shown here) in the central section of the backbone.

Cartilage
Attaches rib to sternum

BLOOD CELL PRODUCTION

The red bone marrow inside bones produces billions of new blood cells daily. In adults it is found in the axial skeleton, the shoulder and hip girdles, and the top ends of each humerus and femur. Within red bone marrow, blood cells arise from unspecialized stem cells called hemocytoblasts. These divide and their offspring follow different maturation pathways to become either red or white blood cells. In the case of red blood cells, progressive generations of hemocytoblast descendants lose their nuclei and fill up with hemoglobin (see p.341), finally becoming red blood cells.

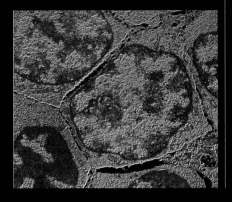

Erythroblasts
At the earlier stages of red blood cell production, these erythroblasts still have a large nucleus (red) and divide rapidly.

MINERAL STORAGE

Bones contain 99 percent of the body's calcium, and store other minerals, including phosphate. Calcium and phosphate ions are released into, or removed from, the bloodstream as required. Calcium ions, for example, are essential for muscle contraction, transmission of nerve impulses, and blood clotting. Calcium salts make teeth and bones hard. Bones are constantly reshaped both in response to stresses and as a result of the antagonistic effects of the hormones calcitonin and parathyroid hormone (PTH); these, respectively, stimulate calcium deposition in bones and calcium release from bones. Collectively, these various influences ensure that calcium withdrawals from, and deposits to, the bony mineral reserve are balanced to keep calcium levels in the bloodstream constant.

BONES

They may appear to be lifeless organs, but bones are composed of active cells and tissues, which enable bones to grow when a fetus is developing and during childhood. They also reshape fully grown bones throughout life to ensure that they are strong and able to withstand the stresses they are exposed to daily.

HOW BONES GROW

The growth and development of the skeleton begins early in the life of an embryo and continues until the late teens. The embryonic skeleton is initially made up of flexible connective tissues, either fibrous membranes or pieces of hyaline cartilage. By the time it has reached 8 weeks old, the process of ossification (bone making) has started to replace these structures with hard bone tissue, and over the ensuing months and years bones grow and develop. Two different methods of ossification replace original connective tissue with bony matrix. Intramembranous ossification forms the bones of the skull from fibrous membranes (see below). Endochondral ossification replaces hyaline cartilage to form the majority of bones, other than those in the skull. The sequence (right) shows the progress of endochondral ossification in a long bone from the cartilage template of a young embryo to the hard, weight-bearing bone of a 6-year-old child that will increase yet more in length and width to enable the child to grow.

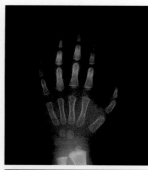
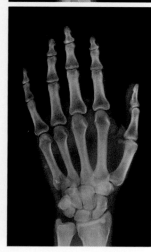

Bone development
An X-ray of a 3-year-old's hand (top) shows large areas of cartilage in the finger joints and wrist, where ossification gradually occurs. In the adult hand (bottom), all the bones of the wrist are present and the joints are fully formed.

SKULL BONES

The flat bones of the skull grow and develop through the process of intramembranous ossification, which begins in the fetus around 2 months after fertilization (see p.413). Fibrous connective tissue membranes form the bone models. Ossification centers develop inside the membranes, lay down a bone matrix, and eventually produce a latticework of spongy bone surrounded by compact bone. At birth, ossification is still incomplete and the skull bones are connected by unossified sections of fibrous membranes at the fontanelles (see p.418). The fontanelles close by around the ages of two. The presence of these flexible, fibrous joints allows changes in the shape of the skull, facilitating the passage of the baby through the birth canal.

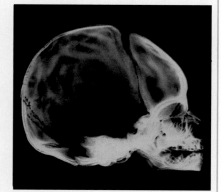

Baby's skull
This X-ray shows the anterior fontanelle (dark zone) between two bones that surround the brain. Fontanelles allow the baby's brain to expand and grow.

7-week embryo
Cartilage cells create the model for a future long bone. It has a clear diaphysis (shaft) with an epiphysis (head) at each end. By dividing and laying down more matrix, cartilage cells make the "bone" grow longer and wider.

Diaphysis (shaft)

Epiphysis (head)

10-week fetus
Cartilage cells in the middle of the diaphysis cause the surrounding matrix to calcify (harden). As a result, small cavities open up and are invaded by nutrient-carrying blood vessels and osteoblasts (bone-making cells), which lay down spongy bone to form the primary ossification center.

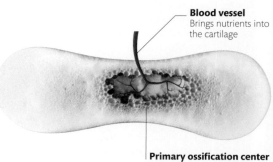

Blood vessel
Brings nutrients into the cartilage

Primary ossification center

12-week fetus
The primary ossification center now occupies most of the enlarged and ossified diaphysis. In the center of the diaphysis, osteoclasts (bone-destroying cells) break down newly formed spongy bone to create a medullary cavity. Cartilage cells in the epiphyses divide to cause bone elongation. At the same time, cartilage at the base of each epiphysis is steadily replaced by bone.

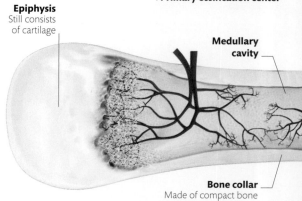

Epiphysis
Still consists of cartilage

Medullary cavity

Bone collar
Made of compact bone

Baby, at birth
Bones continue to lengthen as the primary ossification center continues its work. In the center of each epiphysis, a secondary ossification center with its own blood supply develops. There, cartilage is replaced by spongy bone that remains there; no medullary cavities are formed in the epiphyses. The medullary cavity in the diaphysis is filled with red bone marrow, which manufactures blood cells.

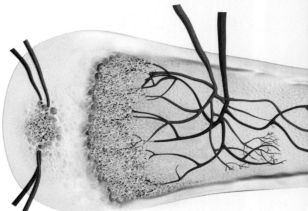

During childhood
Hyaline cartilage is now found in only two locations: covering the epiphysis as articular cartilage and between the epiphysis and diaphysis as the epiphyseal growth plate. Cartilage cells in the epiphyseal plate divide, pushing the epiphysis away from the diaphysis, making the bone grow lengthwise. At the same time, cartilage in the epiphyseal plate adjacent to the diaphysis is replaced by bone. This process continues until the late teens, when the epiphyseal plate disappears, the epiphysis and diaphysis fuse, and bone growth is complete.

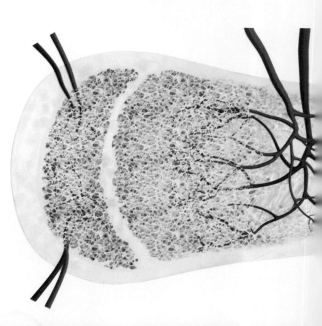

BONE REMODELING

Throughout life bones are remodeled, a reshaping process in which old bone tissue is removed and new tissue is added. Remodeling maximizes the strength of bones in response to changing mechanical demands or forces. Up to 10 percent of an adult's skeleton can be replaced annually. Remodeling has two distinct stages—bone resorption and bone deposition—performed by bone cells called osteoclasts and osteoblasts, which have opposing actions. Osteoclasts break down and remove old bone matrix and then a team of osteoblasts lays down new bone matrix. Remodeling is controlled by two mechanisms. Firstly, osteoclasts and osteoblasts respond to the mechanical stresses put on bones by gravity and muscle tension. Secondly, two hormones, parathyroid hormone (PTH) and calcitonin respectively stimulate and inhibit osteoclast activity in order to regulate the release of calcium ions from bone matrix. This maintains constant levels of calcium, essential for muscle contraction and many other processes, in the blood.

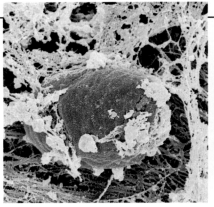

Osteoblast
An osteoblast (red) secretes and is surrounded by the organic part of bone matrix. This is then mineralized by calcium salts to form hard matrix.

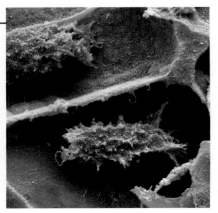

Osteoclasts
Osteoclasts (purple) move along the bone surface excavating spaces by using enzymes and acid to break down both organic and mineral matrix.

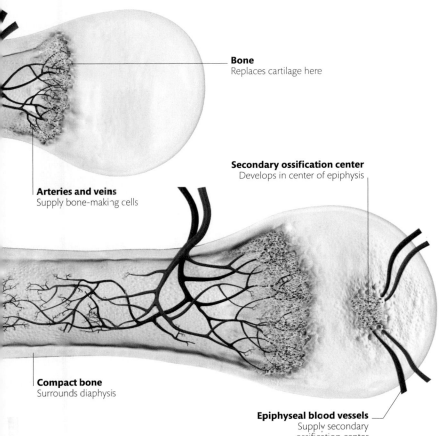

Bone
Replaces cartilage here

Arteries and veins
Supply bone-making cells

Secondary ossification center
Develops in center of epiphysis

Compact bone
Surrounds diaphysis

Epiphyseal blood vessels
Supply secondary ossification center

EXERCISE

Bones are subject to two main mechanical stresses: the weight that bears down on them as a result of the downward pull of gravity and the force of tension exerted by muscles as they move bones. These stresses increase during weight-bearing exercises such as walking, running, dancing, or tennis. Performed several times weekly, such exercises stimulate bone cells to reshape bones and make their strength and mass significantly greater than the bones of an inactive person.

Bone mass peaks in our 20s and 30s, a time when regular exercise and a healthy diet pay dividends. After the age of 40, bone strength and mass decrease, but if they were elevated by regular exercise in young adulthood, age-related bone loss is slowed. Weight-bearing exercise in older people can reverse decreases in bone strength and mass, reducing the risk of osteoporosis (see p.441).

EXTREME HUMAN
EXERCISE IN SPACE

An astronaut onboard an orbiting space shuttle exercises on a rowing machine in an attempt to counteract the effects of weightlessness. On Earth, bones maintain their strength and mass by resisting body weight created by the downward pull of gravity. In space, bones have little gravity to pull against, and as a result weaken, losing up to 1 percent of their mass monthly. Although exercise in space reduces loss of bone mass, it does not prevent it.

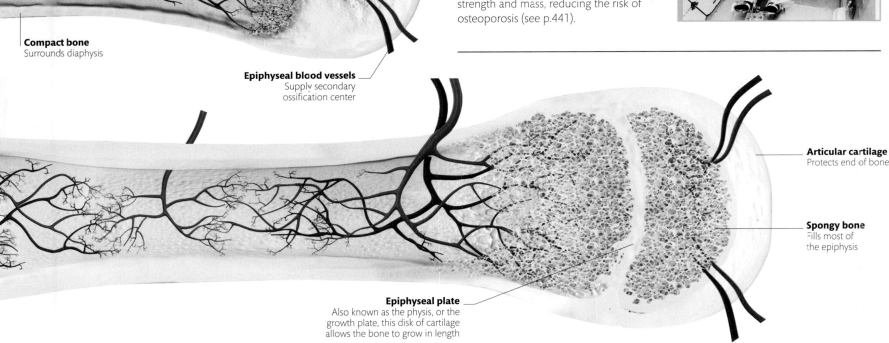

Articular cartilage
Protects end of bone

Spongy bone
Fills most of the epiphysis

Epiphyseal plate
Also known as the physis, or the growth plate, this disk of cartilage allows the bone to grow in length

JOINTS

Wherever in the skeleton two or more bones meet they form a joint or articulation. This gives the skeleton its flexibility and, when bones are pulled by muscles across joints, the ability to move. Joints are classified according to their structure and the amount of movement they allow.

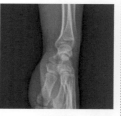

HOW JOINTS WORK

The majority of the body's 320 or so joints, including those in the knee and shoulder, are free-moving synovial joints. They allow the body to perform a wide range of movements including walking, chewing, and writing. In a synovial joint, bone ends are covered and protected by articular cartilage made from glassy, smooth hyaline cartilage. The most common type of cartilage in the body, hyaline cartilage is strong but compressible. Articular cartilages reduce friction between bones when they move, and absorb shocks during movement to prevent jarring. A capsule surrounding the joint contains fibrous tissue that, aided by ligaments, helps hold the joint together. Its innermost layer, the synovial membrane, secretes oily synovial fluid into the cavity between the articular cartilages, making them even more slippery, and allowing the joint to move with less friction than two ice cubes sliding over each other. There are six types of synovial joint (see right). Each allows a different range of movement according to the shape of their articular surfaces.

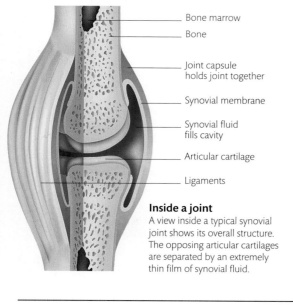

Bone marrow

Bone

Joint capsule
holds joint together

Synovial membrane

Synovial fluid
fills cavity

Articular cartilage

Ligaments

Inside a joint
A view inside a typical synovial joint shows its overall structure. The opposing articular cartilages are separated by an extremely thin film of synovial fluid.

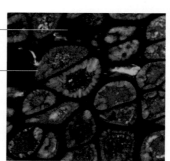

Matrix
Contains collagen fibers

Chondrocytes
Secrete cartilage matrix

Hyaline cartilage
This consists of cells separated by a nonliving matrix (purple), as shown in this micrograph.

1 ELLIPSOIDAL

This type of joint is formed where an egg-shaped end of one bone moves within the oval recess of another. Found in the wrist, between the radius and the carpals, it allows bending and straightening and side-to-side movements.

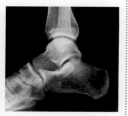

WRIST

2 GLIDING

Articular surfaces between the bones in these joints are almost flat and facilitate short, sliding movements, which are further limited by strong ligaments. Gliding joints are found between the tarsals in the heel (below) and carpals in the wrist.

FOOT

SEMIMOVABLE AND FIXED JOINTS

Some joints are either semimovable or fixed. What they lack in mobility, relative to synovial joints, they make up for with strength and stability. In semimovable joints, such as the pubic symphysis in the pelvic girdle, bones are separated by a disk of fibrocartilage. Resilient and compressible, this allows limited movement. In fixed joints, notably the sutures between skull bones, fibrous tissue anchors the wavy edges of adjacent bones so that they are locked together. In younger people this arrangement still allows growth to occur at the edges of skull bones.

Pubic symphysis
This semimovable joint is found at the junction between the two pubic bones, the anterior portions of the two bones that, with the sacrum, make up the pelvic girdle.

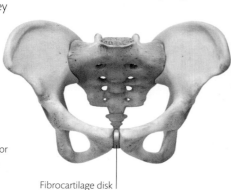

Fibrocartilage disk

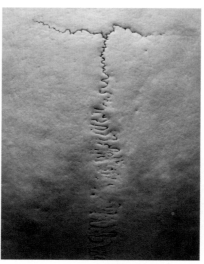

Sutures
This view of the adult skull shows sutures between bones. By middle age, fibrous tissue within sutures has ossified so that adjacent bones fuse together.

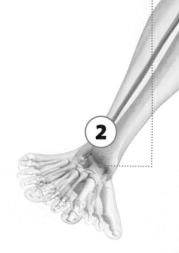

Moving joints
Here are shown the main types of synovial joint, the range of movement associated with each, and examples of the different types in various parts of the body.

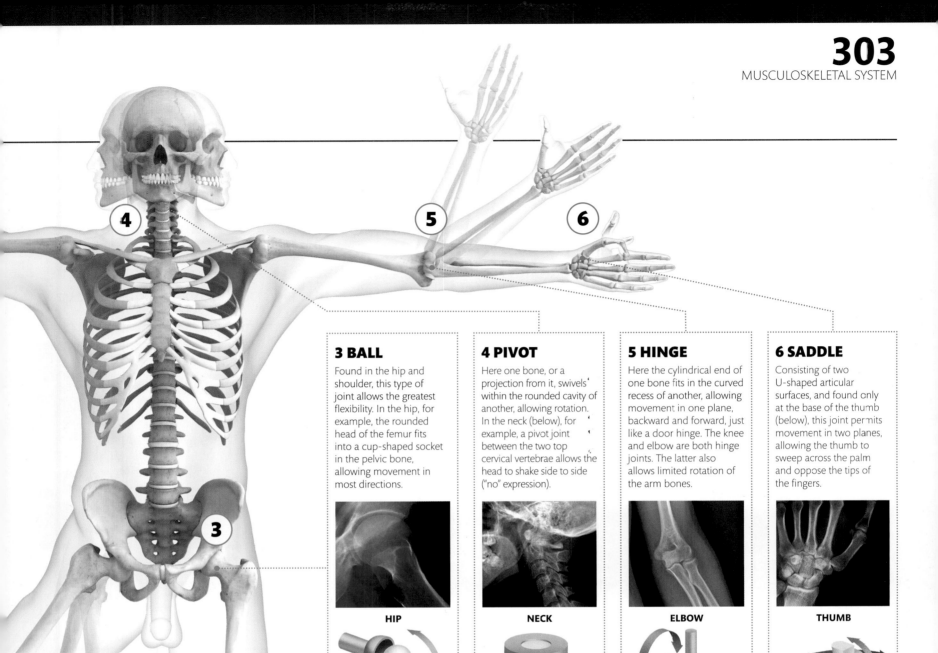

3 BALL

Found in the hip and shoulder, this type of joint allows the greatest flexibility. In the hip, for example, the rounded head of the femur fits into a cup-shaped socket in the pelvic bone, allowing movement in most directions.

HIP

4 PIVOT

Here one bone, or a projection from it, swivels within the rounded cavity of another, allowing rotation. In the neck (below), for example, a pivot joint between the two top cervical vertebrae allows the head to shake side to side ("no" expression).

NECK

5 HINGE

Here the cylindrical end of one bone fits in the curved recess of another, allowing movement in one plane, backward and forward, just like a door hinge. The knee and elbow are both hinge joints. The latter also allows limited rotation of the arm bones.

ELBOW

6 SADDLE

Consisting of two U-shaped articular surfaces, and found only at the base of the thumb (below), this joint permits movement in two planes, allowing the thumb to sweep across the palm and oppose the tips of the fingers.

THUMB

SPINE FLEXIBILITY

In the spine, two types of joint allow limited movements between adjacent vertebrae. Fibrocartilage intervertebral disks form semimovable joints that allow bending and twisting movements, and absorb shocks created during running and jumping. Synovial joints between articular processes allow limited gliding movements. Collectively, however, these joints give the backbone considerable flexibility.

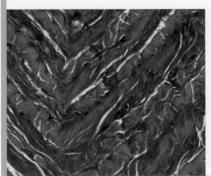

Facet joint
Gliding joints between articular processes limit twisting and slippage

Springy ligament
Ligaments between spinous processes limit movement and store energy for recoil

Fibrocartilage
Consisting of alternate layers of matrix and collagen (pink), fibrocartilage resists tension and heavy pressure.

Intervertebral disk
Composed of tough, flexible fibrocartilage with jellylike core

Spinal joints
Limited by ligaments, the joints between two vertebrae permit small movements, but added to those of other vertebrae they allow the spine to bend and twist.

HOW MUSCLES WORK

Muscles have the unique ability to contract and exert a pulling force. They do this by using stored chemical energy obtained from food to power an interaction between protein filaments inside their cells, in order to generate movement. In skeletal muscles contraction is triggered by nerve impulses that arrive from the brain when we make a conscious decision to move.

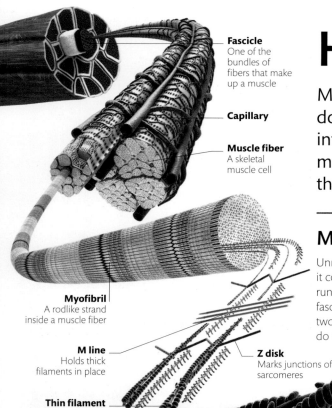

Fascicle
One of the bundles of fibers that make up a muscle

Capillary

Muscle fiber
A skeletal muscle cell

Myofibril
A rodlike strand inside a muscle fiber

M line
Holds thick filaments in place

Z disk
Marks junctions of sarcomeres

Tropomyosin

Thin filament
Consists chiefly of coiled strands of the protein actin

Thick myofilament
Made of the protein myosin

Myosin head
Forms cross bridge with actin during contraction

MUSCLE CONTRACTION

Unraveling the structure of a skeletal muscle is key to understanding how it contracts. A muscle consists of long, cylindrical cells called fibers, which run lengthwise in parallel and are bound together in bundles called fascicles. Each muscle fiber is packed with rodlike myofibrils that contain two types of protein filament, called myosin and actin. These filaments do not run the length of the myofibril but are arranged in overlapping patterns in "segments" called sarcomeres that give the myofibril, and muscle fiber, a striped appearance. Thin actin filaments extend inward from a "Z disk," which separates one sarcomere from the next, and surround and overlap thick myosin filaments in the sarcomere's center. When the muscle receives a nerve impulse instructing it to contract, small "heads" extending from each myosin filament interact with actin filaments to make the myofibril shorten.

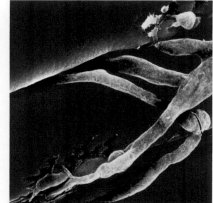

Neuromuscular junction
Motor neurons (green) transmit the nerve impulses to muscle fibers (red) that instruct them to contract. Neurons end in axon terminals that form nerve–muscle junctions with muscle fibers.

CONTRACTION CYCLE

A nerve impulse triggers a cycle of events inside a muscle fiber that causes contraction. Binding sites on the actin filaments become exposed, allowing myosin heads, already activated by the energy molecule adenosine triphosphate (ATP), to repeatedly attach, bend, detach, then reattach. This pulls thin filaments toward the center of sarcomeres, contracting the muscle fiber.

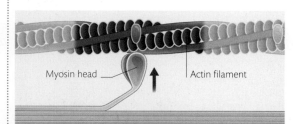

Myosin head — Actin filament

1 Attachment
In its high-energy configuration, the activated myosin head attaches to an exposed binding site on the actin filament to form a cross bridge between the filaments.

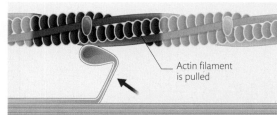

Actin filament is pulled

2 Power stroke
During what is known as the "power stroke" the myosin head pivots and bends, pulling the actin filament toward the center of the sarcomere.

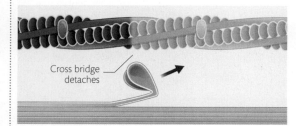

Cross bridge detaches

3 Detachment
A molecule of ATP binds to the myosin head causing it to loosen its hold on the binding site on the actin filament so that the cross bridge detaches.

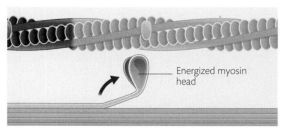

Energized myosin head

4 Energy release
ATP releases energy to convert the myosin head from its bent, low-energy position to its high-energy configuration, ready for the next cycle.

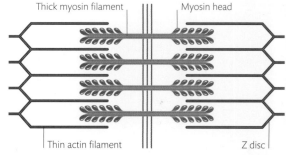

Thick myosin filament — Myosin head

Thin actin filament — Z disc

Relaxed muscle
This diagram shows a longitudinal section through a sarcomere (the section between Z disks) in a relaxed muscle. The thick and thin filaments overlap only slightly. The myosin heads are "energized" and ready for action but they do not interact with the actin filaments.

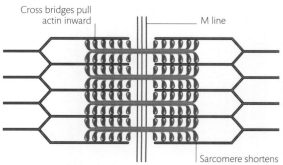

Cross bridges pull actin inward — M line

Sarcomere shortens

Contracted muscle
During muscle contraction, repeated cycles of cross bridge attachment and detachment pull actin filaments inward so that they slide over the thick filaments, shorten the sarcomere, and increase the overlap between filaments. As a result, muscles become significantly shorter than their resting length.

TYPES OF CONTRACTION

When a muscle is activated it exerts a force called tension on the object it is moving or supporting. If the muscle tension balances that of the load the muscle does not shorten, giving an isometric ("same length") contraction, such as when a book is held steady for reading. Isometric contractions of neck, back, and leg muscles will maintain posture, holding the body upright. If the muscle force exceeds the load, movement occurs. A steady speed of movement requires a steady force called isotonic ("same force") contraction. Day-to-day actions, such as picking up a book, are a complex mixture of accelerative, isotonic, and isometric contractions.

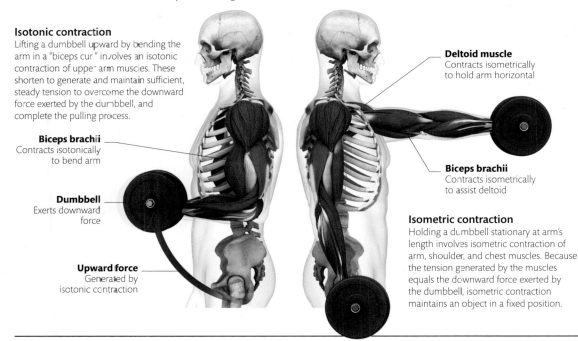

Isotonic contraction
Lifting a dumbbell upward by bending the arm in a "biceps curl" involves an isotonic contraction of upper arm muscles. These shorten to generate and maintain sufficient, steady tension to overcome the downward force exerted by the dumbbell, and complete the pulling process.

Biceps brachii
Contracts isotonically to bend arm

Dumbbell
Exerts downward force

Upward force
Generated by isotonic contraction

Deltoid muscle
Contracts isometrically to hold arm horizontal

Biceps brachii
Contracts isometrically to assist deltoid

Isometric contraction
Holding a dumbbell stationary at arm's length involves isometric contraction of arm, shoulder, and chest muscles. Because the tension generated by the muscles equals the downward force exerted by the dumbbell, isometric contraction maintains an object in a fixed position.

EXTREME HUMAN
BODYBUILDERS

Weight lifters increase muscle size by using exercises that increase the number of myofibrils inside muscle fibers, in order to increase strength. However, bodybuilders also aim to increase the amount of liquid sarcoplasm inside muscle fibers so that their muscles increase in bulk. Coupled with a protein-rich diet and aerobic exercise to reduce body fat, this produces the bodybuilder's characteristic physique.

Overdeveloped muscles
A bodybuilder flexes her muscles to show off her highly defined muscles.

MUSCLE GROWTH AND REPAIR

Skeletal muscle fibers do not increase in number through cell division but retain the ability to grow during childhood, and to hypertrophy in adulthood. Muscular hypertrophy is the increase in size—but not number—of muscle fibers through strength training. One cause of hypertrophy is microtrauma: tiny muscle tears produced by strenuous exercise. Satellite cells in the muscle repair torn tissue and as a result fibers—and muscles—increase in size.

MUSCLE METABOLISM

Energy-rich "fuels" such as glucose cannot be used directly for muscle contraction. First they must be converted into ATP (adenosine triphosphate), a substance that stores, carries, and releases energy. During contraction ATP enables myosin and actin to interact (see opposite). ATP is generated inside a muscle fiber by two types of cell respiration—aerobic or anaerobic. A muscle fiber holds enough ATP to power a few seconds of contraction. Thereafter, concentrations of ATP need to be maintained at a steady level.

Amino acids

Fatty acids

Oxygen

Long-distance runner
During a prolonged aerobic exercise, such as long-distance running, sufficient oxygen is delivered by the bloodstream to muscles to break down glucose and especially fatty acids to make ATP.

Glucose → Glycolysis → Pyruvic acid → Aerobic respiration in mitochondria → 36 ATP molecules

Aerobic respiration
When a person is resting, or performing light or moderate exercise, aerobic respiration provides most of the ATP for muscle contraction. During aerobic respiration glucose, other fuels including fatty acids and amino acids, are [ ] completely to water and carbon dioxide by a [ ] f reactions that take place inside mitochondria. [ ] s requires the input of oxygen.

2 ATP molecules
This initial stage of aerobic respiration happens in the cytoplasm. Glucose is broken down to pyruvic acid, generating a little ATP. ATP moves inside mitochondria for the next stage of aerobic respiration.

Carbon dioxide

Water

Waste product
The reactions of respiration in the mitochondria release waste carbon dioxide, which is then expelled by the lungs.

36 ATP molecules
After pyruvic acid enters the mitochondrion it is processed in a cycle of chemical reactions. This releases carbon dioxide, which is removed, and hydrogen. The hydrogen passes along an electron transport chain that uses energy stored in the hydrogen to make up to 36 ATP molecules for each molecule of glucose. At the end of this process, hydrogen combines with oxygen to make water.

Glucose → Glycolysis → Pyruvic acid → Fermentation → Lactic acid

Sprinter
This sprinter's race will be over in just a few seconds. During that short burst of strenuous activity, anaerobic respiration "burns" huge amounts of glucose without oxygen to supply the ATP needed for muscle contraction.

Anaerobic respiration
During bursts of strenuous exercise, when muscles contract to maximum possible effect, blood vessels supplying oxygen to muscle fibers are squeezed, limiting the delivery of oxygen. Under these circumstances muscle fibers switch to anaerobic respiration, which does not require oxygen, to meet their energy needs. It frees far less energy than aerobic respiration, but happens much more rapidly.

2 ATP molecules
Glycolysis during anaerobic respiration is the same as during aerobic respiration and releases two ATP molecules for every glucose molecule broken down. This is the total energy yield of anaerobic respiration.

Muscle fatigue
Fermentation breaks down pyruvic acid to lactic acid, which causes muscle fatigue and, if allowed to accumulate, cramps. It is therefore converted back to pyruvic acid and recycled.

MUSCLE MECHANICS

In order for them to work effectively, muscles are organized in very specific ways. They are attached to bones by tough, compact tendons. They operate lever systems to move body parts efficiently. And muscles work as antagonists with opposing effects to create a wide range of controlled movements.

MUSCLE ATTACHMENT

Tough cords called tendons attach muscles to bones, and transmit the force of their contraction. Tendons are endowed with enormous tensile strength because they contain parallel bundles of tough collagen fibers. These extend through the periosteum, the bone's outer membrane, to make firm anchorage in the bone's outer layer. Muscles are attached, by their tendons, at one end to one bone, then having stretched across a joint, at the other end to another bone. When a muscle contracts, one of the bones it is attached to moves, while the other does not. A muscle's attachment to an immovable bone is called its origin; its attachment to a movable bone is called its insertion (see pp.56–57).

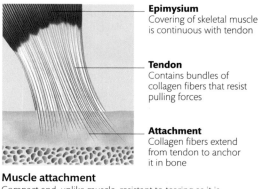

Epimysium
Covering of skeletal muscle is continuous with tendon

Tendon
Contains bundles of collagen fibers that resist pulling forces

Attachment
Collagen fibers extend from tendon to anchor it in bone

Muscle attachment
Compact and, unlike muscle, resistant to tearing as it is moved over bony projections, a tendon provides a strong connection between muscle and bone.

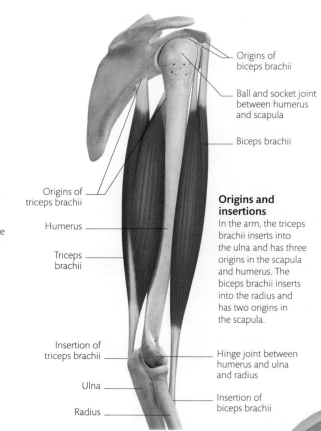

Origins of biceps brachii

Ball and socket joint between humerus and scapula

Biceps brachii

Origins of triceps brachii

Humerus

Triceps brachii

Insertion of triceps brachii

Ulna

Radius

Origins and insertions
In the arm, the triceps brachii inserts into the ulna and has three origins in the scapula and humerus. The biceps brachii inserts into the radius and has two origins in the scapula.

Hinge joint between humerus and ulna and radius

Insertion of biceps brachii

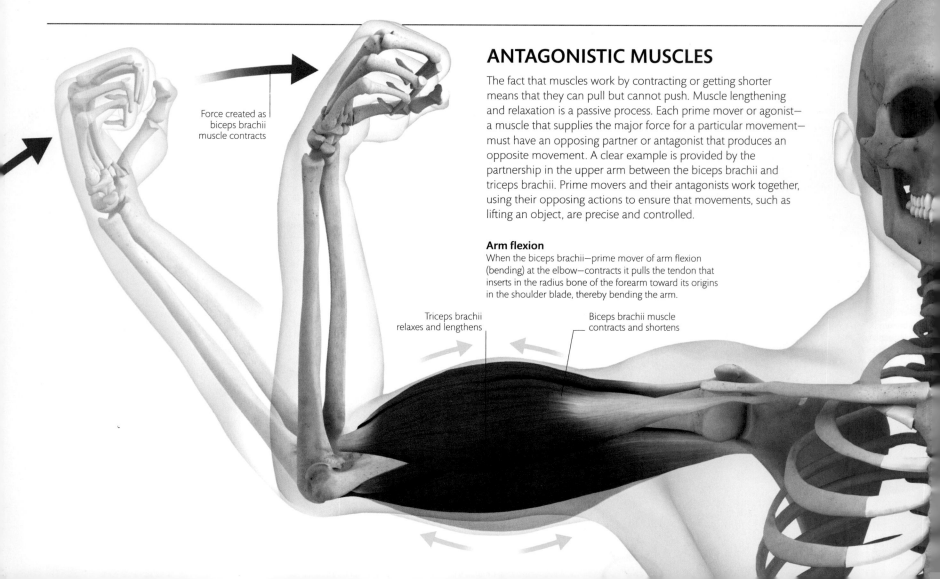

Force created as biceps brachii muscle contracts

ANTAGONISTIC MUSCLES

The fact that muscles work by contracting or getting shorter means that they can pull but cannot push. Muscle lengthening and relaxation is a passive process. Each prime mover or agonist— a muscle that supplies the major force for a particular movement— must have an opposing partner or antagonist that produces an opposite movement. A clear example is provided by the partnership in the upper arm between the biceps brachii and triceps brachii. Prime movers and their antagonists work together, using their opposing actions to ensure that movements, such as lifting an object, are precise and controlled.

Arm flexion
When the biceps brachii—prime mover of arm flexion (bending) at the elbow—contracts it pulls the tendon that inserts in the radius bone of the forearm toward its origins in the shoulder blade, thereby bending the arm.

Triceps brachii relaxes and lengthens

Biceps brachii muscle contracts and shortens

BODY LEVERS

The simplest kind of machine, a lever is a rod that tilts on a pivot, or fulcrum. When a force, or effort, is applied to one point on the rod it swings around the fulcrum to perform work by moving a load at another point. Levers have a multitude of everyday uses including cutting, as in scissors, and prying objects apart, as in a crowbar. Exactly the same mechanical principles of lever action apply to the interaction of bones, joints, and muscles to generate movement. Bones act as levers, joints are the fulcrums, and muscles contract to apply the force that moves the body part, or load. The body's various lever systems allow a wide range of movements, including lifting and carrying. Like all levers, body levers fall into three classes, according to the relative position of force, fulcrum, and load.

Examples of each are shown here; red arrows show the direction of force and blue arrows the movement of load.

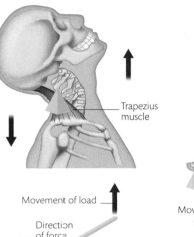

Trapezius muscle

Movement of load

Direction of force

Fulcrum

First-class lever
As in a seesaw, here the fulcrum lies between force and load. For example, muscles in the back of the neck and shoulder pull the rear of the skull, pivoted on neck vertebrae, to lift the face upward.

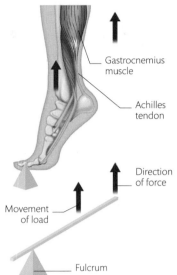

Gastrocnemius muscle

Achilles tendon

Direction of force

Movement of load

Fulcrum

Second-class lever
Here, as in the case of a wheelbarrow, the load lies between the fulcrum and the force. For example, using toes as a fulcrum, calf muscles contract to raise the heel and lift the body.

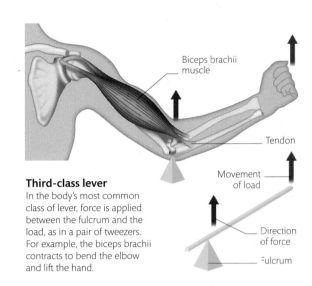

Biceps brachii muscle

Tendon

Movement of load

Direction of force

Fulcrum

Third-class lever
In the body's most common class of lever, force is applied between the fulcrum and the load, as in a pair of tweezers. For example, the biceps brachii contracts to bend the elbow and lift the hand.

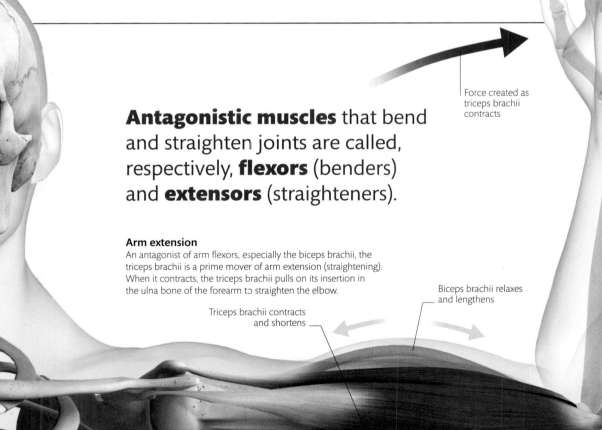

Antagonistic muscles that bend and straighten joints are called, respectively, **flexors** (benders) and **extensors** (straighteners).

Force created as triceps brachii contracts

Arm extension
An antagonist of arm flexors, especially the biceps brachii, the triceps brachii is a prime mover of arm extension (straightening). When it contracts, the triceps brachii pulls on its insertion in the ulna bone of the forearm to straighten the elbow.

Triceps brachii contracts and shortens

Biceps brachii relaxes and lengthens

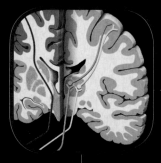

BRAIN

Packed with 100 billion nerve
cells, the brain works in tandem
with the spinal cord to control
everything we sense and do.

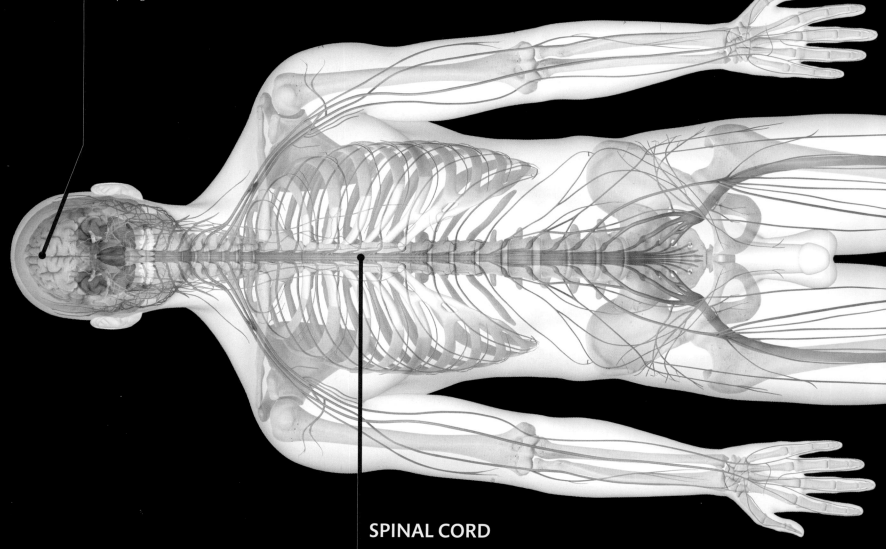

SPINAL CORD

This highly organized bundle
of nerves relays information
and performs basic processing
en route to the brain.

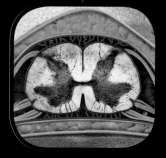

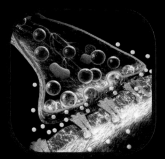

NERVE

Information travels along nerves to and from the brain and spinal cord in the form of a "language" of tiny electrical impulses.

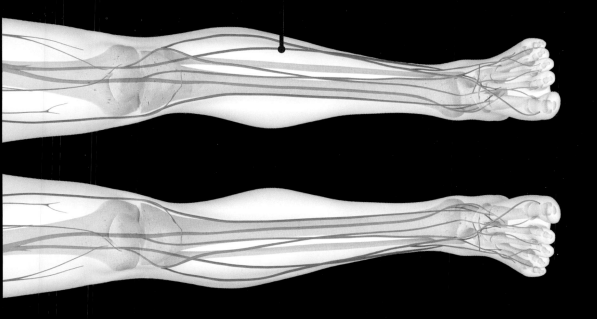

Command, control, and coordination—the nervous system is at the very core of the body's existence. It enables us to adapt to our surroundings as we sense the world around us and engage with it accordingly.

NERVOUS SYSTEM

HOW THE BODY IS WIRED

The human nervous system is composed of three main parts: central, peripheral, and autonomic. Their definitions are partly anatomical and partly functional. Some nerves are under our conscious control while the activity of others is automatic and designed to maintain our body's status quo.

NERVOUS SYSTEM SUBDIVISIONS

The central nervous system (CNS) consists of the brain in the skull and the main nerve from it—the spinal cord, which extends along the inside of the backbone. The peripheral nervous system (PNS) includes all of the nerves branching from the CNS—12 pairs of cranial nerves from the brain and 31 pairs of spinal nerves from the spinal cord. The third main subdivision is the autonomic nervous system (ANS), which shares some structures with the CNS and the PNS as well as having unique features of its own.

THE SOMATIC DIVISION

The somatic division of the PNS is concerned with voluntary movements—that is, conscious actions under free will that we make and control by choice. The brain sends out instructions (motor information) to the skeletal muscles to control their contraction and relaxation in precise ways. Meanwhile, this division also receives and deals with all the data (sensory information) arriving from the skin and other sense organs.

Power of touch
The somatic division of the PNS mediates the intimate feelings from touch as well as coordinating delicate finger movements.

THE ENTERIC DIVISION

The PNS's enteric division controls most of the abdominal organs, chiefly the gastrointestinal tract (stomach and intestines) and to some extent the urinary system. These work mainly under automatic control, without stimulation from or monitoring by the brain. Contractions of the muscles in the tract walls must be coordinated carefully so that digested food moves along the tract in the correct sequence, with suitable timing. The enteric division has its own sensory and motor nerve cells with information-processing interneurons between them. Parts of the enteric division work alongside the ANS (see opposite).

Brain
The body's ultimate control center, encased in the skull

Cranial nerves
These 12 pairs of nerves (yellow) control functions in the head and neck (see pp.116–17).

Spinal nerves
Branch in pairs with each vertebra of the spine, carrying information between the brain and the rest of the body

Spinal cord
The pattern of gray matter (nerve cell bodies) amid the white matter (nerve cell axons) in the spinal cord is butterfly-shaped in cross section.

Sensory rootlet

Ganglion
Lumplike junction area with many nerve cell bodies

Motor rootlet

Spinal nerve root
The motor and sensory rootlets converge to give rise to one spinal nerve.

Sacral plexus
Several nerves join and branch at multijunctions called plexi

Inside a nerve
A nerve contains bundles of strongly wrapped axons (nerve fibers).

Bodywide system
The nervous system seems concentrated in the brain and spinal cord, or CNS. Yet in terms of its basic units—the nerve cells—such parts are greatly outnumbered by the network of nerves in the PNS. Gradually dividing from finger-thick to thinner than a hair, nerves snake into, around, and between almost every tissue and organ—from the scalp to the tips of the toes.

THE AUTONOMIC NERVOUS SYSTEM

We are not aware of the vast amounts of nervous system activity that occur below the brain's level of consciousness. Such activity is mainly the province of the ANS (along with the enteric division, see opposite). We can think of the ANS as our "automatic pilot": it monitors internal conditions such as temperature and the levels of chemical substances and keeps these within narrow limits; it also controls processes we rarely think about such as heart rate, breathing, digestion, and excretion, by stimulating muscles to contract and glands to release their products. There are two divisions —the sympathetic and the parasympathetic— of the ANS, the complementary actions of which are shown below.

Out of our control

When overwhelming emotions such as sudden grief sweep through the body, this is mainly the result of ANS activity. It takes time and mental effort for the brain to reassert conscious control.

BREAKTHROUGHS
JOHN NEWPORT LANGLEY

In 1921, Part 1 of what was to become the influential book *The Autonomic Nervous System* was published. Its author, John Newport Langley, was based at Cambridge University, England. He coined the term "autonomic" to describe what had been called the "vegetative nervous system." In its pages, he discussed antagonistic subsystems, established that the ANS had central and peripheral components, and suggested synapses with glands and smooth muscle.

THE SYMPATHETIC DIVISION

The sympathetic part of the ANS is mainly stimulatory; that is, it raises the activity of its target tissues and organs. Heartbeat, breathing, and various hormone levels all increase and prepare the body for stressful situations (the "fight or flight" response). Information flows from the brain to the spinal cord and then to two chains of ganglia that lie along either side of the backbone, before traveling on to muscles, such as those in the stomach that churn food, and glands, such as the adrenal gland, which releases epinephrine.

THE PARASYMPATHETIC DIVISION

Within the parasympathetic division, information flows from the brain and spinal cord along major nerves directly to the targets, where ganglia-like sets of nerve cells integrate the activity. This division counteracts the sympathetic's stimulation by reducing the activity of target tissues and organs, thereby inducing a calming effect (often referred to as "rest and digest"). For instance, after the heartbeat races, it gradually settles back to normal based on parasympathetic activity. Between them, the two divisions exert close control of the body with a "push-pull" balance.

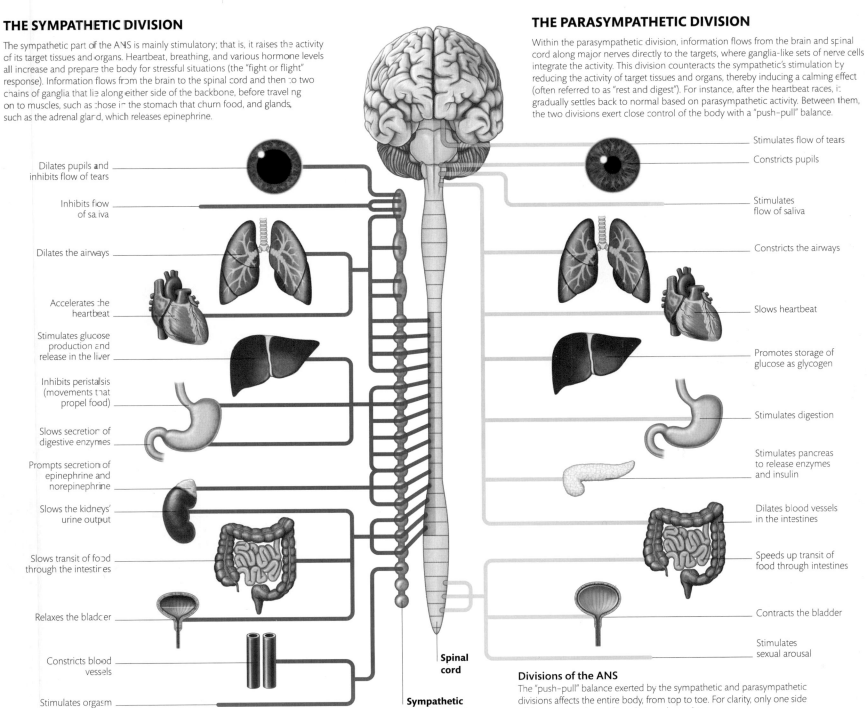

Dilates pupils and inhibits flow of tears

Inhibits flow of saliva

Dilates the airways

Accelerates the heartbeat

Stimulates glucose production and release in the liver

Inhibits peristalsis (movements that propel food)

Slows secretion of digestive enzymes

Prompts secretion of epinephrine and norepinephrine

Slows the kidneys' urine output

Slows transit of food through the intestines

Relaxes the bladder

Constricts blood vessels

Stimulates orgasm

Stimulates flow of tears

Constricts pupils

Stimulates flow of saliva

Constricts the airways

Slows heartbeat

Promotes storage of glucose as glycogen

Stimulates digestion

Stimulates pancreas to release enzymes and insulin

Dilates blood vessels in the intestines

Speeds up transit of food through intestines

Contracts the bladder

Stimulates sexual arousal

Spinal cord

Sympathetic ganglion chain

Divisions of the ANS

The "push-pull" balance exerted by the sympathetic and parasympathetic divisions affects the entire body, from top to toe. For clarity, only one side of the sympathetic ganglion chain is shown here.

NERVE CELLS

All body parts are made of cells. The nervous system's main cells are called neurons. The brain has at least 100 billion of them, and they communicate using a language of tiny electrical pulses or nerve signals.

HOW NEURONS WORK

The basic parts inside a neuron are similar to those in other cells (see pp.20–21). What makes neurons among the most delicate and specialized of all body cells is a combination of their shape and the way the outer cell membrane carries or conducts nerve signals. Each signal travels along the membrane as a pulse, or peak, of electricity, caused by the movement of electrically charged particles called ions (see opposite). Every neuron has its own individual shape, usually with many short branches called dendrites, and one longer, thinner, wirelike extension, the axon (see also pp.64–65). Dendrites collect nerve signals from other neurons. The cell body combines and integrates these signals, and sends its outgoing signal along the axon and then on to other neurons, or to muscle or gland cells.

SUPPORTING ROLES

Less than half of the cells in the brain are neurons. Most of the rest are glial (glue) cells, or neuroglia, of several types. Together these support, nourish, maintain, and repair the delicate neurons. Astrocyte glial cells form a framework through which dendrites and axons snake as they grow and make new extensions. Astrocytes are also important in repairing damage caused by temporary lack of blood, toxins, or infecting microbes. Oligodendrocyte glial cells manufacture the myelin sheath for certain axons in the central nervous system; in the peripheral nervous system, this is carried out by Schwann cells. Ependymal glial cells form coverings and linings and produce cerebrospinal fluid (see pp.316–17).

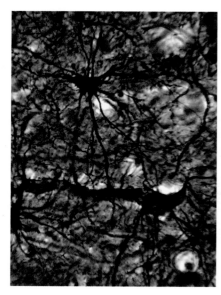

Cell body (soma)
Mainly a soupy liquid, cytoplasm, in which other parts float or move

Nucleus
Cellular control center containing the genetic material, DNA

Dendrite
Branched extension that receives signals from other neurons

Axon hillock
Where the cell body narrows to form the axon; nerve signals are generated here

Astrocyte
Provides neuron with physical support and nourishment

Astrocyte framework
Star-shaped astrocytes signal to each other using the mineral calcium, which helps coordinate their growth and their support of neurons.

SPECIALIZED INSULATION

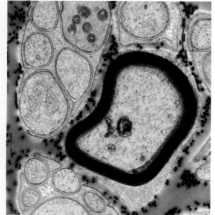

The fatty substance myelin forms a barrier to electrical impulses and to chemical movements. In the brain and spinal cord it is manufactured by oligodendrocytes. These extend their cell membranes to wrap in a spiral fashion around the axons of certain neurons, forming a multilayered myelin covering known as the myelin sheath. The covering is not continuous but exists in sections about $1/32$ in (1 mm) long, with breaks between them called nodes of Ranvier. Myelin's insulation prevents the electrical pulses of nerve signals from leaking away into the surrounding fluids and cells. It also speeds conduction of an impulse by forcing it to "jump" from one node to the next—a process called saltatory conduction. As a result, nerve signals are faster and stronger in myelinated axons than in those without myelin.

Super-fast signals
The insulating layers of myelin (brown) around this axon, compared with the other nonmyelinated axons (green), result in super-fast nerve signals.

THE ELECTRICAL NATURE OF NERVE SIGNALS

Nerve signals are pulses of electricity caused by the mass movement of tiny particles called ions. Electrical charge is a fundamental property of matter. Minerals such as potassium and sodium dissolve in bodily fluids and exist as ions, each with a positive charge. The more ions in a certain place, the higher the charge. The fluids inside and outside of cells are electrically neutral, but there is a polarizing shell of charge coating every cell's membrane, and this creates the resting potential. When ions move across the membrane, the associated move of charge creates a pulse of electricity or action potential. An action potential measures about 100 mV from peak to trough and is over in 1/250 th of a second.

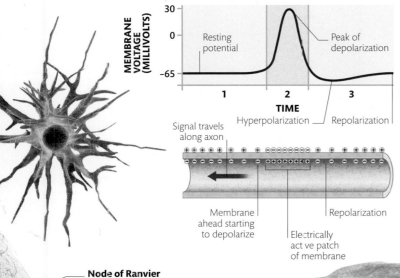

MEMBRANE VOLTAGE (MILLIVOLTS)

30
0
-65

Resting potential

Peak of depolarization

TIME

1   2   3

Hyperpolarization   Repolarization

Action potential
Ions move in and out of small patches of the axon's membrane to generate an action potential by changing the cell's voltage.

Signal travels along axon

Membrane ahead starting to depolarize

Electrically active patch of membrane

Repolarization

Traveling signals
The region of reversed charge "fizzes" along the length of the axon, much like a lit fuse, before passing the message on at a synapse (see p.314). Charges across the membrane are disrupted ahead of and behind the depolarization.

Node of Ranvier
Slight gap between neighboring sections of myelin sheath

Oligodendrocyte
Makes the myelin sheaths in the CNS; it can extend "arms" to more than 30 neurons

Myelin sheath
Wrap-around covering that insulates the axon and speeds signal conduction

Axon
The neuron's longest and thinnest projection; nerve signals travel from the cell body along the axon to the synapse

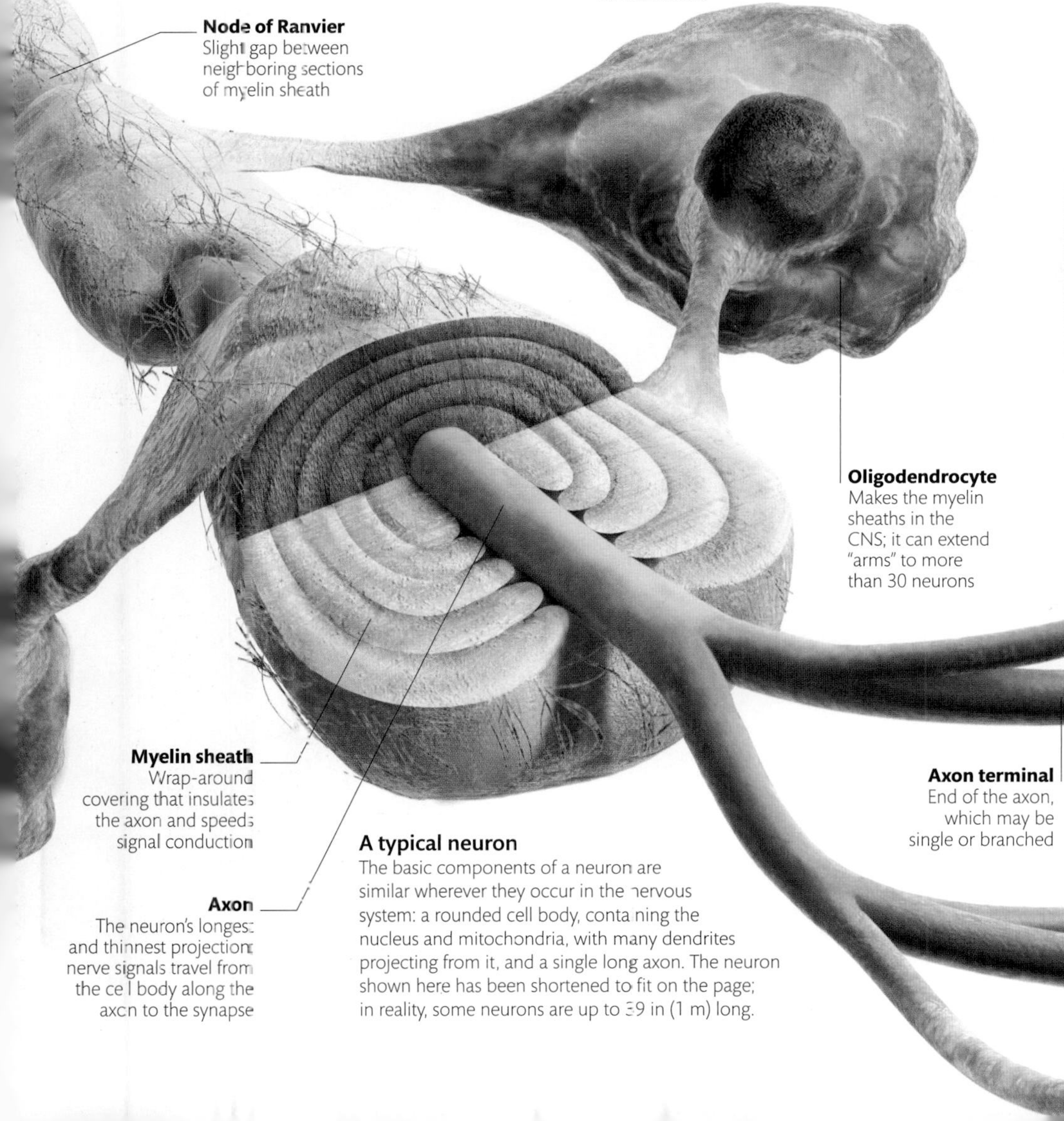

A typical neuron
The basic components of a neuron are similar wherever they occur in the nervous system: a rounded cell body, containing the nucleus and mitochondria, with many dendrites projecting from it, and a single long axon. The neuron shown here has been shortened to fit on the page; in reality, some neurons are up to 39 in (1 m) long.

Axon terminal
End of the axon, which may be single or branched

Synaptic knob
Conveys nerve signals to other cells across a tiny gap or synapse (see pp.314-15)

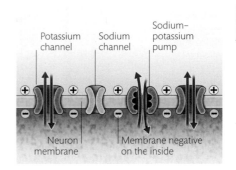
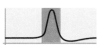

Potassium channel   Sodium channel   Sodium-potassium pump

Neuron membrane   Membrane negative on the inside

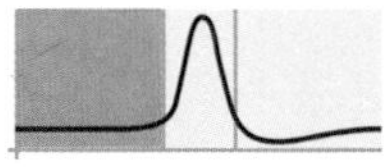

1 Resting potential
Every nerve cell's sodium–potassium pump distributes sodium and potassium across the cell membrane, which creates differences in concentration and a polarization of electrical charge at the membrane—the resting potential—with the inside of the cell negatively charged.

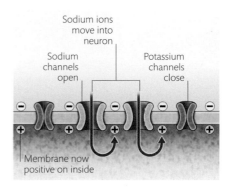

Sodium ions move into neuron

Sodium channels open   Potassium channels close

Membrane now positive on inside

2 Depolarization
A stimulus arrives and triggers voltage-gated sodium channels to open. Sodium ions flood into the neuron, causing a movement of positive charge. If this depolarization (reversal of the polarity of the membrane) achieves a critical level (called threshold) the membrane generates an action potential.

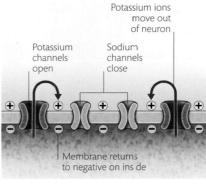
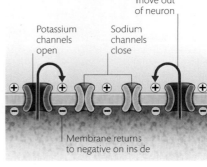

Potassium ions move out of neuron

Potassium channels open   Sodium channels close

Membrane returns to negative on inside

3 Repolarization
The depolarizing change in voltage causes sodium channels to snap shut and voltage-gated potassium channels to open. Now, potassium ions move out of the neuron, removing the positive charge brought in by the sodium ions. In fact, a brief hyperpolarization occurs (inside is even more negative) before returning to its resting potential.

PASSING ON THE MESSAGE

Nerve messages travel along individual neurons as tiny pulses of electricity. They change into chemical form, as molecules of neurotransmitters, to cross the tiny gaps at the junctions, or synapses, between neurons.

At the synapse
Neurons do not quite touch at their main communication points, the synapses. Their cell membranes are separated by a synaptic cleft just 20 nanometers wide. As a nerve impulse in the sending neuron arrives at the synapse, it triggers the release of neurotransmitter molecules. These "jump the gap" and set off a nerve impulse in the receiving neuron.

1 Neurotransmitter ready
Vesicles travel from the sending neuron's cell body to the presynaptic membrane. An impulse arrives and makes them fuse with the membrane and release their contents.

2 Crossing the gap
Neurotransmitter molecules cross the cleft in a few thousandths of a second and attach to receptor sites in the postsynaptic membrane of the receiving neuron.

3 The message continues
Neurotransmitter molecules bind to receptors on ion channels in the postsynaptic membrane, causing them to open. Positive ions then flow into the receiving neuron. If enough channels open, a new wave of depolarization is triggered.

Postsynaptic membrane
Part of the receiving neuron

Microtubule
Microscopic conveyor belt that carries vesicles to the synapse

Axon of neuron
Nerve impulses travel along this to the synapses at its end

Vesicle
Membrane bag of neurotransmitter molecules

Ion
This electrically charged particle floats in the fluid on either side of the cells' membranes

Neurotransmitter molecule
Relatively large chemical "messenger" units; there are several main kinds, including GABA, acetylcholine, and dopamine

Presynaptic membrane
The end part of the sending neuron

Synaptic cleft
Fluid-filled gap less than $1/5000$th the width of a human hair

HOW NERVE CELLS COMMUNICATE

The basic "language" of the nervous system is nerve signals or impulses. This language is frequency based—that is, it "talks" in digital and not analogue terms. The precise information nerves carry depends on how many impulses there are, how close together, where they come from, and where they go.

Resting or quiet neurons, for instance, might send an impulse every second or two. A highly stimulated neuron—for example, dealing with sudden pressure on the skin—might send 50 impulses per second. These signals are passed onward to other neurons with which it has synaptic connections. The pattern of connections between neurons changes over time, through natural body development and also through learning (see p.321).

In the brain's cortex, one neuron may have synapses with more than 200,000 others, so that a piece of cortex the size of this "o" contains more than 100 billion synapses. The way that each neuron processes its incoming signals, and what it sends onward, is shown below.

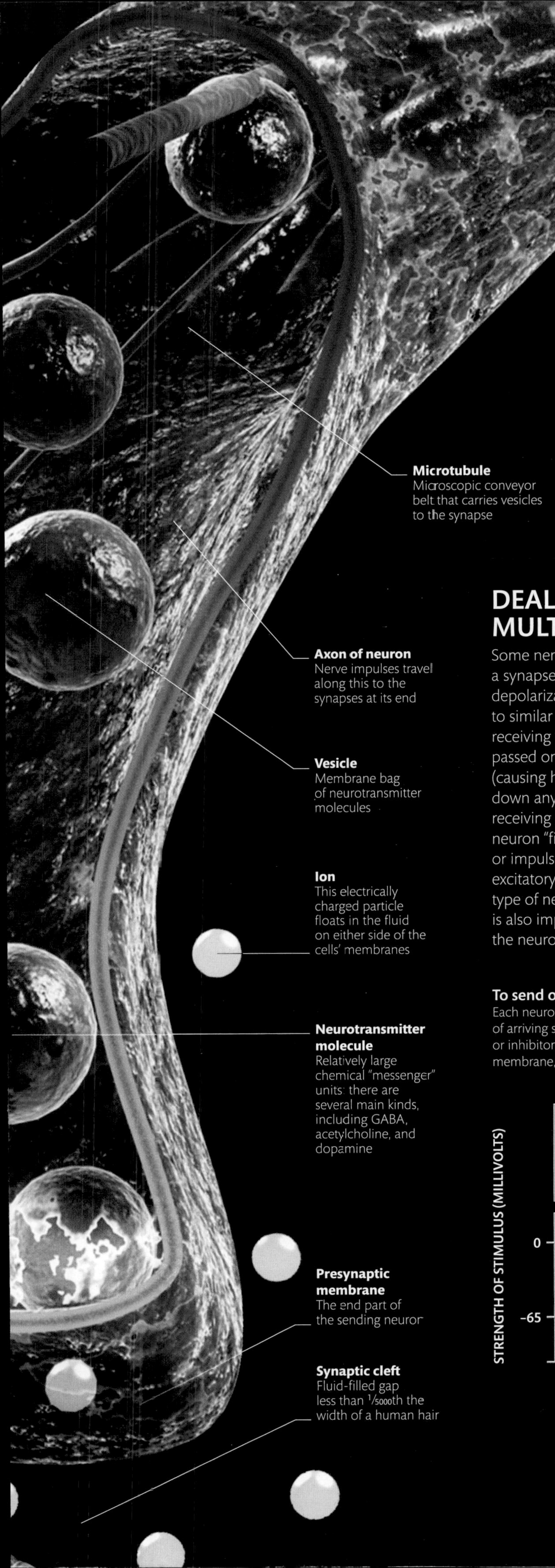

DEALING WITH MULTIPLE SIGNALS

Some nerve impulses arriving at a synapse are excitatory (causing depolarization) and thereby contribute to similar impulses being formed in the receiving neuron and the message being passed on. Other inputs are inhibitory (causing hyperpolarization), damping down any impulse formation in the receiving neuron. Whether the receiving neuron "fires off" an action potential, or impulse, depends on the sum of its excitatory and its inhibitory inputs. The type of neurotransmitter at the synapse is also important, as is the structure of the neurotransmitter receptor site.

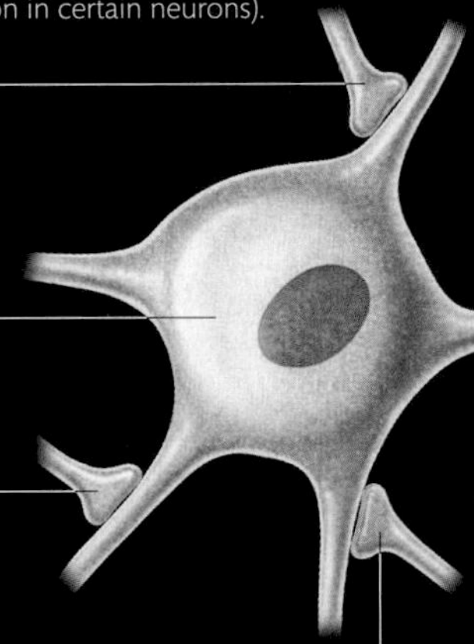

To send or not to send?
Each neuron's inputs (A, B, or C) vary depending on the frequency of arriving signals, their synapse positions, and whether they are excitatory or inhibitory. As a complex web of electricity ripples around the neuron's membrane, it may send its own signals onward—or not.

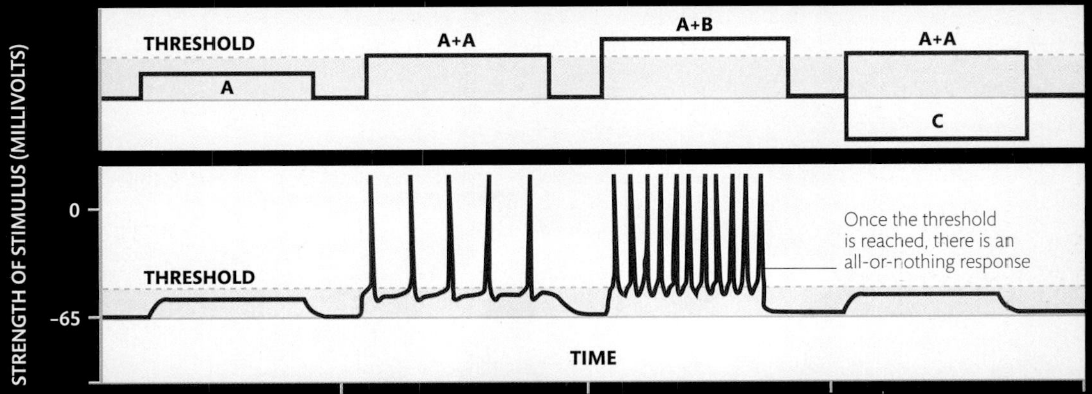

Signal summation
At any instant, a neuron's activity is affected by "summing" the numbers and types of signals it receives and by their positions on its dendrites and cell body (and perhaps the axon in certain neurons).

Excitatory input (A)
This input comes a short distance from a neighboring neuron

Neuron cell body
The cell body receives inputs, as do dendrites

Excitatory input (B)
This axon terminal is from a neuron many inches away

Inhibitory input (C)
Information received here works against the excitatory inputs

THRESHOLD   A+A   A+B   A+A

A

C

STRENGTH OF STIMULUS (MILLIVOLTS)

0

THRESHOLD

-65

Once the threshold is reached, there is an all-or-nothing response

TIME

Subthreshold stimulation
The depolarization of this excitatory input (A) is too small to reach the threshold level, and so the neuron doesn't "fire" an action potential.

Threshold stimulation
The greater the excitatory input (A+A), the greater the chance of exceeding the threshold; here a series of action potentials results for the duration of excitation.

Hyperstimulation
When even greater stimulatory impulses arrive (A+B), far exceeding the threshold level, they result in a higher-frequency sequence of outgoing signals.

Inhibition
The inhibitory input (C) cancels out the stimulatory impulses (A+A), which would normally depolarize to the threshold, so here no signal is generated.

THE BRAIN AND SPINAL CORD

The central nervous system—the brain and spinal cord—receives information from all body parts and replies with instructions to all tissues and organs. These nerve centers are protected and nourished by an elaborate system of membranes and fluids, including blood.

INFORMATION PROCESSING

The spinal cord gathers messages from the torso and limbs and relays them to the brain. But the cord is not just a passive conveyor of signals; it also carries out basic body "housekeeping," receiving and sending messages without involving the brain. In general, the "higher" the information goes—heading up to the top of the brain—the nearer it gets to our conscious awareness. As the cord merges with the brain it leads to the brain stem, where centers monitor and adjust vital functions, such as heartbeat and breathing, usually without bothering the upper brain. Higher still is the thalamus, a "gatekeeper" that selects which information to allow into the uppermost area, the cerebral cortex. Many of the highest mental functions occur in the cortex—thoughts, imagination, learning, and conscious decision-making.

PROTECTING THE BRAIN

Around most of the brain is the rigid, curved case of the upper skull, the cranium. Bone and brain are separated by a set of three sheetlike membranes —the meninges—and two layers of fluid. Outermost is the tough dura mater membrane lining the inside of the skull. Next is the spongier, blood-rich arachnoid. Spaces called venous sinuses between the dura and the arachnoid contain the outer cushioning liquid—slow-flowing venous blood leaving the brain to return to the heart. Within the arachnoid is an inner cushioning layer of cerebrospinal fluid (see opposite). Below this is the innermost and thinnest membrane, the pia mater, which closely follows the brain's contours directly beneath it.

Between brain and skull
Cerebrospinal fluid circulates in a thin gap, the subarachnoid space (see opposite), between the arachnoid and the pia mater. The meninges and fluid work together to absorb and disperse excessive mechanical forces so they don't result in injury.

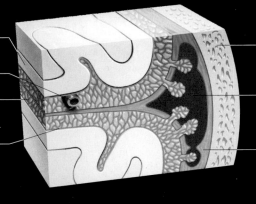

Cerebral cortex
Outermost layer of the brain

Blood vessel

Arachnoid
Weblike layer rich in blood vessels and fluid

Pia mater
Thin membrane around the surface of the brain

Skull bone

Dural venous sinus
Venous blood drains away from the brain

Dura mater
Outermost and strongest membrane

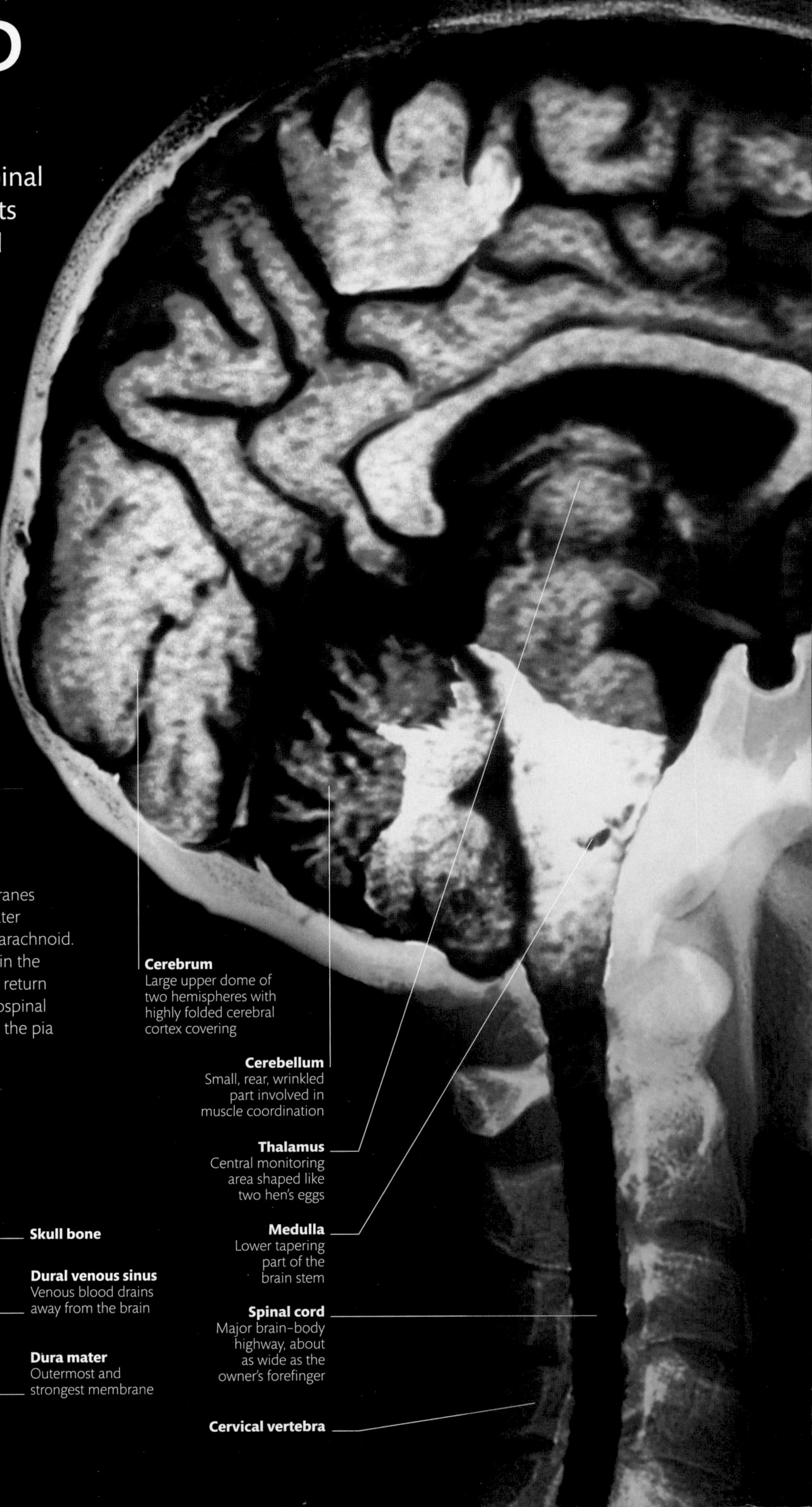

Cerebrum
Large upper dome of two hemispheres with highly folded cerebral cortex covering

Cerebellum
Small, rear, wrinkled part involved in muscle coordination

Thalamus
Central monitoring area shaped like two hen's eggs

Medulla
Lower tapering part of the brain stem

Spinal cord
Major brain–body highway, about as wide as the owner's forefinger

Cervical vertebra

FEEDING THE BRAIN

The brain has two main sources of nourishment and waste disposal. One is blood, brought mainly by the carotid and vertebral arteries in the neck to the Circle of Willis at the brain's base. The second system involves a liquid derived from blood, cerebrospinal fluid (CSF). This fluid is made at a slow, steady rate by the linings of two chambers inside the brain's hemispheres called the lateral ventricles, and it flows within and around the brain. About 17 fl oz (half a liter) of CSF is produced every day, with up to 5 fl oz (150 milliliters) present at any time. It transports glucose, proteins, and other materials to brain tissues, and takes away waste substances; it also carries infection-fighting white blood cells. In addition to metabolic functions, CSF provides physical comfort for the brain and spinal cord since they "float" in it.

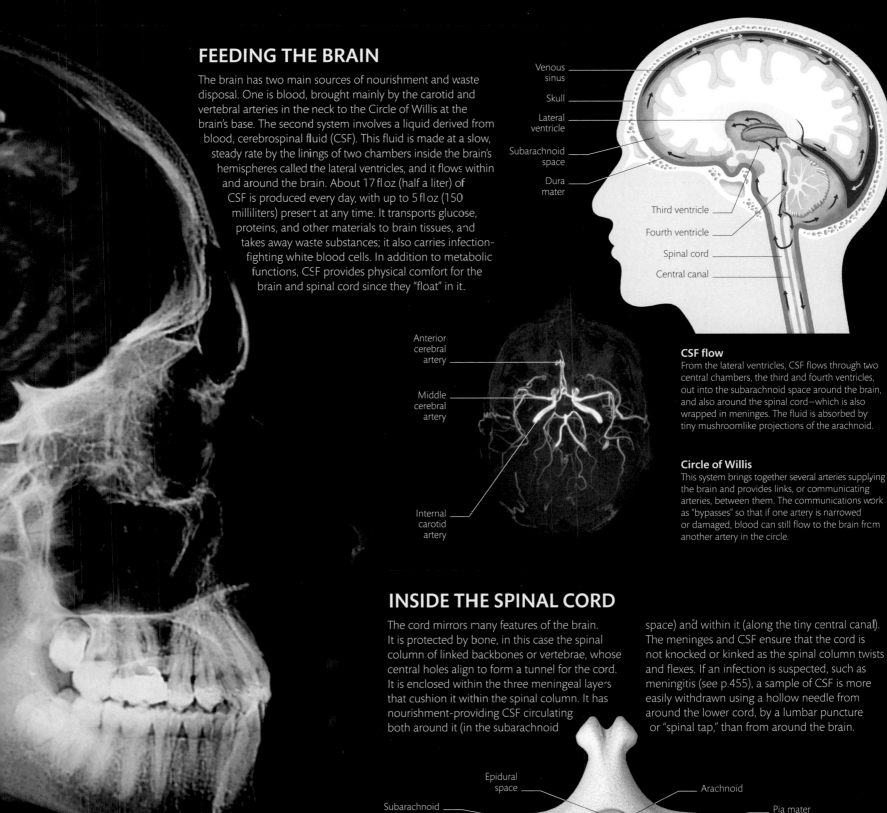

Venous sinus
Skull
Lateral ventricle
Subarachnoid space
Dura mater
Third ventricle
Fourth ventricle
Spinal cord
Central canal

Anterior cerebral artery
Middle cerebral artery
Internal carotid artery

CSF flow
From the lateral ventricles, CSF flows through two central chambers, the third and fourth ventricles, out into the subarachnoid space around the brain, and also around the spinal cord—which is also wrapped in meninges. The fluid is absorbed by tiny mushroomlike projections of the arachnoid.

Circle of Willis
This system brings together several arteries supplying the brain and provides links, or communicating arteries, between them. The communications work as "bypasses" so that if one artery is narrowed or damaged, blood can still flow to the brain from another artery in the circle.

INSIDE THE SPINAL CORD

The cord mirrors many features of the brain. It is protected by bone, in this case the spinal column of linked backbones or vertebrae, whose central holes align to form a tunnel for the cord. It is enclosed within the three meningeal layers that cushion it within the spinal column. It has nourishment-providing CSF circulating both around it (in the subarachnoid space) and within it (along the tiny central canal). The meninges and CSF ensure that the cord is not knocked or kinked as the spinal column twists and flexes. If an infection is suspected, such as meningitis (see p.455), a sample of CSF is more easily withdrawn using a hollow needle from around the lower cord, by a lumbar puncture or "spinal tap," than from around the brain.

Epidural space
Subarachnoid space
Dura mater
Cerebrospinal fluid
Arachnoid
Pia mater
Central canal
Vertebral bone
FRONT OF BODY

The spinal cord in section
The cord is encased within the central space of the vertebral column; its nerve roots (yellow) pass out through gaps between adjacent vertebrae.

A slice through the brain
This MRI scan through the middle of the brain and cord (from front to back) shows their major features. The darker areas of the brain are fluid-filled spaces and internal chambers known as ventricles. In blue around the brain are the protective bones of the skull and, on either side of the cord, the bones of the neck (cervical vertebrae).

THE CNS IN ACTION

Our brain and spinal cord are always active—in constant communication with each other and the rest of our bodies. Messages stream in from the peripheral nervous system (PNS), and are channeled to the central nervous system (CNS), which processes the signals and sends instructions back out.

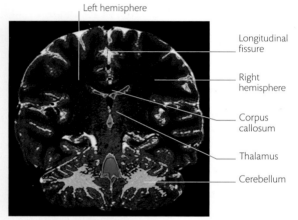

Two sides working as one
This vertical "slice" through the brain shows the longitudinal fissure as a deep furrow between the left and the right cerebral hemispheres. At its base the corpus callosum, a bridge of more than 200 million nerve fibers, links the hemispheres.

Labels on image: Left hemisphere, Longitudinal fissure, Right hemisphere, Corpus callosum, Thalamus, Cerebellum

KNOWING LEFT FROM RIGHT

Anatomically, the nervous system shows left–right symmetry (see pp.60–63); but in terms of function, it's not as simple. The brain's wrinkled cerebrum is almost completely divided by a deep front-to-back groove into two cerebral hemispheres, left and right. Although these may look outwardly similar, each hemisphere dominates for certain mental functions (see table, right). The two hemispheres "talk" constantly via a straplike collection of nerve fibers—the corpus callosum.

Information from the body swaps sides on its way to the brain. Nerve signals travel within organized bundles of nerve fibers called tracts, which cross over from the left side of the body to the right side and vice versa. So, for example, sensory information from the body's left side ends up in the right hemisphere, and motor instructions sent from the left hemisphere control muscles on the right side of the body.

LEFT SIDE OF BRAIN	RIGHT SIDE OF BRAIN
Breaks up a whole into constituent parts	Intuitively combines parts into a whole
Analytical activity, with progressive sequencing	Tends to make random leaps and links
Tends to be objective, impartial, detached	More subjective and individualistic
More active with words and numbers	More active with sounds, sights, and items in space
Deals more with logic and implication	Deals more with ideas and creativity
Leads in rational problem-solving	Jumps with insight to possible solutions
Location of speech and language centers	Rarely dominates speech and language
Stores literal meanings of words, grammar	Gives language context and accentuation
More active in recalling names	More active in facial recognition
Controls right side of the body	Controls left side of the body

Which side takes charge?
Brain scans and studies of brain injury or disease reveal that the "take-apart" left side is more concerned with logic and reasoning, while the "put-together" right side is more intuitive and holistic; although each side assists the other.

TO THE BRAIN AND BACK AGAIN

Information from the world around us reaches the brain via the major sense organs (see p.324). An external stimulus is converted into nerve impulses by specialized receptor cells. The impulses begin a journey through the sensory nerves of the peripheral nervous system and on to the higher centers in the brain; the route to the cerebral cortex may involve a series of up to 10 neurons linked by synapses (see p.314). At each relay station in the sequence, additional messages are sent out along other pathways, like branches diverging from a tree trunk. In the cortex, we become aware of the stimulus and decide to act. The result is a cascade of outgoing or motor messages that travel in the reverse direction, out to various muscles and glands.

Dorsal root
Carries sensory nerves into the spinal cord

Dorsal column–medial lemniscus tract
Sensory information (other than pain) diverges in the spinal cord: one branch stays within the cord to synapse with another neuron; the other branch ascends the spinal cord to the medulla

Dorsal root ganglion
Neuron cell bodies and synapses relay the signals into the spinal cord

Myelinated axon
The myelin sheath speeds the nerve impulse transmission

CROSS SECTION OF THE SPINAL CORD

Sensory receptor
Responds to activation by sending impulses along its axon

White and gray matter
White matter (axons) surrounds the central gray matter (neuron bodies, interconnecting dendrites, and synapses)

Motor messages
Motor nerve impulses descend the corticospinal tract and relay along more axons to the arm and hand muscles

Spinothalamic tract
Information about pain synapses with the next neuron and crosses over within this level of the spinal cord before ascending to the brain

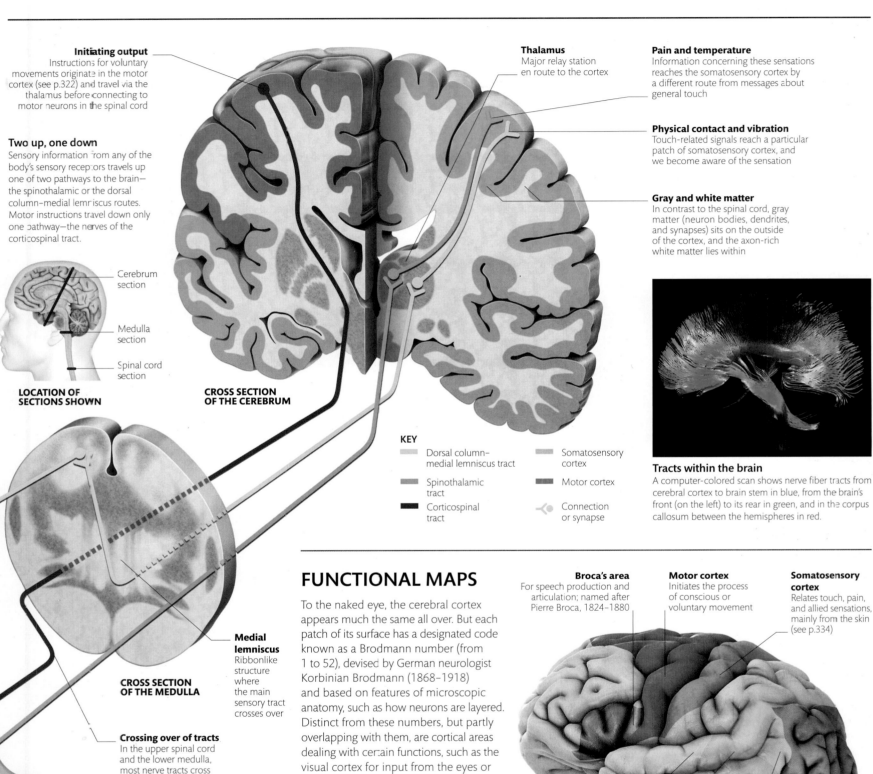

Initiating output
Instructions for voluntary movements originate in the motor cortex (see p.322) and travel via the thalamus before connecting to motor neurons in the spinal cord

Thalamus
Major relay station en route to the cortex

Pain and temperature
Information concerning these sensations reaches the somatosensory cortex by a different route from messages about general touch

Two up, one down
Sensory information from any of the body's sensory receptors travels up one of two pathways to the brain—the spinothalamic or the dorsal column-medial lemniscus routes. Motor instructions travel down only one pathway—the nerves of the corticospinal tract.

Physical contact and vibration
Touch-related signals reach a particular patch of somatosensory cortex, and we become aware of the sensation

Gray and white matter
In contrast to the spinal cord, gray matter (neuron bodies, dendrites, and synapses) sits on the outside of the cortex, and the axon-rich white matter lies within

Cerebrum section

Medulla section

Spinal cord section

LOCATION OF SECTIONS SHOWN

CROSS SECTION OF THE CEREBRUM

KEY

	Dorsal column-medial lemniscus tract		Somatosensory cortex
	Spinothalamic tract		Motor cortex
	Corticospinal tract		Connection or synapse

Tracts within the brain
A computer-colored scan shows nerve fiber tracts from cerebral cortex to brain stem in blue, from the brain's front (on the left) to its rear in green, and in the corpus callosum between the hemispheres in red.

Medial lemniscus
Ribbonlike structure where the main sensory tract crosses over

CROSS SECTION OF THE MEDULLA

Crossing over of tracts
In the upper spinal cord and the lower medulla, most nerve tracts cross over (decussate) to the other side of the body

Ventral root
Motor axons leave the cord here to take instructions to the muscles

FUNCTIONAL MAPS

To the naked eye, the cerebral cortex appears much the same all over. But each patch of its surface has a designated code known as a Brodmann number (from 1 to 52), devised by German neurologist Korbinian Brodmann (1868–1918) and based on features of microscopic anatomy, such as how neurons are layered. Distinct from these numbers, but partly overlapping with them, are cortical areas dealing with certain functions, such as the visual cortex for input from the eyes or Broca's and Wernicke's areas for language. "Live" brain scans using methods such as PET (positron emission tomography) and fMRI (functional magnetic resonance imaging) are revealing ever more details about how the cortex works.

Cortical brain map
Major mental functions are localized in certain areas of the cerebral cortex. These areas do not work alone, they communicate constantly with each other and with inner brain parts. Some are named for their function, while others reference the scientists who discovered their function.

Broca's area
For speech production and articulation; named after Pierre Broca, 1824–1880

Motor cortex
Initiates the process of conscious or voluntary movement

Somatosensory cortex
Relates touch, pain, and allied sensations, mainly from the skin (see p.334)

Auditory cortex
Processes sound information (see p.330)

Wernicke's area
For understanding spoken words; named after Carl Wernicke, 1848–1905

Geschwind's territory
Connects Wernicke's and Broca's areas; named after Norman Geschwind, 1926–1984

Visual cortex
Analyzes what we see (see p.329)

MEMORY AND EMOTION

Memory is not just the storage and recall of facts. It encompasses all kinds of information, events, experiences, and contexts—from names to faces and places—and references our emotional state at the time.

Brain areas involved in memory
There is no single "memory center." Information is processed, selected for memorizing, and stored in various brain parts. For the memory of a roller-coaster ride, for example, what we saw resides in the visual areas, sounds in the auditory areas, and so on. These are pulled together to recall the whole experience.

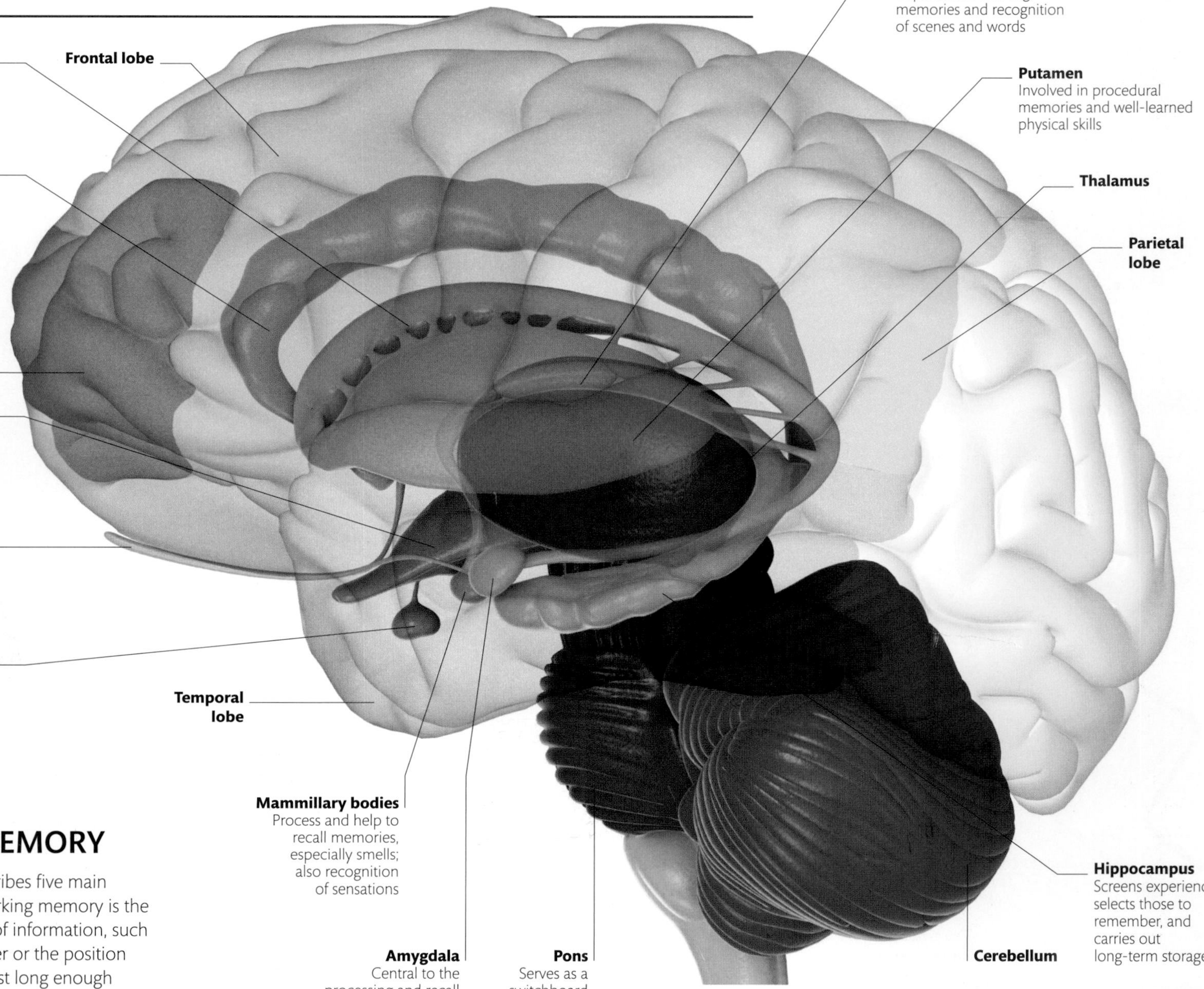

Caudate nucleus
Involved in learning and especially feedback to modify procedural memories for actions

Frontal lobe

Fornix
Important in forming memories and recognition of scenes and words

Putamen
Involved in procedural memories and well-learned physical skills

Thalamus

Parietal lobe

Cingulate gyrus
Deals with learning and memory processing; suppresses overly powerful reactions and behaviors

Central executive
Coordinating area that calls up information from other parts and formulates action plans

Hypothalamus
Links brain to hormonal system; center for major drives, instincts, emotional reactions, and feelings

Olfactory bulb
Preprocesses smells (which are closely tied to emotions) ahead of olfactory areas

Pituitary gland
Chief hormonal gland; responds to instructions from the hypothalamus, just above

Temporal lobe

Mammillary bodies
Process and help to recall memories, especially smells; also recognition of sensations

Amygdala
Central to the processing and recall of the emotional components of memories

Pons
Serves as a switchboard connecting the cortex and the cerebellum

Cerebellum

Hippocampus
Screens experiences, selects those to remember, and carries out long-term storage

TYPES OF MEMORY

Current thinking describes five main kinds of memory. Working memory is the short-term retention of information, such as a telephone number or the position of doors in a room, just long enough to be useful, before rapidly fading away. Semantic memory is for detached facts, independent of our personal existence, such as the date of a famous historical event. Episodic memory recalls episodes and events from our personal perspective, including our sensations and emotions, such as a happy birthday party. Procedural memory is for learned, well-practiced physical skills, such as walking, bicycling, and tying shoelaces. Implicit memory affects us without our awareness, for example being more likely to believe something is true if we've heard it before.

Memory-processing areas
For the four best-understood types of memory, several brain areas work in a coordinated fashion. The thalamus is a general gatekeeper and the frontal lobe, in particular, has an overall executive capacity in both learning and recalling most kinds of memories.

MEMORY TYPE	THALAMUS	PARIETAL LOBE	CAUDATE NUCLEUS	MAMMILLARY BODY	FRONTAL LOBE	PUTAMEN	AMYGDALA	TEMPORAL LOBE	HIPPOCAMPUS	CEREBELLUM	CINGULATE GYRUS	OLFACTORY BULB	FORNIX	CENTRAL EXECUTIVE
WORKING	■	■	■		■	■		■			■		■	■
SEMANTIC	■				■		■	■	■				■	
EPISODIC	■			■	■		■	■	■		■	■	■	■
PROCEDURAL	■		■		■	■			■	■				

HOW EMOTIONS AFFECT MEMORY

The "emotional brain" is a term often applied to the limbic system, a group of parts nestling on top of the brainstem, under and within the overarching dome of the cerebrum. They include the amygdala, thalamus, hypothalamus, fornix, and mammillary bodies (see opposite), plus inward-facing (medial) areas of the cerebral cortex and the cingulate gyrus that form a collar-shape around them.

The limbic system takes the lead in deep-seated feelings and instinctive reactions that seem to well up inside us during times of great emotion, and which the rational-thinking parts of the brain may have trouble controlling. In particular, the fingertip-sized hypothalamus—almost at the anatomical center of the brain —plays vital roles in powerful basic drives for survival such as hunger, thirst, and sex, and the strong emotions that may accompany them, for instance rage or ecstatic joy. The hypothalamus sends out nerve signals to various brain parts that then convey their own nerve signals to various muscles, often through the autonomic nervous system (see p.311). For example, in response to a sudden scare, the hypothalamus takes control and tells the heart to beat faster, the skeletal muscles to tense, and the adrenal glands to release epinephrine, ready for sudden action—the "fight or flight" response. The hypothalamus also links via a thin stalk to the pituitary gland (see p.400) below it. This gland secretes various hormones and other substances that affect other hormonal glands, to complement and reinforce the nervous system's actions.

Several limbic parts are also intimately involved in memory formation, especially episodic memory (see opposite). This fact explains why being in a state of high emotion helps form strong memories at the time, and why we feel emotional again when we recall such memories.

Lasting memories
Events such as our first day at school, first time riding a bicycle, and getting married involve strong emotional components, such as anxiety mixed with achievement, so the memories persist and stay "real."

Average working memory holds five words, six separate letters, or seven single numbers. **Training memory**, such as reordering to assign a meaning, can usually **double** this.

FORMING MEMORIES

Each memory is formed by a unique pattern of connections between the billions of neurons in various parts of the brain, especially the cerebral cortex. The event to be memorized—from reading a number to meeting a celebrity—occurs as a particular set of neurons sending impulses to each other during the initial experience. Activating this set of signals again, by remembering the experience, strengthens its pattern of links so they are more likely to occur together—a process known as potentiation. After several activations the links become semipermanent. Triggering a few of them, by a new thought or experience, activates the pattern's whole network and recalls the memory.

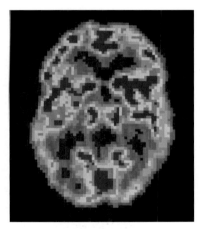

Sleep and memories
Electrical traces and scans show the brain is very active during sleep. With no distraction from conscious thoughts the memory circuits may sift through recent events, move some to longer-term storage, and consolidate established memories while we sleep.

KEY

Brain activity levels, based on the uptake of glucose

HIGHEST   **LOWEST**

EXTREME HUMAN
PEOPLE WHO CANNOT FORGET

Total recall, or hyperthymestic syndrome, is a rare condition in which people can remember vast amounts of information, from incredibly significant to numbingly trivial, for many decades. Even if they try to forget, they cannot. But the memories tend not to be "total" in that, when questioned about a past event, they may recall the date, place, and what people said, but not what they were wearing. Similarly, most of their memories are centered on their personal life and experiences, and less on what was happening in the wider world. Hyperthymestic people show tendencies to obsessive-compulsive traits, such as collecting memorabilia and keeping diaries.

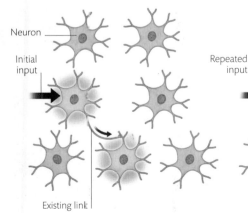

Neuron

Initial input

Existing link

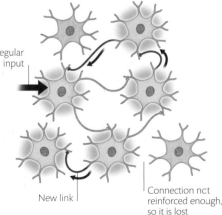

Repeated input

New link

Regular input

New link

Connection not reinforced enough, so it is lost

Hyperthymestic syndrome
One of the first people with hyperthymestic syndrome studied by scientists in the US, Jill Price can recall every day since she was 14.

1 Initial experience
A stimulus causes one neuron to "fire" and send a particular string of nerve signals to the next one. This is part of the process of thinking and being aware of a fact, experience, or learned skill.

2 Further modification
Repeating the stimulus strengthens the initial link, or synaptic communication, and also recruits other neurons into the network. In reality, this occurs with thousands of neurons.

3 Consolidation, or not
Regular use of connections both maintains them structurally and increases the strength of synaptic signaling between the neurons. Links that are not refreshed regularly tend to fade and are lost.

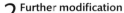

HOW WE MOVE

Every split second, the brain coordinates the precise tensing and accurate contraction of more than 600 muscles all around the body, from full-speed running to the blink of quick an eye. Such a huge task would be impossible with every muscle under conscious control, so the brain has a hierarchy of delegation.

VOLUNTARY MOVEMENT

Moving—part of everyday life
The motor cortex works intimately with other areas of the brain involved in movement, such as the cerebellum (see opposite), so that we can move around almost without thinking.

A voluntary action is one we plan with awareness and carry out with purpose. We may hardly be aware of turning a book's pages, or we might concentrate on its every detail, but both are intentional. Central to these voluntary movements is the motor cortex—a strip of gray matter arching "ear to ear" on the brain's outer surface (see also p.319). It sends and receives millions of nerve impulses every second—even when we do not move, because muscles are still needed to hold the stationary body in position or it would simply flop in a heap.

Different patches of motor cortex deal with instructions to certain parts of the body—it's a similar "map" of size-related specialization to that in the somatosensory cortex (see p.335). Parts that need intricate muscle control, such as the lips and fingers, have a correspondingly larger patch of motor cortex dedicated to them, compared with those needing less refined control, such as the thigh.

Making a move
These views show with arrows which parts of our brains are "talking to each other" during the execution of a simple sequence—Ready, Get Set, Go!

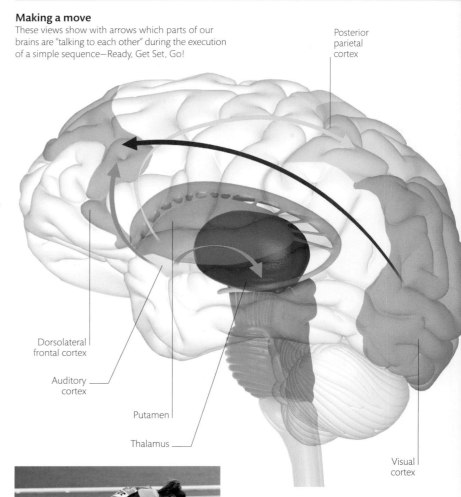

Posterior parietal cortex

Dorsolateral frontal cortex

Auditory cortex

Putamen

Thalamus

Visual cortex

READY ...
The visual and auditory brain centers relay sensory information to the dorsolateral frontal cortex, which continually assesses the start time. The putamen feeds its memories and preparations for well-rehearsed movement patterns to the posterior parietal cortex, whose activity is largely subconscious.

INVOLUNTARY MOVEMENTS—REFLEXES

Most involuntary actions begin not at the conscious level, but unintentionally. They happen automatically, although even as they start, we become aware of them and can start to modify them. Many involuntary actions are reflexes—set patterns of movements in response to a specific situation or stimulus. Reflexes such as lifting the foot up after having stepped on a sharp object have survival value. They protect the body by carrying out a fast reaction to danger, even if we are not paying attention. Reflexes receive sensory nerve messages about a stimulus, "short-circuit" these through the spinal cord or the subconscious parts of the brain, and then send out motor signals to initiate muscle action, without "permission" of the conscious mind. As these nerve circuits quick-fire their impulses, they also send signals up to the brain's higher centres where, a fraction of a second later, they register in our awareness. We can then take over voluntary control.

Duck and dive
Protective reflexes, such as ducking to avoid a fast-approaching object, are rooted deep in our evolutionary past. Ducking is a cascade of four reflexes (see right) that are "learned" as one; the order reflects the journey the motor signals take from the lower brain down the spinal cord to the body.

Sense danger
Long-term training and real-time vision warn that a blow to the head is on the way.

Subconscious processing
Sensory information alerts lower levels of conscious, especially the thalamus.

Motor output begins
Motor areas organize all aspects of the action a split second before awareness clicks in.

Eyes blink
Reflex 1: eyelids blink and screw up to shield the eyes.

Face turns
Reflex 2: neck muscles twist the head to the side.

Head jerks back
Reflex 3: upper body muscles draw the head back.

Hands throw up
Reflex 4: arm muscles raise hands for extra protection.

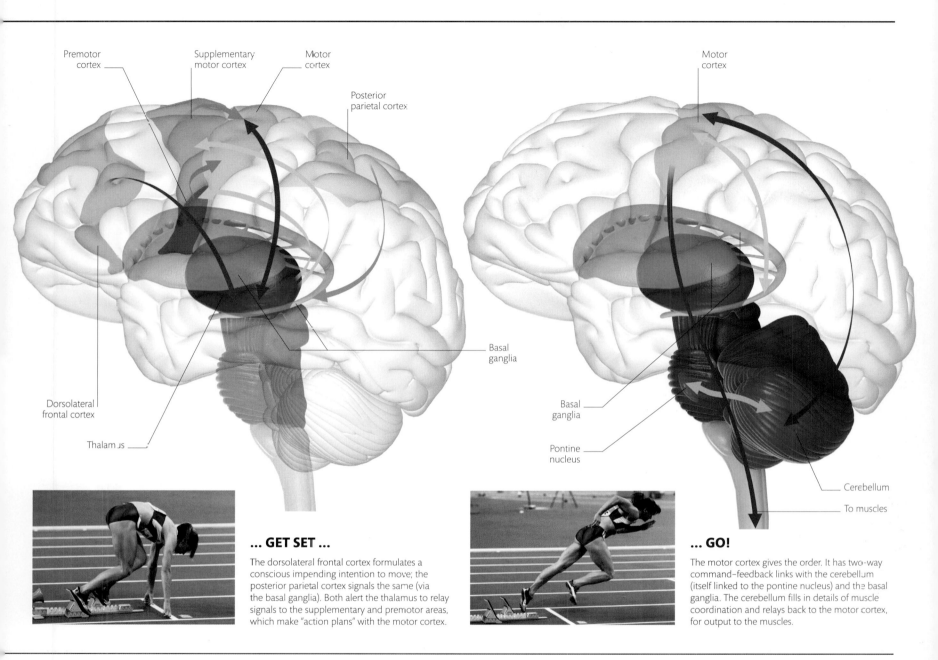

Premotor cortex

Supplementary motor cortex

Motor cortex

Posterior parietal cortex

Basal ganglia

Dorsolateral frontal cortex

Thalamus

Motor cortex

Basal ganglia

Pontine nucleus

Cerebellum

To muscles

... GET SET ...

The dorsolateral frontal cortex formulates a conscious impending intention to move; the posterior parietal cortex signals the same (via the basal ganglia). Both alert the thalamus to relay signals to the supplementary and premotor areas, which make "action plans" with the motor cortex.

... GO!

The motor cortex gives the order. It has two-way command–feedback links with the cerebellum (itself linked to the pontine nucleus) and the basal ganglia. The cerebellum fills in details of muscle coordination and relays back to the motor cortex, for output to the muscles.

THE "LITTLE BRAIN"

In some ways, the rounded, grooved cerebellum ("little brain") at the brain's lower rear mirrors the dominating domed cerebrum above. Like the cerebrum, it has gray matter formed of neuronal cell bodies, dendrites, and synapses in its outer layer, or cortex, with an inner medulla of mainly nerve axons (fibers), arranged in tracts or bundles linking it to many other brain parts. The cerebellar cortex is even more highly folded than the cerebral cortex.

Its anatomical location allows the cerebellum to "see" all the sensory information on its way to the brain as well as all the motor instructions on their way from the brain

to the spinal cord and then the body. The cerebellum also has intimate relationships with other movement-controlling brain zones, such as the basal ganglia. Its chief role is to fill in fine details of the broad instructions for movements coming from the motor cortex, send these back to the motor cortex for detailed output to muscles, and monitor feedback to ensure that all movements are smooth, skilled, and coordinated.

Recent research shows that the cerebellum is also active in focusing attention onto a situation, and in speaking and understanding language.

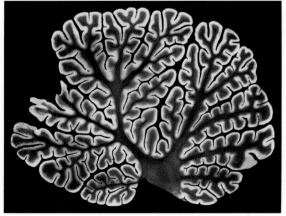

Cerebellum in cross section
The cerebellar cortex (palest yellow) is intricately folded around a multiple treelike branching system of nerve fiber tracts (red). At the thickest "trunks" of the trees are clusters of neurons, or gray matter, known as cerebellar nuclei, which are coordinating centers for the massive inputs and outputs of motor nerve messages.

The **cerebellum** is only **10 percent** of the brain's volume, yet it contains **more than twice the number of neurons** than the other **90 percent** put together.

HOW WE SENSE THE WORLD

The brain itself is surprisingly insensitive. With hardly any sensory nerve receptors of its own, it is incapable of feeling that it is being touched or injured. However, it is highly attuned to what happens in the rest of the body—and in the world outside—through the work of sense organs as they respond to many kinds of stimuli.

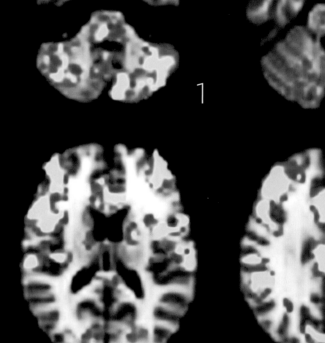

1

2

5

6

OUR MAIN SENSES

The idea of five senses is oversimplified. Four of them and their stimuli are well defined: vision using light rays (see p.326), hearing and sound waves (see p.330), smell involving airborne odor molecules (see p.332), and taste from waterborne flavor molecules (see p.332).

Other modes of sensation are more complex. Balance (see p.330) is less of a discrete sense and more of an ongoing process involving several senses simultaneously as well as the muscular system. Touch is based in the skin, but not exclusively, and is a multifactored sense that responds not just to physical contact but also to vibration and to temperature (see p.334). The sensation

of pain is handled differently by the nervous system compared with other sensations (see opposite).

The body also has internal sensory receptors in muscles, joints, and other parts (see Inner Sense opposite). But at the simplest level, all sensory parts do the same thing. Scientifically, they are transducers, changing energy from their specific stimuli into the nervous system's common "language" of nerve impulses.

A sensational world
We can imagine the main sensory inputs in these situations (clockwise from top left: ears, balance, tongue, nose, skin, and eyes), yet the only actual stimulus here is light for vision.

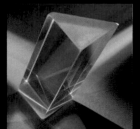

SYNESTHESIA

In normal sensory nerve pathways, messages travel from a sense organ to specific regions of the brain, especially to the cerebral cortex, where they enter conscious perception. Signals from the eyes, for instance, end up in the visual cortex, and so on. Rarely, these pathways diverge and connect to other sensory brain regions. In such cases a person may experience more than one kind of sensation from a single type of stimulus. For example, seeing the color blue may

also bring on a taste of cheese, while sardines are tasted while listening to certain instruments play. This condition is known as synesthesia and affects about 1 person in 25, although to varying degrees. Synesthesia can also be brought on by certain chemicals, especially perception-altering or psychedelic drugs.

Painting by music
British artist and synesthete David Hockney said, of designing the sets for the LA Opera, that the colors and shapes "just painted themselves" when he listened to the music.

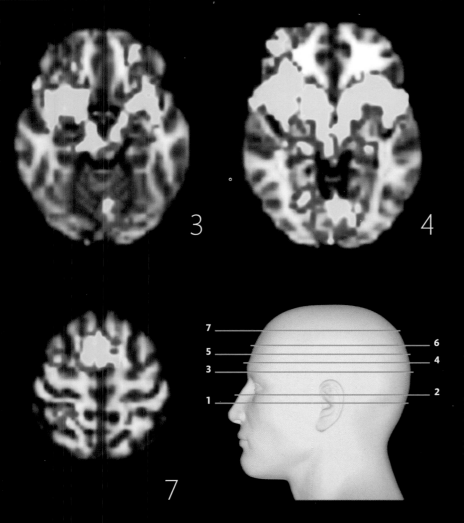

3   4

7   5   7
3
1

7   6
5   4
3
1   2

HOW WE FEEL PAIN

Pain is a sensation that is very difficult to measure objectively. We have a set of terms to describe it, such as aching, stabbing, burning, and crushing. Pain begins in specialized nerve endings—nociceptors—in the skin and in many other body parts. When nociceptors or tissues are damaged they release substances such as prostaglandins, adenosine triphosphate (ATP), and bradykinin. These stimulate the nociceptors to transmit pain signals. The signals follow a different pathway from touch or other sensations from that body part (see p.318), especially in the spinal cord. Most end up in the cortex of the cerebral hemispheres, where we perceive them as pain related to a particular body part.

Whole-brain pain
Left: These fMRI scans show sequential horizontal "slices" up through the brain of a healthy person being subjected to a painful stimulus. The yellow areas show brain activity, reflecting how widely pain is dealt with by different parts of the brain.

Pain pathways
Right: In all sensations, nerve signals take time to travel from their receptors to the brain and enter our conscious awareness. In the time gap of a second or so, damage could already be advanced.

Initiation of pain
Injury causes the release of chemicals such as prostaglandins and bradykinin, which prompt nociceptors to initiate pain signals.

Spinal cord
Nerve signals travel in pain-related axons (fibers) into the dorsal horn of the spinal cord for onward transmission.

Brain stem
The signals pass via the medulla and activate the sympathetic division of the autonomic system (see p.311).

Midbrain
Pain-registering regions monitor the signals and trigger the release of the body's own analgesics in the brain stem and spinal cord.

Cerebral cortex
Signals reach several areas of the cerebral cortex. The pain is felt consciously and regionalized to a body part.

INNER SENSE

Without looking or touching, we know where our arms and legs are, if we are upright or lying down, what our posture is like, and how we are moving through space. This body sense is known as proprioception; it makes us aware of our position and movements.

Proprioception relies on internal sensory parts, mostly microscopic, known as proprioceptors. There are many thousands spread throughout the body, being especially numerous in muscles and tendons, and in the ligaments and capsules of joints. They respond to changes in tension, length, and pressure in their particular area, such as when a relaxed muscle is stretched. Such information is integrated with signals concerning changes of orientation and position in space, for example, via hair cells in the vestibule and the semicircular canals in the inner ear (see p.330).

As the proprioceptors are stimulated, they send streams of nerve signals through the peripheral nervous system to the brain. For example, messages coming from proprioceptors in the biceps muscle of the upper arm inform the brain that they are being compressed and shortened, meaning that the elbow is bending

BLOCKING PAIN AND SENSATIONS

Despite its unwanted nature, pain has survival value as it warns us that a part of our body is in trouble, that any potential cause of the pain should be spotted and removed, and that the part should be protected and rested so it can heal. The body has its own pain-reducing or analgesic substances, principally the endorphins group, which are released by the brain's hypothalamus and pituitary gland and spread in the blood and nervous system. They affect transmission of nerve signals carrying pain information by interfering, for example, at the level of synapses (see p.314) by preventing the production of certain neurotransmitter chemicals or blocking receptor sites, so that impulses do not continue in the receiving neuron.

Levels of relief
Pain messages travel to the higher brain centers along a series of neurons and their synapses. So, there are several opportunities to block these pathways and lessen the perceived pain.

PAINKILLERS	HOW THEY WORK
OPIOIDS (for example, morphine)	Like endorphins, these work mostly within the central nervous system and inhibit the brain's conscious ability to perceive pain.
ACETAMINOPHEN	This analgesic is similar to a weak opioid. It inhibits prostaglandin formation and also affects formation of the neurotransmitter AEA (anandamide), mainly within the central nervous system.
NSAIDS (nonsteroidal anti-inflammatory drugs)	Ibuprofen and other NSAIDs suppress the formation of certain prostaglandins that would otherwise produce pain sensations. They work mainly in the peripheral nervous system.

ANESTHETICS	HOW THEY WORK
GENERAL ANESTHETICS	Act primarily on the brain but also affect the spinal cord, causing muscle relaxation and producing loss of consciousness; precise mechanisms are unclear.
LOCAL ANESTHETICS	Impede peripheral nerve impulses in a specific part, for example, by blocking sodium channels in neuron membranes (see p.313) to reduce all sensory information.
EPIDURAL ANESTHETICS	Injected into the cerebrospinal fluid around the dura mater (the outermost of the meninges surrounding the spinal cord) to quash all sensations felt from below the site of injection.

HOW WE SEE

For most people, vision is the most important sense. Using information in the form of light rays, gathered by our eyes, the brain creates clear images of the world allowing us to experience our surroundings.

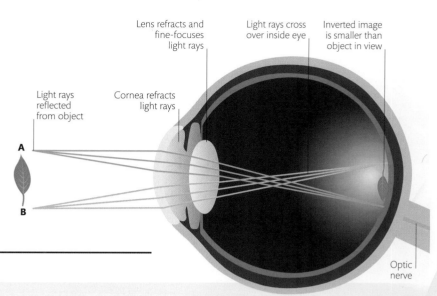

Light rays reflected from object

Cornea refracts light rays

Lens refracts and fine-focuses light rays

Light rays cross over inside eye

Inverted image is smaller than object in view

Optic nerve

THE VISUAL SYSTEM

Cushioned within sockets in the skull, their surfaces washed by tears and wiped by blinking of the eyelids, the eyes relentlessly scan the surroundings to collect light rays reflected or generated by objects in view. Those rays enter the eye through a clear, bulging window, the cornea. Aided by the adjustable lens behind it, the cornea focuses light rays onto the retina, the thin layer of light-sensitive receptors that lines the inside of the rear part of the eyeball. As in a modern camera, the process of focusing is automatic, as is the adjustment of the size of the iris, which controls the amount of light entering the eye. When light hits the retina's photoreceptors, they generate billions of nerve impulses that stream along the optic nerve to the visual areas at the back of the brain. Here signals are analyzed to give a mental impression of what we are looking at, where it is, and whether or not it is moving.

Image production
Refracted by the cornea and lens, light rays cross over and create on the retina a sharply focused, upside-down, and back-to-front image of the object in view.

BENDING LIGHT

Light rays usually travel between objects in a straight line. When they pass through both the cornea and the transparent lens they are bent, or refracted. As a result of refraction, a clear, inverted view of the outside world is projected onto the retina. The cornea does most of the light bending, but its shape and, therefore, refractive powers, cannot be altered. It is the elastic lens that changes shape to fine-focus light (see opposite).

Cornea
Domed transparent membrane that covers front of eye and refracts light

LIGHT REFRACTION

When light rays pass from one transparent medium to another they bend, or refract. This is the case when light enters and leaves the eye's lens, which is convex—curving outward on both surfaces. The greater the angle at which light hits the surface of the convex lens, the more it is refracted inward.

Convex lens
Light rays refracted by a convex lens are focused on a single focal point. The thicker the lens, the more the light rays are refracted.

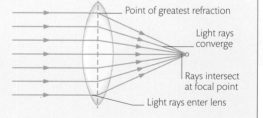

Point of greatest refraction

Light rays converge

Rays intersect at focal point

Light rays enter lens

LIGHT CONTROL

The eyes can operate in most light conditions because of a control system that automatically and unconsciously regulates the amount of light entering through the hole at the center of the iris, the pupil. The iris, the colored part of the eye, has two layers of muscle fibers: concentric circular fibers, and radial fibers arranged like the spokes of a wheel. These muscles contract on signals from the autonomic nervous system (see p.311). The system's opposing parasympathetic and sympathetic branches ensure that the pupil shrinks in bright light to avoid dazzling, and expands in dim light to allow enough light into the eye to make vision possible.

Inner iris
This colored electron micrograph shows the inner surface of the iris (pink). To the right (dark blue) is the edge of the pupil, and the folded structures in the center (red) are the ciliary processes.

BRIGHT LIGHT

Pupil is constricted

Circular muscle fibers contract

NORMAL LIGHT

DIM LIGHT

Pupil is dilated

Radial muscle fibers contract

Narrow pupil
Stimulated by parasympathetic nerves, circular muscle fibers in the iris contract to make the pupil narrow—less light enters the eye.

Normal pupil
In normal light conditions both circular and radial muscle fibers partially contract. The pupil is neither too wide nor too narrow.

Wide pupil
Stimulated by sympathetic nerves, radial muscle fibers in the iris contract to make the pupil wider—more light enters the eye.

Under normal conditions, the pupils of both eyes **respond identically** to a light stimulus, **regardless of which eye** is being stimulated.

VISUAL PATHWAY

Although the eyes are in front of the brain, the cerebral areas that process their information are located at the rear. Nerve impulses from the eyes pass along the million or so axons (nerve fibers) of each optic nerve. These two nerves converge in the underside of the brain at the optic chiasma, where about half of the fibers from each cross to the other side. Next, each set of fibers passes to a dedicated area known as the lateral geniculate nucleus in the thalamus (see p.316). This screens the information for relevance to what is going on in the conscious mind and for links to other senses. Axons from each nucleus then fan out through the brain tissue, as the optic radiation, to the primary visual cortex at the lower rear of the brain. Here the information is initially processed, sorted, and then partitioned to other areas of the brain. These include zones of secondary visual cortex around the primary cortex, which discriminate features such as lines, angles, colors, shapes, and movements, and the temporal lobe on the side of the brain for recognition of familiar objects.

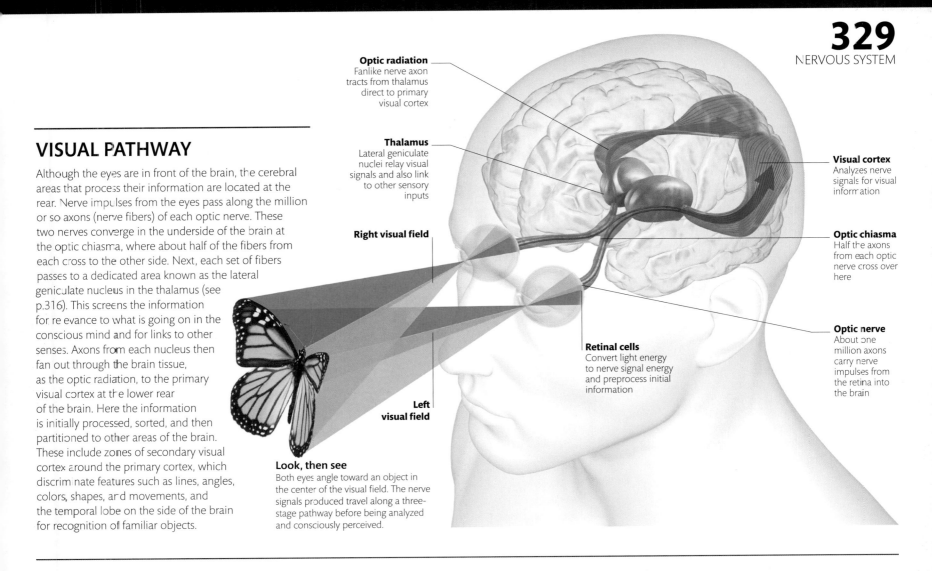

Optic radiation
Fanlike nerve axon tracts from thalamus direct to primary visual cortex

Thalamus
Lateral geniculate nuclei relay visual signals and also link to other sensory inputs

Right visual field

Left visual field

Visual cortex
Analyzes nerve signals for visual information

Optic chiasma
Half the axons from each optic nerve cross over here

Optic nerve
About one million axons carry nerve impulses from the retina into the brain

Retinal cells
Convert light energy to nerve signal energy and preprocess initial information

Look, then see
Both eyes angle toward an object in the center of the visual field. The nerve signals produced travel along a three-stage pathway before being analyzed and consciously perceived.

DEPTH AND DIMENSION

We experience the visual field in three dimensions, with depth, and can determine whether one object in a scene is closer than another. The brain achieves this by combining information from many varied sources.

Memory is important. We recall that mice are small and elephants are big. Linking this to relative size in the visual field, we expect a mouse we see as large to be closer than an elephant that appears smaller. Movements in and around the eye when viewing objects also supply information on their distance. The more the two eyes angle inward as detected by sensors in the eyeball-moving muscles, and the more the lens bulges, due to ciliary muscle contraction, the closer the object.

The fact that we have two eyes and the visual pathways swap information left to right also plays a part. Each eye has its own visual field, which overlap in the middle to form the binocular visual field. Nerve fibers cross at the optic chiasma, so the left part of the visual field of each eye ends up in the left visual cortex, and the right half in the right visual cortex. The brain then compares the differing views from each eye, known as spatial binocular disparity.

17,000

The average number of times the human **eye blinks each day**— that is **once every five seconds**.

Seeing 3-D
An object in the binocular visual field is seen by each eye at a slightly different angle. This means that the view of the image received by each side of the visual cortex from each eye, is different. By combining and comparing the views the brain can judge depth.

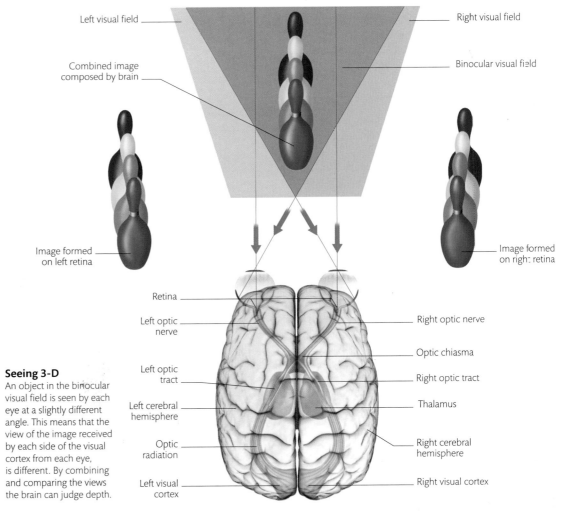

Left visual field

Right visual field

Combined image composed by brain

Binocular visual field

Image formed on left retina

Image formed on right retina

Retina

Left optic nerve

Right optic nerve

Left optic tract

Optic chiasma

Left cerebral hemisphere

Right optic tract

Thalamus

Optic radiation

Right cerebral hemisphere

Left visual cortex

Right visual cortex

HEARING AND BALANCE

Our ears greatly complement our eyes in providing vast amounts of information about the world around us—indeed, we can often hear what we cannot see. Balance is anatomically adjacent to hearing, and employs similar physiological principles, but has no direct connection.

The cochlea
Three fluid-filled ducts spiral within the cochlea and carry sound vibrations. The outer scala vestibuli and scala tympani connect at the apex, or point,ß of the spiral. Between them is the cochlear duct, divided from the scala tympani by the basilar membrane bearing the organ of Corti.

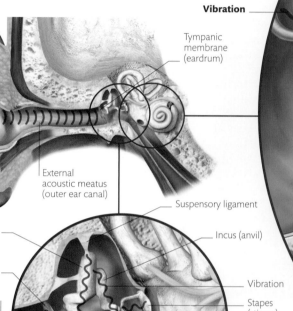

HOW WE HEAR

Sounds consist of areas of alternating high and low pressure, called sound waves, propagating through air. The auditory sense allows us to perceive sounds in the mind through a series of conversions. The first occurs when sound waves hit a skinlike sheet, the tympanic membrane (eardrum). These pressure waves then pass from the eardrum through the middle ear, causing vibrations along a chain of the three smallest bones in the body, called the ossicles. The last ossicle butts against another flexible membrane, the oval window, set into a fluid-filled chamber in the inner ear. The vibrations change into waves of fluid pressure rippling through the snail-shaped cochlea. Within the cochlea lies the organ of Corti, containing a fine membrane in which hair cells are embedded. The vibrations distort these hairs, causing them to produce nerve signals. These signals pass along the cochlear nerve, which becomes part of the auditory nerve, to the brain's auditory cortex—just under the skull, almost alongside the ear itself. Here the nerve impulses are analyzed to gauge the frequency (pitch) and intensity (loudness) of the original air pressure waves—and we hear.

Sound waves arrive
Air pressure waves are funneled by the outer ear flap, or pinna, into the slightly S-shaped external acoustic meatus (canal). They bounce off the tympanum, which is about the size of the little fingernail, causing it to vibrate.

Middle-ear vibrations
The tympanum is connected to the first ossicle, the malleus. Vibrations proceed from here through the air-filled middle ear cavity, along the incus, and then to the stapes. The base of the stapes presses against the membrane of the oval window, and as it vibrates, it pushes and pulls against the window.

Vibration

Tympanic membrane (eardrum)

Sound waves

External acoustic meatus (outer ear canal)

Suspensory ligament

Malleus (hammer)

Tympanic membrane (eardrum)

Sound wave

Incus (anvil)

Vibration

Stapes (stirrup)

Oval window

BALANCE

Balance is an ongoing process, coordinating many sensory inputs. It does this largely at subconscious levels, with outputs to muscles all over the body, enabling us to retain our poise and adjust our posture. For example, vision monitors the head's angle to horizontals such as the ground, the skin registers pressure as we lean, and muscles and joints detect levels of strain (see proprioception, p.325). Balance information comes from the fluid-filled organs in the inner ear, via the vestibular nerve.

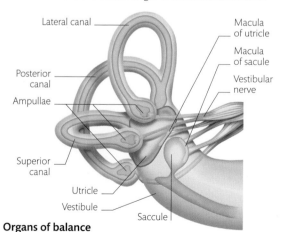

Lateral canal

Macula of utricle

Macula of sacule

Vestibular nerve

Posterior canal

Ampullae

Superior canal

Utricle

Vestibule

Saccule

Organs of balance
Three semicircular canals, each at right angles to the others, detect head movements. Two neighboring chambers, the utricle and saccule, are more specialized for the head's static position.

Responding to movement
The utricle and saccule have a patch of hair cells, the macula, the hair tips of which are set into a membrane bearing mineral crystals. The pull of gravity on the membrane depends on the position of the head. At one end of each semicircular canal is a wide area, the ampulla, with hair cells set into the cupula.

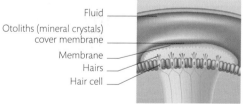

Fluid

Otoliths (mineral crystals) cover membrane

Membrane

Hairs

Hair cell

Hairs deflected

Macula rotated

Gravity pulls membrane

Ampulla

Cupula

Hairs

Fluid swirls

Cupula bends

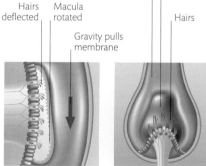

Utricule and saccule
With the head level, gravity pulls evenly on the membrane. As the head nods, gravity tugs it and distorts the hairs, whose cells produce nerve signals.

Semicircular canals
A head movement makes the fluid in at least one canal swirl around. This disturbs the cupula and bends the hair cells, generating nerve impulses.

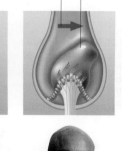

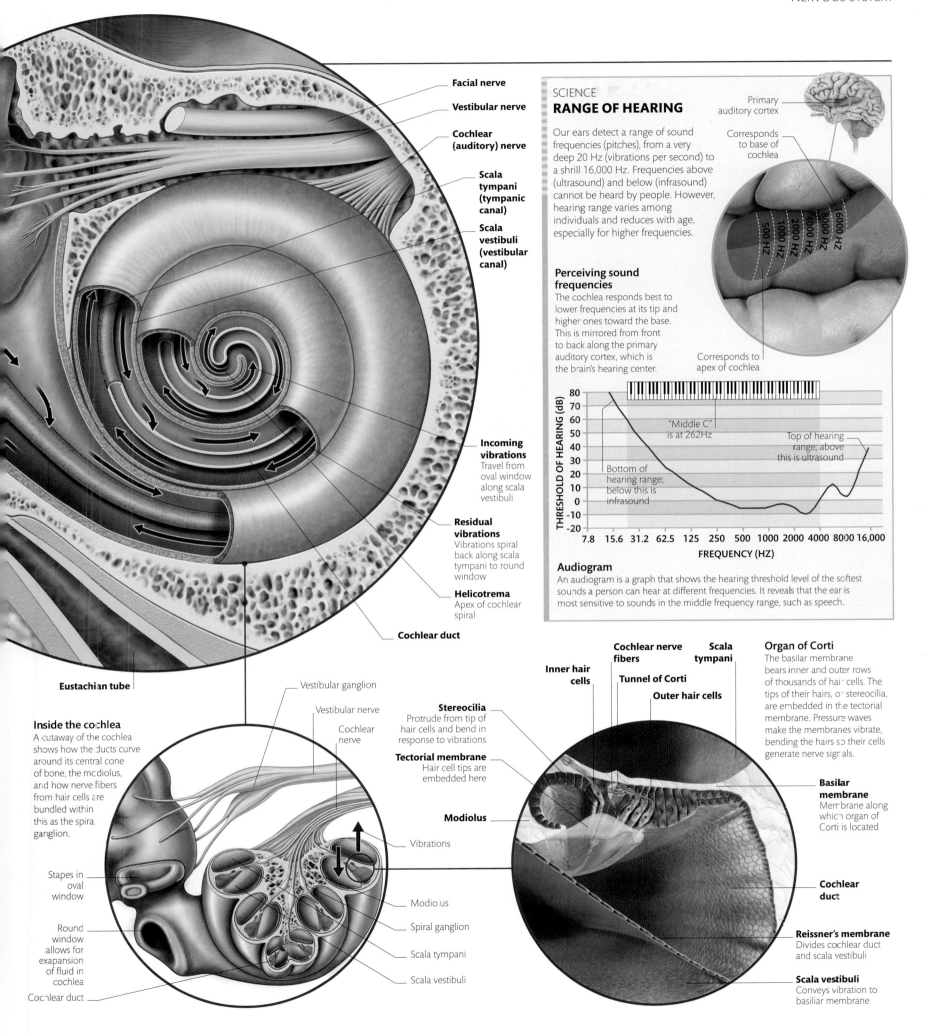

Facial nerve

Vestibular nerve

Cochlear (auditory) nerve

Scala tympani (tympanic canal)

Scala vestibuli (vestibular canal)

Incoming vibrations
Travel from oval window along scala vestibuli

Residual vibrations
Vibrations spiral back along scala tympani to round window

Helicotrema
Apex of cochlear spiral

Cochlear duct

Eustachian tube

SCIENCE
RANGE OF HEARING

Our ears detect a range of sound frequencies (pitches), from a very deep 20 Hz (vibrations per second) to a shrill 16,000 Hz. Frequencies above (ultrasound) and below (infrasound) cannot be heard by people. However, hearing range varies among individuals and reduces with age, especially for higher frequencies.

Primary auditory cortex

Corresponds to base of cochlea

500 HZ
1000 HZ
2000 HZ
4000 HZ
8000 HZ
16000 HZ

Perceiving sound frequencies
The cochlea responds best to lower frequencies at its tip and higher ones toward the base. This is mirrored from front to back along the primary auditory cortex, which is the brain's hearing center.

Corresponds to apex of cochlea

"Middle C" is at 262Hz

Top of hearing range; above this is ultrasound

Bottom of hearing range; below this is infrasound

THRESHOLD OF HEARING (dB)

80
70
60
50
40
30
20
10
0
-10
-20

7.8 15.6 31.2 62.5 125 250 500 1000 2000 4000 8000 16,000

FREQUENCY (HZ)

Audiogram
An audiogram is a graph that shows the hearing threshold level of the softest sounds a person can hear at different frequencies. It reveals that the ear is most sensitive to sounds in the middle frequency range, such as speech.

Inside the cochlea
A cutaway of the cochlea shows how the ducts curve around its central cone of bone, the modiolus, and how nerve fibers from hair cells are bundled within this as the spiral ganglion.

Vestibular ganglion

Vestibular nerve

Cochlear nerve

Stereocilia
Protrude from tip of hair cells and bend in response to vibrations

Tectorial membrane
Hair cell tips are embedded here

Modiolus

Stapes in oval window

Round window allows for expansion of fluid in cochlea

Cochlear duct

Vibrations

Modiolus

Spiral ganglion

Scala tympani

Scala vestibuli

Inner hair cells

Cochlear nerve fibers

Tunnel of Corti

Outer hair cells

Scala tympani

Organ of Corti
The basilar membrane bears inner and outer rows of thousands of hair cells. The tips of their hairs, or stereocilia, are embedded in the tectorial membrane. Pressure waves make the membranes vibrate, bending the hairs so their cells generate nerve signals.

Basilar membrane
Membrane along which organ of Corti is located

Cochlear duct

Reissner's membrane
Divides cochlear duct and scala vestibuli

Scala vestibuli
Conveys vibration to basiliar membrane

TASTE AND SMELL

The senses of taste and smell both detect chemical substances, are adjacent, work in similar ways, are fine-tuned for survival value, and seem inextricably linked as we enjoy a meal. Yet until their sensations reach the brain, there is no direct connection between them.

HOW WE SMELL

Smell particles, or odorant molecules, are detected by the olfactory epithelia—two patches, each thumbprint-sized, one in the roof of each nasal cavity, left and right. These epithelia contain several million specialized olfactory receptor cells, whose lower ends project into the mucus lining the nasal cavity and bear hairlike processes, called cilia, on which are located receptor sites. When suitable odorants dissolve in the mucus and stimulate receptor sites, the cells fire nerve impulses. This may happen when an odorant fits onto a site like a key in a lock. But there is also a "fuzzy coding" component that is less understood, where each odor produces a variable pattern or signature of impulses. Smell information is analyzed by the brain's olfactory cortex, which has close links with limbic areas, including emotional responses. This is why smells can provoke powerful recollections and feelings (see p.321).

Epithelial cells
Separated by smooth supporting cell ends, tufts of cilia, each from an olfactory receptor cell, dangle from the surface of the olfactory epithelium.

Cilia

Dura mater
Glomerulus
Mucus-secreting gland
Olfactory bulb

Ethmoid bone
Nerve fiber (axon)
Basal cell
Receptor cell
Supporting cell
Cilia

Air flow
Mucus
Odor molecule

Olfactory epithelium
Receptor cells send signals along their axons, through holes in the skull's ethmoid bone, to the olfactory bulb. This outgrowth of the brain processes signals at ball-like groups of nerve endings (glomeruli) and sends them along the olfactory tract.

HOW WE TASTE

Like smell, taste or gustation is a chemosense. Its stimuli are chemical substances, in this case taste molecules dissolved in food juices and the saliva that coats the tongue and the inside of the mouth. The main organ for taste is the tongue, which has several thousand tiny cell clusters called taste buds distributed mainly on its tip and along its upper sides and rear. The buds detect different combinations of five main tastes—these being sweet, salty, savory (umami), sour, and bitter. Most of these are detected equally in all the parts of the tongue furnished with taste buds. A similar "lock and key" system to smell (see above) probably works for gustation, with receptor sites for different taste molecules located on the hairlike processes of gustatory receptor cells in each taste bud.

Up to three-quarters of what we think of as taste is a **combination of taste and smell** perceived simultaneously—blocking off the nose makes foods taste very bland.

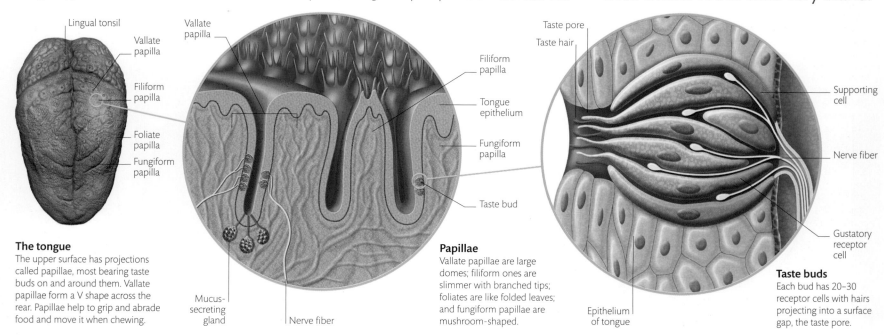

Lingual tonsil
Vallate papilla
Filiform papilla
Foliate papilla
Fungiform papilla

Vallate papilla

Filiform papilla
Tongue epithelium
Fungiform papilla
Taste bud

Taste pore
Taste hair

Supporting cell

Nerve fiber

Gustatory receptor cell

The tongue
The upper surface has projections called papillae, most bearing taste buds on and around them. Vallate papillae form a V shape across the rear. Papillae help to grip and abrade food and move it when chewing.

Mucus-secreting gland
Nerve fiber

Papillae
Vallate papillae are large domes; filiform ones are slimmer with branched tips; foliates are like folded leaves; and fungiform papillae are mushroom-shaped.

Epithelium of tongue

Taste buds
Each bud has 20–30 receptor cells with hairs projecting into a surface gap, the taste pore.

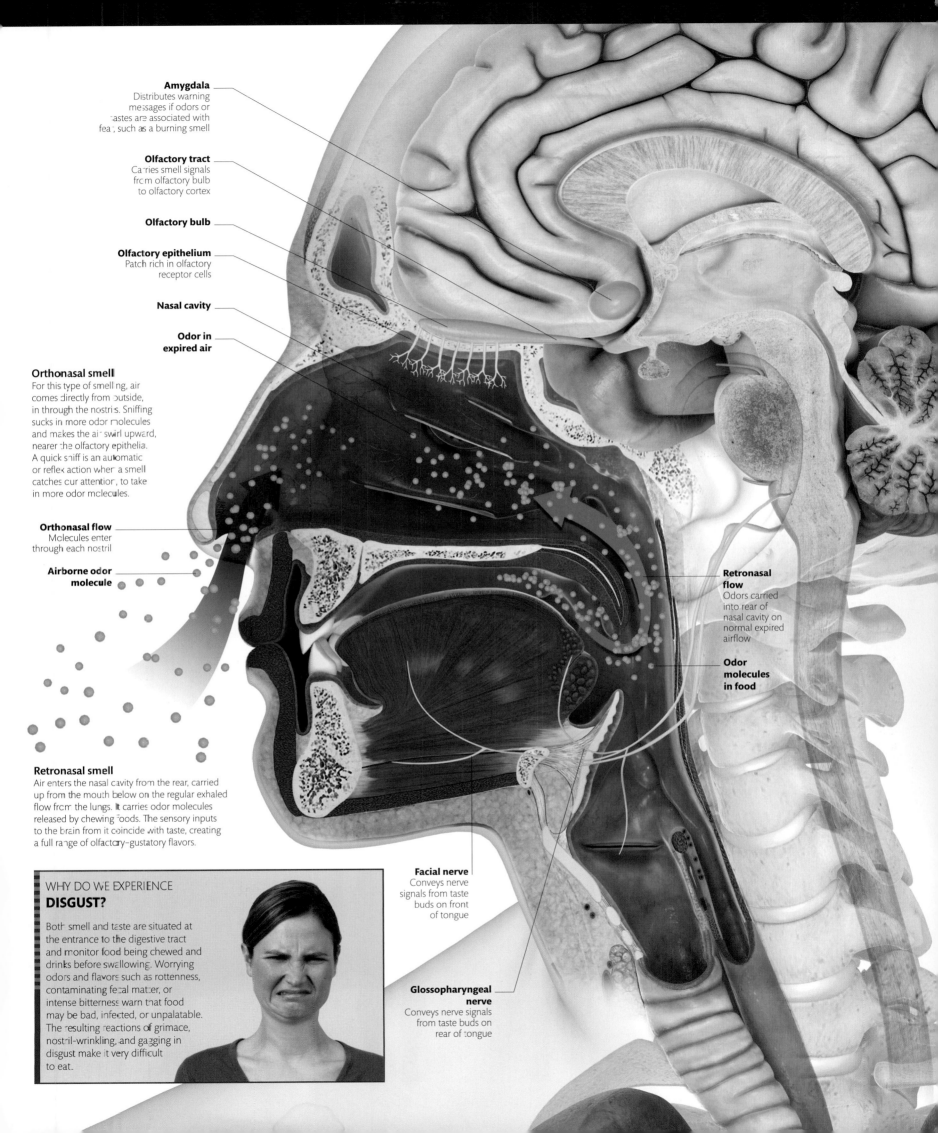

Amygdala
Distributes warning messages if odors or tastes are associated with fear, such as a burning smell

Olfactory tract
Carries smell signals from olfactory bulb to olfactory cortex

Olfactory bulb

Olfactory epithelium
Patch rich in olfactory receptor cells

Nasal cavity

Odor in expired air

Orthonasal smell
For this type of smelling, air comes directly from outside, in through the nostrils. Sniffing sucks in more odor molecules and makes the air swirl upward, nearer the olfactory epithelia. A quick sniff is an automatic or reflex action when a smell catches our attention, to take in more odor molecules.

Orthonasal flow
Molecules enter through each nostril

Airborne odor molecule

Retronasal flow
Odors carried into rear of nasal cavity on normal expired airflow

Odor molecules in food

Retronasal smell
Air enters the nasal cavity from the rear, carried up from the mouth below on the regular exhaled flow from the lungs. It carries odor molecules released by chewing foods. The sensory inputs to the brain from it coincide with taste, creating a full range of olfactory-gustatory flavors.

Facial nerve
Conveys nerve signals from taste buds on front of tongue

Glossopharyngeal nerve
Conveys nerve signals from taste buds on rear of tongue

WHY DO WE EXPERIENCE **DISGUST?**

Both smell and taste are situated at the entrance to the digestive tract and monitor food being chewed and drinks before swallowing. Worrying odors and flavors such as rottenness, contaminating fecal matter, or intense bitterness warn that food may be bad, infected, or unpalatable. The resulting reactions of grimace, nostril-wrinkling, and gagging in disgust make it very difficult to eat.

TOUCH

Touch does far more than detect physical contact. It tells us about temperature, pressure, texture, movement, and bodily location. Pain seems to be part of touch, but it has its own dedicated receptors and sensory pathways.

TOUCH PATHWAYS

The skin contains millions of touch receptors of different kinds, including Merkel's disks, Meissner's and Pacinian corpuscles, and free nerve endings (see p.293). Although most receptors show at least some reaction to most kinds of touch, each kind is specialized to respond to certain aspects of touch. Meissner's corpuscles, for example, react strongly to light contact. The more a receptor is stimulated, the faster it produces nerve impulses. These travel along peripheral nerves into the central nervous system at the spinal cord, then along the dorsal column–medial lemniscus tract (see p.318) to the brain, which figures out the type of contact from the pattern of impulses.

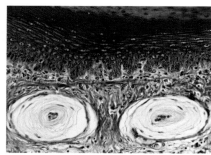

Under pressure
The largest skin receptors are Pacinian corpuscles, about 1/32 in (1 mm) long. They register changes in pressure and fast vibrations in particular.

SPINAL NERVES

Snaking out from the spinal cord, through the narrow gaps between adjacent vertebrae, are 31 pairs of spinal nerves (see pp.150–51 and 180–81). They divide into smaller peripheral nerves that extend to all organs and tissues, including skin. Most of these nerves carry both sensory nerve signals about touch on the skin to the cord, and motor signals from the cord to muscles.

Dermatomes
Each spinal nerve carries sensory information via its dorsal root into the spinal cord from a specific skin area or dermatome. Facial skin (V1–3) is served by cranial nerves (see p.116).

Cervical region
Eight pairs of cervical nerves serve skin covering the rear head, neck, shoulders, arms, and hands

Thoracic region
Twelve pairs of thoracic nerves connect to skin on chest, back, and underarms

Lumbar region
Five pairs of lumbar nerves serve skin on the lower abdomen, thighs, and fronts of the legs

Sacral region
Six pairs of sacral nerves connect to skin on the rear of the legs, feet, and anal and genital areas

Spinal regions
Each pair of spinal nerves, from the upper neck to the lower back, links to one of four specific regions of the body.

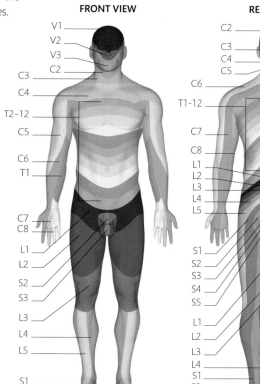

FRONT VIEW

V1
V2
V3
C2
C3
C4
T2–12
C5
C6
T1
C7
C8
L1
L2
S2
S3
L3
L4
L5
S1

REAR VIEW

C2
C3
C4
C5
C6
T1–12
C7
C8
L1
L2
L3
L4
L5
S1
S2
S3
S4
S5
L1
L2
L3
L4
S1
S2
L5

Somatosensory cortex
Left side receives touch signals from right side of body

Medial lemniscus
Fibers cross over to other side here

Spinal cord
Carries signals up ascending tracts into brain stem

Foot to brain
A touch on the foot sends nerve signals along peripheral fibers in the leg to the spinal cord, then up to the brain stem. Here the fibers cross over, right to left, in the medial lemniscus and continue up to the thalamus and the brain's somatosensory cortex (see opposite).

Ganglion
Concentration of neuronal (nerve cell) bodies

Sacral plexus
Nerve junction where information is shared and coordinated

Lateral branch of tibial nerve
Carries nerve impulses up leg

Stimulus
Light touch on skin of outer heel

THE FEELING BRAIN

The main "touch center" of the brain is the primary somatosensory cortex. It arches over the outer surface of the parietal lobe, just behind the motor cortex. It has two parts, left and right. Because of the way nerve fibers cross to the other side in the brain stem (see opposite), the left somatosensory cortex receives touch information from the skin and eyes on the body's right side, and vice versa. Touch information starting as nerve signals from a particular body region, such as the fingers, always ends up at a corresponding dedicated region of the somatosensory cortex. Skin areas with more densely packed touch receptors, giving more sensitive feeling—as in the fingers—have proportionately larger regions of cortex.

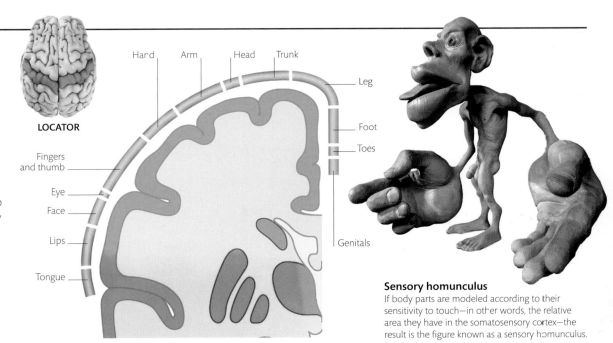

LOCATOR

Hand · Arm · Head · Trunk · Leg · Foot · Toes · Genitals · Fingers and thumb · Eye · Face · Lips · Tongue

Touch map
The surface of the somatosensory cortex has been mapped to skin areas. The order of these, from the lower outer side, up and over to its medial or inner surface, reflects body parts from head to toes.

Sensory homunculus
If body parts are modeled according to their sensitivity to touch—in other words, the relative area they have in the somatosensory cortex—the result is the figure known as a sensory homunculus.

EXPERIENCING PAIN

Pain information comes from a class of receptors, called nociceptors, present not just in skin but throughout the body. However, the skin has the highest numbers, so we can localize a pain here more easily—in a fingertip, for example—whereas pain within organs and tissues is vague and difficult to pinpoint. Nociceptors respond to many kinds of stimuli, such as temperature extremes, pressure, tension, and certain chemical substances, especially those released from cells when the body

suffers physical injury or microbial infection (see p.325). The nociceptors send their nerve signals into the spinal cord along specialized nerve fibers (axons) of two main kinds, A-delta and C. Instead of crossing to the opposite side up in the brain stem, as for touch (see opposite), pain information moves to the opposite side at its entry level in the cord (see pp.318–19). The signals then pass up the spinal cord to the medulla and thalamus, where automatic reactions such as reflexes are triggered.

Inflammatory "soup"
An "insult" to the body breaks tissues and damages cells, which release various substances into the general extracellular fluid to cause inflammation and begin repair. Several of these substances, such as bradykinin, prostaglandins, and ATP, stimulate nociceptors.

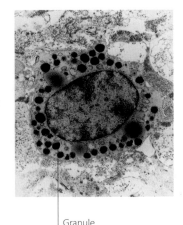

Mast cell with histamine
Mast cells are scattered throughout tissues and play roles in inflammation following injury, and in the allergic response. When damaged or involved in fighting microbes, they release granules (dark purple in this micrograph) containing heparin and histamine. Heparin prevents blood clotting and histamine increases blood flow and swelling.

Granule

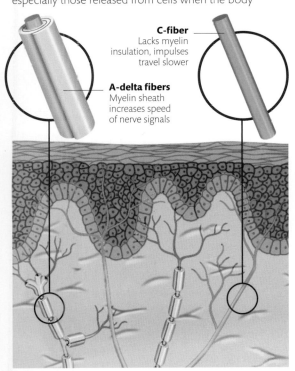

C-fiber Lacks myelin insulation, impulses travel slower

A-delta fibers Myelin sheath increases speed of nerve signals

Pain fibers
Dedicated sensory nerve fibers convey pain information toward the brain. A-delta fibers have myelin sheath insulation, carry impulses fast and serve a small area, usually a 1 mm² patch of skin. C-fibers are more widespread and diffuse and their impulses are slower.

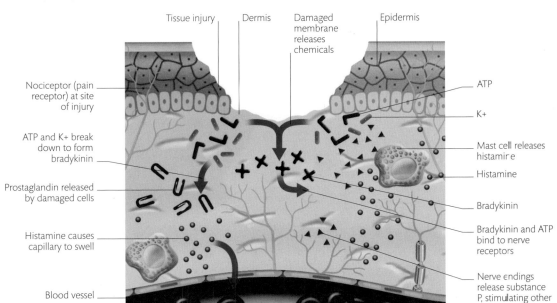

Tissue injury · Dermis · Damaged membrane releases chemicals · Epidermis · ATP · K+ · Mast cell releases histamine · Histamine · Bradykinin · Bradykinin and ATP bind to nerve receptors · Nociceptor (pain receptor) at site of injury · ATP and K+ break down to form bradykinin · Prostaglandin released by damaged cells · Histamine causes capillary to swell · Blood vessel · Red blood cell · Nerve endings release substance P, stimulating other nerves to do the same, causing redness at site of injury

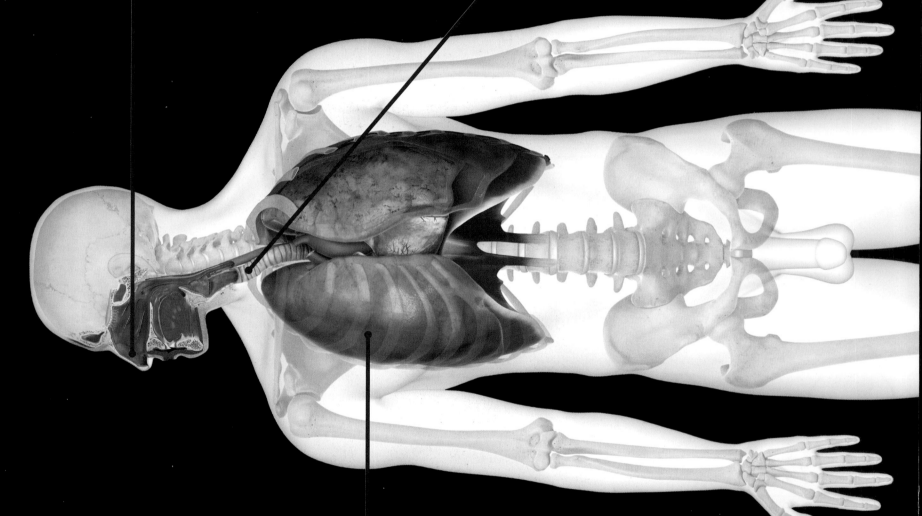

NOSE

Air usually enters the body via the nostrils, which open into the nasal cavity. The linings of both help filter out dust particles.

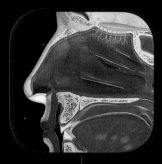

TRACHEA

This main airway, also known as the windpipe, channels air from the nose and throat to deep within the lungs.

LUNG

The highly branched "tree" of tubes in each lung end at millions of balloonlike alveoli where gas exchange takes place.

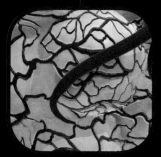

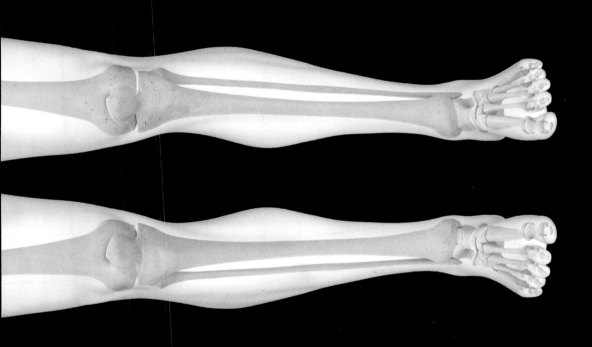

RESPIRATORY SYSTEM

Every living cell in our bodies requires a constant supply of oxygen and the removal of waste carbon dioxide. The respiratory system brings air from the atmosphere into the body so that this vital exchange of gases can occur.

JOURNEY OF AIR

The respiratory tract is responsible for transporting air into and out of the lungs, and for the essential exchange of oxygen and carbon dioxide between the blood and the air in the lungs. It also protects the entire body by providing key lines of defense against potentially harmful particles that are inhaled.

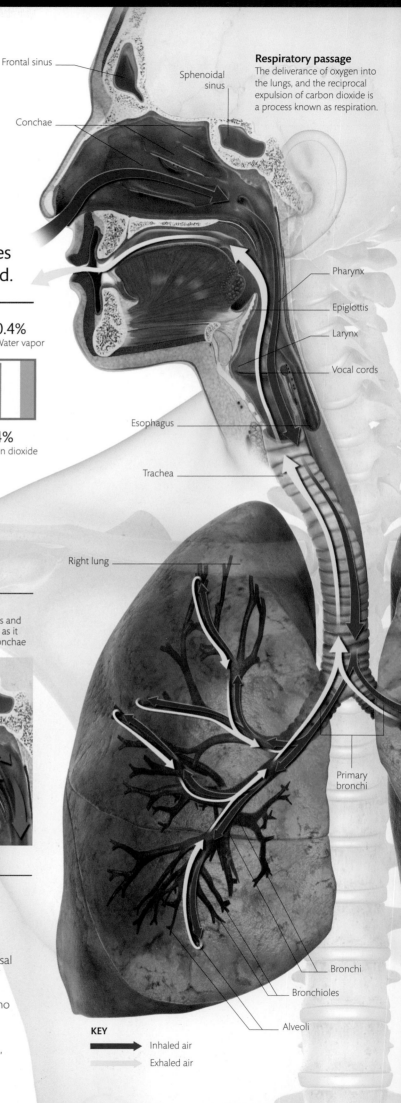

Frontal sinus

Sphenoidal sinus

Conchae

Respiratory passage
The deliverance of oxygen into the lungs, and the reciprocal expulsion of carbon dioxide is a process known as respiration.

Pharynx

Epiglottis

Larynx

Vocal cords

Esophagus

Trachea

Right lung

Primary bronchi

Bronchi

Bronchioles

Alveoli

AIR FLOW

With every breath, air is drawn into the alveoli of the lungs via the respiratory tract. It travels from the nose or mouth, past the pharynx, through the larynx, and enters the trachea. This splits into two smaller tubes, one entering each lung, called the primary bronchi, which in turn branch into increasingly smaller bronchi and then into bronchioles attaching to the alveoli (tiny air sacs). During this long journey, the air is warmed to body temperature and has any particles filtered out. Used air makes the same journey in reverse, but as it passes though the larynx it can be employed to produce sound.

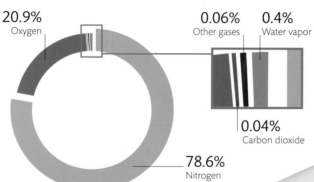

20.9%
Oxygen

0.06%
Other gases

0.4%
Water vapor

0.04%
Carbon dioxide

78.6%
Nitrogen

Breathable air
Nitrogen is the gas that occupies the largest part of atmospheric air, yet at the pressure at sea level, very little dissolves in human blood, so it is able to pass harmlessly into and out of the body.

NASAL CONCHAE

Three shelflike projections in the nasal cavity provide an obstruction to inhaled air, forcing it to spread out as it passes over their surfaces. This fulfills several roles. The moist, mucus-lined conchae humidify passing air and entrap inhaled particles, while their many capillary networks warm the air to body temperature before it reaches the lungs. Nerves within the conchae sense the condition of the air and, if needed, cause them to enlarge—if the air is cold, for example, a larger surface area helps warm it more effectively. This is what gives a feeling of nasal congestion.

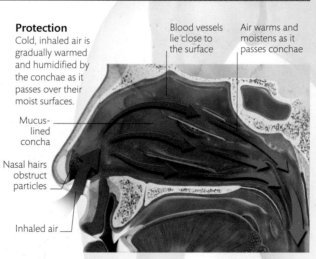

Protection
Cold, inhaled air is gradually warmed and humidified by the conchae as it passes over their moist surfaces.

Blood vessels lie close to the surface

Air warms and moistens as it passes conchae

Mucus-lined concha

Nasal hairs obstruct particles

Inhaled air

Frontal sinus

Ethmoid sinus

Maxillary sinus

Sphenoidal sinus

Continuous space
The paranasal sinuses are filled with air that moves into and out of them from the nasal passageways.

PARANASAL SINUSES

Four pairs of air-filled cavities called paranasal sinuses sit within the facial bones of the skull. They are lined with cells that produce mucus, which flows into the nasal passageways through very small openings. The roles of the sinuses are to lighten the heavy skull bones and to improve the resonance of the voice by acting as an echo chamber. Their effectiveness becomes obvious during a cold, when the small openings into the nose become blocked, giving a nasal quality to the voice.

KEY

→ Inhaled air

→ Exhaled air

TRACHEA

The trachea (or windpipe) acts as a conduit for air from the larynx to the lungs. It is kept open by rings of C-shaped cartilage, which encircle it at intervals along its length. The ends of these rings are connected by muscles that contract to increase the speed of air expelled during coughing. In order to swallow, the trachea closes against the epiglottis, a cartilage flap, and the vocal cords close tightly shut. Cells that line the trachea either produce mucus or display cilia (see below), which transport mucus up to the mouth.

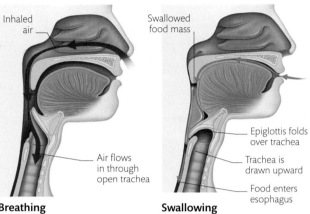

Inhaled air

Air flows in through open trachea

Breathing
The trachea remains open, allowing air to flow freely into and out of the lungs.

Swallowed food mass

Epiglottis folds over trachea

Trachea is drawn upward

Food enters esophagus

Swallowing
The trachea is pulled upward so that it is closed off by the epiglottis. Food passes down the esophagus.

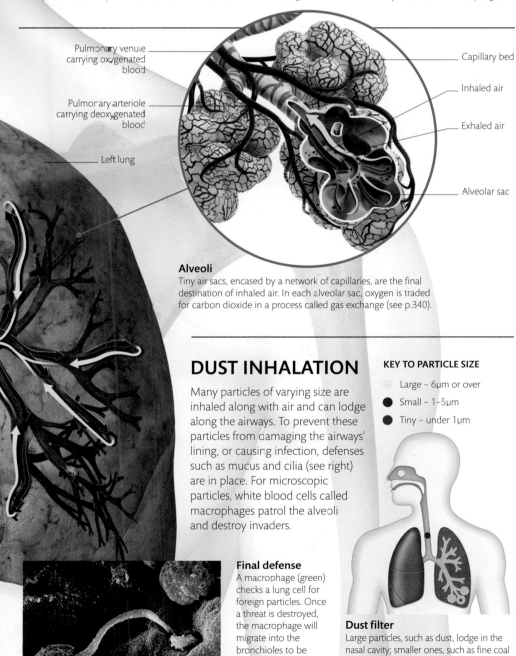

Pulmonary venule carrying oxygenated blood

Pulmonary arteriole carrying deoxygenated blood

Left lung

Capillary bed

Inhaled air

Exhaled air

Alveolar sac

Alveoli
Tiny air sacs, encased by a network of capillaries, are the final destination of inhaled air. In each alveolar sac, oxygen is traded for carbon dioxide in a process called gas exchange (see p.340).

DUST INHALATION

Many particles of varying size are inhaled along with air and can lodge along the airways. To prevent these particles from damaging the airways' lining, or causing infection, defenses such as mucus and cilia (see right) are in place. For microscopic particles, white blood cells called macrophages patrol the alveoli and destroy invaders.

KEY TO PARTICLE SIZE
- Large – 6μm or over
- Small – 1–5μm
- Tiny – under 1μm

Final defense
A macrophage (green) checks a lung cell for foreign particles. Once a threat is destroyed, the macrophage will migrate into the bronchioles to be expelled from the airways via mucus.

Dust filter
Large particles, such as dust, lodge in the nasal cavity; smaller ones, such as fine coal dust, in the trachea; and the tiniest, such as cigarette smoke particles, reach the alveoli.

SNORING

Over one third of people snore. The incidence is higher in older people and those who are overweight. The noise is produced by the vibration of soft tissues in the airways as air is breathed in and out. When a person is awake, the soft tissues at the back of the mouth are kept out of the way of the airflow by the tone of the surrounding muscles. During sleep these muscles relax and the soft tissues flop into the air stream and cause it to vibrate, producing the snoring noise.

Sleepless nights
Severe snoring can cause "obstructive sleep apnea", a condition where the snorer stops breathing during sleep.

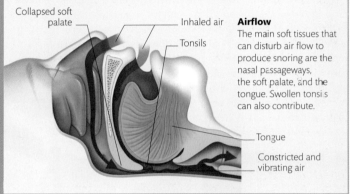

Collapsed soft palate

Inhaled air

Tonsils

Tongue

Constricted and vibrating air

Airflow
The main soft tissues that can disturb air flow to produce snoring are the nasal passageways, the soft palate, and the tongue. Swollen tonsils can also contribute.

CILIA

The air passages from the nose through to the bronchi are lined with two types of cells: epithelial cells and goblet cells. The more numerous epithelial cells have tiny, hairlike projections called cilia on their surface. Cilia continually beat toward the upper airways. The goblet cells produce mucus, which they secrete into the lining of the airways where it can trap inhaled particles, such as dust. The cilia then act as a conveyor belt, transporting the mucus, along with any trapped particles, away from the lungs into the upper airways, where it can be coughed or blown out, or swallowed.

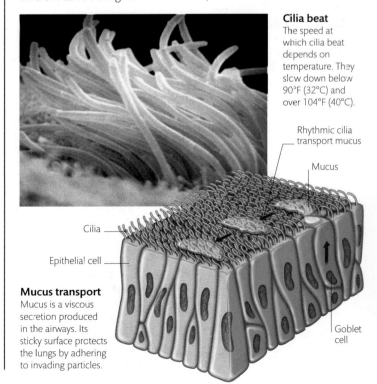

Cilia beat
The speed at which cilia beat depends on temperature. They slow down below 90°F (32°C) and over 104°F (40°C).

Rhythmic cilia transport mucus

Mucus

Cilia

Epithelial cell

Mucus transport
Mucus is a viscous secretion produced in the airways. Its sticky surface protects the lungs by adhering to invading particles.

Goblet cell

GAS EXCHANGE

Cells need a continual supply of oxygen that they combine with glucose to produce energy. Carbon dioxide is continually generated as a waste product of this process and is exchanged for useful oxygen in the lungs.

Hundreds of **millions** of **alveoli** provide a total **surface area** of 750 sq ft (70 sq m), over which gas exchange can take place.

PROCESS OF GAS EXCHANGE

The respiratory tract acts as a transport system, taking air to millions of tiny air sacs (alveoli) in the lungs where oxygen is traded for carbon dioxide in the bloodstream. This exchange of gases can take place only in the alveoli. However, during normal breathing, air is only drawn into and out of the respiratory tract as far down as the bronchioles. This means that the alveoli are not regularly flushed with fresh air and stale, carbon dioxide-rich air remains in them. Carbon dioxide and oxygen in the alveoli therefore have to change places by moving down a concentration gradient—the oxygen molecules migrate to the area where oxygen is scarce, while the carbon dioxide molecules migrate to the area where carbon dioxide is scarce. Using this process, known as "diffusion," oxygen enters the alveoli, and from there diffuses into the blood (see below), while carbon dioxide moves out of the alveoli and into the bronchioles, and is exhaled normally.

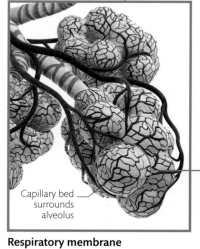

Lung tissue
A color-enhanced micrograph of a section of a human lung clearly displays the numerous alveoli, which form the site of gas exchange.

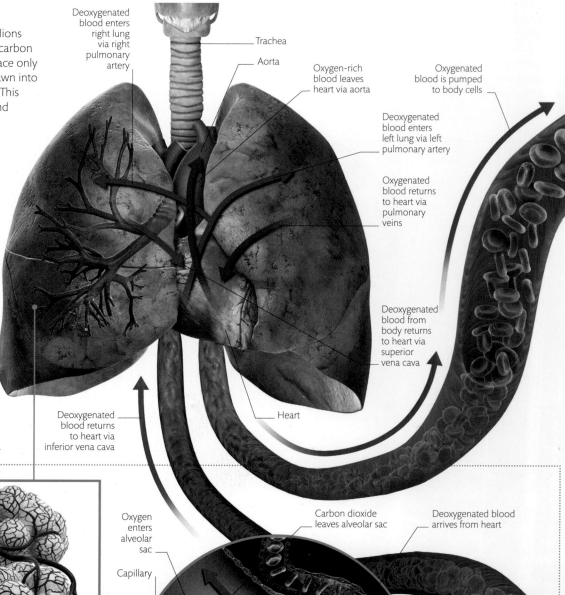

Deoxygenated blood enters right lung via right pulmonary artery

Trachea

Aorta

Oxygen-rich blood leaves heart via aorta

Oxygenated blood is pumped to body cells

Deoxygenated blood enters left lung via left pulmonary artery

Oxygenated blood returns to heart via pulmonary veins

Deoxygenated blood from body returns to heart via superior vena cava

Heart

Deoxygenated blood returns to heart via inferior vena cava

DIFFUSION FROM ALVEOLI

In human lungs there are nearly 500 million alveoli, each of which is around $1/128$ in (0.2 mm) in diameter. Taken together, the alveoli represent a large surface area over which gas exchange can take place. To move between the air and the blood, oxygen and carbon dioxide have to cross the "respiratory membrane," which comprises the walls of the alveoli and their surrounding capillaries. Both of these are just one cell thick, so the distance that molecules of oxygen and carbon dioxide must travel to get into and out of the blood is tiny. The exchange of gas through the respiratory membrane occurs passively, by diffusion, where gases transfer from areas of a high concentration to a low concentration. Oxygen dissolves into the surfactant (see p.343) and water layers of the alveoli before entering the blood, while carbon dioxide diffuses the opposite way, from the blood into the alveolar air.

Capillary bed surrounds alveolus

Respiratory membrane
The vast number of capillaries that surround the alveoli mean that up to 32 fl oz (900 ml) of blood can take part in gas exchange at a given time.

Oxygen enters alveolar sac

Capillary

Carbon dioxide leaves alveolar sac

Deoxygenated blood arrives from heart

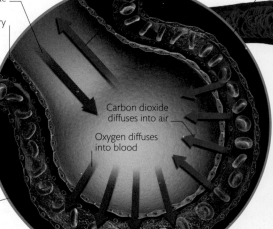

Carbon dioxide diffuses into air

Oxygen diffuses into blood

Oxygenated blood returns to heart

Exchange of gas
Capillaries alongside alveoli give up their waste carbon dioxide and pick up vital oxygen across the respiratory membrane.

HEMOGLOBIN

Hemoglobin s found in red blood cells and is a specialized molecule for transporting oxygen. It is made up of four ribbon-like protein units, each containing a heme molecule. Heme contains iron, which binds oxygen to the hemoglobin and therefore holds it within the red blood cell (oxygenating the blood). When oxygen levels are high, for example in the lungs, oxygen readily binds to hemoglobin; when oxygen levels are low, for example in working muscle, oxygen molecules detach from hemoglobin and move freely into the body cells.

No oxygen molecules

Oxygen molecules

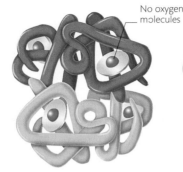

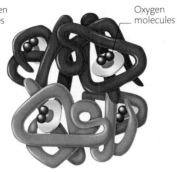

Deoxyhemoglobin
Deoxyhemoglobin is hemoglobin without oxygen. Once it has lost one oxygen molecule, the hemoglobin changes its shape to make it easier to release its remaining oxygen.

Oxyhemoglobin
Oxygen binds to deoxyhemoglobin in the lungs to form oxyhemoglobin. Once one oxygen molecule has been picked up, the structure changes so more oxygen will quickly attach.

DIFFUSION INTO CELL TISSUES

Body cells constantly take in oxygen from hemoglobin (see left) and excrete their waste into the bloodstream. As a result, the concentration of oxygen in the capillaries is low, and the concentration of waste products is high; a situation that prompts hemoglobin to give up its oxygen. The free oxygen then diffuses into the cells, where it is used to create energy, while carbon dioxide diffuses out of the cells and into the blood. Hemoglobin picks up around 20 percent of this carbon dioxide, yet most returns to the lungs dissolved in plasma.

Oxygenated red blood cells enter capillary

Carbon dioxide diffuses out of tissue cells, through the capillary wall, and into the blood plasma

Oxygenated red blood cell

Essential supply
Oxygen absorbed in the lungs is taken in the blood to the left side of the heart, which pumps it through the body. When it reaches the capillaries, oxygen is exchanged for carbon dioxide. Carbon dioxide is then transported in the blood to the right side of the heart, which pumps it to the lungs to be exhaled.

Body cells

Capillary bed

Oxygen is released by hemoglobin within the red blood cells

Capillary gas exchange
Blood flows through the capillaries, where hemoglobin releases oxygen, and carbon dioxide dissolves in plasma to be taken back to the lungs.

Smoke inhalation
Inhaled smoke particles travel deep into the lungs. They damage the alveolar walls and cause them to thin and stretch. This results in the individual air sacs fusing, which reduces available surface area for gas exchange. Breathing difficulties can then arise in later life.

Deoxygenated red blood cell

Deoxygenated blood is carried back to the heart

THE BENDS

Divers breathe pressurized air, which forces more nitrogen than usual to dissolve into the blood (see p.338). If they ascend too fast, nitrogen forms gas bubbles in their blood, blocking the vessels and causing widespread damage, known as "the bends." Treatment is to redissolve the bubbles in a decompression chamber until nitrogen levels return to normal.

MECHANICS OF BREATHING

The movement of air into and out of the lungs, known as respiration, is brought about by the action of muscles in the neck, chest, and abdomen, which work together to alter the volume of the chest cavity. During inhalation fresh air is drawn into the lungs, and during exhalation stale air is expelled into the atmosphere.

MUSCLES OF RESPIRATION

The diaphragm is the main muscle of respiration. It is a dome-shaped sheet of muscle that divides the chest cavity from the abdominal cavity, attaching to the sternum at the front of the chest, the vertebrae at the back of the chest, and to the lower six ribs. Various accessory muscles are located within the rib cage, neck, and abdomen, but these muscles are used only during forced respiration. For normal, quiet respiration, the diaphragm contracts and flattens to inhale, increasing the depth of the chest cavity and drawing air into the lungs. Normal, quiet exhalation is passive and brought about by the relaxation of the diaphragm as well as the elastic recoil of the lungs. If extra respiratory effort is required, for example during exercise, when the body's cells need a greater supply of oxygen to function efficiently, then contraction of the accessory muscles bolsters the action of the diaphragm to allow deeper breathing. Different accessory muscles are used for inhalation and exhalation.

PLEURAL CAVITY

The pleural cavity is a narrow space between the lining of the lungs and the lining of the chest wall. It contains a small amount of lubricating fluid (pleural fluid) that prevents friction as the lungs expand and contract within the chest cavity. Pleural fluid is held under slight negative pressure. This creates a suction between the lungs and the chest wall that holds the lungs open and prevents the alveoli from closing at the end of exhalation. If the alveoli were to close completely, an excessive amount of energy would be needed to reinflate them during inspiration.

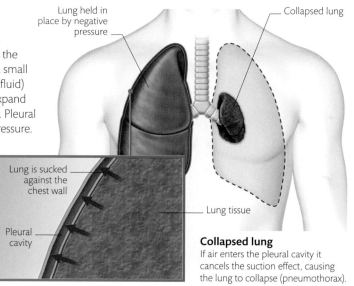

Lung held in place by negative pressure

Collapsed lung

Lung is sucked against the chest wall

Pleural cavity

Lung tissue

Collapsed lung
If air enters the pleural cavity it cancels the suction effect, causing the lung to collapse (pneumothorax).

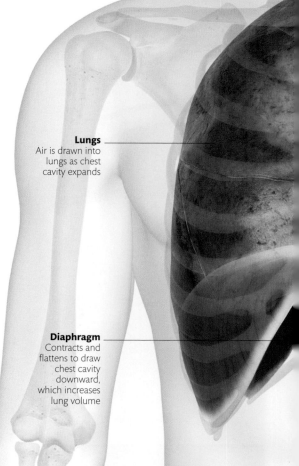

Lungs
Air is drawn into lungs as chest cavity expands

Diaphragm
Contracts and flattens to draw chest cavity downward, which increases lung volume

Circular breathing enables a **single continuous exhalation** by inhaling while exhaling air stored **in the cheeks**—the longest exhalation on record has exceeded **1 hour**.

NEGATIVE AND POSITIVE PRESSURE

The generation of "pressure gradients" is what causes air to move into and out of the lungs. When the muscles of inhalation contract to increase the volume of the chest cavity, the lungs, which are sucked onto the chest wall by the effect of pleural fluid, expand. This reduces the pressure in the lungs relative to that of the atmosphere and air flows down the pressure gradient into the lungs. For exhalation, the elastic recoil of the lungs compresses the air within them, forcing it out into the atmosphere.

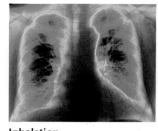
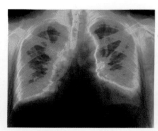

Chest cavity expands

Inhalation
Enlarging the chest cavity creates a negative pressure in the lungs, causing air to be drawn into them.

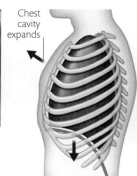
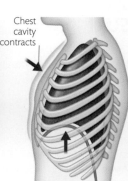

Chest cavity contracts

Exhalation
Reducing the chest cavity volume exerts a positive pressure on the lung tissue and forces the air out.

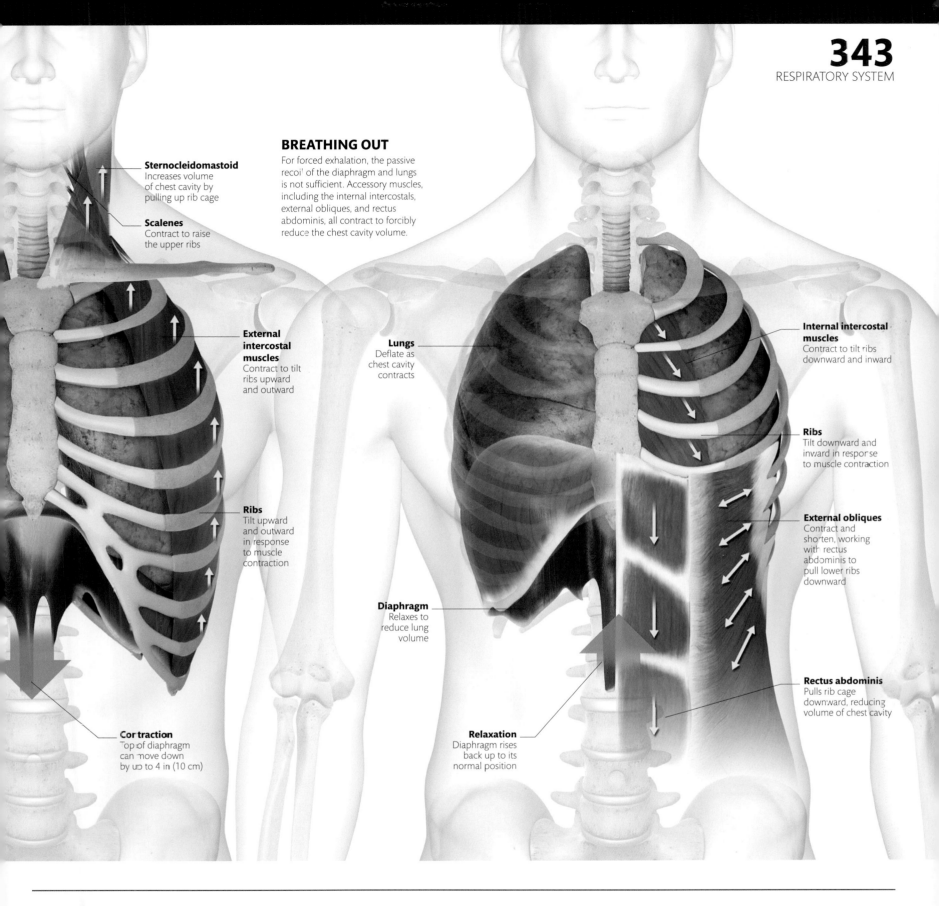

Sternocleidomastoid
Increases volume
of chest cavity by
pulling up rib cage

Scalenes
Contract to raise
the upper ribs

BREATHING OUT

For forced exhalation, the passive
recoil of the diaphragm and lungs
is not sufficient. Accessory muscles,
including the internal intercostals,
external obliques, and rectus
abdominis, all contract to forcibly
reduce the chest cavity volume.

**External
intercostal
muscles**
Contract to tilt
ribs upward
and outward

Lungs
Deflate as
chest cavity
contracts

**Internal intercostal
muscles**
Contract to tilt ribs
downward and inward

Ribs
Tilt downward and
inward in response
to muscle contraction

Ribs
Tilt upward
and outward
in response
to muscle
contraction

External obliques
Contract and
shorten, working
with rectus
abdominis to
pull lower ribs
downward

Diaphragm
Relaxes to
reduce lung
volume

Rectus abdominis
Pulls rib cage
downward, reducing
volume of chest cavity

Contraction
Top of diaphragm
can move down
by up to 4 in (10 cm)

Relaxation
Diaphragm rises
back up to its
normal position

SURFACTANT

Cells lining the alveoli are coated with a layer of water molecules. These
have a high affinity for each other, meaning that the water layer tries to
contract and pull the alveolar cells together, like a purse string. To prevent
the alveoli from closing under this pressure, a layer of surfactant spreads
over the water surface. Oil-based surfactant
molecules have a very low affinity for each other
and can therefore counteract the pull of the water
molecules, ensuring the alveoli remain open.
Alveoli are made of two types of cell: Type I form
the alveolar walls and Type II secrete surfactant.

Oily layer
A surfactant molecule's
water-loving end
dissolves in water; its
fat-loving end forms a
boundary with the air.

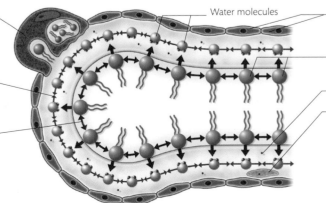

Type II
alveolar cell
produces new
surfactant
molecules

Water molecules

Type I cells form
alveolar wall

Water molecules
pull toward
each other

Surfactant
molecules

Dust particle

Low-affinity
surfactant
molecules
resist the pull
of the water

Alveolar
macrophage
engulfs tiny
dust particles
that enter
alveolar sac
(see p 339)

INSTINCTIVE BREATHING

The aim of respiration is to maintain the necessary blood levels of oxygen and carbon dioxide for the corresponding level of activity. The trigger to breathe, as well as breathing itself, is subconscious, but the rate and force of breathing can be consciously modified.

RESPIRATORY DRIVE

Oxygen is vital for cells to function, yet the drive to breathe is mainly determined by levels of carbon dioxide in the blood. Hemoglobin, the oxygen-carrying molecule (see p.341), has a built-in reserve, and can continue to donate oxygen to cells even when blood levels of oxygen are low. However, carbon dioxide readily dissolves in plasma and is converted to carbonic acid, which quickly damages the cells' ability to function properly. Therefore, breathing is triggered by rising levels of carbon dioxide or acid, and only very low oxygen levels stimulate breathing.

Specialized cells called chemoreceptors measure blood levels and send nerve impulses to the respiratory center of the brain stem within the medulla oblongata. Corresponding messages from the brain then activate the respiratory muscles.

TRIGGER

Clusters of specialized cells, known as chemoreceptors, located in the aortic and carotid bodies (peripheral chemoreceptors) and the brain stem (central chemoreceptors), monitor levels of carbon dioxide and oxygen in the blood. They then send signals to the brain to trigger a response.

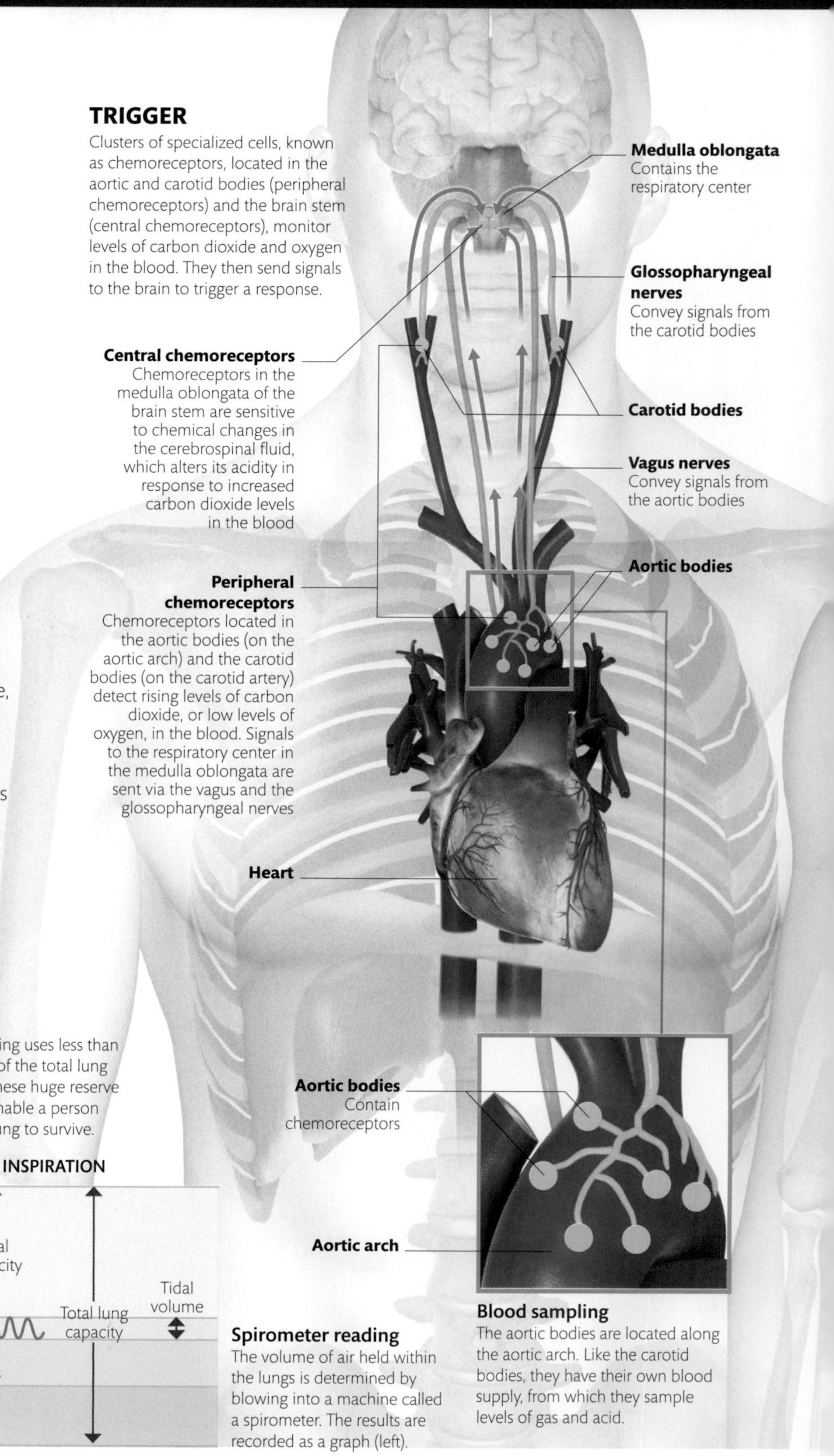

Medulla oblongata
Contains the respiratory center

Glossopharyngeal nerves
Convey signals from the carotid bodies

Carotid bodies

Vagus nerves
Convey signals from the aortic bodies

Central chemoreceptors
Chemoreceptors in the medulla oblongata of the brain stem are sensitive to chemical changes in the cerebrospinal fluid, which alters its acidity in response to increased carbon dioxide levels in the blood

Aortic bodies

Peripheral chemoreceptors
Chemoreceptors located in the aortic bodies (on the aortic arch) and the carotid bodies (on the carotid artery) detect rising levels of carbon dioxide, or low levels of oxygen, in the blood. Signals to the respiratory center in the medulla oblongata are sent via the vagus and the glossopharyngeal nerves

Heart

PATTERNS OF BREATHING

During normal breathing, only 18 fl oz (500 ml) of air flows into and out of the lungs. This is known as the tidal volume. The lungs have extra, reserve capacity (the vital capacity) for both inhalation and exhalation so that they can increase the amount of air they take in during exercise.

The maximum amount of air that the lungs are able to hold is around 204 fl oz (5,800 ml), but about 35 fl oz (1,000 ml) of this remains within the respiratory passages after each out breath. This is called the residual volume and cannot be displaced voluntarily.

90%
Excess space

10%
Used space

Overbuilt
Quiet breathing uses less than 10 percent of the total lung capacity. These huge reserve volumes enable a person with one lung to survive.

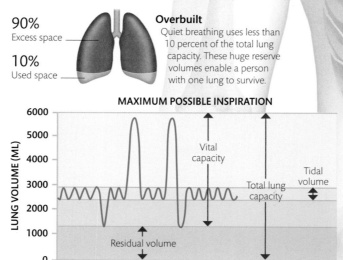

MAXIMUM POSSIBLE INSPIRATION

LUNG VOLUME (ML)

Vital capacity

Tidal volume

Total lung capacity

Residual volume

Aortic bodies
Contain chemoreceptors

Aortic arch

Spirometer reading
The volume of air held within the lungs is determined by blowing into a machine called a spirometer. The results are recorded as a graph (left).

Blood sampling
The aortic bodies are located along the aortic arch. Like the carotid bodies, they have their own blood supply, from which they sample levels of gas and acid.

Divers often **exceed** depths of 328 ft (100 m), which involves them **not breathing** for several minutes at a time.

EXTREME HUMAN
FREE DIVING

Some forms of free diving involve divers competing to go as deep as possible without using breathing apparatus. They train by exercising on land while holding their breath to get their muscles used to working without oxygen. Prior to the dive, some divers hyperventilate in an effort to rid their blood of as much carbon dioxide as possible—high levels would normally tell their brain of the need to stimulate inhalation. This allows them to dive for longer without feeling they need to breathe. However, this is highly dangerous because their cells may run out of oxygen before their brain realizes they need to take a breath. They risk blacking out under water and drowning.

Into the deep
Free diving with fins, or flippers (as shown here), provides extra propulsion and allows divers to reach depths beyond their usual capabilities.

RESPONSE

If carbon dioxide levels rise or oxygen levels fall, the respiratory center signals to the muscles of respiration, via the nerves, to trigger breathing, increasing both its rate and depth. These signals are sent continually so that respiration always matches the demands of the body.

Respiratory center

Cervical vertebrae

Phrenic nerves
Messages from the respiratory center pass down the phrenic nerves, which originate from the spinal cord in the neck, and stimulate the diaphragm to contract and expand the thoracic cavity

Intercostal nerves
The intercostal nerves take impulses from the respiratory center to the intercostal muscles and cause them to contract. Each nerve leaves the spinal cord at the same level of the muscle that it supplies

Intercostal muscles
Contract to expand the rib cage

Diaphragm
Contracts via innervation by the phrenic nerves

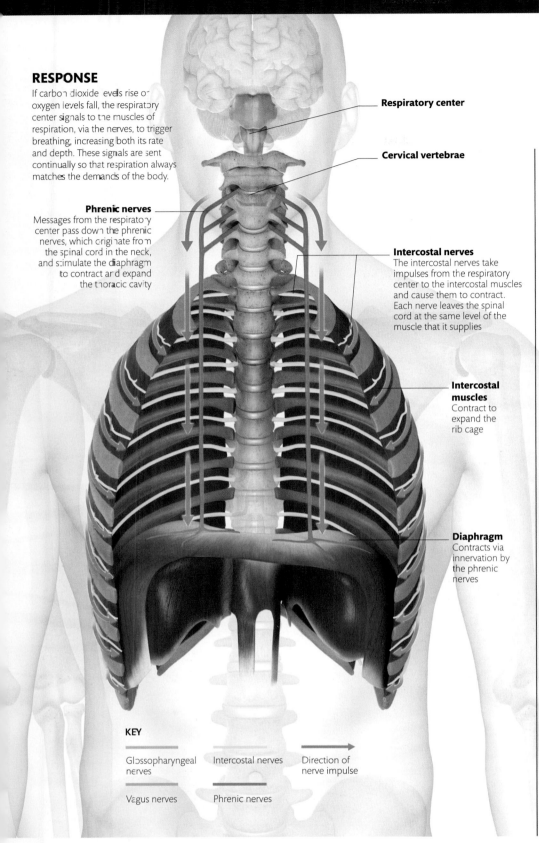

KEY

Glossopharyngeal nerves	Intercostal nerves	Direction of nerve impulse
Vagus nerves	Phrenic nerves	

REFLEXES

Inhaled air often contains particles of dust or corrosive chemicals that could damage the surfaces of the lungs and reduce their ability to function. Cough and sneeze reflexes exist to detect and expel such irritants before they reach the alveoli. Nerve endings in the respiratory tract are very sensitive to touch and chemical irritation and, if stimulated, send impulses to the brain to initiate a sequence of events that causes the offending object or chemical to be coughed or sneezed out.

Forcible explusion
Schlieren photography, which registers density changes, reveals the air turbulence from a cough.

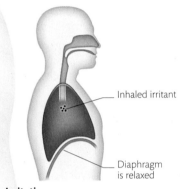

Inhaled irritant

Diaphragm is relaxed

1. Irritation
Inhaled particles or chemicals irritate sensitive nerve endings, which send signals to alert the brain to the intrusion.

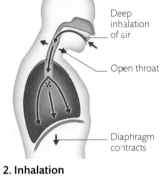

Deep inhalation of air

Open throat

Diaphragm contracts

2. Inhalation
The brain signals to the respiratory muscles to contract, causing a sudden intake of breath (88 fl oz/2,500 ml).

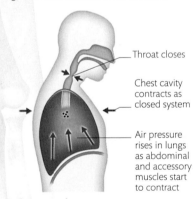

Throat closes

Chest cavity contracts as closed system

Air pressure rises in lungs as abdominal and accessory muscles start to contract

3. Compression
The vocal cords and the epiglottis shut tightly and the abdominal muscles contract, raising air pressure in the lungs.

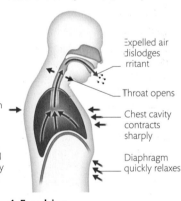

Expelled air dislodges irritant

Throat opens

Chest cavity contracts sharply

Diaphragm quickly relaxes

4. Expulsion
The epiglottis and vocal cords open suddenly, expelling the air at high velocity and taking the irritant with it.

VOCALIZATION

Speech involves a complex interaction between the brain, vocal cords, soft palate, tongue, and lips. When air passes against the vocal cords they vibrate to produce noise. Muscles attaching them to the larynx can move the cords apart for normal breathing, together to create sound, or stretch them to increase pitch. Vibrations are articulated into words by the soft palate, lips, and tongue. Higher air pressure beneath the vocal cords will increase volume. The voice itself finds resonance in the paranasal sinuses (see p.338).

Vocal cords vibrate at a variety of speeds depending on how tightly they are stretched: faster vibrations create high-pitched sound. For example, the vocal cords of a bass singer vibrate at around 60 times per second, whereas those of a soprano can vibrate at up to 2,000 times per second.

Back of tongue

Epiglottis

Open vocal cords

Air passes through trachea

Back of throat

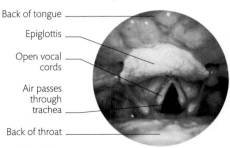

Breathing
The vocal cords are held fully open during breathing. Air passes easily between them without causing any vibration and no sound is made.

Vocal cords press together

Restricted air flow causes vocal cords to vibrate

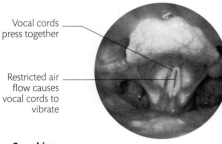

Speaking
During normal speech, the muscles of the larynx move the vocal cords close together so that air passing through them causes them to vibrate.

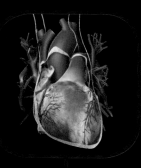

HEART

Sitting at the center of the circulation, the muscular heart pumps all of the blood around the body once every minute.

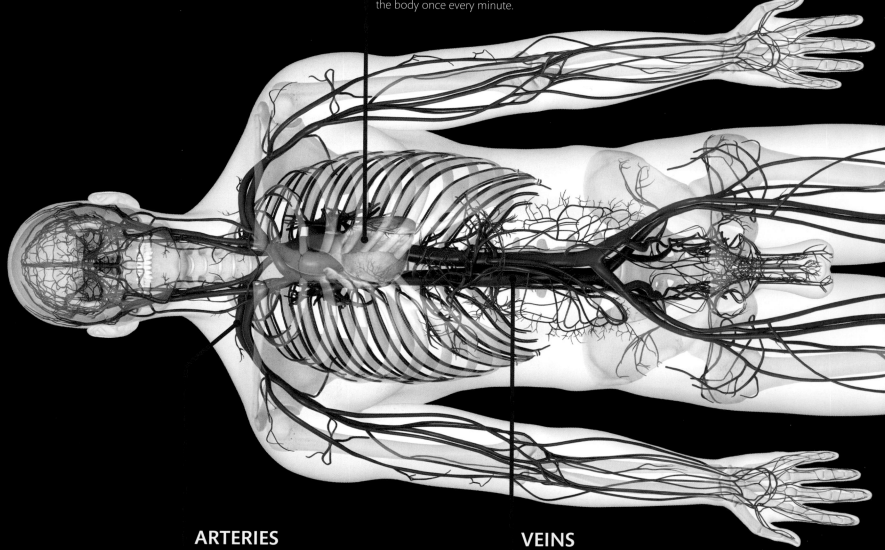

ARTERIES

Blood vessels that carry blood away from the heart have thick, muscular, elastic walls that cope with the high pressures generated by a heartbeat.

VEINS

Blood vessels that bring blood back to the heart have thinner, expandable walls and one-way valves that prevent backflow.

CAPILLARIES

Oxygen diffuses out of these
minute, thin-walled vessels
to supply body cells, while waste
carbon dioxide diffuses in.

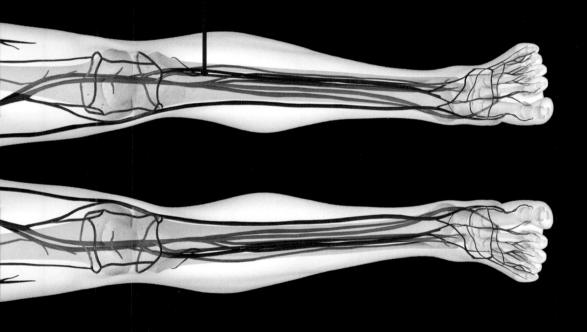

CARDIOVASCULAR SYSTEM

The heart is a pumping engine, powering the
transport of life-giving blood around the body.
Blood carries oxygen, nutrients, and immune
cells to every part of the body via arterial
vessels, and carries away waste via the veins.

BLOOD

Adults have approximately 11 pints (5 liters) of blood, which consists of specialized cells suspended in plasma. It supplies cells with nutrients and oxygen and removes their waste. Blood also transports hormones, antibodies, and cells that fight infection.

Constant supply
Blood flows to every cell in the human body. Throughout the body, the cells continually release chemicals to ensure that they get enough blood to supply them with nutrients and remove any waste.

BLOOD AS TRANSPORT

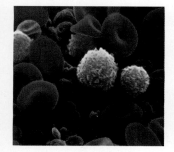

Blood is the main transport system of the body. The heart pumps all 11 pints (5 liters) of a resting adult's blood around the body every minute. Components of the blood pick up nutrients absorbed from the gut as well as oxygen from the lungs and deliver these to the body's cells. The blood also removes the cells' waste chemicals, such as urea and

In the stream
This magnified image reveals the cells and platelets in blood.

lactic acid, and transports them to the liver and kidneys, which break down or excrete them from the body. Carbon dioxide is taken from the cells and excreted by the lungs.

Blood also transports hormones (see p.398) from the glands in which they are produced to the cells they affect. Cells and other substances involved in healing and fighting infection circulate in the blood stream, only becoming active when they are needed.

Blood vessel

COMPONENTS OF BLOOD

The liquid component of blood (plasma) is 92 percent water, but also contains glucose, minerals, enzymes, hormones, and waste products, including carbon dioxide, urea, and lactic acid. Some of these substances, such as carbon dioxide, are just dissolved within the plasma. Others, such as the minerals iron and copper, are attached to specialized plasma transport proteins. Plasma also contains antibodies that fight infection.

Mainly water
Blood is made up of around 46 percent solids (cells), suspended in 54 percent liquid plasma.

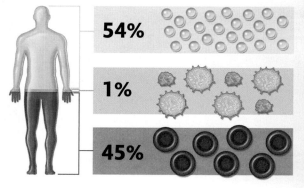

54%

1%

45%

Plasma
Plasma is a straw-colored liquid that forms the largest portion of blood.

White blood cells and platelets
These cells play a vital role in immunity and clotting.

Red blood cells
Each milliliter of blood contains around 5 billion red blood cells.

Capillary network

Blood vessel wall

BLOOD CLOTTING

When a blood vessel is damaged, platelets rush to the site to plug the gap. As they adhere to the damaged area, they release chemicals. These trigger what is called the clotting, or coagulation, cascade. This results in the formation of strands of a protein called fibrin, which cross-link to form a robust plug, or clot, with platelets and red blood cells trapped within.

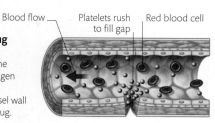

Blood flow — Platelets rush to fill gap — Red blood cell

Platelet plug
Platelets are attracted to the exposed collagen fibers in the damaged vessel wall and form a plug.

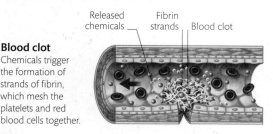

Released chemicals — Fibrin strands — Blood clot

Blood clot
Chemicals trigger the formation of strands of fibrin, which mesh the platelets and red blood cells together.

PRODUCTION OF CELLS

Red and white blood cells, as well as platelets, are produced in the bone marrow, and pass from here into the circulation. White blood cells, involved in immunity, can also pass into the lymphatic system (see p.358–63). Red blood cells, which lack a nucleus, remain in the blood circulation, where they can live for up to 120 days.

Waste product of blood cell — Useful products returned — Cells form in bone marrow

Waste is excreted from body

Life of a red blood cell
After about 120 days of life, red blood cells are broken down by white blood cells called macrophages. Waste products are excreted while useful ones return to the bone marrow.

Macrophage in liver or spleen engulfs red blood cell

Tired red blood cell

New red blood cell

Enters circulation

Protein

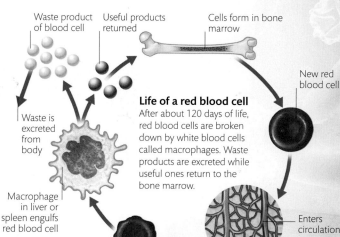

BLOOD TYPES

Blood type is hereditary. It is determined by proteins, called antigens, on the surface of red blood cells. The main antigens are called A and B, and cells can display A antigens (blood group A), B antigens (group B), both together (AB), or none (O). Antigens are triggers for the immune system. An individual's immune system ignores antigens on their own red blood cells, but produces antibodies to recognize and help destroy foreign cells that display new antigens. So, in blood group A, cells display the A antigen, which the immune system ignores, but it produces antibodies to the B antigen, and destroys foreign cells displaying this antigen.

Antigens
There are 30 different antigens that red blood cells can display, but the ABO antigens, illustrated here, are the most well known.

	GROUP A	GROUP B	GROUP AB	GROUP O
BLOOD GROUP				
ANTIGENS	A antigen	B antigen	A and B antigens	None
ANTIBODIES	Anti-B	Anti-A	None	Anti-A and Anti-B

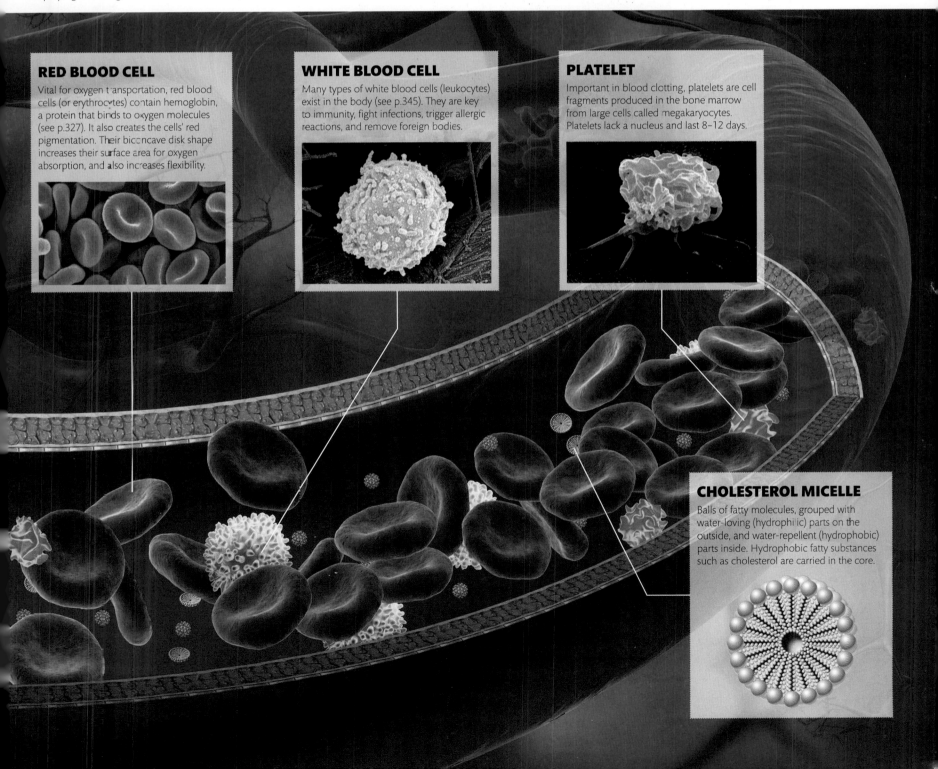

RED BLOOD CELL

Vital for oxygen transportation, red blood cells (or erythrocytes) contain hemoglobin, a protein that binds to oxygen molecules (see p.327). It also creates the cells' red pigmentation. Their biconcave disk shape increases their surface area for oxygen absorption, and also increases flexibility.

WHITE BLOOD CELL

Many types of white blood cells (leukocytes) exist in the body (see p.345). They are key to immunity, fight infections, trigger allergic reactions, and remove foreign bodies.

PLATELET

Important in blood clotting, platelets are cell fragments produced in the bone marrow from large cells called megakaryocytes. Platelets lack a nucleus and last 8–12 days.

CHOLESTEROL MICELLE

Balls of fatty molecules, grouped with water-loving (hydrophilic) parts on the outside, and water-repellent (hydrophobic) parts inside. Hydrophobic fatty substances such as cholesterol are carried in the core.

CARDIAC CYCLE

The heart is a two-sided muscular pump. The right side of the heart receives deoxygenated (oxygen-poor) blood from the body and pumps it to the lungs, where it is topped up with oxygen. The left side receives oxygenated (oxygen-rich) blood from the lungs and pumps this around the body.

PUMPING HEART

The heart combines two separate pumps within a single organ—one for oxygenated blood (left), and one for deoxygenated (right). When at rest, it beats on average 100,000 times per day. Every heartbeat involves the coordinated contraction (systole) and relaxation (diastole) of the heart's four chambers. These regulated muscular pulses transfer blood from the upper two chambers (atria) into the lower two (ventricles) via a system of valves, and from there eject it from the heart through the aorta and the pulmonary artery. Known as the cardiac cycle, this process divides into five key stages (see opposite).

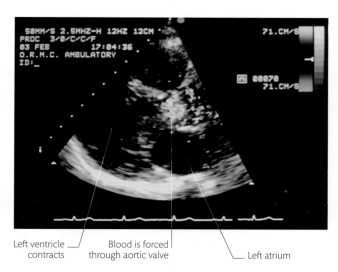

Left ventricle contracts

Blood is forced through aortic valve

Left atrium

Cardiac echo
Echocardiography (or echo) produces an ultrasound of the heart, visually recording the real-time movement of blood through its four chambers. Echo reveals any abnormalities of the valves or of the pumping ability of the heart.

Pulmonary veins carry blood from the lungs

Cardiac cycle
Contraction of the heart muscle occurs in response to electrical activity within the cardiac conducting system (see p.352). Under normal circumstances this electrical activity follows a strict pattern, with contractions of the heart chambers following suit. Despite this regulation, the heart can easily respond to the demands of the body by altering the rate, as well as the force, of its contractions.

CARDIAC MUSCLE

Cardiac muscle (myocardium) can be distinguished from the other types of muscle (skeletal and smooth) by its appearance and behavior. Apart from being branched, cardiac muscle fibers look similar to skeletal muscle, yet they behave very differently.

Striated muscle
A colored micrograph shows pink muscle fibers and oval mitochondria.

The divisions between cardiac muscle cells are highly permeable, allowing electrical impulses (action potentials) to flow quickly and easily between cardiac muscle cells so that all of the cells in an area of muscle can contract as one. Cardiac muscle also contains large numbers of energy-producing mitochondria, meaning that it doesn't fatigue, unlike skeletal muscle.

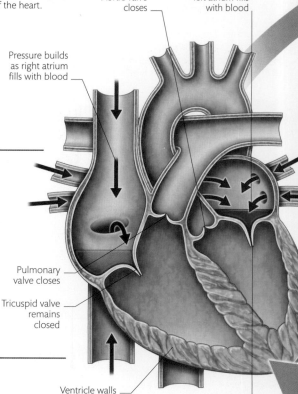

Aortic valve closes

Pressure builds as left atrium fills with blood

Pressure builds as right atrium fills with blood

Pulmonary valve closes

Tricuspid valve remains closed

Ventricle walls relax

Mitral valve remains closed

HEART VALVES

Four heart valves, two at the exit of the atria and two at the exit of the ventricles, prevent blood from flowing backward into the heart chambers. They open or close passively depending on the pressure of the blood surrounding them. If the blood pressure behind the valves is greater than that in front of them they will open; if the pressure in front is greater, they will close—the closing of the valves is what creates the familiar "lub-dub" sound of a heartbeat. The mitral and tricuspid valves located between the

Heart

atria and the ventricles have specialized attachments called papillary muscles and chordae tendineae. These prevent the valves from opening backward into the atria when ventricular pressure rises.

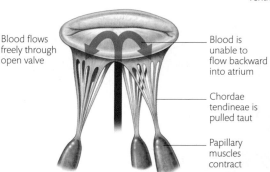

OPEN VALVE

Blood flows freely through open valve

CLOSED VALVE

Blood is unable to flow backward into atrium

Chordae tendineae is pulled taut

Papillary muscles contract

Held tight
Papillary muscles contract along with the ventricle, pulling taut the chordae tendineae (attached to the valve) in order to keep the valve tight shut.

5 ISOVOLUMIC RELAXATION
Isovolumic relaxation is the earliest phase of diastole. The ventricles start to relax and the pressure of blood within them falls to below that of the blood in the aorta and pulmonary artery; therefore the aortic and pulmonary valves both close. However, the pressure in the ventricles is still too high to allow the mitral and tricuspid valves to open.

Valves and pressure
Ventricular pressure decreases so the pulmonary and aortic valves close, yet it is not low enough for the mitral and tricuspid valves to open.

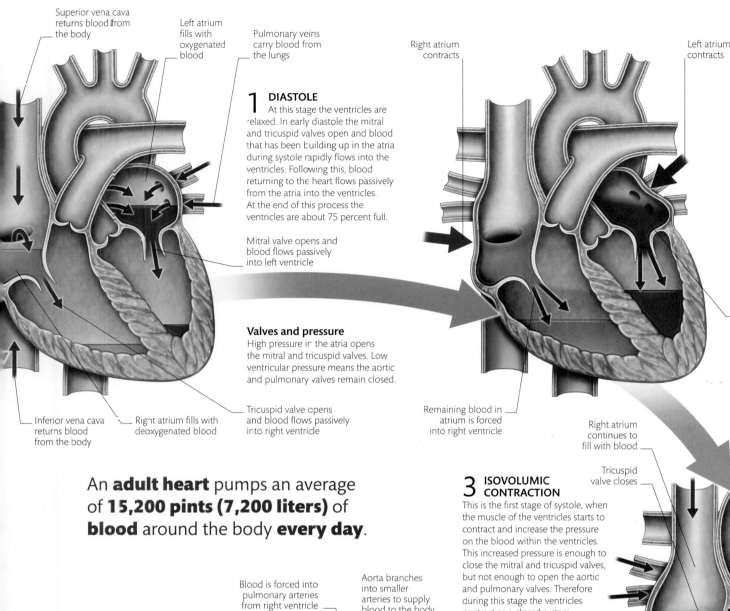

Superior vena cava returns blood from the body

Left atrium fills with oxygenated blood

Pulmonary veins carry blood from the lungs

Right atrium contracts

Left atrium contracts

1 DIASTOLE
At this stage the ventricles are relaxed. In early diastole the mitral and tricuspid valves open and blood that has been building up in the atria during systole rapidly flows into the ventricles. Following this, blood returning to the heart flows passively from the atria into the ventricles. At the end of this process the ventricles are about 75 percent full.

Mitral valve opens and blood flows passively into left ventricle

2 ATRIAL SYSTOLE
The right and left atria contract simultaneously, forcing any remaining blood into the ventricles, which are still relaxed, through the mitral and tricuspid valves. After atrial systole the ventricles are full, yet the contraction of the atria has only contributed to 25 percent of this volume.

Valves and pressure
Even higher pressure in the now contracting atria keeps the mitral and tricuspid valves open. The aortic and pulmonary valves remain closed.

Valves and pressure
High pressure in the atria opens the mitral and tricuspid valves. Low ventricular pressure means the aortic and pulmonary valves remain closed.

Inferior vena cava returns blood from the body

Right atrium fills with deoxygenated blood

Tricuspid valve opens and blood flows passively into right ventricle

Remaining blood in atrium is forced into right ventricle

Remaining blood in atrium is forced into left ventricle

An **adult heart** pumps an average of **15,200 pints (7,200 liters)** of **blood** around the body **every day**.

Pulmonary valve remains closed

Left atrium continues to fill with blood

Right atrium continues to fill with blood

Tricuspid valve closes

Mitral valve closes

3 ISOVOLUMIC CONTRACTION
This is the first stage of systole, when the muscle of the ventricles starts to contract and increase the pressure on the blood within the ventricles. This increased pressure is enough to close the mitral and tricuspid valves, but not enough to open the aortic and pulmonary valves. Therefore during this stage the ventricles contract as a closed system.

Valves and pressure
Increased ventricular pressure means the mitral and tricuspid valves close, yet it is not high enough to open the pulmonary and aortic valves.

Blood is forced into pulmonary arteries from right ventricle

Aorta branches into smaller arteries to supply blood to the body

Blood is forced into aorta from left ventricle

Pulmonary arteries carry blood to lungs

4 EJECTION
Eventually the ventricular contraction causes the pressure of the blood within the ventricles to exceed the pressure of the blood in the aorta and pulmonary arteries. At this point the aortic and pulmonary valves are forced open and blood is powerfully ejected from the ventricles. The papillary muscles prevent the mitral and tricuspid valves from opening.

Pulmonary arteries carry blood to lungs

Aortic valve remains closed

Right ventricle begins to contract

Left ventricle begins to contract

Valves and pressure
The aortic and pulmonary valves are forced open by high pressure in the contracting ventricles. The mitral and tricuspid valves remain closed.

Right atrium continues to fill with blood

Left atrium continues to fill with blood

Pulmonary valve opens

Aortic valve opens

Right ventricle contracts fully

Descending aorta

Left ventricle contracts fully

SCIENCE
ARTIFICIAL HEART
Many people die while waiting for heart transplants because there are not enough donors to satisfy demand. Artificial hearts were therefore developed to help these people survive until a heart became available. They may eventually replace transplanted hearts altogether, and allow more patients to live a normal life.

CONTROLLING THE HEART

The heart beats around 70 times per minute, although this varies dramatically throughout the day. Heart rate is finely tuned by nerves and circulating hormones that work to ensure the speed is just right to provide all the cells in the body with the blood that they need.

CARDIAC CONDUCTING SYSTEM

The cardiac conducting system consists of specialized cells that transport electrical impulses through the cardiac muscle in order to trigger its contraction. The impulse for each heartbeat starts in the sinoatrial (SA) node, which is located in the right atrium. It flows rapidly through the atria and causes them to contract (atrial systole). Electricity cannot pass directly between the atria and ventricles; instead it is channeled into the atrioventricular (AV) node, where it is delayed slightly to ensure that the atrial contraction is over before the ventricles start to contract. After leaving the AV node, the electrical impulse rushes through the bundle of His and Purkinje fibers, which are conducting fibers that run through the ventricle walls, to stimulate contraction of the ventricles.

ELECTRICAL ACTIVITY

The heart's electrical activity can be recorded using an electrocardiogram (ECG). Electrodes are positioned on the chest and limbs in such a way that electrical currents in all areas of the heart can be monitored. The recording displays the voltage between pairs of electrodes. In a typical ECG, each heartbeat produces three distinctive waves (P, QRS, and T), showing a regular beat. In addition to recording the heart's rhythm, an ECG can pinpoint the site of any damage that disturbs the flow of electricity, as the waves will form an unusual pattern.

Sinoatrial node
Also called the pacemaker of the heart, the SA node emits an electrical impulse that runs through the atrial walls and stimulates atrial systole. This is what instigates a heartbeat

Right atrium

Currents
Electrical impulses rush through the atrial walls

Atrioventricular node
The electrical current cannot breach the fibrous tissue dividing the atria and ventricles. It enters the AV node and is delayed there for 0.13 seconds, before being quickly propelled through the ventricle walls

Tricuspid valve

Right ventricle

Purkinje fibers

Papillary muscle

Electrical rhythm
Each heartbeat is triggered by the flow of electricity through the muscle in an exact sequence that can be detected using an ECG. Deviations from the horizontal line on the ECG tracing are caused by electrical activity resulting in specific actions within the heart.

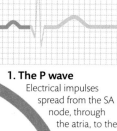

Electrical activity in the SA node instigates atrial systole

SA node prepares for next heartbeat

Electrical impulse

AV node forward electrical impulse to contract ventricles

Electrical impulse recedes as heart resets itself

1. The P wave
Electrical impulses spread from the SA node, through the atria, to the AV node.

3. The T wave
Represents the electrical recovery (repolarization) of the ventricles. Both atria and ventricles relax completely.

2. The QRS complex
Electrical activity continues from the AV node through the ventricles to produce ventricular contraction.

Conductors of the heart
Both the SA and the AV nodes are capable of self-excitation, meaning that the heart will beat without input from the nervous system—nerves regulate, rather than instigate, the heartbeats (see opposite). The SA node sets the heart's rhythm, but if the impulse from the atria is blocked, the AV node can stimulate the ventricles to contract.

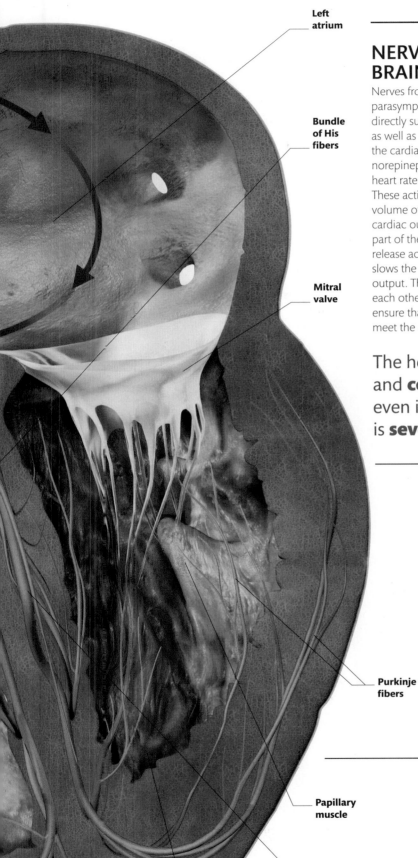

Left atrium

Bundle of His fibers

Mitral valve

Purkinje fibers

Papillary muscle

Bundle of His and Purkinje fibers
These specialized conducting fibers transport electrical impulses extremely rapidly throughout the ventricle walls to ensure that all the muscle cells in the ventricles contract almost simultaneously

Left ventricle

NERVE AND BRAIN CONTROL

Nerves from both the sympathetic and parasympathetic nervous systems (see p.311) directly supply the cardiac conducting system, as well as being widely distributed throughout the cardiac muscle. Sympathetic nerves release norepinephrine, which can increase both the heart rate and the force of muscle contraction. These actions considerably increase the volume of blood that the heart ejects (the cardiac output). The vagus nerves, which form part of the parasympathetic nervous system, release acetylcholine, a chemical that conversely slows the heart rate, thus reducing the cardiac output. These opposing systems complement each other to regulate the heart muscle and ensure that sufficient blood is pumped to meet the demands of the body.

The heart is **self-excitable** and **continues to beat** even if its nerve supply is **severed** completely.

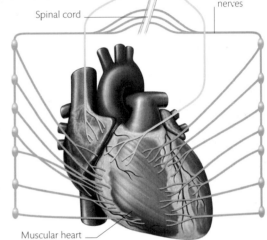

Nerve supply
Parasympathetic nerve supply to the heart, from the vagus nerves, begins in the medulla oblongata (brain stem). Sympathetic supply is from the spinal cord.

Medulla oblongata

Vagus nerves (parasympathetic)

Spinal cord

Sympathetic nerves

Muscular heart

BLOOD SUPPLY

The heart is the most active muscle in the body and needs a constant supply of blood to deliver oxygen and nutrients to its cells and remove their waste. Although the heart chambers are always full of blood, this cannot reach all the cells of its thick walls, so the heart has its own blood vessels: the coronary circulation. The coronary arteries that supply the heart are forced shut under the pressure of the contracting muscle. They therefore can only fill when the heart is relaxed, during diastole.

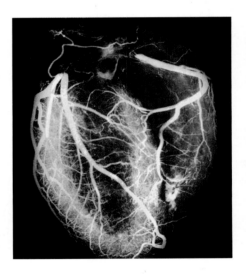

Vital supply
A colored angiogram shows large coronary arteries branching into a network of smaller blood vessels that supply the heart.

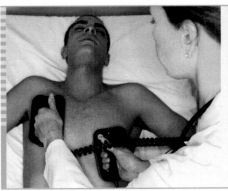

SCIENCE
DEFIBRILLATOR

Defibrillators can deliver electric shocks to kick-start a heart that has stopped beating properly. They are also used to treat abnormal heart rhythms, where the heart cells contract in a haphazard way. The external dose of electricity causes all the heart cells to contract at once, which resets them and allows them to resume working in a coordinated manner. These machines can be external, as shown, but they can also be implanted into patients who are susceptible to abnormal heart rhythms.

BLOOD VESSELS

Blood vessels are a network of branching tubes that join together to form part of the circulatory system. They can dilate or contract to adjust blood flow and in this way finely tune the blood supply to organs, as well as assist with thermoregulation.

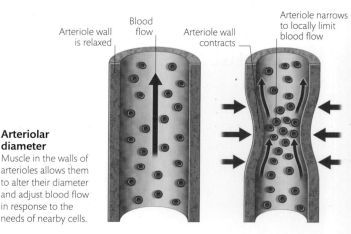

Arteriolar diameter
Muscle in the walls of arterioles allows them to alter their diameter and adjust blood flow in response to the needs of nearby cells.

Arteriole wall is relaxed

Blood flow

Arteriole wall contracts

Arteriole narrows to locally limit blood flow

BLOOD VESSELS

Great variation in the size and structure of blood vessels allows each to perform a specific task. Arteries (the largest) carry oxygenated blood away from the heart. They expand to fill with blood and then propel it forward as they return to their normal diameter. Less muscular, veins return deoxygenated blood to the heart, via a series of valves. Capillaries, the smallest vessels, are the site of gas exchange (see pp.340–41). Their walls are just one cell thick to allow easy gas diffusion. The smallest is just 7μm in diameter, whereas the diameter of the aorta (the largest artery) is 1 in (2.5 cm), with walls so thick they require their own blood supply.

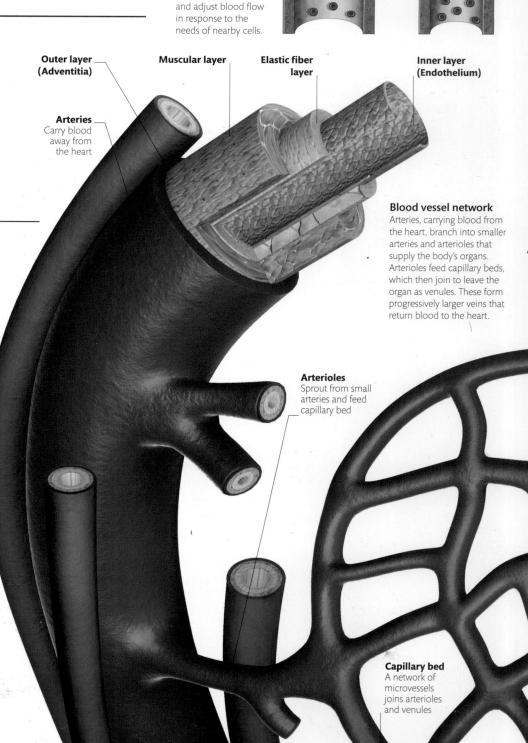

Outer layer (Adventitia)

Muscular layer

Elastic fiber layer

Inner layer (Endothelium)

Arteries
Carry blood away from the heart

Blood vessel network
Arteries, carrying blood from the heart, branch into smaller arteries and arterioles that supply the body's organs. Arterioles feed capillary beds, which then join to leave the organ as venules. These form progressively larger veins that return blood to the heart.

Arterioles
Sprout from small arteries and feed capillary bed

Capillary bed
A network of microvessels joins arterioles and venules

DOUBLE CIRCULATION

The circulation has two main divisions: pulmonary (lungs) and systemic (body). The pulmonary circulation takes blood from the right side of the heart to the lungs, where it is oxygenated and releases carbon dioxide. Blood is then returned to the left side of the heart. The systemic circulation takes the oxygen-rich blood to the body's cells, picks up carbon dioxide and waste products, and returns to the right side.

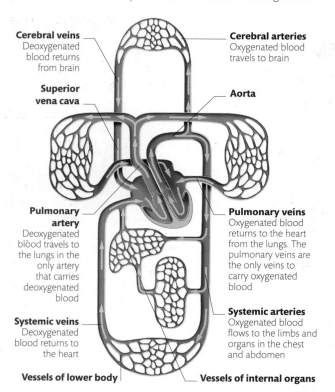

Cerebral veins
Deoxygenated blood returns from brain

Cerebral arteries
Oxygenated blood travels to brain

Superior vena cava

Aorta

Pulmonary artery
Deoxygenated blood travels to the lungs in the only artery that carries deoxygenated blood

Pulmonary veins
Oxygenated blood returns to the heart from the lungs. The pulmonary veins are the only veins to carry oxygenated blood

Systemic veins
Deoxygenated blood returns to the heart

Systemic arteries
Oxygenated blood flows to the limbs and organs in the chest and abdomen

Vessels of lower body

Vessels of internal organs

Multiple blood supplies
The pulmonary and systemic circulatory systems ensure a constant supply of blood to the lungs and to the body. A third system—the coronary circulation—supplies blood directly to the heart itself (see p.353).

THERMOREGULATION

When ambient temperature increases, circulating chemicals signal to blood vessels in the skin to dilate (widen). In this way, warm blood is diverted to the skin, where it can lose its heat to the surrounding air, thus cooling the body. When the temperature falls, blood vessels constrict so the skin loses less heat, and therefore essential warmth is retained in the core of the body, where the vital organs are. This mechanism helps to keep the body temperature at a constant level of around 98.6° F (37° C).

Thermal imaging
On the far right, a thermal scan shows a hot hand that radiates red heat, as warm blood flows through its vessels. On the near right, the hand is cold, blood flow through the vessels is reduced, and less heat is radiated (blue).

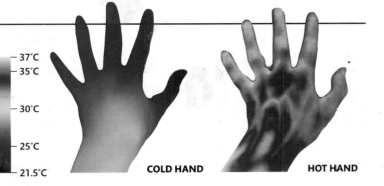

COLD HAND   HOT HAND

Vein valves
Vein pressure only reaches 5–8mmHg (millimeters of mercury); therefore a one-way valve system is in place to keep blood from flowing backward under the force of gravity.

Open valve · Blood flows upward · Closed valve · Blood cannot flow back

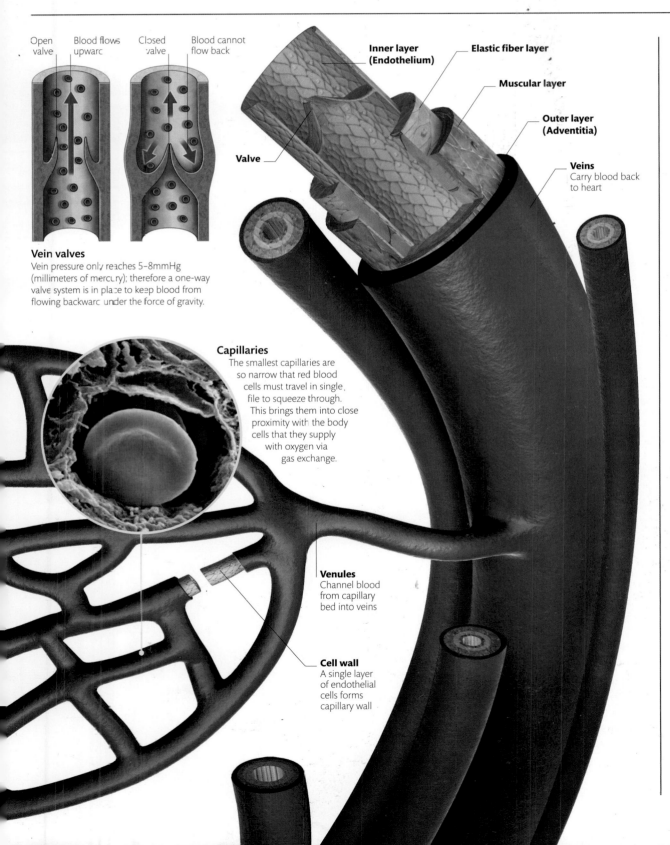

Inner layer (Endothelium)

Elastic fiber layer

Muscular layer

Outer layer (Adventitia)

Valve

Veins
Carry blood back to heart

Capillaries
The smallest capillaries are so narrow that red blood cells must travel in single file to squeeze through. This brings them into close proximity with the body cells that they supply with oxygen via gas exchange.

Venules
Channel blood from capillary bed into veins

Cell wall
A single layer of endothelial cells forms capillary wall

SKELETAL MUSCLE PUMP

Pressure in the veins is too low to actively pump blood back to the heart against gravity. Therefore, veins have to rely on pressure from their surrounding tissues to squeeze blood back toward the heart. In the chest and abdomen, organs such as the liver perform this task. In the limbs, the contraction and relaxation of muscles during movement effectively "pumps" blood toward the heart.

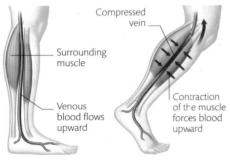

Compressed vein

Surrounding muscle

Venous blood flows upward

Contraction of the muscle forces blood upward

RELAXED MUSCLE   **CONTRACTED MUSCLE**

Pumping muscles
When the muscle contracts, blood in the vein is squeezed upward. When it relaxes, the one-way valves prevent blood from flowing back down.

BLOOD PRESSURE

Blood pressure, measured in millimeters of mercury (mmHg), refers to the pressure within the arteries. It peaks (systolic pressure) as blood pumps into the arteries. As the heart relaxes, pressure in the vessels falls, but the tone of the artery walls never allow it to reach zero, so blood always flows. This lower pressure is called the diastolic pressure.

Peaks and troughs
A heartbeat has a systolic (peak) and a diastolic (minimum) pressure.

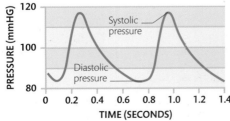

Systolic pressure

Diastolic pressure

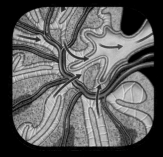

LYMPH NODE

Lymph flows slowly through nodes, where it is filtered. Antibodies are made in nodes, which enlarge during infection.

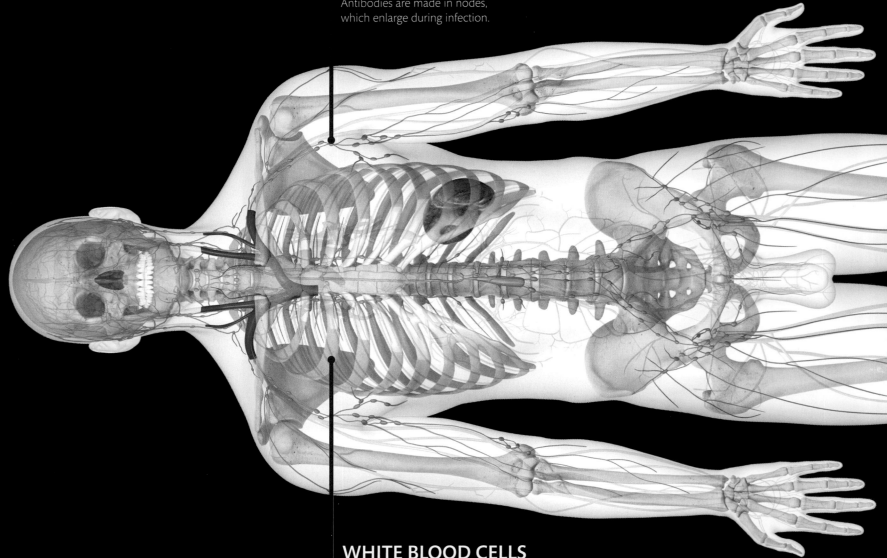

WHITE BLOOD CELLS

White blood cells are produced in bone marrow. The chief immune cells, lymphocytes, are stored in the spleen and lymph nodes.

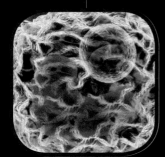

VESSEL

Thin-walled lymph vessels are valved and work as a s milar way as veins, transporting clear lymph fluid around the body.

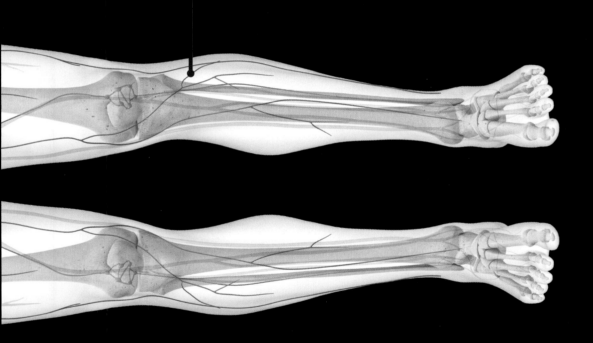

LYMPHATIC AND IMMUNE SYSTEM

Running in parallel with the blood's circulation, the lymphatic system collects excess tissue fluid from the body (via a network of lymph nodes and lymph vessels) and returns it to the blood. This system has vital immune functions.

LYMPHATIC SYSTEM

The lymphatic system is a network of vessels and ducts, with associated lymph nodes, that collects and drains fluid from body tissues. It has important roles in maintaining tissue fluid balance, dietary fat absorption, and the functioning of the immune system.

LYMPHATIC CIRCULATION

The lymphatic circulation, closely linked to the blood circulation, plays a key role in draining fluid from body tissues. Delivery of nutrients to body cells and the elimination of waste products via the blood is not a direct process, but occurs by means of the interstitial fluid, which is derived from blood plasma (see below) and bathes the cells of the tissues. The lymphatic system prevents a buildup of this fluid by collecting and returning it to the blood, via a series of vessels found throughout the body. Once it has entered the lymphatic circulation it is referred to as lymph. Lymph re-enters the blood via ducts that drain into the left and right subclavian veins (see right).

The lymphatic system also forms the basis of an effective surveillance network for the body's immune cells (white blood cells) that monitor tissues for signs of infection. These cells move, via lymph, through lymph nodes located throughout the body (see opposite).

Right lymphatic duct
Lymph drains into blood at junction of right internal jugular and subclavian veins

Thoracic duct
Lymph drains into blood at junction of left internal jugular and subclavian veins

Drainage of the body
The right lymphatic duct drains fluid from the right side of the head and neck, the right arm, and part of the thorax. The remainder of the body is drained by the thoracic, or left lymphatic, duct.

MOVEMENT OF LYMPH

Fluid components of blood plasma, containing nutrients, hormones, and amino acids, filter out of the blood through the capillary walls, and enter the interstitial spaces of body tissues. This interstitial fluid is secreted faster than it can be reabsorbed. Blind-ended channels, called initial lymphatics, allow the excess fluid to drain into the lymphatic system, via one-way valves, forming lymph. White blood cells also migrate into the system in this way.

The initial lymphatics drain into the main lymphatic vessels, which carry the lymph around the body. These vessels have contractile walls that aid the forward movement of lymph, and bicuspid valves that prevent reversal of flow as lymph circulates around the body.

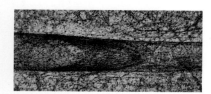

Vessel valves
A bicuspid valve (left) permits one-way fluid flow. Reverse lymph flow causes it to shut.

Fluid pressure
When the pressure of fluid outside the initial lymphatic is greater than the pressure of fluid within it, the valve in the vessel wall opens, allowing interstitial fluid to drain through, forming lymph.

Right subclavian vein

Right internal jugular vein
Left internal jugular vein

Left subclavian vein

Drainage of right lymphatic duct

Drainage of thoracic duct

Vessels of head and upper body

Blood and lymph
This schematic diagram of the body shows the close links between the blood vessels and their associated lymphatic vessels that enable drainage of body tissues.

Right lymphatic duct

Thoracic (left lymphatic) duct

Left lung

Heart

Vessels of abdominal cavity

Right lung

Vessels in gut permit absorption of fat and fat-soluble vitamins from small intestine

Valve
Allows fluid to enter initial lymphatic

Body cell

Vessels of lower body

Body cells

Interstitial space

Initial lymphatic

Initial lymphatic
Entry point of lymph into lymphatic system

Lymph moves into circulation

Interstitial fluid enters initial lymphatic, carrying white blood cells

Plasma filters out of capillary

LYMPHOID TISSUES AND ORGANS

The primary lymphoid tissues are the thymus and bone marrow, both associated with immune cell generation and maturation. Secondary lymphoid tissues—lymph nodes, spleen, adenoids, tonsils, and gut-associated lymphoid tissue (GALT)—are where adaptive immune responses originate (see pp.362–63). Lymph nodes are integrated with the lymphatic system, while the spleen acts as a lymph node for the blood. Adenoids, tonsils, and GALT are key for generating immune responses at mucosal surfaces.

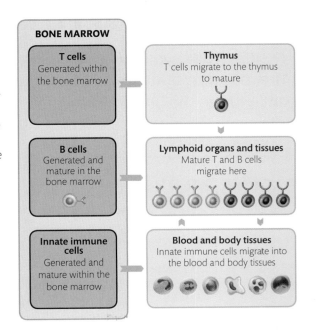

Adenoids
Tonsils
Thymus
Lymph nodes
Bone marrow
Nodes in lungs
Spleen
Gut-associated lymphoid tissue

KEY
- Primary lymphoid tissues
- Lymph nodes and spleen
- Mucosa-associated lymphoid tissue

Guarding the body
The main locations of lymphoid structures show their close links with entry points for infection.

GENERATION OF IMMUNE CELLS

White blood cells, or immune cells (see below), are all produced in the bone marrow. Cells involved in innate immunity (see pp.360–61) migrate to the blood and tissues after maturation. Adaptive immune cells are T and B lymphocytes: T cells mature in the thymus, while B cells mature in the bone marrow. Maturation results in their collective ability to recognize a huge range of specific pathogens (see pp.362–63). Mature lymphocytes migrate to secondary lymphoid tissues, and circulate and scan for infection.

Sites of production
Blood cell generation initially takes place in most bones, but by the time of puberty it is centered on the sternum, vertebrae, pelvis, and ribs.

BONE MARROW

T cells
Generated within the bone marrow

B cells
Generated and mature in the bone marrow

Innate immune cells
Generated and mature within the bone marrow

Thymus
T cells migrate to the thymus to mature

Lymphoid organs and tissues
Mature T and B cells migrate here

Blood and body tissues
Innate immune cells migrate into the blood and body tissues

LYMPH FILTERING

Lymph nodes are small, encapsulated structures that filter passing lymph. They are home to cells of the immune system, primarily T and B lymphocytes but others, such as dendritic cells, are also present. B cells are concentrated in the outer cortex, while T cells are found more centrally in the inner (paracortical) region. Lymph enters through afferent lymphatic vessels, and exits via efferent vessels. As lymph travels through the node it is screened for signs of infection by immune cells. A pathogen may simply flow into the node via the lymph, or it may be actively carried in by another immune cell and presented to resident lymphocytes. Recognition of infection will result in an adaptive immune response (see p.362–63). Numerous lymph nodes are positioned at intervals along draining lymphatic vessels, enabling them to monitor particular regions of the body.

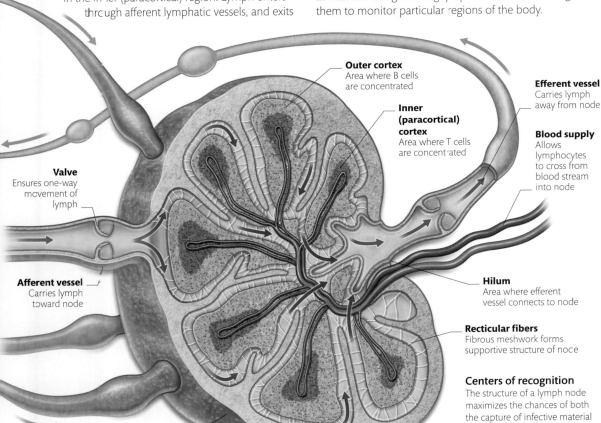

Outer cortex
Area where B cells are concentrated

Inner (paracortical) cortex
Area where T cells are concentrated

Efferent vessel
Carries lymph away from node

Blood supply
Allows lymphocytes to cross from blood stream into node

Valve
Ensures one-way movement of lymph

Afferent vessel
Carries lymph toward node

Hilum
Area where efferent vessel connects to node

Recticular fibers
Fibrous meshwork forms supportive structure of node

Capsule
Fibrous casing for lymph node

Centers of recognition
The structure of a lymph node maximizes the chances of both the capture of infective material carried in the lymph, and also of its exposure to immune cells —in particular T and B cells.

IMMUNE CELLS

White blood cells carry out immune responses. The many different types reflect their varied roles in combating infection. Immune cells broadly divide into two groups: innate cells respond similarly to all infections; adaptive cells respond to specific pathogens toward which they generate immunity.

Monocyte (innate)
Precursor immune cell, found in the blood. Migrates to the tissues where it differentiates into both macrophages and dendritic cells.

Neutrophil (innate)
Phagocytic cell. Often the first immune cell to reach an infection site, these are short-lived and engulf microbes via phagocytosis (see p.361).

Macrophage (innate)
Phagocytic cell, often resident in tissues. Able to promote adaptive immune responses via interactions with lymphocytes.

Natural killer cell (innate)
Cytotoxic cell. Specialized for targeting intracellular pathogens (those living inside body cells) as well as malignant tumor cells.

Mast cell / Basophil (innate)
Inflammatory cells. When activated they release inflammatory factors that promote an immune response. Also responsible for allergic reactions.

Eosinophil (innate)
Inflammatory cell. Specialized for targeting larger pathogens such as parasitic worms. Associated with allergic reactions.

Dendritic cell (innate)
Primary antigen-presenting cell (see p.362). They present material linked to infection to lymphocytes to promote adaptive immune responses.

T and B lymphocytes (adaptive)
Key cells of the adaptive system. T cells target body cells infected with specific pathogens. B cells secrete antibodies that mark microbes for destruction.

INNATE IMMUNITY

The specialized cells and molecules of the innate immune system, supported by barrier immunity, respond rapidly to the typical signs of infection produced when pathogens gain entry to the body. Although highly effective, innate immunity relies upon the recognition of generalized pathogen characteristics and may not be effective against all infections.

BARRIER IMMUNITY

A key strategy in keeping the body free from infection is to prevent the entry of harmful organisms in the first place. Barrier, or passive, immunity acts as a first line of defense against pathogens, providing protection via the physical and chemical barriers presented by the various surfaces of the body. These include both external surfaces, for example, the skin, as well as mucus-lined internal surfaces, for example the airways and the gut.

Each body surface forms a physical barrier to infection, and this is then supplemented by a variety of substances secreted at these barriers that exhibit antimicrobial properties, such as enzymes, which break down bacteria. Additional mechanisms function to expel or flush out microbes from the body, for example, coughing, sweating, and urination.

Tears
Flush the eyes and associated membranes and contain the enzyme lysozyme, which disrupts bacterial cell walls.

Saliva
Flushes the oral cavity, trapping microbes. Contains lysozyme and lactoferrin (antimicrobial agents).

Mucous membranes
Secrete mucus to trap microbes. Cilia (see p.339) line the airways and transport microbes up to the mouth.

Skin
Physically blocks pathogens. Sebaceous secretions contain fatty acids that disrupt microbial membranes.

Stomach acid
Produces very low pH in the stomach that helps to kill many (but not all) microbes present in ingested food.

Urine
Flushes the vessels of the genitourinary system, helping to keep them free of infection.

First line of defense
The body's physical, chemical, and mechanical barriers are maintained constantly and, as such, are a passive means of defense. If they are unable to keep pathogens out of the body, an active immune response takes over.

ACTIVE IMMUNITY

If barrier immunity is breached, for example by a skin wound, and pathogens enter the body, the innate immune system then becomes actively involved. Key to this is the activation of an inflammatory response and the deployment of immune cells (see p.359).

Tissue damage results in inflammation, which helps to prevent microbes from spreading. The capillary walls in the affected area become more permeable, enabling immune cells to easily enter the interstitial fluid and access the infected tissue. Damaged cells release chemicals that attract the immune cells once they have migrated from the blood stream. The first cells to arrive are usually phagocytes (predominantly neutrophils), but other elements, including Natural killer cells (see below) and the complement system (see opposite) may also be engaged. If innate immunity cannot resolve the infection, the adaptive immune system may be set in motion (see pp.362–63).

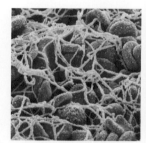
Micrograph of a blood clot
Blood clots (see p.348), seal broken tissues and prevent the entrance of harmful microbes.

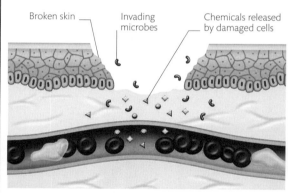

Broken skin — Invading microbes — Chemicals released by damaged cells

Breaching the barrier
Injury to a body surface results in bacteria gaining access to internal tissues. To minimize damage, a defensive inflammatory response is immediately activated as the injured cells release chemicals that attract phagocytes to the scene. Inflammation of body tissue is characterized by four key features: swelling, heat, pain, and redness.

Swollen, red tissue — Phagocytes exit from capillary wall — Phagocytes attack microbes

Inflammatory response
Local blood vessels dilate, allowing more blood to pass through the area. Tissue permeability to blood plasma increases, and the now more-porous capillary enables phagocytes to access the interstitial fluid. The "chemical trail" produced by the damaged tissue then leads them to the site of infection where they attack invading microbes.

INTRACELLULAR INFECTIONS

Natural killer (NK) cells target body cells infected with pathogens. Body cells display surface receptors, called the major histocompatibility complex (MHC), that provide information about the cell's internal environment and indicate when it is infected. NK cells closely monitor these receptors, as infected body cells may avoid displaying them to evade detection. However, NK cells become activated when they detect reduced numbers of MHC on a cell surface and will target such cells for destruction.

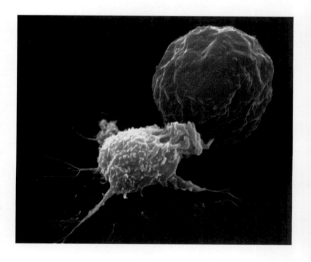

Malignant targets
NK cells are also able to identify and attack malignant cancer cells, as shown in this electron micrograph. The NK cell (white) extends long projections to wrap around the cancer cell (pink).

EXTRACELLULAR INFECTIONS

Fundamental to the innate immune response are cells known as phagocytes (macrophages and neutrophils) that "eat," or engulf, microbes that have infected tissue fluids. This process is known as "phagocytosis." The cell surfaces of bacteria are composed of materials that are different from those of human tissues, and this fact has allowed a system of contact recognition to evolve. Once identified, an invading bacterium is enveloped, absorbed, and then digested by the phagocyte.

Phagocytosis
This series of time-lapse, microscopic images illustrates the process of phagocytosis. The bacterium (green) is identified by the phagocyte (red) via surface contact and has been completely ingested within 70 seconds.

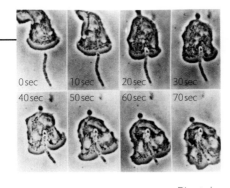

Phagocyte extends pseudopods

Phagolysosome encases bacterium

Phagocyte expels waste products

Digested cellular fragments

Bacterium

Bacterium is gradually digested

Recognition
Recognition of a target bacterium by the phagocyte is achieved on contact of the two cells' surfaces. The phagocyte then extends projections (pseudopods) that engulf and absorb the bacterium.

Digestion
The bacterium is contained within a specialized vesicle called the phagolysosome, in which it is neutralized and broken down by the internal molecular killing mechanisms of the phagocyte.

Expulsion
Aggressive chemical reactions ensure that the bacterium is killed quickly. Digested cellular fragments that cannot be broken down further by the phagocyte are then expelled.

COMPLEMENT SYSTEM

Specialized proteins, together known as the complement system, circulate freely in blood plasma where they target microbes. They are ordinarily present as separate molecules, yet once activated the proteins act together as a "cascade," initiating a complementary chain reaction that attacks and destroys microbes. Like phagocytes, complement proteins can be activated by bacterial surface features, allowing them to easily respond to infections throughout the body, accessing tissues via inflammation (see opposite). They also react to pathogens that have been bound by antibodies (see p.363).

Approach
Bacterial surface proteins activate the complement system, causing the individual proteins to assemble at the cell surface.

Membrane attack
The proteins combine to form the "membrane attack complex" —a structure that punches a hole in the bacterium's surface.

Perforation
The resultant hole allows extracellular fluid to enter the bacterium. This process occurs repeatedly over the cell surface.

Rupture
The combined fluid influx causes the bacterium to swell and eventually rupture.

INFECTIOUS AGENTS

Causes of infection and disease are often microscopic, and broadly divide into five categories. Bacteria and viruses, the smallest and most prevalent, cause many well-known illnesses. Fungi infect the skin and internal mucosa, causing systemic disease in the immunocompromized. Protozoa (single-celled animals with nuclei) cause serious diseases, such as malaria. Parasitic worms infect areas such as the gut, causing debilitating, or even fatal, diseases.

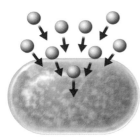

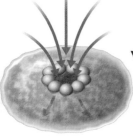
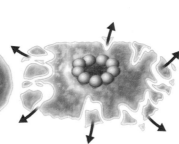
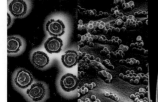

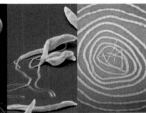

VIRUS **BACTERIUM** **FUNGUS** **PROTOZOAN** **PARASITIC WORM**

FRIENDLY BACTERIA

The human gut represents a huge surface area that is vulnerable to infection. A large population of harmless bacteria that colonize the gut wall form another key barrier to infection. These "friendly" bacteria prevent harmful bacteria from gaining a foothold, and subsequently infecting the body.

ADAPTIVE IMMUNITY

The adaptive immune system provides the body with the means to develop highly specific immune responses to particular pathogens encountered during its life span. Crucially, such responses may be quickly redeployed if a pathogen reinfects.

AGENTS OF SPECIFIC RESPONSE

T and B lymphocytes are the key agents of the adaptive immune response. Unlike innate immune cells, they can recognize and target specific pathogens that enter the body, and are capable of remembering a specific pathogen and acting quickly to eliminate it if it should ever reinfect. T and B cells can attack particular pathogens through their ability to recognize specific molecular targets, called antigens, as foreign. Antigens are recognized via cell-surface receptors displayed by lymphocytes. These receptors are individually programed to recognize a specific antigen.

Two types of T cell—killer, or cytotoxic (attack cells) and helper (coordinating cells)—respond to cellular infections; B cells respond to fluid infections (see opposite). These cells circulate through the body, via the secondary lymphoid tissues, in search of their target antigen.

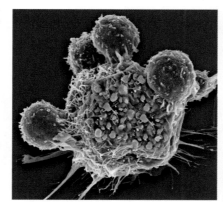

Multiple attack
T cells are able to target body cells that have become malignant, as seen in this micrograph, where four T cells (red) attack a cancer cell (gray).

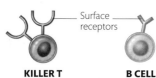

Surface receptors

HELPER T   **KILLER T**   **B CELL**

Maturation of T and B cells
As they mature T and B cells gain receptors that enable them collectively to recognize a huge range of specific antigens. During maturation, any cells that recognize, and may therefore attack, body tissues are eliminated. This usually ensures that antigens that are recognized are foreign in origin.

ANTIGEN PRESENTATION

T cells are only able to recognize an antigen if it is "presented" to them by other immune cells—most commonly dendritic cells, but also macrophages. These are known as antigen-presenting cells (APCs) and are widespread in body tissues. During infection, APCs absorb antigen fragments and migrate, via lymphatic vessels, to local lymph nodes. Here they present the fragment to resident T cells, enabling any with a corresponding receptor to recognize the antigen and launch an attack (see opposite). B cells can interact directly with antigens carried in the lymph, independently of APCs. For adaptive immune cells, the lymphatic system therefore forms a comprehensive surveillance network for the entire body.

Interaction
An electron micrograph captures the remarkable interaction between a T cell (pink) and a dendritic cell (green) that occurs during antigen presentation.

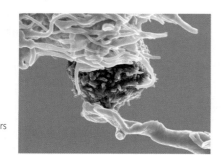

Uptake of antigen
A virally infected body cell bursts, releasing microbial antigen. APCs absorb this antigen for presentation to T cells in the lymph node.

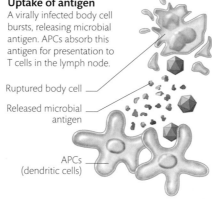

Ruptured body cell

Released microbial antigen

APCs (dendritic cells)

APC presents antigen fragment

T-receptor interacts with antigen

Antigen

T-receptor

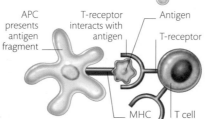

MHC   T cell

Presentation of antigen
An APC presents an antigen to a T cell via a receptor called the major histocompatibility complex (MHC). If the antigen is recognized, the T cell will become activated (see opposite).

CELL-MEDIATED RESPONSE

This immune response targets pathogens that infect body cells, for example viruses. It occurs when an APC bearing a microbial antigen derived from the infected tissue migrates to a lymph node and presents the antigen to a T cell that is able to recognize it. Recognition results in activation of the T cell and triggers a series of reactions that create a swift, coordinated attack. Killer T cells target the infected body cell, while helper T cells produce key signaling molecules that shape the immune response. Only a few T cells of each specificity exist within the body, yet their rapid circulation maximizes their chances of encountering target antigens.

T CELL RECOGNITION
Presentation by the APC in the lymph node results in recognition of the antigen by the killer T cell. If that recognition is confirmed, via signals, by an activated helper T cell nearby, the killer T cell then becomes activated.

APC
Presents antigen to killer T cell

Antigen fragment

Killer T cell
Recognizes antigen

CLONAL EXPANSION
Once activated, the killer T cell undergoes a process of division called "clonal expansion." This involves the production of multiple effector cells and memory cells. Effector cells exit the lymph node to locate and attack the pathogen—the APC will have imprinted the original killer T cell with information about the site of infection, and this is transferred to effector cells. Memory cells stay in the lymph node, but may be activated subsequently to provide a rapid response if the same pathogen reinfects.

Activated killer T cell
Undergoes clonal expansion to produce hundreds of clone T cells

Memory cells
Remain in lymph node to recognize future infections

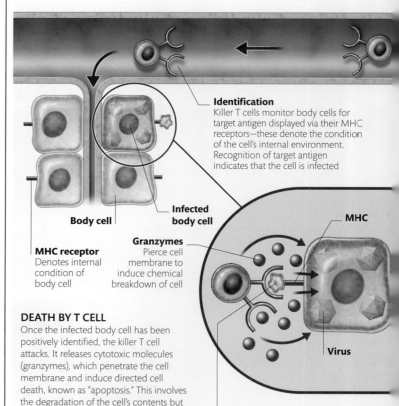

Identification
Killer T cells monitor body cells for target antigen displayed via their MHC receptors—these denote the condition of the cell's internal environment. Recognition of target antigen indicates that the cell is infected

Body cell

Infected body cell

MHC receptor
Denotes internal condition of body cell

Granzymes
Pierce cell membrane to induce chemical breakdown of cell

MHC

Virus

DEATH BY T CELL
Once the infected body cell has been positively identified, the killer T cell attacks. It releases cytotoxic molecules (granzymes), which penetrate the cell membrane and induce directed cell death, known as "apoptosis." This involves the degradation of the cell's contents but without the release of the components, limiting the possible spread of virus particles to neighboring cells.

Microbial antigen
Displayed on cell surface via MHC, and indicates that cell is infected

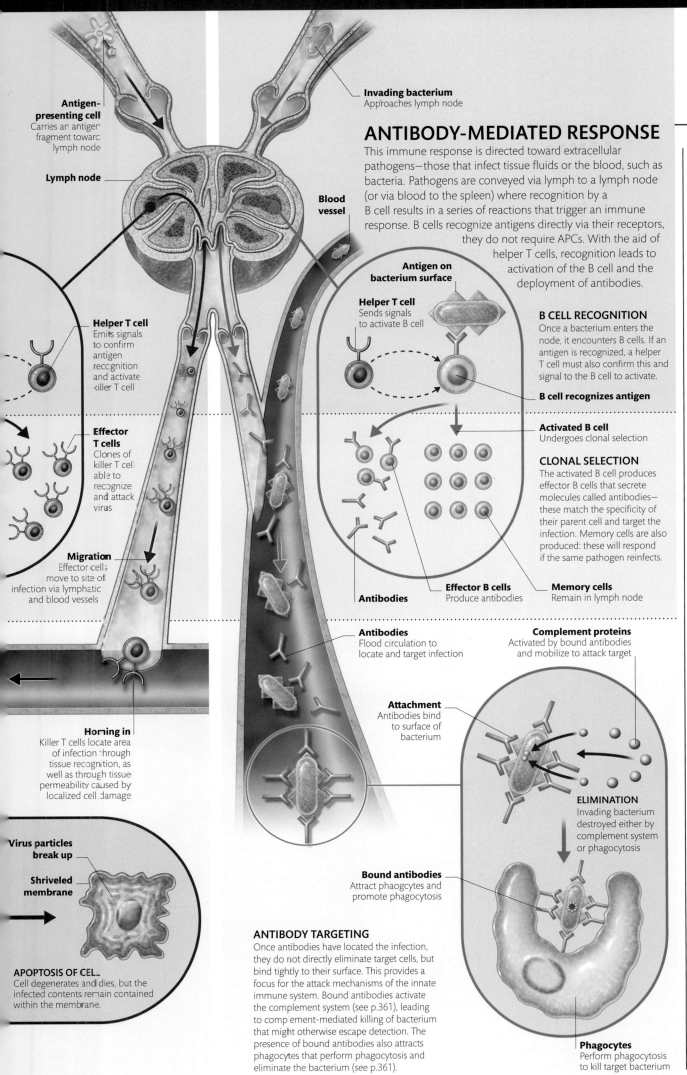

Antigen-presenting cell
Carries an antigen fragment toward lymph node

Lymph node

Invading bacterium
Approaches lymph node

Blood vessel

Helper T cell
Emits signals to confirm antigen recognition and activate killer T cell

Effector T cells
Clones of killer T cell able to recognize and attack virus

Migration
Effector cells move to site of infection via lymphatic and blood vessels

Homing in
Killer T cells locate area of infection through tissue recognition, as well as through tissue permeability caused by localized cell damage

Virus particles break up

Shriveled membrane

APOPTOSIS OF CELL
Cell degenerates and dies, but the infected contents remain contained within the membrane.

ANTIBODY-MEDIATED RESPONSE

This immune response is directed toward extracellular pathogens—those that infect tissue fluids or the blood, such as bacteria. Pathogens are conveyed via lymph to a lymph node (or via blood to the spleen) where recognition by a B cell results in a series of reactions that trigger an immune response. B cells recognize antigens directly via their receptors, they do not require APCs. With the aid of helper T cells, recognition leads to activation of the B cell and the deployment of antibodies.

Antigen on bacterium surface

Helper T cell
Sends signals to activate B cell

B CELL RECOGNITION
Once a bacterium enters the node, it encounters B cells. If an antigen is recognized, a helper T cell must also confirm this and signal to the B cell to activate.

B cell recognizes antigen

Activated B cell
Undergoes clonal selection

CLONAL SELECTION
The activated B cell produces effector B cells that secrete molecules called antibodies—these match the specificity of their parent cell and target the infection. Memory cells are also produced: these will respond if the same pathogen reinfects.

Antibodies

Effector B cells
Produce antibodies

Memory cells
Remain in lymph node

Antibodies
Flood circulation to locate and target infection

Complement proteins
Activated by bound antibodies and mobilize to attack target

Attachment
Antibodies bind to surface of bacterium

ELIMINATION
Invading bacterium destroyed either by complement system or phagocytosis

Bound antibodies
Attract phagocytes and promote phagocytosis

ANTIBODY TARGETING
Once antibodies have located the infection, they do not directly eliminate target cells, but bind tightly to their surface. This provides a focus for the attack mechanisms of the innate immune system. Bound antibodies activate the complement system (see p.361), leading to complement-mediated killing of bacterium that might otherwise escape detection. The presence of bound antibodies also attracts phagocytes that perform phagocytosis and eliminate the bacterium (see p.361).

Phagocytes
Perform phagocytosis to kill target bacterium

IMMUNOLOGICAL MEMORY

The retention of memory cells during adaptive immune responses is central to the development of immunological memory for T and B cells. The disadvantage of initial responses by these lymphocytes is that they are relatively slow to develop, reflecting the time needed for adaptive cells to proliferate and differentiate into effector cells and memory cells. Innate immunity is thus of key importance during an initial infection. If a pathogen reinfects the body, however, it will activate a preformed population of specific cells (the memory cells), which results in a far more rapid secondary response.

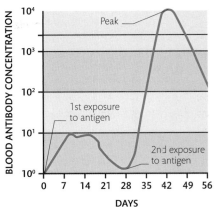

Primary and secondary immune response
This graph illustrates the difference between initial and subsequent exposure to the same pathogen. The secondary response is markedly quicker to develop and much greater in magnitude.

IMMUNIZATION

A vaccine provides an individual with immunity to a disease that has not yet been encountered. It works by mimicking an infection, but doing so safely in order to generate memory cells that are specific. This may involve utilizing microbes that have been killed or attenuated (rendered harmless), or an antigen derived from component parts of the pathogen. These may be given with other chemicals (adjuvants) to make the immune response stronger. This ensures that the primary response develops without the other less desirable aspects of natural infection. If the pathogen is subsequently encountered, then a ready-made memory response, equivalent to a secondary response, is generated, and rapidly clears the infection, often before symptoms develop.

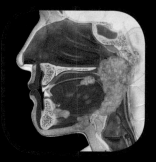

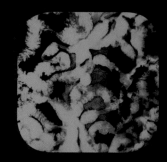

MOUTH

Three pairs of salivary glands secrete 3.1 pints (1.5 liters) of saliva every day, which helps moisten food and makes it easier to swallow.

STOMACH

Acid and enzymes make an environment hostile to bacteria but perfect for the physical and chemical breakdown of food.

SMALL INTESTINE

The highly folded interior of this tube provides a huge surface area of about 3,100 ft^2 (290 m^2), ideal for absorbing nutrients.

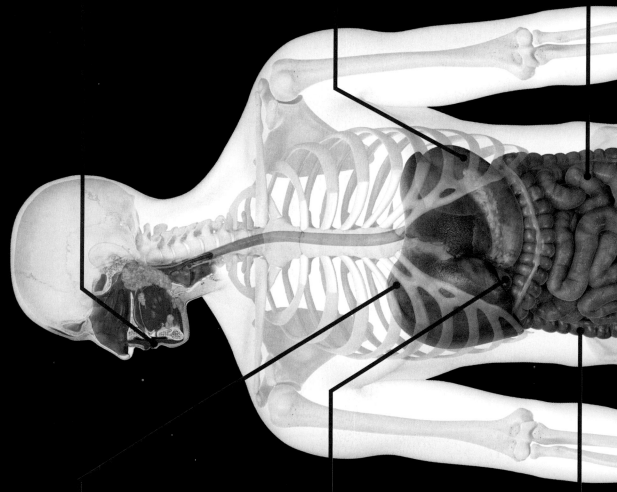

LIVER

This wedge-shaped organ stores certain nutrients and regulates the levels of nutrients in the blood, so that cells receive uninterrupted supplies.

GALLBLADDER AND PANCREAS

Secretions from these organs help break down foods during the first part of digestion in the small intestine.

LARGE INTESTINE

The colon transports indigestible waste from the small intestine—removing water and salts along the way—to the rectum, ready for defecation.

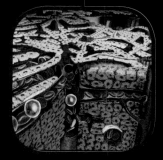

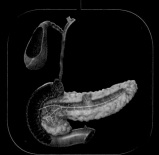

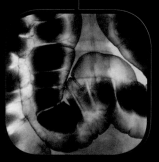

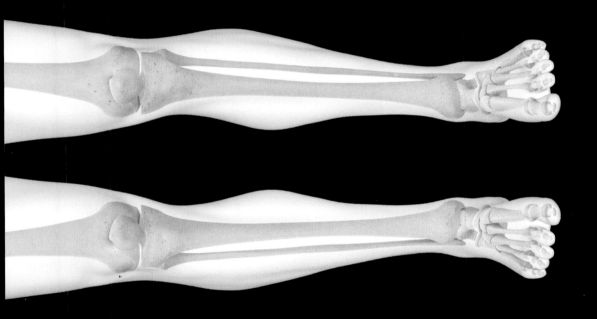

DIGESTIVE SYSTEM

Hunger and thirst prompt us to eat and drink, but after that our digestive system takes care of everything else automatically. As food travels on its journey of digestion, which takes up to two days, it is broken down to release essential nutrients,

MOUTH AND THROAT

Unlike some other animals, humans cannot swallow large chunks of food. It must first be chewed into smaller pieces, an activity that takes place in the mouth. Once chewing has turned food into a slippery pulp, it is pushed into the throat and swallowed, an action that propels it to the stomach.

BITING AND CHEWING

Anchored in sockets in the upper and lower jaws, four types of teeth grasp food by biting it, then chew it into pieces small enough to be swallowed. Chisel-shaped incisors bite and slice; more pointed canines grip and pierce; broad-crowned premolars chew and crush; and broad molars with four cusps (raised edges) bite with great force to grind food into small particles. Biting and chewing is made possible by powerful muscles that elevate the lower jaw to bring opposing sets of teeth into contact.

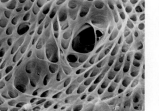

Dentine
This bonelike tissue forms the inner framework and roots of each tooth, and supports the outer enamel.

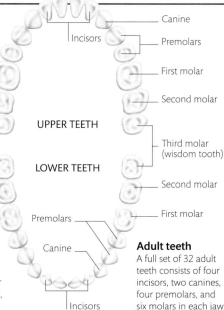

Canine
Incisors
Premolars
First molar
Second molar

UPPER TEETH

Third molar
(wisdom tooth)

LOWER TEETH

Second molar

Premolars
First molar

Canine

Adult teeth
A full set of 32 adult teeth consists of four incisors, two canines, four premolars, and six molars in each jaw.

Incisors

MANEUVRING FOOD

Occupying the floor of the mouth, the tongue is a highly flexible, muscular organ that can change shape and also be protruded, retracted, and moved from side to side. During chewing, the tongue maneuvres food between the teeth, without—usually—being bitten itself, and mixes food particles with saliva. The tongue's upper surface is covered with tiny bumps called papillae that enable the tongue to grip food and contain receptors that detect tastes, heat, cold, and touch. When food has been thoroughly chewed, the tongue compacts it into a mass, or bolus, by pushing it against the roof of the mouth. The tongue then initiates swallowing by pushing the bolus backward into the throat.

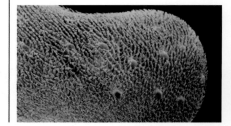

Surface of the tongue
Spiky papillae on the tongue's surface grip food; rounded papillae house taste buds that detect sweet, sour, salty, bitter, and umami (savory) tastes.

10

The **number** of seconds it takes food to travel from the **mouth** to the **stomach**.

This involves introducing a sword at least 15 in (38 cm) long into the upper digestive tract, and requires years of practice. Although the sword takes the same path as food traveling from mouth to stomach, this is different from swallowing food. Practitioners learn to suppress the natural gag reflex that prevents anything, apart from food, entering the throat. They also inhibit involuntary contractions of muscles that push food down the throat and esophagus and learn how to extend the neck to align the mouth, throat, esophagus, and stomach entrance.

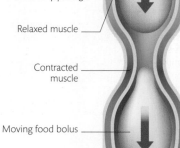

Art of the sword swallower
This X-ray of the upper body shows that there is no trickery involved in legitimate sword swallowing. The head is tilted backward as the sword passes down the throat and esophagus.

SALIVARY GLANDS

Three pairs of salivary glands—parotid, sublingual, and submandibular—are connected to the mouth cavity by ducts through which they release saliva. This is also produced, in small amounts, by tiny glands in the mouth's lining. Saliva is 99.5 percent water, but also contains mucus, the digestive enzyme salivary amylase, and bacteria-killing lysozyme. It is released continuously in amounts sufficient to moisten and clean the mouth and teeth. The taste, smell, sight, or thought of food triggers the release of copious amounts of saliva when hungry. Water and mucus in saliva moisten and lubricate food, making it easier to chew and swallow. Salivary amylase breaks down starch in food into the sugar maltose.

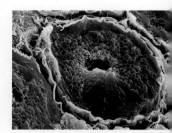

Inside a salivary gland
This acinus inside a salivary gland is a cluster of glandular cells that release saliva into a central duct.

PERISTALSIS

In the last part of swallowing, food is pushed actively down the esophagus, from the throat to the stomach, by a wave of muscular contraction called peristalsis. This is the main means of propulsion in the digestive tract. The wall of the esophagus contains layers of smooth muscle that are under involuntary control. During peristalsis, alternate waves of contraction and relaxation pass down the esophagus to squeeze the bolus of food toward its destination. So powerful is peristalsis that it will propel food to the stomach even if someone is standing on his or her head. At the lower end of the esophagus, the lower esophageal sphincter, normally closed to prevent backflow of food, relaxes to allow food into the stomach.

Movement of food
Smooth muscle in the esophagus wall contracts behind the food bolus to push it downward, and relaxes around and in front of it to allow easy passage.

Relaxed muscle

Contracted muscle

Moving food bolus

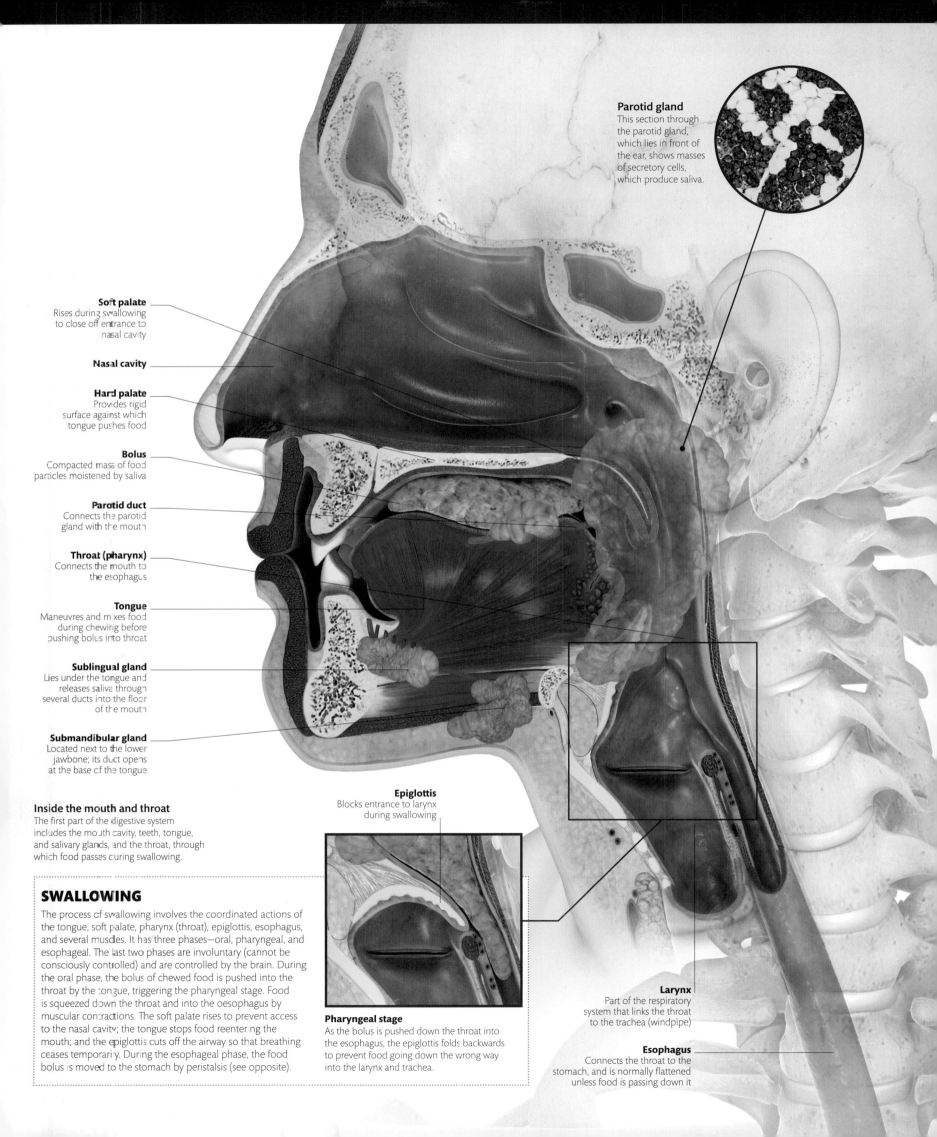

Parotid gland
This section through the parotid gland, which lies in front of the ear, shows masses of secretory cells, which produce saliva.

Soft palate
Rises during swallowing to close off entrance to nasal cavity

Nasal cavity

Hard palate
Provides rigid surface against which tongue pushes food

Bolus
Compacted mass of food particles moistened by saliva

Parotid duct
Connects the parotid gland with the mouth

Throat (pharynx)
Connects the mouth to the esophagus

Tongue
Maneuvres and mixes food during chewing before pushing bolus into throat

Sublingual gland
Lies under the tongue and releases saliva through several ducts into the floor of the mouth

Submandibular gland
Located next to the lower jawbone; its duct opens at the base of the tongue

Epiglottis
Blocks entrance to larynx during swallowing

Inside the mouth and throat
The first part of the digestive system includes the mouth cavity, teeth, tongue, and salivary glands, and the throat, through which food passes during swallowing.

SWALLOWING

The process of swallowing involves the coordinated actions of the tongue, soft palate, pharynx (throat), epiglottis, esophagus, and several muscles. It has three phases—oral, pharyngeal, and esophageal. The last two phases are involuntary (cannot be consciously controlled) and are controlled by the brain. During the oral phase, the bolus of chewed food is pushed into the throat by the tongue, triggering the pharyngeal stage. Food is squeezed down the throat and into the oesophagus by muscular contractions. The soft palate rises to prevent access to the nasal cavity; the tongue stops food reentering the mouth; and the epiglottis cuts off the airway so that breathing ceases temporarily. During the esophageal phase, the food bolus is moved to the stomach by peristalsis (see opposite).

Pharyngeal stage
As the bolus is pushed down the throat into the esophagus, the epiglottis folds backwards to prevent food going down the wrong way into the larynx and trachea.

Larynx
Part of the respiratory system that links the throat to the trachea (windpipe)

Esophagus
Connects the throat to the stomach, and is normally flattened unless food is passing down it

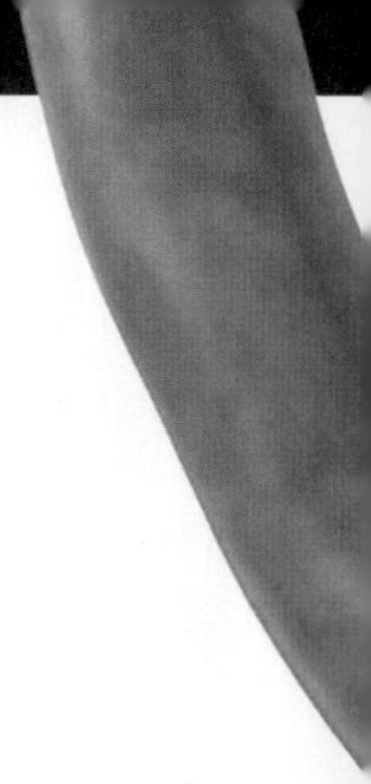

STOMACH

The widest part of the alimentary canal, the stomach is a J-shaped bag linking the esophagus to the first part of the small intestine. It begins the digestive process, churning food and dousing it in gastric juice that contains protein-digesting enzymes.

STOMACH FUNCTIONS

The stomach expands by a considerable amount as soon as food enters it. Two types of digestion happen at the same time in the stomach and together produce a soupy mix of part-digested food called chyme. Chemical digestion is carried out by the enzyme pepsin, contained in acidic gastric juice, initiating the breakdown of protein. Mechanical digestion is carried out by three layers of smooth muscle in the stomach wall, which contract to create waves of peristalsis (see right). This process mixes food with gastric juice, churns it into a liquid, and pushes it toward the pyloric sphincter (muscular opening) at the stomach's exit. The stomach also stores food, releasing chyme through the pyloric sphincter in small amounts to avoid overwhelming the small intestine's digestive processes (see pp.370–71).

A healthy stomach
This colored, contrast X-ray of the stomach shows its upper and lower curves, and the duodenum (top left).

Inside the stomach (below)
The highly elastic stomach wall has three muscle layers arranged at angles to each other. Deep folds in its lining appear when the stomach is shrunken and empty.

GASTRIC JUICE

The gastric mucosa, or stomach lining, is dotted with millions of deep gastric pits that lead to gastric glands. Different types of cells within these glands secrete the various components of the digestive liquid, gastric juice. Mucous cells in the neck of the gland release mucus. Parietal cells release hydrochloric acid, which makes the stomach contents very acidic, activates pepsin, and kills bacteria ingested with food. Zymogenic cells release pepsinogen, the inactive form of pepsin. Enteroendocrine cells release hormones that help control gastric secretion and contraction.

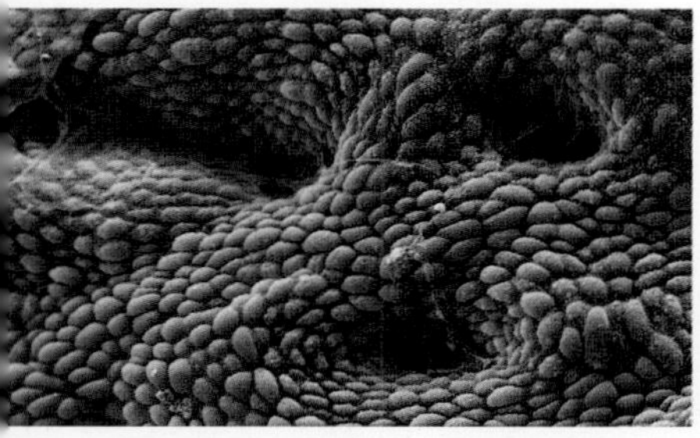

Pyloric sphincter
Ring of muscle that controls the exit to the duodenum

Duodenum
The first short section of the small intestine

Mucus
Coats mucosa and protects it from acidic gastric juice

Mucous cell
Secretes mucus

Gastric mucosa (stomach lining)

Zymogenic cell
Secretes pepsinogen

Stomach lining
This magnified view of the stomach lining, or mucosa, shows its closely packed epithelial cells and the gastric pits (dark holes) that lead to gastric glands.

Parietal cell
Secretes hydrochloric acid

Enteroendocrine cell
Secretes hormones

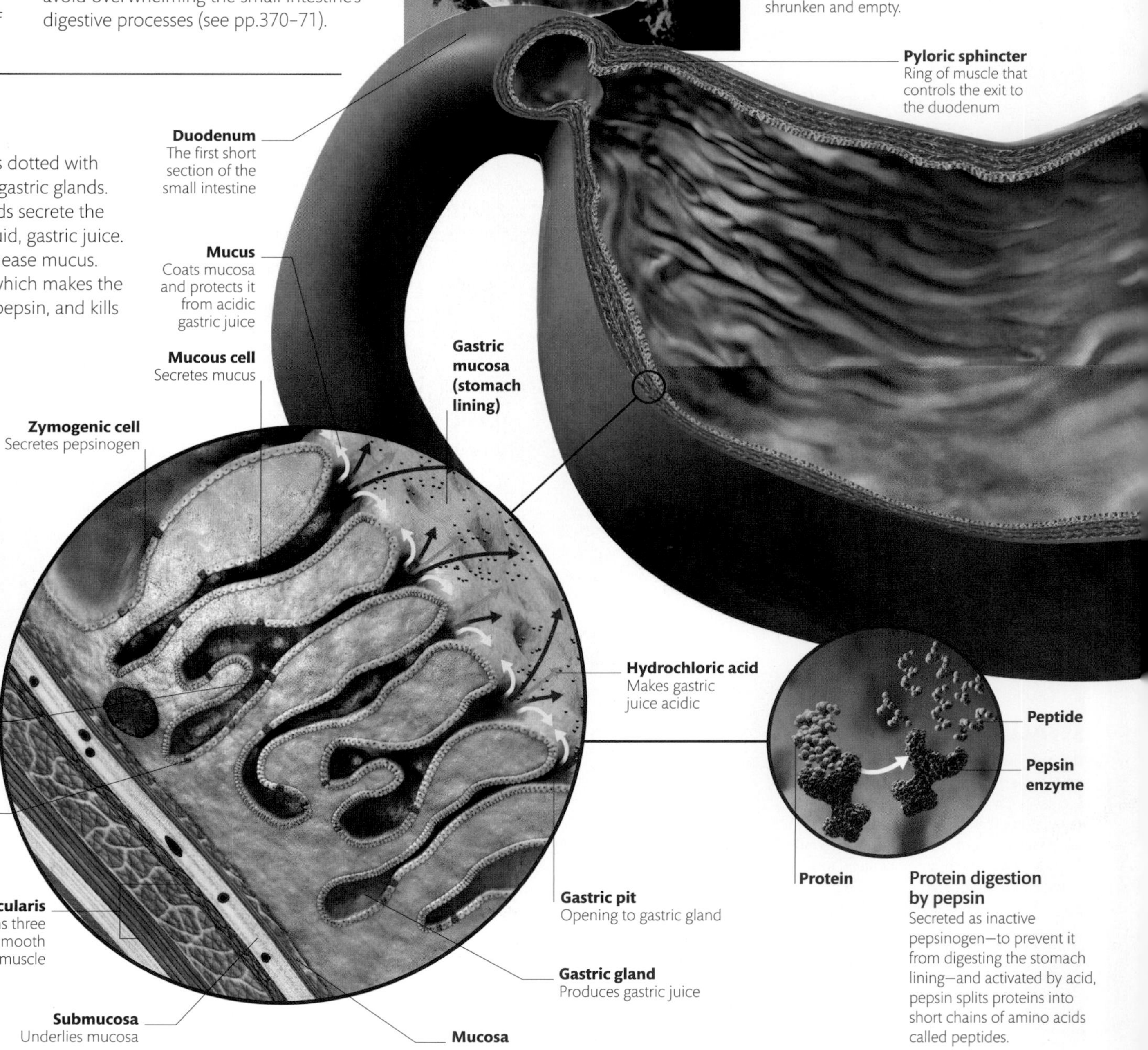

Gastric glands
A section through the stomach wall shows deep gastric glands in the mucosa lining and different secretory cells within those glands. The submucosa connects the three-layered muscularis to the mucosa.

Muscularis
Contains three layers of smooth muscle

Submucosa
Underlies mucosa

Hydrochloric acid
Makes gastric juice acidic

Gastric pit
Opening to gastric gland

Gastric gland
Produces gastric juice

Mucosa

Peptide

Pepsin enzyme

Protein

Protein digestion by pepsin
Secreted as inactive pepsinogen—to prevent it from digesting the stomach lining—and activated by acid, pepsin splits proteins into short chains of amino acids called peptides.

Cardiac sphincter
Prevents the backflow of gastric juice into the esophagus

Longitudinal muscle layer
Runs the length of the stomach

FILLING AND EMPTYING

The stomach expands enormously as it fills with recently chewed food arriving through the esophagus. This food is mixed with gastric juice by peristaltic waves of contractions generated by the three smooth muscle layers in the stomach wall. These waves of contractions gather strength as they push food toward the closed pyloric sphincter, where they become powerful enough to churn food into creamy chyme. Once chyme is liquid and lump-free, the stomach gradually releases it in squirts through the relaxed pyloric sphincter.

Circular muscle layer
Wraps around stomach

Oblique muscle layer
Runs diagonally around stomach

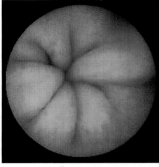

Closed pyloric sphincter
This endoscopic view shows a pyloric sphincter tightly closed to prevent the exit of food into the duodenum while digestion takes place inside the stomach.

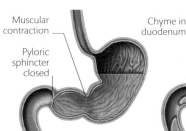

1 During a meal
As the stomach fills, waves of muscular contraction mix food with gastric juice released by gastric glands.

2 1–2 hours after a meal
Food churned by powerful muscular contractions and part-digested by gastric juice is turned into chyme.

3 3–4 hours after a meal
The pyloric sphincter opens slightly at intervals to allow small quantities of chyme into the duodenum.

Rugae
Folds that disappear as the stomach expands with food

Chyme
Creamy liquid produced by digestion of food in the stomach

3
The number of **hours** food spends in the **stomach** before entering the **small intestine**.

REGULATION

The release of gastric juice and the contraction of the stomach wall are regulated by the autonomic nervous system and by hormones released by the alimentary canal. Regulation happens in three overlapping phases: cephalic (head), gastric (stomach), and intestinal. Before eating and during chewing the cephalic phase gives the stomach advance warning that food is on its way. The sight, thought, smell, and taste of food stimulates gastric glands to release gastric juice and triggers peristalsis. When food arrives in the stomach, the gastric phase begins. Gastric juice secretion increases greatly and the waves of peristalsis become much stronger. When semidigested food is released into the duodenum, the intestinal phase inhibits the release of gastric juice and the muscular contractions of the stomach wall.

WHY DO WE VOMIT?

Vomiting can be caused by many factors but is often the result of the stomach being irritated by bacterial toxins. Irritants are detected by receptors in the stomach's lining that send impulses to the vomiting center in the brain stem (the base of the brain). This triggers the vomiting reflex in order to forcibly remove the irritant. During vomiting, the diaphragm and abdominal muscles contract, compressing the stomach so that semidigested food is forced up the esophagus and throat and out of the mouth.

Vomit reflex
The closed pyloric sphincter, soft palate, and epiglottis ensure that food is vomited out through the mouth and does not enter the esophagus or small intestine.

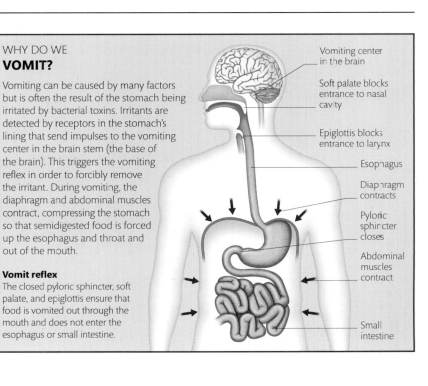

Vomiting center in the brain

Soft palate blocks entrance to nasal cavity

Epiglottis blocks entrance to larynx

Esophagus

Diaphragm contracts

Pyloric sphincter closes

Abdominal muscles contract

Small intestine

SMALL INTESTINE

The longest and most important part of the digestive system, the small intestine's coiled tube fills much of the abdomen. This is where, with the help of the pancreas and gallbladder, food digestion is completed, and where simple nutrients are absorbed into the bloodstream.

HOW THE SMALL INTESTINE WORKS

Extending from the stomach to the large intestine, the small intestine has three parts. The short duodenum receives food from the stomach. The jejunum and ileum, together the longest section of the small intestine, is where the final stages of digestion occur and food is absorbed. Digestion occurs in two phases in the small intestine. First, pancreatic enzymes work inside the small intestine, digesting nutrient molecules as intestinal wall muscles contract to propel food onward by peristalsis. Then enzymes attached to the surface of villi, the millions of fingerlike structures that project from the intestinal lining, complete digestion before the villi absorb digested nutrients.

Muscularis
Contains two muscle layers

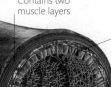

Mucosa
Lining of small intestine

Small intestine wall
The wall of the small intestine has two layers of smooth muscle that mix and propel food along it. Its lining is covered with tiny, fingerlike projections called villi.

23feet

The **length** of the small intestine.

GALLBLADDER AND PANCREAS

These two organs play a key part in digestion in the duodenum, the first part of the small intestine, when semidigested chyme arrives from the stomach. Tucked under the much larger liver, the gallbladder is a small, muscular bag that receives, stores, and concentrates bile, produced by the liver, then releases it along the bile duct into the duodenum where it aids fat digestion. The pancreas produces pancreatic juice, which contains a number of digestive enzymes, and is released along the pancreatic duct that merges with the bile duct before emptying the enzymes into the duodenum.

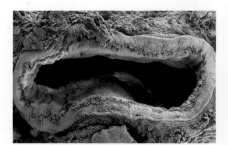

Bile duct
This micrograph image shows a section through the bile duct that carries bile from the gallbladder to the duodenum, absorbing water from the bile.

Pancreas
Secretes pancreatic juice and releases it into the duodenum

Duodenum

Gallbladder
Stores bile and releases it into the duodenum when food arrives from the stomach

Jejunum
The middle section of the small intestine between the duodenum and the ileum

Middle digestive tract
The small intestine, pancreas, and gallbladder make up the central part of the alimentary canal —also known as the middle digestive tract.

Ileum
The longest section of the small intestine

DIGESTION AND ABSORPTION

As food is moved along the jejunum and ileum, digestion continues by the enzymes on the surface of the villi. These tiny projections increase the inner surface area of the small intestine for digestion and absorption by thousands of times. Embedded enzymes such as maltase and peptidase break down, respectively, maltose and peptides to their simplest units, glucose and amino acids. These are absorbed into blood capillaries inside the villi and carried to the liver. Meanwhile, fatty acids and monoglycerides, the result of pancreatic enzyme digestion, are passed into a lacteal or lymph capillary, and despatched to the liver by way of the lymphatic duct and circulatory systems.

Villus projecting from the intestinal wall

Lacteal (lymph capillary)

Capillary network

Artery

Vein

Wall of intestine

Direction of blood flow

Absorption across the villi
The villi of the small intestine provide a massive surface area for the absorption of digestive products. These are shown accumulating in the bloodstream from left to right.

PANCREATIC ENZYMES

Acidic, semidigested liquefied food called chyme arrives in the duodenum, causing the intestinal wall to secrete hormones. These trigger the release of pancreatic juice and bile through a common opening into the duodenum. Alkaline pancreatic juice contains over 15 enzymes, including lipase, amylase, and proteases, that catalyze the breakdown of a range of food molecules. Bile contains bile salts that emulsify large fat and oil droplets into tiny droplets that present a bigger surface area for digestion by lipase. After digestion by pancreatic enzymes, nutrients move to the surface of villi for further digestion and absorption.

Monoglyceride
Lipase Fatty acid

Amylase
Maltose
Starch

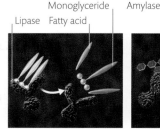

Protease
Protein Peptide

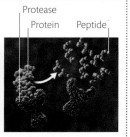

Fat breakdown
After "treatment" with bile salts, fats (triglycerides) are broken down by pancreatic lipase into free fatty acids and monoglycerides (a fatty acid joined to glycerol).

Carbohydrate breakdown
Pancreatic amylase breaks down complex long-chain carbohydrates, such as starch, into disaccharide sugars, such as maltose (two linked glucose molecules).

Protein breakdown
Pancreatic proteases break down proteins into short chains of amino acids called peptides. Peptidases break down peptides into individual amino acids.

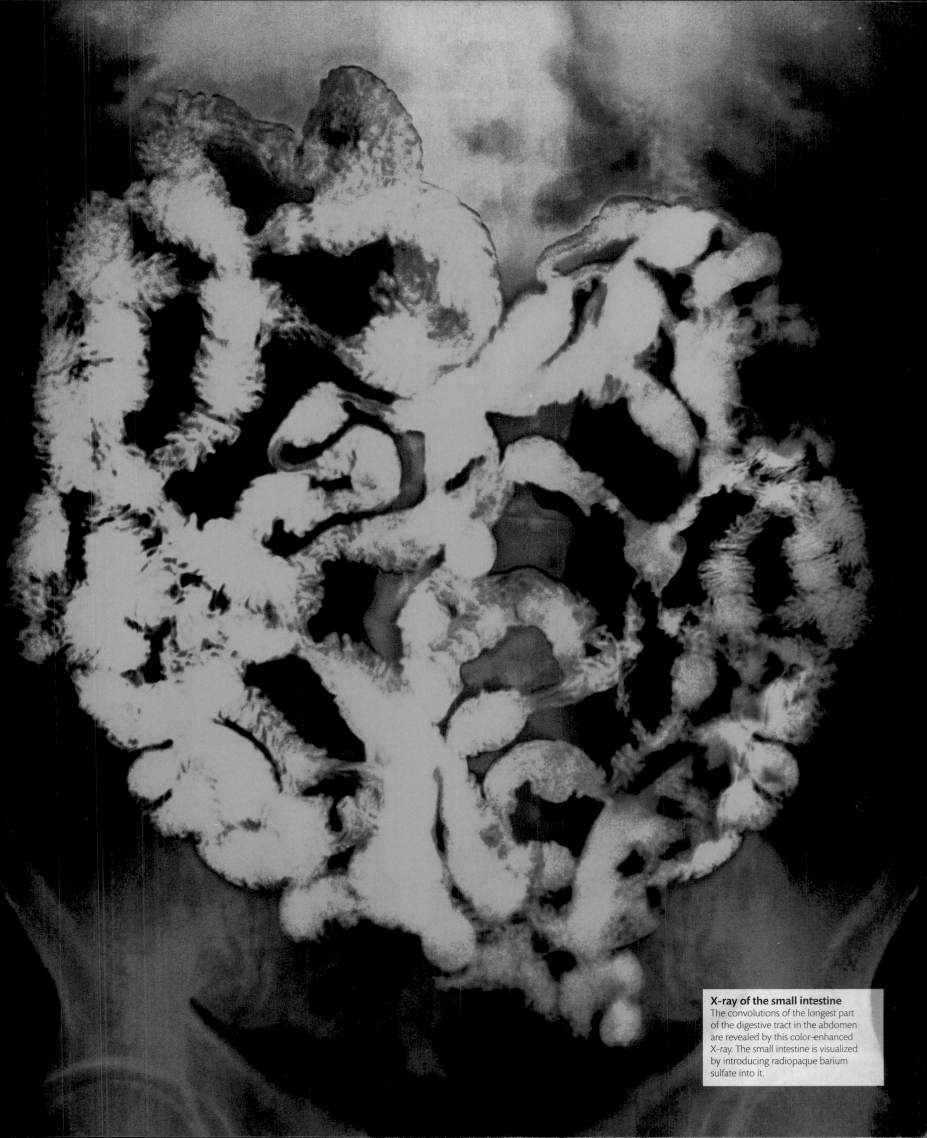

X-ray of the small intestine
The convolutions of the longest part of the digestive tract in the abdomen are revealed by this color-enhanced X-ray. The small intestine is visualized by introducing radiopaque barium sulfate into it.

LIVER

The liver is the body's largest internal organ. It plays a key role in maintaining homeostasis—a stable environment inside the body—by carrying out many metabolic and regulatory functions that ensure the constancy of the blood's composition.

ROLE OF THE LIVER

The deep red color of the liver is an external indicator of what it does—process large volumes of blood to control its chemical composition. Most of the liver's functions, apart from the work carried out by debris-removing Kupffer cells, are performed by hepatocytes, the multitasking cells that are the workhorses of the liver. As blood flows past hepatocytes, they take up nutrients and other substances to be stored, used in metabolic processes, or broken down, and also empty into the blood secretory products and nutrients released from storage. The liver's only direct role in digestion is the manufacture of bile, which is stored in the gallbladder and released into the duodenum. However, once digestion is complete, it "intercepts" nutrients arriving from the intestines and processes them.

SOME LIVER FUNCTIONS

Apart from making bile, controlling the metabolism of carbohydrates, fats, and proteins from food, and storing minerals and vitamins, the liver also, among other things, makes a range of proteins that circulate in blood plasma; breaks down drugs and other dangerous chemicals from the bloodstream; destroys worn-out red blood cells, recycling the iron inside them (see p.348); and removes pathogens and debris in the blood.

Bile production
Hepatocytes produce up to 2 pints (1 liter) of this greenish fluid daily. Bile contains a mixture of bile salts, and wastes, such as bilirubin (from the breakdown of hemoglobin), which are excreted with the feces. Bile salts aid fat digestion in the duodenum, after which they are returned to the liver and secreted again in bile.

Protein synthesis
Liver cells secrete most of the plasma proteins found in blood plasma, using amino acids from digested food or hepatocytes. These proteins include albumin, which helps maintain water balance in the blood; transport proteins, which carry lipids and fat-soluble vitamins; and fibrinogen, for blood-clotting.

Hormone production
The body's chemical messengers, hormones work by changing the activities of target tissues. Once a hormone has exerted its effect it is destroyed; otherwise, it would continue to operate out of control. Many hormones are broken down by liver cells. Their breakdown products are usually excreted by the kidneys in urine.

Heat generation
The vast numbers of metabolic processes occurring in hepatocytes generate, as a by-product, a considerable amount of heat. This heat, together with that from working muscles, is distributed around the body by the blood, keeps the body warm, and enables it to maintain a constant temperature.

STRUCTURE AND BLOOD SUPPLY

Hepatocytes, the functioning units of the liver, are arranged into highly ordered functional units called lobules, each the size of a sesame seed. Within a lobule, sheets of hepatocytes radiate from a central vein. The liver is unusual in having two blood supplies. Oxygen-rich blood delivered by the hepatic artery makes up around 20 percent of its supply. The rest consists of oxygen-poor blood, rich in nutrients and other substances, including drugs, absorbed during digestion, which are transported to the liver along the hepatic portal vein. Inside each liver lobule, blood from both supplies mixes together and is processed as it flows past the massed hepatocytes.

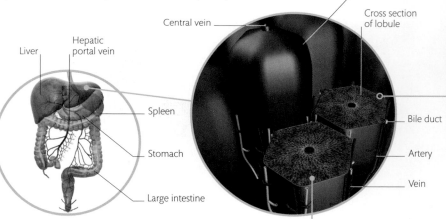

Kupffer cell
Removes bacteria, debris, and old red blood cells from the blood

Liver
Hepatic portal vein
Central vein
Exterior of lobule
Cross section of lobule
Spleen
Stomach
Large intestine
Bile duct
Artery
Vein

Hepatic portal system
A portal system consists of blood vessels with capillary networks at each end. Here, veins from digestive organs, including the intestines and stomach, converge to form the hepatic portal vein that enters the liver.

Inside a liver lobule
Blood flows along sinusoids past hepatocytes to the central vein; bile travels in the opposite direction.

Structure of liver lobules
In section, the tiny liver lobules appear to be six-sided. Running vertically up each corner of the lobule is a threesome of vessels—a tiny vein, artery, and bile duct—that either deliver blood to, or remove bile from, the lobule.

Sinusoid
Receives blood from hepatic portal vein and hepatic artery

Hepatocytes
Process blood and make bile

Central vein
Carries away processed blood to be returned to the heart

Branch of portal vein
Supplies nutrient-rich blood to lobule

Branch of bile duct
Carries bile away from the hepatocytes that make it

Branch of hepatic artery
Supplies oxygen-rich blood to lobule

500
The number of **different** chemical functions the **liver** performs.

KEY

→ movement of nutrient-rich blood

→ movement of oxygen-rich blood

→ movement of bile

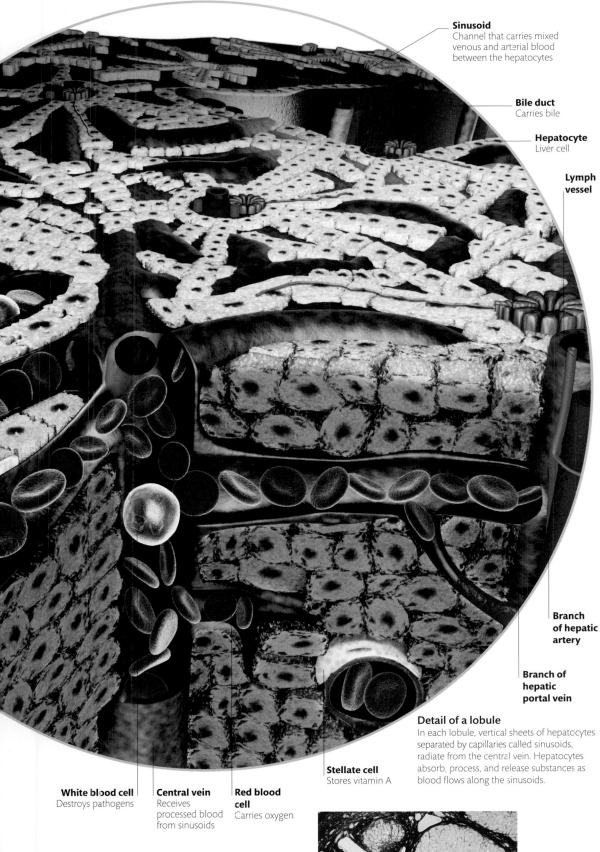

Sinusoid
Channel that carries mixed venous and arterial blood between the hepatocytes

Bile duct
Carries bile

Hepatocyte
Liver cell

Lymph vessel

Branch of hepatic artery

Branch of hepatic portal vein

Detail of a lobule
In each lobule, vertical sheets of hepatocytes separated by capillaries called sinusoids, radiate from the central vein. Hepatocytes absorb, process, and release substances as blood flows along the sinusoids.

Stellate cell
Stores vitamin A

White blood cell
Destroys pathogens

Central vein
Receives processed blood from sinusoids

Red blood cell
Carries oxygen

PROCESSING NUTRIENTS

When nutrients—particularly glucose, fatty acids, and amino acids—flood into the bloodstream following digestion, the liver processes them. Glucose is the body's main fuel source, and its level in the blood must be kept steady. Liver cells gather glucose; they store it as glycogen if blood glucose levels rise and release it from store if levels drop. They also convert excess glucose to fat. The liver breaks down fatty acids to release energy or stores them as fat. It also manufactures packages called lipoproteins to transport fats to and from body cells. It breaks down excess amino acids, using them to release energy and converting their nitrogen into waste urea, which is excreted in urine.

STORING VITAMINS AND MINERALS

Several vitamins, notably vitamin B_{12} and the fat-soluble vitamins A, D, E, and K, are stockpiled by the liver and released when required. The liver can store up to 2 years' supply of vitamin A, and 4 months' worth of vitamins D and B_{12}. Since they are stored, and any excess cannot be excreted, it is important not to overdose on vitamin supplements because the presence of excess fat-soluble vitamins can damage the liver. The liver stores iron, needed to make hemoglobin (see p.341) and copper, which plays a part in many metabolic reactions.

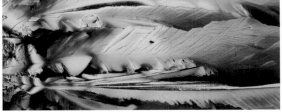

Crystals of vitamin D
This is one of the vitamins stored by liver cells. It is essential for normal absorption of calcium ions, which is needed for bone-building and many other functions, from the small intestine.

RED BLOOD CELL REMOVAL

Defunct red blood cells are destroyed by Kupffer cells, which are macrophages that form part of the lining of sinusoids (red cells are also destroyed in the spleen). Iron is retrieved from one part of the blood cells' hemoglobin molecules, stored by hepatocytes, and reused when required; another part of the hemoglobin molecule is broken down into the bile pigment bilirubin and excreted in bile (see opposite). Kupffer cells also remove bacteria and other debris from blood, and intercept some toxins.

Kupffer cell
This micrograph shows a Kupffer cell (yellow) trapping and "eating" worn-out red blood cells (red) contained in blood (blue) flowing between liver cells (brown).

DETOXIFICATION

While ingested or injected drugs may be helpful to the body in the short term, they are harmful if they remain in the bloodstream. The liver plays a vital role in detoxification by breaking down drugs, bacterial toxins, manmade poisons, and pollutants. Hepatocytes detoxify these harmful substances by converting them into safer compounds that can then be excreted. However, over time, excessive detoxification may, as in the case of alcohol, cause fibrous tissue to develop, which stops the liver from working properly.

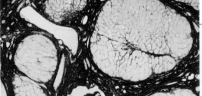

Liver cirrhosis
This section through the liver of an alcoholic person with cirrhosis, shows in liver lobules (white) surrounded by fibrous scar tissue (red) caused by excessive detoxification.

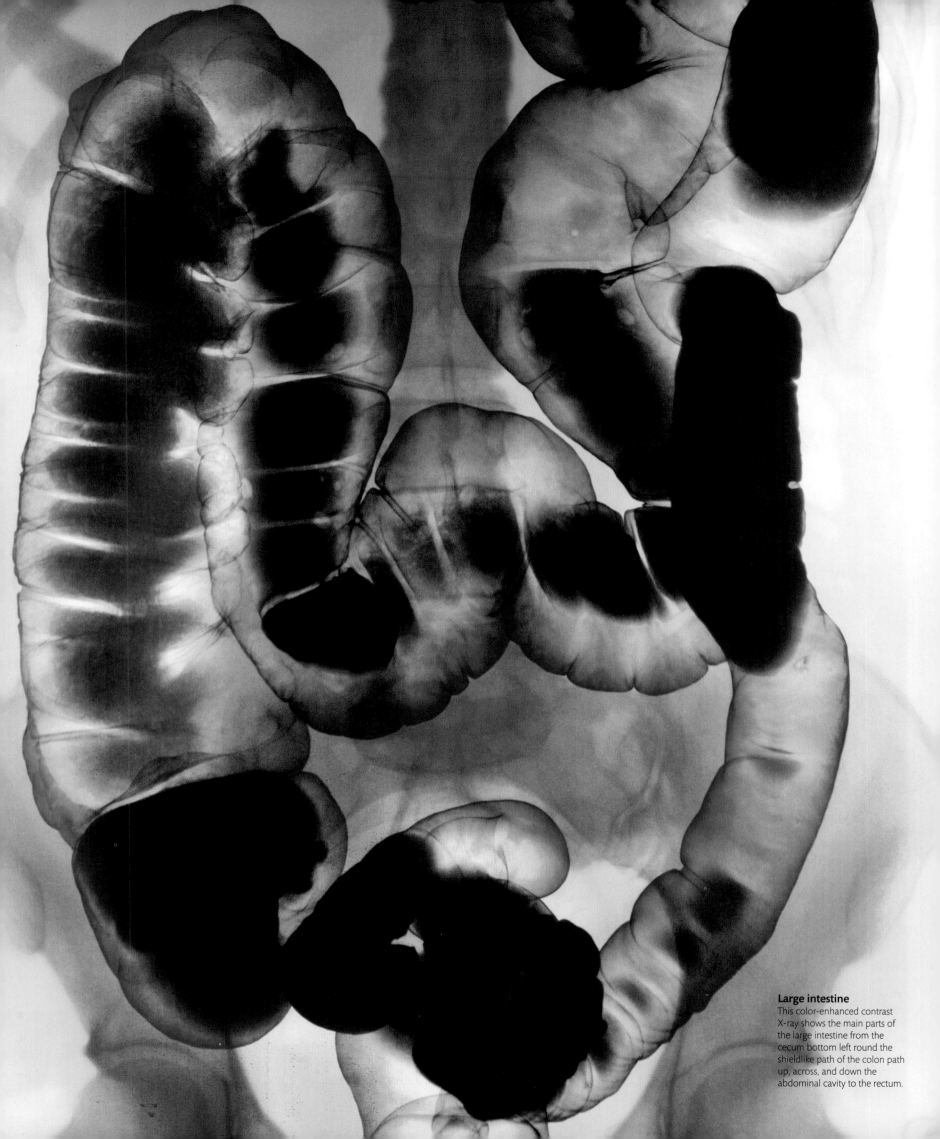

Large intestine
This color-enhanced contrast X-ray shows the main parts of the large intestine from the cecum bottom left round the shieldlike path of the colon path up, across, and down the abdominal cavity to the rectum.

LARGE INTESTINE

This final stretch of the digestive tract is twice the width of the small intestine, although only one-quarter the length. Consisting of the cecum, colon, and rectum, the large intestine processes indigestible waste to form feces.

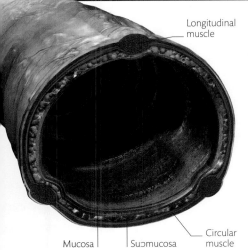

Longitudinal muscle

Mucosa | Submucosa | Circular muscle

FUNCTION OF COLON AND RECTUM

At 5 ft (1.5 m) long, the colon is the longest part of the large intestine. Every day it receives around 3 pints (1.5 liters) of watery, undigested waste from the small intestine. The colon's primary functions are to move this waste so that it can be eliminated from the body, at the same time reabsorbing water and salts—mainly

sodium and chloride ions—through its lining into the bloodstream. This reabsorption of water helps the body maintain its normal water content and avoid dehydration, and also converts the watery waste into solid feces that are easier to move and dispose of. In addition to food waste, feces also contain dead cells, scraped from the intestinal lining, and bacteria, which can make up to 50 percent of fecal weight. At the end of the colon, the rectum stores feces and then contracts to expel them through the anus.

Layers of the colon wall
This section shows the longitudinal and circular muscle layers that produce movements. The mucosa releases mucus to lubricate the passage of feces.

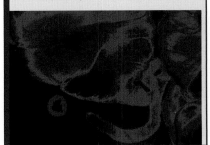

WHY DO WE HAVE AN APPENDIX?

The worm-shaped appendix projects from the cecum, the baglike pouch that is located beneath the point where small and large intestines connect. For many years it was assumed that the appendix was a vestigial organ, one that had a function in our ancient ancestors but is now without purpose, apart from becoming inflamed during appendicitis. More recent research suggests that it contains lymphoid tissue that forms part of the immune system, and that it contains a reservoir of "good" bacteria to repopulate the colon's gut flora should it be flushed away or otherwise destroyed.

COLONIC MOVEMENT

Three types of colonic movement—segmentation, peristaltic contractions, and mass movements—occur during the 12 to 36 hours it takes indigestible waste to travel from the small intestine to the rectum. These movements are produced by the contractions of a layer of circular muscle and of the three bands of longitudinal muscle. They are generally much more sluggish and short-lived than those found in other parts of the digestive tract, giving time for water to be reabsorbed effectively. The strength and efficiency of colon contractions increases when the diet contains more fiber or roughage.

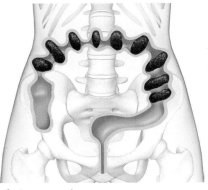

1 Segmentation
When its bands of longitudinal muscle contract, the colon forms pouches that churn and mix fecal material but generate little propulsion. Segmentation happens around every 30 minutes.

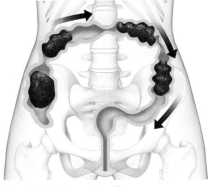

2 Peristaltic contractions
These contractions are similar to peristaltic movements elsewhere in the digestive tract. Small waves of muscular contraction and relaxation pass along the colon, pushing feces toward the rectum.

3 Mass movements
Around three times per day, stimulated by the arrival of food in the stomach, these slow-moving, powerful waves of peristalsis force feces from the transverse and descending colon into the rectum.

ROLE OF BACTERIA

The colon is colonized by microorganisms, principally bacteria, known as the gut flora. They are harmless unless allowed to spread elsewhere in the body. Bacteria digest nutrients, such as cellulose in plant fiber, that cannot be digested by human enzymes. Bacterial digestion releases fatty acids, as well as B complex vitamins and vitamin K, that are absorbed through the colon wall and used by the body. It also releases waste gases including odorless hydrogen, methane, and carbon dioxide, and odorous hydrogen sulfide. Colon bacteria control pathogenic bacteria that enter the large intestine by preventing their proliferation. They aid the immune system by promoting the production of antibodies against pathogens and the formation of lymphoid tissues in the intestinal lining.

DEFECATION

Normally, the rectum is empty and the internal anal sphincter, under involuntary control, and external sphincter, under voluntary control, are contracted to keep the anus closed. When a mass movement pushes feces into the rectum, its walls are stretched. This is detected by stretch receptors, which initiate the defecation reflex by sending impulses along sensory nerve fibers to the spinal cord. Motor signals from the spinal cord instruct the internal sphincter to relax and make the rectal wall contract, building up pressure inside the rectum. Sensory messages to the brain make a person aware of the need to defecate, and a conscious decision is made to relax the external sphincter so that feces can be pushed out through the open anus.

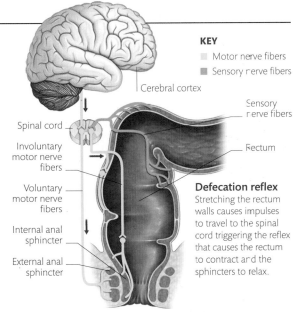

KEY
▨ Motor nerve fibers
▧ Sensory nerve fibers

Cerebral cortex

Spinal cord

Involuntary motor nerve fibers

Voluntary motor nerve fibers

Internal anal sphincter

External anal sphincter

Sensory nerve fibers

Rectum

Defecation reflex
Stretching the rectum walls causes impulses to travel to the spinal cord triggering the reflex that causes the rectum to contract and the sphincters to relax.

NUTRITION AND METABOLISM

The process of digestion produces a range of simple nutrients that provide the raw materials for metabolism, the collection of chemical reactions that together bring cells to life. Before they can be used, however, most nutrients are processed by the liver.

FATE OF NUTRIENTS

During digestion, complex carbohydrates, fats, and proteins are broken down by enzyme action into, respectively, glucose, fatty acids, and amino acids. These simple molecules, along with vitamins and minerals, are nutrients—food substances that are essential to the body to provide energy and building materials, or to make the metabolism work efficiently. Nutrients are absorbed from the small intestine and most travel through the hepatic portal vein to the liver; fatty acids reach the liver by way of the lymph system and then the bloodstream. According to the body's immediate needs, and in order to maintain constant levels of nutrients in the blood, the liver stores some nutrients, breaks others down, or simply allows them to continue their onward journey to be used by body cells.

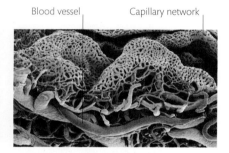

Blood vessel   Capillary network

Blood vessels of the small intestine
This cast shows the fine networks of blood capillaries that infiltrate the wall of the small intestine and collect newly absorbed nutrients.

CATABOLISM AND ANABOLISM

Thousands of chemical reactions take place inside every body cell at any one time, most of them catalyzed by enzymes. These reactions make up the body's metabolism. This has two closely interlinked components: catabolism and anabolism. Catabolism involves the breaking down of complex molecules to simpler ones, often to release energy. In the digestive tract, catabolic reactions break down foods. Anabolism is the opposite of catabolism. It involves processes where smaller molecules are used as building blocks to construct larger ones, such as linking together amino acids to make proteins.

Breaking down and building up
During metabolism, nutrients such as glucose, amino acids, and fatty acids that are absorbed following digestion are broken down or built up.

```
          ┌──────────────────────────────────────────┐
          │   Simple molecules from digested food     │
          └──────────────────────────────────────────┘
              │                              │
              ▼                              ▼
┌────────────────────────────┐  ┌────────────────────────────┐
│ Catabolic processes         │  │ Anabolic processes          │
│ Many catabolic processes    │  │ The enzyme-catalyzed        │
│ involve breaking down fuel  │  │ reactions involved in       │
│ molecules such as glucose   │  │ anabolic processes use      │
│ to release their energy.    │  │ energy to join simple       │
│ Catabolism provides energy  │  │ molecules to construct      │
│ for other chemical          │  │ larger ones, such as        │
│ reactions.                  │  │ multipurpose proteins or    │
│                             │  │ glycogen.                   │
└────────────────────────────┘  └────────────────────────────┘
              │                              │
              ▼                              ▼
       ┌─────────────┐              ┌──────────────────┐
       │   Energy    │              │ Complex molecules │
       └─────────────┘              └──────────────────┘
```

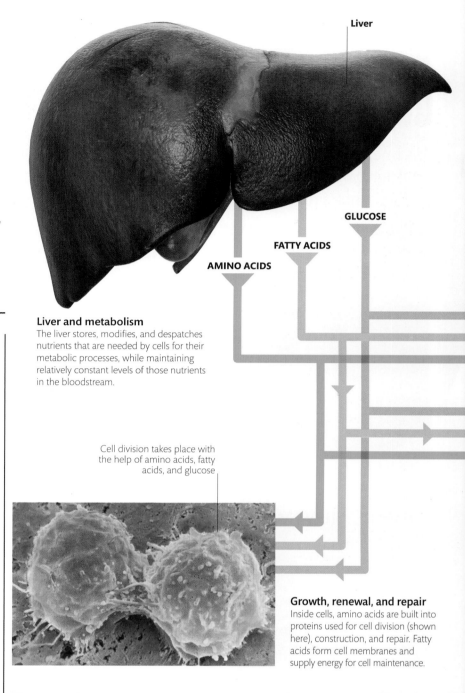

Liver

GLUCOSE

FATTY ACIDS

AMINO ACIDS

Liver and metabolism
The liver stores, modifies, and despatches nutrients that are needed by cells for their metabolic processes, while maintaining relatively constant levels of those nutrients in the bloodstream.

Cell division takes place with the help of amino acids, fatty acids, and glucose

Growth, renewal, and repair
Inside cells, amino acids are built into proteins used for cell division (shown here), construction, and repair. Fatty acids form cell membranes and supply energy for cell maintenance.

ENERGY BALANCE

The chart below shows energy requirements in kilocalories (kcal) and kilojoules (kJ) for different ages, genders, and activity levels. The amount of energy each person needs depends on age, gender, and level of activity. A teenage boy, for example, requires large amounts of energy because his body is growing rapidly. Food energy obtained should balance energy expended because any excess is stored as fat.

AVERAGE DAILY ENERGY REQUIREMENTS

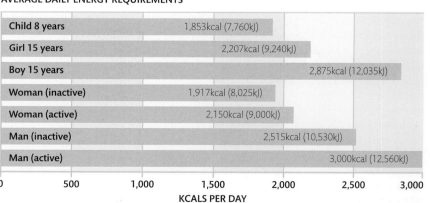

Child 8 years	1,853kcal (7,760kJ)
Girl 15 years	2,207kcal (9,240kJ)
Boy 15 years	2,875kcal (12,035kJ)
Woman (inactive)	1,917kcal (8,025kJ)
Woman (active)	2,150kcal (9,000kJ)
Man (inactive)	2,515kcal (10,530kJ)
Man (active)	3,000kcal (12,560kJ)

0   500   1,000   1,500   2,000   2,500   3,000
KCALS PER DAY

HOW FOOD IS USED IN THE BODY

Glucose is either taken up by liver cells inside the liver (see pp.372-73) and stored as the complex carbohydrate glycogen, or it remains in the bloodstream to provide body cells with a ready source of energy. Fatty acids may be stored in the liver, used by liver and muscle cells to supply energy, or picked up by cells to construct the membranes inside and around them. However, most fatty acids are despatched to adipose tissue (body fat) for storage as fat, providing the body with both an energy reserve and insulation. Some amino acids are broken down by liver cells; others are used by the liver to manufacture plasma proteins, such as fibrinogen, which is involved in blood clotting. Most amino acids, however, remain in the bloodstream to be used by cells throughout the body to build the wide range of proteins needed for growth and maintenance. Excess amino acids cannot be stored and are converted by liver cells to glucose or fatty acids.

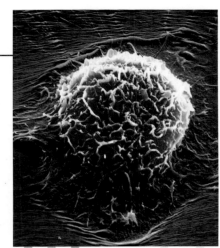

Energy release
Like all body cells, this skin cell needs energy to make it work. The primary source of energy is glucose, although muscle fibers and liver cells also use fatty acids. Under starvation conditions, amino acids may be used.

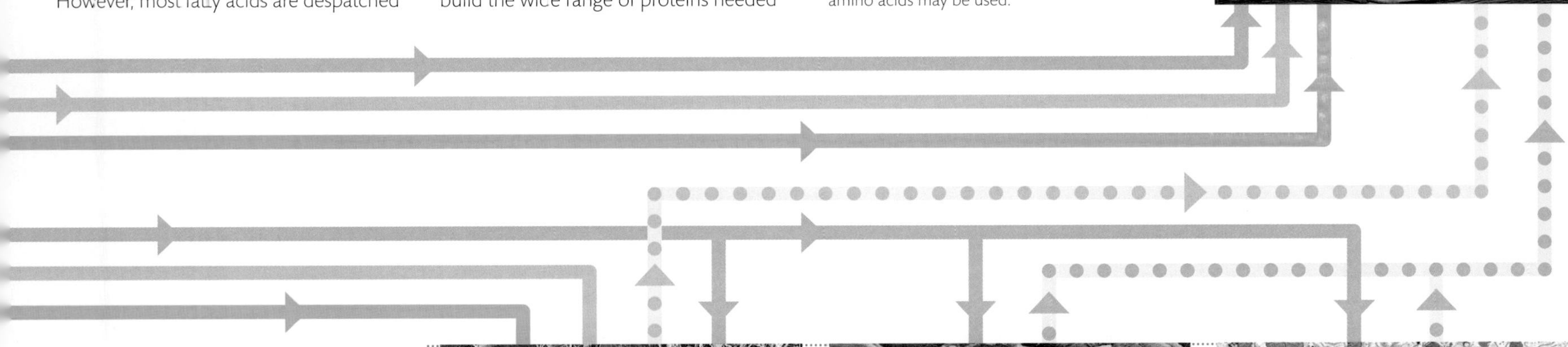

KEY

— Glucose leaves the liver to be used

••• Glucose released from storage

— Fatty acids leave the liver to be stored

••• Fatty acids released from storage

— Amino acids leave the liver to be used

Fat cells
Energy-rich fatty acids are stored as fat inside fat cells, then released when required into the bloodstream and used by some cells as an energy source. Excess glucose is also converted to fat.

Muscle cells
Like liver cells, muscle cells can store glucose as glycogen. Glucose is released from store to provide energy for muscle contraction, or released into the bloodstream if blood glucose levels fall.

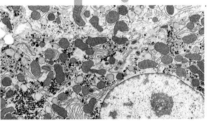

Liver cells
Inside liver cells, surplus glucose is stored as glycogen granules (brown), then released as required. Multiple mitochondria (green) generate the energy needed to power the cell's functions.

VITAMINS AND MINERALS

Essential for normal body functioning, most vitamins and all minerals can only be obtained from food. Vitamins are organic (carbon-containing) substances that act as co-enzymes, which assist many enzymes that control metabolic processes. They are classified according to whether they dissolve in fat (A, D, E, and K) or water (B complex and C). Minerals are inorganic substances needed for enzyme function and in roles such as bone formation. Some, including calcium and magnesium, are needed in larger amounts; trace minerals, including iron and zinc, in tiny amounts.

Use of vitamins and minerals in the body
Some key roles played by vitamins and minerals are shown here. A persistent dietary lack of certain vitamins or minerals impairs body function, resulting in deficiency diseases.

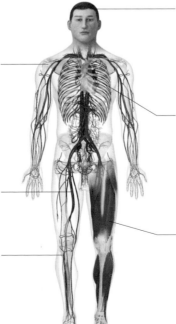

Bone formation
Vitamin A
Vitamin C
Vitamin D
Fluorine
Calcium
Copper
Phosphorus
Magnesium
Boron

Blood clotting
Vitamin K
Calcium
Iron

Blood cell formation and functioning
Vitamins B₅ and B₁₂
Vitamin E
Folic acid
Copper
Iron
Cobalt

Healthy hair and skin
Vitamin A
Vitamin B₂
Vitamin B₃
Vitamin B₆
Vitamin B₁₂
Biotin
Sulphur
Zinc

Heart functioning
Vitamin B₁
Vitamin D
Inositol
Calcium
Potassium
Magnesium
Selenium
Sodium
Copper

Muscle functioning
Vitamin B (Thiamine)
Vitamin B₆
Vitamin B₁₂
Vitamin E
Biotin
Calcium
Potassium
Sodium
Magnesium

WHY DO WE FEEL HUNGRY?

The feeling of hunger, which motivates us to eat, is generated by the brain's hypothalamus in response to a range of signals received from the body, including those delivered by various hormones. For example, the hormone ghrelin, released by an empty stomach, activates parts of the hypothalamus that make a person feel hungry. The hormone leptin, released after eating by the body's fat stores, causes the hypothalamus to inhibit hunger and create a feeling of satiety (fullness).

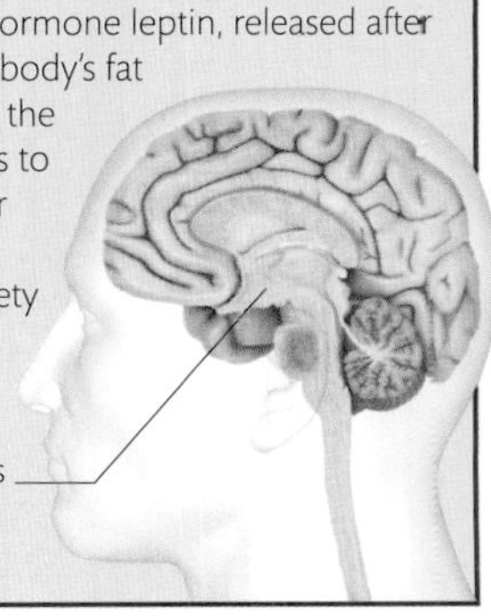

Hypothalamus

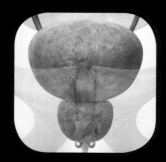

KIDNEY

This bean-shaped organ cleans and filters all of our blood every 25 minutes. All the waste products are excreted in urine.

BLADDER

As it fills with urine, this muscular, elastic bag stretches and expands. The muscles in its wall contract during urination.

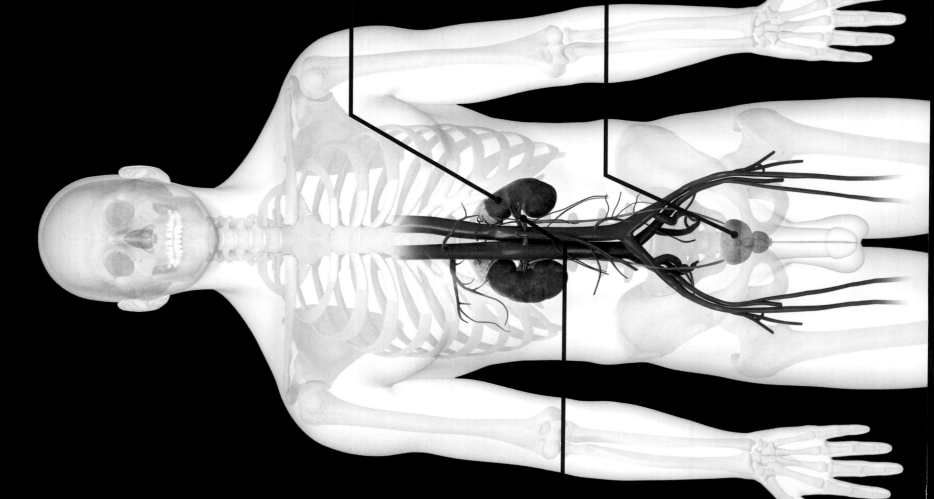

URETER

This urine duct originates in the kidney and channels urine to the bladder, where it is stored for a while.

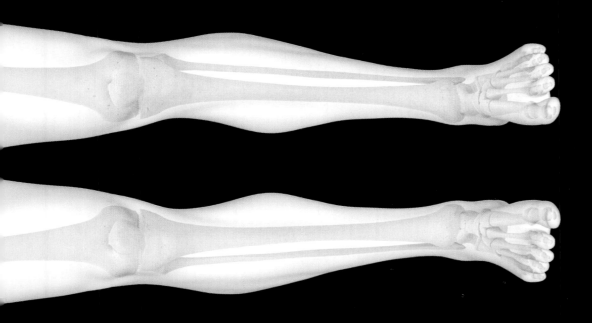

The removal of waste produced by body cells and maintenance of the body's chemical balance are performed by the urinary system. Blood is filtered by the kidneys to remove toxins and any excess substances, ready to be expelled in urine.

URINARY SYSTEM

KIDNEY FUNCTION

The urinary system plays a vital role in keeping the body's fluid and chemical composition in balance and in detoxifying the blood. The kidneys control fluid balance, "rinse" the blood by removing waste products and toxins, and regulate blood pH, or acidity.

INSIDE A KIDNEY

The cortex (outer part) of each kidney contains about one million nephrons. These are filtration units, each made up of a glomerulus and a tubule. The glomerulus consists of a capillary network surrounded by the glomerular (Bowman's) capsule. The tubule is a looped tube connected to the glomerulus. Together, they filter up to 380 pints (180 liters) of blood plasma each day, reabsorbing most of the water and valuable chemicals from the filtrate and producing 2⅛–4¼ pints (1–2 liters) of urine as an excretory product. Loops from the nephrons dip down into the medulla (inner part of the kidney), where the amount of salt and water in the urine is controlled. About 85 percent of nephrons are cortical (short-looped), the rest are juxtamedullary (long-looped). Collecting ducts carry the outflow of the nephrons to the renal pelvis, from where urine flows into the ureter and the bladder for excretion. In addition, the kidney has secondary hormonal functions (see p.405).

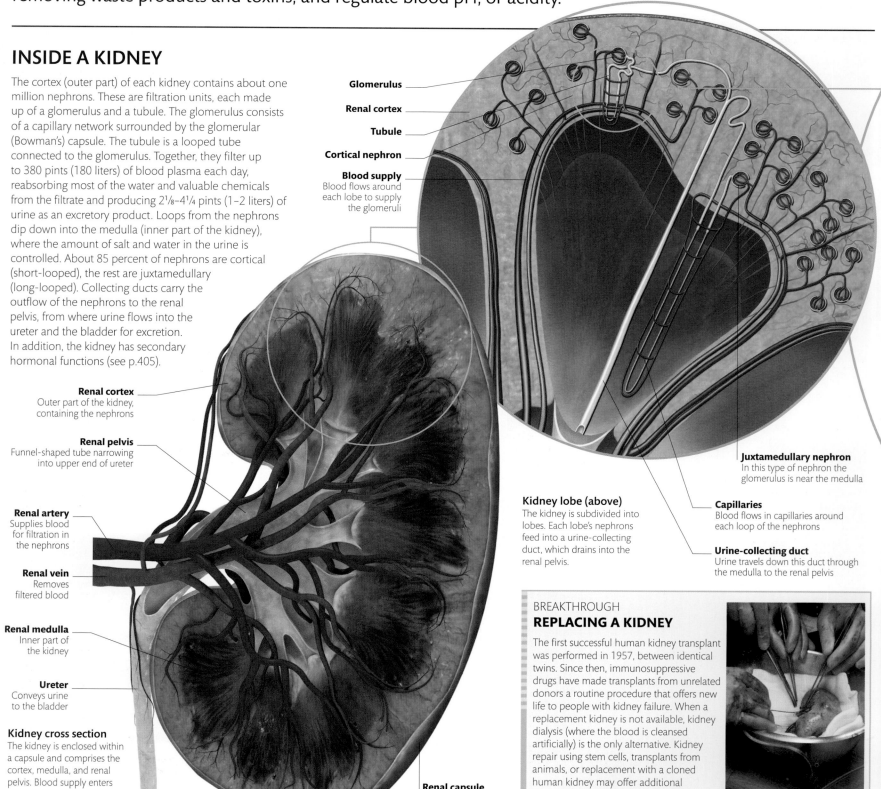

Glomerulus

Renal cortex

Tubule

Cortical nephron

Blood supply
Blood flows around each lobe to supply the glomeruli

Renal cortex
Outer part of the kidney, containing the nephrons

Renal pelvis
Funnel-shaped tube narrowing into upper end of ureter

Renal artery
Supplies blood for filtration in the nephrons

Renal vein
Removes filtered blood

Renal medulla
Inner part of the kidney

Ureter
Conveys urine to the bladder

Kidney cross section
The kidney is enclosed within a capsule and comprises the cortex, medulla, and renal pelvis. Blood supply enters through the renal artery and leaves via the renal vein.

Kidney lobe (above)
The kidney is subdivided into lobes. Each lobe's nephrons feed into a urine-collecting duct, which drains into the renal pelvis.

Juxtamedullary nephron
In this type of nephron the glomerulus is near the medulla

Capillaries
Blood flows in capillaries around each loop of the nephrons

Urine-collecting duct
Urine travels down this duct through the medulla to the renal pelvis

Renal capsule
Outer shell of white, fibrous tissue

BREAKTHROUGH
REPLACING A KIDNEY

The first successful human kidney transplant was performed in 1957, between identical twins. Since then, immunosuppressive drugs have made transplants from unrelated donors a routine procedure that offers new life to people with kidney failure. When a replacement kidney is not available, kidney dialysis (where the blood is cleansed artificially) is the only alternative. Kidney repair using stem cells, transplants from animals, or replacement with a cloned human kidney may offer additional treatment options in the near future.

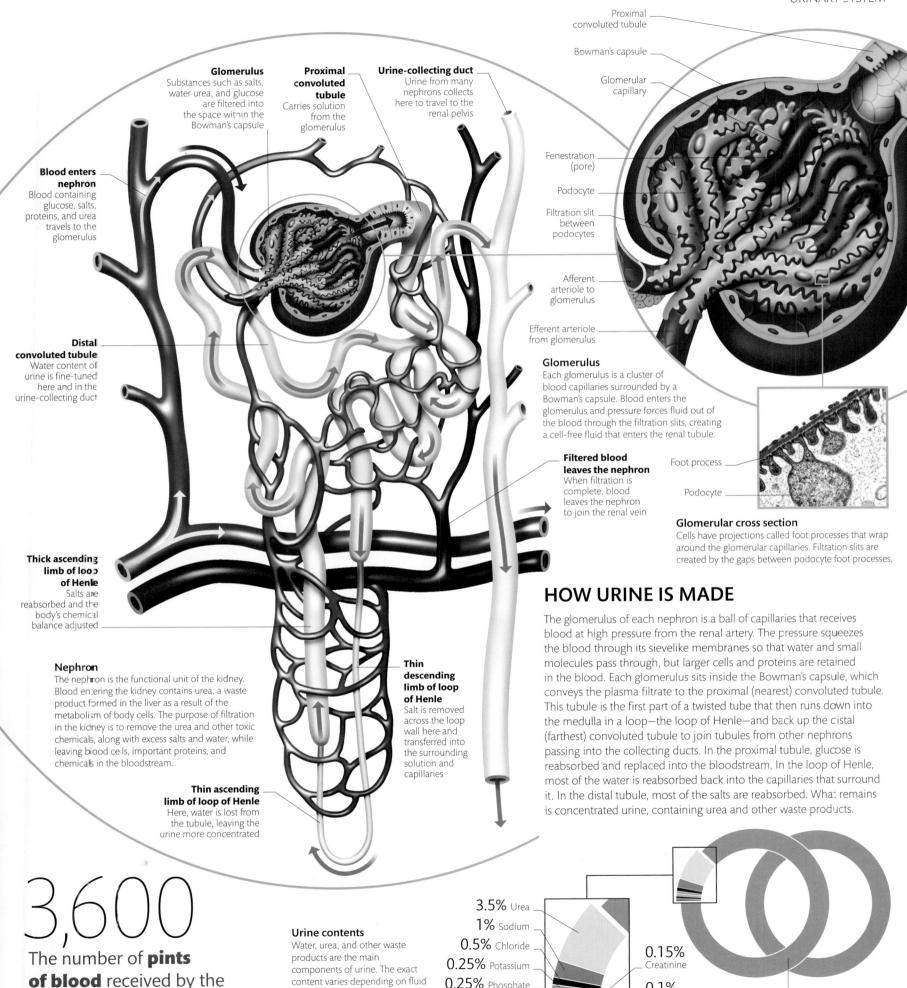

Proximal convoluted tubule

Bowman's capsule

Glomerular capillary

Fenestration (pore)

Podocyte

Filtration slit between podocytes

Afferent arteriole to glomerulus

Efferent arteriole from glomerulus

Foot process

Podocyte

Glomerulus
Substances such as salts, water urea, and glucose are filtered into the space within the Bowman's capsule

Proximal convoluted tubule
Carries solution from the glomerulus

Urine-collecting duct
Urine from many nephrons collects here to travel to the renal pelvis

Blood enters nephron
Blood containing glucose, salts, proteins, and urea travels to the glomerulus

Distal convoluted tubule
Water content of urine is fine-tuned here and in the urine-collecting duct

Glomerulus
Each glomerulus is a cluster of blood capillaries surrounded by a Bowman's capsule. Blood enters the glomerulus and pressure forces fluid out of the blood through the filtration slits, creating a cell-free fluid that enters the renal tubule.

Filtered blood leaves the nephron
When filtration is complete, blood leaves the nephron to join the renal vein

Glomerular cross section
Cells have projections called foot processes that wrap around the glomerular capillaries. Filtration slits are created by the gaps between podocyte foot processes.

Thick ascending limb of loop of Henle
Salts are reabsorbed and the body's chemical balance adjusted

Nephron
The nephron is the functional unit of the kidney. Blood entering the kidney contains urea, a waste product formed in the liver as a result of the metabolism of body cells. The purpose of filtration in the kidney is to remove the urea and other toxic chemicals, along with excess salts and water, while leaving blood cells, important proteins, and chemicals in the bloodstream.

Thin descending limb of loop of Henle
Salt is removed across the loop wall here and transferred into the surrounding solution and capillaries

Thin ascending limb of loop of Henle
Here, water is lost from the tubule, leaving the urine more concentrated

HOW URINE IS MADE

The glomerulus of each nephron is a ball of capillaries that receives blood at high pressure from the renal artery. The pressure squeezes the blood through its sievelike membranes so that water and small molecules pass through, but larger cells and proteins are retained in the blood. Each glomerulus sits inside the Bowman's capsule, which conveys the plasma filtrate to the proximal (nearest) convoluted tubule. This tubule is the first part of a twisted tube that then runs down into the medulla in a loop—the loop of Henle—and back up the distal (farthest) convoluted tubule to join tubules from other nephrons passing into the collecting ducts. In the proximal tubule, glucose is reabsorbed and replaced into the bloodstream. In the loop of Henle, most of the water is reabsorbed back into the capillaries that surround it. In the distal tubule, most of the salts are reabsorbed. What remains is concentrated urine, containing urea and other waste products.

3,600
The number of **pints of blood** received by the kidney every **24 hours**.

Urine contents
Water, urea, and other waste products are the main components of urine. The exact content varies depending on fluid and salt intake, environmental conditions, and health.

3.5% Urea
1% Sodium
0.5% Chloride
0.25% Potassium
0.25% Phosphate
0.25% Sulfate

0.15% Creatinine
0.1% Uric acid

94% Water

BLADDER CONTROL

The bladder is a muscular bag that expands to store urine and contracts to expel it. The ability to inhibit spontaneous urination is acquired in early childhood and is vital to maintaining continence. This can be lost as a result of damage to the pelvic floor or to the nerves supplying it.

Bladder lining
Colored micrograph showing the internal surface folds of the wall of the bladder when empty. The bladder expands and contracts as it fills and empties.

DISCHARGE OF URINE

Waves of muscular contractions in the walls of the ureters help propel the urine to the bladder from the kidneys. At the point where they enter the bladder, valves prevent urine reflux back up the ureters. This is important in preventing microbes from traveling up the ureters and infecting the kidneys. At the exit to the bladder there are two sphincters that prevent the urine from draining into the urethra. The internal sphincter at the bladder neck opens and closes automatically but the external sphincter, located lower down, is under voluntary control. When the bladder is empty, the detrusor muscle in its walls is relaxed and both sphincters are closed. As the bladder fills, the walls become thinner and stretch, prompting a small reflex contraction in the detrusor muscle and triggering the urge to urinate. This can be resisted voluntarily by keeping the external sphincter closed until an appropriate time. When it is convenient to urinate, the external sphincter and pelvic floor muscles are consciously relaxed, and the detrusor muscle contracts, propeling urine out of the bladder.

Bladder fills
As urine flows into the bladder the detrusor muscle in the wall relaxes and the bladder stretches. The sphincters remain closed.

Two ureters carry urine from the kidneys to the bladder

Openings of the ureters have valves

As the bladder fills the detrusor muscles relax, allowing the bladder to stretch

External urethral sphincter remains closed

Bladder empties
The sphincters relax and open and the detrusor muscle contracts, squeezing the urine out through the urethra.

Internal sphincter remains closed

The urethra leads from the bladder to the outside of the body

Both internal and external sphincters relax, allowing urine to exit

Detrusor muscles in the bladder walls contract, voiding the bladder

BLADDER SIZE

The size and shape of the bladder changes with the amount of urine it is storing. When empty, the bladder is flattened into a triangular shape. As it fills, the wall thins and it gradually distends and expands upward into a more spherical shape protruding out of the pelvis into the abdominal cavity. Its length may increase from 2 in (5 cm) to 5 in (12 cm) or more.

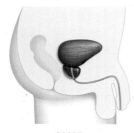

FEMALE   **MALE**

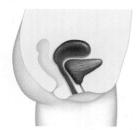

Different bladder sizes
The female bladder is generally smaller than the male with less room to expand on filling.

KEY

■	Bladder	■	Prostate
■	Urethra	■	Uterus

NERVE SIGNALS

Control of micturition (urination) involves nerve centers in the brain and spinal cord, and peripheral nerves supplying the bladder, sphincters, and pelvic floor. As the bladder fills, its internal pressure increases. Stretch receptors in the wall transmit signals to the sacral micturition center in spinal cord segments S2 to S4, which triggers reflex contraction of the detrusor muscle. Signals sent to the micturition center in the brain allow voluntary control, so the need to urinate is consciously recognized, but the sacral reflex is inhibited. When the decision to urinate is made the detrusor muscle in the bladder wall contracts, the internal sphincter relaxes, and the external sphincter is relaxed voluntarily. Once urination begins, further reflexes from the urethra also cause detrusor muscle contraction and sphincter relaxation.

Spinal cord segments S2, S3, and S4
Spinal reflexes travel from here to the bladder where they trigger bladder contraction and sphincter relaxation to allow urination

Pudendal nerve fibers
Control external sphincter

Pelvic nerve fibers
Have both parasymathetic and sympathetic components (see p.311)

Bladder nerve impulses
This schematic shows the connection between segments S2–S4 of the spinal cord with the bladder via the pudendal and pelvic nerves.

S2

S3

S4

Control in the brain
The micturition center in the brain inhibits the sacral micturition center until a conscious decision is made to urinate. The pontine micturition center, lower in the brain, enables the internal sphincter to relax at the same time.

17 fl oz

The capacity of the **average bladder** of an adult male.

FLUID BALANCE

The body's fluid content is maintained by balancing intake with excretion. The osmolarity (concentration) of body fluids is detected in the brain by nerve cells called osmoreceptors. If osmolarity rises, signaling dehydration, antidiuretic hormone (ADH) is secreted from the pituitary gland and acts on the kidney to increase reabsorption of water and decrease urine output. If water intake is increased, osmolarity falls and ADH output is reduced, leading to decreased fluid reabsorption in the kidney and increased urine volume. When the body is sufficiently hydrated, urine is a pale straw color. Darker urine signals a need for increased water intake.

The process of thirst
Although the kidney can conserve body water, it cannot replace it. Thirst, prompted by increased osmolarity, reduced body fluid volume, and symptoms such as a dry mouth, signals the need to increase fluid intake.

Fluid balance upset by loss of water
Water is lost from the body through urination, respiration, sweating (shown here), vomiting, diarrhea, burns, or bleeding. This affects the balance of fluids, setting in motion a series of events.

Osmoreceptors in the hypothalamus activated

Concentration of body fluids
As the body loses fluid, plasma osmolarity (concentration of body fluids) increases, triggering thirst and activation of osmoreceptors

Thirst

ADH released

Water is retained and reabsorbed

Dilution of body fluids
As fluid levels in the body increase, plasma osmolarity (concentration of body fluids) decreases

Increased intake of water

Release of ADH inhibited

Inhibition of thirst

Loss of water and return to fluid balance

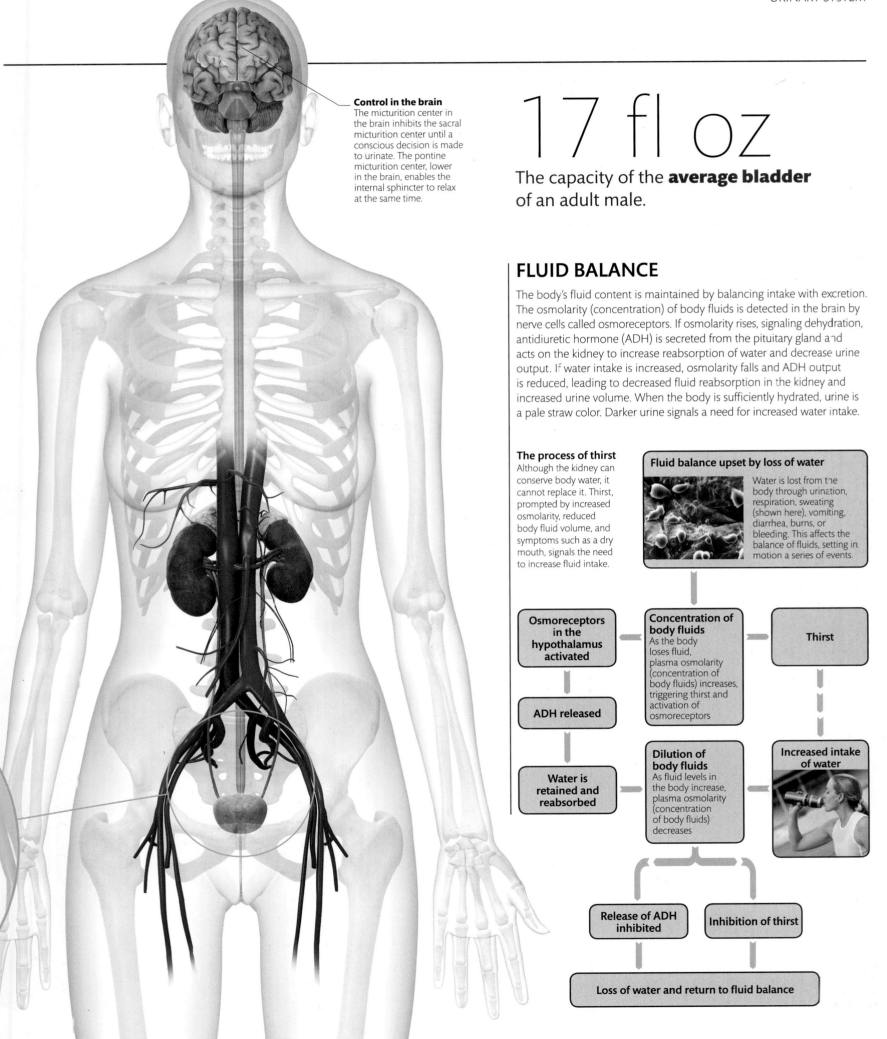

BREAST

Both men and women have breasts containing mammary glands. In women these are larger, and produce milk after childbirth.

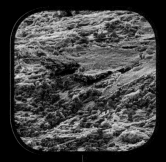

UTERUS

A muscular sac that sheds its lining during menstruation. Inside the uterus, a fertilized egg can develop into a fetus.

OVARY

Two organs, one either side of the uterus, house and mature eggs (ova). One egg is released each month during ovulation.

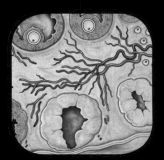

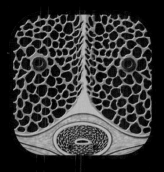

PENIS

The structure and blood supply of the penis allow it to become engorged and remain firm enough to deliver sperm during intercourse.

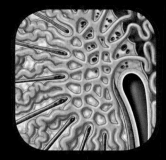

TESTIS

Sperm grow, develop, and mature in a maze of tubules in each of a man's two testes before traveling to, and then out of, the penis during ejaculation.

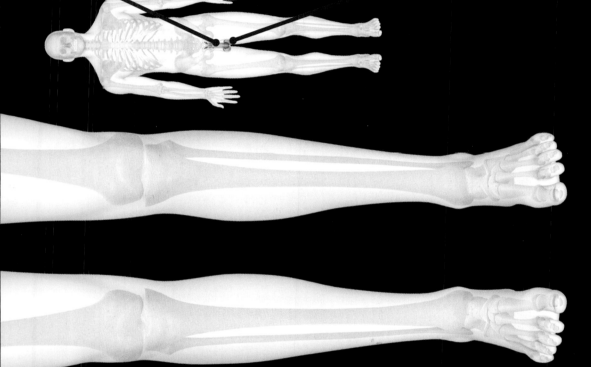

REPRODUCTIVE SYSTEM

The only system that differs greatly between the male and female bodies, the reproductive system is designed to fulfill the purpose of producing offspring—the ultimate biological goal of the human body and all living things.

MALE REPRODUCTIVE SYSTEM

The reproductive organs of an adult male manufacture and supply sperm (spermatozoa), together with the secretions of various glands that make up the semen, or ejaculate. In addition the testes, which are the site of sperm production and storage, produce the male sex hormone testosterone.

SPERM PRODUCTION

The production of sperm cells (spermatozoa) in the testes is known as spermatogenesis. Each testis contains about 500 tightly packed tubes called seminiferous tubules, containing the immature male germ cells (spermatogonia). The germ cells initially multiply by normal cell division, or mitosis (see p.21), to produce spermatocytes. These undergo a special reproductive division called meiosis (see p.410), in which the number of chromosomes in each cell is halved from 46 to 23. These cells, carrying half the genetic material needed to create a new human, are called haploid cells (all other body cells are diploid). Further divisions form sperm precursors (spermatids), which develop into mature spermatozoa, completing the process. Sperm are produced at a rate of several hundred million per day, from puberty into old age.

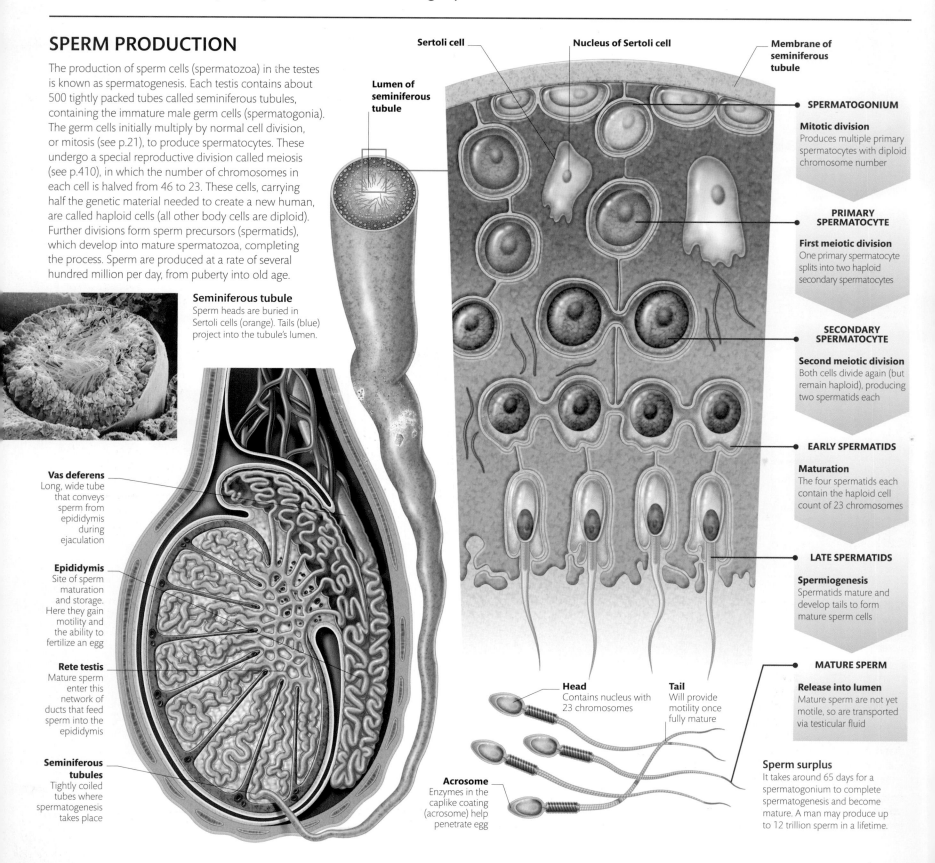

Seminiferous tubule
Sperm heads are buried in Sertoli cells (orange). Tails (blue) project into the tubule's lumen.

Vas deferens
Long, wide tube that conveys sperm from epididymis during ejaculation

Epididymis
Site of sperm maturation and storage. Here they gain motility and the ability to fertilize an egg

Rete testis
Mature sperm enter this network of ducts that feed sperm into the epididymis

Seminiferous tubules
Tightly coiled tubes where spermatogenesis takes place

Sertoli cell

Nucleus of Sertoli cell

Membrane of seminiferous tubule

Lumen of seminiferous tubule

SPERMATOGONIUM

Mitotic division
Produces multiple primary spermatocytes with diploid chromosome number

PRIMARY SPERMATOCYTE

First meiotic division
One primary spermatocyte splits into two haploid secondary spermatocytes

SECONDARY SPERMATOCYTE

Second meiotic division
Both cells divide again (but remain haploid), producing two spermatids each

EARLY SPERMATIDS

Maturation
The four spermatids each contain the haploid cell count of 23 chromosomes

LATE SPERMATIDS

Spermiogenesis
Spermatids mature and develop tails to form mature sperm cells

MATURE SPERM

Release into lumen
Mature sperm are not yet motile, so are transported via testicular fluid

Head
Contains nucleus with 23 chromosomes

Tail
Will provide motility once fully mature

Acrosome
Enzymes in the caplike coating (acrosome) help penetrate egg

Sperm surplus
It takes around 65 days for a spermatogonium to complete spermatogenesis and become mature. A man may produce up to 12 trillion sperm in a lifetime.

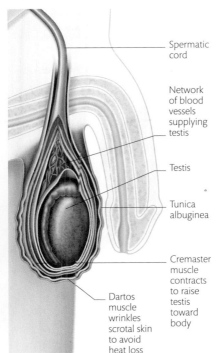

Spermatic cord

Network of blood vessels supplying testis

Testis

Tunica albuginea

Cremaster muscle contracts to raise testis toward body

Dartos muscle wrinkles scrotal skin to avoid heat loss

TESTES AND SCROTUM

The seminiferous tubules make up about 95 percent of testicular volume. They contain male germ cells, from which sperm develop, and Sertoli cells, which provide the developing sperm with nourishment. Fibrous tissue between the tubules contains Leydig cells, which produce testosterone. Each testis has a tough coat called the tunica albuginea and sits within a pouch of skin and muscle called the scrotum. Scrotal muscles are vital for thermoregulation of sperm, which must stay 3.5–5.5° F (2–3° C) below core body temperature to survive. The scrotum moves the testes to and away from the body in response to fluctuations of air temperature, to promote fertility.

Temperature regulation
When it is cold, scrotal muscles contract to wrinkle the skin and elevate the testes, conserving temperature. When warm, they relax, smoothing scrotal skin and lowering the testes to cool them.

SPERM PROTECTION

Tight connections between the Sertoli cells in the seminiferous tubules form what is known as a "blood–testis barrier." This separates the tubules from the blood vessels to prevent harmful substances in the blood from damaging developing sperm. If this barrier is breached, sperm cells can seep into the blood and may provoke an immune response if the body mistakes them for foreign invaders. Antibodies may then enter the tubules and attack the sperm, impairing fertility.

Sertoli cells
Sertoli cells (blue) nourish developing sperm in the coiled seminiferous tubules, and offer them protection via the vital blood–testis barrier.

HORMONAL CONTROL

The hypothalamus (a gland in the brain) secretes gonadotropic-releasing hormone (GnRH). This triggers the pituitary gland (also in the brain) to release luteinizing hormone (LH) and follicle-stimulating hormone (FSH), which both act on the testis. LH stimulates Leydig cells to produce testosterone (responsible for spermatogenesis and male secondary sexual characteristics). FSH prompts Sertoli cells to support developing spermatozoa. Feedback loops reduce GnRH secretion in response to rising levels of testosterone.

Micrograph of testosterone
Testosterone promotes spermatogenesis in the testes, and maintains male sexual characteristics, such as a deep voice, and facial and body hair.

PATH OF SPERM

Seminal vesicle

Bladder

Vas deferens

Cowper's gland

Prostate gland

Sperm leave epididymis

Urethra

Toward ejaculation
Sperm are propelled through the vas deferens into the ejaculatory duct, where added secretions form semen. This continues into the urethra, aided by contractions of the muscular prostate gland.

Sperm make up less than 5 percent of semen volume. As they pass from the seminiferous tubules into a long duct called the epididymis, they undergo further maturation to become motile and fertile before entering the vas deferens, a muscular tube that joins the duct of the seminal vesicle (behind the bladder) to form an ejaculatory duct. The seminal vesicle adds a fructose-rich solution that provides energy and nutrients for the sperm, and contributes around two-thirds of the total semen volume. It is highly alkaline (to counteract vaginal acidity) and contains prostaglandins, which dampen vaginal immune responses to semen. As semen enters the urethra, the prostate gland contributes a slightly alkaline fluid that makes up around a quarter of the seminal fluid. Finally, Cowper's gland secretes a fluid (comprising less than 1 percent of the total volume) to lubricate the urethra and flush out any urine before ejaculation.

ERECTILE FUNCTIONS

The penis has a dual role in the urinary and reproductive systems, by conveying both urine and semen through the urethra. The urethra is contained within a tube called the corpus spongiosum, which runs the length of the penis. On either side are two larger tubes called the corpora cavernosa, each of which has a large central artery surrounded by an expansile, spongy tissue that fills with blood during erections, prompted by nerve impulses that cause the blood vessels to dilate. This usually occurs due to sexual arousal, but can be unprompted.

Prior to ejaculation, contractions within the duct system drive the semen into the urethra. Rhythmic contractions of perineal muscles during male orgasm then eject the semen from the body.

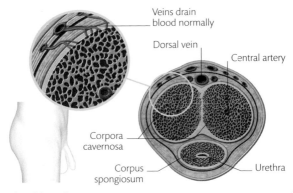

Veins drain blood normally

Dorsal vein

Central artery

Corpora cavernosa

Corpus spongiosum

Urethra

Flaccid penis
In the nonerect penis, the corpora cavernosa have minimal blood flowing through them, while the veins of the penis are wide open and full. The penis droops forward and is soft and flexible.

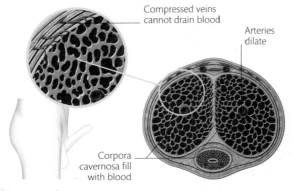

Compressed veins cannot drain blood

Arteries dilate

Corpora cavernosa fill with blood

Erect penis
During an erection, the corpora cavernosa fill with blood, and as a result the veins become compressed, hindering outflow. The engorgement results in enlargement and elevation of the penis.

FEMALE REPRODUCTIVE SYSTEM

The female reproductive organs release a stored egg (or ovum) at monthly intervals, with two possible outcomes each time: to allow shedding of the uterine lining at menstruation, or to enable fertilization, implantation, and nurture of a developing embryo.

OVULATION

The ovaries are paired, oval organs, each one about the size of an almond, that sit at the ends of the fallopian tubes. Female germ cells (eggs, or ova) mature in the ovaries and are regularly released in a process known as ovulation.

Each month 10 or more follicles, the protective casings surrounding each egg (see below), start to ripen, but usually just one releases its egg from either the right or the left ovary—right is favored 60 percent of the time. The egg travels down the fallopian tube to the uterus and is shed from the body along with the uterine lining during the woman's next menstrual period. If, however, the egg is fertilized in the fallopian tube, the resulting cell mass may implant in the wall of the uterus.

An **unfertilized** egg stays in the reproductive tract for between **12 and 24 hours** after ovulation.

Cilia
The fallopian tube lining has cells bearing tiny hairs or cilia (yellow) that help transport the egg to the uterus.

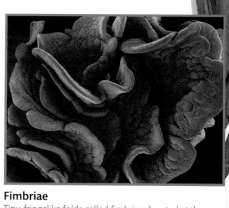

Fimbriae
Tiny, fringelike folds called fimbriae, located at the junction of the fallopian tube with each ovary, pick up the egg and guide it into the tube after ovulation.

Egg travels down fallopian tube

Fallopian tube
Provides egg with a 4 in (10 cm) pathway to the uterus

Egg to uterus
An egg is released from the ovary midway through each reproductive cycle, and reaches the uterus 6–12 days later. Only a tiny minority of eggs, if any, will be fertilized.

Released egg

Fimbriae
Help direct egg into fallopian tube

FOLLICULAR DEVELOPMENT

Immature ova are protected within layers of cells called ovarian follicles. The smallest, primordial follicles, have just a single layer of cells. Each month, some of these develop to become mature (Graafian) follicles. Just before ovulation, one mature follicle moves toward the surface of the ovary and bursts through to release its egg. Its remnants form a body called the corpus luteum and, if the egg is not fertilized, this shrinks to a small, white body called the corpus albicans. At birth, girls have around 1 million follicles per ovary. These will degenerate to about 350,000 by puberty, and 1,500 by menopause.

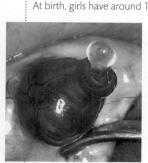

Ovulation
A magnified image of an egg (in reality the size of a period) shows its release from a follicle.

Cyclical development
Each month, some primordial follicles enlarge to become primary then secondary follicles, until they are fully mature. These follicles continually develop in each ovary.

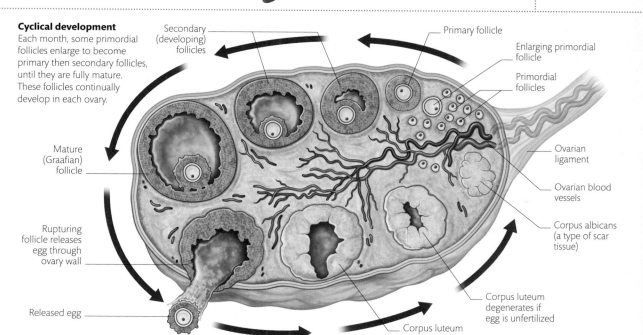

Secondary (developing) follicles

Primary follicle

Enlarging primordial follicle

Primordial follicles

Ovarian ligament

Ovarian blood vessels

Corpus albicans (a type of scar tissue)

Mature (Graafian) follicle

Rupturing follicle releases egg through ovary wall

Released egg

Corpus luteum degenerates if egg is unfertilized

Corpus luteum

UTERUS AND MENSTRUATION

A menstrual cycle is counted from the first day of menstruation and usually lasts 28–32 days. Just prior to ovulation, which usually occurs on day 14, the uterine lining (endometrium) gradually thickens in preparation for a possible pregnancy. If fertilization does not occur, the outer endometrial layer (functionalis) is shed as menstrual blood. The inner layer (basalis) remains and regenerates the functionalis with each new cycle. If an egg is fertilized, the whole endometrium remains to protect the embryo.

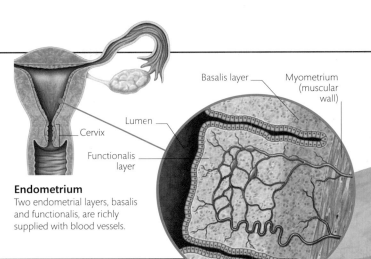

Endometrium
Two endometrial layers, basalis and functionalis, are richly supplied with blood vessels.

Shedding the uterine lining
An electron micrograph shows the process of menstruation: the endometrium (red) breaks away from the uterus wall and is released as blood.

Egg reaches uterus opening

Ovarian ligament

Myometrium
Muscular wall of uterus

Endometrium
Uterine lining, part of which sheds during menstruation

Path of egg
An unfertilized egg is expelled from the uterus during menstruation

HORMONAL CONTROL

The reproductive cycle is controlled by two hormones from the pituitary gland in the brain (see p.386). Follicle-stimulating hormone (FSH) causes ovarian follicles to ripen and produce estrogen. When estrogen levels are high enough, a surge of luteinizing hormone (LH) from the pituitary prompts final maturation of the egg and its release from the ovary. After ovulation, as estrogen levels fall, FSH production increases to repeat the cycle.

Endometrial responses
Estrogen stimulates endometrial thickening. This is temporarily maintained by progesterone from the corpus luteum, but it sheds as levels fall.

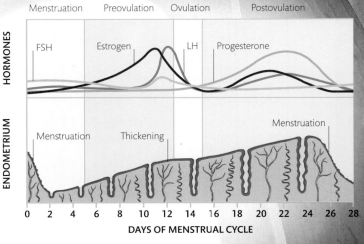

CHANGES DURING MENSTRUAL CYCLE

Menstruation | Preovulation | Ovulation | Postovulation

HORMONES

FSH   Estrogen   LH   Progesterone

ENDOMETRIUM

Menstruation   Thickening   Menstruation

0 2 4 6 8 10 12 14 16 18 20 22 24 26 28

DAYS OF MENSTRUAL CYCLE

FUNCTION OF THE CERVIX

The cervix connects the uterus with the vagina and forms a vital barrier to the outside. It secretes mucus that varies in form and function throughout the reproductive cycle. For most of the cycle and during pregnancy, the mucus is thick and sticky to protect the uterus from infection. It also forms an impenetrable barrier to sperm. During a woman's fertile period, rising levels of estrogen make the mucus thin and stretchy (sort of like egg white), to enable sperm to pass through the cervix and reach the ovulated egg.

Healthy cervix
The tight cervical entrance can be clearly seen in this image. Fertile cervical mucus protects sperm from the acidic vagina.

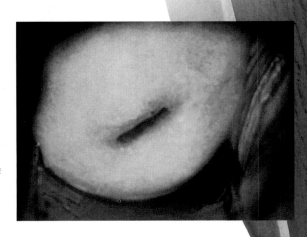

CREATION OF LIFE

Human reproduction involves the fusion of male and female germ cells (spermatozoa and ova), each containing half of the genetic information required to create a fetus that will develop into a new human being.

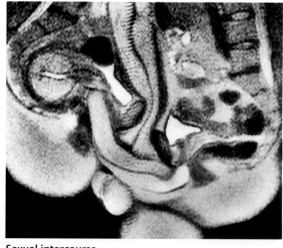

SEX

Sexual arousal in both sexes leads to progressive engorgement of the genital organs as blood flow increases, along with muscle tension, heart rate, and blood pressure. The penis becomes erect and the woman's clitoris and labia increase in size. The vagina lengthens and its walls secrete lubricating fluid to enable the penis to enter and ejaculate semen high up in the vagina, near the opening of the cervix.

Arousal
Sexual responsiveness passes through various phases, and timing differs for men and women.

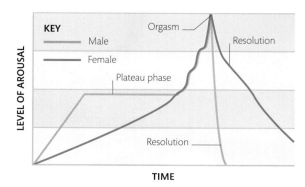

KEY
— Male
— Female

LEVEL OF AROUSAL

Orgasm
Resolution
Plateau phase
Resolution
TIME

Sexual intercourse
This remarkable MRI scan shows a couple having sexual intercourse. The penis (blue) is bent like a boomerang. The uterus is shown in yellow.

After **sex,** males have a **refractory period**, during which they cannot have another orgasm. Women may experience **multiple orgasms**.

SPERM RACE

Male fertility depends on a vast overproduction of sperm compared with the single sperm cell required to fertilize an egg. An average ejaculate contains 280 million sperm per $^1/_{16}$–$^1/_6$ fl oz (2–5 ml) of semen. Only around ovulation will any sperm survive the vaginal acidity and cervical mucus barrier to take part in the competitive race to reach the released egg.

200 sperm enter both fallopian tubes

Fallopian tube

Egg and sperm meet

Egg is released

Ovary

100,000 enter uterus

Uterus

60–80 million pass the cervix

Cervix

Vagina

100–300 million sperm enter vagina at ejaculation

KEY
→ Path of sperm
→ Path of egg

Against the odds
Even during a woman's fertile period, of the 300 million sperm that can enter the vagina, only about 200 reach the fallopian tubes.

Strong swimmers
Sperm swim the 4 in (10 cm) fallopian tube towards the egg at about $^1/_8$ in (3 mm) per hour.

Cervical mucus
During ovulation, cervical mucus becomes clear, slippery, and stretchy, making it easier for sperm to pass through. Mucus at this time dries in a "fern leaf" pattern.

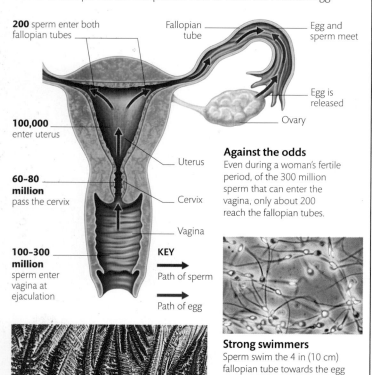

FERTILIZATION AND IMPLANTATION

The first sperm to reach the egg in the fallopian tube binds to its surface, releasing enzymes from the acrosome surrounding its head (see p.386) that help it to break through the egg's protective coating. The egg responds by releasing its own enzymes to block any other sperm from entering, and the rest fall away. The successful sperm is then absorbed into the egg and loses its tail. The nuclei of the egg and sperm fuse, enabling their genetic material to join together: conception has occurred. The newly fertilized egg then continues to travel down the fallopian tube, undergoing various stages of cell division to become a ball of cells called a blastocyst that implants in the uterus.

3 Morula
Cell division continues—the cells are confined within the original egg cell membrane so get progressively smaller. By around day 4 there is a ball of about 30 cells called a morula.

4 Blastocyst
A fluid-filled core forms. The outer cells (the trophoblast) invade the uterine lining and develop into the placenta.

2 Zygote
The single cell that results from fusion carries the complete amount of human DNA and is called a zygote. About 24 hours after fertilization, the cell divides into two.

The journey
The fertilized egg undergoes progressive cell division, at first just increasing the number of cells in the mass. After implantation, these cells start to specialize, to create the different tissues of the embryo.

1 Fertilization
A single sperm burrows into the egg, and they fuse. The egg is about 20 times the size of the sperm.

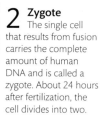

Uterus

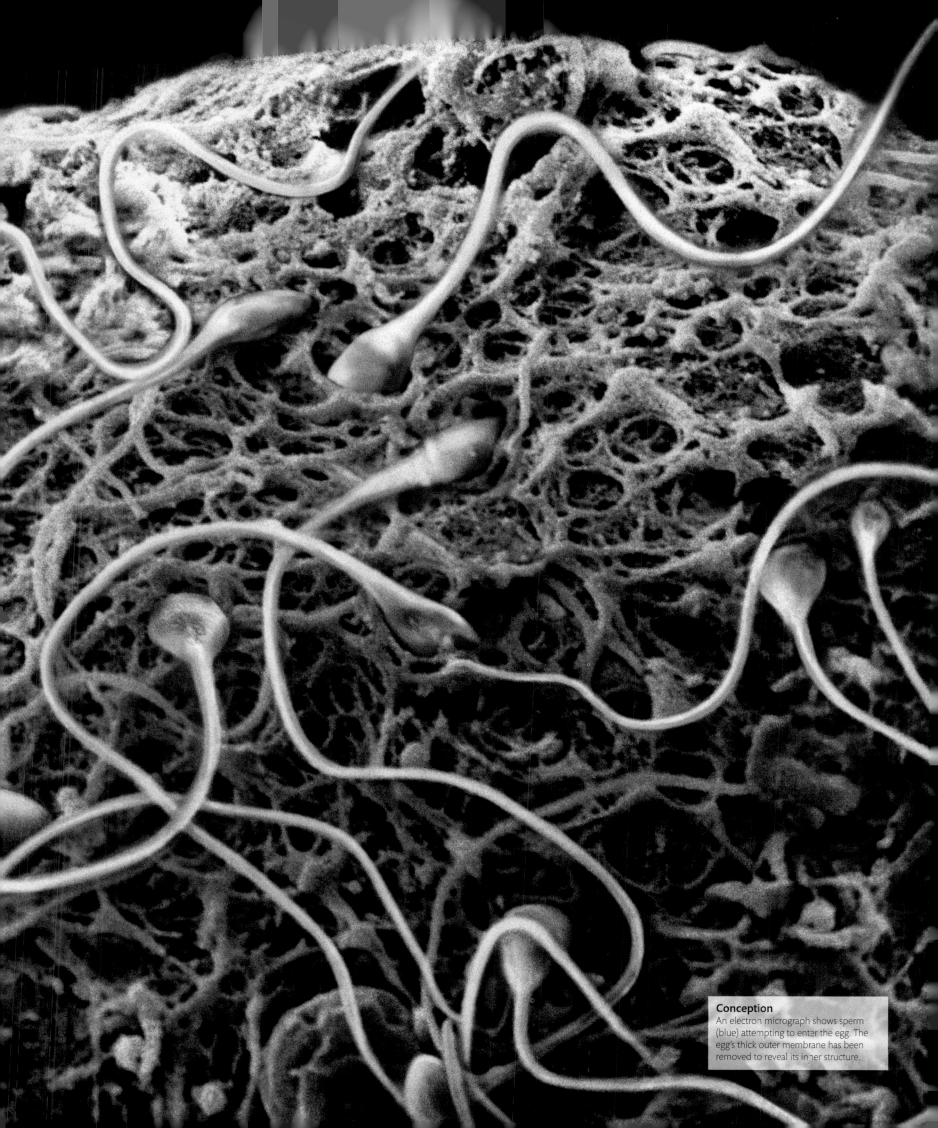

Conception
An electron micrograph shows sperm (blue) attempting to enter the egg. The egg's thick outer membrane has been removed to reveal its inner structure.

THE EXPECTANT BODY

Pregnancy is a time of remarkable physical change in the body, when hormonal surges and metabolic demands affect every tissue and organ, not just the uterus. The blood, cardiovascular and respiratory systems, gastrointestinal organs, and kidneys are all involved in this process.

MEASURING PREGNANCY

Weeks of pregnancy are dated from the first day of the woman's last menstrual period, since the actual date of conception is rarely known. Pregnancy usually lasts for 40 weeks, and is arbitrarily divided into three 12-week periods known as trimesters. The first signs of pregnancy are cessation of menstruation (or sometimes irregular bleeding), nausea or vomiting, breast tenderness, urinary frequency, and fatigue. As pregnancy progresses, the uterus gradually rises up out of the pelvis, and the level at which its top can be felt (the fundal height) is an important guide to fetal growth and development.

Weight gain (right)
A healthy woman will gain 24–35 lb (11–16 kg) during pregnancy, only a quarter of which is the weight of the baby.

7% Breast
7% Uterus
26% Body fluids
7% Amniotic fluid
5% Placenta
23% Fat and protein
25% Baby

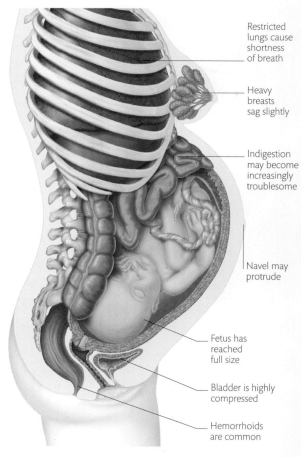

Pregnancy posture
The weight of the enlarged uterus throws a pregnant woman's center of gravity forward, causing her to lean backward and arch her back. Backaches are common.

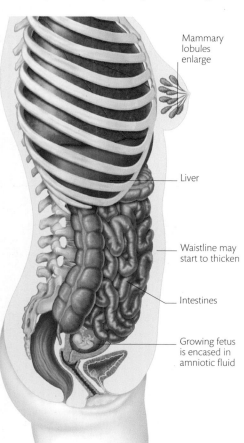

Mammary lobules enlarge

Liver

Waistline may start to thicken

Intestines

Growing fetus is encased in amniotic fluid

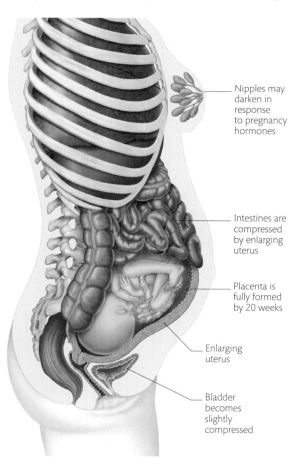

Nipples may darken in response to pregnancy hormones

Intestines are compressed by enlarging uterus

Placenta is fully formed by 20 weeks

Enlarging uterus

Bladder becomes slightly compressed

Restricted lungs cause shortness of breath

Heavy breasts sag slightly

Indigestion may become increasingly troublesome

Navel may protrude

Fetus has reached full size

Bladder is highly compressed

Hemorrhoids are common

First trimester
Nausea is common, breasts may enlarge and feel tender, and there is an increased need to urinate. Heart rate rises and the woman often feels unusually tired. Food transit through the gut slows and heartburn or constipation may result.

Second trimester
Any sickness usually subsides and food cravings may be experienced. The woman gains weight rapidly. Back pain is common, as are stretch marks on the abdomen. Increased circulation may cause nosebleeds and bleeding gums.

Third trimester
The abdomen reaches maximum protrusion and the navel may bulge outward. Leg cramps and swelling of hands and feet may occur. Irregular Braxton-Hicks contractions ("false labor") often begin in the weeks leading up to labor.

0–12 WEEKS

13–24 WEEKS

25–40 WEEKS

SUPPORTING THE FETUS

The placenta develops from the trophoblast (cells within the blastocyst, see p.390) and draws a blood supply from the uterus lining to nourish the fetus as it develops, dispose of its waste products, and protect it from microorganisms. Clear amniotic fluid surrounds the fetus, offering protection and allowing movement and lung development. As it grows, the uterus increases its blood flow and its suspensory ligaments stretch. The woman's whole body increases its blood and body fluid volume and fat reserves, to prepare for labor and feeding. A healthy diet, including calcium, iron, vitamins, and minerals is also crucial.

Life support system
The placenta is richly supplied with blood vessels, which provide essential oxygen and nutrients to the fetus.

Safe haven
The fetus is protected within the sac of warm amniotic fluid and is nourished by the placenta via the umbilical cord.

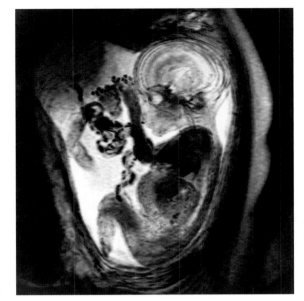

NON-PREGNANT — Pear

8 WEEKS — Orange

14 WEEKS — Cantaloupe melon

20 WEEKS — Honeydew melon

FULL TERM — Watermelon

Relative size of uterus
The above guide to uterine growth during pregnancy indicates the vast change that occurs. The uterus may never return to its previous size.

HORMONE CHANGES

After fertilization, progesterone from the corpus luteum in the ovary prompts endometrial thickening in readiness to receive the fertilized egg. A few days after implantation, the trophoblast produces human chorionic gonadotropin (hCG), a hormone that stimulates the corpus luteum to produce more progesterone, and estrogen. Estrogen keeps the uterus growing, stimulates fetal development and breast enlargement, and boosts blood circulation. It also prompts uterine contractions, along with the hormone oxytocin. Progesterone, which maintains the uterine lining and placenta, tends to relax the uterus. In the second trimester, progesterone is produced by the placenta, and acts with the hormone relaxin to soften cartilage and loosen joints and ligaments, aiding pelvic expansion, in preparation for birth. Human placental lactogen (HPL) and prolactin both prompt milk production.

Chemical surge
The huge surge of the hormone human chorionic gonadotropin (hCG) during early pregnancy is what causes a pregnancy test to register as positive.

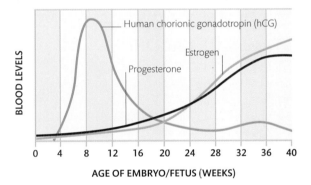

CHANGES IN THE CERVIX

In order for the muscular cervix to dilate before birth, it must first soften and then efface, a process where the tissue thins, or shortens. During pregnancy, the cervix also produces extra-thick mucus that forms a plug in the cervical canal. This helps protect the fetus from infection.

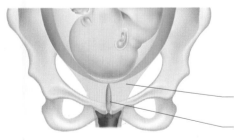

Cervical softening
In late pregnancy, substances called prostaglandins in the blood cause the cervical tissue to soften and become malleable (like the lips).

Cervical tissue forms a necklike canal

Mucus plug

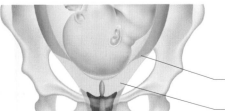

Cervical effacement
As it softens, the cervix begins to thin (efface) and is drawn in toward the lower part of the uterus.

Cervix gradually retracts and fuses with the uterus

Softening cervical tissue begins to thin (efface)

BREAST CHANGES

During pregnancy, the breasts gradually expand and may feel tender. The nipples and areolae (the surrounding circles) enlarge and darken due to pregnancy hormones, and small bumps called Montgomery's tubercles appear around the areolae. Increased blood supply can make veins under the skin more prominent. As birth approaches, the nipples may leak a yellowish fluid called colostrum, or "pre-milk," that is rich in minerals and antibodies to nourish and protect the baby. Breast-feeding after birth stimulates the release of oxytocin, which promotes uterine contractions and helps to birth the placenta.

Milk production
Milk glands and ducts multiply and expand from early pregnancy, and are able to produce milk even during the second trimester.

Mammary lobules

MULTIPLE PREGNANCIES

Twin pregnancies may result from a single fertilized egg that splits in half early in cell division, resulting in monozygotic, or identical, twins. The fetuses have exactly the same DNA and are genetically identical. More often twins are nonidentical (dizygotic), resulting from the fertilization of two separate eggs by two different sperm. They are no more alike than any two siblings. Multiple pregnancies place a greater strain on the woman's body and there is a higher risk of adverse outcomes.

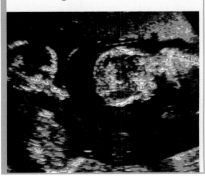

LABOR AND BIRTH

Labor, the process by which a baby is delivered, can be both a joyful and painful experience. The mother undergoes huge physiological and emotional stress, from the first contractions of the latent phase through to the delivery of the placenta.

Oxytocin
This light micrograph shows crystals of oxytocin, the hormone secreted by the pituitary gland to instigate labor. The trigger for its release is still unknown.

CONTRACTIONS

Labor involves strong contractions of the uterine muscle that open up the cervix and expel the baby through the birth canal. Irregular, short-lived "tightenings" known as Braxton-Hicks contractions may be felt much earlier in the pregnancy. As labor progresses, contractions become stronger, last longer, and occur at regular, increasingly short, intervals—most women require analgesics. Contractions and fetal response are monitored by a cardiotocograph (see right) via sensors on the abdomen and on the baby's head as it presents through the opening cervix.

Cardiotocograph (CTG)

The CTG shows two corresponding lines: the strength of uterine contractions and the correlating fetal heart rate. Normal fetal heart rate is 110–160 beats per minute, and abnormal patterns, such as deceleration, indicate fetal distress during contractions.

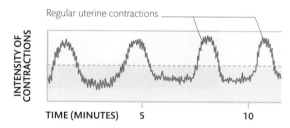

Regular uterine contractions

INTENSITY OF CONTRACTIONS

TIME (MINUTES)   5   10

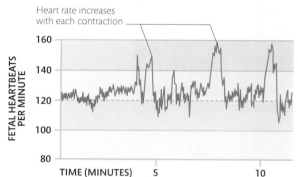

Heart rate increases with each contraction

FETAL HEARTBEATS PER MINUTE

160
140
120
100
80

TIME (MINUTES)   5   10

STAGES OF LABOR

Labor begins in response to the release of oxytocin hormone, which stimulates uterine contractions. It divides into three stages: the latent stage occurs when the cervix starts to dilate; the first stage is defined by dilation of the cervix from 1½ to 4 in (4 to 10 cm); the second stage, from full cervical dilation to delivery of the baby; and the third stage ends with delivery of the placenta. During the second stage, pushing, or bearing down, by the mother is synchronized with the contractions to help expel the baby. Maternal pain, particularly during the second and third stages, may be managed by oral or injected analgesics or epidural anesthesia. Common problems include failure to progress, abnormal presentations such as "breech," tearing of the birth canal and perineum, and difficult placental delivery (see pp.492–93). Forceps or vacuum suction may be used to help pull out the baby, while cesarean section (delivery through the abdominal wall) is used when either the baby or the mother is at risk.

Placenta
Attached to uterine wall

Uterus
Strong contractions push baby forward

Bladder
Compresses as baby moves through birth canal

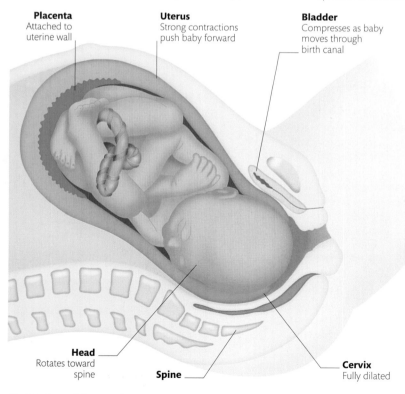

Head
Rotates toward spine

Spine

Cervix
Fully dilated

1 Dilation of the cervix
In the first stage of active labor, the cervix dilates from 1½ to 4 in (4 to 10 cm), which can take hours. Delivery can only begin when the cervix is fully dilated. The baby usually faces its mother's back, so the widest part of its head passes through the widest axis of the pelvis.

Umbilical cord

Contracting uterus
Contractions are combined with active pushing

Presenting part
Crowning head flexes backward as it emerges

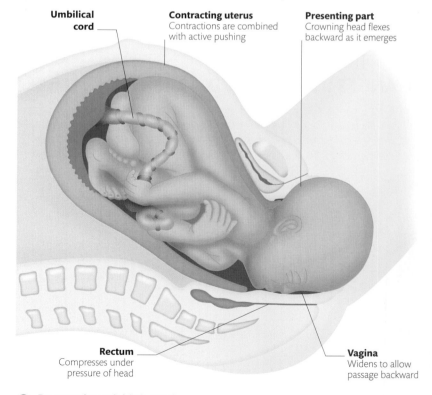

Rectum
Compresses under pressure of head

Vagina
Widens to allow passage backward

2 Descent through birth canal
The presenting part, usually the head, is pushed forward by repeated contractions and pushing. The head progresses from the open cervix, through the vagina, until visible at the perineum ("crowning"). It begins to flex backward to allow the rest of the body to follow.

DILATION OF CERVIX

Once labor has begun, cervical effacement (see p.393) gives way to dilation, when the cervix begins to open in order for the baby to be delivered. Dilation usually begins during the latent phase of labor. Contractions in the upper part of the uterus cause it to shorten and tighten, consequently pulling up the lower part of the uterus and retracting the cervix. In the latent phase, dilation does not exceed 1½ in (4 cm), but it can be long and uncomfortable, with irregular contractions.

Eventually, uterine activity continues into active labor, where regular, increasingly powerful contractions lead to the progressive dilation of the cervix up to a maximum of 4 in (10 cm), at which stage it is wide enough to accommodate the baby. The cervix moves from a posterior to an anterior position, and once it is fully dilated the fetal head rotates, flexes, and molds, before descending into the birth canal.

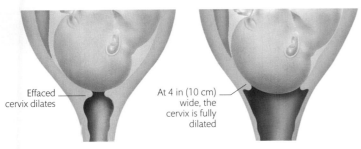

Effaced cervix dilates

At 4 in (10 cm) wide, the cervix is fully dilated

Beginning to dilate
The effaced cervix begins to dilate in response to uterine contractions. For first-time mothers the cervix dilates at an average speed of ⅓ in (1 cm) per hour. The rate is faster for subsequent births.

Fully dilated
As the contractions become stronger and more painful, their frequency and regularity also increase. The cervix dilates further under this strain as well as under the pressure of the fetus's head.

RUPTURE OF MEMBRANES

Shortly before labor is due to begin, the membrane of the amniotic sac that surrounds the fetus ruptures, allowing amniotic fluid to leak out into the birth canal. This is known as the water breaking and most women go into spontaneous labor within 24 hours. If it occurs before 37 weeks, it is considered premature rupture of the membranes, and may put the fetus at risk of infection or premature delivery. Conversely, if the membranes have not ruptured naturally, or if labor is being induced, they may be ruptured artificially to speed up labor and allow a fetal monitor to be attached to the baby's scalp.

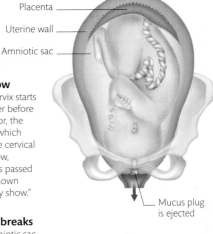

Placenta

Uterine wall

Amniotic sac

1 The show
As the cervix starts to open, either before or during labor, the mucus plug, which has sealed the cervical canal until now, loosens and is passed out. This is known as the "bloody show."

Mucus plug is ejected

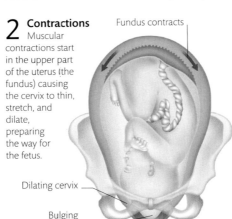

2 Contractions
Muscular contractions start in the upper part of the uterus (the fundus) causing the cervix to thin, stretch, and dilate, preparing the way for the fetus.

Fundus contracts

Dilating cervix

Bulging amniotic sac

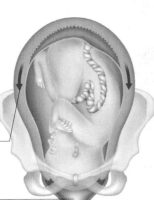

3 Water breaks
The amniotic sac stretches and eventually ruptures under the pressure of the contractions, releasing the amniotic fluid and allowing further descent of the fetus's head.

Continuing contractions

Amniotic fluid drains out through the birth canal

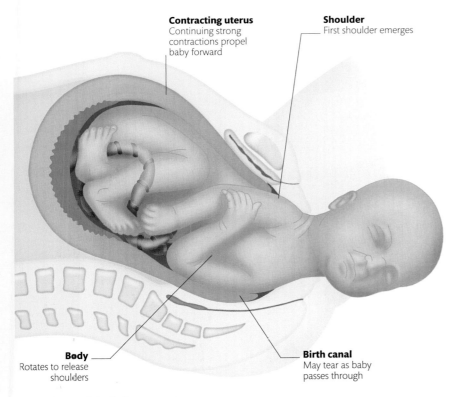

Contracting uterus
Continuing strong contractions propel baby forward

Shoulder
First shoulder emerges

Body
Rotates to release shoulders

Birth canal
May tear as baby passes through

3 Delivery of the baby
As the head is delivered, the doctor ensures that the baby's airway is clear of mucus, and that the umbilical cord is not wrapped around its neck. The baby turns in the birth canal to allow the shoulders to be delivered. The rest of the body then slips out easily.

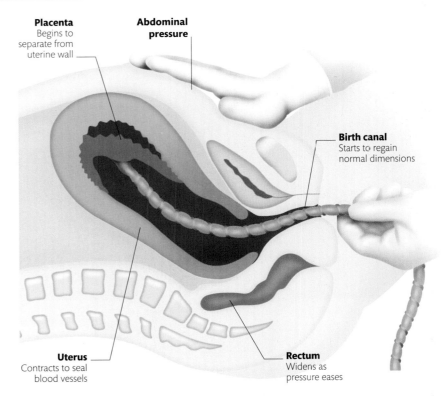

Placenta
Begins to separate from uterine wall

Abdominal pressure

Birth canal
Starts to regain normal dimensions

Uterus
Contracts to seal blood vessels

Rectum
Widens as pressure eases

4 Delivery of the placenta
Further contractions compress the uterine blood vessels, preventing blood loss. The doctor eases the placenta out by pulling the umbilical cord and applying pressure to the lower abdominal wall, or an injection of oxytocin hormone may be given to induce delivery.

HYPOTHALAMUS

The hypothalamus links the nervous and endocrine systems; it secretes hormones that spur the pituitary into action.

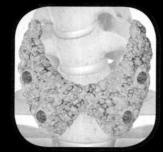

THYROID GLAND

The butterfly-shaped thyroid produces hormones that help to regulate the body's metabolism and heart rate.

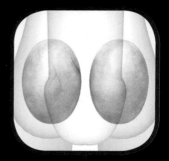

TESTIS

The testes produce sex hormones, which stimulate sexual development and sperm production.

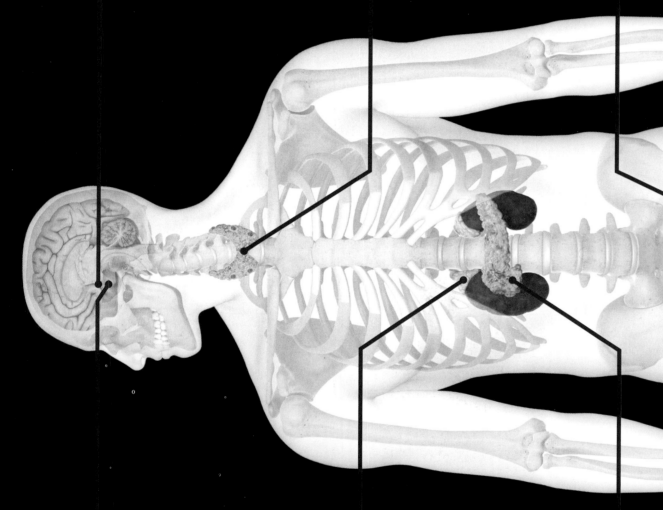

PITUITARY GLAND

Often known as the "master gland," the pituitary controls the activities of many other glands. It is closely connected to the hypothalamus.

ADRENAL GLAND

The distinct parts of this gland (medulla and cortex) produce hormones that help us deal with stress and that attain homeostasis.

PANCREAS

This gland has a dual purpose: secreting the hormones insulin and glucagon as well as digestive enzymes.

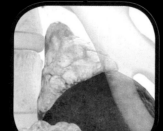

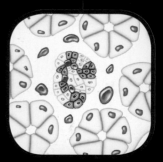

OVARY

Each ovary makes the sex hormones progesterone, which thickens the uterine wall, and estrogen, which ripens eggs.

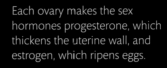

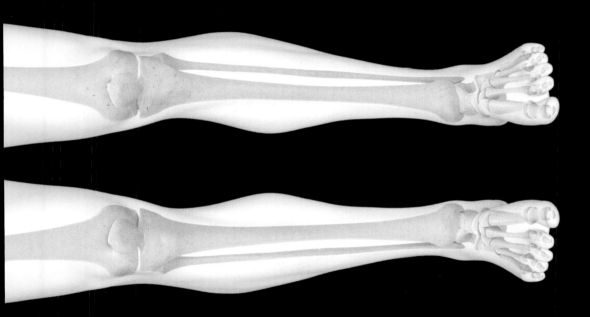

ENDOCRINE SYSTEM

The body's internal environment is monitored and regulated by a chemical communication network. Working alongside the nervous system, endocrine glands produce hormones that control and coordinate many bodily functions.

HORMONES IN ACTION

Hormones are powerful chemicals that work by altering the activity of their target cell. A hormone does not initiate a cell's biochemical reactions, but adjusts the rate at which they occur. Endocrine cells secrete their hormones into the fluid surrounding them; hormones then travel through the bloodstream and affect cells and tissue in distant parts of the body.

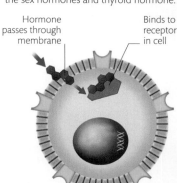

Traveling hormones
Hormones are secreted into the bloodstream by endocrine glands, such as the thyroid in this example, and travel to their target cells—which may be at some distance from the gland.

Endocrine tissue

THYROID GLAND

Fat-soluble hormone in bloodstream, such as thyroid hormone

Water-soluble hormone in bloodstream, such as calcitonin

Blood vessel

HOW HORMONES WORK

Although hormones come into contact with essentially all cells in the body, they produce an effect on only certain cells, called target cells. These target cells have receptors that the hormone recognizes and binds to, triggering a response inside the cell. Each hormone can only affect specific target cells that possess the right kind of receptor for that hormone. For example, thyroid-stimulating hormone only binds with receptors on cells of the thyroid gland. The mechanism is similar to the way a radio broadcast works—

although the signal reaches everyone within range, you need to be tuned to the right frequency to be able to hear it.

A hormone can have several different target cells. However, these do not all react in the same way to the hormone. For example, insulin stimulates liver cells to store glucose but prompts adipose cells to store fatty acids. Once hormones reach their target cell, there are two different mechanisms by which they bind to the cell's receptors and produce a reaction, depending on whether a hormone is water soluble or fat soluble (see right). Water-soluble hormones are built from amino acids (the building blocks of proteins), while most fat-soluble hormones are made from cholesterol.

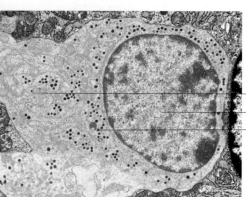

Cytoplasm

Cell nucleus

Secretory granule

Endocrine cell
This micrograph shows a parafollicular cell in the thyroid, which produces and secretes the hormone calcitonin. Dots in the cytoplasm (colored red) are secretory granules, where calcitonin is stored.

SCIENCE
PROSTAGLANDINS

Chemicals called prostaglandins act in a similar way to hormones, by stimulating activity in target cells. However, they act locally, near where they are produced, rather than traveling in the blood. Prostaglandins are released by nearly all cell membranes and have many different effects, including lowering blood pressure and increasing uterine contractions during labor. They are also involved in inflammation, and their release contributes to the sensation of pain.

Prostaglandin crystals
Crystals of prostaglandin BI are seen in this micrograph, taken in polarized light. There are more than 20 types of prostaglandin.

WATER-SOLUBLE HORMONES

These hormones are unable to pass through the cell membrane, which has fatty layers. Therefore, to have an effect on target cells, they bind to receptors on the surface of the cell. Most hormones are water-soluble.

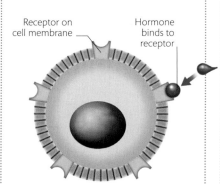

Receptor on cell membrane

Hormone binds to receptor

1 Receptor binding
The hormone recognizes a receptor protruding from the surface of the target cell and binds to it. The mechanism works in a similar way to that of a key in a lock.

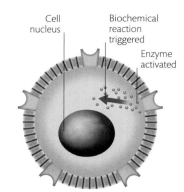

Cell nucleus

Biochemical reaction triggered

Enzyme activated

2 Activation
Enzymes inside the cell are activated, altering the biochemical activity of the cell —either increasing or decreasing the rates of normal cell processes.

FAT-SOLUBLE HORMONES

Hormones that are fat soluble are able to pass through the cell membrane. They produce their effects by binding with receptors in the cell. Fat-soluble hormones include the sex hormones and thyroid hormone.

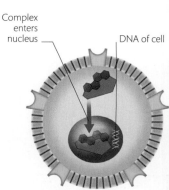

Hormone passes through membrane

Binds to receptor in cell

1 Binding in cell
The hormone diffuses through the cell membrane and binds to a mobile receptor within the cell itself, which is activated by the process of binding.

Complex enters nucleus

DNA of cell

2 Genes triggered
The hormone–receptor complex makes its way to the nucleus, where it binds to a region of DNA. This triggers genes to switch on or off enzymes that alter the cell's biochemical activity.

TRIGGERS FOR HORMONE RELEASE

Factors stimulating the production and release of hormones vary. Some endocrine glands are stimulated by the presence of certain minerals or nutrients in the blood. For example, low blood levels of calcium stimulate the parathyroid glands (see p.402) to release parathyroid hormone, while insulin, made in the pancreas, is released in response to rising glucose levels.

Many endocrine glands respond to hormones produced by other endocrine glands. For example, hormones produced by the hypothalamus stimulate the anterior pituitary gland to produce its hormones. These pituitary hormones in turn stimulate other glands; for example, adrenocorticotropic hormone stimulates the cortex (outer part) of the suprarenal gland to produce corticosteroid hormones.

Hormonal stimulation leads to the rhythmic release of hormones, with hormone levels rising and falling in a particular pattern. In a few cases, release of hormones is triggered by signals from the nervous system. An example is the medulla (inner part) of the suprarenal gland, which releases epinephrine (also called adrenaline) when stimulated by nerve fibers from the sympathetic nervous system. With this type of stimulation, hormone release occurs in bursts rather than rhythmically.

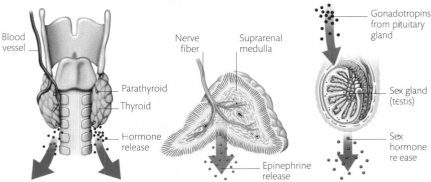

Blood level response
Low blood calcium prompts the parathyroid to release parathyroid hormone, which raises calcium levels. The release of calcitonin from the thyroid is also inhibited.

Nervous stimulation
Nerve fibers of the sympathetic nervous system, signaled by the hypothalamus, stimulate the suprarenal medulla to release epinephrine in times of stress.

Response to hormones
Gonadotropin hormones from the pituitary gland stimulate the sex glands (ovaries and testes) to secrete more sex hormones. In the testes, this is testosterone.

HORMONE REGULATION

Hormones are powerful and affect target organs at low concentrations. However, the duration of their action is limited—from seconds to several hours—so blood levels need to be kept within limits, tailored to the specific hormone and the body's needs. Many hormones are regulated by negative feedback mechanisms. These work like a thermostat-controlled heating system. The thermostat is set at the desired temperature and its sensor monitors the air. If the temperature drops, a control unit in the thermostat triggers the boiler to go on.

When the desired temperature is reached, the control unit triggers the boiler to go off. In a hormonal feedback system, the blood levels of a hormone (or chemical) are equivalent to the air temperature and the thermostat is often the hypothalamus–pituitary complex. If the blood levels of a hormone (or chemical) drop lower than is optimal, this triggers the endocrine gland to "turn on" and release hormones. Once blood levels have risen, the endocrine gland is triggered to "turn off."

Negative feedback loop
Hormone blood levels are kept within an optimal range (known as homeostasis) by negative feedback mechanisms. Levels are monitored and if they get too high or low production switches off or on.

Hormone secretion
Thyroid hormone (yellow) is secreted from the thyroid gland, following stimulation by hormones from the pituitary. The hormones enter the capillaries (blue) and travel in the bloodstream.

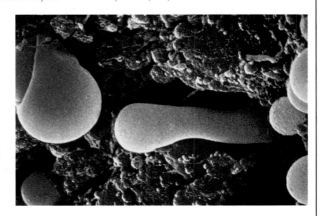

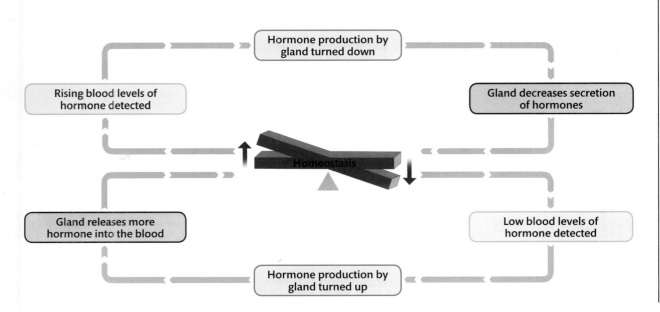

Rising blood levels of hormone detected

Hormone production by gland turned down

Gland decreases secretion of hormones

Homeostasis

Gland releases more hormone into the blood

Low blood levels of hormone detected

Hormone production by gland turned up

HORMONAL RHYTHM

The blood levels of some hormones vary according to the time of the month or day. Levels of female sex hormones follow a monthly cycle (see p.389), regulated by the rhythmic release of gonadotropin-releasing hormone (GnRH) from the hypothalamus. GnRH regulates release of hormones from the pituitary gland: follicle-stimulating hormone, which causes egg follicles to develop, and luteinizing hormone, which triggers egg release. Growth hormone (GH), cortisol from the suprarenal gland, and melatonin from the pineal gland follow diurnal (daily) cycles. GH and melatonin are highest at night, while cortisol peaks in the morning. Diurnal hormone rhythms are linked with sleep–wake or light–dark cycles.

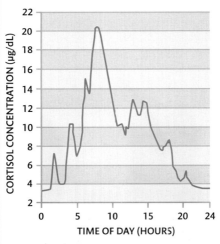

Cortisol levels
The hormone cortisol affects the metabolism and is controlled on a 24-hour cycle. Maximum concentration is achieved between 7 and 8 am each day, with a nadir at about midnight.

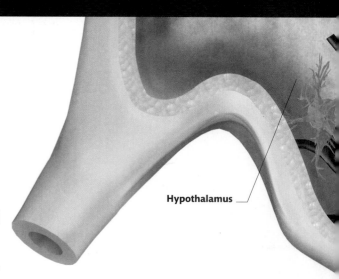

Hypothalamus

THE PITUITARY GLAND

The tiny pituitary gland, at the base of the brain, secretes hormones that stimulate other glands to produce their own hormones. It is often called the master gland because of its wide-ranging influences, but the real master is the hypothalamus, linking the endocrine and nervous systems.

HORMONE CONTROLLERS

The pituitary gland consists of two anatomically and functionally different parts: an anterior lobe and a posterior lobe. The anterior lobe forms the bulk of the pituitary, and consists of glandular tissue that manufactures hormones. The posterior pituitary is really part of the brain and is derived from hypothalamic tissue. It does not make hormones itself, but stores and releases hormones produced by the hypothalamus.

The two lobes link to the hypothalamus differently. The anterior lobe is linked by a system of interconnected blood vessels called a portal system. In a portal system, blood from arteries and veins connects directly rather than traveling through the heart first. This system allows hormones from the hypothalamus to be delivered to the anterior pituitary rapidly. The posterior lobe is linked to the hypothalamus by a nerve bundle, the hormone-producing neurons of which originate in the hypothalamus. The axons of these neurons extend into the posterior lobe and carry their hormones there for storage. Nerve signals from these neurons prompt release of their hormones "on demand."

Anterior lobe
Secretory cells, which manufacture hormones, can be seen around the edge of this color scanning electron microscope picture. Controlling hormones from the hypothalamus reach the secretory cells through capillaries, one of which is visible toward the bottom of the image. The inside of the capillary contains a macrophage, a type of cell that helps fight infection.

9
The number of **hormones** made by the **pea-sized pituitary** gland.

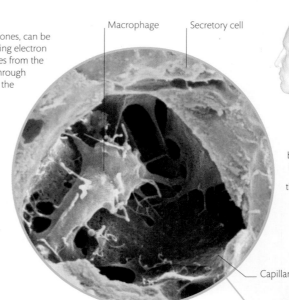

Macrophage   Secretory cell

Capillary wall

LOCATOR

Pituitary gland

Portal system
The system of blood vessels that carries regulatory hormones from the hypothalamus to the anterior pituitary

ANTERIOR LOBE HORMONES

Seven hormones are produced in the anterior pituitary. Four of these, known as tropic hormones, target other glands, prompting them to release their hormones. They are thyroid-stimulating hormone (TSH), adrenocorticotropic hormone (ACTH), follicle-stimulating hormone (FSH), and luteinizing hormone (LH). The others—growth hormone (GH), prolactin, and melanocyte-stimulating hormone (MSH)—act directly on target organs.

The release of hormones from the anterior pituitary is regulated by the hypothalamus, which secretes releasing or inhibiting hormones. Although different hormones from the hypothalamus reach the anterior lobe, secretory cells recognize those directed at them and secrete or release their specific hormones accordingly. The hormones are secreted into capillaries that drain into veins and into the general circulation to reach their target organs.

Capillary
Hypothalamic hormones enter the anterior lobe via capillaries

Secretory cell
Cells of the anterior lobe make and release hormones

Anterior lobe

Adrenal gland

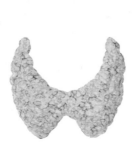

Testis

Ovary

Skin
MSH targets skin cells called melanocytes, which produce the hormone melanin. If produced in excess MSH can cause the skin to darken.

Adrenal glands
ACTH stimulates the cortex of the adrenal glands to secrete steroid hormones that help the body resist stress; they also affect the metabolism.

Thyroid gland
TSH stimulates the thyroid to secrete hormones that affect metabolism and body heat production, and promote normal development of many body systems.

Bone, skeletal, muscle, and liver
GH promotes the enlargement of bones, increase of muscle mass, and tissue building and renewal.

Sex glands
LH and FSH trigger the sex glands to make hormones. In females, they cause egg cells to ripen and stimulate ovulation; in males, they prompt sperm production.

Breast
Prolactin helps stimulate milk production by the mammary glands. Levels rise before menstruation, which may account for breast tenderness.

Venule
Small veins called venules carry hormones from the lobes of the pituitary gland into the blood stream

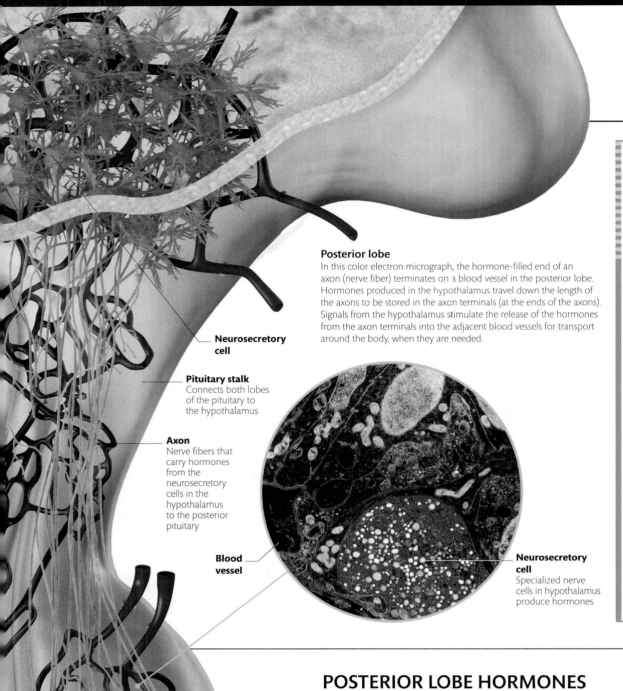

Neurosecretory cell

Pituitary stalk
Connects both lobes of the pituitary to the hypothalamus

Axon
Nerve fibers that carry hormones from the neurosecretory cells in the hypothalamus to the posterior pituitary

Blood vessel

Posterior lobe
In this color electron micrograph, the hormone-filled end of an axon (nerve fiber) terminates on a blood vessel in the posterior lobe. Hormones produced in the hypothalamus travel down the length of the axons to be stored in the axon terminals (at the ends of the axons). Signals from the hypothalamus stimulate the release of the hormones from the axon terminals into the adjacent blood vessels for transport around the body, when they are needed.

Neurosecretory cell
Specialized nerve cells in hypothalamus produce hormones

Axon terminal
Hormones made by the hypothalamus are stored and released here

Posterior lobe

GROWTH HORMONE

During childhood and the teenage years, growth hormone (GH) is essential for normal growth. In adults, it is needed to maintain muscle and bone mass and for tissue repair. If too much GH is produced during childhood, the actively growing long bones are affected and the person becomes abnormally tall, but with relatively normal body proportions. Too little GH during childhood results in slowed growth of long bones and short stature. An overabundance of GH after the growth of the long bones is complete results in enlarged extremities because bones of the hands, feet, and face remain responsive to the hormone. Too little GH in adulthood does not usually cause problems. If a lack of GH is identified before puberty, treatment with synthetic growth hormone means that affected children will reach a nearly normal height.

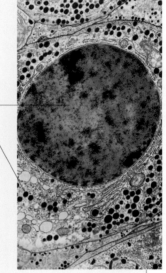

Nucleus

Granule

Somatotroph
Growth hormone is produced in cells called somatotrophs in the anterior lobe of the pituitary gland. This color electron micrograph shows numerous hormone-containing granules within the cell cytoplasm.

POSTERIOR LOBE HORMONES

Two hormones—oxytocin and antidiuretic hormone (ADH)—are stored in the posterior lobe of the pituitary gland. These hormones are not made in the gland but by the cell bodies of neurons located in two different areas of the hypothalamus. After production, the hormones are packaged in tiny sacs and transported down the axons (nerve fibers) of the neurons to the axon terminals, where they are stored until needed. Nerve impulses from the same hypothalamic neurons where they were produced trigger the release of the hormones into capillaries. From the capillaries, they pass into veins for distribution to their target cells. Oxytocin and ADH are almost identical in structure: each is made of nine amino acids, only two of which differ between them. However, each has a different effect. Oxytocin stimulates smooth muscle to contract, especially that of the uterus, cervix, and breast. ADH influences the balance of water in the body (see p.383).

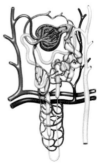

Muscle stretches

Pituitary gland anatomy
The pituitary gland consists of two lobes and a stalk, or infundibulum, which connects the lobes to the hypothalamus. Traveling through the stalk are blood vessels and nerve fibers that transport hormones from the hypothalamus.

Breast
Oxytocin prompts the release of milk from the mammary glands in breast-feeding. The baby's suckling triggers this hormonal response.

Uterus
Oxytocin stimulates contractions in labor. Stretching of the uterus triggers the hypothalamus to make oxytocin, which the posterior lobe releases.

Kidney tubules
ADH causes water to be returned to the blood by the kidney's filtering tubules, making urine more concentrated. ADH also affects blood pressure.

Cuddle hormone
Oxytocin is produced naturally during childbirth and is thought to play an important role in promoting nurturing maternal behavior. Oxytocin may also be responsible for feelings of satisfaction after intercourse.

HORMONE PRODUCERS

The thyroid, parathyroid, adrenal glands, and pineal gland are all organs of the endocrine system that exclusively produce hormones. Other organs and tissues also considered part of the endocrine system, but which are not exclusively endocrine organs, are discussed on pages 404–405.

THYROID GLAND

The butterfly-shaped thyroid gland is composed mainly of spherical sacs called follicles, the walls of which produce two important hormones, T3 (triiodothyronine) and T4 (thyroxine), collectively known as thyroid hormone (TH). Almost every cell in the body has receptors for TH, and it has widespread effects in the body. The thyroid gland is unusual among endocrine glands as it can store large quantities of hormones—maintaining about 100 days' supply of TH. The thyroid gland also produces calcitonin from parafollicular cells located between the follicles. An important effect of this hormone is to inhibit the loss of calcium from bones into the blood. It is most important in childhood, when skeletal growth is rapid.

Thyroid hormone regulation
Thyrotropin-releasing hormone (TRH) from the hypothalamus and thyroid-stimulating hormone (TSH) from the anterior pituitary stimulate the production and release of thyroid hormones (TH). Blood levels of TH feed back to the pituitary and hypothalamus to stimulate or inhibit activity.

PROCESSES INVOLVING TH	EFFECTS
Basal metabolic rate (BMR)	Increases BMR by stimulating the conversion of fuels (glucose and fats) to energy in cells; when BMR increases, metabolism of carbohydrates, fats, and proteins increases
Temperature regulation (calorigenesis)	Stimulates cells to produce and use more energy, which results in more heat being given off, raising body temperature
Carbohydrate and fat metabolism	Promotes use of glucose and fats for energy; enhances cholesterol turnover, thus reducing cholesterol
Growth and development	Acts with growth hormone and insulin to promote normal development of nervous system in fetus and infant, and normal growth and maturation of skeleton
Reproduction	Necessary for normal development of male reproductive system; promotes normal female reproductive ability and lactation
Heart function	Increases heart rate and force of contraction of heart muscle; enhances sensitivity of cardiovascular system to signals from the sympathetic nervous system (see p.311)

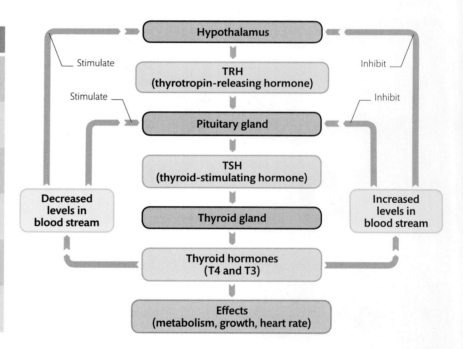

PARATHYROID GLANDS

The four tiny parathyroid glands at the back of the thyroid gland produce parathyroid hormone (PTH), the major regulator of calcium levels in blood. The correct balance of calcium is essential for many functions, including muscle contractions and the transmission of nerve impulses, so it needs to be controlled precisely. When blood calcium levels fall too low, PTH stimulates the release of stored calcium from bone into the blood and reduces calcium loss from the kidneys into urine. It indirectly increases the absorption of calcium from ingested food in the small intestine. In order for the intestine to absorb calcium, vitamin D is needed, but the ingested form is inactive: PTH stimulates the kidneys to convert vitamin D from its precursor form into its active form, calcitriol.

Effects of parathyroid hormone
Parathyroid hormone acts on the bone, kidneys, and (indirectly) the small intestine in order to increase the amount of calcium in the blood.

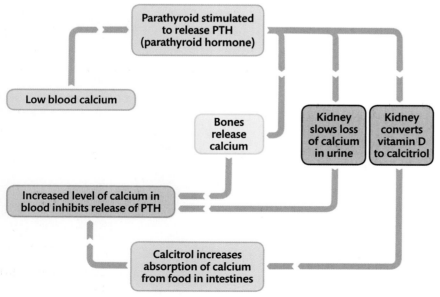

Parathyroid hormone has a relatively **short life span** in the blood stream, its levels falling by **50 percent every 4 minutes**.

ADRENAL GLANDS

The outer and inner regions of the adrenal glands differ from each other in structure, and each produces different hormones. The outer adrenal cortex is glandular tissue, while the inner medulla is part of the sympathetic nervous system and contains bundles of nerve fibers.

The adrenal cortex produces three groups of hormones: mineralocorticoids, corticosteroids, and androgens. An important mineralocorticoid is aldosterone, which regulates the sodium–potassium balance in the body and helps adjust blood pressure (see p.405) and volume. The main glucocorticosteroid is cortisol, which controls the body's use of fat, protein, carbohydrates, and minerals.

It also helps the body to resist stress, including from exercise, infection, extreme temperatures, and bleeding. The androgens produced by the adrenals are relatively weak in their effects, compared with those produced by the ovaries and testes during late puberty and adulthood. However, they probably play a role in the appearance of underarm and pubic hair in both sexes. In adult women, they are linked to the sex drive. The adrenal medulla produces epinephrine and norepinephrine. In stressful situations, when the sympathetic nervous system becomes activated, the hypothalamus stimulates the adrenal medulla to secrete these hormones, which augment the stress response (see right).

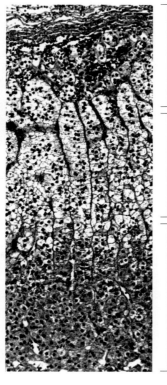

Adrenal anatomy
Each adrenal gland sits on a fatty pad on top of the kidney. The cortex forms the bulk of the gland. The medulla contains nerve fibers and blood vessels.

Adrenal cortex **Adrenal medulla** **Blood vessel**

Adrenal cortex zones
The adrenal cortex has three layers, or zones. Each consists of a different cell type and makes its own hormones. The outer zone, zona granulosa, is located just under the fibrous capsule that encloses the gland. The middle zone, zona fasciculata, is the widest and has columnar cells. Cells of the inner zone, zona reticularis, are cordlike.

Zona granulosa
Secretes mineralocorticoids, mainly aldosterone, which is important for regulating mineral balance and blood pressure

Zona fasciculata
Secretes corticosteroids, mainly cortisol, which regulates metabolism and helps the body cope with stress

Zona reticularis
Secretes weak androgens, which prompt growth of pubic and underarm hair at puberty and are responsible for the female sex drive

STRESS RESPONSE

When stress is detected, nerve impulses from the hypothalamus activate the sympathetic nervous system, including the adrenal medulla. These nerves start a fight-or-flight response, preparing the body for action. Hormones from the adrenal medulla prolong the response. Next, the body tries to respond to the emergency. This reaction is initiated mainly by hypothalamic-releasing hormones, which trigger the anterior pituitary to release growth hormone and other hormones that prompt the thyroid and adrenal cortex to secrete their hormones. These mobilize glucose and proteins for energy and repair.

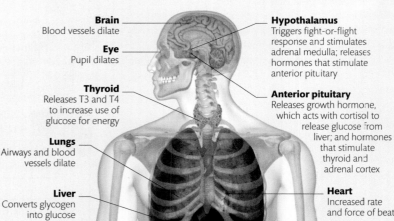

Brain
Blood vessels dilate

Eye
Pupil dilates

Thyroid
Releases T3 and T4 to increase use of glucose for energy

Lungs
Airways and blood vessels dilate

Liver
Converts glycogen into glucose

Adrenal cortex
Releases cortisol, which prompts liver to release glucose, adipose tissue to release fatty acids

Adrenal medulla
Secretes epinephrine and norepinephrine, which supplement the effects of the sympathetic nervous response

Skeletal muscle
Blood vessels dilate

Hypothalamus
Triggers fight-or-flight response and stimulates adrenal medulla; releases hormones that stimulate anterior pituitary

Anterior pituitary
Releases growth hormone, which acts with cortisol to release glucose from liver; and hormones that stimulate thyroid and adrenal cortex

Heart
Increased rate and force of beat

Stomach
Digestive activity decreases

Spleen
Contracts

Kidney
Urine output decreases

Intestines
Movement of food slows

Bladder
Sphincter muscle constricts

Skin
Blood vessels constrict, hair stands on end, and sweat pores open

PINEAL GLAND

The tiny pinecone-shaped pineal gland is located near the center of the brain, behind the thalamus. It secretes the hormone melatonin, which is involved in the body's sleep–wake cycle. Pineal activity lessens in bright light, so melatonin levels are low during the day. They rise at night, increasing about tenfold, making us sleepy. Bright light does not directly affect the pineal gland; instead, input from the visual pathways stimulates the suprachiasmatic nucleus (part of the hypothalamus), which

sends signals to the pineal gland via nerve connections near the spinal cord. The suprachiasmatic nucleus also controls other diurnal biological rhythms, such as body temperature and appetite, and it is likely that melatonin cycles influence these processes. Melatonin is also an antioxidant and may protect against damage from free radicals in the body. In animals that breed seasonally, melatonin inhibits reproductive function but it is not known whether melatonin affects reproduction in humans.

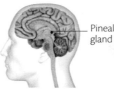

Pineal gland

LOCATOR

Melatonin levels
The level of circulating melatonin rises at night or when it is dark, creating a daily rhythm of rising and falling hormone levels.

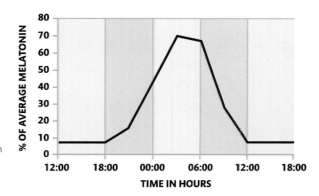

PANCREAS

The pancreas is a dual-purpose gland with both digestive and endocrine functions. The bulk of the gland consists of acinar cells, which produce enzymes used in digestion (see pp.376–77). Scattered among these cells are about a million pancreatic islets, or islets of Langerhans, cell clusters that produce pancreatic hormones. There are four different types of hormone-producing cell. Beta cells make insulin, which enhances transport of glucose into cells, where it is used for energy or converted into glycogen for storage. In this way, beta cells lower blood glucose levels. Alpha cells secrete glucagon, which has the opposite effect of insulin, stimulating release of glucose from the liver and raising blood glucose levels. Somatostatin, secreted by delta cells, regulates alpha and beta cells. There are only a few F cells. They secrete pancreatic peptide, which inhibits secretion of bile and pancreatic digestive enzymes.

Pancreatic islets
Surrounded by enzyme-producing acinar cells, the islets contain four types of cell: alpha, beta, delta, and F.

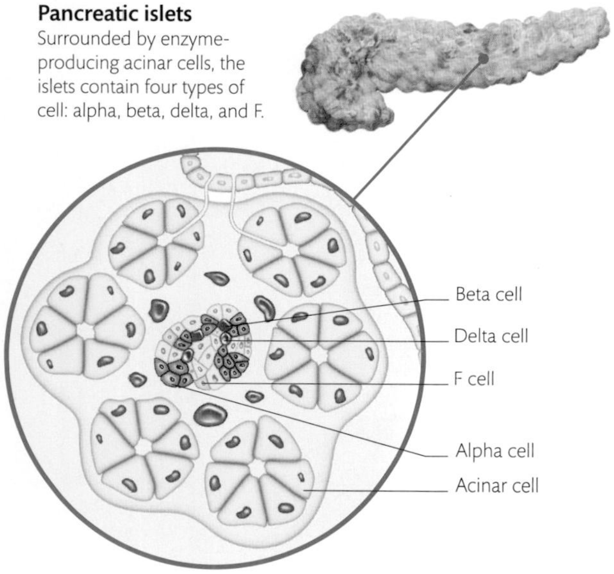

Beta cell
Delta cell
F cell
Alpha cell
Acinar cell

Blood sugar regulation
The body needs to regulate blood glucose levels so that cells receive enough energy to meet their needs. The main source of fuel is glucose, which is carried in the blood stream—any excess glucose is stored in liver, muscle, and fat cells. The pancreatic hormones insulin and glucagon prompt storage or release of glucose from cells, keeping blood levels stable.

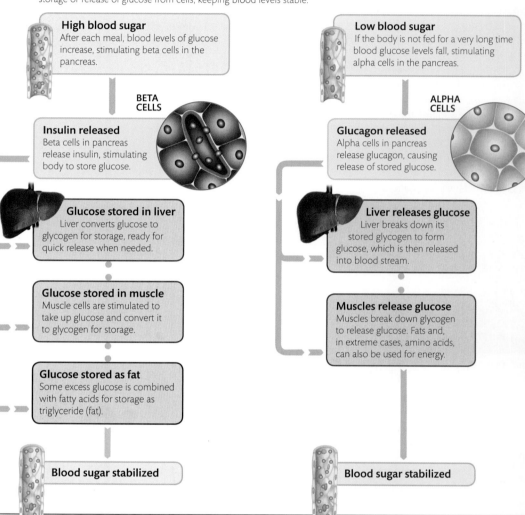

High blood sugar
After each meal, blood levels of glucose increase, stimulating beta cells in the pancreas.

BETA CELLS

Insulin released
Beta cells in pancreas release insulin, stimulating body to store glucose.

Glucose stored in liver
Liver converts glucose to glycogen for storage, ready for quick release when needed.

Glucose stored in muscle
Muscle cells are stimulated to take up glucose and convert it to glycogen for storage.

Glucose stored as fat
Some excess glucose is combined with fatty acids for storage as triglyceride (fat).

Blood sugar stabilized

Low blood sugar
If the body is not fed for a very long time blood glucose levels fall, stimulating alpha cells in the pancreas.

ALPHA CELLS

Glucagon released
Alpha cells in pancreas release glucagon, causing release of stored glucose.

Liver releases glucose
Liver breaks down its stored glycogen to form glucose, which is then released into blood stream.

Muscles release glucose
Muscles break down glycogen to release glucose. Fats and, in extreme cases, amino acids, can also be used for energy.

Blood sugar stabilized

OVARIES AND TESTES

The female ovaries and male testes, also known as gonads, produce eggs and sperm respectively. They also produce sex hormones, the most important of which are estrogens and progesterone in females, and testosterone in males. Release of these sex hormones is stimulated by follicle-stimulating hormone (FSH) and luteinizing hormone (LH) from the anterior pituitary gland. Before puberty, FSH and LH are almost absent from the blood stream, but during puberty they begin to rise, causing the ovaries and testes to increase hormone production. As a result, secondary sexual characteristics develop and the body is prepared for reproductive functions. The hormone inhibin inhibits release of FSH and LH. In males it regulates sperm production and in females it plays a role in the menstrual cycle. The ovaries also produce relaxin, which prepares the body for childbirth.

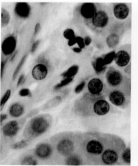

TESTICULAR TISSUE

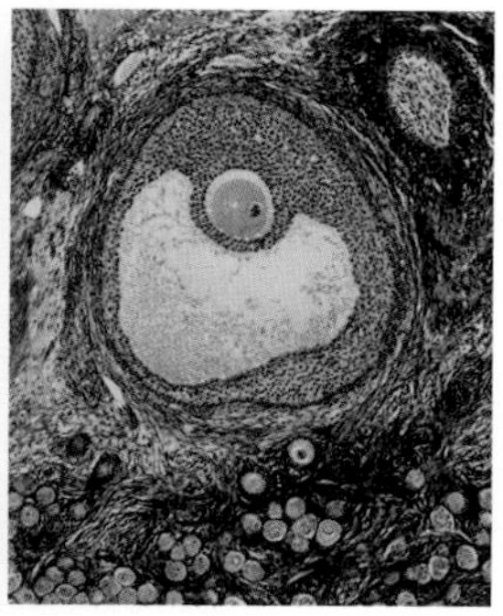

OVARIAN TISSUE

Hormone-producing cells
In the testes, interstitial cells (dark circles) secrete testosterone. In the ovaries, granulosa cells (dark purple dots), shown here surrounding an egg follicle, produce estrogen.

OVARIAN HORMONES	TESTICULAR HORMONES
Estrogens and progesterone Stimulate egg production; regulate menstrual cycle; maintain pregnancy; prepare breasts for lactation; promote development of secondary sexual characteristics at puberty	**Testosterone** Determines "sex" of brain in fetus; stimulates descent of testes before birth; regulates sperm production; promotes development of secondary sexual characteristics at puberty
Relaxin Makes the pubic symphysis more flexible during pregnancy; helps cervix to widen during labor and delivery	**Inhibin** Inhibits secretion of follicle-stimulating hormone from the anterior pituitary
Inhibin Inhibits secretion of follicle-stimulating hormone from the anterior pituitary	

OTHER HORMONE PRODUCERS

Many organs in the body that primarily have another function also produce hormones, including the kidneys, heart, skin, adipose tissue, and gastrointestinal tract. Although not as well known as hormones from purely endocrine glands such as the thyroid, they are just as important in controlling vital functions. Hormones from the kidneys and heart help control blood pressure and stimulate production of red blood cells. Skin is responsible for supplying the body with much of its vitamin D by producing cholecalciferol, a precursor form of the vitamin. Endocrine cells lining the gastrointestinal

tract secrete a number of different hormones, most of which play a role in the digestive process. Some of these hormones, called incretins, have sparked particular interest as they affect many different body tissues. Incretins stimulate insulin production in the pancreas, enhance bone formation, help promote energy storage, and, by targeting the brain, suppress appetite. Researchers hope that in the future incretins may be useful in treating diabetes mellitus and obesity. The hormone leptin, produced by adipose tissue, also affects appetite, and has provoked interest as a possible aid to weight control.

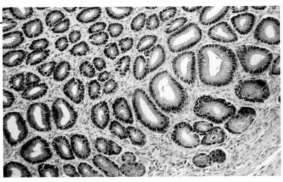

Stomach pylorus glands
This micrograph shows a section through gastric glands (pink) in the stomach. These glands contain endocrine cells that produce gastrin.

Adipose tissue is not just a passive energy reserve, but an active endocrine organ that may hold the **key to controling obesity** and its damaging effects.

HORMONAL CONTROL OF BLOOD PRESSURE

The nervous system responds to sudden changes in blood pressure, but longer term control is managed by hormones. Low blood pressure prompts the kidneys to secrete renin. Renin generates angiotensin, which constricts arteries and raises blood pressure. The adrenal glands, pituitary gland, and heart also respond to low or high blood pressure by secreting aldosterone, ADH (antidiuretic hormone), and natriuretic hormone respectively. These hormones alter the amount of fluid excreted by the kidneys, which affects the volume of blood in the body and hence blood pressure.

Hormone-producing tissues
Various body organs not classified as endocrine glands contain isolated cell clusters that release hormones. These hormones regulate many important processes in the body.

Kidney
Hormone: erythropoietin
Trigger: low level of oxygen in blood
Effects: stimulates bone marrow to increase production of red blood cells

Hormone: renin
Trigger: low blood pressure or blood volume
Effects: initiates mechanism for release of aldosterone from adrenal cortex; returns blood pressure to normal

Stomach
Hormone: gastrin
Trigger: response to food
Effects: stimulates gastric acid secretion

Hormone: ghrelin
Trigger: long period without eating
Effects: appears to stimulate appetite and eating; stimulates growth hormone secretion

Duodenum
Hormone: intestinal gastrin
Trigger: response to food
Effects: stimulates gastric acid secretion and movements of gastrointestinal tract

Hormone: secretin
Trigger: acid environment
Effects: stimulates release of bicarbonate-rich juice from pancreas and bile ducts; inhibits production of gastric acid in stomach

Hormone: cholecystokinin
Trigger: response to fats in food
Effects: stimulates secretion of enzymes in pancreas, and contraction and emptying of gallbladder to allow bile and pancreatic enzymes to enter duodenum

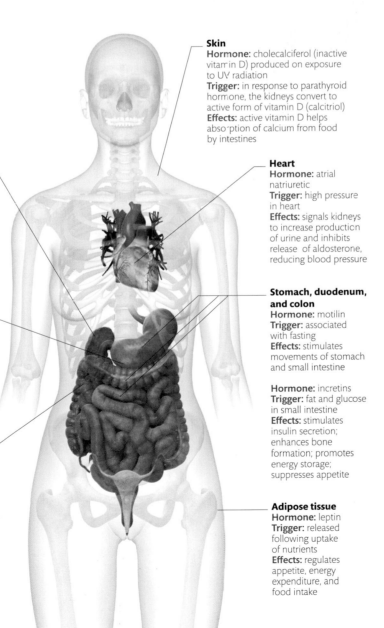

Skin
Hormone: cholecalciferol (inactive vitamin D) produced on exposure to UV radiation
Trigger: in response to parathyroid hormone, the kidneys convert to active form of vitamin D (calcitriol)
Effects: active vitamin D helps absorption of calcium from food by intestines

Heart
Hormone: atrial natriuretic
Trigger: high pressure in heart
Effects: signals kidneys to increase production of urine and inhibits release of aldosterone, reducing blood pressure

Stomach, duodenum, and colon
Hormone: motilin
Trigger: associated with fasting
Effects: stimulates movements of stomach and small intestine

Hormone: incretins
Trigger: fat and glucose in small intestine
Effects: stimulates insulin secretion; enhances bone formation; promotes energy storage; suppresses appetite

Adipose tissue
Hormone: leptin
Trigger: released following uptake of nutrients
Effects: regulates appetite, energy expenditure, and food intake

Pituitary gland
ADH produced by hypothalamus is stored here and secreted when blood pressure falls

ADH
Promotes water retention by kidneys, which raises blood pressure

Natriuretic hormone
Acts on kidneys to lower blood pressure by inhibiting renin secretion and promoting excretion of sodium and water

Heart
Elevated blood pressure stretches atria of heart, stimulating atrial endocrine cells to produce natriuretic hormone

Adrenal glands
Produce aldosterone when stimulated by angiotensin, which is activated by renin from kidneys

Kidney
Low blood pressure reduces blood flow through kidneys and stimulates them to produce the hormone renin

Aldosterone
Causes kidneys to retain sodium and water, increasing amount of fluid in body and raising blood pressure

Renin
Activates angiotensin in arteries

Hormonal action
The hormones that raise or lower blood pressure become effective over a period of several hours. Their effects may last for days.

life cycle

Each human is unique, with an individual genetic makeup. This section tracks the changes that take place over each person's life cycle, from what characteristics are inherited from their parents, through to childhood, puberty, old age, and eventually death.

LIFE'S JOURNEY

Like all living organisms, every human is created out of elements from its parents. Having grown from infancy to a mature state, where reproduction of the next generation is possible, a gradual aging precedes the eventual decline toward death.

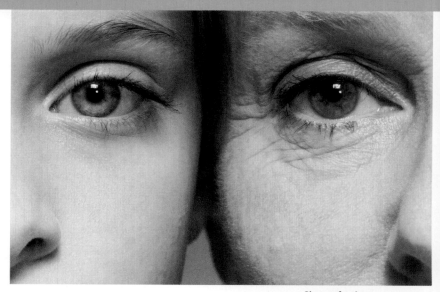

Signs of aging
Wrinkles form with age as the skin becomes drier, thinner, droopier, and less elastic.

CONCEPTION TO DEATH

From the moment of fertilization, through the resulting development of a ball of cells that contains a new combination of genetic material, the human fetus grows in size and complexity. By birth, its organs are functioning, yet size and proportion continue to change as the infant grows. Major changes occur at puberty when, under new hormonal influences, the secondary sexual characteristics develop, preparing the body for potential reproduction. Fertility is time-limited for women, and at menopause the female reproductive system becomes less responsive to hormonal stimulation and eventually ovulation ceases. Men produce sperm until the end of their lives, although less efficiently. As the body ages, its tissues become less able to repair and regenerate and disease develops, leading to death.

By 2020, for the first time in human history, the number of people in the world aged **65 years** and older will **exceed** the number of children **under five**.

DEVELOPMENT AND AGING

Little is understood about the aging process, including why and how it occurs. During development there is evidence of degenerative change affecting many cellular components. Cells are the fundamental structures that comprise organs; factors known to affect ongoing cell function, division, and repair, such as free radicals and UV radiation, have been shown to reduce cellular longevity and hence organ function. On a macroscopic level, disease processes can be found to have started even in children, for example the fatty deposits that occur within blood vessel walls in atherosclerosis.

The multiplication, regeneration, and death of cells is a necessary part of life, but at some point their ability to regenerate successfully fails. Cancers develop when cell regeneration is uncontrolled and cells multiply rapidly and abnormally; organ failure occurs when the cells cannot regenerate at all.

Death rates rise after the age of 30, with women often surviving longer than men, probably due to the protective effects of female hormones prior to the menopause. Age-related deterioration of cell function relates to many factors, but eventually death occurs as a result of organ failure.

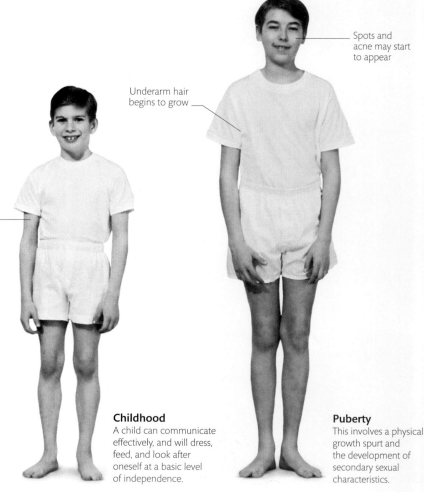

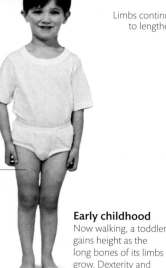

Young and old
The hands of babies and adults are similar in shape and structure, yet size, muscle bulk, skin color, texture, and surface markings can identify the individual's age.

Spots and acne may start to appear

Underarm hair begins to grow

Limbs continue to lengthen

Stages of man
All the organs and tissues in the body continue to grow until the end of puberty. Brain development generates early motor skills, such as walking and dextrous tool use, as well as higher functions, such as speech and logical thought. After middle age, these skills decline as the brain deteriorates and body tissues, including muscles, become weaker and less able to respond to cerebral command.

Skeletal and muscular proportions start to change

Infancy
During the first year, an infant develops many motor skills, including mobility: from crawling, to shuffling, then walking.

Early childhood
Now walking, a toddler gains height as the long bones of its limbs grow. Dexterity and language develop.

Childhood
A child can communicate effectively, and will dress, feed, and look after oneself at a basic level of independence.

Puberty
This involves a physical growth spurt and the development of secondary sexual characteristics.

LIFE EXPECTANCY

There are many factors affecting life expectancy. Women usually survive longer than men, probably due to the protective effects of hormones released before the menopause. Around the world, average life expectancy varies, from less than 50 years in parts of Africa to more than 80 in Japan, Canada, Australia, and parts of Europe. This is due to genetic tendencies, lifestyle factors, sanitation, and the prevalence of infectious diseases. Historically, lifespans have been increased by improvements in sanitation, healthcare, and nutrition, among other factors.

Life expectancy around the world
This chart shows the life expectancy of people living in the world's 25 most highly populated countries. Life expectancy is lowest in poor countries and those affected by war, and highest in the developed world.

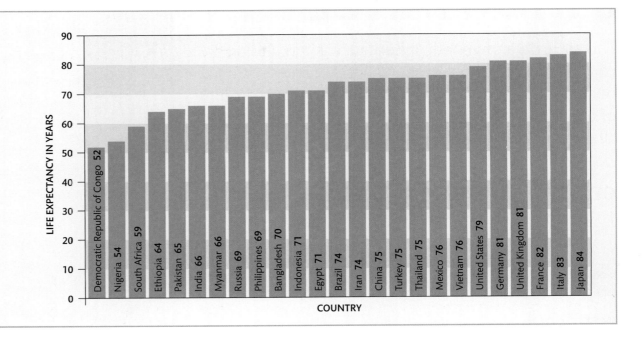

LIFE EXPECTANCY IN YEARS

Country	Life expectancy
Democratic Republic of Congo	52
Nigeria	54
South Africa	59
Ethiopia	64
Pakistan	65
India	66
Myanmar	66
Russia	69
Philippines	69
Bangladesh	70
Indonesia	71
Egypt	71
Brazil	74
Iran	74
China	75
Turkey	75
Thailand	75
Mexico	76
Vietnam	76
United States	79
Germany	81
United Kingdom	81
France	82
Italy	83
Japan	84

COUNTRY

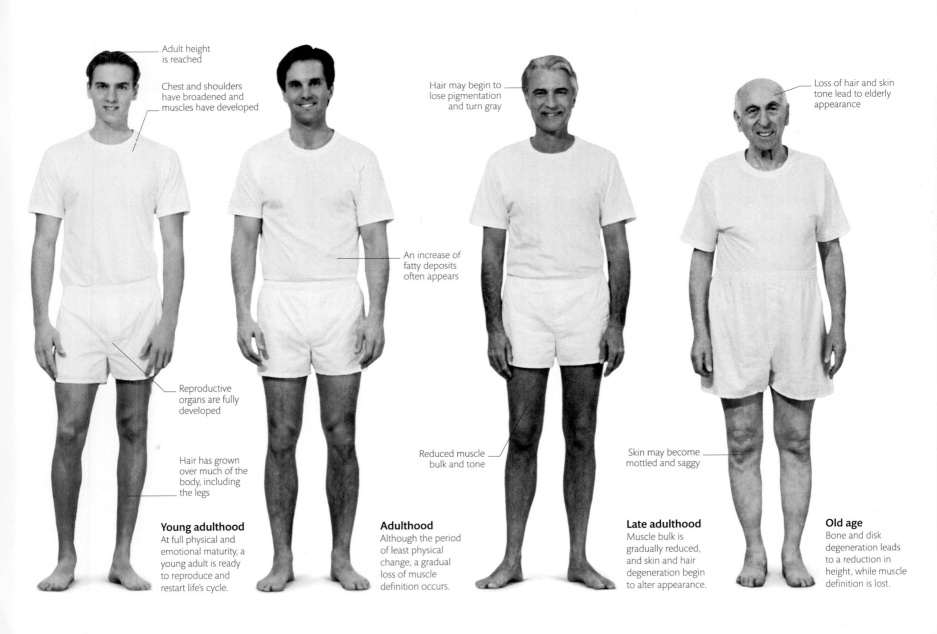

Adult height is reached

Chest and shoulders have broadened and muscles have developed

Reproductive organs are fully developed

Hair has grown over much of the body, including the legs

Hair may begin to lose pigmentation and turn gray

An increase of fatty deposits often appears

Reduced muscle bulk and tone

Loss of hair and skin tone lead to elderly appearance

Skin may become mottled and saggy

Young adulthood
At full physical and emotional maturity, a young adult is ready to reproduce and restart life's cycle.

Adulthood
Although the period of least physical change, a gradual loss of muscle definition occurs.

Late adulthood
Muscle bulk is gradually reduced, and skin and hair degeneration begin to alter appearance.

Old age
Bone and disk degeneration leads to a reduction in height, while muscle definition is lost.

INHERITANCE

The basic data of genetic inheritance is the unique combination of genes lying in chromosomes within our cells. Created from our parents' genes at the point of conception, this combination forms a template for all cellular forms and functions throughout the body.

GENERATION TO GENERATION

Chromosomes are inherited as a unique parental combination. Most tissues are comprised of cells that contain two sets of 23 chromosomes (diploid cells). These divide by mitosis (see p.21) to make replica cells with the same chromosomal content. However, sex cells (the egg or sperm), or gametes, form with only one set of chromosomes. When an egg and a sperm fuse at conception, the resulting embryonic cells contain two sets again, combining 23 chromosomes each from the mother and father. Traits from both parents may or may not be expressed, depending on what has been inherited, and whether genes are recessive or dominant (see opposite). The physical expression of a gene (its phenotype), such as hair color, can be obvious, but unseen tendencies to disease may also be inherited. Mutations that occur during cell division can be passed down through generations.

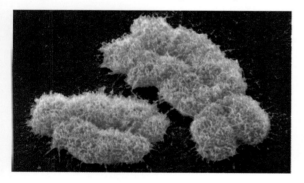

X and Y chromosomes
The sex chromosomes provide data for sexual development and function. Females have two X chromosomes (right) in each cell; males have one X and one Y (left), named because of their basic shapes.

MAKING SEX CELLS

Sex cells divide in a different way from normal mitotic cell division (see p.21). This process, called meiosis, is distinct from mitosis and also includes a further division, so that the chromosomal content of the resulting gametes is halved and also mixed.

Duplicated chromosome

Nuclear membrane

1 Preparation
The cell's DNA strands divide to form two identical sets of each chromosomal pair. The nuclear membrane starts to break down.

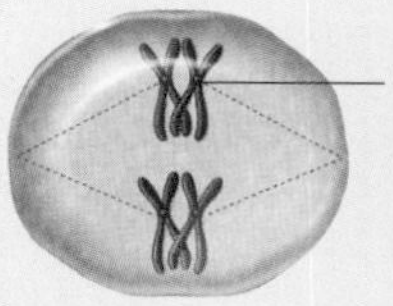

Matching pair of chromosomes

2 Pairing
The two sets then pair up and part again; genetic material may cross over within the pairs, giving a new mix for the daughter cells.

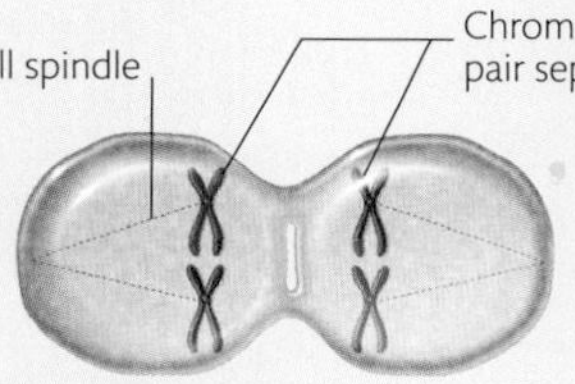

Cell spindle

Chromosome pair separates

3 First separation
The cell spindles pull the chromosomes apart so that there is one set of each pair in each of the two cells that form.

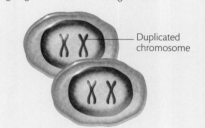

Duplicated chromosome

4 Two offspring
There are now two daughter cells, each with a pair of the 23 chromosomes (but these are slightly different from the original ones).

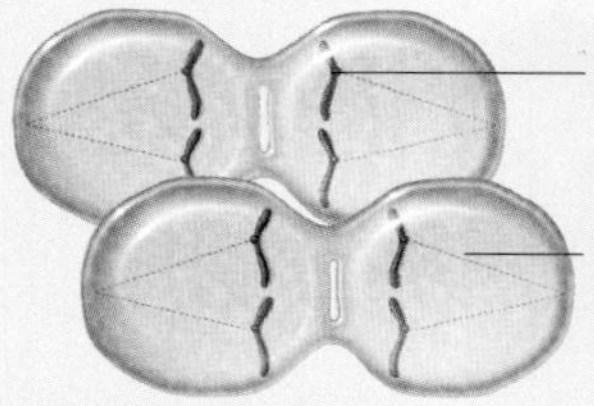

Single chromosome

Spindle

5 Second separation
The chromosomal content is divided again so that each sex cell contains just one set of 23 chromosomes.

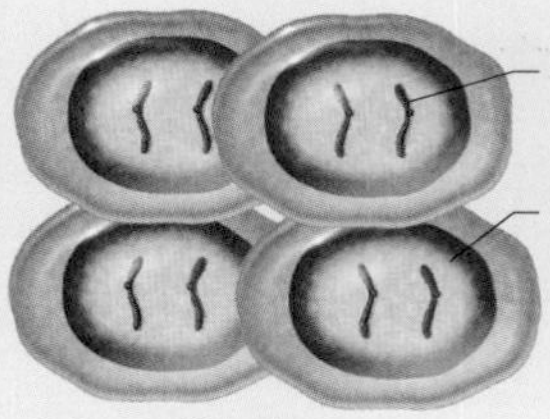

Chromosome

Nucleus

6 Four offspring
The resulting four cells all have a single set of 23 chromosomes, each set containing a mix of the genes from the original pair of chromosomes.

EPIGENETIC PROCESSES

Although the human genome has been mapped and partly explains patterns of disease inheritance, environmental factors also play a part. Epigenetics is the science of all modifications to genes other than changes to the DNA sequence itself. Various intracellular changes, called epigenetic processes, alter gene activity—in effect they can switch particular genes on or off. Although every cell contains a full set of DNA, each cell epigenetically silences some genes, leaving active only those it needs to do its specialized function. However, when this process is affected by external, environmental factors, abnormal cells may develop and grow uncontrollably as a tumor. As the understanding of epigenetics increases, scientists are learning more about how genes are affected by their environment, and how resultant conditions may eventually be prevented or treated.

Twin studies
Studies of genetically identical (monozygotic) twins have shown that, over time, environmental factors affect genetic expression.

MIXED GENES

Sophisticated technology allows the study of gene sequences within several generations of a family. This enables scientists to understand the origin of a particular gene as well as to predict the risk of a feature, or disease, linked to that gene developing within current and future generations. A child's genetic material is inherited from both parents. They, in turn, will have inherited genetic material from their own mother and father, and so on back through the generations.

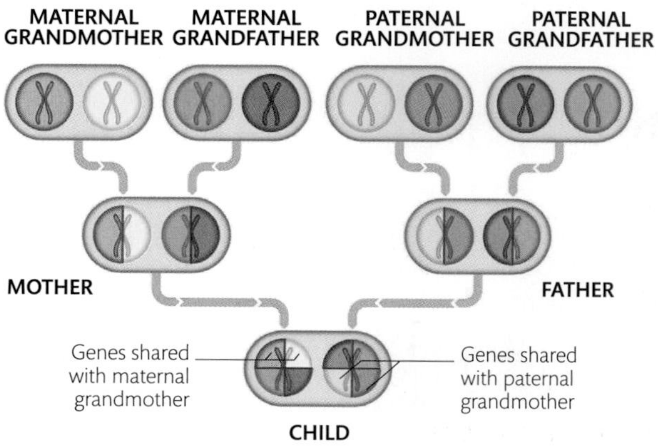

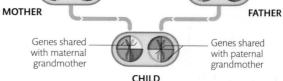

MATERNAL GRANDMOTHER MATERNAL GRANDFATHER PATERNAL GRANDMOTHER PATERNAL GRANDFATHER

MOTHER

FATHER

Genes shared with maternal grandmother

Genes shared with paternal grandmother

CHILD

Units of inheritance
This diagram shows how genes are passed down generations and shuffled—not blended—to create new combinations.

RECESSIVE AND DOMINANT GENES

Whether the effects of the message held in a gene on one of the chromosomal pairs is expressed or not depends on whether it is recessive or dominant. If both genes are the same, the individual is said to be homozygous for that gene, but if they are different the person is described as heterozygous. Dominant genes overwhelm the message in recessive genes, so that only one of the pair needs to be dominant to see its effects. Recessive genes may show their effects if both of the pair are recessive, but if there is only one recessive gene it is suppressed by the presence of the dominant gene.

Recessive and recessive
When both parents are homozygous for a recessive gene, here the gene for blue eyes, the phenotype will be expressed because there is no dominant gene to overwhelm it. This means that all offspring will have blue eyes.

Recessive gene for blue eyes — BLUE EYE   BLUE EYE

ALL INDIVIDUALS HAVE BLUE EYES

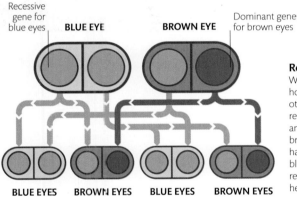

Recessive gene for blue eyes — BLUE EYE   BROWN EYE — Dominant gene for brown eyes

BLUE EYES   BROWN EYES   BLUE EYES   BROWN EYES

Recessive and mixed
When one parent is homozygous recessive and the other heterozygous (has one recessive gene for blue eyes and one dominant gene for brown eyes), the offspring have an equal chance of being blue-eyed homozygous recessive, or brown-eyed heterozygous.

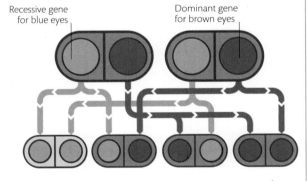

Recessive gene for blue eyes   Dominant gene for brown eyes

Mixed and mixed
When both parents are brown-eyed heterozygous, the offspring have a one in two chance of being brown-eyed heterozygous; a one in four chance of being homozygous blue-eyed; or a one in four chance of being homozygous brown-eyed.

BLUE EYES   BROWN EYES   BROWN EYES   BROWN EYES

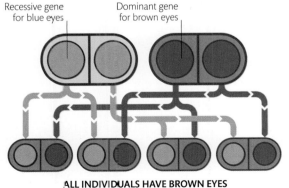

Recessive gene for blue eyes   Dominant gene for brown eyes

ALL INDIVIDUALS HAVE BROWN EYES

Dominant and recessive
With two homozygous individuals, where one is homozygous recessive blue-eyed and the other is homozygous dominant brown-eyed, all the offspring will be heterozygous brown-eyed.

SEX-LINKED INHERITANCE

Because males have only one X chromosome, if recessive genetic phenotypes are carried on the sex chromosomes they will show a sex-linked pattern of inheritance. Women have two X chromosomes, so recessive phenotypes may be hidden by a dominant gene on the other, and she will "carry" the gene. However, in males, the presence on their single X chromosome allows that gene to be expressed whether recessive or dominant.

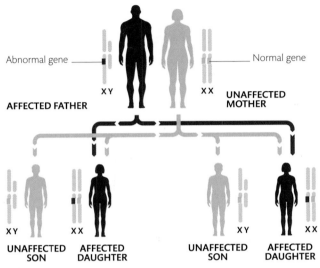

Abnormal gene   X Y — AFFECTED FATHER   X X — UNAFFECTED MOTHER — Normal gene

XY — UNAFFECTED SON   XX — AFFECTED DAUGHTER   XY — UNAFFECTED SON   XX — AFFECTED DAUGHTER

X-linked dominant inheritance
The "abnormal" gene is on the father's X chromosome. This example shows an abnormal gene inherited in a dominant fashion. The gene is expressed even if there is also a normal gene present.

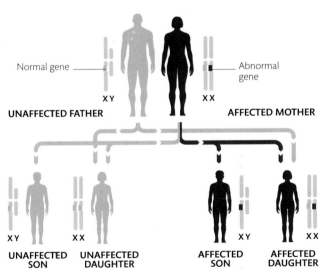

Normal gene — X Y — UNAFFECTED FATHER   X X — AFFECTED MOTHER — Abnormal gene

XY — UNAFFECTED SON   XX — UNAFFECTED DAUGHTER   XY — AFFECTED SON   XX — AFFECTED DAUGHTER

Affected mother, unaffected father
In this case, the mother is affected. There is a 50 percent chance that a daughter or son would inherit the faulty gene and have the condition.

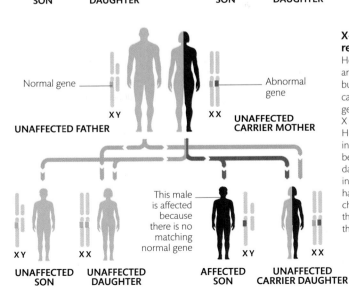

Normal gene — X Y — UNAFFECTED FATHER   X X — UNAFFECTED CARRIER MOTHER — Abnormal gene

This male is affected because there is no matching normal gene

XY — UNAFFECTED SON   XX — UNAFFECTED DAUGHTER   XY — AFFECTED SON   XX — UNAFFECTED CARRIER DAUGHTER

X-linked recessive gene
Here, both parents are unaffected, but the mother carries the abnormal gene on one of her X chromosomes. Her sons have a one in two chance of being affected. Her daughters have a one in two chance of having one affected chromosome and, therefore, carrying the condition.

DEVELOPING EMBRYO

From fertilization to the end of the eighth week of pregnancy, the embryo grows rapidly from a ball of cells into a mass of distinct tissue areas and structures, which develop into organs within a recognizable human form.

EMERGING BODY STRUCTURES

The cell mass, or embryo, that results from fertilization undergoes cell division (cleavage) within 24–36 hours to become two cells. About 12 hours later, it divides into four cells, and continues to divide until it becomes a ball of 16–32 cells, which is called a morula. During cell division, the embryo progresses down the fallopian tube to the uterine cavity. Around day six, the morula develops a hollow central cavity, after which it is described as a blastocyst. The blastocyst then implants into the richly vascular endometrium (uterus lining).

The embryonic cells have started to differentiate into specific cell types as genes within its chromosomes are switched on or off. Within the inner cell mass of the blastocyst, an embryonic disk forms, consisting of three primary germ layers: endoderm, mesoderm, and ectoderm. These layers are the origins of all the structures in the body. The endoderm cells will form linings of systems such as the gastrointestinal, respiratory, and urogenital tracts, as well as some glands and ductal parts of organs such as the liver; mesoderm cells develop into the skin dermis, the connective tissues of muscle, cartilage and bone, the blood and lymphatic systems, as well as some glands; ectoderm cells form the skin epidermis, tooth enamel, sensory organ receptor cells, and other parts of the nervous system.

Fertilization
Sperm approach the zona pellucida (the outer layer, or shell, that surrounds the egg), which must be pierced by a single sperm in order for the egg to be fertilized.

Embryo at 5 weeks
Already the embryo's external features, including the eyes, spine, and limb buds, are clearly visible, as is the umbilical cord. Scans can detect a pumping heart, and rudimentary major organs are in place, although not developed.

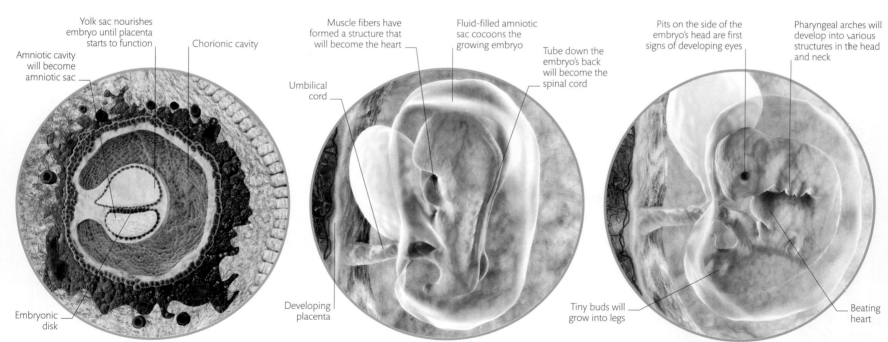

Amniotic cavity will become amniotic sac

Yolk sac nourishes embryo until placenta starts to function

Chorionic cavity

Embryonic disk

Muscle fibers have formed a structure that will become the heart

Umbilical cord

Fluid-filled amniotic sac cocoons the growing embryo

Tube down the embryo's back will become the spinal cord

Developing placenta

Pits on the side of the embryo's head are first signs of developing eyes

Pharyngeal arches will develop into various structures in the head and neck

Tiny buds will grow into legs

Beating heart

Differentiation
Having embedded into the maternal endometrium, the embryo at 2 weeks has already started to differentiate into various cellular types. The outer layers are forming the placenta, to provide nutrition via the maternal blood, but the main source of energy comes from the yolk sac, which has developed alongside the rapidly changing embryo.

Neural tube formation
Attached by the umbilical cord to the placenta, and suspended in the fluid of the amniotic sac, the $\frac{1}{8}$ in- (3 mm-) long embryo has formed a neural tube that will become the spinal cord. An enlarged area at one end will form the brain, while the other end curls under in a tail-like shape. Heart muscle fibers begin to develop in a simple tubal structure that pulsates.

Major organ formation
By 4 weeks, the $\frac{1}{5}$ in- (5 mm-) long embryo has formed rudimentary major organs. The heart has reorganized into four chambers, and now beats to pump blood through a basic vessel system. The lungs, gastrointestinal system, kidneys, liver, and pancreas are all now present, and a basic cartilaginous skeletal system has developed to provide a supportive structure.

2 WEEKS

3 WEEKS

4 WEEKS

DEVELOPMENT OF THE PLACENTA

The placenta develops from the outer layer of the blastocyst—the ball of cells that results when the sperm fertilizes the egg. The placenta has several functions. It provides a barrier to protect the baby from harmful substances and even foreign matter such as bacteria in maternal blood, while being a membrane across which it can bring in nutrients and oxygen from maternal blood and expel waste products. It also produces hormones essential for the continuation of the pregnancy.

1 Trophoblast proliferates
The outer layer of blastocyst cells become the trophoblast, which taps into the blood vessels of the maternal endometrium. This forms the placental bed across which nutrients and oxygen cross into the fetal blood system and waste products flow out.

2 Chorionic villi form
The flat trophoblastic layer develops fingerlike projections, called chorionic villi, growing out into the tissue of maternal blood sinuses to increase the surface area and augment nutrient transfer. Fetal blood vessels then grow into the chorionic villi.

3 Placenta established
By the fifth month, the placenta has become established, with a large network of villi protruding deep into maternal blood-filled chambers called lacunae. After implantation the placenta produces the human chorionic gonadotropin (hCG) hormone.

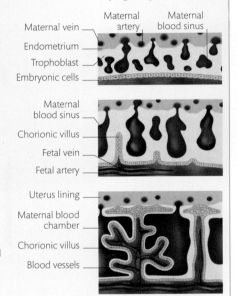

Maternal vein
Maternal artery
Maternal blood sinus
Endometrium
Trophoblast
Embryonic cells

Maternal blood sinus
Chorionic villus
Fetal vein
Fetal artery

Uterus lining
Maternal blood chamber
Chorionic villus
Blood vessels

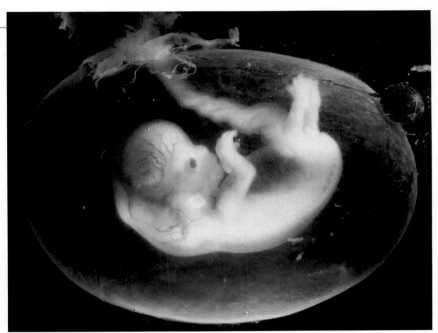

All the basic **organs** have formed and the **skeletal cartilage** starts to turn into **bone**. Spontaneous **movements** are occurring.

Cocooned fetus
An 8-week-old fetus is shown suspended by the umbilical cord within an intact amniotic sac. The shriveled yolk sac (red) can be seen separately to the right, hanging from the placental root of the umbilical cord.

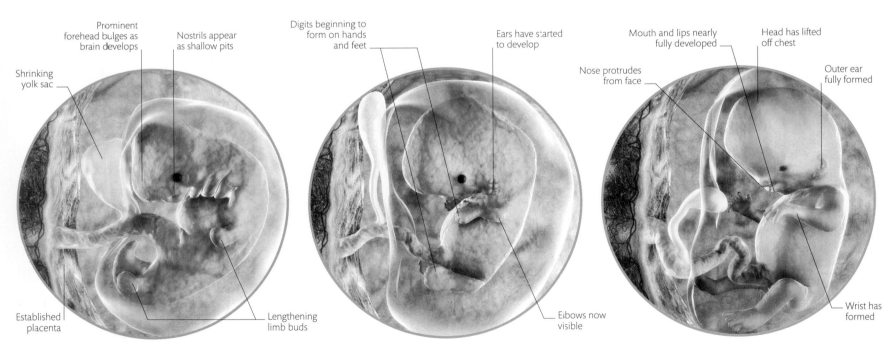

Prominent forehead bulges as brain develops
Nostrils appear as shallow pits
Shrinking yolk sac
Established placenta
Lengthening limb buds

Digits beginning to form on hands and feet
Ears have started to develop
Elbows now visible

Mouth and lips nearly fully developed
Head has lifted off chest
Nose protrudes from face
Outer ear fully formed
Wrist has formed

Limb development
The embryo starts to show a recognizably human form as the limb buds develop and lengthen and the early "tail" is reabsorbed. Neural tissue rapidly evolves into specialized sensory areas, such as the eye and the cochlear structure of the inner ear. An increasing amount of nutrition now comes via the placenta as the yolk sac starts to shrink.

Structural details
At 1 in (25 mm) long, the embryo is growing rapidly and its finer structural detail is forming. By 6 weeks, the hands will have formed fingers, the feet will have developed toes, and the basic eyes will have differentiated into structures including a lens, retina, and eyelids. Electrical brain activity is established and sensory nerves are developing.

Basic human shape
Now 1½ in (40 mm) long, the embryo has an obvious human shape, including a recognizable face and even the early detail of fingerprints. All the basic internal organs have formed and the skeletal cartilage starts to develop into bone. Spontaneous movements are occurring. After the end of the 8th week, the embryo is referred to as a fetus.

5 WEEKS   **6 WEEKS**   **8 WEEKS**

FETAL DEVELOPMENT

From 8 weeks until delivery, the fetus grows rapidly in size and weight. During this time, its body systems develop until it has reached a stage when it is sufficiently mature to sustain itself once separate from its mother after birth.

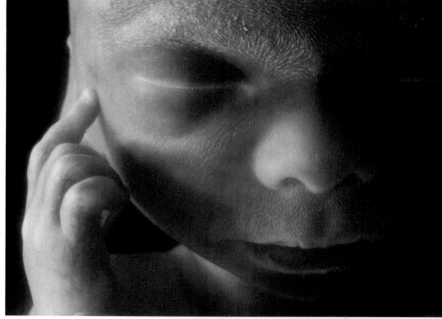

THE GROWING BABY

By the time an embryo has become a fetus, it has developed a clearly human form. From this point, measuring 1in (2.5 cm) long or roughly the size of a grape, it has 32 weeks to grow to an average birth weight of around 6½–8¾ lb (3–4kg) in

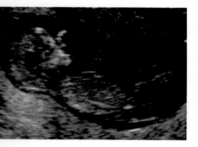

Fetus at 12 weeks
Ultrasound imaging shows the fetal heartbeat, spine, limbs, and even recognizable details such as facial features.

developed countries (less in developing countries, where maternal health can be less certain). Growth will depend on many factors, including maternal health, nutrition and lifestyle, fetal or placental disease or abnormalities, and also ethnic or familial trends in size and weight. Generally, the fetus is protected from minor or transient maternal illness, but more serious illnesses can affect its growth. Initially floating free in the amniotic fluid, as the fetus grows its movement becomes increasingly restricted until it fills the stretched uterine cavity. During the early period, growth is focused on the organs gaining size, body length, and structure, while fat deposition occurs later. Bones grow by cell division from the growth plates at either end of the long bones. Specialized cells of the nervous system, such as the retinal cells, become more refined and the brain cells gather detailed information as sensory input increases.

Fetus at 20 weeks
The skin is coated in a greasy substance called vernix, which protects it from prolonged contact with amniotic fluid.

Limbs are lengthening rapidly

Eyes have moved to the front of the face, but remain closed

Body has no underlying fat and bones appear prominent

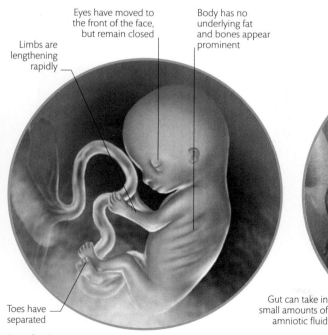

Toes have separated

Developing sensation
Weighing around 1½ oz (45 g) and measuring 3½ in (9 cm) long, the fetus is now active and is able to stretch out and test its muscles. Its eyes are shut but the brain and nervous system are both sufficiently developed for the fetus to sense pressure on its hands and feet, and it can open and close its fists and curl its toes in response to such stimuli.

11 WEEKS

Greater hand mobility means that the baby is able to suck its thumb

In the brain, nerve cells are growing from central to outer areas

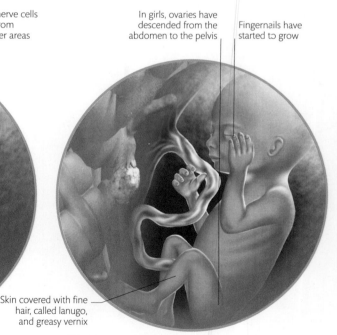

Gut can take in small amounts of amniotic fluid

Sucking, breathing, and swallowing
By this stage, the fetus has developed a swallowing action and will ingest amniotic fluid, which is then absorbed by the body. The kidneys are functioning, cleansing the blood and passing urine back into the amniotic fluid via the bladder and urethra. Breathing movements are occurring and the fetus will have discovered its mouth with its hands and may suck its thumb.

14 WEEKS

In girls, ovaries have descended from the abdomen to the pelvis

Fingernails have started to grow

Skin covered with fine hair, called lanugo, and greasy vernix

Making its presence felt
At 6 in (15 cm) long and weighing 11–14 oz (300–400 g), the fetus is highly active and the mother begins to feel fluttering sensations through the uterine wall. (The top of the uterus can now be felt above the pubic bone.) Unique fingerprints are now fully established on the fingers and toes of the fetus, and its heart and blood vessel systems are fully developed.

19 WEEKS

HOW THE PLACENTA WORKS

The placenta supplies the growing fetus with nutrients, such as glucose, amino acids, minerals, and oxygen, and removes waste products such as carbon dioxide. It does this by acting as a barrier between adjacent maternal and fetal blood flows, allowing these molecules to cross while protecting the fetus from maternal waste variation in her metabolism, and bacteria. The placenta secretes hormones, including estrogen, progesterone, and human chorionic gonadotropin (hCG). Maternal antibodies can cross the placenta in late pregnancy, giving the fetus passive immunity to infections, but the placenta also has several mechanisms to keep the mother's immune system from recognizing the fetus as foreign and attacking it.

- Uterine muscle
- Maternal blood vessels
- Flow of wastes
- Fetal blood vessels
- Maternal blood in intervillous space
- Flow of nutrients

Umbilical cord

Direction of blood flow from the fetus

Direction of blood flow to the fetus

Exchange of nutrients
Nutrient and waste exchange occurs across the walls of the placental blood vessels.

CONNECTED AND NOURISHED

The 6 in- (15 cm-) long umbilical cord connects the blood vessels of the placenta to the blood system of the fetus, allowing the flow of nutrients and return of waste. Unlike most adult blood vessels, the umbilical vein supplies oxygenated blood and nutrients, while the two arteries carry deoxygenated blood and waste products to the placenta. Abnormalities of the cord, such as being unusually short, long, or having only one artery, are associated with a variety of fetal malformations. The cord has few sensory nerves and is clamped and cut after birth.

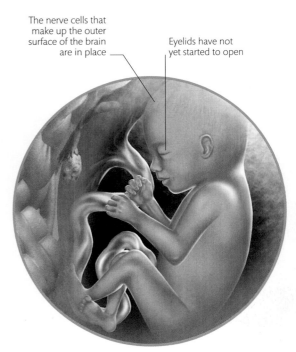

Umbilical lifeline
The blood vessels of the umbilical cord are protected and insulated within a gelatinous substance called Wharton's jelly.

From **22 weeks,** the fetus begins to stand a small but increasing **chance of survival** should it be born **prematurely**.

Hands are very active, touching the face, body, and umbilical cord

Inner ear organs have matured enough to send nerve signals to the brain

The nerve cells that make up the outer surface of the brain are in place

Eyelids have not yet started to open

Fluid-filled lungs are not quite ready for the outside world

A chance of survival
From 22 weeks, the fetus begins to stand a small but increasing chance of survival should it be born prematurely. Most body systems are sufficiently developed to cope with independence from the mother, although the biggest problem at this stage is the immature respiratory system. Although the breathing reflex is in place, the lungs are unable to secrete the vital surfactant needed to keep them open.

Every bone in the body now contains bone marrow, which produces red blood cells

Layers of body fat are being stored beneath the skin; fats contribute to the development of the nervous system

Responsive to sound and motion
Surrounded by constant maternal internal noise—heartbeat, bloodflow, and intestinal gurglings—the fetus is responsive to external noise or movement, quickening its heartbeat and increasing its own movement (felt by the mother as "kicks"), or, conversely, slowing when soothed. Now with developed balance mechanisms, it is aware of positional change.

TOWARD FULL TERM

Development during the final 3 months is mostly a process of consolidation as the fetus's organs have all formed but need to mature. The fetus continues to refine its various activities and functions, including movement, breathing, swallowing, and urination. The bowels show rhythmical activity, but contain a plug of sterile contents called meconium (comprising amniotic fluid, skin cells, lanugo hairs, and vernix) that is not usually passed until delivery. (However, if the fetus becomes stressed, for example by falling oxygen levels, some meconium may get passed into the amniotic fluid.) The fetus is rapidly gaining fat stores, and its growing lungs will have reached a stage of maturity at which they may be able to cope with breathing if premature delivery occurs.

Sensations become more acute—the eyes (already detecting simple light levels) will open, the ears pick up familiar sounds—and the fetus displays a sense of its surroundings and also of the state of its mother. If the mother relaxes, increasingly the fetus will too; if she is anxious or restless, it will respond to this.

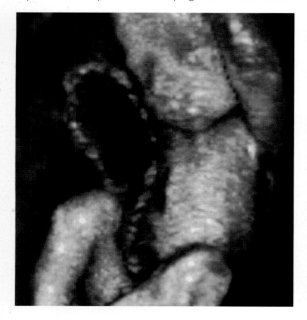

Fetus at 26 weeks
This 4D ultrasound image gives an all-round view of the fetus, showing head, torso, and limbs together with the umbilical cord and placenta. When the baby moves (time being the 4th dimension), its movement and structural development can also be assessed.

Brain waves revealing electrical activity have been detected from 6 weeks, and by 26 weeks **rapid eye movement** sleep occurs— usually associated with **dreaming**.

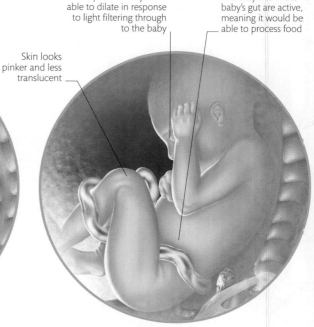
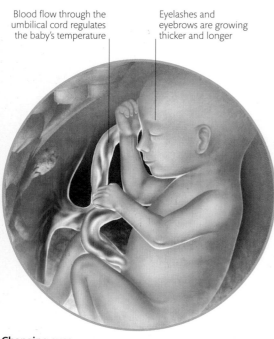

Blood flow through the umbilical cord regulates the baby's temperature

Eyelashes and eyebrows are growing thicker and longer

Changing eyes
Measuring 13 in (33 cm) long and weighing around 2 lb (850 g), the fetus has full sets of eyelashes and eyebrows, but will not open its eyes for another week or two, when the upper and lower lids have separated. The initial eye color will be blue, as true pigmentation does not occur until later, often not until after birth.

26 WEEKS

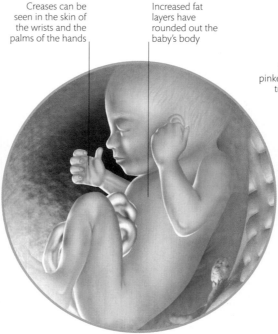

Creases can be seen in the skin of the wrists and the palms of the hands

Increased fat layers have rounded out the baby's body

Maturing lungs
The heart rate will have begun to slow slightly from its previous rate of 160 beats per minute (bpm) to 110–150 bpm. The cells that line the lungs are by now starting to secrete a substance (surfactant) that will help them to inflate when the baby takes its first breath. In boys, the testes will have moved down from the abdomen and will descend into the scrotum.

30 WEEKS

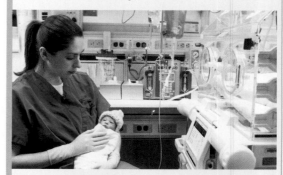

Pupils of the eyes are able to dilate in response to light filtering through to the baby

The enzymes in the baby's gut are active, meaning it would be able to process food

Skin looks pinker and less translucent

Skin changes and space restrictions
At a weight of about 4lb (1.9kg), increased fat deposition fills out early wrinkles. The vernix and lanugo begin to disappear and the skin loses its translucency. The fetus wriggles but there is little space for vigorous movement. Its eyes blink and breathing movements may result in hiccoughs—harmless spasms of the diaphragm.

35 WEEKS

Ready for birth
By 40 weeks, the baby's organs are mature, and it now fills the entire uterine space. It is ready to leave the womb and face the outside world.

THE NEWBORN

The first four weeks of a baby's life, known as the neonatal period, are a time of immense change and adaptation. This is also one of the most dangerous stages of life, with a higher risk of death than at any other time until retirement age.

STARTING OUT IN THE WORLD

At birth a baby has a head that is large in proportion to its body, and often misshapen due to molding of the skull during passage through the birth canal. The abdomen is relatively large, with the appearance of a pot belly, whereas the chest is bell-shaped and about the same diameter as the abdomen, so it appears small. The breasts may be swollen as a result of maternal hormones, and sometimes a pale, milky fluid leaks out. Most newborn babies appear somewhat blue, but turn pink as they start to breathe. Some have a fine covering of pale, downy hair called lanugo, which will disappear within a few weeks or months. More than 80 percent of babies have some kind of birthmark, an area of pigmented skin that usually fades or disappears as the child gets older.

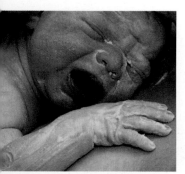

Skin protection
At birth a baby's delicate new skin is protected by a waxy, cheeselike coating known as vernix caseosa, formed from skin oils and dead cells.

SIGN	SCORE: 0	SCORE: 1	SCORE: 2
HEART RATE	None	Below 100	Over 100
BREATHING RATE	None	Slow or irregular; weak cry	Regular; strong cry
MUSCLE TONE	Limp	Some bending of limbs	Active movements
REFLEX RESPONSES	None	Grimace or whimpering	Cry, sneeze, or cough
COLOR	Pale or blue	Blue extremities	Pink

Apgar score
A newborn baby's health is assessed at one minute and five minutes after birth, based on five characteristics. A perfect score is 10. A score of 3 or less shows the baby needs immediate resuscitation.

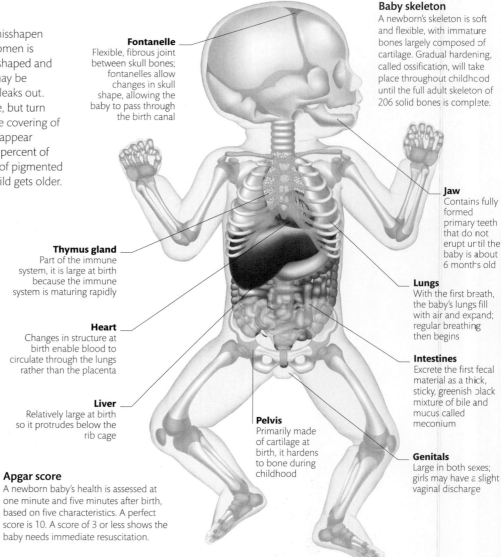

Fontanelle
Flexible, fibrous joint between skull bones; fontanelles allow changes in skull shape, allowing the baby to pass through the birth canal

Baby skeleton
A newborn's skeleton is soft and flexible, with immature bones largely composed of cartilage. Gradual hardening, called ossification, will take place throughout childhood until the full adult skeleton of 206 solid bones is complete.

Jaw
Contains fully formed primary teeth that do not erupt until the baby is about 6 months old

Thymus gland
Part of the immune system, it is large at birth because the immune system is maturing rapidly

Heart
Changes in structure at birth enable blood to circulate through the lungs rather than the placenta

Lungs
With the first breath, the baby's lungs fill with air and expand; regular breathing then begins

Intestines
Excrete the first fecal material as a thick, sticky, greenish black mixture of bile and mucus called meconium

Liver
Relatively large at birth so it protrudes below the rib cage

Pelvis
Primarily made of cartilage at birth, it hardens to bone during childhood

Genitals
Large in both sexes; girls may have a slight vaginal discharge

Just arrived
In developed countries the average weight of a newborn baby is 7½ lb (3.4 kg) and the average length, from crown to heel, is 20 in (50 cm).

CHANGING CIRCULATION

While in the womb and unable to breathe or eat for itself, the fetus receives nourishment and oxygen, via the umbilical cord, from the blood flowing through the placenta, and gets rid of waste products, including carbon dioxide, in blood flowing back to the placenta. The fetal circulation is adapted to make this arrangement work by having specialized blood vessels that convey blood to and from the umbilical cord and enable most of the blood to take a route that bypasses the immature liver and lungs. At birth, the lungs start to inflate with the first breath, causing pressure changes that increase blood flow through the lungs and close off these special channels. The baby has made the transition to breathing air.

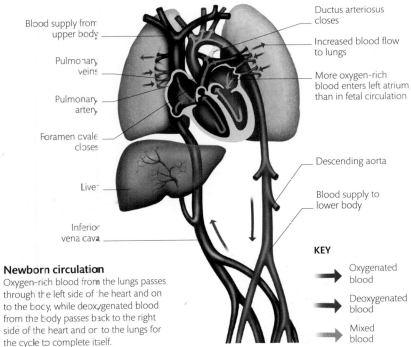

Fetal circulation
Oxygen- and nutrient-rich blood is supplied through the placenta, and deoxygenated blood containing waste products flows back through it to be enriched again.

Blood supply from upper body

Pulmonary artery

The foramen ovale, a window between atria, is a short-cut for blood passing from placenta to fetus

Ductus venosus connects umbilical vein to inferior vena cava

Umbilical vein carries all nourishment and dissolved gases

Placenta links blood supplies of mother and baby

Blood supply to upper body

Ductus arteriosus allows umbilical blood to bypass lungs

Left atrium

Left lung

Heart

Descending aorta

Inferior vena cava

Umbilical arteries take waste products and deoxygenated blood back to placenta

Blood supply to lower body

Newborn circulation
Oxygen-rich blood from the lungs passes through the left side of the heart and on to the body, while deoxygenated blood from the body passes back to the right side of the heart and on to the lungs for the cycle to complete itself.

Blood supply from upper body

Pulmonary veins

Pulmonary artery

Foramen ovale closes

Liver

Inferior vena cava

Ductus arteriosus closes

Increased blood flow to lungs

More oxygen-rich blood enters left atrium than in fetal circulation

Descending aorta

Blood supply to lower body

KEY

→ Oxygenated blood

→ Deoxygenated blood

→ Mixed blood

CUTTING THE CORD

Unless it has already been cut, the umbilical cord will continue to pulse for up to 20 minutes after a baby is born, maintaining the baby's oxygen supply and keeping the placental blood supply flowing until it is no longer needed. After this, the cord can be safely clamped or tied and cut—this is painless because there are few nerves in the cord. At birth, the average umbilical cord is about 20in (50cm) long and usually a stump 1–1½in (2–3cm) long is left attached to the baby's umbilicus. The placenta will be expelled naturally around 20 minutes to an hour after the baby is born, although this may be accelerated by an injection given during the birth. Meanwhile the baby can be put to the breast.

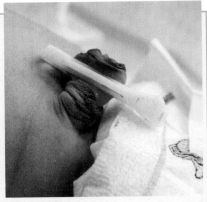

Umbilical stump
The umbilical stump will gradually shrivel and dry out. It will fall off by itself in 1 to 3 weeks, leaving a "belly button" that may be inverted or protrude outward.

FOOD FOR LIFE

A newborn baby instinctively attempts to find its mother's breast and suckle. Thanks to an automatic response called the rooting reflex, babies turn their head toward a touch on their cheek or lips and make sucking motions. If put to the breast, the mouth will automatically open and the baby will latch on, taking the whole areola into its mouth, and begin to suck. After a few seconds, the mother's let-down reflex comes into play and milk starts to flow. Sweet, pre-milk colostrum helps guard against infection and contains beneficial "good bacteria" to protect the baby's immature gut. Breast milk proper is nutritionally ideal and contains antibodies that defend against infection. Breastfed babies are also less likely to develop allergies later in life.

Suckling instinct
The suckling instinct is strongest for about half an hour after birth, when feeding also stimulates maternal hormones that help the uterus to contract down and the placenta to be expelled.

LIFE OUTSIDE THE WOMB

Most newborn babies sleep for much of the day and night, but wake to feed every few hours. An average baby will cry for between 1 and 3 hours a day. Within the first 24 hours a baby should urinate and have a first bowel movement, although for the first few days this will be meconium, a green-black, sticky substance representing the fetal bowel contents. Once the baby is settled into a feeding routine, its stools will become grainy and brown, then yellowish. In the first week or two of life, babies actually lose weight, up to 10 percent of their birth weight, before starting a steady gain.

Looking and touching
Babies soon start to explore the world through looking and touching. Young babies focus best at about 8–14 in (20–35 cm) from an object and love to gaze at faces. The mouth and the hands are important for touch sensations.

CHILDHOOD

Childhood is a time of continual physical change and developmental progress on a scale that does not occur again in life. Along with growth in height and weight comes the acquisition of physical and mental skills, social understanding, and growing emotional maturity.

GROWTH AND DEVELOPMENT

The first two years of a child's life are marked by extremely rapid physical growth, after which the rate slows until puberty. The size and weight of all body tissues and organs increase during childhood, with the exception of lymphatic tissue, which shrinks. Both growth rate and final stature are largely dependent on genetic inheritance, so that, to an extent, a child's final height can be predicted from the height of the parents. However, growth and development are also influenced by the child's environment, so health or illness, nutrition, intellectual stimulation, and emotional support all contribute to physical and mental outcome.

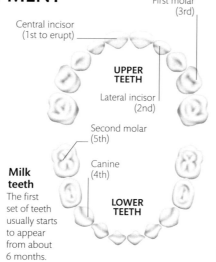

First molar (3rd)

Central incisor (1st to erupt)

UPPER TEETH

Lateral incisor (2nd)

Second molar (5th)

Milk teeth
The first set of teeth usually starts to appear from about 6 months.

Canine (4th)

LOWER TEETH

The cartilaginous joints in a baby's skull facilitate rapid brain growth. The newborn's brain is about a quarter of the size it will reach at adulthood, but by its third year it will have enlarged to 80 percent of its eventual size. While almost all the brain's neurons are present at birth, their links are limited and interconnections will continue to develop until adulthood. Dental development during childhood is marked by the succession of the primary or "milk" teeth by permanent adult teeth, which erupt through the gums below.

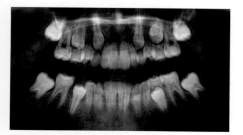

Erupting teeth
Permanent adult teeth begin to erupt and baby teeth fall out at about 6 years of age. By the age of 13 a full set of adult teeth (except for the wisdom teeth) has grown.

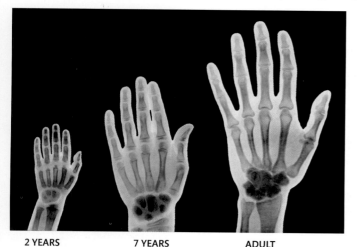

2 YEARS   7 YEARS   ADULT

Developing bones
As a child grows, the cartilage in the skeleton gradually turns to bone. In adults, the wrist consists of eight bones, which gradually develop from cartilage during childhood.

Once a child has reached a **particular milestone**, practice and enthusiasm **spur progress** toward the next.

Exploring the world
Every child has innate curiosity about the world and will learn from whatever catches the attention.

CHANGING PROPORTIONS

At birth, a baby's head is relatively large, representing one quarter to one third of its total body length—compared with just one eighth for an adult's head. In addition, a baby's skull is quite large compared with its face. The trunk of a baby is about three-eighths of its total height—about the same as in an adult—although its shoulders and hips are fairly narrow and its limbs are relatively short. Thus, as a child grows, its height and weight gains are accompanied by distinct changes in body proportions. The trunk grows steadily throughout childhood but the head does not enlarge very much, although the face gets bigger relative to the skull, while the limbs grow proportionately very much longer, often in spurts. The growth of the long bones of the legs is largely responsible for the increase in height during childhood. The first two years of life are the time of maximum growth. An average infant gains around 10 in (25 cm) in height and triples its birthweight in the first year. However, after the age of two, growth usually settles down to a steady 2½ in (6 cm) per year until puberty (see p.422), and eventually ceases at about 18–20 years.

Body-head proportions
A newborn baby's head is already almost adult-sized, whereas its limbs are relatively short. As the child grows, increases in height and weight are therefore accompanied by changes in body proportions.

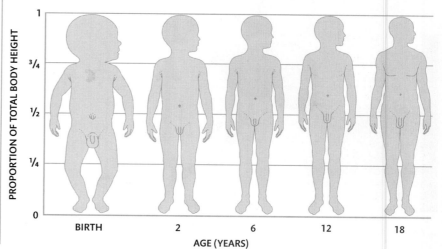

PROPORTION OF TOTAL BODY HEIGHT

BIRTH   2   6   12   18

AGE (YEARS)

STAGES OF DEVELOPMENT

A child's acquisition of skills and abilities in different spheres is marked by certain achievements known as developmental milestones. These may be seen as stepping stones to future development—children must be able to walk before they can run, and to understand and vocalize simple words before they can start to construct sentences. Once a child has reached a particular milestone, practice and enthusiasm spur progress toward the next. Children are individuals and develop at different rates, so even siblings may

vary enormously in the age at which they achieve these stages or learn certain skills. Some children will miss out on certain stages and go straight on to the next, and a child who is "ahead" in one area may lag behind in others. New circumstances, especially stress and changes at home (such as a new baby or moving house) may delay the achievement of milestones, but most children will adapt readily given time and support. Below is a guide to the average ages at which children reach developmental milestones.

THE IMPORTANCE OF PLAY

Play is far from a trivial activity—it is crucial to the acquisition of physical, mental, and social skills. Unlike passive entertainment, play requires involvement, imagination, and resourcefulness. Pretend play stimulates creativity and understanding, while playing with other children boosts communication and social skills. For a parent, playing with children on their level is one of the best ways to give them emotional security and cement the bond with them.

Manual dexterity
Children develop the ability to grasp and manipulate objects very early on. Gradually, they learn to perform increasingly complex movements.

AGES (YEARS)

| 0 | 1 | 2 | 3 | 4 | 5 |

PHYSICAL ABILITIES
Many of a baby's physical responses at birth are involuntary and largely reflex actions, such as the suckling reflex. Gradually, but steadily, a child will make the transition to more purposeful and active motions, learning in sequence to hold its head up, turn over, crawl, stand, and walk. Balance and coordination improve in parallel, and eventually children learn the highly complex motor skills needed for sophisticated activities such as riding a bicycle or writing.

- Lifts head and chest
- Brings hand to mouth
- Grasps objects with hands

- Reaches for objects
- Rolls over
- Supports own weight on feet

- Crawls
- Walks holding furniture
- Bangs objects together
- Eats finger foods unaided

- Crawls up stairs
- Squats to pick up objects
- Jumps with both feet

- Walks unaided
- Carries or pulls toys
- Starts to run
- Can kick a ball

- Walks up and down stairs
- Can hold and use pencil
- Shows hand preference
- Gains control of bowels

- Runs easily
- Can pedal and steer tricycle
- Turns pages in a book
- Controls bladder by day

- Turns handles and jar lids
- Draws straight lines and circles
- Can build a tower to six blocks

- Hops
- Can dress and undress unaided
- Climbs and descends stairs unaided
- Can catch and throw a bounced ball
- Draws basic shapes and figures
- Uses scissors

- Holds pencils with precision
- Can write some words
- Feeds self using utensils
- Uses bathroom unaided

THINKING AND LANGUAGE SKILLS
Speech and language development are vital to a child's ability to interact with their surroundings. An infant starts to understand basic words and commands long before being able to speak, and verbal skills are readily learned by imitation. The more parents and others involved in an infant's care talk to the child, the more vocal and verbal the child is likely to become. Along with growing understanding of the world, language helps the child develop thinking, reasoning, and problem-solving skills.

- Smiles at parent's voice
- Starts to imitate sounds

- Begins to babble
- Investigates with hands and mouth
- Reaches for out-of-reach objects
- Understands "no," "up," and "down"

- Recognizes own name
- Responds to simple commands
- Uses first words
- Imitates behavior

- Starts to drink from cup

- Points to named objects
- Sorts shapes and colors
- Says simple phrases
- Follows simple instructions
- Engages in fantasy play

- Uses simple sentences
- Can state name, age, and gender
- Uses pronouns ("I," "you," "we," "he," "they")
- Understands spatial location ("in," "on," "under")
- Begins to understand numbers

- Understands basic grammar
- Starts to count
- Starts to understand time
- Tells stories
- Follows three-part commands

- Understands future tense
- Can state name and address
- Names four or more colors
- Can color in shapes
- Can count more than 10 objects
- Able to distinguish reality from fantasy
- Understands concept of money
- Aware of gender

SOCIAL AND EMOTIONAL DEVELOPMENT
Almost from birth, a baby recognizes its mother and shows a marked preference for her over other people. Many children go through phases of shyness with strangers but most are enthusiastic for interactions with others. Soon they grow in independence and show a capacity to control their behavior, understand social rules, cooperate, and display empathy for others.

- Makes eye contact
- Recognizes familiar people
- Cries when needing attention
- Smiles at mother, then socially
- Watches faces intently
- Recognizes parents' voices

- Responds to own name

- Cries when parent leaves
- Shows preferences for people and objects

- Imitates others' behavior
- Enjoys company of other children
- Demonstrates defiant behavior

- Peak separation anxiety

- Shows affection for other children
- Takes turns when playing
- Understands possession ("mine," "yours")

- Interested in new experiences
- Cooperates and negotiates with other children
- May imagine threats such as "monsters"

- Wants to please and to be like friends
- Increasingly independent
- Likes to demonstrate skills, such as singing, dancing, acting
- Shows empathy for others

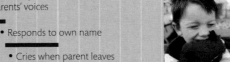

| 0 | 2 | 4 | 6 | 8 | 10 | 12 | 14 | 16 | 18 | 20 | 22 | 24 | 26 | 28 | 30 | 32 | 34 | 36 | 38 | 40 | 42 | 44 | 46 | 48 | 50 | 52 | 54 | 56 | 58 | 60 |

AGE (MONTHS)

ADOLESCENCE AND PUBERTY

Adolescence is the period of transition between childhood and adulthood, during which puberty is marked by a great physical transformation in both boys and girls and the onset of sexual maturity.

TRANSITION TO MATURITY

During adolescence, increasing physical maturity is accompanied by behavioral changes that mark the start of growing up. As teenagers seek to develop their own sense of identity, interactions with friends and peer groups gain increased importance, and their social skills expand. Adolescents are attracted to peer group interests, such as music and fashions, and may become increasingly distanced from their parents. They need to discover their individuality and prove their independence in thought and actions, so may start to take their values more from their peers, making them vulnerable to peer pressure. Without a strong sense of identity and self-confidence developed in childhood, they may be at risk from experimenting with alcohol, drugs, smoking, and sexual relationships. Many teenagers have mixed emotions as they try to establish their own values, which may lead to rebellion and negative effects such as

Growth spurts
Puberty marks a time of rapid growth in hormone-driven spurts. Boys usually start later but grow more during peak periods.

family disharmony, falling school grades, or trouble with authority. In addition to coping with the physical changes and hormone surges of puberty, teenagers are often anxious about their body development, changing appearance, and attractiveness to the opposite sex.

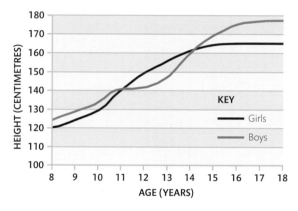

KEY
— Girls
— Boys

This may lead to body image problems, which may spiral into eating disorders. With all these pressures, including academia and future work, it is perhaps not surprising that adolescents may come across as moody and volatile.

Girls and boys
On average, girls reach puberty two years before boys. The age difference in sexual maturity is paralleled by a similar gap in physical and mental development.

RAGING HORMONES

The hormonal surges that occur at puberty are responsible for some of the most dramatic changes that ever occur in the human body. In both sexes, the trigger of puberty is the release from the hypothalamus, a gland in the brain, of a hormone called gonadotropin-releasing hormone (GnRH). This stimulates the nearby pituitary gland to release two more hormones called luteinizing hormone (LH) and follicle stimulating hormone (FSH). These in turn travel through the bloodstream to trigger the production of the sex hormones—primarily estrogens and progesterone from the ovaries in girls and testosterone from the testes in boys. These hormones are responsible for all the developments underlying puberty in both sexes. Female sex hormones stimulate the ovaries to start releasing eggs and the body to prepare for a possible pregnancy. Male sex hormones prompt the testes to start producing sperm.

Feedback loops
Hormone production is regulated by feedback, when the amount of a substance in the system controls how much is produced.

The **physical changes** associated with puberty are initially triggered by **hormones in the brain**.

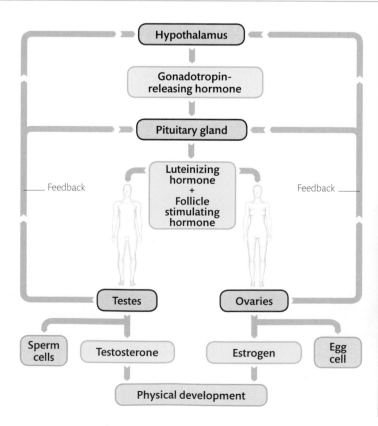

Hypothalamus

Gonadotropin-releasing hormone

Pituitary gland

Luteinizing hormone + Follicle stimulating hormone

Feedback   Feedback

Testes   Ovaries

Sperm cells   Testosterone   Estrogen   Egg cell

Physical development

MIXED EMOTIONS

Surges in hormone levels have traditionally been linked to fluctuations in mood and emotions during puberty. However, sex hormones are now not thought to play the major role. Instead, social and environmental influences, coupled with physical changes in the brain as it matures, are believed to have a greater effect on the emotions.

Appearance anxiety
The physical changes that take place during puberty provoke anxiety about appearance and attractiveness to other adolescents.

PHYSICAL DEVELOPMENT

The age of onset of the physical changes marking the start of puberty is highly variable, but will often be around the age that the same-sex parent made the transition. Most girls enter puberty between age 8 and 13; most boys from age 10 to 15. In both sexes, the sequence of physical changes that culminates in physical maturity lasts 2 to 5 years. It will be complete in most girls by age 15 and most boys by age 17.

Both genders have a remarkable growth spurt associated with puberty, at its peak resulting in height increases of up to 3½ in (9 cm) in a year in boys, and 3 in (8 cm) in girls (see opposite). Although on entering puberty boys are generally ¾ in (2 cm) shorter than girls of the same age, at full adult height they are, on average, 5 in (13 cm) taller.

In addition to boosting height, puberty marks the onset of sexual development, with growth and maturation of the sex organs (testes and ovaries) to enable fertility and secondary sexual characteristics. In both sexes these include increased genital size, the appearance of underarm and pubic hair, and skin changes that may promote acne. In addition, girls undergo breast development, their hips widen, and they lay down an extra layer of insulating body fat. Menstruation begins, usually preceding the onset of ovulation. In boys, the

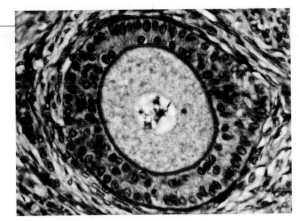

Ripening egg
A girl is born with a full complement of half a million eggs in her ovaries. After puberty, several start to ripen each month, but usually only one is released.

Adam's apple enlarges, the vocal cords stretch and the voice deepens, muscle bulk increases, and additional body and facial hair appears. Most boys will experience spontaneous nocturnal ejaculations (also known as wet dreams) during and after puberty.

Sperm production
Puberty triggers sperm production in the testes. It takes 72 days to produce a mature sperm capable of movement.

Puberty marks the onset of **sexual development**, with **growth and maturation** of the sex organs (testes and ovaries) to **enable fertility**.

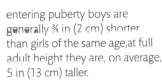

Facial hair
The appearance of facial hair is one of the last changes to occur during puberty in boys, occurring on average around age 15.

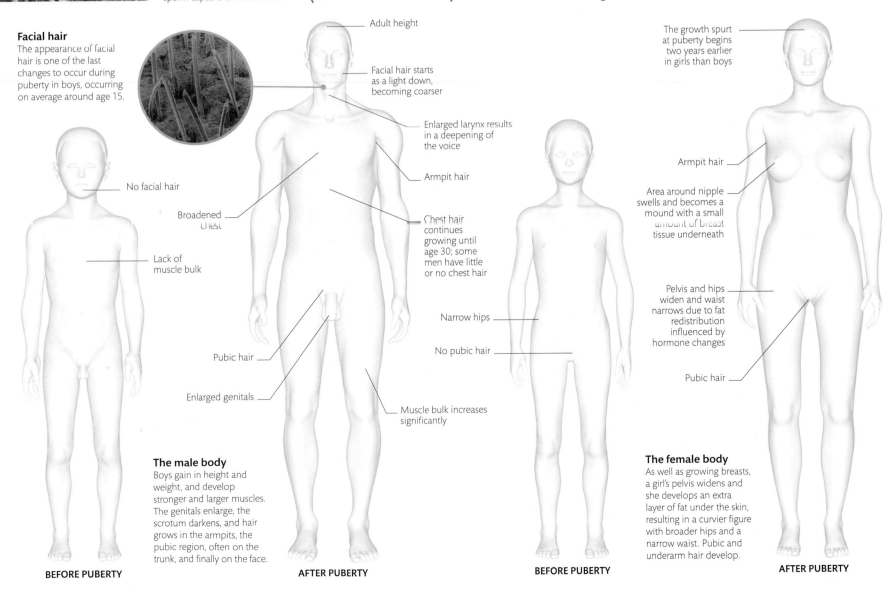

Adult height

Facial hair starts as a light down, becoming coarser

Enlarged larynx results in a deepening of the voice

Armpit hair

Chest hair continues growing until age 30; some men have little or no chest hair

No facial hair

Broadened chest

Lack of muscle bulk

Pubic hair

Enlarged genitals

Muscle bulk increases significantly

Narrow hips

No pubic hair

The growth spurt at puberty begins two years earlier in girls than boys

Armpit hair

Area around nipple swells and becomes a mound with a small amount of breast tissue underneath

Pelvis and hips widen and waist narrows due to fat redistribution influenced by hormone changes

Pubic hair

The male body
Boys gain in height and weight, and develop stronger and larger muscles. The genitals enlarge, the scrotum darkens, and hair grows in the armpits, often on the trunk, and finally on the face.

The female body
As well as growing breasts, a girl's pelvis widens and she develops an extra layer of fat under the skin, resulting in a curvier figure with broader hips and a narrow waist. Pubic and underarm hair develop.

BEFORE PUBERTY

AFTER PUBERTY

BEFORE PUBERTY

AFTER PUBERTY

ADULTHOOD AND OLD AGE

The inevitable progression from adulthood through middle age to old age is accompanied by gradual changes in all body systems. Although there are many possible contributors to the aging process, scientists still do not fully understand why we age as we do.

Signs of aging
Perhaps the most visible outward signs of aging are wrinkling and discoloration of the skin, and graying hair, which results from fading pigment.

THE AGING PROCESS

As we get older, all the cells in our bodies undergo progressive changes that inevitably affect the tissues and organs they comprise. During their lives, cells accumulate internal debris, enlarge, and become less efficient. They are less able to take on board essential nutrients and oxygen, or to get rid of the waste products of metabolism.

As their function is impaired, cells become less capable of reproducing and replacing themselves. Gradual effects include stiffening of connective tissues, leading to loss of elasticity in the walls of the arteries, along with skin thinning, lowered immunity, and loss of organ function.

As people age they become less able to cope with increased physical demands. For example, as heart muscle ages the heart may be less able to increase its pumping capacity during exercise or stress. Similarly, lung and kidney capacities are gradually reduced. Also, the body becomes less able to detoxify harmful substances, meaning that older people are more at risk from the side-effects of drugs.

Because immune function is reduced, the body becomes more vulnerable to illness and less able to cope with it. Gradually, the body's repair and renewal functions wind down until a point is reached at which the body may be unable to recover from the onset of a disease.

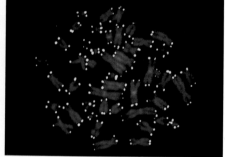

Telomeres
DNA strands at the end of each chromosome get shorter every time a cell divides, limiting the number of possible divisions and perhaps holding a clue to the mechanisms of aging.

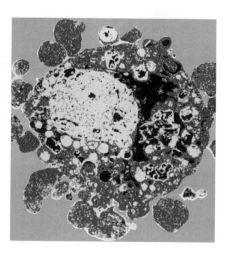

20-35

Between these ages the body's **biological functioning** and **physical performance** reach their peak.

Dying cell
The repair and renewal of tissues depends on a process of programmed cell death called apoptosis. Normally, cells die in a controlled manner, to be replaced by new cells. With age, apoptosis is less well regulated, contributing to disease.

METABOLISM AND HORMONES

Aging affects both the production of the body's hormones and the way in which target organs respond to them. Output and responses to thyroid hormones, which control the body's metabolism, may decline with age alongside a loss of muscle tissue, which uses more energy than fat. This means that metabolic rate decreases with age, so the body burns fewer of the calories in food. Unless this is counteracted by exercise, to increase muscle mass, older people can develop a susceptibility to a rise in body fat levels. From middle age, body cells become less sensitive to the effects of insulin, produced in the pancreas, with the result that blood glucose levels tend to rise

slowly, so older people are more likely to develop diabetes. Reduced parathyroid hormone levels affect levels of calcium in the body and this may contribute to bone thinning or osteoporosis. Reduced secretion of aldosterone, a hormone from the adrenal glands that regulates body fluid and chemical balance, may impair blood pressure regulation. Another hormone from the adrenals, called cortisol, is produced in response to stress, and high levels seem to accelerate age-related changes. Estrogen levels in women decrease markedly after the menopause, whereas testosterone levels in men decline slowly, so male fertility can continue into old age.

MENOPAUSE

The decline in estrogen production from a woman's ovaries eventually leads to cessation of ovulation and loss of fertility, along with menopause, when periods stop. The transition may take several years, with the last period on average at age 51 in developed countries. After menopause, a woman is more vulnerable to osteoporosis, cardiovascular disease, and breast and endometrial cancers.

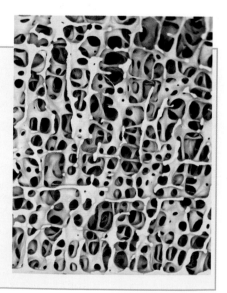

Osteoporosis
In the brittle bone disease osteoporosis, bones gradually lose density and strength, and fractures of weakened bones, especially in the hip or spine, may occur (see p.441).

SKIN

With age the outer layer of the skin gets thinner, as does the underlying fat layer. Aging skin becomes less elastic and more fragile, with reduced sensitivity, so it not only sags but also is more easily damaged. Blood vessels in the subcutaneous tissue become more fragile, so skin is more susceptible to bruising. The sebaceous glands produce less oil, making the skin more prone to dryness and itching.

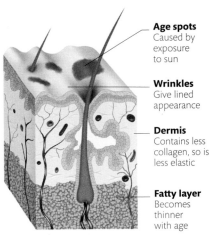

Age spots
Caused by exposure to sun

Wrinkles
Give lined appearance

Dermis
Contains less collagen, so is less elastic

Fatty layer
Becomes thinner with age

Aging skin
Older skin has less subcutaneous fat and elastic tissue and its glands produce less oil. Pigment cells reduce in number but may get larger. The skin appears paler but age spots may appear.

MUSCULOSKELETAL AND ORGAN CHANGES

Multiple changes occur in the musculoskeletal system with age, including loss of bone density, joint stiffening, and loss of muscle mass and tone. Older people become more liable to osteoporosis, in which calcium and other important minerals are lost from the skeleton. This makes bones more porous and brittle, reducing their strength and increasing the risk of fractures. A good intake of calcium and vitamin D, along with weight-bearing exercise, can strengthen bones and ameliorate some of these changes. Exercise also mitigates loss of muscle bulk with age and may partly compensate for less flexible joints and age-related arthritic changes. Even so, older age is often accompanied by stooped posture, muscle weakness, loss of agility, and slower movements, leading to changes in gait, made worse by impaired sensation and balance. With age the heart's pumping ability progressively decreases and loss of elasticity in the arteries may increase blood pressure, putting further strain on a weakened heart. Heart rhythm abnormalities become more common as the heart's electrical conducting system is disrupted. Lung capacity decreases as the elastic support of the airways weakens, and, especially after age 65, this reduces the amount of oxygen available to the tissues.

Loss of cartilage in hip joint

Osteoarthritis
Wear and tear gradually erode joint cartilage and may produce osteoarthritis where joint surfaces rub together. Pain and stiffness become increasingly common as people age.

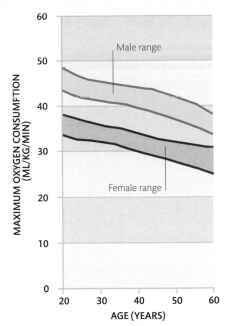

Heart and lung performance
Both heart and lung function progressively decrease with age, so there is less reserve capacity to cope with additional demands.

Exercise mitigates loss of **muscle bulk** and may partly compensate for less flexible joints and age-related **arthritic changes**.

BRAIN, NERVES, AND SENSES

Like other body cells, those of the nervous system function less well as people get older. The brain and spinal cord lose nerve cells, and those that remain may accumulate waste products that can slow nerve impulses, reduce reflexes and sensation, and blunt cognitive abilities. Vision and hearing also tend to become less acute, and the senses of touch, taste, smell, balance, and proprioception may be impaired. While a healthy lifestyle with good nutrition, physical exercise, and mental stimulation can ameliorate many of these changes, older people remain more vulnerable to accidents, memory loss, dietary impairment, and general reductions in quality of life. Senility and dementia are not normal or inevitable, although older people are more likely to develop Alzheimer's disease. Most people become farsighted with age, and need reading glasses. Sharpness of vision and color perception may be dimmed and various eye problems, including cataracts, become more common. Reduced taste and smell can diminish enjoyment of eating and contribute to nutritional deficiencies.

Decline in hearing
Loss in hearing with age especially affects higher frequencies, such as women and children's voices, or ringing telephones. Hearing is more likely to be impaired with age among people who were exposed to loud noises earlier in life.

KEY

— Age 20
— Age 30
— Age 50
— Age 70

Brain of 27-year-old
A brain scan in a young person shows little atrophy—the shrinkage that represents loss of brain cells with aging—and normal-sized ventricles and subarachnoid spaces.

Ventricle | Subarachnoid space

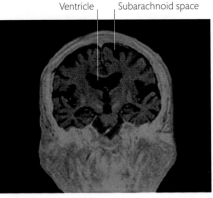

Brain of 87-year-old
This scan shows considerable shrinkage and loss of brain tissue, with expanding ventricles and enlarged subarachnoid spaces. There are also fewer cells in the hippocampus, the area where memory is processed.

Ventricle | Subarachnoid space

END OF LIFE

Death is the cessation of all biological functions. It may result from disease, trauma, or lack of vital nutrients. Unless one of these events occurs, all people will eventually die of senescence—simple old age.

DEFINING DEATH

Traditionally death has meant the cessation of heartbeat and respiration, almost inevitably followed by irreversible bodily deterioration and decomposition. Modern medical technology has made it possible to maintain vital body functions artificially, so that the boundary between life and death has become increasingly blurred. We can now intervene in events that were previously irreversible—such as cardio-respiratory arrest—and, as a result, death is now seen as a process, rather than an event, with varying definitions. Clinical death accords with the traditional definition of the absence of vital signs of heartbeat and breathing—but from which individuals may now be resuscitated. Brain death, a criterion developed to enable removal of

viable organs for transplantation, may be pronounced when it is judged that brain failure is permanent and irreversible, even if heart and lung function is maintained artificially. Similarly, brainstem death occurs when the brain is judged no longer capable of sustaining vital functions. Legal death is simply when a doctor pronounces death, which may be contemporaneous with pronouncement of brain death or some time after clinical death.

Intensive care
With advances in medical technology, failure of vital body functions can now be overridden by maintaining the patient artificially, especially by ventilators or "life support" machines.

122
The age of Jeanne Calment, the **longest-living human**.

Death mask
In past centuries, death masks were often made to record a person's appearance. They were cast in wax or plaster immediately after death, before facial features could become distorted. This is the death mask of Austrian writer Adalbert Stifter.

NEAR-DEATH EXPERIENCE

Some individuals who have been pronounced clinically dead and then revived, or who have undergone resuscitation after a cardiac arrest, report a set of strikingly similar perceptions known as near-death experiences. These include out-of-body sensations, moving through a tunnel toward a bright light, and encountering familiar figures from their past. Usually these sensations are experienced as positive. Some people believe they represent physiological changes in the dying brain; others think that they are evidence of an afterlife, through reincarnation or other spiritual phenomena.

Common visions
Near-death experiences are often characterized by a feeling of floating out of the body and moving through a tunnel toward bright light.

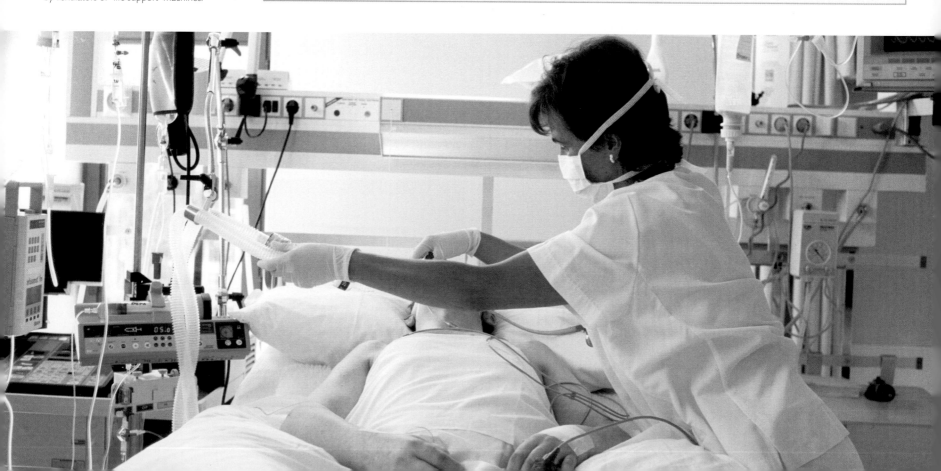

CAUSES OF DEATH

Worldwide, the leading causes of death are linked with cardiovascular disease, which is, to a large extent, preventable. For example, scientists have shown that nine potentially modifiable lifestyle factors, including smoking and obesity, account for more than 90 percent of the risk of having a heart attack. Compared with high-income countries, low-income countries have a much greater occurrence of death from infectious diseases. This is largely due to the effects of poverty, including inadequate nutrition, poor hygiene, and lack of health provisions.

The most common causes
These tables show the top 10 causes of death worldwide, and compare the leading causes of death in developing and developed countries.

WORLDWIDE	LOW-INCOME COUNTRIES	HIGH-INCOME COUNTRIES
Coronary artery disease **12.2 %**	Lower respiratory infections **11.2 %**	Coronary artery disease **16.3 %**
Stroke and other cerebrovascular diseases **9.7 %**	Coronary artery disease **9.4 %**	Stroke and other cerebrovascular diseases **9.3 %**
Lower respiratory infections **7.1 %**	Diarrheal diseases **6.9 %**	Trachea, bronchus, lung cancers **5.9 %**
Chronic obstructive pulmonary diseases **5.1 %**	HIV/AIDS **5.7 %**	Lower respiratory infections **3.8 %**
Diarrheal diseases **3.7 %**	Stroke and other cerebrovascular diseases **5.6 %**	Chronic obstructive pulmonary diseases **3.5 %**
HIV/AIDS **3.5 %**	Chronic obstructive pulmonary diseases **3.6 %**	Alzheimer's and other dementias **3.4 %**
Tuberculosis **2.5 %**	Tuberculosis **3.5 %**	Colon and rectum cancers **3.3 %**
Trachea, bronchus, lung cancers **2.3 %**	Neonatal infections **3.4 %**	Diabetes mellitus **2.8 %**
Traffic accidents **2.2 %**	Malaria **3.3 %**	Breast cancer **2.0 %**
Prematurity and low birth weight **2.0 %**	Prematurity and low birth weight **3.2 %**	Stomach cancer **1.8 %**

AFTER DEATH

The human body undergoes many changes after death, which may be useful to establish a time of death if this is unknown. Usually, after an initial lag period of 30 minutes to 3 hours, the body progressively loses heat at an average rate of about 2.7°F (1.5°C) per hour until it reaches the same temperature as its surroundings. Muscles undergo chemical changes that make them stiffen. This process, called rigor mortis, begins with the small facial muscles and works down the body toward the larger muscles of the arms and legs. Rigor mortis happens more quickly at higher temperatures and in thinner people. After around 8 to 12 hours, the body has become stiff and fixed in the position of death. Thereafter the tissues begin to decompose and the stiffness is lost during the following 48 hours. As blood flow ceases, it pools in various parts of the body, creating a purple hue known as lividity. Initially, the position of the discoloration is affected by moving the body, but after 6 to 8 hours it becomes

Physical changes
After death, the body slowly cools to the same temperature as its environment and becomes temporarily stiff, with the joints fixed in the position at death.

fixed. Finally, bacteria and enzymes start to decompose the tissues, and the body will start to smell after 24 to 36 hours. The skin takes on a green-red hue, body orifices may leak, and the skin may split as gas forms in the putrefying flesh and body cavities. The various procedures undertaken by mortuaries are designed to prevent this until after the funeral.

Bodies that are **buried in the ground** after death turn to skeletons within about **10 years**.

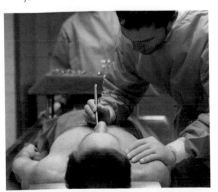

Post-mortem
A body may undergo a medical examination by a pathologist to discover or further investigate the cause of death.

CHEATING DEATH

In future, new techniques to repair the damage done by the aging process may hold out the hope of extending the healthy human lifespan. One promising line of research is the use of stem cells, which can reproduce indefinitely and develop into any new body cell. These might regenerate worn out or diseased organs and so avert or delay many leading causes of death. This might involve using a

Living longer
Japanese women have the world's highest life expectancy (87 years). Studies suggest that a combination of good diet, low stress, and high levels of physical activity is responsible.

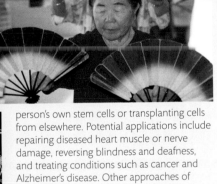

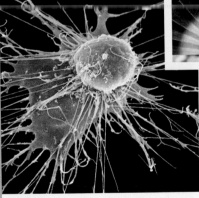

Stem cell research
Adult stem cells become increasingly inefficient with age. Scientists hope to find a way to replace or rejuvenate them, to repair age-related damage to worn-out organs and tissues.

person's own stem cells or transplanting cells from elsewhere. Potential applications include repairing diseased heart muscle or nerve damage, reversing blindness and deafness, and treating conditions such as cancer and Alzheimer's disease. Other approaches of regenerative medicine include manipulating the genetic influences underlying aging or the major diseases of older age, targeting body metabolism or hormones to delay age-related changes, and learning more about the factors that contribute to natural longevity. For example, studying the lifestyle of centenarians may provide clues to how we could all perhaps live a little bit longer.

Although **family history** influences how long a person will live, many of the factors that **affect lifespan** are within people's **own control**.

diseases and disorders

The body is a complex construction, vulnerable to disease and malfunction. This section catalogs major diseases and disorders, starting with those that are not specific to any single body system, such as infectious diseases and cancer, and then moving on to look at each system of the body in turn.

428
DISEASES AND DISORDERS

INHERITED DISORDERS

Defective genes and chromosome disorders are usually passed from parent to child. Chromosome disorders are caused by a fault in the number or structure of the chromosomes. Gene disorders are due to a fault on one or more of the genes that are carried on the chromosomes.

CHROMOSOME DISORDERS

Chromosomes are strands of coiled DNA, the genetic material arranged in a double helix that instructs our cells how to grow and behave. Humans have 23 pairs of chromosomes—one in each pair from the father and one from the mother. Major chromosomal abnormalities can produce serious defects and disease. There may be errors on any of the chromosomes, such as breakages, missing pieces, extra pieces, or translocations (pieces that are incorrectly swapped). These usually result from mistakes during meiosis (cell division to form egg or sperm cells).

DOWN SYNDROME

A partial or complete extra copy of chromosome 21 causes Down syndrome. The extra genetic material causes abnormalities in many systems.

Down syndrome is the most common chromosomal abnormality in which the fetus can survive. It is caused by a fault in the normal parental production of eggs and sperm (90 percent or more are eggs rather than sperm), giving rise to one that contains extra genetic material. This fault is more common in older women. In about 3 percent of cases, however, Down syndrome is due to one parent having a translocation, which means that a piece of one chromosome 21 is attached to another chromosome. This pattern of inheritance does not increase with parental age.

Down syndrome can be diagnosed through tests in early pregnancy and also after birth with a blood test. It causes learning difficulties and affects physical appearance, causing characteristics such as floppy limbs, round face, and eyes that slant up at the outer corners. Children with the condition may require long-term medical support, and life expectancy is shortened to about 50 years.

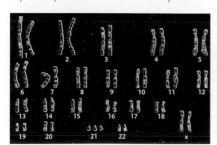

Chromosome set
This set of chromosomes from a child with Down syndrome shows the extra copy of chromsome 21 that causes the condition.

TURNER SYNDROME

In this condition, girls are born with only one active X chromosome in each cell instead of two. It does not affect boys.

Girls with Turner syndrome share certain physical characteristics—they are of short stature, and they have an abnormal or absent uterus and ovaries and are infertile. They may have abnormalities of other organs such as the heart, thyroid, and kidneys, but the condition varies among individuals. It is often only detected when a girl does not reach puberty at the normal age. The underlying genetic defect probably results from a fault when the egg or sperm is made. In some cases, mosaicism occurs (both X chromosomes are present in some cells but not in others).

KLINEFELTER SYNDROME

Klinefelter syndrome only affects boys. It is caused by the inheritance of an extra X chromosome in each cell, in addition to the normal X and Y chromosome.

Individuals with Klinefelter syndrome are physically male due to the presence of the Y chromosome. About 1 in 500 males have an extra X chromosome. The XXY status results from an abnormality during sex cell division, leading to a sperm or egg with an extra X chromosome. This leads to boys being born with two active X chromosomes in each cell instead of the normal one. The presence of the Y chromosome allows some of the genes on the extra X to be expressed. These are called triploid genes and are thought to cause the syndrome. The condition causes a number of physical and behavioral characteristics, including infertility with absence of sperm. Individuals have low testosterone levels and are often shy and lack muscularity, but in many cases the condition is not detected. Some men with Klinefelter syndrome do produce sperm, and assisted conception may be possible.

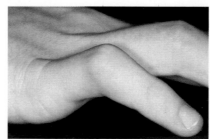

Clinodactyly of the little finger
This abnormal curving of the little finger toward the ring finger is often found in people with Klinefelter syndrome. However, it may also occur without any genetic abnormality.

Around 98 percent of fetuses affected by Turner syndrome are not viable and are miscarried. The condition affects about 1 in 2,500 live births. It is not fatal, although it can cause medical problems. It cannot be inherited, since affected individuals cannot reproduce.

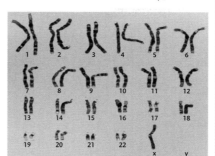

Turner syndrome chromosomes
This set of chromosomes from a female with Turner syndrome shows only one X chromosome rather than the usual two.

AMNIOCENTESIS

One of the tests that can be done to detect inherited abnormalities is amniocentesis. At around 16–18 weeks of pregnancy, a small amount of the amniotic fluid that surrounds the baby is extracted using a long needle guided by ultrasound. Cells from the baby, found in the amniotic fluid, can be examined for simple genetic information such as the presence of too many or too few chromosomes.

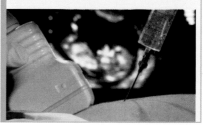

BIRTH DEFECTS

Genetic and chromosomal abnormalities may be relatively minor, or incompatible with successful development so that the fetus never reaches birth.

Birth defects are relatively uncommon and may be caused by inherited factors or by behavior. Many affected fetuses are lost early in pregnancy, due to abnormalities in the chromosomes that are incompatible with further successful growth and development. Miscarriage is extremely common, probably affecting at least 1 in 4 fertilized eggs and possibly many more at a very early stage. This may be due to interruptions and problems in the complex series of genetic maneuvers that take place when an egg is fertilized. We may never know what proportion of egg-sperm interactions are faulty.

GENE DISORDERS

Chromosomes are made up of thousands of genes. Each gene provides the blue-print for making a particular protein that the body needs to function. Abnormalities in these genes result in faulty instructions being sent to dividing cells. Abnormal genes may be passed on through inheritance. There are around 4,000 recognized inherited disorders caused by defects of single genes. Recessive diseases occur when both parents pass on a faulty gene. Dominant diseases are expressed, or partially expressed, if only one abnormal gene is inherited.

HUNTINGTON'S DISEASE

An abnormal gene on chromosome 4 causes Huntington's disease, a brain disorder that causes personality changes, involuntary movement, and dementia.

This is a dominant genetic disorder—if a person inherits the abnormal gene from either parent, he or she will develop Huntington's disease. Children of an affected parent have a 50 percent chance of inheriting the disease, which does not usually become apparent until the fifth decade. Huntington's is a degenerative brain disorder, which causes a progressive loss of brain function, often resulting in abnormal movements and dementia.

Diagnosis is made through CT scan and physical examination. Treatment may be given to relieve the symptoms. Those at risk can be tested, but many choose not to have the test, because the condition has no cure, and may only affect them far into the future.

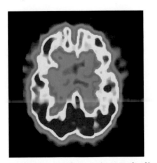

Enlarged ventricles

Brain scan of Huntington's disease
This scan of sections through the brain shows enlarged lateral ventricles typical of Huntington's disease, leading to loss of brain function.

ALBINISM

This name is used for a group of genetic disorders causing a lack of the pigment that gives color to skin, eyes, and hair.

Albinism is a recessive disorder, meaning that both parents need to have the affected genes in order to pass on the condition. If both parents are carriers, a child has a 25 percent chance of inheriting the condition and a 50 percent chance of being a carrier. No prenatal test is possible unless parents have previously had a child with albinism so that the particular genetic abnormality can be identified. Usually the genes instructing the body to make pigment are abnormal. Individuals with albinism have poor vision and little or no pigment in eyes, skin, or hair, resulting in pale skin, fair hair (which can be white), and eyes that are usually blue or violet but with a thin iris that tends to

Recessive inheritance
If both parents carry the genes for albinism but do not have the disorder, there is a 1 in 4 chance that their child will inherit both affected genes.

give back a red reflection in bright light. There is no cure, but those with the condition are advised to stay out of the sun. Problems with vision can be corrected to some degree.

COLOR BLINDNESS

Color blindness is a difficulty in distinguishing between colors. It is a genetic condition more common in males.

Most color blindness is due to abnormal genes on the X chromosome (where many genes that are concerned with color vision lie) that lack a matching opposite number on the Y. It causes a defect in the cones of the eye which are sensitive to different colors. Because the abnormal gene is recessive, a female will be affected only if she has two abnormal genes. A male will be affected if he has one abnormal gene from his mother; his father, who gave him his Y chromosome, will not have bequeathed him a matching gene. This is termed X-linked recessive inheritance: it is carried by women but expressed in men. It can also be expressed in a female who has two abnormal genes (from an affected father and a carrier mother).

About 8 percent of males, but only 0.5 percent of females, are color-blind. Most commonly, red and green are confused, but there are many other variations, some of which increase in severity through life, and others of which remain stable and cause few problems.

CYSTIC FIBROSIS

The gene for this inherited disorder is carried by 1 in 25 people. It produces thick secretions in the lungs and pancreas.

Cystic fibrosis (CF) is one of the most common life-affecting genetic diseases in the West. The child of two carrier parents has a 25 percent chance of having CF and a 50 percent chance of being a carrier for CF. Testing for carrier status is possible, as is testing of the fetus.

The gene responsible normally creates the cystic fibrosis transmembrane regulator protein, important in the regulation of sweat, digestive juices, and mucus. CF is characterized by thick, dehydrated mucus in the lungs, which accumulates, attracting infection and causing lasting damage. The secretion of pancreatic juices is also affected, impairing the absorption of nutrients from food. The severity of the condition is variable, and modern medical techniques have contributed enormously to the health and life expectancy of those affected.

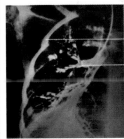

Ribcage

Mucus in bronchi

Cystic fibrosis lungs
This colored chest X-ray shows the bronchi in a lung of a person with cystic fibrosis. They are filled with mucus, causing recurrent chest infections.

ACHONDROPLASIA

Defective bone growth caused by an abnormal gene, achondroplasia is the most common cause of dwarfism, or extreme short stature.

Achondroplasia affects around 1 in 25,000 people. Affected people are typically not much over 4 ft (131 cm) in height, due to a mutation in the gene that affects the growth of bones. Altered body proportions also result from the condition. People with achondroplasia have one abnormal gene, but the matching gene in the pair is normal. A combination of two abnormal genes is fatal before or soon after birth. If both parents have achondroplasia there is a 1 in 4 chance that the baby will not survive and a 1 in 2 chance that the baby will also have dwarfism. There is also a 1 in 4 chance that the baby will be of normal stature. However, most cases of achondroplasia are due to new mutations of genes, with neither parent being affected. It is not possible to carry the gene without showing its effects. There is no cure, and treatment is rarely needed.

MULTIFACTORIAL INHERITANCE

Most inherited diseases are multifactorial, which means that they result from a combination of genetic and environmental factors. Genes may cause the condition or increase the chances of it developing, and the condition may vary widely. Such inheritance can be difficult to trace through families. Autism is one example of multifactorial inheritance, and it may be caused by a number of genes.

Autistic child
Usually diagnosed in childhood, individuals with autism generally have unusual or problematic social and communication skills, sometimes with other unusual abilities.

CANCER

Cancer is most often a growth or lump caused by the abnormal multiplication of cells that spread beyond their natural space. It is not a single disease but a large group of disorders with different symptoms, and may be caused by faulty genes, aging, or cancer-causing agents such as cigarettes.

BENIGN AND MALIGNANT TUMORS

A tumor is a growth or lump. Malignant tumors can invade normal tissue and spread to other parts of the body. Benign tumors do not spread.

A tumor is a mass of cells that divide abnormally quickly and fail to carry out their usual function. These growths can be benign (noncancerous) or malignant (cancerous), depending on the behavior of the cells.

Generally speaking, malignant tumors have the greatest potential to cause harm—but not all do so. Rapid growth and fast cell division, more structurally abnormal cells, and a pattern of spread all suggest greater malignancy. Benign growths are also caused by changed cells that multiply abnormally and do not carry out their proper functions. Unlike malignant cancers, they grow slowly and do not spread.

Treatment may be given for benign tumors if they bleed or press on important structures, but generally benign tumors are less likely to progress and cause harm. It is important to detect whether a tumor is benign or malignant because cancerous cells can spread through the body. Malignancy is usually tested by taking a sample of the affected tissue and checking its behavior microscopically. Some cancers produce specific chemicals, and measuring the levels of these substances can also help in diagnosing the type of cancer.

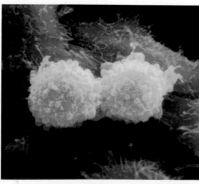

Cancer cells dividing
This magnified image shows a cancerous cell dividing to form two cells containing damaged genetic material. Untreated cancer cells multiply uncontrollably and spread through the body.

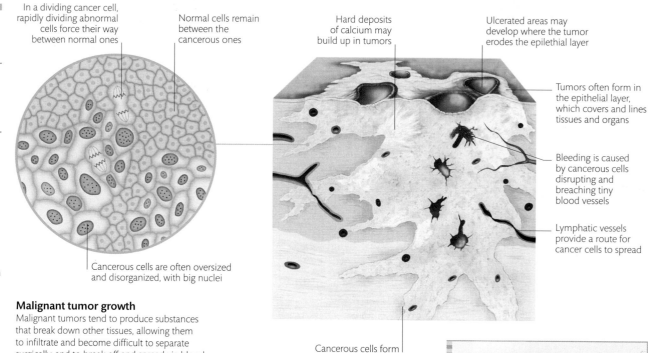

In a dividing cancer cell, rapidly dividing abnormal cells force their way between normal ones

Normal cells remain between the cancerous ones

Hard deposits of calcium may build up in tumors

Ulcerated areas may develop where the tumor erodes the epilethial layer

Tumors often form in the epithelial layer, which covers and lines tissues and organs

Bleeding is caused by cancerous cells disrupting and breaching tiny blood vessels

Lymphatic vessels provide a route for cancer cells to spread

Cancerous cells are often oversized and disorganized, with big nuclei

Malignant tumor growth
Malignant tumors tend to produce substances that break down other tissues, allowing them to infiltrate and become difficult to separate surgically, and to break off and spread via blood and lymph to seed in distant parts of the body.

Cancerous cells form tendril-like outgrowths that infiltrate surrounding tissues

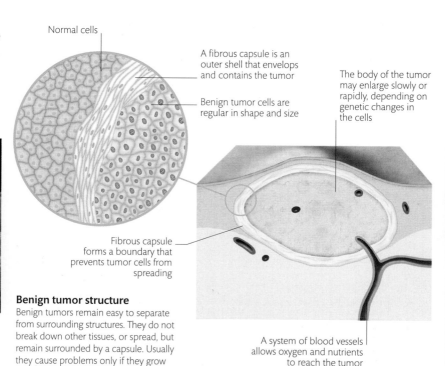

Normal cells

A fibrous capsule is an outer shell that envelops and contains the tumor

Benign tumor cells are regular in shape and size

The body of the tumor may enlarge slowly or rapidly, depending on genetic changes in the cells

Fibrous capsule forms a boundary that prevents tumor cells from spreading

Benign tumor structure
Benign tumors remain easy to separate from surrounding structures. They do not break down other tissues, or spread, but remain surrounded by a capsule. Usually they cause problems only if they grow too big or press on surrounding organs.

A system of blood vessels allows oxygen and nutrients to reach the tumor

SCREENING FOR CANCER

Some cancers can be detected before they cause symptoms; screening for these cancers looks for changes in cells before they become cancerous (such as in colon, cervical, and prostate cancers). This allows detection of conditions that may progress to cancer but have not yet done so, enabling intervention and prevention. Other cancers may be detected at an early stage, which may be asymptomatic (have no symptoms). This is commonly done for breast cancer. If caught early, curative treatment is more likely to be possible.

Mammogram
Testing for breast cancer is done using a mammogram. This is a special X-ray technique that shows tissue in the breast and allows cancer to be detected at an early stage.

HOW CANCER STARTS

Cancer is often triggered by carcinogens (cancer-causing agents) such as tobacco. Faulty genes may increase the risk of developing the disease.

Cellular damage occurs all the time, but the body's DNA usually repairs itself. Several things have to occur for a cancer to begin. The initial trigger is usually damage to the DNA of genes called oncogenes, which program cell behavior. If mutated or damaged, oncogenes may prevent the normal processes of natural cell death (apoptosis), and instead encourage cells to keep dividing.

Various substances can damage DNA and are carcinogenic (cancer-causing). They include radiation such as sunlight, toxic chemicals such as alcohol, and many of the byproducts of tobacco. Sex hormones may provoke cancers by overstimulating cell growth, and chemotherapy, which damages cellular DNA, can actually cause cancer. Viruses including hepatitis C can also damage DNA. Successful repair requires a functioning immune system, so cancer risk is increased when a person has a condition that leads to weakened immunity (such as AIDS). Cancer is also more likely to result if the damage is repeated, or severe, and sustained, or if the person has inherited defective oncogenes. In these cases the damage becomes permanent, and key cell functions are irreparably affected.

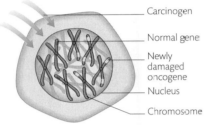

- Carcinogen
- Normal gene
- Newly damaged oncogene
- Nucleus
- Chromosome

1 Damage from carcinogens
Carcinogens damage the DNA of oncogenes, which manage the normal restrictions on cell growth. Toxins, radiation, and viruses can all damage DNA, which is under constant attack.

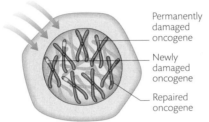

- Permanently damaged oncogene
- Newly damaged oncogene
- Repaired oncogene

2 Permanent damage
While DNA can repair itself, if damage is severe or sustained or if the repair system fails, oncogenes may be permanently damaged and their cancer-preventing function switched off.

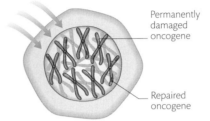

- Permanently damaged oncogene
- Repaired oncogene

3 Cell becomes cancerous
If the oncogene is permanently damaged, then abnormal cell growth can begin. The malignancy depends on the nature of the affected cells and the manner in which they grow.

HOW CANCER SPREADS

Cancer spreads by local growth, and when cells break off from the tumor and are carried via the blood or lymphatic system to other parts of the body.

Local cancer growth occurs through the growth and multiplication of cancer cells in their original site. If the cells look and behave normally and spread neatly, pushing at local tissues rather than growing into them, the cancer is behaving in a benign way—even though it may grow rapidly. Malignant cancer cells produce substances that allow them to break into other structures, growing through other tissues (local invasion) and potentially breaching the walls of blood vessels, lymphatic vessels, and important structures.

The main routes of spread are through the blood and lymph systems, the body's main ways of distributing nutrients and collecting waste. Once the walls of blood or lymph vessels are breached, cancer cells can enter those vessels and be transported to other sites in the body—often the liver, brain, lungs, or bones. When they lodge in these distant areas, more aggressive cancers can become established and start growing independently of the original tumor. This is called metastasis, and the distant growths are called metastases. Particular cancers tend to spread to characteristic places; for example, bowel cancer typically spreads to the liver, because the blood vessels of the bowel travel from there to the liver for processing products of digestion.

SPREAD BY LYMPH

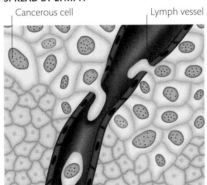

Cancerous cell — Lymph vessel

1 Lymph vessel breached
As the primary tumor grows, its cells invade adjacent tissues. The lymphatic vessels form a suitable transport system for abnormal cells to move around the body.

SPREAD BY BLOOD

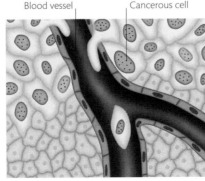

Blood vessel — Cancerous cell

1 Blood vessel wall ruptured
The rupture of a blood vessel wall as a tumor expands may cause bleeding and allow tumor cells to enter into the blood system. In this way they can be transported to virtually anywhere in the body.

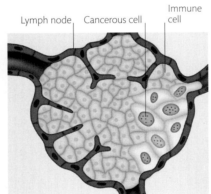

Lymph node — Cancerous cell — Immune cell

2 Tumor in lymph node
Cancerous cells entering a local lymph node can start to divide and grow into a secondary tumor (metastasis). Immune cells here may halt the spread of the disease temporarily.

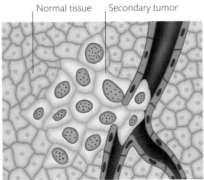

Normal tissue — Secondary tumor

2 Secondary tumor formed
Cancerous cells may be bigger than red blood cells and can become lodged in narrow vessels. As the cells divide, they push into surrounding tissues, establishing a secondary tumor.

TREATMENT OF CANCERS

Cancer may be treated with surgery to remove a tumor, with radiation therapy, or with anticancer drugs known as chemotherapy, which kill cancer cells.

Some cancers—particularly early cancers and benign tumors—are cured by surgery to remove the tumor. Surgery is also used to reduce the size of tumors prior to other treatment, or to prevent them from damaging the surrounding tissue. Radiation therapy destroys cancer cells using high-intensity radiation. It can cure the disease or slow or prevent its growth, and it can be accurately focused on surgically inaccessible tumors. Side effects include fatigue, loss of appetite, nausea and vomiting, and painful skin at the site of the treatment. It may be used along with other treatments.

Chemotherapy includes different chemical agents that target damaged or mutated oncogenes (genes that have mutated and cause tumors), growth factors, and the division of cancer cells. Some agents work against all dividing cells, and side effects such as hair loss or nausea are caused by the normally rapid division of hair follicles and gut cells. Others pick on specific characteristics of certain cancers and target all cells with that characteristic. The treatment may cure the disease or relieve its symptoms, and can be given orally or into the bloodstream or spinal fluid. The success of the treatment depends on the age and general health of the person being treated and the type of cancer.

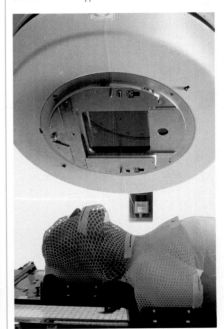

Radiotherapy treatment
Radiation is used to destroy cancer cells. During the treatment, high-intensity radiation is carefully focused on the cancerous area to destroy it or slow its growth.

INFECTIOUS DISEASES

Infection is the invasion of the body by pathogens (harmful microorganisms) that multiply in the body tissues. Organisms that can produce infectious disease include viruses, bacteria, fungi, protozoa, parasites, and aberrant proteins called prions.

ROUTES OF INFECTION

The body is constantly exposed to infection, but disease only occurs when an organism overwhelms the immune system's attempts to overcome it.

Infectious organisms can enter the body via any breach of its natural defenses: through the skin, by puncture or other injury or through

the mucous membranes of the eyes, nose, ears, digestive tract, lungs, and genitals, by inhalation, absorption, or ingestion. From there they may spread in the bloodstream (as with HIV), along nerves (like rabies), or by invading body tissues (as in invasive gastroenteritis). Most pathogens, apart from prions, are living organisms, and when they enter the body the immune system typically mounts a response to fight them off. This response produces the symptoms of illness, such as fever, inflammation, and increased production of mucus. The severity of the disorder depends on the strength and numbers of the invading organism and the immune response of the host. Some infections last only a short time before either defeat by the host's defenses or the death of the host. Others become chronic.

Airborne infections
Many viruses and bacteria spread by airborne droplets, expelled from the nose or mouth when people cough or sneeze, then entering a new host, via the mucous membranes.

VIRAL INFECTIONS

Viral pathogens range from the relatively harmless, such as those causing warts and the common cold, to the life-threatening, as in HIV (which causes AIDS).

Viruses are the smallest type of infectious organisms, made of genetic material inside a coating of protein. They cannot multiply alone, but invade body cells and use their replication mechanisms to multiply. The new particles then burst out of the cell and destroy it, or bud through the surface, and travel to infect further cells. Infections are usually systemic, involving many parts of the body at one time.

Many of the symptoms they cause, such as swollen glands and nasal congestion, are in part due to the activation of the immune system to fight the invasion. The immune response commonly begins with a fever, which is in essence an attempt to slow viral replication by increasing the body temperature

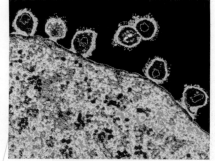

HIV virus budding from cell
Once the virus has used the body cell's DNA and reproductive mechanisms to replicate itself, the daughter organisms bud out from the cell and are each free to infect further cells.

above the optimum level for replication. Inflammation occurs when the immune system directs disease-fighting white blood cells and chemicals to the affected area. Viruses can affect any organ or body system. They commonly cause rashes, but do not often produce pain. An exception are the *Herpes zoster* virus 3 , which causes chickenpox and shingles, and the *Herpes simplex* viruses 1 and 2, which cause cold sores and genital herpes.

BACTERIAL INFECTIONS

Bacteria can cause illness by multiplying so fast that the immune system cannot control them, or by releasing toxins that damage body tissues.

Bacteria are single-celled organisms, much larger than viruses and capable of reproducing independently. They exist everywhere in the environment. The human body contains many types, largely on the skin and in the gut. Most coexist harmlessly with us, and many are beneficial. However, if the immune system is weakened by an injury such as a burn, or by illness, some can become infective; for example, *Staphylococcus aureus* lives on the skin, but in people with reduced immunity can cause boils or even invade the bloodstream.

Other disorders are caused by bacterial pathogens that invade the body and spread via the bloodstream, body fluids, or tissues. They may infect one area, as in meningitis (affecting the membranes of the brain and spinal cord), or the whole body, as in septicemia (blood

poisoning). Symptoms vary according to the site of infection, and include pain, fever, sore throat, vomiting or diarrhea (as the body tries to expel the infection), inflammation, and pus (a buildup of white blood cells and dead material). Bacterial infection can follow viral infection: tissues inflamed by a virus allow bacteria to multiply. Many infections can now be treated by antibiotics, which kill bacteria, but some bacteria have evolved to become resistant to these drugs (see right).

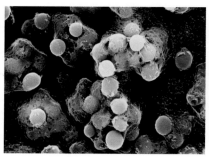

Streptococcus bacteria
The enhanced electron micrograph shows *Streptococcus pyogenes*, the bacteria that can cause scarlet fever. Sufferers have a sore, pus-coated throat, red tongue, fever, and a scarlet rash.

ANTIBIOTIC RESISTANCE

All organisms adapt to cope with changes in their environment. Since humans started using antibiotic drugs, bacteria have evolved many mechanisms to withstand them, such as plasmids. Once a method of withstanding a drug has been randomly generated by one of millions of dividing bacteria, it is coded onto a piece of genetic material, the plasmid, and transferred between bacteria, rendering the antibiotic useless.

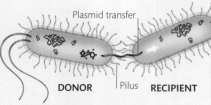

Plasmid transfer

DONOR   Pilus   RECIPIENT

2 Spread of plasmids
Plasmid transfer takes place during a process known as conjugation. The plasmid copy is passed from the donor through a tube called a pilus to the recipient bacterium.

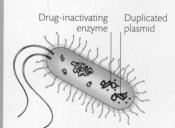

Drug-inactivating enzyme   Duplicated plasmid

Drug-inactivating enzymes

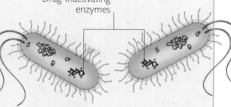

1 Activity of plasmid
Plasmids may cause the bacterium to make enzymes against antibiotic drugs, or to alter its surface receptor sites, to which antibiotics bind. Then the plasmids duplicate themselves.

3 Drug-resistant strains
Whole populations of bacteria become resistant to a range of antibiotics; some types can cause serious illnesses, such as methicillin-resistant *Staphylococcus aureus* (MRSA).

FUNGAL INFECTIONS

Infections caused by fungi or yeasts rarely result in harm unless the immune system is weakened, in which case overwhelming infection is possible.

Yeasts and fungi are simple organisms that grow as colonies of round single cells (yeasts) or in long threads (filamentous fungi). Many live on moist areas of the skin, where they cause only minor symptoms such as flaky skin or rashes. They can also inhabit mucous

Candida organisms
Candida albicans is a yeast infection that lives naturally in the bowel of many healthy people but can be an opportunistic pathogen of other parts of the body in people with weakened immunity.

membranes such as those lining the mouth or the vagina; for example, *Candida albicans* can cause oral thrush, with a thick white coating, itching, and soreness, or vaginal yeast, with a vaginal discharge. Infective fungi can also enter the body from soil or decaying material.

Some may enter via broken skin as in sporotrichosis which causes a skin infection; others may be breathed into the lungs and spread through the body, as in aspergillosis. Fungal infections do little harm to healthy people, and most can be cured with antifungal drugs. People with a weakened immune system, such as those with AIDS, may develop serious illness, even from normally harmless fungi.

Athlete's foot
Also called tinea pedis, athlete's foot is a fungal skin infection on the feet, usually between the toes. The tinea fungus favors warm, moist spaces; it can also occur on the scalp or in the groin.

PROTOZOAL INFECTIONS

Particularly common in tropical regions or in areas with poor sanitation, protozoa enter the body via vectors (carriers) such as mosquitoes, or from food or water.

Protozoa are single-celled organisms. Many live in water or other fluids, and they tend to flourish in warmer climates. The best known protozoal infection is malaria, caused by *Plasmodium* parasites, which kills more than a million people each year. The parasites spend some stages of their life cycle in mosquitoes, which transmit the infection to humans via bites. They enter the bloodstream and multiply in the liver, then penetrate and destroy red blood cells. This causes a malarial attack, with high fever, chills, headache, and confusion.

There is no vaccine, but infection spread can be reduced by mosquito control measures, nets, and repellents. Other protozoal infections, such as amebiasis and giardiasis, are spread via contaminated food and water, and cause digestive symptoms such as abdominal pain and diarrhea. Toxoplasmosis is a worldwide protozoal infection and can be contracted via contact with cat feces or undercooked meat.

Blood cell — *Plasmodium* vivax protozoan

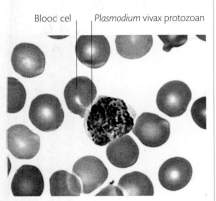

Malaria protozoa
The *Plasmodium* parasites spend part of their life cycle within human red blood cells. The parasites multiply inside the cells, causing them to rupture, thus releasing the parasites to invade new cells.

WORM INFESTATIONS

Worms interfere with the body's supply of nutrients, hijacking it for their own benefit. Most are passed on through poorly cooked food, water, and feces.

Worms, also called helminths, live inside and feed off living hosts, usually attaching via a mouth structure within the gut to drink the blood. They are sequential hermaphrodites; in other words, they may be male or female at

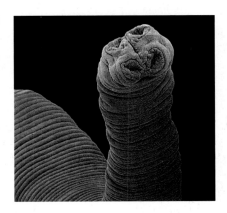

different times. Worms enter the body through ingestion, reproduce in the digestive tract, and emerge from the anus to lay eggs, which can then be transferred to a new host. Millions of people are affected worldwide—in developing countries, helminthic infection is widespread and a common cause of anemia. In the West, pinworms is the most common infestation.

Tapeworm
Tapeworms live in the gut of a host, classically causing weight loss despite increased food intake. Humans become infested by eating traces of contaminated meat or ingesting traces of feces.

ZOONOSES

Zoonoses are diseases caught from other animal species. Many are extremely serious, and some cause widespread illness in human populations.

As pathogens evolve, they occasionally mutate (change) and cross the species barrier. This is true of bacteria (for example, plague), viruses (such as rabies), protozoans (such as toxoplasma), abnormal proteins (for example, Creutzfeldt-Jakob disease), or worms. Many

human diseases began as zoonoses, including influenza, measles, smallpox, and HIV. The common cold probably came from birds, and tuberculosis may also have begun in animals. In the early stages of the encounter, the organisms have not yet adapted well to their new host, which is likewise not adapted to them with an immune response. Catastrophic infection then results as the host dies quickly.

To survive and reproduce successfully, an infective organism needs to stay alive in a living host. In severe zoonotic

illnesses, the human is a dead-end host, often infected accidentally, as in anthrax, rabies, and HIV. These diseases made the "species leap" recently in evolutionary terms. Over time, a pathogen adapts to its new host, which in turn acquires immunity, so zoonoses become milder over time.

Lyme disease
Spread by ticks, the bacterium causes a rash and flulike symptoms and, if untreated, heart and joint problems.

IMMUNIZATION

The body normally becomes immune to infections only after it has overcome them, but immunization allows immunity to develop without exposure to the disease. Most immunization is done by vaccination: the injection of either an "attenuated" form of the disease-causing organism (which is alive but not dangerous) or a dead vaccine (made from the protein coat of an organism), to provoke the immune system into attacking the organism. Alternatively, antibodies (immune system proteins) from other humans or from animals may be given. Immunization is available against

many common bacterial and viral diseases, including tetanus, diphtheria, polio, hepatitis B, and seasonal flu. It has effectively eradicated smallpox worldwide. Other infectious organisms, such as HIV, have proved to be more of a challenge because they change their form rapidly and frequently.

Measles vaccination
Measles used to be a common infectious disease of childhood. However, immunization of whole populations of children has enabled the disease to be relatively well controlled in the West.

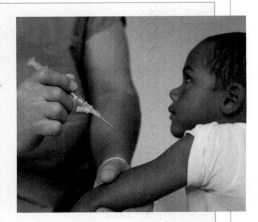

SKIN, HAIR, AND NAIL DISORDERS

The skin is frequently exposed to irritants and microorganisms and can become inflamed and infected. Skin cancers are usually caused by excessive exposure to sunlight. Nail and hair disorders may be due to localized disease or general health problems.

ATOPIC ECZEMA

Eczema is a common long-term condition that causes itching, redness, dryness, and cracking of the skin, usually in children prone to allergies.

Around one fifth of children develop eczema, but most grow out of it by adulthood. Very rarely it begins in adulthood. The condition runs in families and often occurs with hay fever and asthma. It affects both sexes equally. Eczema may come and go, with flare-ups triggered by ingesting allergens such as dairy products or gluten; contact with allergens such

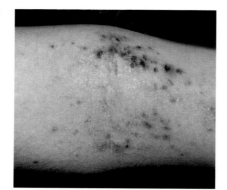

Eczema on the arm
The affected skin is reddened and thickened, with prominent skin creases and markings, crusting, and fissuring. It is very itchy and can be painful.

as house dust mites, pollen, and pet skin and saliva; and stress and fatigue. Typically, the condition occurs in the creases of the skin around elbows, knees, ankles, wrists, and neck.

A patch of eczema begins as itchy red skin. This progresses to dry scaling, and the skin may eventually thicken further, with accentuated skin lines and severe dryness, cracking, and fissuring. Eczema has no cure, and it may cause considerable emotional distress. Treatment includes avoiding the triggers, using anti-itch medications, and using topical emollients to reduce dryness of the skin. Topical corticosteroids or immunosuppressives are used either during flare-ups or more regularly, depending on the severity of the condition. Infected eczema requires antibiotics.

BIRTHMARKS

Birthmarks are colored marks on the skin that commonly develop before or soon after birth. These include café au lait spots (permanent oval, light brown patches) and port wine stains (permanent red or purple patches). A strawberry nevus (pictured below) is caused by abnormal distribution of blood vessels and usually diminishes by 6 years of age. Stork bites (pink patches) and Mongolian blue spots (large blue bruises) usually fade in childhood.

CONTACT DERMATITIS

Contact dermatitis is an inflammation of the skin due to an allergic reaction or caused by direct irritation to the skin.

Irritant contact dermatitis is more common than allergic contact dermatitis and may be due to a wide variety of chemical or physical irritants. Common chemical causes include solvents, abrasives, acids and alkalis, and soaps. Physical causes include prolonged friction from clothing and certain plants. Allergic contact dermatitis is most commonly caused by metals (such as nickel jewelry), adhesives, cosmetics, and rubber. Symptoms include a burning, itchy, or painful red rash, blisters, and hives. If due to allergy, dermatitis may take up to 3 days to develop; with an irritant, the inflammation is

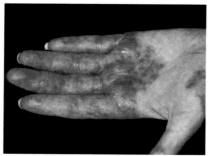

Skin affected by dermatitis
Work-related contact dermatitis is common in certain occupations such as hairdressing, where the hands are repeatedly exposed to the mild chemicals in shampoos.

often immediate. Affected skin may become dry, thickened, and cracked over time. Treatment includes avoiding trigger factors and using emollients and topical corticosteroids.

IMPETIGO

A highly contagious superficial bacterial infection of the skin, commonly on the face, impetigo rarely causes complications.

Two types of impetigo are recognized, depending on whether large blisters (bullae) form. Nonbullous impetigo is most common. It typically starts as a painless red fluid-filled blister that rapidly bursts, causing weeping and crusting, typically around the mouth and nose. In bullous impetigo, blisters are larger and may take days to burst and crust; they are most common on the arms, trunk, or legs. Impetigo heals in a few days without scarring. It is common in children, people living in confined environments, or contact sport players. Topical (applied externally) or oral antibiotics are

needed to treat the infection and prevent it spreading to others. It is highly contagious through direct contact with lesions or sharing linen and towels. Complications are rare but include cellulitis and septicemia.

Impetigo infection
An infected fluid-filled vesicle or pustule ruptures and then develops a golden-yellow crust. Touching affected areas may transfer the infection to other areas of the body and other people.

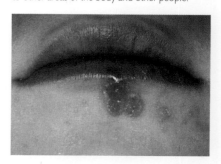

PSORIASIS

Psoriasis is a long-term skin disorder in which the skin cells reproduce too rapidly, causing itchy, flaky patches.

Psoriasis affects around 1 in 50 people. Men and women are affected equally, and it runs in families. It begins between the ages of 10 and 45 and can be triggered by a throat infection,

skin injury, drugs, and physical or emotional stress. Around 80 percent of those with the disorder have plaque psoriasis, where red, flaky patches (plaques) covered in silver scales appear usually on the elbows, knees, and scalp, which are itchy and sore. In flexural psoriasis, less scaly patches occur in skin folds such as the groin and armpit. In guttate psoriasis, smaller scaly red patches occur all over the body in a young person, following a throat infection. Guttate psoriasis usually clears up completely. Psoriasis may affect only the scalp.

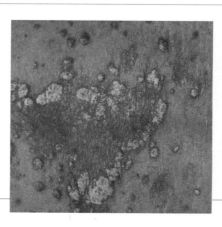

The condition is diagnosed on its appearance. Psoriasis responds well to phototherapy (UV light) but is usually a long-term condition. Topical (external) treatments include emollients, coal tar-based preparations, corticosteroids, dithranol, and vitamin D and A analogues.

Plaque psoriasis
Patches (plaques) of the skin are thickened, red, flaky, and covered in silvery-white scales, and have a sharp border. They usually itch and may burn.

RINGWORM

"Ringworm" is an umbrella term for a variety of common fungal infections of the nails, scalp, and skin.

Ringworm (tinea) infections are classified by the site of infection; usually warm, moist areas that allow fungi to thrive. In tinea corporis, an enlarging, red, itchy, slightly raised, ring-shaped skin rash develops on exposed body areas (for example, face and limbs). It is contagious by direct contact or via contaminated items such as clothing, animals, carpets, and bathing surfaces. In tinea capitis, which mainly affects children, scaly patches appear on the scalp and the local hairs break off.

In tinea cruris ("jock itch"), an itchy, red, raised rash develops in the skin folds of the groin and enlarges with a redder, more raised advancing edge. In tinea pedis (athlete's foot), scaling, flaking, and itching of the feet occurs, especially in the webs between toes. Onychomycosis (fungal infection of the nails) causes the nails to become thick, yellow, friable, and deformed. Fungal infections are diagnosed by their appearance and microscopic analysis of skin scrapings or nail cuttings. Treatment is with oral or topical antifungals, depending on the site and severity of the infection.

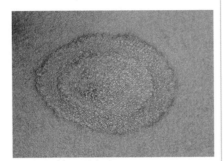

Ring-shaped rash of tinea corporis
A raised red ring with healing within the center is characteristic of ringworm. Scales, crusts, and papules may develop especially on the advancing edge. Ringworm is most common in children.

URTICARIA

Itchy red raised bumps on the skin, urticaria ("hives") is commonly caused by an allergic reaction and lasts a few hours.

Urticaria is caused by the release of histamine and other inflammatory substances from skin cells. These substances cause small blood vessels in the lower layer of the skin to leak fluid. About 1 in 4 people develop urticaria in their life, usually as children or young adults, and it is more common in women. Acute urticaria lasts less than 6 weeks; most cases last only a few hours.

Allergic urticaria is commonly due to food or drug allergies or direct skin contact with substances. Nonallergic causes of urticaria include certain foods (such as rotten fish), stress, and an acute viral illnesses. In the rarer physical urticarias, pressure, exercise, heat and cold, vibration, and sunlight may cause hives.

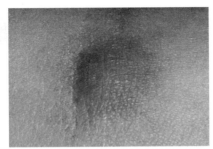

Red swelling caused by urticaria
The red, itchy, raised areas of skin due to urticaria can vary in shape and size. Typically, they are round but can form into rings or large patches.

In chronic (long-lasting) urticaria, the hives last more than 6 weeks (sometimes years), usually no cause can be found, and it can be difficult to treat. Investigations include allergy testing and searching for triggers. Treatment involves avoiding the triggers and taking oral antihistamines during attacks or to prevent them. Oral corticosteroids may be used to treat chronic urticaria.

ACNE VULGARIS

Blockage and inflammation of the sebaceous glands leads to spots on the face, upper chest, and back. Acne affects nearly all teenagers.

Acne may last for many years with repeated flare-ups but typically disappears by the age of 25. Acne is more common in boys and may run in families. Adult acne occurs mainly in women and may worsen a few days before menstrual periods or during pregnancy. Drugs such as corticosteroids or phenytoin may cause acne. The condition is neither infectious nor due to poor hygiene but does cause much psychological distress.

The skin itself appears greasy. The lesions that develop include open comedones (blackheads), closed comedones (whiteheads), papules (red bumps), and pustules (pus-filled bumps). Severe cases may include nodules (painful, deep, large, hard lumps) and cysts (painful, large, pus-filled lumps that look like boils). These may scar when they rupture, leaving "ice-pick" scars that look like holes punched into the skin, or keloid scars that are red and lumpy. It is important not to squeeze or pick lesions, to prevent scarring. Acne is diagnosed by its typical appearance. Treatment depends on the severity of the condition, but includes combinations of oral antibiotics for many months with topical treatments such as benzoyl peroxide, retinoids, topical antibiotics, and azelaic acid. Visible improvement may take 2–3 months. Severe acne may require 4–6-months of an oral retinoid, which is a powerful drug used by specialists. Acne scarring may require dermabrasion or laser therapy.

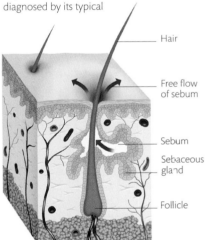

Normal hair follicle
The pilosebaceous unit consists of a hair follicle, a sebaceous gland, and, a sebaceous duct. The gland produces oil called sebum that flows out of the skin pore to lubricate the skin and hair.

Hair
Free flow of sebum
Sebum
Sebaceous gland
Follicle

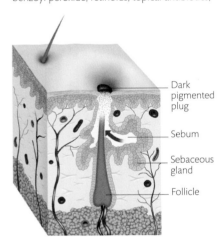

Blackhead
In acne, excessive amounts of sebum are produced and a large plug of sebum and dead skin cells blocks the follicle, forming a blackhead (comedone), which is dark due to pigmentation.

Dark pigmented plug
Sebum
Sebaceous gland
Follicle

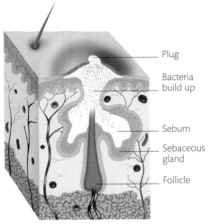

Infected follicle
Harmless bacteria that live on the skin that contaminate the plugged follicle, causing inflammation and infection, which leads to papules, infected pustules, nodules, and cysts.

Plug
Bacteria build up
Sebum
Sebaceous gland
Follicle

ROSACEA

Rosacea is a long-term skin condition that primarily affects the face of fair-skinned people, causing flushing and redness.

Rosacea is twice as common in females as males and starts after the age of 30. It causes facial flushing that may spread to the neck and chest and typically lasts a few minutes. Rosacea has a variety of triggers including caffeine, alcohol, sunlight, wind, spicy foods, and stress. Persistent facial redness on the cheeks, nose, forehead, and chin may develop. Spots and pustules may appear, and small red blood vessels (telangiectasia) can become prominent on the skin. The skin may thicken and, rarely, the nose can become bulbous and disfigured (a condition known as rhinophyma).

Rosacea is diagnosed by its characteristic appearance. Treatment includes avoiding triggers and, if severe, using topical or oral antibiotics. Camouflage creams may be used to cover the rash. Telangiectasia can be treated with laser therapy. Cosmetic treatment for rhinophyma may require surgery.

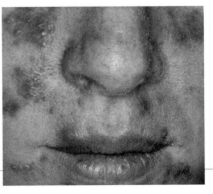

Redness on the face caused by rosacea
The face is red and liable to flush easily. There are red bumps (papules) and some pus-filled spots (pustules), which can be mistaken for acne vulgaris.

BURNS AND BRUISES

Burns are skin injuries due to heat, cold, electricity, friction, chemicals, light, or radiation. Bruises are caused by internal bleeding into tissues from capillaries.

Superficial-thickness burns affect only the epidermis (outer layer of skin), leading to mild swelling, redness, and pain, and rarely scar. Superficial partial-thickness burns involve the epidermis and the superficial dermis, leading to pain, dark red or purple coloration, marked swelling, blisters, and the weeping of clear fluid. Deep partial-thickness burns involve the epidermis and whole dermis, look whiter or mottled, and are less painful due to nerve

damage. Full-thickness burns involve the epidermis, dermis, and the subcutaneous fat layer and cause no or minimal pain. The burn may be charred and black, leathery and brown, or white and pliable. Subdermal burns reach down even further, to the underlying tissues and structures. Treatment of a burn depends on its site, depth, and extent. Full-thickness and subdermal burns often require skin grafts. Extensive burns may easily become infected and can cause massive fluid loss.

A bruise is called an ecchymosis; red or purple bruises 3–10mm in size are called purpura, and ones smaller than that, petechiae. Treatment for bruises includes analgesics and protection, rest, ice, compression, and elevation ("PRICE"). Unexplained bruising can signal an underlying disorder such as a blood clotting problem, septicemia, or leukemia.

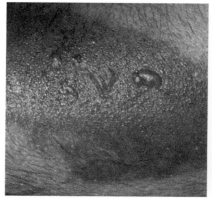

Scald
A scald is a burn caused by hot liquid or steam, often boiling water from a tap. As shown here, it results in a well-demarcated area of swelling and redness with some blistering.

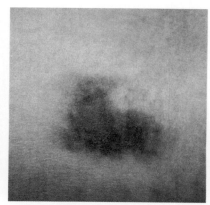

Bruise
Bruises change color due to hemoglobin from red blood cells being broken down to form chemicals of various colors including green, yellow, and golden-brown.

SKIN CANCER

Skin cancers are the most often diagnosed cancers worldwide. The most common forms are basal cell carcinoma, squamous cell carcinoma, and malignant melanoma.

Basal cell and squamous cell carcinoma are both usually caused by cumulative ultraviolet (UV) light exposure (often from sunshine and tanning beds). They are most common in people with light skins in countries with high levels of UV light. They affect males more often, perhaps due to differing lifetime sunlight exposures.

Basal cell carcinoma (BCC) arises from the basal cell layer and is rare before the age of 40. It accounts for around 80 percent of skin cancers. The lesion appears as a raised, smooth, pink or brown-gray bump with a pearly border, which may have visible blood vessels. It is not painful or itchy. The center may be pigmented or ulcerate. It grows slowly and only very rarely metastasizes (spreads to other organs or parts of the body). Diagnosis of basal cell carcinoma is by skin biopsy, and it can usually be cured by surgical excision (removal).

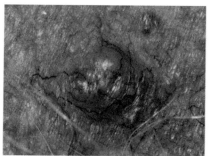

Basal cell carcinoma
The typical smooth pink bump of a basal cell carcinoma. The center may crust and bleed and is often described as a sore that does not heal.

Squamous cell carcinoma (SCC) arises from the squamous cell layer. It may rarely be due to exposure to chemical carcinogens (such as tar) or ionizing radiation as well as UV light. It usually occurs from the age of 60 onward, but this varies. SCC accounts for about 16 percent of skin cancers. The lesion is a raised, hard, scaling, pinkish patch that may ulcerate, bleed, and crust. It slowly enlarges, sometimes developing into a large mass, and it rarely metastasizes. It is diagnosed by skin biopsy, and the usual treatment is surgical excision.

Malignant melanoma arises from the melanocytes (pigment-producing cells) in the skin. Sunlight exposure especially in childhood, episodes of blistering sunburn, using sunbeds, and a family history increase the risk. It is most common in light-skinned people and those with many moles. Melanoma may arise from a preexisting mole or appear as a new, enlarging black or brown mole (see below), and is treated by complete surgical excision. The prognosis depends on the depth and spread of the tumor. Melanomas often metastasize and are fatal in around 1 in 5 cases. All people—not just those with already diagnosed skin cancer—should have yearly screening and should avoid the sun by wearing protective clothing, applying sunscreen regularly, and staying out of the sun in the middle of the day.

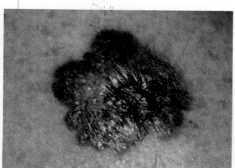

Melanoma on the skin
Warning signs of malignant change in a mole include a change in size, shape, color, or height; bleeding; itching; ulceration; irregular shape; variable color, and asymmetric border.

SKIN BIOPSY

During a skin biopsy, a small sample of a skin lesion is removed so that it may be examined under a microscope. This may be done to diagnose infections or cancers of the skin and other skin conditions. In an excisional biopsy, the lesion and a margin of normal skin around it are completely removed. In a punch biopsy, a small cylindrical core is taken from the lesion, leaving the rest of it behind if it is large. In a shave biopsy, a very thin slice of the top part of a lesion is removed. This may be sufficient to completely remove a superficial skin lesion.

Melanocyte

Melanoma skin biopsy
This microscopic view of a tissue sample shows cancerous melanocytes, containing brown melanin pigment, that have invaded the epidermal (uppermost) skin layer.

PIGMENTATION DISORDERS

Loss of normal skin color is usually due to the skin's inability to produce the pigment melanin. This may be hereditary or develop later in life.

Melanin is the pigment that gives the skin its color. Abnormal pigmentation is caused by several conditions, including albinism and vitiligo. Albinism (see p.431) is a genetic disorder resulting in a lack of melanin pigment. This may just affect the eyes (ocular albinism) or the eyes, skin, and hair (oculocutaneous albinism).

Vitiligo affects up to 1 in 50 people. It is an autoimmune disorder that is caused by the immune system's antibodies reacting against its own tissues, destroying the cells that produce melanin. White or pale skin patches appear, commonly on the face and hands, and then enlarge. New patches then develop, usually all over the body. There is no cure, but phototherapy or laser therapy may help repigment areas. Camouflage cosmetics can hide smaller areas. Topical treatments may be used.

Vitiligo
Depigmented patches of skin typically occur symmetrically on the extremities, appearing after childhood usually before the age of 30. Psychologically distressing, vitiligo may also be associated with other autoimmune disorders.

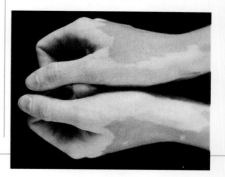

MOLES, WARTS, CYSTS, AND BOILS

Local overgrowth of certain skin cells leads to a mole or a wart. A sebaceous cyst or a boil causes a lump in the skin.

Common warts are small, raised, rough lumps usually found on the hands or knees. Plantar warts occur mainly on pressure points on the sole of the foot, forming painful, hard lumps. Warts (verrucae) are diagnosed by their appearance. They often disappear by themselves but may be treated by cryotherapy (the use of cold to freeze them) or topical treatments containing salicylic acid.

Sebaceous cysts vary in size, are smooth and round, freely move under the skin, grow slowly, and are painless unless they become infected. Usually harmless, they are diagnosed by their appearance. They can be left alone, or can be surgically removed if they cause distress or become infected. A boil is a bacterial infection—a warm, painful lump that develops a central yellow or white head of pus before it discharges the pus and then heals. Clusters of boils may interconnect to form a carbuncle. Recurrent boils may occur in diabetes, or those with a weakened immune system. Large boils may need incision and drainage.

Moles are dark, pigmented lesions that may be raised from the surface of the skin. They vary in size and can develop anywhere on the body. Most moles occur before the age of 20 and disappear after middle age. They may be removed if malignant change or melanoma (see opposite) is suspected. Warning signs can be easily remembered by "ABCDE": Asymmetry; Bleeding; Color change or variability of color; Diameter (if bigger than a pencil eraser); Elevation. Some inherited conditions lead to a large number of moles.

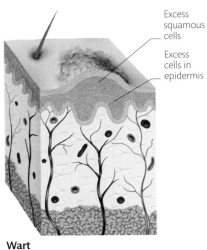

Mole
Localized overproduction and build-up of melanocytes leads to a (sometimes raised) pigmented area. As the cells are not cancerous, they do not invade beneath the epidermis.

Raised pigmented area
Pigment cells

Wart
Overgrowths of epidermal cells in a small area, warts are caused by the human papilloma virus (HPV). Warts are passed on by direct contact or from objects used by affected people.

Excess squamous cells
Excess cells in epidermis

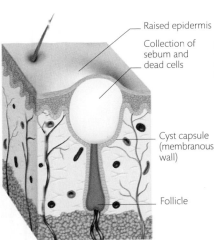

Sebaceous cyst
This is a closed sac under the skin surface, filled with accumulated sebum and dead cells. These most commonly occur on the hairy areas of the scalp, face, trunk, and genitals.

Raised epidermis
Collection of sebum and dead cells
Cyst capsule (membranous wall)
Follicle

Boil
Collections of pus in hair follicles, sometimes including the sebaceous gland, boils are commonly caused by infection by *Staphylococcus* bacteria and usually clear up within 2 weeks.

Head of boil
Swollen area
Pus-filled sebaceous gland
Pus-filled follicle

NAIL DISORDERS

Localized infection, inflammation, and deformity of the nails is common. The nails may also show evidence of diseases that occur elsewhere.

Onycholysis (loosening of the nail from the nail bed) may be caused by infection, drugs, or trauma. Nail trauma can also result in blood collecting under the nail, leading to pain. The blood is released by making a hole in the nail. Onychomycosis (fungal nail infection) causes thickened, friable, discolored nails. They are diagnosed by examining nail clippings for fungi and treated with local or oral antifungals.

Paronychia (bacterial infection where the nail and skin meet at the side or the base of a nail) leads to a painful, throbbing, red, hot swelling of the area. It responds to antibiotics but may need to be drained if there is pus. Koilonychia (spoon-shaped nails), where the nails curve upward, is seen in people with iron-deficiency anemia (see p.472). Pale nails occur in all anemias and may also be due to kidney or liver disease. The skin condition psoriasis (see p.436) can cause pitted nails. Leukonychia punctata (white flecks on the nails) is common and usually due to injury to the base of the nail, disappearing as the nail grows out.

With neglect, nails may thicken, develop grooves, and discolor (onychogryphosis). In clubbing of the nail, the nails become curved and bulbous and the ends of the fingers eventually thicken. This can occur with chronic heart and lung disease, malabsorption, inflammatory bowel disease, and cirrhosis.

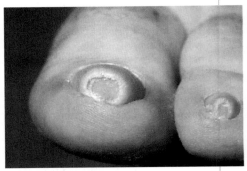

Ingrown toenail
The nails cut into the sides of the nail bed, often leading to localized redness, swelling, warmth, and pain, sometimes with pus and bleeding. Minor surgery may be needed.

HIRSUTISM

Increased or excessive hair growth in areas where hair is usually absent or minimal, hirsutism can lead to distress and may have a serious cause.

Around 1 in 10 women develop dark, coarse hairs on the chin, upper lip, chest, around the nipples, or on the back, abdomen, and thighs. In most cases, there is no underlying disorder. Serious causes include polycystic ovary syndrome, hypothyroidism, Cushing's syndrome, anabolic steroid use, and tumors that produce male hormones. Investigations for hirsutism may include measuring hormone levels and assessing the menstrual cycle. Drug therapies include certain combined oral contraceptive pills.

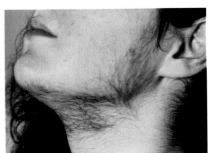

Excessive hair growth
Shaving, waxing, plucking, electrolysis, depilatory creams, and bleaching may help the appearance of excessive hair growth, especially on the face.

ALOPECIA

The temporary or permanent loss of hair from the head or body may be in one area or all over the body, and may indicate an underlying medical condition.

Androgenic alopecia (male-pattern baldness) causes hair to recede and is most common in males. Alopecia areata is caused by autoimmune attack on the hair follicles. Scalp skin disorders such as tinea capitis, burns, and chemicals can also cause hair loss. Iron deficiency and hypothyroidism (underactive thyroid gland) may cause general hair loss. Physical or psychological stress can cause telogen effluvium (diffuse general hair loss) by interrupting the normal life cycle of the hair. Chemotherapy can cause the loss of all body hair.

Alopecia areata
Hair lost from the scalp in patches usually regrows over several months, but the condition can be permanent and body-wide.

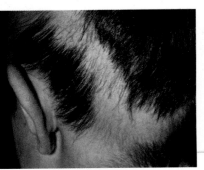

BONE AND JOINT DISORDERS

Bones and joints can be damaged by injury or by disease. Many conditions become more common with increasing age as the bones become weaker. Some disorders may be inherited or associated with poor nutrition and lifestyle.

FRACTURE

A fracture can be a complete break, a crack, or a split part of the way through a bone anywhere in the body.

Bones can normally withstand most strong impacts, but they may fracture if subjected to violent force. A sustained or repeated force can also cause a fracture; long-distance runners are particularly prone to this kind of injury. Bone diseases such as osteoporosis (see opposite) can make the bones more fragile and less able to withstand impacts. There are two main types of fracture. A simple or closed fracture is a clean break through a bone, but the bone ends stay inside the overlying skin. In a compound or open fracture, the broken bone can pierce the skin, and there is an increased risk of bleeding and infection. Bones may also crack without breaking apart; this is known as a hairline fracture. If there are more than two fragments, the break is known as a comminuted fracture.

In children and adolescents, the long bones of the arms and legs grow from areas near the bone ends known as growth plates; these areas can be damaged in a fracture, which may affect the development of the bone. Young children's long bones are less brittle and can sometimes bend and crack without breaking in two; this is known as a greenstick fracture. As long as the broken parts have not become displaced or abnormally angled, a fracture will usually heal if the pieces are held in position; otherwise, it will need to be reset first. Fractures are always extremely painful. Broken bones bleed, sometimes with considerable blood loss, and movement will provoke further pain. The bone is usually set in a cast, to relieve pain and aid healing. The healing process varies from a few weeks to several months, depending on the person's age, the type of break, whether it is open or closed, and whether it has to be reset.

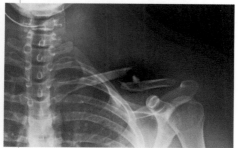

Fractured collar bone
This color-enhanced X-ray shows a collar bone that has fractured into three separate pieces. The fragments need to be realigned before healing begins.

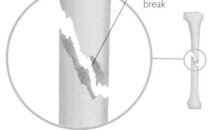

Diagonal break

Spiral fracture
A sharp, twisting force may break a long bone diagonally across the shaft. The jagged ends may be difficult to reposition.

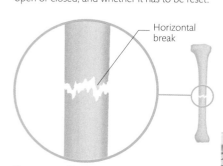
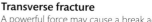

Horizontal break

Transverse fracture
A powerful force may cause a break across the width of a bone. The injury is usually stable; the broken surfaces are unlikely to move.

HOW BONES HEAL

Bone has its own self-repair process. This begins just after a fracture, when blood leaks from severed blood vessels and clots. Over the next few weeks, the broken bone ends generate new tissue. The bone will be immobilized, usually in a cast or splint, to keep the ends aligned as they heal.

Network of fibrous tissue

The first few days
Specialized cells called fibroblasts form a fibrous web across the break. White blood cells destroy damaged cells and debris, and osteoclast cells absorb damaged bone.

After 1–2 weeks
Bone cells called osteoblasts multiply and make callus (new woven bone tissue). The callus grows from each bone end to fill the gap.

New woven bone (callus)

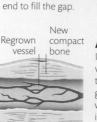

Regrown vessel

New compact bone

After 2–4 months
In time, the blood vessels rejoin across the break. The callus gradually reshapes, while new bone tissue is remodeled into dense, compact bone.

PAGET'S DISEASE

This abnormality affects bone growth, causing bones to become deformed and to be weaker than normal.

Normally, bone is continually being broken down and replaced by new bone, to keep the skeleton strong. In Paget's disease, the cells that break down bone (osteoclasts) are overactive, which makes the cells that produce new bone (osteoblasts) work faster than normal. The resulting new bone is weak and of poor quality. The condition sometimes runs in families, but the cause is not known. The most common sites for Paget's disease are the skull, spine, pelvis, and legs, but it can affect any bone. The disease most commonly causes bone pain, which may be mistaken for arthritis, and can lead to fractures of the long bones. In the skull, it can cause headache, pain in the teeth, and deafness resulting from affected small bones in the ear compressing the hearing nerves; it may also cause pressure on nerves in the neck or spine. Rarely, cancerous changes may develop in affected areas. Paget's disease cannot be cured, but it can be controlled with medication.

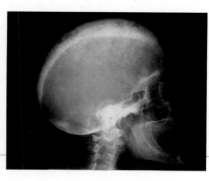

Thickened skull
This color-enhanced X-ray shows abnormalities due to Paget's disease. The bone is overly thick and dense (white areas), and the skull appears enlarged.

ABNORMAL SPINAL CURVATURE

The spine normally has gentle curves along it, but it can become excessively bent due to disease or poor posture.

The spinal column has two main curves: the thoracic curve, in the chest area, and the lumbar curve, in the lower back. Excessive thoracic curvature is called kyphosis; curvature in the lower back is called lordosis. Sideways curvature is called scoliosis. A curved spine is common in children, especially girls, and in most cases there is no obvious cause, although the condition often runs in families. In adults, the excessive curvature may result from weakening of the vertebrae, obesity, or poor posture. In most children, the curvature corrects itself as the child grows, but in severe cases a corrective brace or surgery may be needed to prevent permanent disability.

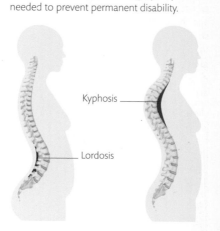

Kyphosis

Lordosis

Types of spinal curvature
A pronounced outward curve in the upper back (thoracic spine) is known as kyphosis. Excessive hollowing of the lower back is called lordosis.

OSTEOPOROSIS

More common in older people, this disorder is a loss or thinning of bone, which increases the risk of fractures.

Bones stay healthy when the cells that form new bone (osteoblasts) work in balance with the cells that eat worn out or damaged bone (osteoclasts). With increasing age this balance is gradually altered, so that less new bone is formed. As a result the bones lose density, becoming more fragile, and are likely to break with only minimal force.

Osteoporosis is common in old age, but in some cases the process starts much earlier. Genetics, poor diet, lack of exercise, smoking, and excessive alcohol use are significant risk factors. Hormones also play a major role: in particular, a lack of estrogen (which is needed to supply minerals for bone replacement) or high thyroid hormone levels can cause more rapid bone loss.

Women may develop osteoporosis after menopause, when their estrogen levels drop rapidly. In addition, long-term treatment with corticosteroids can cause the condition, and people with chronic kidney failure or rheumatoid arthritis are at increased risk of developing it. The most common problem associated with osteoporosis is fracture due to the fragility of the bones; typical sites are the radial bone at the wrist, the femoral neck (hip bone), and the lumbar vertebrae, where crush fractures weaken the spine. The disorder can be diagnosed with a bone density test (see right), and drugs are available to slow the progression. Osteoporosis can be prevented by eating a healthy diet, rich in calcium and vitamin D, and by doing regular weight-bearing exercise, not smoking, and limiting alcohol intake.

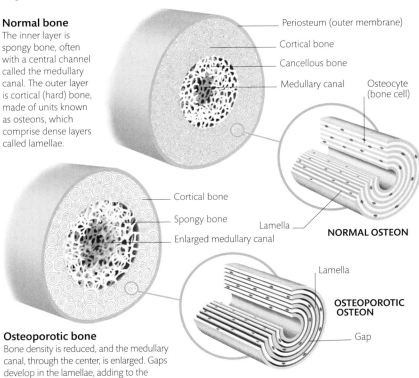

Normal bone
The inner layer is spongy bone, often with a central channel called the medullary canal. The outer layer is cortical (hard) bone, made of units known as osteons, which comprise dense layers called lamellae.

Periosteum (outer membrane)
Cortical bone
Cancellous bone
Medullary canal
Osteocyte (bone cell)

Cortical bone
Spongy bone
Enlarged medullary canal
Lamella
NORMAL OSTEON

Lamella
OSTEOPOROTIC OSTEON
Gap

Osteoporotic bone
Bone density is reduced, and the medullary canal, through the center, is enlarged. Gaps develop in the lamellae, adding to the fragility of the bone.

BONE DENSITOMETRY

A bone density scan, also called a DEXA scan, uses X-rays to measure bone density. Such scans are used to reveal evidence of bone loss and help doctors diagnose osteoporosis. The varying absorption of X-rays as they pass through the body is interpreted by a computer and displayed as an image. The computer calculates the average density of the bone and compares it to that of women in their 30s, when the density is the greatest. The scan is usually perfomed on the lower spine and hips.

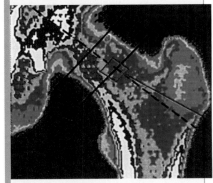

Hip bone density scan
Bone density is shown as a color-coded image, like this scan of a hip joint. In the scan, the least dense areas are blue or green. The most dense areas are white.

OSTEOMALACIA

In this painful condition, known as rickets in children, the bones become softened and may bend and crack.

Osteomalacia is due to a deficiency in vitamin D, which the body needs to absorb calcium and phosphate. These minerals give bone strength and density. In healthy people, vitamin D is made in the skin. Small amounts come from oily fish, eggs, vegetables, fortified margarine, and milk. Deficiency commonly occurs in people who follow a restricted diet or cover their skin, and absorption is reduced in darkly pigmented skin. Symptoms include painful, tender bones, fractures after minor injuries, and difficulty in climbing stairs. Treatment depends on the underlying cause and may include calcium and vitamin D supplements.

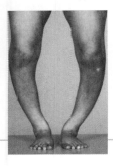

Rickets
This child has rickets, which is caused by vitamin D deficiency. This causes the bones to become softer and weaker, leading to pain and deformity.

HIP DISORDERS IN CHILDREN

The most common hip disorder in children is known as irritable hip, and often relates to viral infection—but more significant problems do also occur.

Serious problems include congenital hip dysplasia, which is evident at birth. This leads to a misalignment of the head of the femur (thigh bone) in the hip socket, and ranges from a mild defect to a complete hip dislocation. Babies are screened for the condition at birth, because it is easy to treat in the first year. If untreated, it can lead to early arthritis of the hip joints. Slipped upper femoral epiphysis occurs in children at times of rapid growth, and is most common in adolescent boys. It involves a slippage between the growth plate of the femur and the shaft, usually following relatively minor trauma. This condition causes symptoms in the hip or knee ranging from mild discomfort to incapacitating pain, and usually needs surgical correction. Perthes' disease develops when the head of the femur dies through lack of a blood supply, following a reduction in blood flow to the joint. The cause is unknown, but it results in hip, knee, or groin pain. It is more common in boys than in girls, affecting mainly prepubertal children.

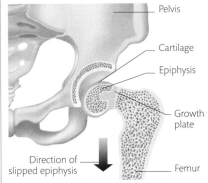

Pelvis
Cartilage
Epiphysis
Growth plate
Direction of slipped epiphysis
Femur

Slipped upper femoral epiphysis
In children, the epiphysis (end of a long bone) is separated from the shaft by a "growth plate." A weakened growth plate in the upper femur can allow the epiphysis to slip out of the hip joint.

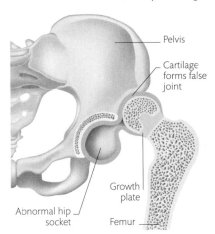

Pelvis
Cartilage forms false joint
Growth plate
Abnormal hip socket
Femur

Congenital hip dysplasia
This picture shows a severe case of hip dysplasia, with the head of the femur failing to engage in the over-shallow socket and instead forming a false socket on the pelvis.

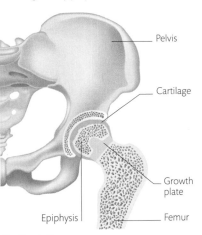

Pelvis
Cartilage
Growth plate
Epiphysis
Femur

Perthes' disease
In this disorder, the blood supply to the epiphysis (head) of the femur is inadequate. As a result, the bone breaks down and cannot engage properly in the socket, causing restriction of movement.

OSTEOARTHRITIS

This degenerative joint condition is the most common type of arthritis, usually affecting people over the age of 50 and largely caused by joint aging.

Osteoarthritis can affect any joint, although it most commonly occurs in the hips, knees, hands, and lower back. In a normal joint the bone ends are protected by a smooth, even layer of cartilage, and fluid is secreted by the synovial membranes (which line the joint capsule) to allow the bones to move easily.

In osteoarthritis, the cartilage becomes frayed or torn. Friction develops, causing inflammation of the membranes and leading to heat, pain, and excess fluid production. Bony growths called osteophytes develop around the joint edges in response to inflammation, further increasing friction and limiting the range of movement. The inflammation may come and go, but eventually the cartilage is so worn that bone grinds on bone. Pieces of cartilage or osteophytes can work loose within the joint, causing sudden locking. Affected joints may also give way suddenly. Exercise can be done to help limit stress on the joints and increase muscle tone to support them. In severe cases, surgery may be needed to remove debris, resurface the bone ends, or replace the joint.

JOINT REPLACEMENT

If a joint is severely damaged by disease or injury, it may be surgically replaced. This procedure, called arthroplasty, involves removing all or part of the joint surface and areas of damaged bone and replacing them with a prosthetic device, which is usually made of metal and hard-wearing plastics or ceramics. Not all joints can be replaced, but the knee and the hip are commonly treated in this way. Arthroplasty is a last resort, used only when pain or limitation of function significantly impair quality of life. It can relieve pain and allow a greater range of movement, but the new joint will last for only 10 to 20 years and will then itself need to be replaced.

Hip replacement
The top of the femur (thigh bone) is removed and the hip socket is hollowed out. A prosthesis is inserted into the shaft of the femur, and a new socket is fitted into the pelvis.

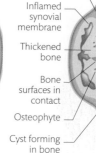

Pelvis

Original hip socket may be hollowed out and replaced

Head of femur is removed and replaced with prosthesis

Skin incision

Shaft of femur

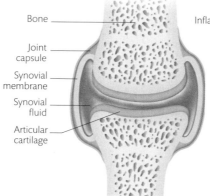

Healthy joint
The healthy bone surfaces are covered in smooth, intact cartilage, and the whole joint capsule (the tissue enclosing the joint) is lined with synovial membrane, which produces lubricating fluid.

Bone

Joint capsule

Synovial membrane

Synovial fluid

Articular cartilage

Early osteoarthritis
Changes begin with damage and degeneration of the cartilage. This leads to narrowing of the joint space, increased friction, and excess synovial fluid production, resulting in swelling, heat, and pain.

Inflamed synovial membrane

Osteophyte

Excess synovial fluid

Thinned articular cartilage

Reduced joint space

Late osteoarthritis
The cartilage is worn away in places, and the bone ends become damaged. Osteophytes and cysts form, the synovial membrane is chronically thickened, and the joint can no longer move freely.

Tight, thickened capsule

Inflamed synovial membrane

Thickened bone

Bone surfaces in contact

Osteophyte

Cyst forming in bone

ANKYLOSING SPONDYLITIS

This is a form of inflammatory arthritis mainly affecting the spine and pelvis, causing pain and stiffness and, in severe cases, making bones fuse together.

Ankylosing spondylitis (AS) is an autoimmune disease, in which the immune system attacks the body's own tissues. It is one of a group of inflammatory disorders called arthropathies, which affect the connective tissue in joints and can cause progressive and irreversible damage. In the case of AS, the damage usually involves the spine and pelvis. In the worst cases, the joints in the spine become fused and the spine loses its flexibility; an affected person will have a rigid gait with permanently impaired mobility.

The tendency to develop AS is inherited. It typically affects men, usually beginning in their 20s, with pain in the lower back and buttocks that is worse during the night and eased by walking around. Almost half of those with the condition have eye symptoms— mainly iritis (inflammation of the iris), which causes pain, redness, and temporary reduction of vision. AS is also associated with psoriasis and Crohn's disease, which share the same predisposing genes. The disorder is incurable, but physical therapy and exercise can help control its course. Nonsteroidal anti-inflammatory drugs (NSAIDs) are used to relieve pain, and immune-modifying drugs given to reduce inflammation.

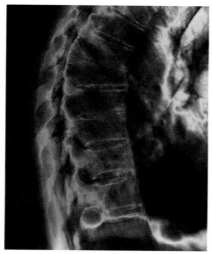

X-ray showing ankylosing spondylitis
This spinal X-ray shows inflammation, destruction of joint spaces, and joint fusion, which produce a flexed deformity of the back. The appearance of late AS on X-ray is referred to as "bamboo spine."

OSTEOMYELITIS AND SEPTIC ARTHRITIS

Osteomyelitis is a bone infection causing damage to surrounding tissue. Septic arthritis is an infection within the joint capsules and can damage joints.

Bones or joints may become infected through injury or surgery, or by the spread of infection from skin and soft tissue or via the blood. Most cases of osteomyelitis in the developed world are due to infection by bacteria such as *Staphylococcus aureus*, but tuberculosis (TB) is a common cause worldwide.

The condition may be acute (develop quickly), with many symptoms, and more common in children, or chronic (longer-lasting). In chronic osteomyelitis, the infection can cause bone tissue to die, and the dead tissue must be surgically removed. The bone marrow can also become infected. Septic arthritis is usually due to *S. aureus* bacteria. It tends to be acute, causing fever with joint pain and restricted movement. If fluid and pus build up inside the joint capsule, the joint may be permanently damaged. Surgery is necessary to drain the affected joint.

PSORIATIC ARTHRITIS

A form of arthritis associated with the inflammatory skin condition psoriasis, this condition can be highly destructive if it is allowed to progress.

This autoimmune condition affects up to 30 percent of people with psoriasis (see p.436). It may occur in both small and large joints, appearing predominantly in the hands, the back and neck, or a mixture of joints. In mild cases, only a few joints are affected—often those at the ends of the fingers or toes. In severe cases, many joints are involved, including those in the spine. Often, the arthritis flares up at the same time as the skin symptoms of psoriasis. If left untreated, psoriatic arthritis can lead to arthritis mutilans, in which the joints are completely destroyed.

The affected joints can no longer move at all, with subluxation (slipping beneath neighboring joints) and telescoping (collapsing in) of the bones. This condition is most often seen in the fingers and feet. Psoriatic arthritis may be treated with analgesics to relieve pain and reduce inflammation, as well as with medications to slow its progress.

RHEUMATOID ARTHRITIS

This connective tissue disorder can cause inflammation in many body systems, but principally attacks the lining of the joints, resulting in progressive damage.

Rheumatoid arthritis (RA) is an autoimmune disorder in which the immune system attacks the body's connective tissues (the fibrous tissues that support and connect body structures). It tends to run in families and affects more women than men. Typically it begins when people are in their 40s, although it can start at any age. The first symptoms are painful, hot swelling and stiffness in the small joints of the fingers and toes, usually worst in the

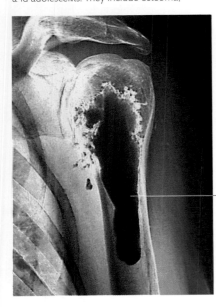

Rheumatoid arthritis
This X-ray shows RA in the joints of the wrist and hand, which has caused deformity of the wrist and finger joints.

morning. RA typically flares up intermittently and unpredictably; flare-ups can be incapacitating, and may last from days to months, sometimes with long symptom-free gaps between them. If left untreated, the disorder can spread to other areas. Joints become damaged by synovitis (inflammation of the membrane lining the joint capsule), leading to erosion of the joint surface. Tendon sheaths become inflamed. As the joint destruction progresses, the fingers may be permanently deformed. Tender nodules can develop in the skin and over joints. The condition can involve the heart, lungs, blood vessels, kidneys, and eyes. General symptoms such as fatigue, fever, and weight loss are common, as is anemia. People with the condition are also at increased risk of developing osteoporosis and heart disease. Blood tests for substances called rheumatoid arthritis "markers" may help doctors to detect RA. There is no cure; treatment involves controlling the symptoms and using "disease-modifying" drugs to slow the progression of the condition.

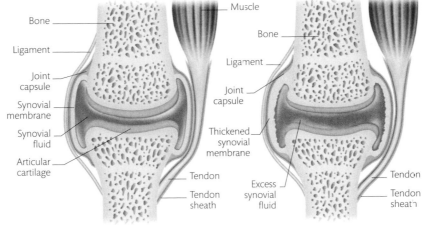

Healthy joint
The bone ends are covered with a smooth, even layer of cartilage. The joint capsule, lined with synovial membranes, is lubricated with synovial fluid, which allows the joint to move freely.

Early rheumatoid arthritis
The synovial membrane becomes inflamed and produces excess synovial fluid. This fluid contains destructive immune cells, which attack the cartilage and distort the joint space.

Late rheumatoid arthritis
Fluid and immune system cells build up to form a pannus—thickened synovial tissue that produces harmful enzymes. These rapidly destroy remaining cartilage and bone and attack other tissues.

BONE TUMORS

Bone can be affected by various kinds of growth, involving the bone tissue itself, the bone marrow, or the joints.

Tumors that originate in the bone can be either benign (noncancerous) or malignant (cancerous). Benign bone growths are fairly common and most often develop in children and adolescents. They include osteoma, osteochondroma, bone cysts (holes that usually form in growing bone), and fibroid dysplasia. Primary malignant tumors (cancers arising in the bone) include osteosarcoma and Ewings' tumor, which develop from the bone itself; chondrosarcomas, which develop from joint cartilage; and myeloma, which develops in the bone marrow.

Secondary bone tumors are caused by cancer that has spread from other areas via the blood or lymph and they are particularly associated with breast, lung, and prostate cancer. They are more common than primary bone cancer. Soft tissue tumors may also spread to invade nearby bone. The most notable symptom of bone tumors is gnawing, persistent pain that becomes worse during movement but can be relieved by anti-inflammatory analgesics. The affected area is often tender, and fractures may occur, in which the abnormal bone breaks and cannot heal.

Tumors may be identified by biopsy (tissue samples). X-rays, CT or MRI scans are also used to study the tumor. Benign tumors often need no treatment, but if they grow very large, press on nerves, or restrict movement, they may need to be removed. Myeloma is treated with chemotherapy, but most other primary bone cancers require surgery as well as chemotherapy. Secondary cancers may be treated with chemotherapy or radiation therapy, depending on their nature and the site of origin.

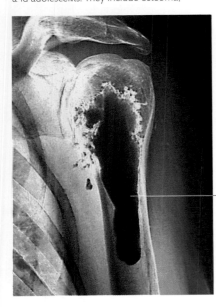

Malignant tumor
Metastases (secondary deposits of cancer) may occur at any site in the skeleton, but most often develop in the axial skeleton—the bones of the skull, chest, pelvis, and spine.

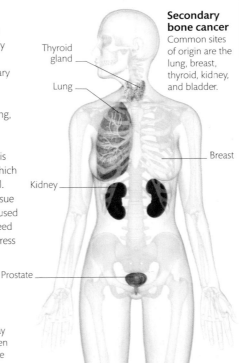

Secondary bone cancer
Common sites of origin are the lung, breast, thyroid, kidney, and bladder.

Thyroid gland

Lung

Breast

Kidney

Prostate

GOUT AND PSEUDOGOUT

In these disorders, crystals formed from chemical substances collect in the joints, causing inflammation and severe pain.

Gout results from excessively high levels of uric acid (a waste product formed by the breakdown of cells and proteins) in the blood. The acid is deposited as crystals in the joint space, causing inflammation and severe pain. Gout may be triggered by foods containing purines, including offal, oily fish, beer, and some drugs. Attacks usually affect middle-aged men and tend to last for about a week. Treatment involves avoiding triggers and taking medication to lower blood levels of uric acid. Pseudogout is caused by deposits of calcium pyrophosphate, and is often seen in older people with joint or kidney disease. Both disorders normally affect single joints, causing severe pain, heat, and swelling.

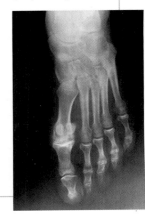

Early gout in foot
This X-ray shows gout as a dense white area in the joint at the base of the big toe; this is the most common site for the condition.

MUSCLE, TENDON, AND LIGAMENT DISORDERS

The muscles enable the skeleton and organs to move. Tendons attach skeletal muscles to bones, while ligaments connect bones to one another. Disorders affecting any of these structures can interfere with conscious movements and other muscle functions.

MYOPATHY

The name myopathy means a disorder of the muscle fibers. Myopathies can lead to cramps, muscle pain, stiffness, weakness, and wasting.

Myopathies range from simple muscle cramps to muscular dystrophy. Some are inherited, including dystrophies (muscle-weakening) and myotonias (abnormally prolonged contraction of muscles). Others are acquired and may be due to autoimmune inflammatory conditions, such as polymyositis. The disorders may also be associated with diabetes or advanced kidney disease. Some myopathies grow worse, and become life-threatening if respiratory muscles are affected. The treatment depends on the cause; for many conditions, only supportive measures are possible.

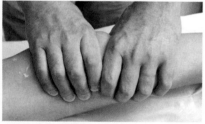

Myopathy treatment
Treatment is mainly given to relieve symptoms of myopathy; it includes physical therapy and exercise programs to make muscles stronger and more mobile and analgesics to manage pain.

MYASTHENIA GRAVIS

A relatively rare autoimmune condition, myasthenia gravis causes fatigue and weakness in muscles under voluntary (conscious) control.

Myasthenia gravis develops when antibodies produced by the immune system attack the receptors in muscles that receive signals from the nerves. As a result, the affected muscles only respond weakly, or fail to respond at all, to nerve impulses. The cause is unknown, but many affected people have a thymoma (a tumour of the thymus, an immune gland in

Myasthenia and the eyes
The condition typically affects the muscles that control the eyelids, causing the eyelids to droop. Other areas of the body may also be affected.

the neck). The condition often develops slowly; it varies in severity as the levels of antibodies fluctuate. Affected muscles still function to some extent but get tired quickly, although they may recover with rest. Myasthenia affects the eye and eyelid muscles in particular. It may also affect the face and limb muscles, causing difficulties in swallowing and breathing, and loss of strength. A severe attack, or myasthenic crisis, can cause paralysis of breathing muscles. There is no cure, but thymectomy (removal of a thymoma) and drugs may relieve symptoms.

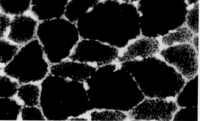

FIBROMYALGIA

This condition, whose cause is unknown, mainly causes muscle pains and tiredness and can last for months or years.

Fibromyalgia develops gradually, over a long time, with widespread muscle pain and tenderness. Muscles appear normal and functional, although affected people experience tiredness, disturbed sleep and memory, mixed sensory symptoms, and anxiety and depression. No specific cause has yet been found, but it has been suggested that the disorder may be due to a problem with the way in which the brain registers pain signals.

Research also suggests some brain abnormalities that may be linked to symptoms. Stress and physical inactivity make the symptoms worse, while programmes that include pain relief, exercise, cognitive behavioural therapy, and education can help.

DUCHENNE MUSCULAR DYSTROPHY

The most common form of muscular dystrophy, this condition mainly affects boys, causing progressive, severe muscle weakness and premature death.

Duchenne muscular dystrophy is an X-linked genetic condition. Females carry the condition on one of their two X chromosomes, but are protected by a normal second X chromosome. Boys, who have one X chromosome and one Y chromosome, may inherit the faulty gene from carrier mothers and develop the disease.

Affected baby boys tend to start walking later than normal, then by the age of 3 or 4 become clumsy and weak, finally losing the ability to walk by the age of 12 years. The progressive weakness and deterioration of the skeletal muscles (those attached to bones) leads to deformities affecting the spine and breathing, but with modern surgical corrective treatments many affected men now live into their 20s and 30s, or sometimes longer.

NORMAL

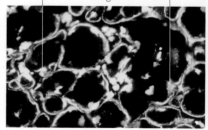

Fat   Damaged membrane

ABNORMAL

Effects of muscular dystrophy
Progressive destruction of muscle is seen here at the cellular level, as muscle cells undergo damage to their outer membranes and are replaced by connective tissue and fat.

CHRONIC UPPER LIMB SYNDROME

This name is used for a group of disorders affecting the hands and arms, such as repetitive strain injury (RSI), which cause pain and restriction of movement.

The cause is often thought to relate to overuse of the arm. Certain inflammatory conditions are also sometimes included within this group. They include carpal tunnel syndrome (see p.448), which affects the hand and forearm due to nerve compression at the wrist, as well as tennis elbow, golfer's elbow, and de Quervain's tenosynovitis, which result from inflammation of tendons through repeated use. Repetitive strain injury (RSI) is often due to occupational overuse. Symptoms include gradual onset of pain, often difficult to pinpoint to one area, and a sensation of swelling, although no swelling can be seen or felt. Numbness and tingling are common, and sleep may be disturbed by the symptoms. The disorders are often relieved by rest, gentle exercise, and modifying the activity that brings on the condition.

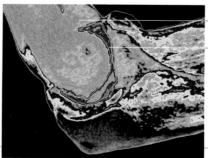

Lower end of humerus

Area of damage to articular surface

Osteoarthritis at the elbow
Abnormal stresses on a joint can predispose to the development of osteoarthritis. Here, stress on the elbow from pneumatic drilling has led to damage of the articular cartilage and underlying bone.

TENDINITIS AND TENOSYNOVITIS

These conditions involve inflammation of the tissues that connect muscles to bones, often due to injury or overuse.

Tendons are fibrous tissues that attach muscles to bones, enabling the bones to move when muscles contract. Inflammation of the tendons is called tendinitis; this often occurs together with tenosynovitis, inflammation of the sheath of tissues enclosing a tendon. Both conditions cause pain on movement, sometimes with a "catch" point during a limb motion, when the affected tendon moves.

Some tendons form pulleys, as seen in the shoulder, where the supraspinatus tendon passes in a groove over the joint; an inflamed tendon that "catches" will cause a painful arc of movement. Tendinitis is generally referred to

anatomically; Achilles tendinitis, for example, affects the back of the heel, causing pain on putting the foot to the floor.

Tenosynovitis can arise as a degenerative disorder or in connective tissue disease, arthritis, or an overuse injury, or with tendinitis. The most common example is de Quervain's tenosynovitis, which affects the tendon sheath enclosing the two tendons that move the thumb outward from the hand. This condition causes pain, swelling, tenderness, and difficulty in gripping. Tenosynovitis may also cause joints to stick, as in "trigger finger." In both conditions, treatment may involve rest or modifying the use of the tendon with braces, splints, or supports, as well as analgesics, anti-inflammatory drugs, and gradual return to exercise.

Humerus — Inflamed supraspinous tendon — Clavicle (collarbone) — Acromial process of shoulder blade

Tendinitis
Tendons transmit the pull of muscles to bones. Injury or overuse can cause inflammation or a tear in the tissues, resulting in pain and sometimes a crackling sensation as the limb moves.

Inflammation

Tendon sheaths — Tendons — Tendon sheath

Tenosynovitis
The synovium, the protective sheet of tissue that covers some tendons, produces fluid to keep the tendon moving smoothly. Inflammation of these tissues causes pain and tenderness.

FIRST AID TREATMENT

Injuries to muscles, tendons, or ligaments can be treated quickly and easily by a technique called PRICE—this stands for Protection, Rest, Ice, Compression, and Elevation. Protection helps prevent further injury; rest relieves the injured area; an ice pack every few hours reduces pain, inflammation, and bruising; compression with an elastic bandage helps reduce swelling; and elevation (raising the limb) also reduces swelling, by allowing excess fluid and waste from the body's repair process to disperse. PRICE reduces blood flow to the injury and therefore reduces bleeding, bruising, and swelling.

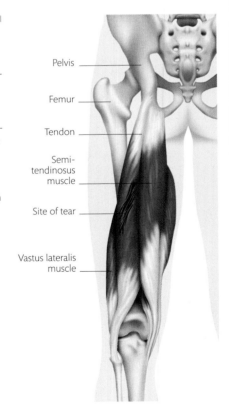

Treatment for strains and sprains
The PRICE technique includes applying an ice pack and elevating (raising) the affected area above the level of the heart.

LIGAMENT SPRAINS AND TEARS

Ligaments are bands of connective tissue that hold bones together; they are tough and thick, but not very stretchy, so are prone to tearing.

Ligaments can stretch gradually under tension, as gymnasts and ballerinas show by stretching their ligaments gradually during their training to achieve extreme body positions. Ligaments also become stretchier in pregnancy, to allow the pelvis a little more "give" during childbirth.

Sprained ankle
The ankle is prone to sprains if the foot twists suddenly. Common injuries are lateral ligament sprains, in which the foot turns inward, or medial ligament sprains, in which it twists outward.

People taking part in sport or exercise are advised to do "warm-up" exercises to protect their ligaments. These tissues are not easily torn owing to their strength, but they can be damaged by a fall or a sudden twisting or wrenching movement. Injuries range from a sprain (minor tear) to a rupture (a complete break in the ligament). The wrists and ankles are common sites of injury. Symptoms come on suddenly and include pain, swelling, and restricted movement in the joint. Damaged ligaments heal relatively slowly because their blood supply is not as rich as that of muscle. Mild sprains can be relieved by PRICE (see right), but severe, incapacitating injuries need medical attention to prevent joint dislocation.

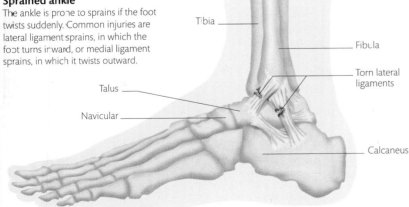

Tibia — Fibula — Torn lateral ligaments — Talus — Navicular — Calcaneus

MUSCLE STRAINS AND TEARS

Excessive stress on muscles can cause strain (sometimes referred to as "pulling a muscle") or even a tear in the muscle.

Muscles contract in order to move joints. They comprise groups of parallel fibers that move relative to one another and grip one another like interlocking ladders. Muscle injuries are common and range from a mild strain, in which fibers are pulled apart lengthwise but without tearing, to a complete tear, which can cause pain, bleeding, and dramatic swelling. Strains are often caused by over-stretching or over-contracting a muscle, in sports or heavy physical work. Some strains are chronic, due to repeated over-stressing of a muscle.

Injury is particularly common during sudden changes in directional force, such as twisting suddenly while running, during falls, and when lifting heavy objects. Injuries need immediate treatment with PRICE (see above), and the affected muscle will have to be kept still for a few days. Muscles have a rich blood supply, so they heal relatively quickly, but recovery time also depends on the severity of the injury, the natural variation in healing time between individuals, and the level of normal activity required of the muscle.

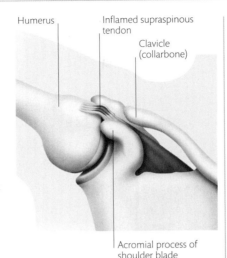

Pelvis — Femur — Tendon — Semi-tendinosus muscle — Site of tear — Vastus lateralis muscle

Torn hamstring
The hamstrings are the muscles at the back of the thigh, which bend the knee and pull the leg back. Hamstring injuries are often seen in athletes who do a lot of sprinting or jumping.

BACK, NECK, AND SHOULDER PROBLEMS

Disorders of the spine and shoulders are common but can be disabling. The lower back is vulnerable to damage since it supports most of the body's weight and is under continual pressure from bending and twisting movements. The shoulder is also prone to problems as the body's most mobile joint.

WHIPLASH

This term is used for a range of injuries caused by sudden back-and-forth movements of the neck.

Whiplash commonly occurs in traffic accidents, due to deceleration: the sudden impact first forcibly flexes the neck as the head is thrown forward, then forcibly extends the neck as the forward momentum of the head is stopped by the body, and the head rebounds backward. The severity of the injury varies from small strains, with tearing of a few muscle fibers, to major trauma, in which neck ligaments are torn. The sudden pull of muscle and tendon on bone may break pieces off the ends of the vertebrae (spinal bones). Nerves may be damaged, causing pain in the neck, shoulders, and arms, and possibly dizziness and disturbed vision; some people also suffer memory problems and depression. In the hours following a whiplash injury, bleeding occurs in the tissues, and tissue swelling and muscle spasm follow; the injury reaches its peak in the first 48 hours. It can take many weeks or months for whiplash to get better. Treatment includes anti-inflammatory drugs and physical therapy.

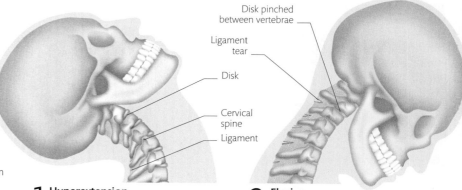

1 Hyperextension
If hit from behind, the head rapidly moves backward then forward. Whiplike backward motion hyperextends the cervical vertebrae.

2 Flexion
Following hyperextension, flexion of the vertebrae occurs as the head's momentum carries it forward and causes the chin to arc down.

Labels: Disk pinched between vertebrae · Ligament tear · Disk · Cervical spine · Ligament

TORTICOLLIS

Also known as wry neck, torticollis usually involves spasm of the muscles in the neck, which pulls the head to one side and results in pain and stiffness.

Torticollis is thought to be due to pulling on the deep ligaments of the neck, causing a muscle spasm. It may occur in babies due to a difficult birth or an awkward position in the womb. In adults, it may be caused by damage to the joint at the skull base, or possibly by a nerve disorder. Often, torticollis can simply be due to sleeping in an awkward position; in this case, it usually improves in 2 or 3 days and can be relieved by anti-inflammatory or antispasmodic drugs, massage, and rest. Further treatment may be needed for more permanent torticollis.

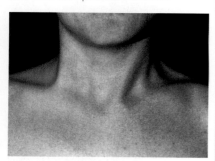

Torticollis
Spasm of the large muscle around the side of the neck causes torticollis, with a resulting tilt of the head to one side.

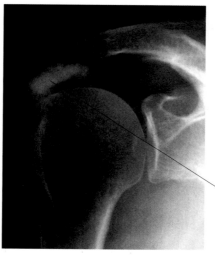

Calcification of rotator cuff

FROZEN SHOULDER

In this condition, the tissue around the shoulder joint becomes inflamed, stiff, and painful, severely limiting movement.

In the shoulder joint, the humerus (upper arm bone) and end of the scapula (shoulder blade) are enclosed in a capsule of fibrous tissue filled with fluid that enables the joint to move easily.

Inflammation of the shoulder joint
Chronic inflammation of the tissues around the shoulder joint can cause calcium deposits to form in the tissues; these show as white areas on X-ray.

Inflammation of the fibrous tissues leads to frozen shoulder, also called adhesive capsulitis. Although the cause is unknown, the disorder is more common in people who also have other inflammatory joint or muscle conditions, and in those with diabetes. It begins gradually, with pain and inflammation in one area or muscle group, but then progresses around the joint, with adhesions (bands of scar tissue) forming between the tissues. The pain can disturb sleep and limit movement. In a typical case, there are three stages: slow, painful "freezing" of the shoulder over several weeks or months; a "frozen" stage lasting for months, when the pain is less but the stiffness is severe; and then several weeks of "thawing." Treatment involves physical therapy, analgesics, and occasionally corticosteroid injections into the shoulder.

DISLOCATED SHOULDER

Dislocation is an injury in which a joint is displaced from its normal position. The shoulder is particularly prone to this problem, usually due to sudden impacts.

The shoulder is a ball-and-socket joint in which the head of the humerus (upper arm bone) sits in a shallow socket at the end of the shoulder blade. The shoulder bones are kept in place by the rotator cuff, a group of strong muscles around the joint. This structure allows the arm a wide range of movement in many directions; however, it also makes the joint unstable, or liable to dislocate under pressure. Dislocation is most commonly due to falling or impacts in sports such as football. It can also be caused by inherited loose joints. A dislocated shoulder is painful and swollen and may look deformed. An X-ray will be needed to confirm and assess the injury. Treatment involves manipulation to move the bones back into place.

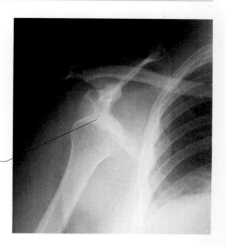

Dislocation of humeral head

X-ray showing shoulder dislocation
This X-ray shows an anterior (forward) dislocation. In most cases the bone is displaced forward, because the rotator cuff is weakest at the front.

LOWER BACK PAIN

Most people suffer back pain at some point, often due to strain on the muscles and ligaments; the lower back is the area most commonly affected.

Although there are many theories about lower back pain, and many studies have been done, the definite cause is unknown. MRI scans can show marked damage to muscles and joints in people who have no back pain, while scans of people with debilitating pain may fail to show any abnormality at all.

It is important to avoid behavior that might lead to back injury, such as poor posture when sitting or standing, or lifting heavy objects in a way that does not safeguard the back. Injury to the structures of the spine and back as a result of excessive strain may be caused by twisting, bending, or lifting. The lumbar area of the back, below the waist, is especially vulnerable to back pain because it already bears much of the body's weight. Pain may arise from the muscles, ligaments, vertebrae (spinal bones) or the disks between the bones, or the nerves, although muscle strain is most common.

In most cases, an affected person can relieve back pain with heat, anti-inflammatory analgesics, and gentle exercise. For pain that is more severe or that lasts for more than a few days, medical treatment or physical therapy may be needed.

MANAGING BACK PAIN

After a back injury, a person should remain physically mobile and resume normal activities as soon as possible. Back pain generally gets better within 2 or 3 weeks with exercise and pain relief. However, a person with chronic back pain may need treatment with physical therapy and back rehabilitation programs in addition to analgesics. Addressing lifestyle issues such as losing weight, and learning a method of using the back muscles safely, such as the Alexander technique, can help ease the pain and prevent recurrences.

Treatment for back pain
Treatments include medication for pain and muscle spasm, physical therapy to strengthen the back, and advice on back health.

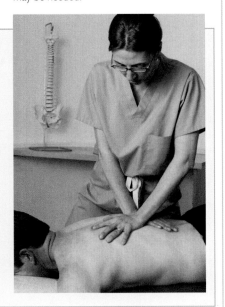

DISK PROLAPSE AND SCIATICA

The vertebrae (spinal bones) are separated by disks of soft tissue; if one of the disks slips out of place or ruptures, this can put pressure on a nerve and cause pain.

The disks that separate the vertebrae are composed of a tough, fibrous coating and a softer, jellylike core. Sometimes, excessive stress on the back can cause a disk to get pushed out of position. If the disk is squeezed, the outer coating may rupture, and the soft core may herniate (protrude) through it; this is often called a slipped disk. The problem is more common in the lumbar disks (those in the lower back), which are subject to the most force—especially if they have started to degenerate due to age. It may occur suddenly or slowly. Sudden slippage may occur after lifting or injury; and it may cause pain or difficulty in moving.

A prolapsed disk may press on the nerves leading from the spinal cord, causing sciatica: burning, tingling pain in the sciatic nerve, which travels via the buttock, down the back of the leg, to the foot. In many cases, gentle exercise and analgesics bring recovery within 6–8 weeks. More serious cases may require physical therapy or surgical repair of the disk.

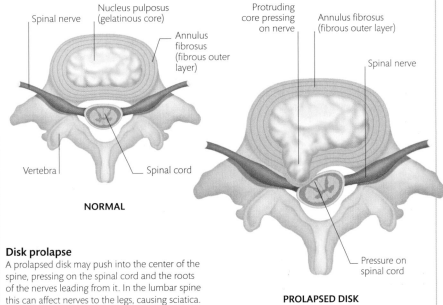

NORMAL

PROLAPSED DISK

Disk prolapse
A prolapsed disk may push into the center of the spine, pressing on the spinal cord and the roots of the nerves leading from it. In the lumbar spine this can affect nerves to the legs, causing sciatica.

SPINAL STENOSIS

Stenosis, or narrowing, of the spinal canal can compress the spinal cord or nerve roots; this condition is usually due to the effects of aging.

Age-related changes to the spine may begin in the mid-30s, but obvious symptoms are unusual before age 60. Stenosis begins with stiffening of the joints between the vertebrae and the formation of bony growths called osteophytes on them. These growths encroach on the spinal canal and the foramina (gaps through which the nerve roots exit the spine), narrowing the spinal canal. Stenosis is most common in the lumbar spine, and can lead to pain, cramping, and weakness in the back, neck, shoulders, legs, or arms. An affected person may be given anti-inflammatory drugs and physical therapy, but serious cases may need decompression surgery, in which bone or tissue is removed to ease pressure on the cord.

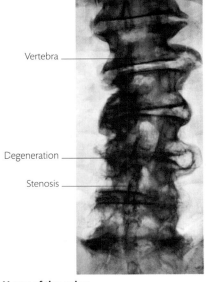

Vertebra

Degeneration

Stenosis

X-ray of the spine
This color-enhanced X-ray shows spinal stenosis caused by severe degeneration of the spine. The red areas are bones distorted by osteophytes, and the greenish area is the spinal canal.

SPONDYLOLISTHESIS

The forward slippage of one vertebra over another is called spondylolisthesis; it usually causes no symptoms, although at worst it can compress the spinal cord.

Spondylolisthesis may result from a congenital spinal deformity (one that is present from birth), or may develop during growth in mid-to-late childhood. However, most cases occur in adults and result from degenerative changes to the joints between vertebrae, which alter the angle of the bones and allow higher vertebrae to slip over lower ones. In most cases there are no symptoms, but some people have pain, stiffness, or sciatica (see above). If there is coexisting spinal stenosis (see left), symptoms may be worsened by the narrowing. Severe cases (in which the upper vertebra is more than 50 percent out of line) may cause significant pressure on the spinal cord, and decompression surgery may be needed.

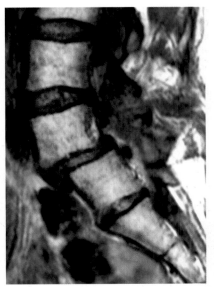

Scan of spine showing spondylolisthesis
Spondylolisthesis most often affects the lumbar vertebrae, in the lower back, as seen in this scan. The overhang where the upper vertebra has slipped is clearly visible.

LIMB JOINT DISORDERS

Problems involving the muscles, tendons, or other soft tissues around the joints are often caused directly by the way we use those joints. They can result in considerable pain, but many of them get better by themselves or need only rest and treatment at home.

EPICONDYLITIS

This condition includes tennis elbow and golfer's elbow and involves inflammation of the epicondyles, the bony protrusions on either side of the elbow joint.

Tennis elbow, affecting the outer epicondyle, and golfer's elbow, which develops in the inner epicondyle, usually result from overuse of the muscles that attach to the bone at these points, or occasionally from direct injury. The damage causes inflammation of the tendons that attach the muscles to the epicondyles. Tennis elbow classically arises from recurrent serving in tennis, and golfer's elbow from the golf swing, but the conditions are far more commonly due to other overuse injuries. The

Tennis elbow
Epicondylitis can affect both sides of the elbow simultaneously. Symptoms include a painful, reddened area around the joint.

area is tender to touch, with pain that is made worse by movement. In golfer's elbow, lifting the arm with the palm upward will worsen pain; in tennis elbow, lifting the arm with the palm downward makes it worse.

Pain moves down one side of the arm, into the hand, with tingling in the forearm and heat, pain, and swelling over the epicondyle. Treatment involves resting the arm and using analgesics. A splint may help, and braces are used to take the strain off the muscles. Physical therapy may be recommended and corticosteroid injections given for severe pain.

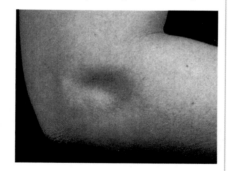

CARPAL TUNNEL SYNDROME

Compression of the median nerve, which passes through the carpal tunnel in the wrist, causes carpal tunnel syndrome.

The median nerve passes down the forearm to the hand, where it operates the muscles at the base of the thumb and controls sensation in the thumb half of the palm. En route it passes through the carpal tunnel, a space between the wrist bones that is enclosed by a ligament. In addition to the nerve, 10 tendons pass through the space. Carpal tunnel syndrome occurs when the nerve is compressed. This may be caused by swelling of the tendons, or by fluid collecting in the carpal tunnel due to wrist arthritis, hormone fluctuations, thyroid problems, diabetes, or overuse. The pressure results in pain, loss of grip, tingling in the thumb, first two fingers, and half the ring finger and, if severe, wasting of the thumb muscles. In mild cases, it is treated with rest, analgesics, and splints. Corticosteroid injections are sometimes given to relieve inflammation. In severe cases, decompression surgery is used to relieve pressure by dividing the ligament.

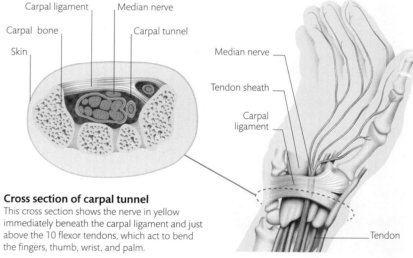

Carpal ligament / Median nerve
Carpal bone / Carpal tunnel
Skin

Cross section of carpal tunnel
This cross section shows the nerve in yellow immediately beneath the carpal ligament and just above the 10 flexor tendons, which act to bend the fingers, thumb, wrist, and palm.

Median nerve
Tendon sheath
Carpal ligament
Tendon

GANGLION

Most often found on the wrist, ganglions are soft, harmless swellings that often disappear by themselves eventually.

Ganglions are cysts that form just under the skin, over a tendon sheath. They often occur close to joints, in which case they tend to be connected to the joint; typical sites are the feet, wrists and hands, most commonly the extensor (upper) side of the wrist. Ganglions contain synovial fluid, a thick, clear, gel-like substance, from inside the joint. If they cause no symptoms they can be left to disappear by themselves. If they cause pain or impede movement, they can be drained or removed.

Ganglion
Like most ganglions, this swelling is situated on the extensor (outer) surface of the thumb joint.

KNEE JOINT EFFUSIONS

Also known as "water on the knee," an effusion on the knee can cause swelling and, sometimes, stiffness and reduced ability to use the joint.

Joints contain bone ends inside the synovial membrane, which produces the synovial fluid that lubricates the joint. Sometimes an effusion—a collection of excess fluid—can accumulate around a joint. An injury, or an infective or inflammatory disorder (such as osteoarthritis or gout), can cause the condition by provoking the membrane to produce excess synovial fluid.

The knee is particularly prone to effusion as it bears considerable downward and rotatory forces and is therefore prone to wear and injury. Knee effusions tend to cause obvious, soft swellings, with pain, and it may be difficult to put weight on the leg. Treatment depends on the cause; it may involve draining the excess fluid and/or giving corticosteroids or anti-inflammatory drugs to reduce the inflammation.

BURSITIS

Inflammation of a bursa, one of the small pads that provide cushioning in joints, can cause pain and obvious swelling.

Bursae are lined with synovial membrane and filled with jellylike synovial fluid. They act as cushions between the moving parts of most joints. Injury or infection of a bursa can cause the lining to produce excess fluid leading to a buildup called bursitis. The area may become red, painful, and swollen. Bursitis is common around the knee and elbow, perhaps because these joints often suffer injury. An olecranon bursa, at the back of the elbow, can become very large because the loose skin allows expansion. Bursae usually settle by themselves; they can be drained but often refill.

Housemaid's knee
Bursitis in the knee, also known as housemaid's knee, often occurs in people who spend a lot of time kneeling, such as gardeners.

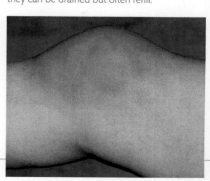

CHONDROMALACIA

Chondromalacia patellae is pain at the front of the knee, which is probably related to overuse and is most often seen in active young people.

The pain in chondromalacia may be caused by chronic friction where the patella (knee cap) passes back and forth over the knee joint as it flexes and extends. In adolescents, the condition can be very painful but is essentially harmless. It may be relieved by rest and physical therapy and, because it usually clears up over a couple of years, most physicians prefer to avoid surgery that could scar the joint. Chondromalacia, which is sometimes known as runner's knee, is also common in adults, particularly in women over 40 years of age. In adults, the condition can usually be relieved only by exercises, rest, ice, and therapy.

OSGOOD-SCHLATTER DISEASE

Usually seen in active teenagers, this condition is caused by inflammation at the front of the tibia (shinbone), just below the knee.

Osgood-Schlatter disease usually occurs during adolescent growth spurts, often in teenagers who do a lot of sports. It develops at the tibial tuberosity, a bony point at the top of the tibia where the quadriceps muscles, at the front of the thigh, attach to the tibia via the patellar ligament (which connects the kneecap to the tibia). It is thought to result from excess strain on tibial tuberosities as the long bones of the leg grow more rapidly than the muscles can lengthen.

Repeated stress from contraction of the overstretched quadriceps is transmitted to the tuberosity, causing pain and swelling. In the most severe cases, this results in the formation

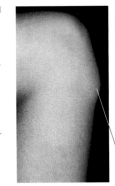

Osgood-Schlatter disease
This photograph shows a prominent tibial tuberosity in a person with Osgood-Schlatter disease. Recurrent small fractures lead to a bony lump, which is usually very painful if it is knocked.

— Bony prominence

of shin splints—stress fractures of the growth plate at the end of the tibia. As the body tries to heal the fractures, it produces new bone growth at the tuberosity, enlarging the point into a prominent lump that is tender to touch and may be so painful that it prevents exercise altogether, particularly when a splint has just formed. However, the condition does clear up after a couple of years, and often needs no treatment except rest, ice, and analgesics.

ACHILLES TENDINITIS

The Achilles tendon connects the muscles of the calf to the ankle, and may often become inflamed in athletes and runners.

Achilles tendinitis results from small tears in the tissue as the foot hits the ground with excessive force—typically when someone is running or jogging on hard or rough ground. Pain and swelling develop at the back of the ankle, and the ankle itself may swell. The inflamed tendon is particularly painful if it has been stretched (as it is when a person flexes the heel to "push off" from the ground during a stride). In many cases, the condition can be relieved by rest, applying an ice pack to the area, and taking analgesics. If it persists, treatment may include physical therapy or temporarily fitting an orthosis into the shoe—a device such as a heel pad or cup that reduces the stress on the tendon as the foot is placed to the floor. The Achilles tendon has a relatively poor blood supply, so healing tends to be slow.

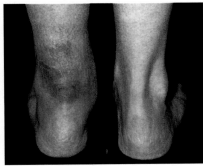

Achilles tendinitis
If an Achilles tendon is severely inflamed, as shown here, excess tissue fluid collects and gravity causes it to move downward, resulting in swelling of the ankle and heel.

PLANTAR FASCIITIS

The plantar fascia is a thick band of tough, fibrous tissue running beneath the sole of the foot and supporting the arch; inflammation here can cause severe pain.

The plantar fascia is the continuation of the Achilles tendon and connects the heel bone to the base of the toes. Inflammation in this tissue

Treating plantar fasciitis
The pain from plantar fasciitis most commonly develops near the back of the sole, where the plantar fascia attaches to the calcaneus (heel bone).

is caused in a similar way to Achilles tendinitis (see right), by repeated overstretching. Common in people who do a lot of walking over rough ground or jogging, it can be degenerative and can accompany inflammatory arthritis, obesity, osteoarthritis, and diabetes. Pain occurs when the sole of the foot is stretched, and is usually felt most severely beneath the heel; it is described as "walking on marbles."

Initial treatment includes rest, ice packs, and analgesics. Exercises may be prescribed to stretch the tissues gently. Some people may be given orthoses, devices that fit in the shoes to relieve the stretch on the fascia when the foot is used. Severe cases may be treated with an injection of corticosteroids and local anesthetic into the affected area.

— Achilles tendon

— Inflammation

— Plantar fascia

FOOT DEFORMITY

Abnormalities in the bones, muscles, and ligaments of the foot can distort the shape and cause problems with function.

The shape of the feet develops as a child grows, and the bones, ligaments, and fascia (connective tissue) form an arch in the sole; this structure gives flexibility and acts as a shock-absorber. Structural disorders can affect the shape of the arch, causing flat or high-arched feet. In flat feet, or pes planus, the arch collapses or may never have developed, and the entire sole contacts the ground on walking. It can lead to pain, but arch supports will help. Pes cavus is a condition in which the arches are abnormally high; it may be inherited or may be acquired in some muscular or nerve

disorders. There are usually no symptoms, but it can cause problems with fitting shoes. Club foot is inward twisting of one or both feet, and is present from birth; the cause is unknown. Most cases are treated with minor surgery, along with physical therapy and special shoes.

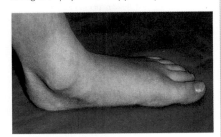

Flat foot
This image shows flattening of the arch due to collapse of the bony structures under the weight of the patient. The entire sole of the foot can be seen in contact with the ground.

HALLUX VALGUS (BUNION)

Some people have a structural deformity of the joint at the base of the big toe, which leads to the formation of a bunion.

The valgus deformity begins with the big toe gradually turning inward, sometimes with the other toes also bending at an angle. As the big toe moves out of position, the joint between its base and the head of the first metatarsal bone (in the body of the foot) is exposed, becoming swollen and painful. Inflammation of the bursa over the joint adds to the enlargement and pressure. The resulting bony lump is called a bunion. The condition tends to run in families. The cause is complex and involves abnormal action of the foot—possibly combined, in some cases, with years of

wearing tight, pointed shoes, which tend to compress the toes into an angled position. The affected toe may develop arthritis. Some people have difficulty finding shoes that fit over the bunion. Pads, orthoses (corrective devices), and comfortable shoes may help relieve the pressure, but if symptoms are severe, surgery will be needed to remove the excess bone and realign the toe.

— Enlarged part of joint

Bunion
Bending of the big toe causes deformity in the bone and thickening of the soft tissues around the toe joint, forming a bunion.

CEREBROVASCULAR DISORDERS

The cerebrovascular system comprises the blood vessels that supply the brain. It is prone to conditions that can affect blood vessels elsewhere, such as blood clots and atherosclerosis, but the effects on the brain are specific and sometimes catastrophic.

STROKE

A stroke causes sudden, irreversible damage to areas of brain tissue due to disturbance of the blood supply—it is the brain's equivalent of a heart attack.

The brain needs a rich supply of oxygen and nutrients from the blood in order to function properly. If the blood supply is interrupted, brain cells can fail and die, interfering with the physical or mental function controlled by the affected part of the brain. This is cerebral infarction. Most strokes are atherotic, occurring when a thrombus (a piece of atherosclerotic material) breaks off from the heart or a large cerebral artery and flows along the blood stream until it lodges in a cerebral artery already narrowed by atherosclerosis. A minority of strokes are hemorrhagic (caused by bleeding), due to a tumor or to blood vessel malformation.

Stroke occurs if the damage caused is not completely reversed within 24 hours. It can involve small or large areas of the brain; it is relatively common for strokes to affect one whole side of the body (hemiplegic stroke) initially, leading to one-sided paralysis. Speech, swallowing, and vision may be affected, as may personality, memory, and mood. The damaged brain swells, and it can be weeks or months before the swelling settles. During this time, function may gradually return, and with rehabilitation it is possible to relearn skills. The risk is reduced by not smoking and by lowering blood pressure and cholesterol levels. Sometimes early treatment with clot-busting drugs can minimize or reverse the damage.

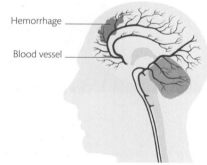

Bleeding within the brain
Rupture of blood vessels in the brain is termed intracerebral hemorrhage. This is the least common type of stroke, usually resulting from a tumor or preexisting blood vessel abnormality.

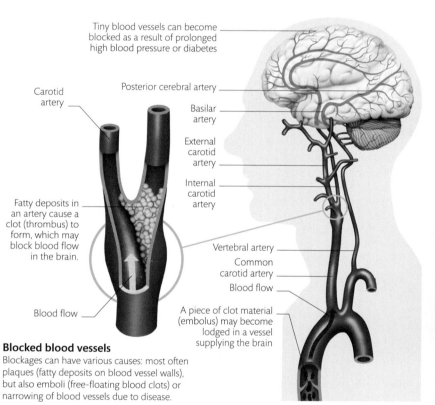

Tiny blood vessels can become blocked as a result of prolonged high blood pressure or diabetes

Carotid artery

Posterior cerebral artery

Basilar artery

External carotid artery

Internal carotid artery

Fatty deposits in an artery cause a clot (thrombus) to form, which may block blood flow in the brain.

Blood flow

Vertebral artery

Common carotid artery

Blood flow

A piece of clot material (embolus) may become lodged in a vessel supplying the brain

Blocked blood vessels
Blockages can have various causes: most often plaques (fatty deposits on blood vessel walls), but also emboli (free-floating blood clots) or narrowing of blood vessels due to disease.

LONG-TERM EFFECTS OF A STROKE

The long-term effects of a stroke depend on which part of the brain is damaged, whether the damage is permanent, and how well the brain learns new pathways to accomplish tasks. Even a major stroke may be followed by gradual but dramatic recovery. Speech is commonly affected, particularly in terms of finding and forming words. Stroke may also alter personality, and increased emotional difficulties and depression are common aftereffects.

Facial paralysis
The facial paralysis sometimes seen with stroke usually affects just one side, preventing the eye and mouth from closing fully.

TRANSIENT ISCHEMIC ATTACK

In this condition, brief interruptions to normal cerebral blood flow result in a sudden, temporary loss of function.

If stroke is the brain's equivalent of a heart attack, transient ischemic attacks (TIAs) are the equivalent of angina. The process of a TIA is like that in a thrombotic (clot-related) stroke, except that in TIA the blockage in the blood vessel is temporary and possibly only partial, and clears itself before permanent damage to brain tissue results. TIAs can last for seconds or hours, and may involve any of the functions that are affected in stroke. It is a warning sign for stroke—more so if prolonged or frequent. TIAs therefore require urgent investigation, including scans of the heart and carotid arteries (which supply the brain), to find the source of the material causing the blockage.

The risk factors for TIA, as for stroke, are high blood pressure, smoking, diabetes (especially if control is poor), and high cholesterol, all of which increase the risk of atherosclerosis and of fatty deposits forming in blood vessels. Treatment aims at reducing the risk factors and thinning the blood with aspirin or warfarin to prevent clots from forming.

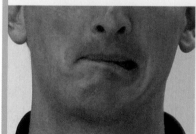

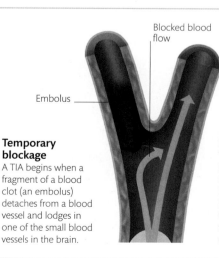

Blocked blood flow

Embolus

Dispersed particles

Blood flow resumes

Temporary blockage
A TIA begins when a fragment of a blood clot (an embolus) detaches from a blood vessel and lodges in one of the small blood vessels in the brain.

Dispersal of blockage
The blockage is moved by the pressure of the blood building up behind it. Oxygenated blood can then reach the area of the brain that has been starved of oxygen.

SUBARACHNOID HEMORRHAGE

This dangerous condition involves blood leaking between the inner two of the three meninges—the layers of membrane covering the brain.

Subarachnoid hemorrhage occurs when an artery near the brain surface suddenly ruptures and blood escapes into the subarachnoid space, between the inner two meninges—the arachnoid mater and the pia mater. In most cases, hemorrhage results from the rupture of a berry aneurysm—a swollen, weakened area at the join between two arteries in the brain—or malformed blood vessels, problems that may be present from birth, but trauma can also damage blood vessels, causing them to hemorrhage. The bleeding causes a sudden, severe "thunderclap headache," with vomiting, confusion, intolerance of light, and, in severe cases, coma and death. Warning headaches may occur before the blood vessel ruptures. A CT scan may be done to find the source of bleeding, and the affected vessels surgically repaired. However, full recovery does not always occur, and almost half of cases are fatal.

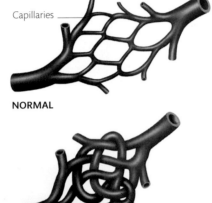

Capillaries

NORMAL

Abnormal knot of vessels

ABNORMAL

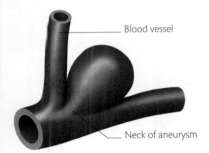

Blood vessel

Neck of aneurysm

Berry aneurysm
A berry aneurysm is a swelling in a blood vessel wall that develops at a weak point in the join between two blood vessels. Berry aneurysms are often found at the base of the brain.

Arteriovenous malformation
Abnormally formed arteries and veins are connected in a tangled knot. High-pressure arterial blood meets low-pressure venous blood at these points, so they are prone to bleeding.

SUBDURAL HEMORRHAGE

A subdural hemorrhage occurs when blood leaks into the space between the outer two of the three meninges—the membranes covering the brain.

Subdural hemorrhage usually results from tears in the veins that cross the subdural space, between the dura mater (the outermost of the meninges) and the arachnoid mater. This leads to acute, severe bleeding or a slow, chronic bleed. The bleeding produces a pocket of blood called a hematoma, which presses on brain tissue. Severe bleeding causes pressure on the brain, with rapid loss of consciousness.

Acute subdural hematomas are usually caused by severe head injuries. They are most often seen in young men, and in babies—possibly as a result of shaking ("shaken baby syndrome"). Chronic subdural hemorrhages cause gradual confusion and decline in consciousness; they are usually seen in older people, in whom they may be mistaken for dementia, or in people who abuse alcohol. This is because age and alcohol are associated with a tendency to cerebral shrinkage, which stretches the veins crossing the meninges and possibly makes them more liable to rupture.

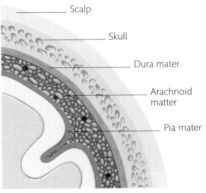

Scalp

Skull

Dura mater

Arachnoid matter

Pia mater

Normal
The brain is enclosed in three layers of membrane called meninges: the dura mater, arachnoid mater, and pia mater. These layers carry sensitive nerves and blood vessels over the brain's surface.

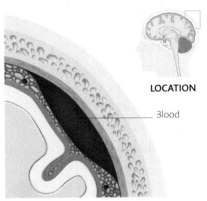

LOCATION

Blood

Subdural hemorrhage
The hematoma, or collection of blood, between the outer two layers exerts pressure on the brain. It may grow rapidly within hours, or may take weeks or months to increase in size.

MIGRAINE

Migraine is a recurrent and often severe headache, usually on just one side of the head, which occurs with disturbed vision, nausea, and other abnormal sensations.

Migraine affects more women than men and tends to run in families. The condition can first appear at any age, although it rarely starts beyond the age of 50. The cause is not fully understood, but one theory suggests that migraine starts with sudden constriction of blood vessels in the meninges (membranes covering the brain), causing a transient slight ischemia, followed by a "flush" or dilation, which stretches the sensitive veins and nerves and leads to pain. It is typically triggered by factors such as stress, hunger, fatigue, and certain foods and drinks including chocolate, red wine, and caffeine. In women, attacks may be associated with hormone fluctuations and often occur before menstrual periods.

Migraine attacks are often disabling and may last for up to 3 days. They typically have four stages—the prodrome (warning signs), the aura, the headache, and the postdrome (recovery stage). Symptoms of the prodrome include loss of appetite and changes in mood

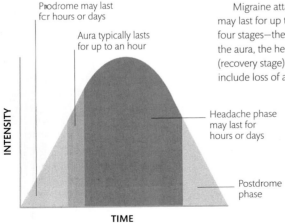

Prodrome may last for hours or days

Aura typically lasts for up to an hour

Headache phase may last for hours or days

Postdrome phase

INTENSITY

TIME

Course of migraine attack
A typical migraine attack comprises four stages, which vary in intensity and duration.

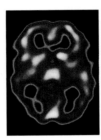

Migraine attack
This scan shows different levels of brain activity during a migraine. The red and yellow areas show high activity; areas of gray and blue indicate low activity.

or behavior. The premigraine aura (if present) often consists of visual disturbances such as blurred vision and seeing flashing lights; abnormal sensations such as numbness or pins and needles; loss of balance or coordination; and difficulty speaking. The headache is typically a throbbing pain on one side of the head, with nausea and vomiting, intolerance of light and noise, and altered scalp sensation.

About 15 percent of sufferers have migraine with aura (previously called classic migraine); migraine without aura is called common migraine. There are also various atypical patterns, such as "ice pick" headache, "cold air" headache, and "hat-band" headache, which tend to recur in certain people. There is no cure, but migraine can be controlled by avoiding triggers and by using drugs that help prevent or limit attacks or relieve headache and nausea.

HEADACHE

Most headaches are tension headaches, resulting from stress; a more painful form is cluster headache, with brief attacks happening several times a day.

Tension headache is a feeling of constriction across the forehead, brought on by tightness in the muscles of the scalp and neck. It is often worse at the end of the day and is increased by fatigue or stress. The pain can usually be relieved with analgesics and relaxation.

Cluster headache affects more men than women, and is an excruciating one-sided pain around one eye or temple, associated with watering red eyes and nasal congestion. The pain is due to dilation (widening) of blood vessels, but the underlying cause is unknown, although temperature changes or drinking alcohol may trigger an attack. Onset is rapid, and the headache is sometimes described as a red-hot poker in the eye. As the name suggests, the headaches occur in clusters. Attacks last from a few minutes to a couple of hours and recur up to several times a day. They may be treated with medication or oxygen therapy.

BRAIN AND SPINAL CORD DISORDERS

The brain and spinal cord process the information coming in from sensory nerves and blood-borne chemicals and formulate responses that are sent to body tissues. Damage to either of these structures can severely impair brain and body functions.

HEAD INJURY

Many bumps and bruises to the head are minor, but a severe blow or other injury can put the brain tissue at risk of damage.

Severe head injuries include open injuries, which expose brain tissue, and closed injuries, in which the brain is shaken inside the head. An open skull fracture may result from a heavy blow or impact—the skull is a strong structure, and only forceful injury will fracture it. The fracture may expose brain tissue and the cerebrospinal fluid (which cushions and protects the brain and spinal cord) to trauma and infection. A fracture at the base of the skull may allow cerebrospinal fluid to leak down the nose or out of the ear. Where fluid can get out, infection can get in. Shaking of the brain in the skull can cause bleeding, and the blood may build up to form a hematoma. The hematoma may be extradural (between the bone and the membranes covering the brain) or subdural (between the brain and the tissues that cover it). The accumulating blood presses against the brain and causes headache and altered consciousness.

The brain may also be bruised in deceleration injuries (which occur when the body is moving fast and is suddenly halted, as in a traffic accident). The shaken brain hits the inner surfaces of the skull, and is bruised both at the site of the impact and then on the opposite side as the brain bounces back. This results in concussion, which can cause vomiting, double vision, and headaches.

The brain may swell, causing symptoms such as confusion, seizures, loss of consciousness, and sometimes death. Urgent treatment is needed to relieve the pressure on the brain, and to treat any bleeding. Care and rehabilitation may be needed for many months afterward.

Movement

Brain

Skull

1 Moving rapidly
In a person moving at speed—for example, in a car—the skull and brain are moving at the same speed as the body and the vehicle.

Brain impact 2   Brain   Brain impact 1

Brain

2 Impact
If movement is suddenly stopped, the brain hits the front of the skull, and then rebounds and hits the back (a "contrecoup" injury).

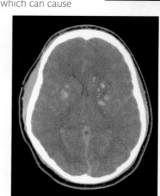

Skull fracture
This three-dimensional CT scan of the skull shows several severe fractures. Injuries like these can cause brain damage or death.

Hematoma
In this scan, the blue area is a hematoma, or a pocket in which blood has accumulated, outside the skull. Severe bleeding within the brain is highlighted in orange.

CEREBRAL PALSY

The name "cerebral palsy" is used for a group of disorders that result from brain damage and cause difficulties with posture and movement.

In many cases, the damage to the developing brain occurs before birth; in others, the brain is starved of oxygen before, during, or just after birth. Cerebral palsy involves damage to the motor cortex of the brain, leading to difficulties in standing and moving. If the disability is severe, there is spasticity (stiffness) of the arms and legs. Mildly affected children may show only slight stiffness and "scissoring" (crossing) of the legs and some alteration in their gait. However, the cognitive (thinking) processes, and therefore the child's intelligence, are not necessarily affected. The child will need physical therapy, to keep the muscles flexible, and possibly help with speech and language. The condition does not worsen over time, and many children adapt well to their difficulties.

HYDROCEPHALUS

This condition results from an excess of cerebrospinal fluid, which puts pressure on brain tissues and can damage them.

Cerebrospinal fluid surrounds the brain and fills the ventricles (spaces) within it. Normally, it cushions and nourishes the brain, and any excess is absorbed into the blood. Excess fluid may build up due to overproduction, or to impairment of the drainage process by a blockage or a structural abnormality. In babies, the skull bones are not yet fused but are held together by stretchy cartilage. As the fluid gathers, the bones separate, causing the skull to become large and translucent.

In adults, hydrocephalus causes increased pressure on the brain, leading to persistent headaches, which tend to be worse in the morning; problems with vision and gait; and drowsiness or lethargy. The excess fluid may be cleared with a shunt (drainage tube) which drains the fluid to another part of the body.

Arachnoid (site of reabsorption)   Choroid plexuses (site of production)

Third ventricle

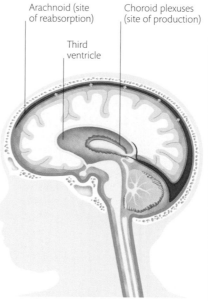

Fluid on the brain
Cerebrospinal fluid is produced by the choroid plexuses that line the ventricles at the center of the brain, and bathes the brain and spinal cord. Excess fluid is reabsorbed via the arachnoid membrane.

BRAIN AND SPINAL CORD ABSCESSES

An abscess is a pus-filled swelling caused by infected material in body tissue; in the brain or spinal cord, it can cause severe or life-threatening damage.

Infected material can reach the brain or spinal cord directly if the blood-brain barrier is breached by injury, or from infections of the sinuses (cavities in the bones of the face) or the meninges. Abscesses of the brain or spinal cord are rare, but the symptoms are severe. In the brain, the pressure caused by the abscess can result in confusion, headache, fever, and possible collapse if the infection is severe. Abscesses around the spine may cause pain and paralysis, and can swiftly lead to meningitis (see p.455) as the cerebrospinal fluid carries infection to the meninges (the tissues covering the brain). Surgery may be needed to drain the abscess, and drugs will be given to kill the infection and prevent seizures.

DEMENTIA

The name "dementia" refers to a gradual loss of cognitive ability—understanding, reasoning, and memory.

Dementia most commonly affects older people. It is usually caused by diseases of the brain or the cerebral blood vessels. The most common form is Alzheimer's disease, in which brain cells degenerate and deposits of protein build up in the tissue. Another form is vascular dementia, in which the small blood vessels supplying the brain are blocked by blood clots, leading to multiple tiny areas of brain damage. Dementia with Lewy bodies is a condition in which tiny round nodules, called Lewy bodies, collect in the brain and impair its function, causing symptoms such as hallucinations.

Dementia can occasionally occur in younger people, as a result of chronic brain injuries, Parkinson's disease, or Huntington's disease. Most forms of dementia grow progressively worse over the years. Typically, the affected person's relatives notice that he or she has become more forgetful, with loss of memory for recent events but with clear, long-lasting memories about distant events. It is initially difficult to distinguish this from the normal aging process. Eventually, however,

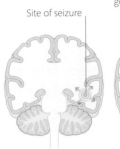

NORMAL   ALZHEIMER'S

Brain activity
This PET scan shows the result of brain stimulation tests made on a healthy person and a person with Alzheimer's disease. The blue areas show reduced brain activity in the person with the disease.

symptoms do become worse and the person begins to forget basic information such as where he or she lives. Problems such as speech difficulties, incontinence, and personality changes may develop. Those with severe dementia may lose all memory of loved ones and friends and need full-time care.

To identify dementia, doctors may carry out scans and and assessments of mental abilities. The risk of dementia can be diminished by mental activity, especially new learning opportunities. Although there is no cure, memory and daily life can be improved with mental exercises, and occasionally with medication.

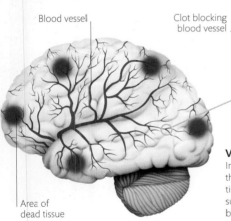

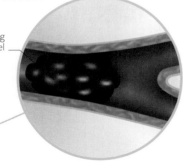

Blood vessel

Clot blocking blood vessel

Area of dead tissue

Vascular dementia
In this form of dementia, tiny blood vessels throughout the brain become blocked, causing tissue death (infarction) in the areas that they supply. The disease becomes worse as further blood vessels are affected.

EPILEPSY

This disorder is typified by recurrent seizures or convulsions as a result of abnormal electrical activity in the brain.

Brain cells send messages to each other, and to the rest of the nervous system, in the form of electrical signals. Seizures occur when these signals are temporarily disrupted. In epilepsy, such abnormal brain activity is recurrent and unprovoked. It can arise spontaneously or result from disease or damage to the brain.

Seizures can be triggered by stress or a lack of food or sleep. Symptoms vary depending on where the abnormal activity arises. Partial seizures involve only one side of the brain. Simple partial seizures, confined to a small area, may just cause twitching of one body part, whereas complex partial seizures, in which the disturbance spreads to nearby areas, produce bizarre movements, confusion, and loss of consciousness.

Generalized seizures, affecting the whole brain, cause loss of consciousness, collapse, and severe muscle spasms, followed by a period of altered consiousness and fatigue. Many sufferers experience a warning "aura" just before a seizure, with abnormal sensations. Epilepsy can be managed with medication to control seizures and lifestyle changes to ensure safety.

ELECTROENCELPHALOGRAPHY

Electroencephalography (EEG) is a recording of electrical activity within the brain. Small electrodes are fixed to the scalp with adhesive gel and record brain activity for several hours. The results are shown as a trace on paper or a computer. EEG is often performed on sleep-deprived patients, in whom abnormalities are more likely to show. During an epileptic seizure an EEG will show areas of abnormal activity, and there may be visible centers of abnormal activity even when the person is not having a seizure.

Site of seizure

Secondarily generalized seizure

Partial seizure

Partial seizure
The abnormal activity originates in one lobe and remains confined to this area. In some cases a partial seizure becomes generalized and spreads (above right).

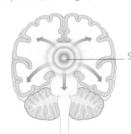

Site of seizure

Generalized seizure
Abnormal activity spreads through the brain. Symptoms vary but typically include uncontrolled movements of the whole body, with loss of consciousness lasting from one to several minutes.

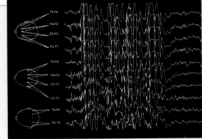

EEG trace during generalized seizure
This EEG shows electrical activity across all areas of the brain, corresponding with a generalized epileptic seizure.

BRAIN TUMORS

Tumors in the brain may be either benign (noncancerous) or malignant (cancerous), but both types can cause severe impairment of brain function.

Most brain tumors are metastatic, meaning that they grow from cancer cells that have spread via the blood from another part of the body. Breast and lung cancers are particularly liable to spread to the brain, and are often a sign that the primary disease is accelerating. Primary brain cancer, originating in the brain,

is far less common. Malignant tumors typically grow fast and spread through the brain. Benign tumors tend to grow more slowly and remain in one area. Any kind of tumor can damage the brain—there is no room in the skull for the tumor to grow, so it puts pressure on the brain tissue. Symptoms vary according to the area affected; they include severe headaches, confusion, blurred vision, paralysis of one body part, difficulty speaking or understanding speech, and changes in personality. If a tumor causes bleeding, there may be sudden pain and loss of consciousness. It may be possible to remove a benign tumor surgically, although this epends on the site. Malignant tumors generally cannot be removed because

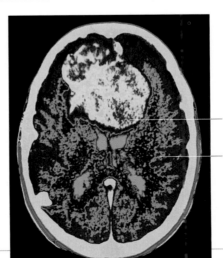

separating them from the surrounding brain tissue would be too destructive, although radiation therapy or chemotherapy can help reduce their size. Many people with benign tumors recover, but those with cancer may face a shortened life expectancy.

Site of tumor

Brain hemisphere

Meningioma
This scan shows a large tumor in the frontal lobes, pushing the healthy brain tissue aside. The frontal lobes affect personality, and changes in this area can lead to abnormal moods and behavior.

GENERAL NERVOUS SYSTEM DISORDERS

The nervous system carries a constant two-way flow of signals from body tissues to the brain and responses returning from the brain to the body. However, certain disorders cause degeneration of brain and nerve tissue, impeding or stopping these signals.

MULTIPLE SCLEROSIS

In multiple sclerosis (MS), nerves in the brain and spinal cord suffer progressive damage, which causes problems with a wide range of body functions.

Electrical signals pass between the brain and the body along the nerves. Healthy nerves in the brain and spinal cord have a protective covering of a fatty substance called myelin, which enables signals to travel faster and more smoothly. MS involves progressive destruction of the myelin sheaths around the nerves. It is an autoimmune disorder in which the immune system attacks the body's myelin tissue. The cause is unknown, although both genetic and environmental factors seem to play a part.

Typically, the disorder first appears between the ages of 20 and 40. Symptoms may include problems with vision or speech, difficulties with balance and coordination, numbness or tingling, weakness, muscle spasms, muscle or nerve pain, fatigue, incontinence, and altered mood. In some people the symptoms come and go, although often with a deterioration after each episode, while in others they grow steadily worse. There is no cure, although various drugs are used to relieve symptoms and delay the progression of the disease.

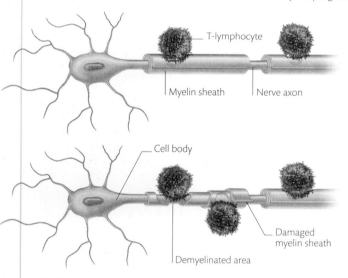

- T-lymphocyte
- Myelin sheath
- Nerve axon
- Cell body
- Damaged myelin sheath
- Demyelinated area

Early stage
T-lymphocytes and macrophages (cells from the immune system) attack the myelin sheaths on the nerves. Some repair may occur in the early stages.

Late stage
Significant nerve injury occurs early in MS. By the late stage this has become irreversible, with death of nerves and scarring and swelling of damaged nervous tissue.

PARKINSON'S DISEASE

This chronic, progressive disorder typically causes tremor, slowing, stiffness, and problems with voluntary movement.

Parkinson's disease results from degeneration of cells in the basal ganglia, a part of the brain involved in initiating movement. Normally, the cells produce a neurotransmitter (a chemical that carries information between nerves) called dopamine, which helps coordinate muscle activity. In Parkinson's disease, these cells produce much less dopamine, and the signals to the muscles become slow and faulty.

The disease is most common in older people, but may also occur in young adults or, rarely, in children. In most people there is no obvious cause, although there is some evidence for a genetic origin. Parkinson's disease can also result from encephalitis, or from damage to the basal ganglia by certain drugs or repeated head trauma. The main symptoms are trembling of one hand, arm, or leg at rest, which may progress to affect the limbs on the opposite side; muscle stiffness, which makes it difficult to begin moving and makes movements slower; and problems with balance. Abnormal head movements are common, and the face may become less expressive as the facial muscles lose their mobility. People may also experience mood disturbance, depression, shuffling gait, problems with speech and cognition, and difficulties with sleep.

Drugs may be given to mimic dopamine production, although over time these can become less effective. Treatment also includes physical therapy and lifestyle changes to help preserve mobility. Some people may be offered surgery such as deep brain stimulation, a procedure in which electrodes are implanted into the basal ganglia to help control tremors.

Brain in Parkinson's disease
This color-enhanced MRI scan of the brain in Parkinson's disease shows generalized shrinking (atrophy) of brain tissue. Other changes are microscopic and cannot be seen on scans.

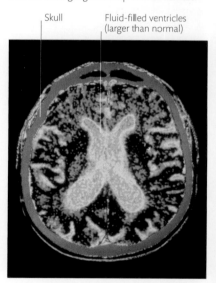

- Skull
- Fluid-filled ventricles (larger than normal)

AMYOTROPHIC LATERAL SCLEROSIS

This incurable condition causes gradual but inevitable loss of function in the motor nerves, which carry signals from the brain to cause conscious movements.

Amyotrophic lateral sclerosis (ALS) typically begins between the ages of 50 and 70. The disorder damages both the nerves and the muscles: as the motor nerves lose the ability to stimulate muscle activity, the muscles weaken and waste away. The cause is unknown, although in a few people there is a genetic susceptibility. Weakness first appears in the hands, arms, and legs. There may be muscle cramps, twitching, or stiffness. Daily activities such as holding objects and climbing stairs may become difficult, and the person may start to stumble. As ALS worsens, it causes spasticity (severe muscle spasms), slurred speech, and difficulty swallowing. Mental abilities are usually unimpaired. Most people with ALS die of respiratory failure only a few years after diagnosis, although there are exceptions to this.

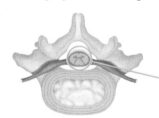

Motor neurons in the spinal cord
ALS destroys the motor nerves in the ventral horns of the spinal cord. The most common form attacks here first, leading to peripheral weakness of the hands, feet, and mouth.

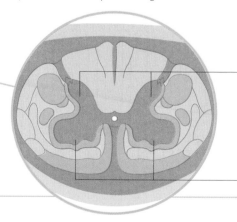

- Neurons (nerve cells) in the dorsal (back) horns receive sensory information from around the body
- Neurons in the ventral (front) horns send motor nerve fibers to skeletal muscles, causing them to contract

NERVOUS SYSTEM INFECTIONS

The brain and spinal cord are extremely well protected from infection, but any infective organisms that do penetrate them can cause problems such as inflammation or tissue abnormalities, which can become serious or even life-threatening.

MENINGITIS

In this disorder, the meninges—the three layers of membranes that surround the brain and spinal cord—become inflamed, usually as a result of infection.

The most common forms in Western countries are bacterial meningitis and viral meningitis. The viral form (caused by organisms such as enteroviruses) is more common but relatively mild; bacterial meningitis (usually caused by *Neisseria meningitides* or *Streptococcus pneumoniae*) is much more serious. Other forms occur in the developing world or in

Meninges
The meninges comprise the dura mater (outermost layer), arachnoid (middle), and pia mater (inner layer)

Dura mater
Arachnoid
Pia mater

people with reduced immunity. Meningitis may also result from certain drug reactions, or from bleeding in the brain. In viral meningitis the symptoms arise gradually; in the bacterial form they come on within hours. Inflammation may spread from the meninges to blood vessels and brain tissue. Symptoms include fever, headache with intolerance of light, stiff neck, vomiting, and altered consciousness. It may be life-threatening and can cause brain damage. Vaccination provides protection; meningitis requires urgent hospital treatment where drugs may be given to kill the infection.

Brain tissue

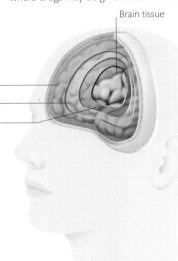

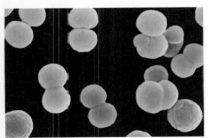

Meningitis-causing bacteria
The bacteria that most commonly give rise to meningitis are meningococci (shown here), *Haemophilus*, and pneumococci, although any bacteria could potentially cause the disease.

Sites of infection
Most bacterial meningitis is caused by bacteria transmitted through the bloodstream. Bacteria can also enter the brain or spinal cord directly in head or spinal trauma, brain abscess, or surgery.

LUMBAR PUNCTURE

In this procedure, a sample of cerebrospinal fluid (the fluid that bathes the brain and spinal cord) is taken from the spine with a needle. Lumbar puncture is used mainly for diagnosing meningitis, by identifying the organisms causing the infection and revealing high numbers of white blood cells (which fight infection). It may also be used to detect abnormal protein and antibody levels if multiple sclerosis is suspected, or to detect bleeding or a tumor in the brain. Lumbar puncture is occasionally used to remove excess cerebrospinal fluid if it is putting too much pressure on the brain.

The procedure
The person lies on one side, curled up as tightly as possible, and the needle is inserted between two of the lumbar vertebrae into the subarachnoid space below the bottom of the spinal cord.

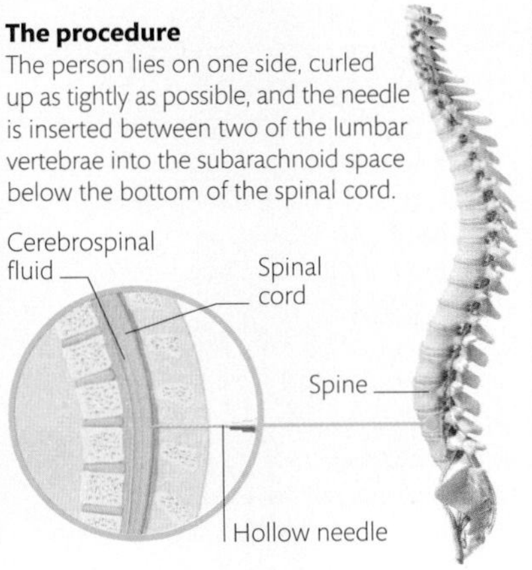

Cerebrospinal fluid
Spinal cord
Spine
Hollow needle

ENCEPHALITIS

Inflammation of the brain, or encephalitis, is usually due to infection but occasionally to autoimmune attack. It is a rare but life-threatening emergency.

Most cases result from viruses, although bacteria and other microorganisms can also cause it. The most common viral causes are herpes simplex (the cold sore virus), measles, and mumps; the incidence in children has greatly reduced since vaccination became widespread. Encephalitis usually results from systemic (whole-body) infection breaching the brain's defenses, but can also occur secondary to meningitis (see left) or brain abscess (see p.452). It causes flulike symptoms, fever, and headache; more severe cases progress rapidly to confusion, seizures, loss of consciousness, and coma. Difficulty speaking and paralysis in part of the body may also be present. The disorder is rare, most often occurring in older people and children under 7. Encephalitis is usually diagnosed by MRI scan and treated with drugs to kill infection. Recovery may be slow and incomplete. Long-term consequences can include epilepsy, memory problems, and personality change.

Infected tissue in temporal lobe

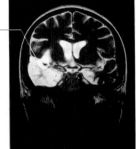

Viral encephalitis
MRI brain scan showing infected tissue due to encephalitis caused by herpes simplex infection.

HERPES ZOSTER

Also known as shingles, herpes zoster is a nerve infection caused by a reactivation of the virus that produces chickenpox.

Chickenpox normally causes a blistering rash and mild illness, and lasts for about a week. However, it can be more serious in adults and older teenagers who did not have the disease as children, in pregnant women, and in people with a weakened immune system. After the illness clears up, the virus lies dormant in the body but can be reactivated, causing shingles (an itchy, blistering rash, with burning or stabbing pain, that follows the path of a nerve). Herpes zoster may also result in inflammation and infection of various organs. In the brain, it can cause loss of coordination, speech disturbance, and encephalitis (see above), and can be life-threatening. It may be treated with antiviral drugs, steroids, and analgesics.

CREUTZFELDT-JAKOB DISEASE

This is a rare brain disease, similar to BSE (mad cow disease) in cattle and scrapie in sheep, that may be contracted by eating contaminated meat or may be inherited.

Creutzfeld-Jakob disease (CJD) is thought to be caused by prions, abnormal proteins that behave as infectious organisms and have a particular affinity for nervous tissue. The new variant form of CJD, first diagnosed in 1996, is acquired from prion-contaminated meat. Similar conditions are seen in some animals.

A rare variant form is inherited. The prion protein triggers a misfolding of normal proteins in the brain. As a result, the brain cells die and are replaced with deposits of prions. This causes rapid loss of body functions, dementia, progressive brain failure, and death within months.

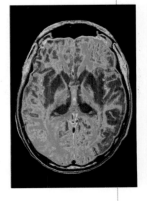

Brain with CJD
In this MRI scan, the red areas show the thalamus, in which the tissue has degenerated as a result of vCJD.

MENTAL HEALTH DISORDERS

Disorders of the mind may involve mood, as in depression; thought, as in OCD; or serious disturbance of brain function. Talking therapies and medications can help managing symptoms, but the more serious relapsing illnesses cannot be cured.

DEPRESSION

This is broadly speaking a condition of depressed mood and feelings of sadness, but it affects people in different ways.

Depression causes lowered mood, drive, and enjoyment, leading to a sense of sadness and hopelessness. It is very common, but is often not diagnosed and treated. Depression is also more than just temporary sadness—it is a medical condition caused by disorders in brain chemistry that can seriously disrupt daily life. Those affected see the whole world, including themselves, as pointless and useless. Some people become flat and down, lacking energy, with excessive eating and sleeping. Others become more anxious, with agitation, poor sleep, and poor appetite. In severe cases, an affected person may consider or attempt suicide, or may develop psychosis (delusional thoughts). It is a chronic condition, typically lasting for several months unless treated, and recurring in most people. Treatment may involve talking therapies, such as cognitive behavioral therapy or psychotherapy, antidepressant medication, or electroconvulsant therapy.

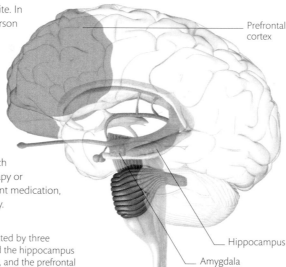

Brain areas and mood
Mood and feelings are regulated by three main areas. The amygdala and the hippocampus produce emotional responses, and the prefrontal cortex generates thoughts about those emotions.

Prefrontal cortex

Hippocampus

Amygdala

ANXIETY DISORDERS

Anxiety is a condition that causes fear, agitation, worrying, poor sleep, loss of appetite, and physical symptoms.

Anxiety is a natural response to stress, which arises in the amygdala and hippocampus—in evolutionary terms, the oldest parts of the brain. It stimulates the fight or flight response, which would have saved our distant ancestors from the many physical dangers that they faced. This primitive but vital response still operates in the modern world but is triggered by stresses such as problems at work or in personal relationships. Some people have a stronger than usual response to stress, which they may have inherited. Otherwise, anxiety can arise as a result of difficult life events such as loss of a job. Chronic anxiety produces physical symptoms such as fast heart rate, sweating, butterflies in the stomach, and heartburn. It also causes feelings of being on edge, anger, sleeplessness, poor concentration, and difficulty coping with simple stresses without feeling overwhelmed. At its worst, anxiety can lead to panic attacks, with shaking, sweating, racing heartbeat, and a feeling that one is about to die.

Treatment may involve learning relaxation techniques or undergoing a "talk treatment" such as cognitive behavioral therapy to help control thought patterns that can lead to stress. Antidepressant drugs may also be given.

BIPOLAR DISORDER

This condition causes extreme mood swings, with alternating periods of profound elevation in mood (hypomania and mania) and depression.

Bipolar disorder, also called manic depression or manic affective disorder, causes episodes of high or euphoric mood, alternating with depression. During a "high" phase, known as mania, the person can feel elated, confident, full of energy, and highly creative. However, the elevated mood can lead to risky behavior such as overspending or unsafe sex; sometimes feelings of being indestructible; disordered thoughts; and delusional beliefs that can render the person a danger to themselves or others. At worst, mania leads to psychosis, disordered perception, or hallucinations. By contrast, in the depressive phase, the person loses all interest in life and hope for the future, and may be low enough to consider suicide.

Most affected people have longer periods of depression and relatively short periods of mania, interspersed with periods of normal mood, and the condition is chronic and recurrent over many years. Treatment is with long-term medication to correct the disorders in brain chemistry and thereby stabilize the moods, together with intensive psychological support through high or low periods. If the symptoms are particularly severe, the person may need treatment in the hospital.

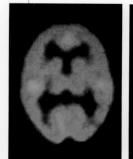

NORMAL   MANIA

Brain activity scan
During a high or manic phase of bipolar disorder the brain shows increased levels of activity as shown in this brain scan. Common symptoms include increased energy and less need for sleep.

SUBSTANCE ABUSE

The main substances of abuse are alcohol; tobacco; and illegal or restricted-use drugs such as heroin, amphetamines, cocaine, cannabis, benzodiazepines, and LSD. Drugs act on the brain's "reward system," which normally responds to pleasurable stimuli, making us want to repeat the activity. Taking a drug overstimulates this system, producing a "high" feeling. The brain can become dependent on the drug and may experience unpleasant withdrawal symptoms if the drug is stopped. The same drugs can also cause mental problems such as paranoia and psychosis.

OBSESSIVE COMPULSIVE DISORDER

The main characteristics of this disorder are repetitive behavior and intrusive thoughts that can interfere with daily life.

Many people have some degree of obsessive or compulsive tendencies. However, in obsessive compulsive disorder (OCD), the need to perform a specific action becomes constant, and the person may become very anxious if unable to carry out the action. The person may also have intrusive or upsetting thoughts, such as fear that their loved ones will die if they do not carry out the action. Many can be helped with antianxiety medications and therapies that help people confront and manage the fear underlying their behavior.

Compulsive handwashing
A common "ritual" action seen in OCD is compulsive washing of hands, due to extreme fear of contact with dirt or germs.

SCHIZOPHRENIA

Schizophrenia is characterized by a loss of contact with reality with hallucinations and delusions.

Schizophrenia involves a mixture of "positive" symptoms such as hallucinations, which tend to predominate early in the course of the condition, and "negative" symptoms such as lack of any pleasure in life, which predominate later as the positive symptoms die down. Hallucinations most commonly involve an affected person hearing voices talking to or about them. A person may also have delusions, such as believing that people on television are speaking directly to them, as well as difficulty distinguishing reality. Other symptoms include disordered thinking and bizarre repetitive movements.

Negative symptoms include loss of emotional expressiveness and social withdrawal. Schizophrenia has some genetic basis, and tends to appear in the late teens or early 20s. Stressful life events may trigger the onset or cause flare-ups. The condition needs long-term treatment with antipsychotic medication, social support, psychotherapy, and rehabilitation, but rates of physical illness, anxiety, and depression are high.

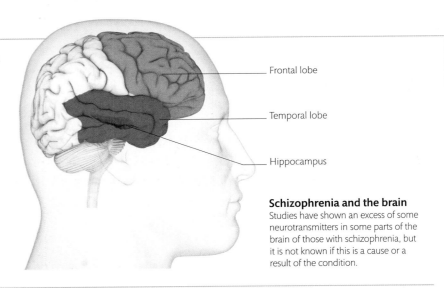

Frontal lobe

Temporal lobe

Hippocampus

Schizophrenia and the brain
Studies have shown an excess of some neurotransmitters in some parts of the brain of those with schizophrenia, but it is not known if this is a cause or a result of the condition.

EATING DISORDERS

Psychological disorders to do with eating cause the affected person to avoid food, induce vomiting, or, conversely, to overeat compulsively.

Anorexia nervosa and bulimia nervosa are the most common eating disorders; many affected people have elements of both. People with anorexia believe themselves to be fat when they are very underweight. The disorder begins with severe calorie restriction but may progress to refusal of all food and fluids. Menstrual periods may stop, and fine, downy hair may grow on the body. Anorexia has a 10 percent fatality rate. Bulimia involves some of the same attitudes to the self, but people alternate short periods of fasting with intense binges of overeating, often of high-calorie "forbidden" foods, followed by self-induced vomiting, sometimes with laxative abuse. People with bulimia may have normal body weight, but they are at risk of salt imbalance, tooth decay, and stomach rupture.

Other disorders include compulsive overeating and eating non-food items such as paper tissues. Eating disorders may be brought on by stress and a need for control in life, but the illness can come to take over a person's life. Treatment involves psychological help and nutritional support.

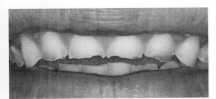

Acid-worn teeth in bulimia
Recurrent vomiting in bulimia causes the teeth to be exposed repeatedly to gastric acid. This wears away the enamel covering the teeth; eventually the enamel is worn through and the teeth decay.

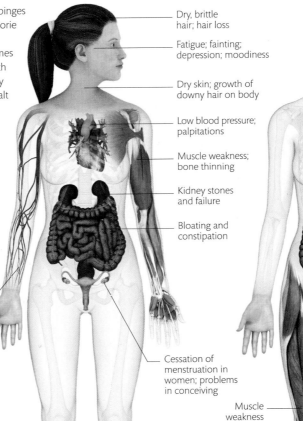

Dry, brittle hair; hair loss

Fatigue; fainting; depression; moodiness

Dry skin; growth of downy hair on body

Low blood pressure; palpitations

Muscle weakness; bone thinning

Kidney stones and failure

Bloating and constipation

Anemia; low levels of electrolytes

Cessation of menstruation in women; problems in conceiving

Muscle weakness

ANOREXIA NERVOSA

Dizziness; depression; low self-esteem

Gum disease; sensitive teeth; tooth erosion and decay

Sore throat; inflammation of esophagus

Low blood pressure; heart muscle disorders

Stomach pain; bloating; ulceration

Anemia; low levels of electrolytes; dehydration

Irregular or absent periods

BULIMIA NERVOSA

Effects on body
Anorexia and bulimia both have widespread effects on the body, affecting almost every system.

PERSONALITY DISORDERS

These disorders involve persistent, fixed dysfunctions in a person's perceptions and the way they relate to others.

Our personality is largely established by the time we reach adulthood. In most people, it continues to develop in response to new experiences. However, people with personality disorders show rigid, dysfunctional patterns of behavior that cause problems for themselves or others. The disorders fall into three main groups. The first group (paranoid, schizoid, schizotypal) involves odd or eccentric thinking. The second (histrionic, borderline, narcissistic, antisocial) is typified by emotional, impulsive, attention-seeking, or cruel behavior. The third (avoidant, dependent, obsessive-compulsive) shows anxious or fearful thinking. Personality disorders cannot be cured, but management involves talking therapies such as cognitive behavioral therapy (CBT), as well as support to help the person gain insight into the way they respond to the world and to help them adapt their behavior and function successfully.

PHOBIAS

A phobia is an intense and persistent fear of certain objects, people, animals, or situations, so that the person feels great anxiety if forced to confront them.

Some fears, such as fear of deadly creatures or fear of heights, are normal, natural survival mechanisms. Phobias, however, are fears that involve nondangerous animals, objects, or situations, or are so intense that they interfere with daily life. Many people manage phobias effectively by avoidance. However, some, such as agoraphobia (a fear of going out), can become disabling, and challenging them can provoke severe anxiety. Phobias can be cured by very gradual phased exposure to the source of their fear, sometimes aided by sedating medication. Alternatively, people may be given "flooding therapy," when massive, sustained exposure is used to demonstrate that the feared object or situation is harmless. A third method is "counter-conditioning," in which the affected person learns relaxation techniques to replace the fear response.

EAR DISORDERS

The ear is a complex structure whose roles include converting sound waves of differing amplitude and frequency into nerve impulses for transmission to the auditory cortex, localization of sound, and the sense of balance and body position.

OUTER EAR DISORDERS

The outer ear comprises the pinna or visible part, and the ear canal, leading to the ear drum. Problems here can cause discomfort but are usually treatable.

The ear canal secretes ear wax to clean and lubricate the canal. Most excess wax comes out by itself. Any build-up can be cleared by warm olive oil or ear drops, to melt the wax and relieve the sensation of blockage. Using objects such as cotton buds to clean the ear, however, interrupts the outward flow of wax, compressing it back against the ear drum and damaging the skin of the ear canal. Infection of the ear canal can occur when the delicate lining of the ear canal is damaged, most often by having objects poked into it, irritation from detergents such as shampoo and chlorinated water, or infection spreading out from the middle ear. It can be extremely painful, but can usually be relieved by ear drops. People with recurrent infection may find that nightly application of olive oil will protect the ear canal and reduce the frequency of attacks.

Infected ear canal
Ear canal infection is prone to cause discharge, usually due to a mixture of pale yellow fluid from the inflamed tissues and wax liquefied by the increase in temperature as a result of infection.

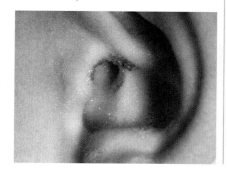

MIDDLE EAR INFECTIONS

The ear drum and the space behind it are highly sensitive structures, so infections in this area can be extremely painful.

The middle ear space, which lies behind the ear drum, contains three small bones (ear ossicles); these bones transmit vibrations from the ear drum to an inner window connecting with the hearing nerves, which turn the vibrations into electrical signals that pass to the brain. Normally this space is filled with air, which enters through the Eustachian tubes. Middle ear infections may arise during an infection such as a cold, when mucus builds up in the middle ear and air can no longer get in. The trapped mucus thickens and becomes infected with viruses and, in some cases, bacteria. This causes pain and reduced hearing. Sometimes the mucus exerts so much pressure on the ear drum that it bursts and lets

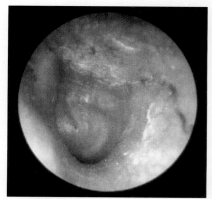

Ear infection
When the ear is infected the normally translucent ear drum looks dull and may bulge under pressure.

the mucus drain out. Ear infections are more common in children under 6, who have shorter, straighter Eustachian tubes than adults, allowing bacteria swift passage to the middle ear.

PERFORATED EAR DRUM

The ear drum, or tympanic membrane, sits between the ear canal and the middle ear. Its job is both to amplify sound and to protect the middle ear from debris.

External or middle ear infections can cause the ear drum to become inflamed. Pressure from fluid in the middle ear can make the ear drum burst; when it does so the ear may discharge a bloody fluid, although pain is often partly relieved. The drum can also be perforated by objects being used to clean the ear canal.

Most perforated drums heal by themselves over a couple of weeks; the ear canal and drum must be kept dry while they recover. If the ear drum does not heal, it may need surgical repair.

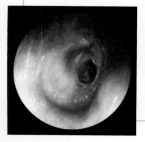

View of perforated ear drum
The eardrum has burst, allowing trapped pus to escape.

CHRONIC OTITIS MEDIA WITH EFFUSION

Common in children, otits media is due to the accumulation of mucus in the middle ear cavity, which normally contains air.

In adults, the condition is commonly due to long-term Eustachian tube blockage (often associated with sinus problems). The Eustachian tubes connect the middle ear space to the back of the throat, keeping the space ventilated and at the correct pressure. If the tubes are blocked, air cannot get into the middle ear. Gluey mucus replaces it and remains trapped there, reducing the ability of the ossicles (tiny bones in the middle ear) to transmit sound. Hearing is reduced, and the ear feels full. A popping sound can occur when the Eustachian tubes intermittently open to let a little air in.

In children, the condition commonly follows ear infection, when the mucus can be slow to clear. If the child has several ear infections in a row, the mucus can become persistent, leading to prolonged hearing reduction that can affect schooling or language development. When this happens, a small ventilation tube, or ear tube, may be fitted into the ear drum to allow air into the middle ear. They do not prevent ear infections, but do help clear mucus and improve hearing. The condition is most common in children under 5 years, who have short, straight tubes, which are more susceptible to viral infections from the throat. As the jaw grows and the adult teeth come through, the tubes lengthen and become less straight.

Ear tube

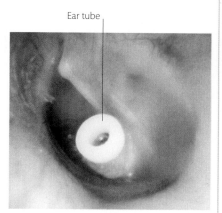

Chronic otitis media with effusion
An ear tube has been inserted into the drum to allow air into the middle ear space, preventing development of chronic otitis media. The condition is often caused by bacteria or viruses.

LABYRINTHITIS

This common condition of dizziness and nausea is caused by inflammation in the inner ear. It is painless but often very unpleasant in its effects.

The labyrinth is a coiled, fluid-filled structure in the inner ear that consists of the cochlea (the hearing apparatus) and the vestibular system (the balance apparatus). The role of the vestibular system is to sense the position of the head in relation to gravity, indicating whether the head is upright or tilted, and to help the eyes stay focused on objects as the head is turned. Inflammation of the labyrinth upsets the balance system, causing vertigo, nausea, and disorientation. If both labyrinths are involved, these symptoms can be severe.

The brain can compensate for the disturbances to the inner ear, but loud sounds and sudden head movements stimulate the labyrinth and make symptoms worse. Viral labyrinthitis is the most common type, and can last from several days to several weeks. Bacterial labyrinthitis is less common but, if untreated, can lead to permanent impairment of hearing.

ADULT HEARING LOSS

Some degree of hearing loss is common as part of the aging process, but hearing can also be damaged by loud noise, injury, or disease.

Hearing loss may be conductive (due to poor sound wave transmission) or sensorineural (due to nerve damage). Conductive hearing loss is often caused by a blockage of earwax

and is usually temporary. It can be resolved by syringing the ear. In children it may be caused by chronic otitis media with effusion (see left). Sensorineural hearing loss most commonly occurs during the aging process as the cochlea deteriorates. This is known as presbyacusis, and it affects many people over the age of 50.

Persistent exposure to loud noise can also cause sensorineural hearing loss as this damages the nerves more rapidly. Ménière's disease (see right) or damage to the cochlea may also cause sensorineural hearing loss. The ability to hear high-frequency (high-pitched) sounds is reduced first, and the problem may initially be noticed when the frequencies of speech become difficult to distinguish. Tests can be done to find the cause and seriousness of the problem. Hearing aids (see below) may help people to cope with loss of hearing.

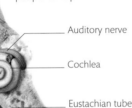

Structure of the ear
Hearing is affected by problems in various structures of the ear, which may result in partial or complete hearing loss. Most adult hearing loss is age-related.

Auditory nerve
Cochlea
Eustachian tube
Outer ear canal
Bones of middle ear

Tumor in left internal auditory canal
This acoustic neuroma (tumor) is growing on the cochlear nerve. These tumors are benign but cause progressive hearing loss, with vertigo and tinnitus, and usually require surgery.

Tumor

HEARING AID

The purpose of a hearing aid is to amplify (increase) the sound reaching the inner ear. The device is an electroacoustic amplifier, consisting of a microphone, amplifier, and speaker. The limitation of hearing aids is that they only amplify sound, but do not clarify it—and much hearing loss involves high-frequency sounds such as consonants. The result can be that speech is less clear rather than not loud enough. To combat this problem, FM listening devices are being developed with wireless receivers integrated with hearing aids.

Wearing hearing aids
Hearing aids are usually worn inside or behind one or both ears. Some have an in-canal receiver with an amplifier behind the ear, while others are surgically implanted.

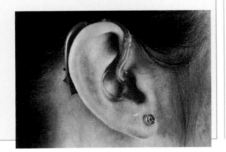

TINNITUS

Damage to the hearing apparatus can cause tinnitus, the perception of sound when no external sound is present.

The sounds associated with tinnitus range from intermittent and quiet to a constant loud noise, and one or both ears may be affected. They include whooshing, hissing, musical sounds, clickings, and buzzings. The origin can be the

pulsating blood vessels in the ear, or false signals from damaged nerves. Causes of temporary tinnitus include ear wax, glue ear, ear infection, and exposure to noise. Permanent tinnitus is usually caused by damage to the auditory (hearing) nerves, including age-related loss (in these cases, the frequency of the sound is often in the range that the person can no longer otherwise hear). The condition can be difficult to tolerate, as people with the condition have to develop strategies to ignore or mask the sound. It can be cured by cutting the auditory nerves, but absolute deafness results.

MENIERE'S DISEASE

This inner ear disorder is common but long-lasting and difficult to treat well and its symptoms can be disabling.

Ménière's disease is a disorder of the fluid in the labyrinth, which contains the organs of hearing and balance. It typically causes tinnitus (see above), hearing loss, vertigo (see below), and a full feeling in the ear, and may affect one or

both ears. The underlying cause is a problem with the drainage of the fluid in the vestibular system (the balance apparatus). This causes an increase in fluid pressure, resulting in damage to the sensitive nerve structures. It usually comes on gradually, but sudden attacks of severe vertigo, lasting less than 24 hours, are common and can cause people to fall to the ground. The underlying trigger is unknown, although the viral infection herpes has been suggested. The vertigo and tinnitus can be cured by cutting the auditory nerves, which causes total deafness, so most sufferers choose to manage the symptoms instead.

LOCATION

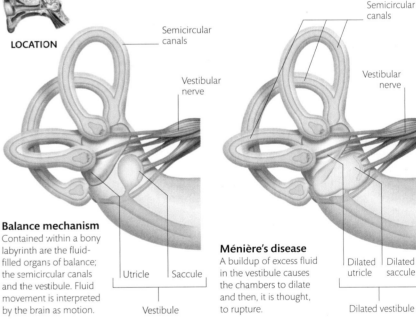

Balance mechanism
Contained within a bony labyrinth are the fluid-filled organs of balance; the semicircular canals and the vestibule. Fluid movement is interpreted by the brain as motion.

Semicircular canals
Vestibular nerve
Utricle
Saccule
Vestibule

Ménière's disease
A buildup of excess fluid in the vestibule causes the chambers to dilate and then, it is thought, to rupture.

Semicircular canals
Vestibular nerve
Dilated utricle
Dilated saccule
Dilated vestibule

VERTIGO

An unsteady feeling due to upset balance, vertigo can be brought on by visual stimuli or by being spun around, or can be a symptom of a balance disorder.

Vertigo gives a sensation of spinning or tilting, sometimes with nausea or vomiting. In some people it is triggered by heights. It can result

from an inner ear disorder, as in benign paroxysmal positional vertigo (BPPV), which is caused by tiny crystals in the balance system being displaced. It can also be caused by poor blood supply to the balance system (often due to atherosclerosis), Ménière's disease, or ear infection. In addition, vertigo can result from problems in the balance centres of the brain; for example, due to migraine or stroke. It is usually worsened by sudden head movements and loud noises, and partially relieved by drugs to relieve nausea or simply closing the eyes.

EYE DISORDERS

The eye collects light by focusing, then converts the light signals into sequences of nerve messages that enable the brain to build an accurate picture of the world, in full color. Disorders can affect all parts of this surprisingly tough structure.

EYELID DISORDERS

The eyelids can be affected by irritation or infections on their surfaces, at the margins, or in their internal structures.

The eyelids protect the eye surface, both directly and by spreading tears and lubricating fluid across it. The most common disorders affecting them are inflammation at the margins, known as blepharitis, styes, and chalazia.

Blepharitis results from infection of the follicles at the roots of eyelashes, usually by staphylococci (the bacteria that commonly cause conjunctivitis) or fungi (often associated with seborrheic dermatitis, a type of eczema). It causes a gritty, irritated feeling, but can be relieved by cleaning the eyelids, ideally with diluted baby shampoo, and warming the eyelid margin to melt and release trapped sebum.

Rosacea, a skin inflammation common in older women, may cause blockage of the eyelid glands, producing a similar result. Styes and chalazia are infected glands in the eyelid, which cause red, painful lumps. Styes develop in the sebaceous glands at the edge of the eyelids. Chalazia develop in Meibomian glands (tiny glands that secrete oily fluid to lubricate the eye) and are bigger and farther from the edge than styes. Both disorders usually get better with the application of warm compresses. Good eyelid care, careful removal of eye makeup, and regularly replacing mascara, if used, may help prevent all of these conditions.

Stye on eyelid
A familiar eyelid disorder, a stye causes pain when blinking, and is sometimes accompanied by a a discharge. Styes are more common in people with seborrheic dermatitis, a type of eczema.

INFLAMMATION OF THE EYE SURFACE

The conjunctiva is a sensitive layer of cells covering the sclera (the white of the eye), inner eyelids, and cornea, and can suffer damage for a variety of reasons.

Infective conjunctivitis may be caused by bacteria (usually staphylococci) or by viruses (often adenoviruses). People who wear contact lenses are particularly vulnerable. Chemical conjunctivitis results from irritants coming into contact with the eye surface. Many chemicals can irritate the eye, such as chlorine, used in swimming pools, and pyruvic acid, released when cutting onions. Allergic conjunctivitis is often caused by pollens, in which case it is seasonal (hay fever), but may occur throughout the year if it results from other forms of allergy.

Atmospheric irritation such as wind, heat, solar radiation, ultraviolet light, and dust may cause progressive damage to the cornea, leading to thickening and degeneration. These changes may result in pinguecula, an area of yellowish thickening, or pterygium, a lumpy growth on the surface of the eye. Surgery may be needed if these areas spread across the cornea.

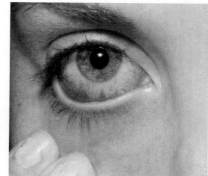

Conjunctivitis
Conjunctival inflammation is common and causes a sore, itchy, red eye, often with sticky or crusty discharge, but without any real impairment of vision or focusing ability.

GLAUCOMA

A common cause of visual loss, glaucoma often runs in families and generally becomes more common with age.

Normally, fluid is secreted into the front of the eyeball by a structure called the ciliary body, to nourish the tissues and maintain the shape of the eye. Excess fluid drains away through a gap called the drainage angle. In glaucoma, the system that allows this movement of fluid becomes blocked, and fluid builds up in the eye. Raised pressure in the eyes is a common risk factor for glaucoma; however, most people with increased pressure in the eye do not go on to develop the condition. Glaucoma may be chronic (long-term) or acute (short- lived). Chronic glaucoma is painless and can go unnoticed for years. Increased pressure within the eye reduces the blood supply to the optic and retinal nerves, causing progressive damage to these nerves and areas of visual loss.

In acute glaucoma, the pressure in the eye rises quickly, because the iris bulges forward and blocks the drainage angle. It causes severe pain and sudden loss of vision, and is a medical emergency, but a small surgical procedure will relieve it. Acute glaucoma is more common in farsighted people because the eyeball is smaller and more prone to structural and functional problems.

LOCATION

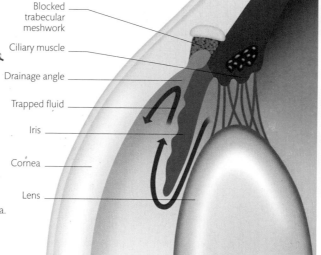

Blocked trabecular meshwork

Ciliary muscle

Drainage angle

Trapped fluid

Iris

Cornea

Lens

Chronic glaucoma
Usually fluid continually flows out through the pupil and drains out of the trabecular meshwork, a sievelike structure located between the iris and the edge of the cornea. In chronic glaucoma, the meshwork becomes blocked and pressure builds up in the eye.

LENS PROBLEMS

The most common disorder of the lens is cataract, in which the lens becomes clouded and cannot focus light properly.

The lens is a clear, rounded structure suspended between the front and rear chambers of the eye, and changes its shape to focus light precisely on the retina. A cataract is a clouding of the lens from clear to milky-white. Symptoms include blurred or distorted vision and "dazzle" around lights, and, if untreated, blindness. Cataracts may be caused by eye trauma, drugs (such as long-term corticosteroid use), overexposure to environmental irritants such as ultraviolet light and solar radiation, or changes due to aging. Most cataracts are treated by surgery, which consists of breaking up and removing the center part of the lens and replacing it with a plastic lens, to allow restoration of useful vision.

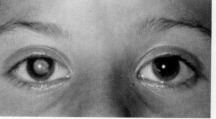

Cataract (cloudy lens)
A cataract may affect just one eye, or both eyes may be affected, one more severely than the other. Here, the right eye has a dense cataract, making the entire pupil appear opaque.

FOCUSING PROBLEMS

The most common disorders of vision are refractive (focusing) errors and can often be corrected by wearing glasses.

The lens is the main structure involved in focusing light, although the cornea and the fluid in the eye also play a part. In particular, the lens is responsible for accommodation, or adjustment of focus between near and far objects. This is achieved by the ciliary body, a ring of muscle that contracts to make the lens rounder, or relaxes to make it flatter.

The ability of the lens to change shape declines with age, partly due to age-related stiffening of the lens and partly due to lessening in the power of the ciliary muscles. By the age of 60 most people cannot achieve near focus (used for reading) without wearing glasses or contact lenses. This condition is called presbyobia. It differs from near- and farsightedness, which affect all aspects of vision. In farsightedness (hypermetropia) either the eyeball is too short, the lens not round enough, or the cornea not curved enough. As a result, light rays are not focused exactly onto the retina, but the "focus point" would actually lie behind the eye, so vision is blurred. In nearsightedness (myopia), the opposite occurs: the light rays converge in front of the retina because the eyeball is too long, the cornea too curved, or the lens too powerful for the length of the eye.

The degree of myopia is measured by the strength of the spectacle lens needed to correct it. In very high myopia there is an increased risk of retinal detachment. Astigmatism, in which there are irregularities in the shape of the lens or the cornea, also impairs focus. These problems can usually be corrected by glasses or contact lenses or through laser eye surgery.

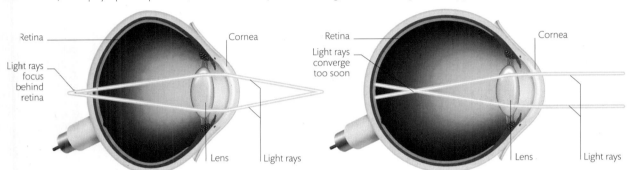

UNCORRECTED FARSIGHTEDNESS

UNCORRECTED NEARSIGHTEDNESS

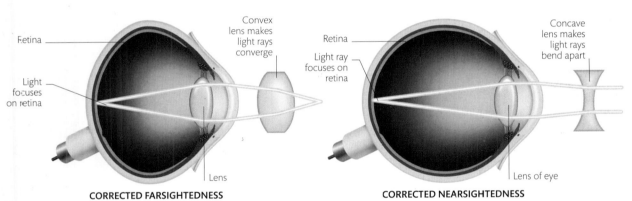

CORRECTED FARSIGHTEDNESS

CORRECTED NEARSIGHTEDNESS

Farsightedness
The eyeball is too short relative to the focusing power of the cornea and lens, so light is focused behind the retina. A convex lens bends the light rays together so that they meet on the retina.

Nearsightedness
The eyeball is too long relative to the focusing power of the lens, so light is focused in front of the retina. A concave lens makes the light rays diverge (bend apart) so that they meet on the retina.

LASER TREATMENT

This procedure is designed to correct nearsightedness, farsightedness, and astigmatism. It involves the use of lasers to reshape the cornea, aiming to eliminate the need for glasses. Laser correction has not previously eliminated the need for reading glasses, since age-related loss of accommodation is not related to lens and corneal curvature. However, new advances mean that this is also becoming possible.

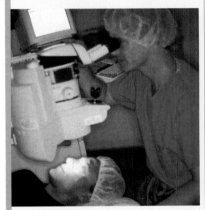

Laser eye treatment
This form of surgery involves opening a flap in the corneal surface and removing some of the tissue from inside, or removing some of the outer layer, to make the cornea flatter.

UVEITIS AND IRITIS

These terms describe inflammation of a group of structures in the eye (the uvea) and the iris, the colored part of the eye.

Both uveitis and iritis cause pain and reduction in vision. There are many possible causes; the most common are inflammatory disorders, such as Crohn's disease, and infection, particularly by herpes viruses, including shingles. Inflammatory joint conditions such as rheumatoid arthritis can also affect the eye. Symptoms include redness, blurring of vision, and aching in the eye. Iritis and uveitis can permanently impair the vision, causing scarring and sticking of the eye structures, and require treatment by an ophthalmologist.

DISORDERS OF THE RETINA

The retina is a delicate, light-sensitive structure lining the back of the eye. It can become damaged through a variety of disorders and injuries.

The retina receives an image of the world from the focusing structures of the eye and converts this into nerve messages that are sent to the brain. It contains light-sensitive cells, as well as a network of blood vessels that supplies it with nutrients. Depending on the area of the retina affected, any problem has the potential to impair vision. Permanent damage results in loss of vision in the corresponding area of the visual field. One possible cause of damage is impaired blood flow, including blockages and bleeding from ruptured blood vessel linings. This condition, often termed retinopathy, is most common in diabetes mellitus and hypertension (high blood pressure).

Chronic glaucoma may also damage the retina through compression of the surface blood vessels, which leads to restriction of the blood supply. Macular degeneration is a common cause of blindness and results from degenerative change in the retina around the macula (the central point of vision). The retina can also become detached from the back of the eye and its blood supply, for example by injury, so that loss of vision results. Reattachment by laser, if done within a few hours, may successfully restore vision.

Normal retina   Leaking blood vessels

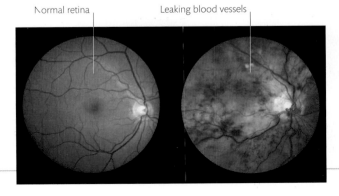

Retinopathy
Magnified retina of a healthy eye (left) and an eye affected by diabetes (right), a common cause of retinopathy. The leaking blood vessels and blockages are evident in the affected eye.

RESPIRATORY DISORDERS

The upper respiratory tract constantly encounters inhaled microbes and often becomes infected. The lower respiratory tract can become irritated and damaged by inhaled agents, especially cigarette smoke—the major cause of lung cancer and chronic obstructive pulmonary disease.

COLDS AND INFLUENZA

Viral upper respiratory tract infections are most common in winter. The common cold is mild and short-lived but influenza can lead to serious complications.

The viruses that cause colds and influenza are airborne, spreading in fluid droplets coughed or sneezed out or in films of moisture transferred to close contacts by sharing objects or shaking hands. Most adults develop the common cold up to four times a year, children more often. It is caused by over 200 different viruses, and, as yet, there is no vaccine. It starts with sneezing and a runny nose with mucus (initially clear, then thicker and darker), then a headache and mild fever may develop along with a sore throat, cough, and sore, reddened eyes. It is relieved by regular intake of fluids and rest.

Influenza is caused by three main types of influenza virus called A, B, and C, and is common. A new vaccine is developed annually to fight the most common strains, and yearly vaccination is important, particularly for those at risk of complications, including people over 65 and those with other health issues. Symptoms include a high fever, muscle aches and pains, coughing, sneezing, sweats, shivers, and exhaustion. It typically lasts a week, but fatigue may persist. Complications include pneumonia, bronchitis, meningitis, and encephalitis. Treatment includes intake of fluids, rest, and antiviral medications.

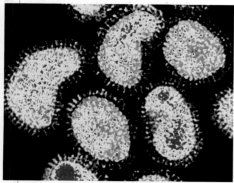

Influenza virus
Colored micrograph of influenza viruses. A core of RNA-genetic material (red) is surrounded by a spiked protein envelope (yellow) that can change its structure to create a new strain of influenza.

RHINITIS AND SINUSITIS

Inflammation of the sinuses and the linings of the nasal cavity may occur together and be acute or chronic. They are due to infection or other causes.

Rhinitis causes a runny nose, sneezing, and nasal congestion. It may be allergic (see p.474), infectious (such as a cold), or vasomotor. In vasomotor rhinitis, blood vessels in the nose are oversensitive and overreact to changes in weather, emotion, alcohol, spicy foods, and inhaled irritants such as pollution. It is treated by avoiding triggers and using nasal sprays.

Sinusitis may be acute—clearing up within 12 weeks, or chronic, lasting for over 12 weeks. Acute sinusitis is most common and typically follows a cold. Symptoms include headache, facial pain, facial pressure when bending forward, a discharge of pus from the nose, and fever. Treatment is with analgesics and decongestants. Antibiotics may be used in bacterial or chronic sinusitis. Chronic sinusitis may sometimes require surgery.

Location of sinuses
There are four pairs of sinuses, which drain through small channels. These channels may become blocked when inflamed, leading to a buildup of fluid and a sensation of pressure.

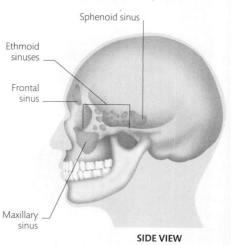

Frontal sinus
Ethmoid sinuses
Sphenoid sinus
Maxillary sinus

FRONT VIEW

Sphenoid sinus
Ethmoid sinuses
Frontal sinus
Maxillary sinus

SIDE VIEW

THROAT DISORDERS

Inflammation in the tonsils or pharynx (throat) leads to a sore throat; in the larynx, to hoarseness; and in the epiglottis, to blockage of the airway.

The pharynx connects the back of the mouth and nose to the larynx (voice box) and the esophagus. Infection of the tonsils (tonsillitis) or pharynx (pharyngitis) can be bacterial or viral. Symptoms include a sore throat, pain and difficulty in swallowing, fever, chills, and enlarged lymph nodes in the throat. Treatment is with rest, fluids, analgesics, and lozenges and sprays. Antibiotics may also be given.

Bacterial infection of the epiglottis (epiglottitis) usually affects children. It causes fever, drooling, hoarseness, and stridor (an abnormal, high-pitched breathing noise). This condition requires urgent medical attention.

Inflammation of the larynx (laryngitis) can be due to infection, overuse of the vocal cords, gastroesophageal reflux disease, or excessive smoking, alcohol, or coughing. Laryngitis causes hoarseness or inability to speak. When due to infection, there may be fever and flu or cold symptoms. Chronic laryngitis is treated by addressing the underlying cause, resting the voice, and voice therapy. Chronic laryngitis can cause white plaques (leukoplakia) to develop on the vocal cords. These may turn cancerous, so specialist treatment is required. Hoarseness or a change in voice lasting for more than 3 weeks needs treatment by a specialist.

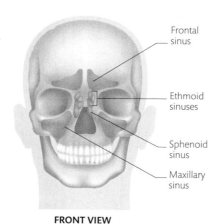

Pharynx
Tonsils
Larynx

Sites of upper respiratory tract infections
Most infections of the nose, sinus, pharynx, and larynx are caused by viruses and do not respond to antibiotics. However, patients with underlying lung disease may be treated with antibiotics.

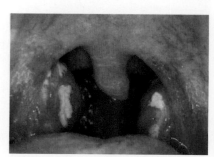

Tonsillitis
The tonsils are swollen and have white, pus-filled spots on them. For recurrent episodes or when swallowing becomes impossible, surgical removal of the tonsils (tonsillectomy) may be performed.

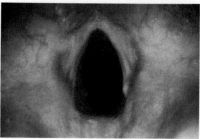

Laryngitis
Endoscopic view of the inside of the larynx, which is inflamed due to acute infection. The vocal cords are the paired central white structures that resonate to produce vocal sounds.

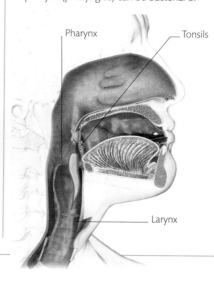

ACUTE BRONCHITIS

Inflammation of the bronchi is usually due to infection by viruses or bacteria, causing a hacking cough, and typically clears up within 2 weeks.

Acute bronchitis typically follows a cold or flu and is more common in smokers. It starts with a dry cough that a few days later becomes "productive," bringing up green, yellow, or gray sputum. Symptoms may include a general feeling of being unwell, fatigue, fever, shortness of breath, and wheeziness. Sometimes a chest X-ray may be needed and sputum sent off for microbiological analysis. Because 90 percent of cases are viral, antibiotics are usually not needed. People with bronchitis are advised to stop smoking, drink fluids, and rest. Noninfective bronchitis may be caused by lung irritants such as smog, tobacco smoke, and chemical fumes.

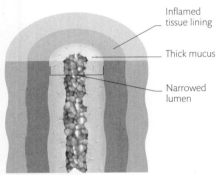

Inflamed tissue lining

Thick mucus

Narrowed lumen

Inflamed bronchus
Infection of the mucosa leads to inflammation that narrows the lumen, along with excessive production of mucus that is filled with white blood cells to fight the infection.

SPIROMETRY

A lung function test called spirometry measures the volume and speed of the air during inhalation and exhalation. Peak expiratory flow rate (PEFR) of air from the lungs gives a measure of obstruction in the airways. Regular monitoring using this test may be carried out in people with asthma (see p.464) and COPD (see right) to measure disease activity and any response to treatment.

CHRONIC OBSTRUCTIVE PULMONARY DISEASE

Chronic obstructive pulmonary disease (COPD) refers to long-term narrowing of the airways, causing obstruction to airflow through the lungs, which leads to shortness of breath. It consists primarily of chronic bronchitis and emphysema, which often coexist in the same person and are usually caused by smoking, or less commonly, by occupational exposure to dusts or fumes (for example, in the mining or textile industry).

CHRONIC BRONCHITIS

Chronic inflammation of the bronchi with excessive mucus production leads to obstruction in the airways of the lungs and a cough that produces sputum.

Chronic bronchitis is defined clinically as a persistent cough that produces sputum for at least 3 months in 2 consecutive years. It is most common in men over 40 years of age who have smoked regularly over a long period of time. Typically, the cough is worst in damp, cold weather, producing a clear white sputum.

Over time, increasing shortness of breath develops and there are frequent and repeated chest infections, with the sputum turning green or yellow, accompanied by worsening of the shortness of breath and wheezing. Eventually there is progressive heart and respiratory failure (which has a poor prognosis) causing weight gain, cyanosis (a blue tinge to lips and fingers), and swollen ankles (edema). Investigations include blood tests, lung function tests, chest X-ray, and analysis of the sputum. Supplemental oxygen may be needed, and inhalers are often prescribed to relax the muscle of the bronchial walls, but airway obstruction is often irreversible. Smoking cessation is vital. Oral corticosteroids may help acute exacerbations. Annual flu vaccination is recommended. Chest infections in chronic bronchitis are usually viral, but antibiotics are used if a bacterial infection is suspected. Many people benefit from disease education, physical training, nutritional assessment, and psychological intervention.

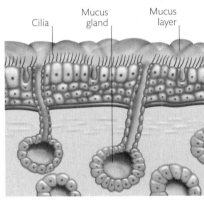

Cilia | Mucus gland | Mucus layer

Normal airway lining
Glands produce mucus to trap inhaled dust and microbes. Tiny hairs (cilia) on the cells move to propel the mucus up into the throat, to be coughed up or swallowed.

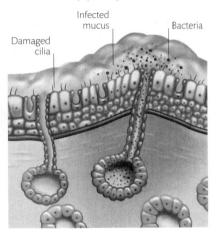

Infected mucus | Bacteria

Damaged cilia

Airway in chronic bronchitis
The mucosa is swollen and there is excessive mucus production, leading to airway obstruction. The cilia are damaged so mucus is not propelled along adequately, thus encouraging infection.

EMPHYSEMA

Destruction of alveolar (air sac) walls, caused by emphysema, reduces the areas for gas exchange and causes the small airways to collapse during exhalation.

Emphysema is usually caused by smoking but can be due to a rare inherited disorder called alpha-1 antitrypsin deficiency. It is most common in men over 40 who have smoked over a long period of time. Emphysema causes progressive shortness of breath. A cough without sputum may occur in the late stage. People with emphysema lose weight; their lungs over-inflate, leading to a characteristic barrel-shaped chest; and they often breathe through pursed lips.

Diagnostic investigations include arterial blood gas analysis, lung function tests, and chest X-ray. A CT scan may show characteristic holes (bullae) in the lungs. To prevent further irreversible progression, smoking cessation and avoidance of cigarette smoke and lung irritants is vital. Treatments include short- and long-acting inhalers that act on bronchial muscles to dilate (widen) the airways, inhaled steroids, and oral corticosteroids. The person may need supplemental oxygen from time to time or continuously. Gastric reflux and allergies may exacerbate the condition. In severe cases, lung volume reduction surgery or lung transplant may be offered. Pulmonary rehabilitation—disease education, advice, and physical training to improve lung function—is often beneficial. Annual flu vaccination is recommended.

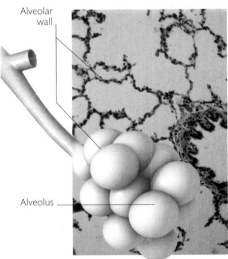

Alveolar wall

Alveolus

Healthy tissue
The alveoli (air sacs) in the lungs are grouped like a bunch of grapes. Each air sac is partly separate from the others. Their elastic walls help to push air out during exhalation.

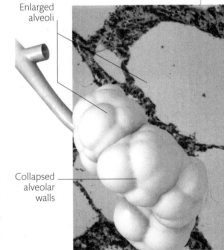

Enlarged alveoli

Collapsed alveolar walls

Damaged tissue
The alveolar walls have been destroyed, with a resulting decrease in elasticity. The alveoli are enlarged and fused together, reducing the available area for gas exchange.

ASTHMA

Reversible narrowing of the airways of the lung, asthma is due to long-term inflammation and leads to episodes of chest tightness and shortness of breath.

Asthma affects around 7 percent of people and often runs in families. It often starts in childhood but can develop at any age. People with asthma have recurrent attacks when the muscle in the walls of the airways contracts, causing narrowing. The narrowing of the airways is reversible, and some people with asthma only rarely experience symptoms, usually in response to the common asthma triggers such as allergens (dust mites, pet

dander, and pollens), medications, exercise, viral upper respiratory tract infections, stress, inhaled dusts or chemicals.

An asthma attack causes the sudden onset of shortness of breath, chest tightness, wheezing, and coughing. Between attacks, some people may have much milder symptoms such as chronic coughing at night, mild chest tightness, and shortness of breath on exertion. Asthma is usually confirmed by spirometry testing and peak flow readings (see p.463) that confirms the reversibility of the airway narrowing. Treatment includes avoiding triggers and use of inhaled medication to relieve symptoms. Mild asthma requires short-acting reliever inhalers that directly dilate the airways. Regular inhaled steroids (preventer inhalers) are used for more persistent symptoms. Oral corticosteroids are used for severe cases.

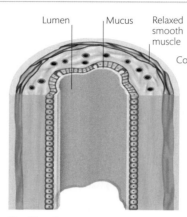

Healthy airway
The smooth muscle is relaxed and does not contract readily in response to triggers. There is a thin coating of mucus covering the lining of the airway. The passageway for air (lumen) is wide.

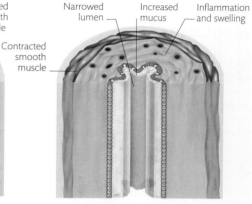

Airway in asthma
The smooth muscle is contracted. The lining of the airway is inflamed and the mucus layer thickened. The lumen is narrowed, causing wheezing and shortness of breath.

PNEUMONIA

Inflammation of the alveoli (tiny air sacs) of the lung, pneumonia is usually due to infection, but it can also be caused by chemical or physical injury.

Infective pneumonia is most common in babies and the very young, smokers, the elderly, and people whose immune system is suppressed. It is caused most commonly by the bacterial infection *Streptococcus pneumoniae* and may affect areas of just one lobe of a lung. Viral pneumonias are commonly due to the viruses that cause colds, flu, and chickenpox. Symptoms include shortness of breath, rapid

breathing, coughing up bloody sputum, fevers, chills, sweating, feeling unwell, and chest pain. A chest X-ray is usually done to confirm the diagnosis. Sputum and blood may be sent for microbiological analysis. Treatment is by taking the appropriate antibiotic. Bacterial pneumonias resolve within a month with treatment, viral pneumonias take longer. Pneumonia that is caused by inhaling any substance into the lung is aspiration pneumonia.

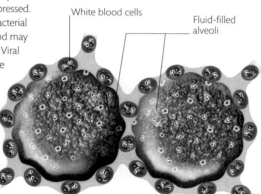

Inflamed alveoli
The air spaces fill with fluid containing white blood cells that kill bacteria. The fluid accumulates and reduces oxygen absorption.

TUBERCULOSIS

A bacterial infection mainly affecting the lungs, tuberculosis (TB) is major global health problem. Around one-third of the world has latent tuberculosis infection.

Tuberculosis (TB) is spread by inhaling tiny droplets of fluid from the coughs or sneezes of an infected person. Most people are able to clear the bacteria, some develop active disease, others develop latent TB with no symptoms but around 10 percent of these will develop the active disease in the future. The bacteria multiply very slowly and may take years to cause symptoms.

Pulmonary TB causes symptoms such as a chronic cough with sputum that may be bloody, chest pain, shortness of breath, fatigue, weight loss, and fever. TB may spread to the lymph nodes, bones and joints, nervous system, and genitourinary tract. It is treated using a combination of antibiotics, over many

months. If left untreated, TB causes the death of half of those infected. Drug-resistant TB is now an increasing problem. Vaccination is rarely used in the US against this condition, and is only 50 percent effective.

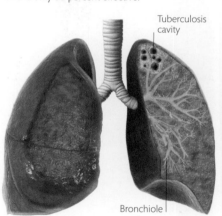

Cavities in the lungs
In active pulmonary TB cavities are often seen in the upper lungs. These are areas of necrosis (cell and tissue death). Passage of air between infected tissue and bronchi releases TB into the airways.

INTERSTITIAL LUNG DISEASE

A variety of diseases can affect the tissue and space around the alveoli and are distinct from obstructive airway diseases.

Most types of interstitial lung disease (ILD) involve fibrosis (development of excess fibrous connective tissue). ILD usually affects adults and may be caused by drugs (such as chemotherapy and some antibiotics), lung infection, radiation, connective tissue disease (for example, polymyositis, dermatomyositis, SLE and rheumatoid arthritis) and environmental or occupational exposure to chemicals such as

silica, asbestos, or beryllium. Sometimes no underlying cause can be found. Symptoms usually develop gradually over many years and include shortness of breath on exertion, a dry cough, and wheezing. The fingernails may become clubbed with increased convexity of the nail fold and thickening of the end part of the finger. Lung function tests, and high resolution CT scan of the thorax are used in the diagnosis. A lung biopsy (tissue sample) may be needed; this is usually done via a bronchoscope (a tube inserted through the airways). Treatment depends on the underlying cause, however the fibrosis is generally irreversible. Specific environmental causes of the disease should be avoided. In occupations where lung disease is a risk, protective clothing and masks should be worn. Smoking cessation is advisable.

SARCOIDOSIS

A multisystem disease, sarcoidosis is characterized by small inflammatory nodules (granulomas) that affect the lungs and lymph nodes.

Sarcoidosis usually affects 20–40 year-olds although it may occur at any age and is most common in northern Europe. It is an autoimmune disease and the exact cause is unknown. Many people with sarcoidosis have no symptoms, some have lung symptoms such as a dry hacking cough and shortness of breath, or eye or skin problems. Typical skin lesions include plaques, erythema nodosum

(reddish, painful, tender lumps), and red or brown papules (raised bumps on the skin). Common eye problems with this condition include uveitis and retinitis (see p.461). General symptoms include weight loss, fatigue, fever, and generally feeling unwell.

Sarcoidosis can affect any organ including the heart, liver, and brain. If the lungs are affected it can lead to progressive lung fibrosis, and around 20–30 percent of those with the condition develop permanent lung damage. Many people do not need any treatment, and the symptoms disappear spontaneously. Severe symptoms are treated with drugs such as corticosteroids. Most people recover fully within 1–3 years, but around 10–15 percent develop chronic sarcoidosis with periods of increased severity of the symptoms and exacerbations.

PLEURAL EFFUSION

Accumulated excess fluid in the pleural cavity, a pleural effusion has a variety of causes and may interfere with lung expansion, causing shortness of breath.

The pleural cavity is the lubricated space between the two pleura (the layers of membrane lining the lungs and innter chest wall). Excess fluid within the cavity causes shortness of breath and, if the pleura is irritated (pleurisy), sharp chest pain typically worse when breathing in. Common causes include heart failure, cirrhosis, pneumonia, lung cancer, pulmonary embolus, TB, and autoimmune diseases such as systemic lupus erythematosus (SLE) and rheumatoid arthritis. The fluid may be removed with a hollow needle, and examined to investigate the underlying cause.

Large effusions may be drained by inserting a tube through the chest wall. Recurrent effusions may be prevented by adhering the pleural surfaces together (pleuradhesis) chemically or surgically.

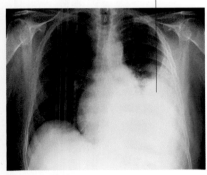

Accumulation of fluid

Pleural effusion
This colour-enhanced chest X-ray shows a large left-sided pleural effusion, which obscures the view of the border of the left side of heart and fills the lower part of the left chest.

PNEUMOTHORAX

A pneumothorax occurs when air or gas enters the pleural cavity and causes the lung to collapse, leading to chest pain and shortness of breath.

A pneumothorax may occur spontaneously (more commonly in tall, thin young men) or following a chest trauma or lung diseases, including asthma, chest infections, tuberculosis, cystic fibrosis, interstitial lung diseases, and sarcoidosis. Penetrating trauma may cause a tension pneumothorax where, with each breath, more air is sucked into the pleural cavity pushing the heart and surrounding structures to the other side of the chest. This can be fatal without urgent treatment and can be confirmed by chest X-ray. Symptoms of the condition include sudden shortness of breath and chest pain. A small pneumothorax may resolve by itself. If a large amount of air has entered the pleural cavity, the lungs need to be decompressed by insertion of a hollow needle through the chest wall or by the insertion of a chest tube.

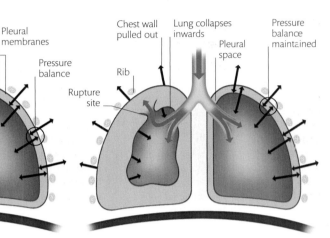

Normal breathing
As the chest wall expands, it lowers the pressure within the pleural space and the lung, acting effectively as a sealed unit, is pulled outwards by the pressure difference.

Collapsed lung
Air from the right lung leaks out into the pleural space and the lung deflates, no longer acting as a sealed unit the lung cannot be pulled outwards by the pressure difference.

PULMONARY EMBOLISM

A blockage to a pulmonary (lung) artery is usually caused by a thrombus (blood clot) breaking away from a deep vein thrombosis (DVT) in the leg.

A pulmonary embolism is a blockage in lung arteries caused by an object not normally found circulating in the blood. Rarely this may be air, fat, or amniotic fluid (in pregnancy) but is usually a clot from a deep vein thrombosis (see p.470). Symptoms include shortness of breath, chest pain worse on breathing in, and the coughing up of blood. Severe cases may cause blueness of the lips and fingers (cyanosis), collapse, and shock. Usually it is diagnosed by specialized CT scanning. Treatment is with anticoagulation ("blood-thinning") drugs (typically heparin and warfarin). Severe cases may require thrombolytics to break up the clot, or the clot may be removed surgically, a procedure known as pulmonary thrombectomy. Untreated, 25–30 percent of people with pulmonary embolism die.

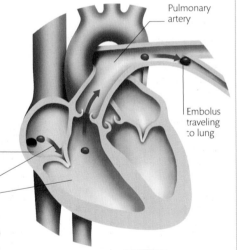

Pulmonary embolism
The clot travels from the deep vein of the legs to the right atrium (chamber of the heart), then into the right ventricle and into the pulmonary artery.

LUNG CANCER

A malignant tumor that develops in the tissue of the lungs is the most common cause of cancer death worldwide.

Primary lung cancer arises from within the lung. There are two main types: small cell lung cancer (SCLC) accounts for 20 percent of all cases, the rest are non-small cell lung cancer (NSCLC). SCLC is more aggressive (spreads faster). Lung cancer occurs mainly in people over 70 and 90 percent of cases are due to smoking. The risk is related to the number of cigarettes smoked and for how long. Breathing in other people's cigarette smoke (passive smoking) is a risk factor for non-smokers. Rarely, lung cancer may be caused by asbestos, toxic chemicals, and radon gas. By the time of diagnosis, most lung cancers have spread elsewhere. Symptoms include a persistent cough or a change in the regular coughing pattern, coughing up blood, chest pain, wheezing, shortness of breath, fatigue, weight loss, loss of appetite, hoarseness, and difficulty swallowing.

Diagnosis initially is made by chest X-ray and a scan of the chest and is confirmed by biopsy (tissue sample) typically taken using bronchoscopy (a tube is passed through the mouth into the lungs). Treatment depends on the type, site, and spread of the tumor. SCLC is usually treated with chemotherapy and radiaton therapy and has the poorer prognosis.

NSCLC is often removed surgically, which may be curative. Only around 25 percent of lung cancer patients survive for more than a year after diagnosis.

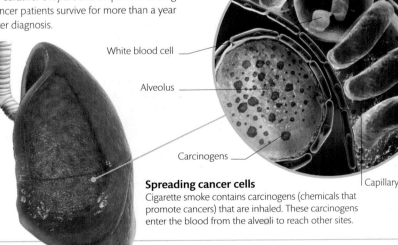

White blood cell

Alveolus

Carcinogens

Capillary

Spreading cancer cells
Cigarette smoke contains carcinogens (chemicals that promote cancers) that are inhaled. These carcinogens enter the blood from the alveoli to reach other sites.

CARDIOVASCULAR DISORDERS

The heart and circulatory system are affected by many diseases, and cardiovascular disease is the leading cause of death in the US. Lifestyle factors such as diet are important risk factors, but some disorders result from structural abnormalities such as defects in the heart valves.

ATHEROSCLEROSIS

Fatty deposits and inflammatory debris, deposited as plaque on artery walls over many years, lead to atherosclerosis or narrowing of the arteries.

Atherosclerosis can begin in childhood, even in healthy people, although risk factors including high cholesterol, smoking, obesity, high blood pressure, and diabetes increase its rate of development. Fatty deposits build up in artery walls, forming clumps, or plaques, known as atheromas. These plaques stimulate inflammation that damages an artery's muscle wall, causing it to thicken. Blood flow is restricted, and tissues beyond that point are starved of oxygen and nutrients. Eventually the plaque may break off in the artery, blocking the blood flow completely. In the coronary arteries (which supply the heart), atherosclerosis can cause angina or a heart attack; in the brain, stroke or dementia; in the kidneys, kidney failure; and in the legs, claudication. The disorder can be slowed, halted, or even reversed by stopping smoking and lowering cholesterol and blood pressure.

Atheromatous plaques
The fatty deposits and inflammatory reaction in the artery lining cause a restriction within the blood vessel before eventually rupturing, blocking the artery completely.

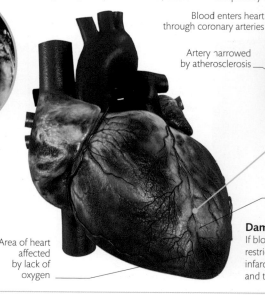

Fatty deposit

Red blood cell

Arterial branch junction

Fatty core of plaque

Fibrous cap

Narrowed arterial channel

Outer protective layer of artery

Muscle layer of artery

Inner lining of artery

Restricted blood flow
Atherosclerosis can often start in a damaged area of artery wall. As a plaque forms and the wall becomes inflamed the area thickens, reducing the space inside and restricting blood flow.

ANGINA

Inadequate blood supply to the heart itself, from the coronary arteries, can lead to angina—pain resulting from too little blood reaching the heart muscle.

Angina is usually caused by narrowing of coronary arteries due to atherosclerosis (see left), but a thrombus (clot), artery wall spasm, anemia, exertion, fast heart rate, and other heart disease may also be factors. Angina is felt in the chest, neck, arms, or abdomen, often with associated breathlessness. It usually comes on with exertion and eases with rest or use of vasodilator drugs (drugs that widen the arteries and let blood flow more easily). Longer-term treatments include lifestyle changes, control of atherosclerosis, nitroglycerine, aspirin, and a beta-blocker drug to protect the heart. Occasionally, surgery or angioplasty are needed to widen or bypass narrowed arteries.

Why angina occurs
Pain arises when part of the lumen (inner channel) of a coronary artery becomes so narrow, due to atheroma and spasm, that the area it supplies is temporarily starved of blood and oxygen.

Blood enters heart through coronary arteries

Artery narrowed by atherosclerosis

Blood supply to heart muscle is reduced

Area of heart affected by lack of oxygen

Damaged heart muscle
If blood flow and oxygen supply are restricted over a long time by myocardial infarction, some heart muscle fibers die, and the pumping action is impeded.

ANGIOPLASTY

This procedure is used to widen narrowed arteries in the heart and elsewhere in the body. Angioplasty is often used to treat severe angina or after a heart attack. Under local anesthetic, a tiny balloon is inserted into the artery to push open the narrowed area. A mesh tube called a stent may also be inserted to hold the artery open.

There are several techniques, and types of stent, used for a variety of atherosclerotic problems. Some stents are coated with drugs to help prevent plaques from forming again. Aspirin or other anti-clotting drugs are given following angioplasty, to reduce the risk of clots.

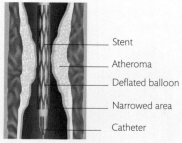

Stent

Atheroma

Deflated balloon

Narrowed area

Catheter

1 Catheter inserted
A guide catheter is fed through an incision in an artery in the leg or arm until its tip reaches the coronary artery. It carries a balloon catheter, covered by a stent, to the narrowed area.

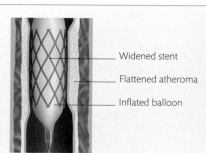

Widened stent

Flattened atheroma

Inflated balloon

2 Balloon inflated
The positioning of the balloon catheter is monitored by X-ray imaging. Once in the correct place the balloon is inflated, expanding the stent and pushing the artery open.

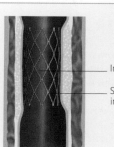

Increased flow

Stent remains in place

3 Catheter removed
Once the stent has been expanded to the correct width, the balloon is deflated and the catheter is withdrawn. The stent remains in place and the catheter is removed from the body.

HEART ATTACK

A myocardial infarction (MI) or heart attack is caused by complete blockage of a coronary artery or one of its branches.

The term myocardial infarction means death of part of the heart muscle. When a coronary artery becomes blocked, usually from a ruptured atheromatous plaque or a thrombus (clot), the area of muscle that it supplies is starved of oxygen and dies. The extent of damage and complications depend on the artery involved; larger arteries supply larger areas of muscle, and MI of large arteries are more likely to cause death.

An MI typically causes central chest pain, although those with diabetes may experience no symptoms at all ("silent MI"). The diagnosis is confirmed by ECG (a trace of the electrical activity of the heart) and raised blood levels of cardiac enzymes—chemicals released by the damaged muscle. Urgent treatment with "clot-busting" drugs or angioplasty can clear the blockage, restoring blood flow. Other treatments include beta-blocker drugs to protect the heart from arrhythmias (see below) and aspirin to prevent further clots.

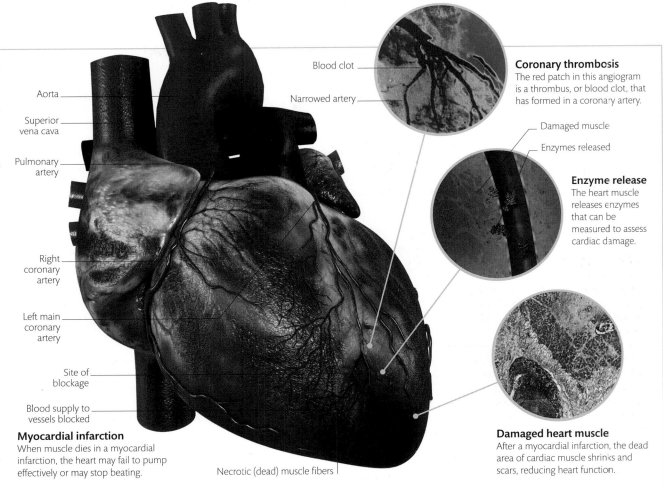

Aorta

Superior vena cava

Pulmonary artery

Right coronary artery

Left main coronary artery

Site of blockage

Blood supply to vessels blocked

Blood clot

Narrowed artery

Coronary thrombosis
The red patch in this angiogram is a thrombus, or blood clot, that has formed in a coronary artery.

Damaged muscle

Enzymes released

Enzyme release
The heart muscle releases enzymes that can be measured to assess cardiac damage.

Myocardial infarction
When muscle dies in a myocardial infarction, the heart may fail to pump effectively or may stop beating.

Necrotic (dead) muscle fibers

Damaged heart muscle
After a myocardial infarction, the dead area of cardiac muscle shrinks and scars, reducing heart function.

HEART RHYTHM DISORDERS

An abnormal heart rate or rhythm is caused by a disturbance in the electrical system that controls the way the heart muscle contracts.

The signal to the heart to contract is driven by electrical pulses from the sinoatrial (SA) node, a natural "pacemaker" in the right atrium. It travels across both atria (upper chambers) via the atrioventricular node, through the septum, and across the ventricles (lower chambers). Arrhythmias (abnormal heart rhythms) occur because of poor signal transmission or abnormal electrical activity. In atrial fibrillation (AF), one of the most common forms of arrhythmia, abnormal "pacemaker" sites override the SA node, producing a contraction pattern that is not effective in pumping blood.

AF may be treated by electrically shocking the heart back into normal rhythm. In ventricular fibrillation, a medical emergency, the very fast, random contractions of different ventricular areas hinder the pumping of blood from the heart, stopping flow to body tissues including the brain. Immediate defibrillation is needed, with drug therapy to stabilize the heart. Problems such as arrhythmia occur when the signal does not transmit through the usual pathway.

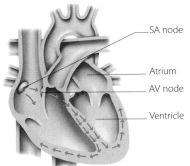

SA node

Atrium

AV node

Ventricle

Sinus tachycardia
In this condition the heart rate of more than 100 beats/minute and a normal rhythm may simply be due to anxiety or exercise, but can also occur in fever, anemia, and thyroid disease.

Irregular impulses through atria

Variable blockage at AV node

Atrial fibrillation
If the sinoatrial node is overridden by random electrical activity in the atria, impulses pass through the atrioventricular node erratically, causing fast, irregular ventricular contractions.

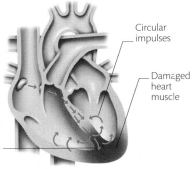

Blockage

Some impulses cross from healthy side

Bundle-branch block
The sinoatrial node impulses are partially blocked, slowing the ventricular contractions. In total heart block, no impulse gets to the ventricles, so they contract at a rate of only 20–40 beats/minute.

Circular impulses

Damaged heart muscle

Slowed conduction through damaged area

Ventricular tachycardia
Abnormal electrical impulses in the ventricular muscle cause the ventricles to contract rapidly, overriding the sinoatrial signal and resulting in a fast regular, but inefficient beat.

HEART FAILURE

Failure of the heart to pump blood effectively can occur as a result of a heart attack, valve damage, or drug therapies used for other medical conditions.

The heart pumps blood to the lungs to pick up oxygen, and to the tissues to deliver oxygen and nutrients. When the heart fails as a pump, it causes symptoms of breathlessness, fatigue, and edema (excess fluid in the tissues). In addition, organs such as the liver and kidneys do not receive enough blood and start to fail. Heart failure can be acute (sudden), often resulting from a heart attack, or chronic (long-term), due to persistent disorders such as atherosclerosis, hypertension, chronic obstructive pulmonary disease, and heart valve disease. It is classified according to the area affected and the phase of the pumping cycle.

In most cases fluid accumulates in the lungs (left-sided ventricular failure). In right-sided failure (which often follows left-sided failure), fluid builds up in the feet, legs, peritoneum, and the abdominal organs. Treatment for acute heart failure includes oxygen and diuretics to remove some of the edema, and medication to help the heart muscles contract. Chronic heart failure is treated with beta-blocker and ACE inhibitor drugs, and by working to control the underlying cause of the problem.

HEART MURMURS

Caused by turbulent blood flow, heart murmurs may signify diseased valves or abnormal blood circulation within the heart.

Unexpected sounds heard when listening to the valves closing or blood flowing through the heart are known as heart murmurs. Common causes include valve defects, such as a valve that is too tight or floppy, or that does not close properly. Congenital defects that produce abnormal blood flow include a hole in the heart (a gap in the wall between two heart chambers) and patent ductus arteriosus (the remnant of a vessel that carries blood in an unborn baby's heart but should close up just after birth). Heart murmurs may also occur in pregnancy or conditions such as anemia, even though the heart is normal. The sound gives a clue to the cause, but echocardiogram (ultrasound of the heart) is done to confirm the type of defect. Most conditions that cause heart murmurs do not need treatment unless the underlying problem causes any symptoms. Then surgery may be done to repair any defects.

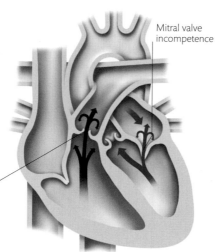

Mitral valve incompetence

Pulmonary valve stenosis

Abnormal flow
Normally, blood flows into and out of the heart via one-way valves. The flow through diseased valves is disturbed, passing through at overly high pressure, or leaking backward through the valve.

INFECTIVE ENDOCARDITIS

A serious infection of the endocardium (the heart's internal lining), endocarditis may occur after valve replacement.

If a heart valve is diseased or has been replaced, bacteria in the bloodstream can stick to its surface, causing an infection that spreads to the endocardium. The area over the valve becomes inflamed, and infected material and blood clots may collect there. Symptoms of endocarditis include persistent fever, fatigue, and breathlessness. Diagnosis is by blood tests, physical examination, and echocardiogram; an ECG might be done to monitor the electrical activity of the heart. Endocarditis can be life-threatening and needs urgent treatment. Antibiotic drugs may be given for six weeks, until the infection has cleared up. If the endocarditis persists, the valve may need to be surgically repaired or replaced (see right).

HEART VALVE DISORDERS

The four heart valves allow blood to flow in the correct direction around the heart, but disease can harden or weaken them.

The valves of the heart are located between the atria (upper chambers) and ventricles (lower chambers), and at the points where blood leaves the ventricles. Their function can be impaired by congenital defects, infections such as rheumatic fever and endocarditis, and atherosclerosis. Stiffness of a valve (stenosis) makes the heart pump harder to push blood past the obstruction, while floppiness (incompetence) forces the heart to do extra work to pump the required volume through, as some leaks back. In both cases, the strain causes the heart to enlarge and become less efficient. This may lead to heart failure (see p.467); valve disease also increases the risk of clots and stroke. The type of valve defect may be identified by ECG, X-ray, or echocardiogram. Drugs to relieve strain on the heart can help, but if symptoms persist, surgery may be needed to repair or replace the valve.

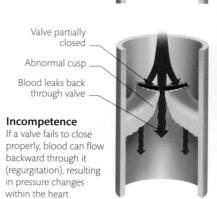

Valve tightly closed

Cusp

Normal valve closed
The pressure outside the closed valve builds, and the valve cusps snap shut so that blood cannot flow backward.

Valve partially closed

Abnormal cusp

Blood leaks back through valve

Incompetence
If a valve fails to close properly, blood can flow backward through it (regurgitation), resulting in pressure changes within the heart.

VALVE SURGERY

There are several procedures for repairing or replacing a damaged heart valve. Repair techniques include valvuloplasty or valvotomy, used to open a stenosed valve. A valve may be replaced with one from a human donor or an animal, or an artificial valve. Another procedure is percutaneous aortic valve surgery, in which a new valve is inserted inside a diseased aortic valve.

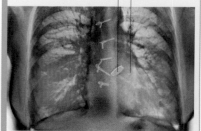

Artificial aortic valve   Heart

Heart valve
Color-enhanced chest X-ray showing an artificial heart valve. The green loops show where the sternum (center of the chest) has been repaired following open-heart surgery.

CONGENITAL HEART DISEASE

Heart abnormalities that are present at birth affect about 8 in 1,000 babies; most of these defects are minor, but some are life-threatening.

The development of the heart in a fetus is complex, and many types of abnormalities can occur. The heart valves may not grow properly, leading to problems such as pulmonary stenosis (narrowing of the valve that allows blood to flow to the lungs). There may be holes in the chamber walls, as in septal defects (hole in the heart), or even absent chambers. The large vessels leading to and from the heart may be abnormal in shape, size, or location, with coarctation of the aorta (narrowing of part of the aorta). In patent ductus arteriosus, a blood vessel that should close off at birth may stay open, causing "shunts" of blood in the wrong direction. Several abnormalities may be present, as in tetralogy of Fallot (pulmonary stenosis, ventricular septal defect, displaced aorta, and thickened right ventricle). Possible causes of developmental problems include chromosomal abnormalities; illness in the mother during pregnancy, which affects the growth of the baby's heart; and the mother's use of medications, drug abuse, alcohol, or tobacco.

Congenital heart disease may be diagnosed during the pregnancy if the fetus is small for its gestational age, or after delivery, if the baby is cyanotic (blue from lack of oxygen). Treatment depends on the defect, the age and condition of the person affected, and the presence of other disease. There is great variation, from extreme defects, which need immediate and possibly repeated surgery, to minor valve defects, which may not become obvious until old age.

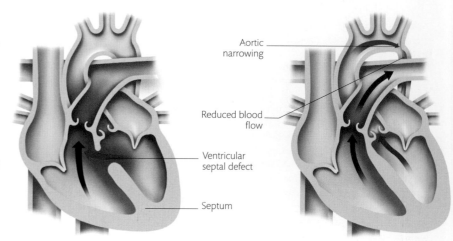

Aortic narrowing

Reduced blood flow

Ventricular septal defect

Septum

Ventricular septal defect
A third of congenital heart defects involve the ventricular septum (the wall between the lower chambers): blood is shunted back from the left to the right ventricle through a hole in the septum.

Coarctation of the aorta
A narrowing of the aorta (major artery from the heart), coarctation causes abnormal circulation patterns with altered blood pressure and flow, including poor blood flow to the lower body.

HEART MUSCLE DISEASE

Many diseases can affect heart muscle, but the cardiomyopathies are four types of disorders.

These disorders are classified by the changes that they produce in the heart muscle: hypertrophic (thickening); dilated (stretching); restrictive (stiffening); and arrhythmogenic (in which fatty and fibrous deposits interfere with the pumping action, causing abnormal beats). There may be a genetic link, or some cases may be associated with specific factors—for example, hypertrophic cardiomyopathy may be linked to high blood pressure, and dilated cardiomyopathy to excess alcohol use. In all forms, the changes lead to inefficient pumping action and heart failure, with symptoms including chest pain, breathlessness, fatigue, and edema (excess fluid in tissues). Treatment includes drugs to reduce fluid and improve heart function. Surgery may help, but the final option is heart transplantation.

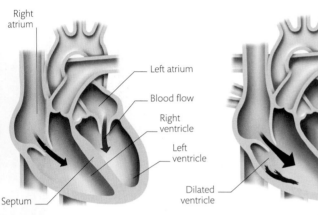

Normal heart
Healthy circulation depends on efficient muscle contractions pumping blood from the right side of the heart to the lungs to be oxygenated, then through the left side to the body tissues.

Dilated cardiomyopathy
If the muscle fibres weaken, the ventricles may expand (dilate) and become floppy. As a result, the heart pumps blood less forcefully, and this loss of efficiency can lead to heart failure.

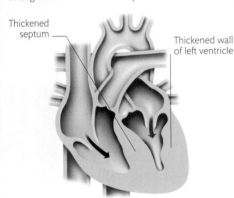

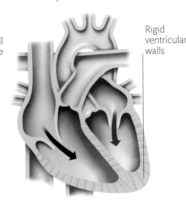

Hypertrophic cardiomyopathy
Thickening (hypertrophy) of the muscle, often around the left ventricle or septum, prevents the chambers from filling as normal and causes the valves to leak, so the heart's output reduces.

Restrictive cardiomyopathy
The heart muscle cannot relax properly between heartbeats, because diseased muscle fibres make the ventricular walls unusually rigid and unable to fill properly or pump effectively.

PERICARDITIS

Inflammation of the pericardium, the twin membrane surrounding the heart, can restrict the heart's pumping action.

Pericarditis is a response to damage, infection, myocardial infarction, or other inflammatory disease such as rheumatoid illness. It can be acute (of sudden onset), or chronic (persistent), causing scarring of the membrane. Fluid may collect between the two layers. Symptoms include chest pain, breathing difficulties, cough, fever, and fatigue. If pericarditis is suspected, ECG, chest X-ray or other imaging, and blood tests may be done. Drugs may be given to reduce inflammation, and excess fluid is drained. If scarring causes constriction, surgery may be needed to release the pericardium.

Pericardial effusion
A buildup of fluid between the two layers of the pericardium can prevent the heart from expanding fully.

HYPERTENSION

Commonly defined as high blood pressure, hypertension slowly damages the heart, blood vessels, and other tissues, but is usually easy to treat.

Normal blood pressure results from the heart forcing blood around the circulatory system. It varies with age, but in hypertension the blood pressure is constantly higher than the recommended level. There are rarely any symptoms, but if it is left untreated the heart becomes enlarged and less efficient as a pump. The long-term effect on other tissues includes damage to the eyes and the kidneys and an increased risk of heart attack and stroke. Causes of hypertension include a genetic tendency to the condition, too much dietary salt, smoking, being overweight and inactive, and drinking too much alcohol. Stress may also be a factor. Secondary hypertension occurs as a result of kidney, hormonal, or metabolic disease, or as a side effect of other medications. Hypertension can be controlled by changes to diet, and by drugs that remove excess fluid or reduce arterial wall tension, to reduce blood pressure. Other treatments, such as cholesterol-lowering medication and aspirin, are used to reduce cardiac risk.

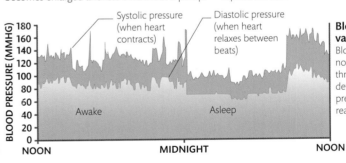

Blood pressure variation
Blood pressure normally varies through the day; to detect persistent high pressure, several readings are taken.

PULMONARY HYPERTENSION

Abnormally high blood pressure in the arteries that carry blood to the lungs is difficult to treat and can be fatal.

Normally, blood passes from the right side of the heart through the pulmonary arteries at low pressure. If the pressure becomes too high, the right side must pump harder, and over time the ventricle thickens and heart failure develops. Pulmonary hypertension can develop after chronic heart or lung disease. There is a genetic link in some families, and in others a link to other disorders, but often the cause is unknown. Symptoms include chest pain, breathlessness, fatigue, and dizziness. Oxygen therapy and drugs to improve blood flow can help improve heart function and reduce clotting problems, but there is no cure. Lung transplantation is an option if medication fails.

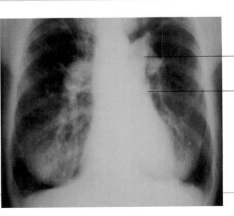

Effects of pulmonary hypertension
Increased pressure in the pulmonary arteries causes them to thicken. In this X-ray the right ventricle is visibly enlarged as a result of working harder to pump blood to the lungs.

PERIPHERAL VASCULAR DISORDERS

The peripheral vascular system includes arteries, which carry blood from the heart to all of the body tissues, and the venous system, returning blood to the heart. Any part of the system can be damaged by disease, which may then affect other organs and tissues.

ANEURYSM

An aneurysm is a swelling in an artery; if it affects the aorta, the body's main artery, it can be life-threatening.

Defects in part of the artery wall weaken the area so that, under pressure from blood flow, it stretches and may burst. Aneurysms can occur in any artery, but the aorta is more prone to problems, and the risk of death from hemorrhage is greatest. Thoracic aneurysms occur near the heart, but aneurysms are more common in the abdominal part.

Underlying causes include atherosclerotic damage (see p.466) or, more rarely, infection or a genetic disorder. In many cases, aneurysms produce no symptoms and are detected only when they burst, or during other investigations or surgery. Small ones can be monitored, but if they grow too large they may need surgery.

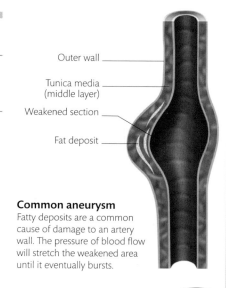

Outer wall
Tunica media (middle layer)
Weakened section
Fat deposit

Common aneurysm
Fatty deposits are a common cause of damage to an artery wall. The pressure of blood flow will stretch the weakened area until it eventually bursts.

Outer wall
Tear in inner wall
Kidney
Swollen wall of abdominal aorta
Blood in false channel
Fatty deposit
Original channel

Dissecting aneurysm
Blood is forced through a tear in the inner wall, creating a false channel between the layers of the wall.

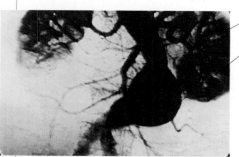

Abdominal aortic aneurysm
In this angiogram (X-ray taken after radiopaque dye has been injected into the bloodstream), the bulging aorta can be seen between the kidneys.

THROMBOSIS

A thrombus, or blood clot, can form in any blood vessel, causing reduced blood flow or blockage, or detach and travel in the circulation as an embolus.

Different types of thrombus (blood clot) may develop anywhere in the body. In veins, thrombi form when blood is flowing sluggishly, if the blood is particularly thick as a result of certain genetic conditions, or where the inner wall of a vein has been damaged and blood sticks to it. In arteries, thrombi usually form where a fatty plaque (atheroma) has damaged the inner wall.

Thrombosis is usually symptom-free until it blocks a blood vessel, when pain, redness, and inflammation occur around the oxygen-starved tissues. Anticoagulant drugs are given to help prevent clotting. If a thrombus is large or cannot be dissolved quickly, surgery to remove it is needed.

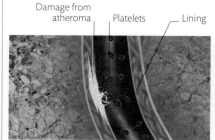

Damage from atheroma
Platelets
Lining

1 How thrombosis begins
Atheromatous plaque forms from a collection of fatty substances, waste products, calcium, and fibrin, a stringy substance that helps blood clot.

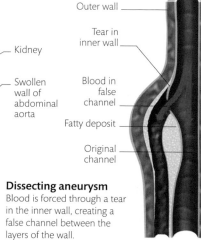

Fibrin strands
Thrombus blocking artery

2 Clot formation
The growing atheroma reduces blood flow and oxygen delivery to the tissues. The plaque ruptures, causing the sudden formation of a clot.

DEEP VEIN THROMBOSIS

Any deep vein can develop a thrombosis (DVT), although it usually occurs in the calf. DVT is due to static or slow-flowing blood and clot formation. The skin over the area is hard, painful, red, and swollen.

Risk factors include clotting diseases; high estrogen levels, as in pregnancy or taking the combined contraceptive pill; and immobility. There is a serious risk that a broken-off piece of clot (embolus) could lodge in an artery in the heart or lung. Treatment includes drugs to limit clotting and perhaps surgery to bypass the clot.

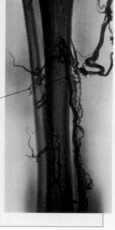

Blood clot

Leg thrombosis
The usual place for a DVT is in the veins deep within the calf. This image shows a clot blocking a vein near the shinbone (tibia).

EMBOLISM

The sudden blockage of an artery by an embolus (a plug of free-floating matter), embolism is serious and can be fatal.

Many emboli are "thromboemboli"—pieces broken off from a blood clot (thrombus) inside a blood vessel. Emboli can also form if fat enters the blood, usually after a fractured pelvis or tibia. Other types include an air embolus, in which air is introduced into the bloodstream

during trauma or surgery, or a foreign body. When an embolus blocks an artery, the tissue supplied by that artery dies. In pulmonary embolism (see p.465), damage to lung tissue results in breathing difficulties, chest pain, and circulatory collapse. Emboli (most commonly thromboemboli) that travel up to the brain can cause a stroke. Fat emboli may affect lung, brain, or skin tissues, while air emboli can be fatal. A suspected embolus requires hospital admission while the type and location of the embolus is determined. Thrombolysis (clot-busting) medication is used to dissolve thromboemboli; surgery may be needed to

remove large blood, fat, or foreign body emboli. Often the embolus is small, but treatment such as anticoagulant drugs to prevent blood clots is given to prevent further emboli occurring from the same source.

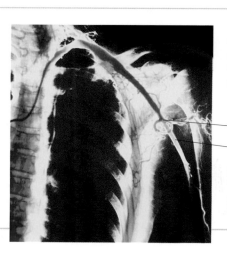

Embolus blocking blood flow
Brachial artery

Embolus blocking an artery
The most common emboli are thromboemboli: pieces from a clot that travel in the bloodstream until they lodge in a smaller artery, as shown here.

LOWER LIMB ISCHEMIA

The lower legs are more prone than other areas to ischemia—oxygen starvation of the tissues—if the blood flow is reduced.

Lower limb ischemia may occur when blood flow in an artery is reduced due to a thrombus (clot) or atheroma (fatty deposit), embolism, or constriction from an injury or local pressure. If ischemia is acute (sudden), as when a large thrombus blocks a major artery, the result is a cold, painful, blue, pulseless leg, which needs emergency treatment to prevent shock and gangrene. Any clot needs to be dissolved by drugs or surgically removed to restore the circulation; if the tissue dies, the only option is amputation.

Chronic (long-term) ischemia may cause intermittent claudication (cramplike pains during exercise), when the muscles do not receive enough oxygen through the narrowed arteries. In these cases atherosclerosis may have partially blocked the arteries, and blood-thinning medication will help blood flow, or angioplasty, stenting, or bypass will dilate the artery again.

RAYNAUD'S DISEASE

The main feature of this condition is Raynaud's phenomenon—constriction of the tiny blood vessels in the extremities.

In Raynaud's phenomenon, the fingers, toes, ears, or nose whiten and cool as the vessels constrict, before turning blue, purple, or black as blood oxygen levels fall. Vessels then dilate again and blood flow increases, turning the tissues red, with pain and throbbing. There may also be joint pain, swelling, rashes, and muscle weakness. Generally the cause is unknown, in which case the condition is defined as Raynaud's disease. In some people, diseases such as rheumatoid arthritis (RA), systemic lupus erythematosus

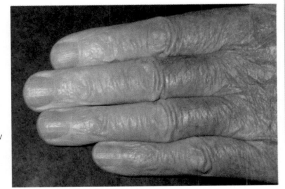

Raynaud's phenomenon
As arteries constrict and blood flow reduces, the extremities turn pale and cool. When the vessels dilate again, pain, numbness, and throbbing are common.

(SLE), scleroderma, or multiple sclerosis cause secondary Raynaud's, or develop after Raynaud's symptoms occur. "Hand–arm vibration syndrome" in workers who use vibrating tools is another cause. In both Raynaud's disease and secondary Raynaud's, attacks may be triggered by cold or stress.

Symptoms can be avoided by keeping the extremities warm, with thermal underwear and heated gloves and socks, and not smoking or using drugs that cause blood vessel constriction. Medication to improve blood flow may be given. Causes of secondary Raynaud's need to be controlled.

VASCULITIS

Inflammation of the blood vessels, or vasculitis, is an uncommon condition but it can affect any organ or body system.

In half of all cases, the cause is unknown, but in the remaining cases the condition results from infection, another inflammatory disease such as rheumatoid arthritis (RA), cancer, some medications, drug use, or contact with chemical irritants. Symptoms depend on the size and location of affected blood vessels. The most common problems are skin lesions, rashes, and ulcers. Internally there may also be bleeding and swelling or blockage of vessels or organs. Blood tests for inflammation and autoimmune diseases, X-rays, and other tests may suggest vasculitis, but it can be confirmed only by tissue biopsy.

Treatment depends on the underlying cause: for example, avoiding any causative medications and treating infection. Further treatment depends on the organs affected and the overall health of the person. In rare cases, surgery is needed to repair damaged large vessels.

VENOUS ULCERS

Usually developing on the lower leg or ankle, venous ulcers are persistent, often painful open sores that are particularly common in older people.

If the walls of the veins become weakened, the circulation will fail to return blood effectively to the heart. As a result, pressure builds up in the veins. This increased pressure causes fluid

to leak out of the veins into the surrounding tissues. The tissues and the skin above them swell, and the skin surface eventually breaks down to form an ulcer. The raw, open tissue may be painful and can become secondarily infected. Without treatment, large areas of skin necrose and die, leaving exposed fat or muscle. Venous ulcers can be identified by their appearance. To assess circulation, the doctor will compare the blood pressure in the ankle to that in the arm, because poor circulation gives a lower ankle pressure. Treatment includes compression bandages applied to

the leg to help blood return to the heart and reduce fluid pressure in the tissues, and elevation of the leg, again to improve blood return. If the ulcer fails to heal, surgery to the vein or the use of skin grafts to cover the ulcer may provide a more permanent solution.

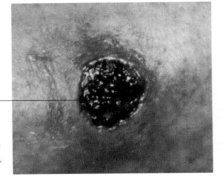

Venous ulcer

Ulceration
Poor blood circulation can result in chronic tissue damage and ulcer formation. Appearing as shallow craters in the skin that expose the underlying tissue, ulcers can be difficult to heal.

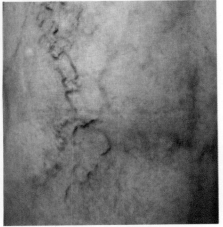

Varicose veins in the leg
Any vein can become varicose, but the most common site is the lower leg, where the swollen, distorted veins may become more prominent if the affected person stands for a long time.

VARICOSE VEINS

Typically visible as lumpy swellings on the legs, varicose veins can run in families and are more common in women.

Normally, muscle contractions in the legs help to push blood through veins back to the heart, and one-way valves in the veins prevent blood from flowing backwards. Varicose veins occur mainly in the legs when the valves fail to close properly, causing backflow to occur and pressure to increase in the veins, making the veins swell. Varicose veins are often caused by increased pressure from abdominal swelling in pregnancy or obesity, or by pressure in the lower legs due to prolonged standing. In rare cases, the vein walls are abnormally elastic or some valves are missing, so the veins are

overstretched by normal blood pressure. Varicose veins may cause no symptoms, or may result in aching, heaviness, itching, and swelling. Diagnosis is usually made by clinical

examination, but specialized ultrasound scanning may also be used to investigate blood flow, especially if there are complications or the problem is recurrent.

TREATING VARICOSE VEINS

Mild varicose veins may need no other treatment apart from surgical stockings to support the vein walls and measures to prevent them from worsening, such as exercise, weight loss, and avoiding standing for long periods. However, varicose veins can be made worse by ulcers, eczema, and swelling of the ankle. Surgery offers some improvement, although the problem may recur. Techniques such as sclerotherapy, radiofrequency, and laser techniques can be used to seal the veins, depending on their severity and location.

Sclerotherapy
During sclerotherapy treatment, veins are injected with a chemical to seal them. They can be highlighted using ultrasound and marked on the skin, as shown here.

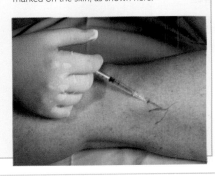

BLOOD DISORDERS

Abnormal numbers and forms of red blood cells, white blood cells, and platelets can occur due to a variety of disorders, including anemia and leukemia. Abnormalities in blood clotting mechanisms result in blood that clots either too readily, leading to thrombosis, or not enough, leading to bleeding and bruising.

ANEMIA

In anemia, there is a reduction in the number of red blood cells or in the concentration of hemoglobin—the pigment in red blood cells that transports oxygen around the body. As a result, anemia can lead to hypoxia (oxygen deprivation) in cells. Different types of anemia are classified by the size of the red blood cells. In microcytic anemia they are smaller than normal, in macrocytic anemia they are bigger, and in normocytic they are normal-sized. Abnormalities in the hemoglobin molecules can cause further variations of the condition.

THALASSEMIA

Genetic defects can cause the formation of abnormal hemoglobin molecules leading to anemia. Beta thalassemia is the most prevalent of these disorders.

Beta thalassemia major is an inherited disorder and is common in the Mediterranean region and southeast Asia. A fault in the production of hemoglobin leads to red cells that are rigid, fragile, and easily destroyed. This leads to severe anemia by the age of 6 months and also to retardation of growth. As the bone marrow expands to produce more red blood cells, the long bones become thin and liable to fracture, and the skull and facial bones become distorted. The liver and spleen enlarge as they try to produce red blood cells as well.

Diagnosis can be made from blood tests that show hemoglobin levels. Frequent blood transfusions with iron chelating treatment (which prevents iron overload) help correct the anemia. Bone marrow transplant is the only cure and may be offered for severe cases.

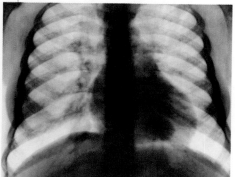

Chest X-ray of person with thalassemia
This color-enhanced chest X-ray shows a deformed ribcage as a result of marrow expansion. The bones become distorted as the body tries to produce more red blood cells.

MICROCYTIC AND MACROCYTIC ANEMIA

Microcytic anemia is often caused by iron deficiency in the diet. The rarer macrocytic anemia is usually due to a deficiency of vitamin B12 or folic acid.

If blood is lost and not replaced by iron in the diet, iron deficiency and microcytic anemia may develop. In this condition the red blood cells are smaller than normal. Causes of bleeding include menstration, parasitic infection, gastritis, peptic ulcers, and colon cancer. Treatment

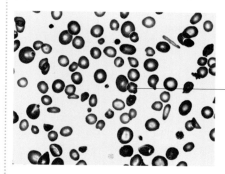

—— Misshapen red blood cell

Severe microcytic anemia
This blood smear shows red blood cells that are smaller and paler than normal, and some misshapen red blood cells. This is characteristic of microcytic anemia.

depends on finding the underlying cause but includes iron replacement. Macrocytic anemia (where the red blood cells are bigger than normal) may be caused by hypothyroidism (see p.496) or alcoholism. A deficiency of vitamin B12 or folic acid causes a type of macrocytic anemia called megaloblastic anemia. Dietary supplements usually help treat this condition.

Pernicious anemia is another type of macrocytic anemia and is caused by a lack of intrinsic factor, produced in the stomach and required to absorb B12 from food. It can be treated with vitamin B12 injections. Normocytic anemia, in which red blood cells are normal-sized but hemoglobin levels are low, occurs in aplastic anemia (see right), chronic diseases, and disorders with increased destruction or loss of red blood cells. Symptoms of anemia include fatigue, shortness of breath on exertion, pallor, and pale nail beds. Treatment depends on the cause.

APLASTIC ANEMIA

In this condition the bone marrow fails to produce sufficient blood cells and platelets to sustain normal function.

The cause of aplastic anemia is often unknown, or it may be due to toxins, radiation, and certain drugs. Lack of platelets in the blood leads to bruising and excessive bleeding. Low levels of white cells lead to unusual and life-threatening infections. Reduction in red blood cells leads to anemia, causing paleness, fatigue, and shortness of breath. Diagnosis is by bone marrow biopsy. Treatment is by bone marrow transplant.

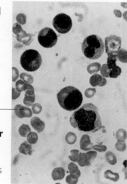

Red blood cell ——

Bone marrow smear
Fewer red and white blood cells than is usual are shown in this bone marrow sample.

SICKLE CELL ANEMIA

A mutation in the hemoglobin gene leads to red blood cells of a fragile, rigid sickle shape that do not pass easily through small blood vessels.

In sickle cell anemia, the red blood cells contain an abnormal type of hemoglobin. It is diagnosed by blood tests, and is usually first detected at the age of 4 months. The abnormal sickle cells restrict blood flow to organs, leading to episodes of severe pain (sickle cell crises) and eventual organ damage. A crisis can be triggered by infection and dehydration; the severity, frequency, and duration varies.

Typical symptoms include painful bones and joints, severe abdominal pain, chest pains, shortness of breath, and fever. Treatment is aimed at prevention through long-term use of hydroxyurea; crises are treated with rehydration, strong analgesics, antibiotics, and transfusions. Bone marrow transplant may be offered in severe cases.

Sickle-shaped cell ——

Deformed red blood cell
Abnormal sickle cells are fragile, have difficulty passing through the blood vessels, and have a reduced lifespan, leading to long-term anemia.

LEUKEMIA

Cancer of the bone marrow and white blood cells leads to bone marrow failure, causing immunosuppression, anemia, and low platelet counts.

In acute leukemia, immature, malignant white blood cells rapidly proliferate and reduce the numbers of normal blood cells. They then spill over into the blood, spreading to other organs in the body. Lack of platelets leads to bruising, excessive bleeding, and petechiae (red or purple spots on the body caused by hemorrhage). Poorly functioning white cells are unable to fight infection, leading to a greater risk of unusual and life-threatening infections. Lack of red blood cells leads to anemia. Leukemia is diagnosed by blood tests and bone marrow biopsy.

Acute leukemia is fatal without treatment, including chemotherapy and bone marrow or stem cell transplant. In children, the prognosis with treatment is excellent. In chronic leukemia, mature malignant white blood cells proliferate slowly over months to years, so bone marrow function is maintained for longer. The cells spread to the liver, spleen, and lymph nodes, causing them to enlarge. Chronic leukemia mainly affects older people and may be treated by chemotherapy or bone marrow transplant.

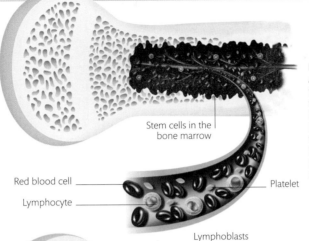

Blood cell production
All blood cells derive from stem cells found in the bone marrow. Red blood cells carry oxygen. Lymphocytes are a type of white blood cell that fights infection. Platelets help the blood clot at injury sites, reducing blood loss.

Stem cells in the bone marrow

Red blood cell

Lymphocyte

Platelet

Lymphoblasts multiply

Acute lymphoblastic leukemia (ALL)
Lymphoblasts (immature malignant lymphocytes) rapidly proliferate in the bone marrow. As a result, the production of normal blood cells is disrupted. Lymphoblasts also spread to the bloodstream and carry the cancer to other organs and tissues in the body.

Fewer red blood cells

Fewer platelets

Lymphoblasts circulating in bloodstream

BONE MARROW TREATMENT

Normal bone marrow may be transplanted into people needing treatment to replace cancerous or defective marrow. This is done for life-threatening conditions such as leukemia or aplastic anemia. First the diseased bone marrow is destroyed through radiation, then healthy bone marrow cells are transfused into the patient's circulation. Cells are harvested (removed) from a large bone such as the pelvis. A donor must have the same tissue type as the patient and so is usually a close relative or even the patient himself. Bone marrow transplants are also done using stem cells taken from a donor or from umbilical cord blood.

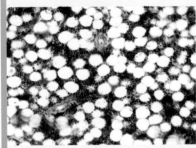

Bone marrow
Microscopic view of healthy bone marrow that can be harvested and used to replaced diseased bone marrow.

LYMPHOMAS

Lymphomas are cancers that develop when the lymphocytes (white blood cells) of the immune system form solid tumours in the lymphatic system.

There are more than 40 different types of lymphoma, classified according to cell type. The major categories are mature B cell neoplasms, mature T cell neoplasms, natural killer cell neoplasms, and Hodgkin's lymphoma. All types may cause swelling of the lymph nodes in the neck, armpits, or groin, and fever, weight loss, night sweats, and fatigue. Hodgkin's lymphoma is a rarer type that affects either adults aged 15 to 35, or people over 50, and runs a very aggressive course. It is easily curable in young people, slightly less so in older adults. The other lymphomas mainly occur in people over 60 and may run an aggressive or indolent (slow) course.

Diagnosis is based on taking a biopsy (tissue sample) from a lymph node and checking for spread by scanning. Treatment includes chemotherapy, radiotherapy, monoclonal antibody therapy, and corticosteroids. Early treatment gives a better outlook.

Lymphoma lymph cells
The stage of the lymphoma can be found by checking whether the cells are confined to one group of nodes or have spread beyond the lymphatic system to the liver, skin, and lungs.

PLATELET DISORDERS

Platelets aid the clotting of blood. Excessive numbers of platelets leads to clots in the blood (thrombosis). A deficiency causes excessive bleeding.

Reduced platelet count (thrombocytopenia) may be due to disorders such as aplastic anemia (see opposite) and leukemia or result from increased destruction of platelets due to conditions including SLE (lupus) and idiopathic thrombocytopenic purpura (low platelet count with no known cause). Certain drugs (such as those used in chemotherapy, and interferon) that suppress the bone marrow also cause a reduced platelet count. This leads to bruising, excessive bleeding, and red or purple spots on the body (petechiae).

Platelet disorders are diagnosed by blood counts or bone marrow biopsy. Platelet count may be raised following inflammation, surgery, bleeding, and iron deficiency, or unknown reasons. This does not usually need treatment. High platelet count causes no symptoms but increases the risk of thrombosis (clotting). Aspirin may be given to reduce this risk. Idiopathic thrombocytopenic purpura may require corticosteroids and specialist drugs.

CLOTTING DISORDERS

Failure of the blood to clot sufficiently may be genetic, autoimmune, or acquired for other reasons and can lead to excessive bruising and bleeding.

Hemophilia A is a rare inherited disorder that causes a deficiency in a blood protein, factor VIII, that is essential for clotting. This leads to prolonged bleeding and rebleeding after trauma or even spontaneously. There may be

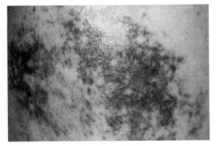

Bruising caused by hemophilia
Extensive bruising occurs after even minor trauma in severe hemophilia. Spontaneous bleeding typically causes nosebleeds and bleeding gums.

bleeding into internal tissues such as the muscles and joints, causing severe pain and joint destruction.

Hemophilia is treated by regular infusions of the deficient clotting factor. Von Willebrand's disease is a common inherited disorder that usually has no symptoms, but can lead to easy bruising, nosebleeds, and bleeding gums; it usually requires no treatment. Other clotting disorders may be caused by liver failure, leukemia, or vitamin K deficiency. Tests may be done to see how long the blood takes to clot. Treatment may be given to keep clotting factors in the blood high enough to prevent bleeding.

ALLERGIES AND AUTOIMMUNE DISORDERS

The basis of an allergy is an inappropriate reaction by the immune system in response to certain substances. In autoimmune disorders, the body's immune system reacts against its own cells and tissues, causing a variety of diseases.

ALLERGIC RHINITIS

Contact with an airborne allergen provokes an immune response in the lining of the nose, causing swelling, itching, and excessive mucus production.

In seasonal allergic rhinitis (hay fever), symptoms occur when certain pollens are in the air. Hayfever is rare before the age of 6 years, usually develops before the age of 30, and affects up to 1 in 5 people. Hay fever is often associated with eczema (see p.436) and asthma (see p.464). Perennial rhinitis can occur throughout the year and is commonly caused by house dust mites or

Dust mites
Millions of dust mites are present in bedding and carpets in the home. Their feces can provoke allergic reactions in many people.

animal saliva and skin flakes (dander). Sneezing, a runny nose, and sometimes runny, itchy eyes and itchy throat occur within minutes of exposure; the nose becomes blocked a few hours later. Allergy testing includes skin prick testing and blood testing. In hay fever, the time of the year may indicate which type of pollen is involved. Allergic rhinitis can be prevented or reduced by avoiding triggers and using oral antihistamines, intranasal corticosteroids, and, if the eyes are affected, cromolyn eyedrops. Immunotherapy and desensitization may be used for severe, chronic cases.

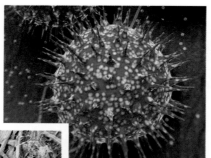

Pollen grain
Grass pollen is a common cause of hay fever. The pollen count is highest from spring to early summer.

ANAPHYLAXIS

A massive immune response to an allergen leads to anaphylaxis, a potentially fatal multisystem reaction, within minutes to hours of the exposure.

Anaphylaxis is a severe, potentially fatal allergic reaction caused by exposure to an allergen (typically nuts, drugs, or insect stings). The allergen may be ingested, injected, touched, or inhaled. The initial sense of anxiety, with itching and flushing, is quickly followed by problems that include a catastrophic fall in blood pressure (anaphylactic shock). This leads to fainting and unconsciousness, wheezing, constriction of the airways, shortness of breath, and respiratory failure. There may also be chest pain and palpitations, nausea and vomiting, diarrhea, angioedema (see right) and skin problems including urticaria (see p.437).

Anaphylaxis comes on suddenly and progresses rapidly. It is a life-threatening emergency because the airways and circulation may become severely impaired within minutes. Affected people should carry epinephrine "pens" for emergency; treatment includes resuscitation and immediate administration of epinephrine to open the airways, stimulate the heart, and constrict the blood vessels. Prevention involves avoiding the cause of the reaction and building up tolerance to the allergen.

ANGIOEDEMA

Angioedema is the local onset of swelling below the surface of the skin, due to the leakage of fluid from blood vessels. This is usually caused by an allergic reaction.

Angioedema usually affects the face and mouth and the mucosa (lining) of the mouth, tongue, and throat, but possibly other areas as well, leading to swelling. This can interfere with breathing, and the airway may have to be kept open with a tube. Common allergic triggers are peanuts, seafood, and insect bites. Drugs may induce nonallergenic angioedema. It is most commonly treated with antihistamines. Known trigger factors need to be avoided, and in severe cases the cause of the reaction may be gradually introduced, to build up tolerance.

Swollen lower lip
In angioedema, swelling occurs beneath the skin around the mouth, rather than on the skin surface. It may continue for hours or even days.

FOOD ALLERGIES

An adverse immune response to a food protein leads to a variety of problems including anaphylactic shock and eczema.

Food allergies affect around 6 percent of children but are slightly less common in adults. The most common triggers are dairy products, egg, nuts, seafood, shellfish, soy, wheat, and sesame products. A food allergy may cause a range of symptoms from itching and rash to nausea, abdominal cramps, and diarrhea. It may also cause wheezing and difficulty swallowing, brought on by swelling of the airways and angioedema (see right). A food

allergy is different from a food intolerance in which symptoms arise from food toxins (for example, bacterial food poisoning), problems with digestive enzymes (such as lactose intolerance), or the direct action of chemicals in the food (such as caffeine causing tremors).

People with a suspected food allergy may be offered blood and skin testing (see right) to find the cause of the problem. Food diaries and dietary exclusion may also help identify the allergen. If this is unsuccessful, the suspected allergen may be given under hospital supervision to provoke the reaction. People with allergies are advised to avoid foods that trigger the condition. Antihistamines may be used to treat mild allergies. People with severe allergies may need to carry an autoinjector of epinephrine (adrenaline) for emergency treatment.

SKIN TESTING FOR ALLERGIES

In a skin prick test, a drop of fluid containing a potential allergen is applied to the skin, which is pierced with a needle or scratched. A positive reaction (itching, redness, and swelling) indicates that the person may be allergic to that substance. In patch testing, used to test for allergic contact dermatitis, the allergen is applied directly onto the skin, covered with adhesive tape, and a reaction is checked for a few days later.

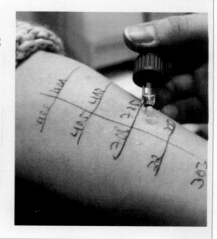

Skin prick testing
A skin prick test is done to diagnose common allergies to pollen, dust, dander (animal skin flakes), saliva (such as cat's saliva), and foods.

SYSTEMIC LUPUS ERYTHEMATOSUS

Commonly called lupus, this condition is an autoimmune disorder of the tissue that provides the structure for the skin, joints, and internal organs.

Systemic lupus erythmatosus (lupus), affects 2–15 people per 10,000 and can run in families. It is more common in women and develops from the teenage years onward. It is caused by the immune system's antibodies reacting against connective tissue in the body. This causes the tissues to become inflamed.

Lupus may be triggered by infections, puberty, menopause, stress, sunlight, and certain drugs. The symptoms vary widely in severity, and come and go. Flare-ups may last for weeks then disappear for months or even years. The progression of the disease ranges from very slow to rapid. The most common symptoms are fatigue, joint pain, fever, and weight loss. Up to half of people with lupus develop the classic "butterfly" rash across the nose and cheeks. Lupus is diagnosed in part by testing the blood for certain antibodies. There is no cure, but immunosuppressive agents, including corticosteroids, can be given to control symptoms, and to help prevent flare-ups and reduce their severity.

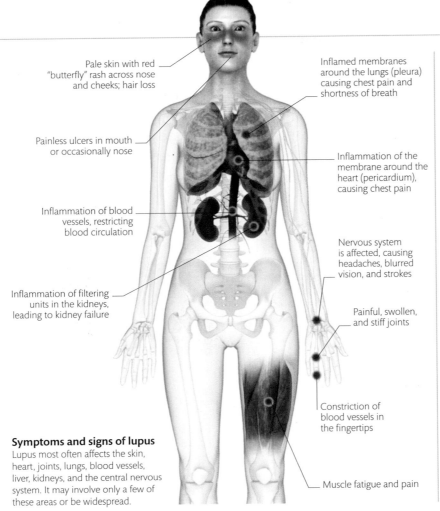

Pale skin with red "butterfly" rash across nose and cheeks; hair loss

Painless ulcers in mouth or occasionally nose

Inflammation of blood vessels, restricting blood circulation

Inflammation of filtering units in the kidneys, leading to kidney failure

Inflamed membranes around the lungs (pleura) causing chest pain and shortness of breath

Inflammation of the membrane around the heart (pericardium), causing chest pain

Nervous system is affected, causing headaches, blurred vision, and strokes

Painful, swollen, and stiff joints

Constriction of blood vessels in the fingertips

Muscle fatigue and pain

Symptoms and signs of lupus
Lupus most often affects the skin, heart, joints, lungs, blood vessels, liver, kidneys, and the central nervous system. It may involve only a few of these areas or be widespread.

POLYMYOSITIS AND DERMATOMYOSITIS

In these two rare related autoimmune disorders, the muscle fibers become inflamed. In dermatomyositis, the skin is also affected.

Polymyositis and dermatomyositis are more common in women than in men, and tend to develop in middle age, but dermatomyositis can occur in children. In both conditions, the arm and leg muscles weaken, typically making it hard to get up from a chair or lift the arms over the head. Other symptoms of polymyositis include fatigue, fevers, and weight loss. If the esophagus is affected, difficulty swallowing also occurs. Weakness of the chest wall muscles and diaphragm can lead to difficulty breathing.

Dermatomyositis also causes skin changes including a red scaly rash on the knuckles, knees, and elbows; rough and cracked skin on the fingertips; swelling and violet discoloration around the eyes; and flat, reddish areas on the face, neck, and chest. These skin changes may appear before any muscle problems occur. Diagnosis is by the presence of certain antibodies in the blood, electrical testing of muscle and nerves, and muscle biopsy (taking a sample of tissue). Treatment includes the use of immunosuppressives, including corticosteroids.

POLYARTERITIS NODOSA

This autoimmune disorder causes inflammation of the walls of small or medium-sized arteries (blood vessels), which restricts blood supply to tissues.

A rare autoimmune disorder that mainly occurs in people aged 40–60 years, polyarteritis nodosa affects the arteries supplying the heart, kidneys, skin, liver, digestive tract, pancreas, testes, skeletal muscles, and central nervous system. Areas of the body supplied by inflamed arteries may ulcerate, die, or atrophy (wither away). The inflamed arteries may dilate and rupture, leading to nodules, mottling, ulcers, and gangrene. People with polyarteritis may feel generally unwell, lose weight, and have fever and loss of appetite. Polyarteritis can lead to kidney failure (see p.483), hypertension (see p.469), and heart attack (see p.467).

Digestive problems caused by the condition include bleeding and perforation of the intestine. In men, the testes may inflame (orchitis). Musculoskeletal involvement causes muscle pain and arthritis. Diagnosis is based on a tissue biopsy of an affected artery or organ. Immunosuppressive agents such as corticosteroids are used to treat the condition.

Weakened artery wall

Artery affected by arteritis
Shown in cross section, the wall of the artery shows marked inflammation, is weakened and may eventually rupture.

SCLERODERMA

In this rare disorder, antibodies damage smaller blood vessels and cause hardening of the connective tissue throughout the body.

Scleroderma runs in families, is more common in women, and typically begins between the ages of 30 and 50. With morphea (limited cutaneous scleroderma), it is mainly the skin that is affected. With diffuse cutaneous scleroderma (systemic scleroderma), large areas of skin and the internal organs are affected, and the condition rapidly progresses. The skin becomes swollen and then thickened, shiny, and tightened, making it hard to move the joints, especially in the hands. Many people with scleroderma develop Raynaud's disease (see p.471). Hardening of the connective tissues elsewhere may affect the lungs, heart, kidneys, and digestive tract. Swallowing problems and gastric reflux are common due to the esophageal muscles becoming stiffened.

Diagnosis is based on a skin biopsy (tissue sample) and, in part, checking for the presence of antibodies (that attack the body's own tissue) in the blood. Immunosuppressive drugs may slow or reverse progression but there is no cure. Other treatments can be given to relieve symptoms. Regular monitoring of the condition is necessary as further complications may arise.

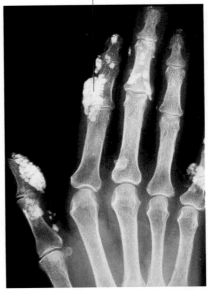

Lump of calcium

X-ray of hand affected by scleroderma
Lumps of calcium can form under the skin on fingers or other areas of the body (calcinosis) in scleroderma. They may require surgical removal.

UPPER DIGESTIVE TRACT DISORDERS

The common disorders of the mouth, esophagus, stomach, and duodenum are often caused by irritation leading to inflammation and problems such as ulcers. Some of these disorders are related to infection with bacteria, such as *Helicobacter pylori* in the stomach.

GINGIVITIS

Inflammation of the gums (gingivitis) is caused by a buildup of dental plaque, usually resulting from poor oral hygiene.

Plaque is a film of bacteria that collects where the teeth meet the gums. The bacteria inflame the gums, causing them to become reddish purple and tender and to bleed easily after brushing. If gingivitis is left untreated, deep pockets may form between the teeth and gums, and the tissues supporting teeth can become inflamed (periodontitis), causing the teeth to fall out. Smoking and alcohol increase the risk of gingivitis, but regular brushing, flossing, and dental checkups help prevent it. Removal of any plaque that develops is important.

MOUTH ULCERS

A break in the mucous membrane of the mouth leads to a painful open sore, or ulcer. Aphthous ulcers are the most common type of mouth ulcer.

Aphthous ulcers or canker sores are painful open sores inside the mouth. Minor ulcers are usually due to injury from vigorous brushing, biting the inside of the cheeks, sharp teeth, braces, and dentures. The ulcer typically forms a small, pale pit, and the area around it may become swollen. Minor ulcers clear up within 2 weeks. Recurrent minor mouth ulcers affect around 1 in 5 people, often appearing in groups of four to six. Major aphthous ulcers are larger (more than 0.5 in / 1 cm wide), deeper,

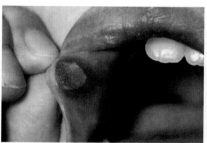

Ulcer inside lip
A minor aphthous ulcer is a small, painful, white, grey, or yellow area, forming an oval-shaped pit, with an inflamed red border.

more painful, take many weeks to heal, and may scar. Treatments include using a saltwater mouthwash, steroid pastes or lozenges, and anesthetic gels. Ulcers persisting for longer than 3 weeks require investigation.

ENDOSCOPY

An endoscope is a thin, flexible or rigid tube containing optical fibers, through which light passes to illuminate internal body structures and relay images back to an eyepiece or monitor. Within the shaft, there are also channels down which instruments or manipulators may be passed to cut out pieces of tissue (biopsy), grasp objects, and allow treatments using laser and electrocautery devices. Irrigating fluids and gases can flow down other channels. Different types of endoscope are used for particular body areas, such as a colonoscope for the large intestine or a gastroscope for the stomach. In most upper digestive tract disorders, endoscopy is replacing barium studies (swallowing a white fluid that shows up on X-rays) as the preferred form of investigation.

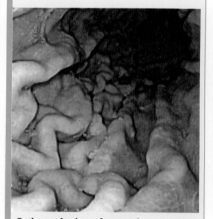

Endoscopic view of stomach
The gastric mucosa (inner lining) of a healthy stomach as seen through an endoscope. This procedure may be carried out to investigate upper digestive tract disorders.

CANCER OF THE ESOPHAGUS

Malignant tumors of the esophagus are often linked with smoking and excess alcohol use, and have a poor prognosis.

Most common in males over 60, this form of cancer usually causes difficulty swallowing solids, and then soft foods, and finally fluids. It commonly leads to substantial weight loss; other symptoms include regurgitation of food, coughing, hoarseness, and vomiting blood. It is diagnosed by barium studies or endoscopy with biopsy, but the cancer has often spread by this time. The tumor will need to be removed, and a tube (stent) may be inserted to keep the esophagus open and allow swallowing.

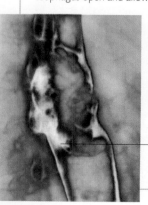

Tumor in the esophagus
This colored barium study shows the large, irregular outline of a tumor protruding into the interior of the esophagus.

Tumor

SALIVARY GLAND STONES

Hard masses formed from calcium phosphate, calcium carbonate, and other minerals can develop in the salivary glands, leading to painful swelling.

Salivary gland stones, also called sialoliths, may be single or multiple. They most commonly form in the submandibular glands, in the lower jaw, and may be associated with chronic infection of the gland, dehydration, poor saliva flow, and injury to salivary ducts. The stone causes a painful swelling that may worsen during meals, when salivary flow increases. It is diagnosed by seeing or feeling a lump in the gland and by imaging the stone on X-ray, ultrasound, or CT scans. Some stones can be removed by just massaging them out of the salivary duct; otherwise, surgery is required. Obstruction of the duct by a stone can cause a bacterial infection of the salivary gland (sialoadenitis), which is treated with intravenous antibiotics and sometimes surgical drainage.

GASTRIC REFLUX

The backflow of acidic stomach contents up into the esophagus causes the painful sensation known as heartburn.

The lower esophagus passes through a hole in the diaphragm before it joins the stomach at the gastroesophageal junction. The hole is normally taut, together with the esophageal sphincter, a ring of muscle at the base of the esophagus, helps prevent acidic stomach contents from flowing up into the esophagus (gastroesophageal reflux). If this structure is weakened and unable to stop the reflux of

stomach contents, it leads to heartburn, a burning sensation behind the breastbone. Common causes of heartburn are overeating, eating fatty foods, excess coffee or alcohol intake, smoking, obesity, and pregnancy. If the reflux is persistent or severe, it may cause the esophagus to become inflamed, leading to ulceration and bleeding. Over time, esophagitis may cause narrowing of the esophagus or cancerous changes. The condition is diagnosed by endoscopy and can usually be relieved by lifestyle changes. For reflux, drugs may be given to reduce acid production in the stomach, tighten the esophageal sphincter, or neutralize the stomach acid. Keyhole surgery can be done to tighten the esophageal sphincter.

Esophagitis
An endoscopic view of the esophagus, showing ulceration and inflammation due to reflux. Over time, inflammation may cause narrowing of the esophagus (stricture) or cancerous changes.

Ulcerated tissue   Inflamed lining

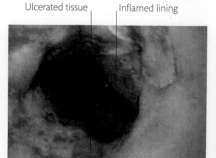

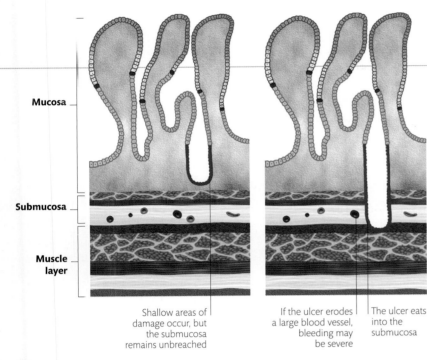

Mucosa

Submucosa

Muscle layer

Shallow areas of damage occur, but the submucosa remains unbreached

If the ulcer erodes a large blood vessel, bleeding may be severe

The ulcer eats into the submucosa

Early ulcer
If the mucus layer protecting the stomach lining is breached, stomach acid can attack and damage mucosal cells.

Progressive ulceration
The ulcer erodes the deeper layers. It may even perforate (break through) the wall of the stomach or duodenum.

PEPTIC ULCER

An erosion in the lining (mucosa) of the stomach or first part of the duodenum, a peptic ulcer can cause pain and bleeding.

The cells lining the stomach and duodenum secrete a layer of mucus that protects them from damage by stomach acid. If this layer is breached, an ulcer can form. Most peptic ulcers are caused by persistent inflammation due to the bacterium *Helicobacter pylori*. The other major cause is the use of aspirin or nonsteroidal anti-inflammatory drugs (NSAIDs) or ibuprofen, which reduce the secretion of mucus. Further contributory factors include smoking, alcohol, family history, and diet.

Symptoms include upper abdominal pain, often related to eating; bloating; and nausea. Ulcers last for days to weeks and may recur every few months. Bleeding ulcers can cause hematemesis (vomiting of blood) or melena (black, tarry stools). Severe ulcers may perforate the stomach or duodenum wall; this is a surgical emergency. Ulcers are detected by endoscopy, and *H. pylori* infection is confirmed by biopsy and blood, stool, or breath tests. Drugs are given to reduce acid production so the ulcer can heal, and to eradicate *H. pylori* infection.

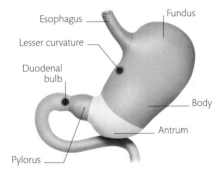

Esophagus

Fundus

Lesser curvature

Duodenal bulb

Body

Antrum

Pylorus

Sites of peptic ulcer
The most common site is the duodenal bulb, the first part of the duodenum, where the stomach empties into the duodenum. In the stomach, most ulcers develop in the lesser curvature.

GASTRITIS

Inflammation of the stomach lining may be acute or chronic and has a variety of causes, often related to irritation or infection of the lining.

The stomach lining normally protects itself from the acidic stomach contents by a layer of mucus, but if this barrier is disrupted, gastritis (inflammation) can occur. Acute (sudden, onset) gastritis is usually caused by excessive alcohol intake, which irritates the stomach lining, or by the use of aspirin or nonsteroidal anti-inflammatory drugs (NSAIDs) such as ibuprofen or naproxen, which reduce the production of mucus by the stomach lining cells. Symptoms can include pain in the upper abdomen, nausea,

vomiting (sometimes with blood), and bloating. Chronic (long-term) gastritis is usually due to infection of the stomach lining by the bacterium *Helicobacter pylori*, which weakens the protective mucus barrier. Gastritis is diagnosed by endoscopy. Treatment includes addressing the underlying cause and using drugs to neutralize stomach acid or reduce acid production.

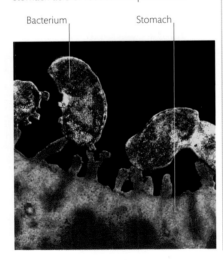

Bacterium

Stomach

Bacteria in the stomach
More than 50 percent of people carry *H. pylori*. It causes long-term, low-level inflammation of the stomach lining and can lead to peptic ulcers, chronic gastritis, and stomach cancer.

HIATUS HERNIA

A tear or weakness in the diaphragm (the large, flat muscle that separates the chest and abdominal cavities) can allow part of the stomach to protrude into the chest.

In the most common "sliding" form of hiatus hernia, the junction between the esophagus and the stomach slides upward through the diaphragm. This form is very common, especially in people over 50. It usually causes no symptoms, but a large hernia may result in gastroesophageal reflux. Measures to relieve this problem include raising the head of the bed, avoiding lying down after meals, losing weight, and using medications to reduce acid production in the stomach and tighten the esophageal sphincter. In the much rarer paraesophageal hernia, the top of the stomach may become constricted in the chest and have its blood supply cut off. This needs urgent surgical treatment. A hiatus hernia is diagnosed by endoscopy or barium study. People with severe symptoms or long-term reflux may have surgery to repair the hernia. During surgery, the upper part of the stomach is wrapped around the lower part of the esophagus, stopping the stomach from protruding through the hiatus.

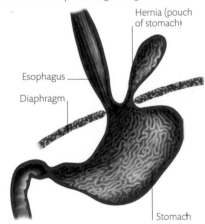

Hernia (pouch of stomach)

Esophagus

Diaphragm

Stomach

Paraesophageal hiatus hernia
A pouchlike part of the upper stomach is pushed upward through the hole (hiatus) in the diaphragm where the esophagus normally passes through to join the stomach.

STOMACH CANCER

A malignant tumor of the stomach is a common form of cancer worldwide but is uncommon in the US.

Males over 40 years of age are most likely to develop stomach cancer. Risk factors include infection with *Helicobacter pylori*; smoking; a family history of this cancer; a diet rich in salted, smoked, or pickled food (as in Japan); disorders such as pernicious anemia; and previous stomach surgery. Symptoms include loss of appetite, unexplained weight loss, nausea, vomiting, bloating, and feeling "full up" after meals. Bleeding from the stomach may cause hematemesis (bloody vomiting), melena (black, tarry stools), or anemia. The diagnosis is made by endoscopy with biopsy or a barium study. Gastrectomy (surgery to remove some or all of the stomach) is the most common treatment; tumors at the top of the stomach may require removal of the esophagus as well (esophagogastrectomy). Typically, stomach cancer has already spread by the time it is detected, so radiation therapy and chemotherapy may be offered too, but the outlook is poor.

Cancer in the lower stomach
This colored barium study shows the large, irregular shape of a tumor in the lower stomach. CT, MRI, and ultrasound scans may be used to find out if the tumor has spread elsewhere.

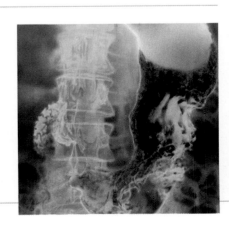

LOWER DIGESTIVE TRACT DISORDERS

Many of the disorders affecting the intestines (bowels) and rectum are caused by inflammation, as in inflammatory bowel disease (IBD). Others may be due to structural changes, as in diverticulosis. Cancers of the colon and rectum are common.

CELIAC DISEASE

A disorder of the small intestine, celiac disease is caused by an immune system reaction to gliadin, a gluten protein found in wheat and some other grains.

The lining of the small intestine has millions of tiny, fingerlike projections called villi, which absorb nutrients from food. In celiac disease, the immune system reacts against gluten in the digestive system. This reaction damages the villi, causing them to flatten and interfering with their normal function. The resulting symptoms vary widely, but include a swollen abdomen, vomiting, diarrhea (typically pale, foul-smelling, and bulky), fatigue, weight loss, and stunted growth. Celiac disease is more common in women and can run in families. It often coexists with other autoimmune disorders such as type 1 diabetes mellitus.

The condition is diagnosed by finding antigliadin antibodies in the blood, endoscopy (see p.476), and taking a biopsy (tissue sample) of the small intestine. Affected people need to follow a strict, lifelong gluten-free diet (avoiding wheat, rye, and barley) to clear up symptoms, and take dietary supplements to correct nutritional deficiencies.

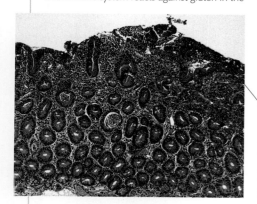

Flat surface due to loss of villi

Celiac disease
This light micrograph of a section through the duodenum of a patient with celiac disease shows the loss of villi from the surface. As a result, the intestine is less able to absorb nutrients effectively.

IRRITABLE BOWEL SYNDROME

A common long-term complaint with no structural or biochemical origin, irritable bowel syndrome (IBS) leads to abdominal discomfort and altered bowel habits.

Occurring mainly in people aged 20–30 years of age, IBS affects up to 1 in 5 people and is two to three times more common in women than men. It causes bouts of recurrent abdominal pain and possibly bloating, associated with changes in the frequency or appearance of stools. The pain is often relieved by defecation. The cause is unknown, but IBS can be triggered by a bout of gastroenteritis. It is a long-term, intermittent condition; flare-ups can be brought on by alcohol, caffeine, stress, and certain foods. It is diagnosed by the symptoms, physical examination, and blood tests. Lifestyle changes, dietary modification, and increasing soluble fiber intake can lessen symptoms. During flare-ups, drugs may help regulate bowel habits and relieve abdominal spasms.

DIARRHEA AND CONSTIPATION

Acute diarrhea (frequent loose or liquid bowel movements) is often due to viral or bacterial infections causing gastroenteritis (inflammation of the stomach and the small intestine). Diarrhea can also have a variety of other causes. Constipation (infrequent or hard stools or difficulty in passing stools) is often due to inadequate intake of dietary fiber and fluid, but may also be caused by a variety of bowel problems, including tumors.

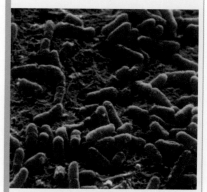

Intestinal bacteria
E. coli bacteria live in the intestines. Most strains are harmless, but some cause severe cramps, vomiting, and bloody diarrhea and may produce toxins that damage the kidneys.

CROHN'S DISEASE

This rare autoimmune disorder does not have a cure, and causes inflammation anywhere along the digestive tract, sometimes in several places at once.

The disease affects both sexes equally and can run in families. It usually appears in teenagers and young adults. The inflammation in Crohn's disease involves all of the intestinal wall and follows two main patterns. In stricturing disease, the affected area narrows, eventually causing blockages. In fistulizing disease, abnormal passageways form between affected areas and nearby structures. Symptoms fluctuate but may include abdominal pain, severe diarrhea (often with blood), loss of appetite and weight, profound fatigue, and anemia. Because it is an autoimmune disorder, Crohn's disease may also cause liver, skin, and eye problems and inflamed joints. Drugs can be taken to reduce inflammation and suppress the activity of the immune system, and the disease may be in remission for years. Often, surgical removal of diseased areas is necessary.

Patches of inflammation
Crohn's disease typically affects the ileum (the last part of the small intestine), but may occur in patches anywhere from the mouth to the anus. Strictures in the bowel can lead to obstruction.

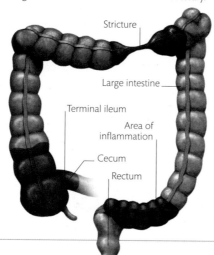

Stricture

Large intestine

Terminal ileum

Area of inflammation

Cecum

Rectum

ULCERATIVE COLITIS

This rare disorder of the large intestine causes inflammation and ulceration (open sores) in the colon and rectum.

Ulcerative colitis (UC) usually affects teenagers and young adults or, less commonly, adults between 50 and 70 years old. The inflammation occurs in the mucosa (lining) of the colon and rectum, leading to ulceration with bleeding and pus. Symptoms, which come and go over months or years, typically include diarrhea mixed with blood and mucus, abdominal pain, fatigue, and weight loss. UC is believed to be an autoimmune disorder; it may also cause skin and eye problems and inflamed joints. People with UC have a greatly increased risk of developing colon cancer. Diagnosis is made by endoscopy (see p.462), barium studies, and blood tests. Treatment involves using drugs to suppress or modulate the immune system and to control inflammation and diarrhea. Up to 40 percent of affected people eventually have surgery to remove the colon and rectum, which cures the condition.

Inflammation and ulceration
In ulcerative colitis, the inflammation is normally continuous, extending from the rectum up the colon to a varying extent, and sometimes reaching all the way to the cecum ("pancolitis").

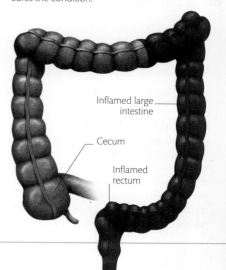

Inflamed large intestine

Cecum

Inflamed rectum

DIVERTICULAR DISEASE

The development of diverticula (pouches) in the colon wall is called diverticulosis. Problems can arise if diverticula become inflamed and infected.

The pea-to grape-sized pouches typically develop from the age of 40 onward and are found in many older people. Risk factors include increasing age, constipation, and a low-fiber and high-fat diet. Diverticula usually produce no symptoms, but in some cases they can cause bloody stools, bloating, abdominal pain, diarrhea, or constipation. The pouches can trap bacteria and become inflamed (acute diverticulitis); this condition typically causes left-sided lower abdominal pain, with fever and later vomiting. Diverticulosis is diagnosed by viewing (colonoscopy) or imaging of the colon (barium studies). Acute diverticulitis is diagnosed by CT scanning. Diverticulosis can, if necessary, be treated with a high-fiber diet and fiber supplements. Acute diverticulitis usually clears up in response to antibiotics and resting of the bowel, but in severe cases surgery may be required to remove the affected area of bowel.

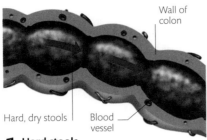

Wall of colon

Hard, dry stools | Blood vessel

1 Hard stools
If stools are small, hard and dry, the smooth muscles in the bowel wall must contract harder to push them along than if they are soft and large.

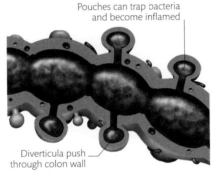

Pouches can trap bacteria and become inflamed

Diverticula push through colon wall

2 Pouches form
Increased pressure from pushing can cause the mucosa and submucosa to push through weak points in the colon wall, forming pouches.

APPENDICITIS

An inflamed appendix (appendicitis) causes severe abdominal pain and requires urgent medical removal.

Infection and blockage in the appendix can cause it to fill up with pus, making it swell. As this swelling grows worse, the appendix starts to die and infected pus forms around it (suppuration). Eventually the appendix ruptures (bursts) and infected material leaks out, causing peritonitis (inflammation of the membrane covering most of the abdominal organs), which can be fatal. Typically, appendicitis begins with sudden, severe pain that starts in the center of the abdomen and shifts down into the lower right area, where the appendix is situated. The condition usually causes a loss of appetite and sometimes fever, nausea, and vomiting. Diagnosis is based on the symptoms and on an examination and blood tests. Treatment may be antibiotics or immediate surgery (removal of the appendix), which can be performed by laparotomy (open surgery) or laparoscopy (keyhole surgery).

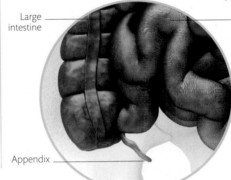

Large intestine | Small Intestine

Appendix

Site of appendix
The appendix is a blind-ended tube connected to the cecum, which is part of the colon. Its removal seems to have no effect on the function of the digestive or immune system.

COLORECTAL CANCER

A malignant tumor of the rectum or colon (bowel) is one of most common forms of cancer in industrialized nations and a leading cause of cancer deaths.

Around 1 in 20 people will have colorectal cancer in their life. It affects both sexes equally, and most cases occur in people over 50 years of age. Risk factors include having a colorectal polyp (a slowly developing overgrowth of the colon or rectum lining), a family history of this cancer, increasing age, smoking, a diet high in red and processed meat and low in fiber, lack of exercise, excess alcohol intake, and a history of inflammatory bowel disease. Symptoms may include a change in bowel habit and stool consistency; mucus or blood in stools; tenesmus (a sensation of not fully emptying the bowels); abdominal pain; anemia; and loss of weight or appetite.

A large tumor may block the bowel, causing abdominal pain and bloating with vomiting and constipation. Tumors may be detected by imaging (barium studies, CT, and PET scans), viewing (endoscopy), and blood tests for chemicals called tumor markers. Treatment of the condition depends on how much the tumor has spread and includes surgery and chemotherapy; early cancers can be cured. Screening programs to detect the disease early exist in many countries.

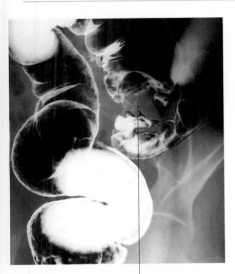

Tumor in the colon

Colon cancer
This color-enhanced X-ray shows a tumor within the colon. The patient was given a barium enema, which highlights the abnormality.

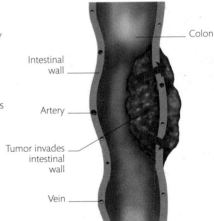

Colon

Intestinal wall

Artery

Tumor invades intestinal wall

Vein

Invasive colonic tumor
Cancers can spread directly, by invasion of local structures such as the colon wall, or indirectly, via the bloodstream and lymphatic system.

HEMORRHOIDS

The veins in the anus and rectum may become varicosed (swollen), causing them to protrude and be liable to bleed.

Hemorrhoids can result from straining to pass stools, so are common in constipation and chronic diarrhea. Internal hemorrhoids occur within the rectum and are painless but may bleed, showing as bright red blood on the stools or toilet paper, or blood dripping into the toilet bowl. Larger internal hemorrhoids may prolapse out of the anus, typically after defecation, but often go back by themselves or may be pushed back in by hand. External hemorrhoids develop outside the anus. Both types can form itchy, tender, painful lumps. Piles can be found by proctoscopy (viewing the anus and rectum). Treatments include increasing fluid and fiber intake, ointments, injections, banding, laser therapy, and surgery.

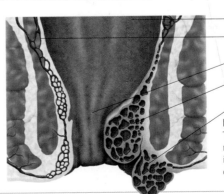

Rectum

Vein network

Anal canal

Internal hemorrhoid

External hemorrhoid

Hemorrhoids
The venous network on the left is normal. On the right, the veins have become swollen, protruding into the anus (internal hemorrhoids) or developing outside it (external hemorrhoids).

LIVER, GALLBLADDER, AND PANCREAS DISORDERS

Producing substances vital to digestion, the liver, gallbladder, and pancreas enable the absorption and metabolism of food, drinks, and medicines and other chemicals. They are vulnerable to infection, cancerous change, and damage by alcohol and other toxins.

ALCOHOLIC LIVER DISEASE

Prolonged, excessive alcohol intake causes increasing damage to liver cells, and can eventually result in permanent harm.

Alcohol is absorbed in the small intestine and enters the liver. There, it is metabolized (broken down) to form fat and chemicals, some of which can damage liver cells. The first sign of damage is fatty liver, in which large droplets of fat collect in liver cells. There are no symptoms, but blood tests may show impaired liver function, and ultrasound scans show the liver to be enlarged and fatty. Continued drinking causes alcoholic hepatitis (liver inflammation). Abstaining from alcohol will stop or delay progression, allowing the liver to recover.

Symptoms include liver enlargement, jaundice, and ascites (fluid within the abdomen). It is diagnosed by liver function blood tests. Mild cases clear up with abstinence, but severe cases may be fatal. In cirrhosis, liver tissue is replaced by fibrous scar tissue, and some of the damaged tissue forms nodules. Symptoms include ascites, jaundice, enlarged breasts and shrunken testes in men, red palms, pruritis, weight loss, confusion, and coma. If the liver fails, transplant is needed.

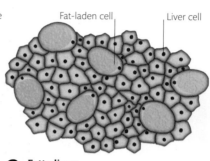

Alcohol
Liver cell
Acetaldehyde
Water

1 How damage occurs
When alcohol (ethanol) is broken down by the liver, it produces fat and a chemical called acetaldehyde, which is toxic to the liver but is itself processed into water and carbon dioxide.

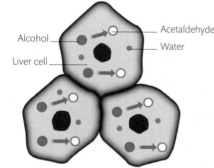

Damaged tissue

3 Alcoholic hepatitis
With continued heavy drinking, liver cells become swollen, damaged, and surrounded by white blood cells. Some cells die and are replaced with fibrous tissue (fibrosis); others regenerate.

Fat-laden cell
Liver cell

2 Fatty liver
Fat builds up in the liver cells, and eventually the deposits become so large that a cell swells and the nucleus is pushed to the side of the cell. The liver becomes enlarged.

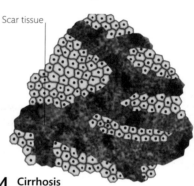
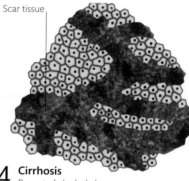

Scar tissue

4 Cirrhosis
Repeated alcohol abuse causes permanent scarring and fibrosis. The liver becomes nodular, shrinks, and cannot function normally. As a result, liver failure and portal hypertension develop.

JAUNDICE

Old red blood cells are broken into bilirubin in the liver and, normally, excreted into the bile. The yellowing of jaundice results from an excess of bilirubin in the blood. In hemolytic jaundice, it is released directly into the bloodstream. In liver disease, bilirubin seeps into the bloodstream because the liver is unable to metabolize or excrete it properly.

Yellow sclera
The sclera (white of the eye) appears yellow because the overlying conjunctiva contains excessive amounts of bilirubin.

VIRAL HEPATITIS

The most common viral causes of hepatitis (liver inflammation) are the hepatitis A, B, and C viruses.

Hepatitis A virus (HAV) is spread by food and water contaminated with infected stools. It produces jaundice, fever, nausea, vomiting, and upper abdominal pain. Most people recover within 2 months. Hepatitis B virus (HBV) and hepatitis C virus (HCV) are spread by infected body fluids such as blood or semen. HBV causes acute hepatitis, which may lead to chronic hepatitis. HCV often has no symptoms initially, but may lead to chronic hepatitis. Chronic viral hepatitis may result in cirrhosis and liver cancer, but antiviral drugs can reduce the risks.

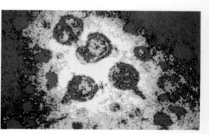

Hepatitis B
This virus is commonly transmitted by sexual contact, blood transfusion, sharing needles for drug use, and nonsterilized tattoo equipment.

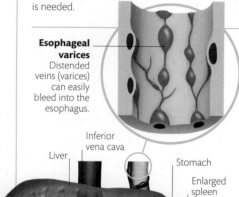

Esophageal varices
Distended veins (varices) can easily bleed into the esophagus.

Inferior vena cava
Liver
Stomach
Enlarged spleen
Blood from stomach
Portal vein
Blood from spleen

PORTAL HYPERTENSION

Raised pressure in the portal vein is usually due to alcoholic cirrhosis, but schistosomiasis (a parasitic worm infection) is a major cause worldwide.

The portal venous system collects blood from the esophagus, stomach, intestine, spleen, and pancreas. The veins merge to form the portal vein, which enters the liver and splits into

Obstructed blood flow
Restricting the blood flow into the portal system raises the pressure behind the blockage, causing the veins to distend and the spleen to enlarge.

smaller branches to supply it. If the liver is scarred and fibrosed, blood flow is impeded, leading to back-pressure in the portal system. This causes the veins to become distended and liable to bleed. Varices (swollen veins) in the esophagus sometimes bleed severely and result in hematemesis (vomiting of blood), which can be life-threatening. Bleeding may be stopped by the use of rubber bands, to seal the veins, or sclerotherapy (injection of a chemical to cause varices to scar).

The spleen may enlarge, and fluid may collect in the abdominal cavity. In addition, poor liver function may cause hepatic encephalopathy, resulting in confusion and forgetfulness. Portal hypertension is treated with beta-blocker drugs, which lower the blood pressure, or sometimes surgery to reduce pressure in the portal venous system. Ultimately, a liver transplant may be needed.

LIVER TUMORS

Growths within the liver are usually benign (non-cancerous), but cancer may spread there from other parts of the body.

Benign liver tumors are most often either a hemangioma (mass of blood vessels) or an adenoma (overgrowth of normal cells). They generally cause no symptoms and require no treatment. Cancerous tumors are usually due to a cancer having spread from another area of the body, most commonly from a cancer in the colon, stomach, breast, ovary, lung, kidney, or prostate. The most common cancer to arise within the liver (primary liver cancer) is a hepatoma, which may result from chronic viral hepatitis cirrhosis, or exposure to toxins. It causes abdominal pain, weight loss, nausea, vomiting, jaundice, and a mass in the abdomen. It is diagnosed by imaging, such as ultrasound or CT scan, and biopsy. Treatment may include surgical removal of the tumor, chemotherapy, radiation therapy, and liver transplant. Prognosis depends on whether the cancer has spread.

LIVER ABSCESS

An abscess, or pus-filled mass, in the liver is most commonly due to bacteria that spread from elsewhere in the body.

A pyogenic (bacterial) abscess is commonly caused by bacteria spreading from an abdominal infection (such as appendicitis, cholangitis, diverticulitis, or perforated bowel) or from the blood. It causes a sudden feeling of illness, loss of appetite, high fever, and pain in the upper right of the abdomen, although it can be present for weeks with few symptoms. The abscess may be detected by ultrasound or CT scanning. It is treated by draining the pus using a needle (either done through the skin or during abdominal surgery), followed by antibiotics. Left untreated, the mortality rate from this condition is high. Abscesses may also result from fungal or amebic infections, especially in the tropics.

Pyogenic abscess
Abscesses may be single or multiple, and usually occur in the right lobe of the liver. They are more common in people with diabetes mellitus or a weakened immune system.

Vein

Pus-filled abscess

Gallbladder

Common bile duct

Liver

GALLSTONES

Hard masses formed from bile, gallstones can occur anywhere in the biliary ducts but usually form in the gallbladder.

Gallstones may be single or multiple and vary in size, some more than an inch wide. Most are made primarily of cholesterol, some are "pigment stones" made of bilirubin (produced from red blood cells) and calcium, and the rest are a mixture of these two types. Gallstones are more common in women, Mexican Americans, Native Americans, those who are overweight, and older people. They take years to form and often cause no symptoms unless they become lodged in the ducts that drain the gallbladder or the pancreas. If this happens when the gallbladder contracts (such as after a fatty meal), it can cause biliary colic: steadily increasing, severe upper abdominal pain, often with nausea and vomiting. The stones may be detected by ultrasound and, if painful, the gallbladder may be surgically removed (cholecystectomy).

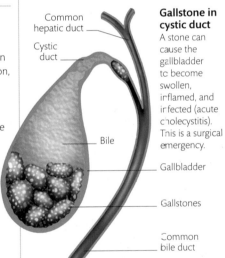

Common hepatic duct

Cystic duct

Bile

Gallbladder

Gallstones

Common bile duct

Gallstone in cystic duct
A stone can cause the gallbladder to become swollen, inflamed, and infected (acute cholecystitis). This is a surgical emergency.

PANCREATITIS

Inflammation of the pancreas, or pancreatitis, is due to enzymes produced by the pancreas damaging the pancreatic tissue itself (autodigestion).

The pancreas produces enzymes to aid the digestion of food in the duodenum. However, if these enzymes are activated within the organ, they digest it. This causes the pancreas to become inflamed. The condition may be acute (sudden-onset) or chronic (long-term). Acute pancreatitis causes severe upper abdominal pain, which penetrates through into the back with severe nausea and/or vomiting and fever, but the pancreas heals without any loss of function. In chronic pancreatitis, recurrent attacks of inflammation cause permanent damage and loss of function, which can lead to diabetes mellitus and reduced ability to digest fats.

The major causes of pancreatitis are gallstones, if they obstruct the drainage of the pancreas, and excessive long-term alcohol intake, which damages the function of pancreatic cells. Other causes include injury to the pancreas, and certain drugs and viral infections. The disorder is diagnosed by finding elevated levels of the pancreatic enzyme amylase in the blood and particular changes on CT scan. It is treated with analgesics and antibiotics, and by addressing the underlying cause.

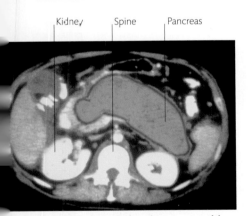

Kidney   Spine   Pancreas

Scan of abdomen showing pancreatitis
The blue area on this CT scan through the upper body shows an enlarged pancreas caused by pancreatitis.

CANCER OF THE PANCREAS

A malignant tumor of the pancreas is a common cause of cancer deaths because it causes no symptoms in its early stages and may not be found until it has spread.

This disease is most common in males over the age of 60. The risk factors for pancreatic cancer include smoking, obesity, chronic pancreatitis, poor diet (excess red and processed meat), and a family history. Symptoms do not appear until late in the disease. They include upper abdominal pain, penetrating through to the back, and severe weight loss. Cancer in the head of the pancreas may block the flow of bile from the gallbladder, leading to jaundice, generalized itching, pale stools, and dark urine. Diagnosis is made by finding tumor markers (chemicals released by cancer) in the blood, CT scanning, and biopsy. Patients may be offered surgery but treatment can only relieve symptoms: only one-fifth of patients survive longer than a year after diagnosis.

Sites of pancreatic cancer
Most tumors occur in the head of the pancreas. Some develop at the ampulla of Vater, where the pancreatic duct and common bile duct join, causing biliary obstruction and jaundice.

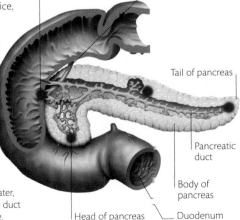

Ampulla of Vater   Common bile duct

Tail of pancreas

Pancreatic duct

Body of pancreas

Head of pancreas   Duodenum

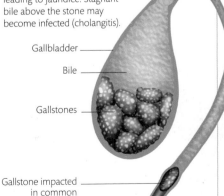

Gallstone in common bile duct
A stone can block the flow of bile into the duodenum, leading to jaundice. Stagnant bile above the stone may become infected (cholangitis).

Cystic duct

Gallbladder

Bile

Gallstones

Gallstone impacted in common bile duct

KIDNEY AND URINARY PROBLEMS

The renal system of kidneys, ureters, bladder, and urethra clears waste from the blood. The kidneys also play a role in the renin-angiotensin system, which regulates blood pressure, and in vitamin D metabolism, and secrete erythropoietin to stimulate red blood cell production. Kidney disease affects all of these functions.

URINARY TRACT INFECTIONS

One of the most common types of infection, urinary tract infections arise when the normally sterile urine is contaminated by bacteria from the bowel. Bacteria may pass up the urethra to the bladder or, less commonly, through the bloodstream to the urinary tract. The presence of sugar in the urine, as in diabetes, or stones in the urinary tract can enable bacteria to take hold, especially where there is any obstruction to urinary flow.

GLOMERULONEPHRITIS

In this complex condition, the glomeruli (the tiny filtering units within the kidneys) are damaged by inflammation.

Glomerular inflammation may occur alone, as a result of an immune system disorder, or due to infection. It may also be caused by other diseases that affect the whole body, such as SLE (see p.475) or polyarteritis nodosa (see p.475). Damaged glomeruli can no longer filter wastes effectively from the blood, so problems include kidney failure, nephrotic syndrome (protein in the urine, high cholesterol, and low protein in the blood), and nephritic syndrome (body tissue swelling, protein and blood in the urine).

The condition is investigated by blood tests, urine analysis, and X-ray, MRI, or biopsy (taking a tissue sample) of the kidney. The management and prognosis depend on the cause of the condition, its severity, and other diseases that may be present.

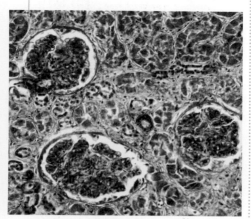

Inflamed glomeruli
Light micrograph of three glomeruli (dark blue areas) in a kidney affected by glomerulonephritis. A sample of the kidney tissue taken in biopsy is analyzed and used to diagnose the condition.

CYSTITIS

Inflammation of the bladder lining, or cystitis, is usually caused by infection, most commonly with bacteria normally found in the bowel.

More frequent in women—in whom the urethra is only 1½ in (4 cm) long, making bacterial access easier—cystitis usually causes symptoms such as pain when urinating, frequent need to urinate, abdominal pain, fever, and blood in the urine. In men, cystitis is rare and usually caused by a disorder of the urinary tract. The immune system can overcome low levels of bacteria, but once cystitis is established, antibiotics may be needed to prevent chronic infection and keep it from spreading to the kidneys. Diagnosis is made from the symptoms and testing urine for white blood cells, nitrites, and blood.

The bacterium causing the infection can be confirmed by analyzing a sample of urine, and tests may be done to find which antibiotic will eradicate it. Drinking plenty of clear fluids and emptying the bladder soon after sexual intercourse helps prevent further infections.

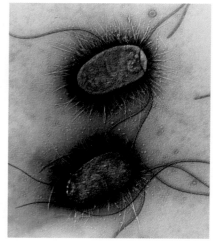

Bacterial cause of infection
E. coli is a bacillus that inhabits the bowel and perineum. It is usually harmless but can migrate to other organs, where it may cause infection. It is responsible for most cases of cystitis.

Other forms of cystitis may be triggered by certain foods or drinks, chlamydia, and urethral syndrome, in which diseases inflaming the urethra and bladder cause cystitis symptoms.

PYELONEPHRITIS

Inflammation of the kidneys due to bacterial infection, is called pyelonephritis. It is usually caused by bacteria entering the urinary tract through the urethra.

Pyelonephritis is a more serious infection than bacterial cystitis (see above), although if treated promptly does not cause permanent damage to the kidneys. Around 80 percent of cases are caused by a virulent subgroup of the *Escherichia coli* bacterium, which has migrated from the bladder to the kidneys via the ureters. More rarely, other organisms such as proteus, staphylococcus, or tuberculosis (TB) may be responsible for the condition. Symptoms include painful or frequent urination, fever, back pain, blood in the urine, nausea, and fatigue. In rare cases, a kidney abscess may form, or the infection can spread in the blood. Diagnosis is made by testing the urine for bacteria. X-ray, ultrasound, or other scans may also be done to show any stones or other damage to the kidneys. Long courses of antibiotics may be needed to clear the infection, and surgery may be required to correct resulting problems such as kidney stones (see right).

KIDNEY STONES

Also known as calculi, these stones are formed from hardened deposits of waste materials that pass through the kidneys; they are most common in young men.

The exact cause of kidney stones is unknown, but predisposing factors include diseases producing high levels of calcium or other compounds, or urinary infection. In some cases, the stones are associated with genetic or metabolic disorders, such as gout. Stones are not usually painful until they pass into the ureter, when they may cause excruciating pain with blood in the urine or infection.

Diagnosis is confirmed by CT scans. About 40 percent of stones pass out in the urine, but some give rise to blockage, infection, backflow, or kidney failure, and need to be removed. Surgical procedures include lithotripsy, external shock waves to break up the stone so it can pass in the urine; ureteroscopy, in which a tube is passed up the urinary tract to reach the stone; or open surgery.

Growth of kidney stones
Most kidney stones are small and pass out of the body in the urine. The biggest stones slowly form in the calyces and renal pelvis at the center of the kidney, developing a horned shape.

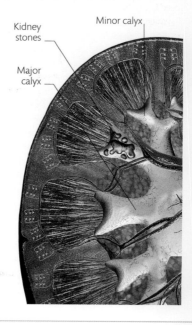

Minor calyx

Kidney stones

Major calyx

KIDNEY FAILURE

Acute loss of kidney function can be immediately life-threatening, while chronic kidney failure involves more gradual, progressive deterioration.

The kidneys' main role in clearing the blood of waste products can be affected suddenly by severe conditions such as shock, burns, blood loss, infection, and heart failure; by diseases of the kidney itself; and by conditions that cause obstruction of urinary flow. Certain drugs, including NSAIDs, anti-inflammatories, some antibiotics, and drugs for heart and cancer disease, may also reduce kidney function. Symptoms of acute kidney failure include nausea, vomiting, low urine output, fluid retention, breathlessness, confusion, and eventually coma. It is treated by dialysis, a system for removing waste products in the blood, until the kidneys recover.

Chronic kidney failure is the progressive loss of renal (kidney) cells and is a feature of long-term disorders including kidney disease, diabetes, hypertension (high blood pressure), and inherited disorders such as polycystic kidneys. Treatment for kidney failure involves dealing with the underlying cause of the condition and supporting production of vitamin D and red blood cells. If the kidneys fail, dialysis and then kidney transplant may be necessary.

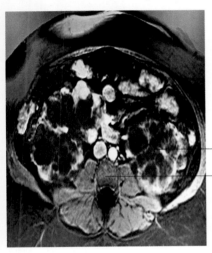

Kidneys

Spine

Polycystic kidneys
Cysts slowly grow within the renal tubules. They can reach a massive size by adulthood, gradually damaging the normal kidney tissue and causing deterioration in kidney function.

DIALYSIS

For people with acute or advanced chronic kidney failure, dialysis may be necessary to replace the kidneys' function of filtering the blood. In hemodialysis, the most common form, blood passes from the patient into a machine via a cannula in a large vein (or a surgically created join between an artery and a vein). In the machine, waste and excess water diffuse into dialysate (dialysis fluid), and the filtered blood is then returned to the body. The process takes several hours and is repeated two or three times a week. Another option is peritoneal dialysis, which makes use of the membrane around the abdominal organs.

Peritoneal dialysis
Dialysate is infused into the abdominal cavity via a catheter. Waste from the blood passes into the fluid through the peritoneal membrane, and later the fluid is replaced with a fresh supply.

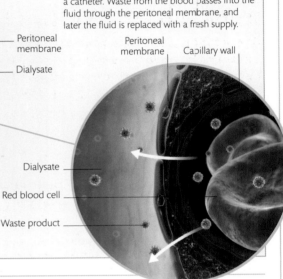

Peritoneal membrane

Dialysate

Peritoneal membrane

Capillary wall

Dialysate

Red blood cell

Waste product

INCONTINENCE

Uncontrollable urinary leakage, or incontinence, is increasingly common with age in both men and women.

There are several forms of urinary incontinence, such as stress incontinence when exercising; urge incontinence, causing an uncontrollable need to urinate; and overactive bladder syndrome, in which there is an urgent need to urinate but no flow. Various illnesses and physical weaknesses can cause incontinence, such as prostate problems in men and poor muscle tone in women. Diagnosis may include urodynamic tests to assess urinary tract function, including flow rates, pressure in the bladder, and urethral sphincter action. Management may involve diet and lifestyle changes, physical therapy, drug treatments, or occasionally surgery.

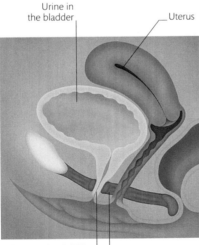

Urine in the bladder

Uterus

Weakened pelvic floor muscle

NORMAL BLADDER

Urethra | Pelvic floor muscle

INCONTINENT BLADDER

Stress incontinence
This results from weakness in the external urethral sphincter and pelvic floor muscles. Coughing or exercise causes pressure in the bladder to exceed that in the urethra, leading to leakage.

KIDNEY TUMORS

Kidney tumors commonly metastasize— tumors that spread to other organs— but cancer may also develop from the kidney tubule cells.

The first signs are usually hematuria (blood in the urine), back pain, abdominal swelling, and anemia. Less often, symptoms relating to the kidneys' other functions, such as hormonal syndromes and high blood pressure, develop. Kidney cancers spread early, particularly to the lungs, liver, and bone, and symptoms of metastasis, such as breathlessness and bone pain, can occur first. Diagnosis is made by ultrasound and CT scans, and biopsy (tissue sampling) to confirm the stage of the tumor. Treatments include removal of the kidney, chemotherapy, and immunotherapy.

BLADDER TUMORS

Most tumors in the bladder arise in the lining cells on the bladder wall, but they can also develop from muscle and other cells within the bladder.

Bladder tumors are more common in smokers; in men; and in people whose jobs involve exposure to carcinogens in the rubber, textile, and printing industries; and those with chronic irritation from bladder stones or the tropical worm infection schistosomiasis. Growth often goes unnoticed; the tumor may only be found when symptoms such as blood in the urine or urinary blockage arise, or when weight loss or anemia develop. Treatments include radiation therapy, removal of the tumor or the bladder, and diversion of urine via the bowel.

Bladder cancer cell
Most bladder cancers develop from the epithelial cells lining the bladder, and can be very advanced before they cause the typical symptoms of blood in the urine or an abdominal swelling.

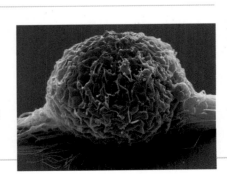

FEMALE REPRODUCTIVE SYSTEM DISORDERS

The functioning of the female reproductive system involves complex physical and hormonal interactions, and disorders can result from disturbances in a number of different tissues. In some cases genetic influences play a role.

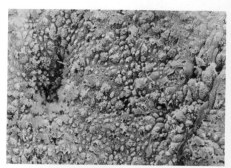

Endometriosis
Endometrial cells, shown in green and yellow on this electron micrograph, are lying on the surface of an ovarian cyst. They respond to cyclical hormones, causing bleeding into the pelvic cavity.

BREAST CANCER

The most common cancer in women, it can develop in part of the breast or in the nearby lymph nodes. It accounts for 15 percent of female cancer deaths.

Breast cancer most often occurs in women aged 45–75 years, and it is rare before the age of 35 years. It affects 1 in 9 women. A small number of cases occur in men. Up to 1 in 10 cases are due to genetic predisposition; the most important genes involved are called BRCA1 and BRCA2. Other risk factors include lack of exercise, obesity, excessive alcohol consumption, hormone therapy, and previous breast cancer).

The most common type of breast cancer is ductal adenocarcinoma, which arises in the milk ducts, but lumps may appear anywhere in the breast tissue or in the nearby lymph nodes. The first symptoms are often a painless lump, skin changes, or a nipple that becomes inverted (turned inward) or develops a discharge. The cancer may be diagnosed by a physical examination, or by ultrasound or mammogram imaging and biopsy (study of

Cancerous tumor

Breast cancer mammogram
A mammogram is an X-ray of the breast. It shows any tumors or other lumps as dense, white areas in the breast tissue. Screening for breast cancer is done by a mammogram.

a tissue sample). Further tests, such as blood tests, X-rays, or CT scans, may be done to find out if the cancer has spread. Possible treatments include surgical removal of the cancer, chemotherapy, and radiation therapy. Symptoms need to be detected as soon as possible for the best chance of treatment, so women aged 50–75 years (who are at highest risk) are offered screening mammograms.

ENDOMETRIOSIS

In this condition, cells belonging to the endometrium (uterus lining) grow in parts of the body outside the uterus.

Abnormal growths of endometrial cells are most commonly found on the ovaries or in the abdominal cavity, but they can also occur in the lungs, heart, bone, and skin. The cause is not known, but theories include reverse menstrual flow or cells spreading via blood and lymph vessels. Some women have no symptoms, but others experience severe

period pain, vaginal or rectal bleeding, pain during intercourse, or reduced fertility. Treatment includes anti-inflammatory drugs, hormones such as progesterone or the contraceptive pill, or surgery to remove the deposits.

FIBROIDS

These noncancerous growths of the smooth muscle inside the uterus often cause no symptoms, but some can grow to an enormous size.

Fibroids affect about 1 in 5 women and are more common in those who have never been pregnant. It is not known why they develop, but they are dependent on the hormone estrogen, so they usually shrink after menopause. They can cause bloating or swelling; abdominal and back pain; heavy, painful periods; and infertility. During childbirth, large fibroids can cause obstruction. Fibroids may be located by an ultrasound scan and may be treated with anti-inflammatory drugs or hormones. Surgery may be needed to remove persistent problematic growths.

BREAST LUMPS

There are several possible kinds of breast lump, of which cancer is only one. The most common cause of breast lumps in women before menopause is fibroadenosis, or fibrocystic disease. In this condition, some of the breast cells become overactive, possibly in response to hormonal changes, producing a thickened but noncancerous area (a fibroadenoma) that can be felt as a lump. Typically, women notice one or more painful lumps, which vary through the menstrual cycle.

Cysts (lumps filled with fluid) are more common in women near menopause, and may cause nipple discharge. Usually, a lump subsides over the next menstrual cycle, but persistent lumps need further investigation to rule out cancer. Another possible problem is nonspecific lumpiness and tenderness, which may occur or get worse before menstrual periods; this may also be related to hormonal changes.

Sites of breast lumps
Lumps can develop anywhere in the breast, but occur most commonly in the upper outer quadrant, near the armpit.

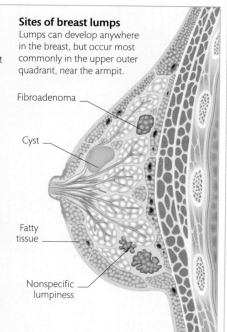

Fibroadenoma

Cyst

Fatty tissue

Nonspecific lumpiness

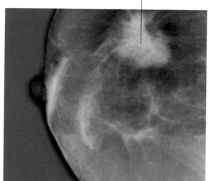

Subserosal — Fallopian tube

Intramural

Ovary — Submucosal

Uterus — Cervical

Sites of fibroids
Fibroids can occur in any part of the uterus wall and are named according to their site: for example in the cervix (cervical), or in the tissue layer in which they occur.

MENSTRUAL DISORDERS

A woman's usual menstrual cycle can be disturbed by a variety of factors, both physical and psychological.

The cycle is controlled by complex hormonal influences from the brain, ovaries, and other tissues. Follicle-stimulating hormone (FSH) stimulates egg release in the first half of the

cycle, and luteinizing hormone (LH) stimulates thickening of the uterus lining in the second half with estrogen, progesterone, and other hormones involved. Short-lived disorders are common because of variation in these hormones, or due to dieting, lowered immune or mental states, medications, or other diseases. Heavy periods, dysmenorrhea (painful periods), or amenorrhea (missing a period), may occur with no serious effects. Bleeding at abnormal times, or recurrent or persistent period problems may require further investigation.

OVARIAN CYSTS

These fluid-filled sacs in the ovary are related to cyclical changes; most are benign but a few can be cancerous.

During the menstrual cycle, a follicle grows around an egg within the ovary, and after the egg is released, the empty follicle (corpus luteum) shrinks away. Both growing and empty follicles can develop into "functional cysts," the most common type, which usually disappear on their own. Between 6 and 8 percent of women have polycystic ovarian syndrome (PCOS), a disorder in which multiple cysts grow. PCOS is associated with hormone imbalance and high testosterone levels, and can cause hairiness, obesity, irregular periods, reduced fertility, and acne. Diet and weight loss may help control it, but some women need hormone treatment. Occasionally, cysts can become cancerous, especially if they grow after menopause.

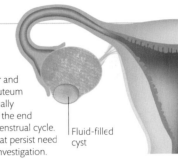

Cyst
Follicular and corpus luteum cysts usually shrink at the end of the menstrual cycle. Those that persist need further investigation.

Fluid-filled cyst

OVARIAN CANCER

Although less common than breast cancer, ovarian cancer can be more dangerous, since it often produces no symptoms until it has already spread.

Most often developing in women aged 40–70 years, ovarian cancer is more common in women with a family or personal history of breast, ovarian, or colon cancer, those who have endometriosis, obese women, and smokers.

Oral contraceptives may give some protection against ovarian cancer, because they suppress ovulation, but hormone therapy (HT) may slightly increase the risk because the cancer is often sensitive to estrogen. Symptoms develop late in the disease. They may include persistent abdominal discomfort and swelling, back pain, weight loss, and, less commonly, irregular vaginal bleeding, trapping of urine in the bladder, and peritonitis (inflammation of the abdominal lining). The cancer can spread to the uterus and intestines, and on through the lymphatic vessels and the blood.

It maybe diagnosed by examination, scans, or biopsy (study of a tissue sample). As much of the tumor as possible may be removed by surgery, and chemotherapy is used to destroy cancer cells before and after surgery.

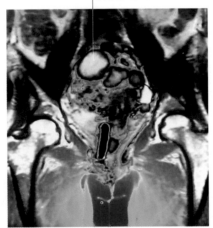

Ovarian tumor

Ovarian cancer
This colored MRI scan of the abdomen shows ovarian cancer (brown, upper center) within the tissues contained inside the pelvic cavity.

CERVICAL CANCER

The development of cancer in the cervix is most common in women aged 30–40. It has been linked to infection with human papillomavirus (HPV).

Cervical cancer is less common than it used to be, mainly because of regular testing programs. It develops slowly and can be detected through screening and treated at an early stage. Risk factors include having multiple sexual partners, smoking, and having many children. The most common symptom of cervical cancer is abnormal vaginal bleeding. Diagnosis is made by colposcopy—an examination of the cervix using a magnifying device—and biopsy of the tissue. Other tests may be carried out to check whether the cancer has spread. The cancer is treated by surgery to remove part or all of the cervix or uterus, and chemotherapy or radiation therapy may also be needed. The outcome depends on how severe the changes are and how far the cancer has spread. The availability of a vaccine against HPV infection should reduce its frequency.

CERVICAL SCREENING

The "smear" or "pap" test is a regular screening for cervical cancer and has been a successful initiative in reducing deaths from the disease. During the test a sample of cells is removed from the cervix and examined for abnormalities. Most cell changes are minor and disappear within 6 months, but more serious or persistent changes may need treatment. Precancerous cells can be detected early; they are most often found in women under 35.

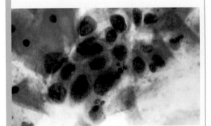

Cervical smear test
The darker areas of this smear test show precancerous cells. Cervical screening can help detect the disease at an early, treatable, stage and prevent cancer from developing.

UTERINE CANCER

Most cancers in the uterus arise from a tumor in the lining (endometrium). In rare cases a sarcoma (cancer of the muscle) can develop.

Endometrial cancer is rare under the age of 50. It usually produces irregular periods, abnormal post-menopausal bleeding, or bleeding after intercourse, sometimes with pain or discharge. The cause is unknown but is linked to excessive estrogen. Risk factors include obesity (fat cells produce some estrogen); early menarche (onset of periods), late menopause, or childlessness; and endometrial hyperplasia (overgrowth of endometrium), or other rare estrogen-producing tumors. The diagnosis is confirmed by ultrasound scan and biopsy. The main treatment is surgery, although radiation therapy, hormone treatment, or chemotherapy are sometimes needed.

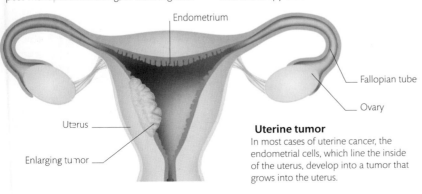

Endometrium

Fallopian tube

Ovary

Uterus

Enlarging tumor

Uterine tumor
In most cases of uterine cancer, the endometrial cells, which line the inside of the uterus, develop into a tumor that grows into the uterus.

PELVIC INFLAMMATORY DISEASE

Inflammation of the uterus and fallopian tubes can cause infertility and an increased risk of ectopic pregnancy.

Pelvic inflammatory disease (PID) most often results from a sexually transmitted disease (STD) that goes unnoticed for weeks or months. Risk factors include a new sexual partner, previous PID or STD, or insertion of an IUD (intrauterine device). There may be abnormal vaginal bleeding, pain, discharge, fever, or back pain, but some women have no symptoms. Left untreated, PID can cause inflammation, thickening, cyst formation and scarring, leading to infertility. Diagnosis is confirmed by swabs of the area, ultrasound, and laparoscopy (a keyhole procedure used to examine the fallopian tubes). PID is treated with antibiotics; partners should also be checked for infection.

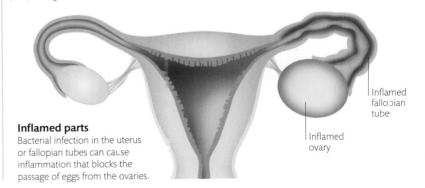

Inflamed fallopian tube

Inflamed ovary

Inflamed parts
Bacterial infection in the uterus or fallopian tubes can cause inflammation that blocks the passage of eggs from the ovaries.

MALE REPRODUCTIVE SYSTEM DISORDERS

The functioning of the male reproductive system involves complex physical and hormonal interactions between the testes, penis, prostate gland, and seminal vesicles; the pituitary and hypothalamus in the brain; and the adrenal glands, liver, and other tissues. Disruption in any of these tissues can result in disorders.

HYDROCELE

A collection of fluid around the testis, a hydrocele may be benign, or it may be a sign of underlying disease that needs further investigation.

Hydrocele is commonly seen in newborn boys. It is thought to arise in the fetus as the testes descend from the abdomen into the scrotum and the passage down which they move then fails to close, allowing abdominal fluid to enter the scrotum. There may be an associated hernia, as part of the bowel can also protrude through the passage to the scrotum. The hydrocele is usually reabsorbed as the baby grows. If it persists after 12–18 months of age, then surgery may be needed to drain it and close the passage. In older men, a hydrocele can develop slowly, often reaching a significant size before a man consults his doctor. There is usually no obvious cause, but occasionally the fluid may come from inflammation of the testis due to infection, injury, or malignancy. An ultrasound scan may be performed to aid detection of any underlying problem. Management may involve drawing off the fluid or treatment of underlying disease.

Swollen testis
The fluid of a hydrocele is contained within a double-layered membrane partially surrounding the testis but not the epididymis, which can be felt above and behind the swelling.

Bladder

Urethra

Epididymis

Scrotum

Testis

Fluid

EPIDIDYMAL CYSTS

These very common, benign, fluid-filled swellings occur in the upper part of the epididymis, the coiled tube that stores sperm from the testis.

Most common in middle-aged and older men, epididymal cysts often occur in both testes, and are painless. They can grow to any size but do not need removal unless they become painful or too large. There is a link to genetic disorders including cystic fibrosis and polycystic kidney disease. A doctor may be able to detect them by physical examination: cysts differ from hydroceles by the fact that a doctor can feel above the swelling, and from testicular cysts because they can be felt as separate from the testis. An ultrasound scan, or very rarely a fluid sample from the cyst, confirms the diagnosis. If painful or bulky, cysts may be removed surgically.

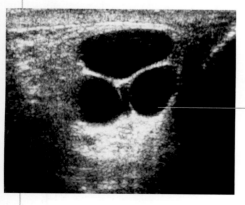

Epididymal cyst

Ultrasound scan of epididymal cysts
This scan shows three fluid-filled epididymal cysts lying within the epididymis at the head of the testis. These develop slowly and are harmless.

TESTICULAR CANCER

The most common cancer in men aged 15–40, testicular cancer usually causes a painless lump within the body of one testis. It is becoming more common.

Risk factors for testicular cancer include having undescended testes, a family history, European ethnicity, and, less commonly, being HIV-positive. There are various types of testicular cancer. Half are seminomas, which arise from the seminiferous tubules (structures responsible for sperm development). The rest, mainly teratomas, grow from other cell types and may require more aggressive treatment.

Diagnosis is confirmed by ultrasound scan and biopsy (removal of cells or tissue for examination) or testis removal if there is a strong possibility of cancer. Chemical markers in the blood may indicate certain tumor types, but a negative result does not rule out all cancers. More than 90 percent of testicular cancers can be cured. Treatment is by surgical removal, and then chemotherapy or radiation therapy. However, these treatments can cause sterility, so semen may be put into storage, to be used later for artificial insemination. Regular self-examination reveals most lumps at an earlier stage, giving a better outlook.

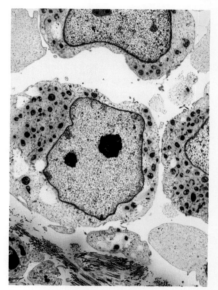

Section through cancer cells
The cells of a malignant teratoma, a cancer of the testis, are shown here as rapidly dividing cancer cells with large, irregular nuclei (pale brown) and green cytoplasm.

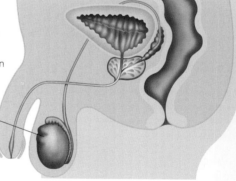

Cancer

Tumor of testis
Testicular tumors of this size are often painless but are noticeable on self-examination as a lump or painful general swelling of the groin or testis.

ERECTION DIFFICULTIES

Difficulty in achieving or maintaining a penile erection is a common problem for men, and can be an indicator of psychological stress or physical disease.

Defined as an inability to achieve or keep an erection, erection difficulties range from insufficient hardness to complete inability to achieve penetration. The simplest causes include fatigue, alcohol, stress, or depression. This experience can then set up performance anxieties that perpetuate the problem. Physical causes are usually due to poor blood supply, as in peripheral vascular disease, or neurological disorders, as in multiple sclerosis, or a combination of both, as in advanced or uncontrolled diabetes. Treatment includes counseling and reassurance, treatment of any underlying disease, and, for more persistent problems, medical therapies such as drugs.

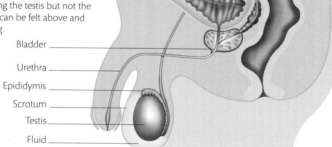

PROSTATE DISORDERS

A walnut-sized gland at the base of the bladder, surrounding the urethra the prostate secretes an alkaline fluid to protect and nourish sperm. The most common prostate gland disorder is benign prostatic hyperplasia (BPH), in which the prostate enlarges with age, sometimes obstructing the flow of urine through the urethra. The cause is unknown, but by the age of 70, about 70 percent of men are affected. The prostate can also become infected or inflamed. Cancer may develop from any of the cell types in the gland.

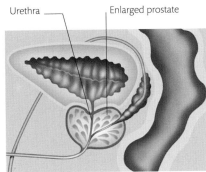

Bladder   Prostate   Urethra   Enlarged prostate

Normal prostate
The prostate gland sits underneath the bladder, surrounding the top of the urethra. It secretes prostatic fluid, which is combined with sperm.

Enlarged prostate
As the prostate enlarges it constricts the urethra, causing a poor, dribbling flow and frequent need to pass urine. Total blockage may require surgery.

ENLARGED PROSTATE

There are several possible causes of an enlarged prostate, including benign prostatic hyperplasia (BPH), prostatitis, and benign or cancerous tumors.

Most men are unaware of their prostate gland, which lies just beneath the bladder, until they reach middle age, when disorders affecting the gland are common. Symptoms include an urgent need to urinate, difficulty passing urine, poor flow, dribbling, erectile dysfunction, or retention of urine. The most common cause of these problems is benign prostatic hyperplasia (BPH), or noncancerous enlargement of the gland. To confirm diagnosis of BPH, and distinguish it from the much rarer prostatic cancer, a physical examination of the prostate, often with an ultrasound scan, biopsy, and PSA test (see below), is done. Urine flow studies and cystoscopy (an internal camera inspection of the bladder) may also be carried out. If the symptoms affect quality of life, medication may be given to relax the smooth muscle of the prostate and bladder neck, or shrink the prostate gland, to improve urine flow. Surgery may be also required to reduce the pressure on the bladder and urethra, or to remove the gland altogether.

PROSTATITIS

Inflammation or infection of the prostate gland, prostatitis can be acute (short-lived) or chronic (long-lasting).

The term prostatitis covers several conditions that have similar symptoms. Acute bacterial prostatitis is a relatively rare but serious condition that may require admission to the hospital but can be treated effectively. Chronic bacterial prostatitis is a long-lasting bacterial infection that can spread to the bladder and kidneys. In some cases no bacteria are found but persistent pain occurs. Symptoms include fever, chills, and pain in the lower back. Chronic nonbacterial prostatitis is the most common type of prostatitis. It is more difficult to treat as its cause is unknown. Symptoms include pain in the groin and penis, and difficulty and pain when urinating. All forms of prostatitis are diagnosed by testing urine or blood for STDs, or by massaging the prostate to obtain samples of prostatic fluid, which is tested for infectious organisms. Chronic and acute bacterial prostatitis can be successfully treated with antibiotics, although the condition can recur. There is no single recommended course of treatment for nonbacterial prostatitis.

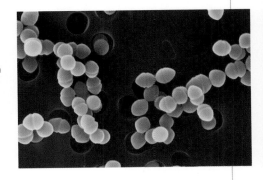

Bacteria associated with prostatitis
Escherichia coli bacteria live in the bowel in large numbers and are the most common infective cause of acute prostatitis.

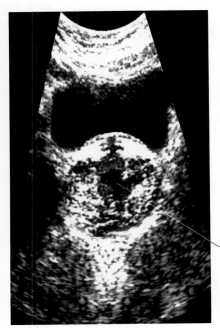

Ultrasound scan of cancerous prostate gland
Rectal ultrasound scan of the prostate can show the type of enlargement and give clues to the cause, such as tumors or inflammation.

Prostate gland

PROSTATE CANCER

The most common cancer in men, prostate cancer is rare before the age of 50, and often grows slowly and silently.

Prostate cancer often causes few symptoms, and is often only revealed late on, after the cancer has spread. Because it is a cancer of older men, who may have other health problems, it is often not the cause of death. It is more common in men with a family history and in African-American men. Cancer can arise from any of the prostate cell types, but most are adenocarcinomas, developing in the gland cells. The diagnosis is confirmed by physical examination, ultrasound scan, PSA test, and biopsy (tissue sample). Bone and liver scans or MRI may show how far the cancer has spread. Treatments depend on the stage of the cancer and the age, health, and wishes of the man, but include removal of the prostate gland, with radiation therapy, chemotherapy, and hormone therapy to block the effect of testosterone and thus limit tumor growth.

PSA TESTING

Prostate-specific antigen (PSA) is a protein produced by cells in the prostate that circulates in the bloodstream. Prostate cancer can cause higher levels of PSA in the blood, so blood samples can be used to help test for this. However, other prostate problems, such as BPH (benign prostatic hypoplasia and prostatitis, can cause raised blood levels of PSA so further testing may be required. For men with prostate disease, PSA levels may be monitored to detect any advance in disease and to plan treatment.

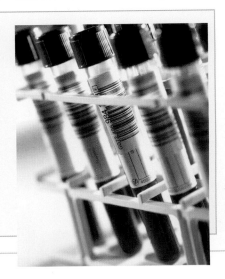

SEXUALLY TRANSMITTED DISEASES

Most sexually transmitted diseases (STDs) may reduce quality of life and lead to chronic health issues, including pain and infertility. More serious infections, such as HIV and syphilis, can be fatal. The incidence of all STDs is increasing despite medical advice about prevention.

CHLAMYDIA

The most common bacterial STD, chlamydia can affect both men and women, causing long-term pain and reduced fertility.

Chlamydial infection is now thought to affect 1 in 10 sexually active young people and many older men and women. The bacterium that causes the infection, *Chlamydia trachomatis*, is carried in semen and vaginal fluids and passed on during sexual contact. It lives in the cells of the cervix, urethra (the tube leading from the bladder to outside the body), and rectum, or in the throat, and, rarely, in the eyes, where it can cause conjunctivitis.

Many of those infected report no or only mild symptoms. As a result, the infection may go undetected for weeks or months, causing inflammation that can reduce both partners' fertility. If symptoms do occur, women tend to notice a slight vaginal discharge, pelvic pain or pain on intercourse, and irregular vaginal bleeding. Men may have urinary pain, urethral discharge, or testicular and prostatic discomfort. In the long term, damage to the fallopian tubes in women causes scarring, increasing the risk of ectopic pregnancy and infertility. The infection can also spread to the liver. Both sexes occasionally suffer an associated inflammation of the joints, urethra, and eyes, called Reiter's syndrome; this is more common in men. During pregnancy chlamydia can be passed to the baby, causing pneumonia or conjunctivitis at birth. The infection can be diagnosed with a urine sample from men and a cervical or vaginal swab from women, and is treated with antibiotics. Condom use and tracing sexual contacts both play an important part in stopping the spread of chlamydia.

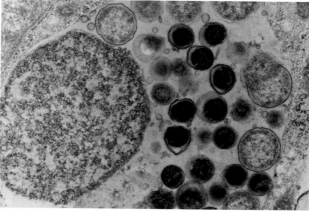

Cell infected with chlamydia
The bacterium multiplies over 48 hours before the cell bursts, releasing new organisms to spread to surrounding cells.

Multicolored condoms
Most condoms are made of latex: this can dissolve on contact with some toiletries, but water- and silicone-based lubricants are safe for use with condoms.

GONORRHEA

A bacterial infection mostly confined to the genital tract, gonorrhea can cause permanent damage and reduced fertility in both men and women.

The bacterium *Neisseria gonorrhoeae* is passed during sexual contact. Infection may cause genital pain, inflammation, a green or yellow discharge from the penis or vagina, and pain on urination in the next few days or even many months later. Women tend to experience recurrent episodes of abdominal pain, irregular bleeding, and heavy periods; men may notice testicular or prostatic pain. The bacterium can live in the cervix, urethra, rectum, and throat, and may spread through the blood to other areas such as the joints, causing arthritis and tenosynovitis, and a rash. During vaginal delivery, an infected mother may pass it to her baby, causing eye and other infections.

Gonorrhea can be detected by a urine sample, or penile, cervical, throat, or eye swabs, and is usually easy to treat with antibiotics. Left untreated, however, chronic inflammation scars women's fallopian tubes, reducing fertility and increasing the risk of ectopic pregnancy, because the egg cannot pass down the tube properly. Chronic infection also puts future sexual partners at risk. Condom use and tracing sexual contacts can help prevent the spread of gonorrhea.

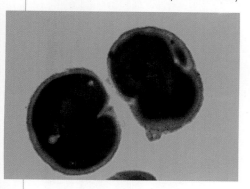

Gonorrhea bacterium
Microscopic view of *Neisseria gonorrhoeae*, the bacterium that causes gonorrhea, which can often be quickly identified under the microscope.

URETHRITIS

Known as nonspecific urethritis or NSU, inflammation of the urethra can be due to infection or a variety of other causes.

NSU can occur in both men and women. Infective causes include STDs such as herpes, chlamydia, and *Trichomonas vaginalis*, as well as nonsexually transmitted infections such as thrush (candida) and bacterial vaginosis. The symptoms of NSU may also occur without infection, possibly due to a chemical sensitivity to soap, spermicide, antiseptics, or latex in condoms. The symptoms depend on the cause but may include discharge, difficulty or pain when urinating, frequent urination, and itchiness or irritation at the end of the urethra. Left untreated, the inflammation may spread, causing testicular and prostatic pain in men, or (with chlamydia) pelvic inflammatory disease (see p.485) in women. Urine tests and swabs help identify infection, and drugs may be used to kill infective organisms. Prevention may include using only nonlatex condoms.

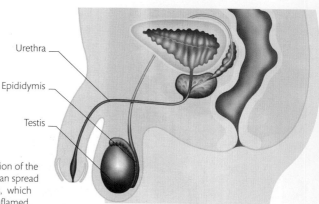

Urethra

Epididymis

Testis

Symptoms of NSU
Urethritis causes inflammation of the urethra. If left untreated it can spread to the testis and epididymis, which can become swollen and inflamed.

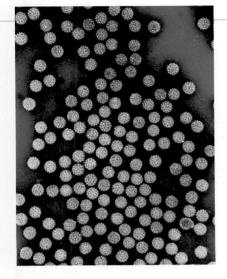

GENITAL WARTS

Some strains of human papillomavirus (HPV) can cause fleshy growths, or warts, in the genital and anal areas.

There are over 100 strains of HPV, although not all cause genital warts. Strain types 6 and 11 are responsible for 90 percent of genital warts. HPV infects the epidermis (skin surface)

Human papillomavirus

The virus that causes genital warts can enter the body through the skin around the genital area, so condom use may not be fully protective.

and mucous membranes, and is spread during genital contact of any sort. In many people, there is no sign of infection and the virus is not carried for long. Sufferers may also be unaware that they carry the virus because warts can take weeks, months, or years to develop. They show as small, painless, fleshy lumps in the genital or anal area, internally or externally. Genital warts do not have serious consequences and most will eventually disappear, although this may take months or years, during which time they remain infectious. Treatment using creams, freezing, electrocautery (removing the wart by burning it with a low-voltage electrical probe), or laser can clear them more quickly: in the meantime, condom use is advised to help prevent spread of the infection.

Anal warts
Also called condyloma acuminata, genital warts are highly contagious. These small, cauliflower-shaped lesions may cause itching, bleeding, and discharge, or may not be noticed.

SYPHILIS

Once rife and untreatable until the development of antibiotics, this infection is now increasing again. Left untreated it can affect many parts of the body.

The bacterium that causes syphilis is passed on during intercourse, or by skin contact with a syphilitic sore or rash. A painless sore called a chancre usually develops on the genitals, but it can occur on the fingers, buttocks, or in the mouth. The chancre may take up to six weeks to heal, and may go unnoticed. The next stage, secondary syphilis, occurs several weeks later, with a flulike illness, a non-itchy rash, and sometimes wartlike patches on the skin. The final stage, tertiary syphilis, may take years to develop. It affects parts such as the blood vessels, kidneys, heart, brain, and eyes, and can cause mental disorder and death. The first and second stages can be treated by antibiotics, but the damage in the third stage is permanent.

GENITAL HERPES

A blistering, painful rash caused by herpes simplex viruses HSV1 and HSV2, this infection may recur repeatedly.

The herpes simplex viruses enter the body via close contact with skin or moist membranes. Both HSV1 and HSV2 can cause genital and oral lesions, either within days of infection or weeks or months later. These small, painful sores can last for several weeks before subsiding. Other symptoms include flulike illness, fatigue, aches, pain on urination, and swollen glands. Many people have only a mild, single infection, but some experience regular relapses. These are often triggered by other illnesses and, although usually less severe each time, can be debilitating. The virus can be passed even by people with no active lesions. Pregnant women with active sores can pass the virus to their baby during pregnancy or in childbirth. Attacks of herpes may be treated with antiviral medication; this is most effective as soon as symptoms start.

Herpes simplex lesion

The lesions of genital herpes are typically painful irregular blisters, which break down to form ulcers, with a raised, reddish outer edge and weeping inner area.

HIV AND AIDS

Infection with human immunodeficiency virus (HIV) is lifelong and can lead to acquired immunodeficiency syndrome (AIDS), a life-threatening condition.

HIV may be passed by contact with bodily fluids including blood, semen, vaginal fluids, and breast milk. (The level of HIV in urine and saliva is thought to be too low to be infectious.) Initially there may be a short flulike illness (called sero-conversion illness), mouth ulcers, or rash for up to 4 weeks, or no symptoms at all. The virus then multiplies in the body over several years, damaging the immune system. This damage can be measured by the reduction in the number of CD4 (T-helper) cells, which are a vital part of the immune system's defense against infection. As the disease progresses, fever, night sweats, diarrhea, weight loss, swollen glands, and recurrent infections may occur. In its late or advanced stage, known as AIDS, the CD4 count drops very low, and a variety of immune system-related conditions develop. These include opportunistic infections caused by organisms that live harmlessly in healthy people, such as *Pneumocystis* pneumonia,

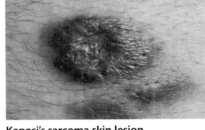

Kaposi's sarcoma skin lesion
These tumors start as small, painless, flat areas or lumps, colored brown, red, blue, and purple, which look like bruises and grow until they merge.

candida, and cytomegalovirus, and a skin cancer called Kaposi's sarcoma.

Those with HIV can be regularly monitored and have opportunistic infections treated promptly. People with HIV remain infectious throughout their life, but can avoid passing it on by practicing safe sex using condoms. Infected mothers, who can pass HIV to their baby before or during birth, and by breastfeeding, may be offered antiretroviral drugs and cesarian delivery. The only way to positively diagnose HIV is an antibody blood test. This can take up to 3 months to become positive after HIV exposure.

Although there is no vaccine or cure for HIV, "HAART" (Highly Active Antiretroviral Therapy) has changed AIDS from a rapidly fatal disease to a chronic condition with many complications, most of which can be managed.

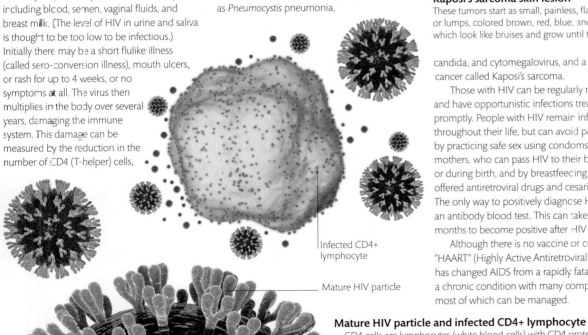

Infected CD4+ lymphocyte

Mature HIV particle

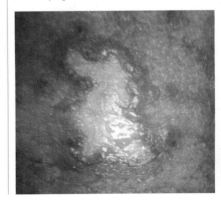

Mature HIV particle and infected CD4+ lymphocyte
CD4 cells are lymphocytes (white blood cells) with CD4 protein molecules on their surface, usually responsible for starting the body's response to invading viruses. HIV binds to CD4 in order to enter the cell, damaging the cell in the process.

INFERTILITY

More than 1 in 10 couples experience infertility—difficulty conceiving a baby. Most male problems center on poor sperm function, but for women fertility depends on a complex interaction between hormonal activity, egg production, and the ability to carry a fetus.

OVULATION PROBLEMS

Ovulation takes place when an egg is released and is ready to be fertilized. Eggs released intermittently or not at all can cause problems with conception.

During the normal 28-day menstrual cycle, many ova (eggs) develop, each in a follicle, in the ovary. Usually one egg is released every month; the other follicles and eggs wither. The process is influenced by many hormones, including the gonadotropins, follicle-stimulating hormone (FSH) and luteinizing hormone (LH), estrogen, and progesterone. Around day 14, the dominant follicle ruptures, and the egg is released into the fallopian tube and travels to the uterus. The control of this process relies on hormonal interaction between the hypothalamus and pituitary gland in the brain, and the ovaries. Factors that can disrupt this process include pituitary and thyroid gland disorders, polycystic ovary syndrome, being under- or overweight, excessive exercise, and stress. Tests are used to determine hormone levels and find out if ovulation is taking place. Treatment may include the use of gonadotropin-releasing hormones, progesterone, and clomiphene to stimulate ovulation.

Cystic follicles

Polycystic ovary
A common condition including multiple ovarian cysts and abnormal hormone levels, polycystic ovary syndrome (PCOS) can cause infertility.

UTERUS ABNORMALITIES

A variety of abnormalities, from defects in development to growths in the uterus, can cause problems with fertilization and the ability to carry a fetus.

As a female fetus develops, the uterus and vagina form from two halves that fuse together. Incomplete fusion can cause abnormalities such as a doubling of the uterus (bicornuate uterus) or the cervix, or a septum (membrane) dividing the vagina. These problems may, in some cases, reduce fertility in adult women. Some problems only become apparent in early pregnancy if an abnormally shaped uterus prevents proper fetal development. Late miscarriage and premature or difficult labor are more likely problems and may arise due to poor implantation of the egg or restricted growth of the fetus and uterus. A minor problem arises, either during development or, more rarely, as a result of scarring, if the hymen, a thin membrane that blocks the entrance to the vagina, is unbroken. This prevents the flow of menstrual fluid, causing a swelling to grow as blood collects every month. It also prevents penetration during intercourse, so fertilization of the egg cannot occur.

Some abnormalities can easily be remedied by surgery—for example, removal of a vaginal septum. Other deformities may need surgical reconstruction. Some abnormalities develop in adulthood. These include tumors and a tightening of the cervix, which can occur after a cone biopsy (used to investigate precancerous cervical changes). The most common tumors that affect uterine shape are fibroids and cervical or endometrial polyps. The risk of fertility problems increases with the size of these growths; it also varies with their position within the uterus. Most of these tumors are noncancerous, but they may need to be removed to improve the chance of conception.

PROBLEMS WITH EGG QUALITY

Egg quantity and quality both decline significantly with age, particularly from the mid-30s.

Poor-quality eggs may not be fertilized, or may be fertilized but not develop properly to achieve implantation in the uterus. If implantation does occur, there is a higher than average chance of miscarriage. Egg quality depends on several factors including normal chromosomes, the ability to combine the chromosomes with those in the sperm, and stored energy to enable cell splitting after fertilization. This energy is held in particles called mitochondria, but levels drop as the eggs age. Smoking is one of the external factors known to reduce egg quality. This condition is hard to treat, although IVF can be used to select good eggs or embryos.

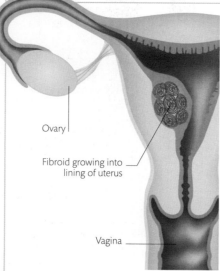

Ovary

Fibroid growing into lining of uterus

Vagina

Fibroid
Benign (noncancerous) smooth muscle tumors of the uterus, fibroids can grow large enough to disturb the interior of the uterus, possibly interfering with egg implantation.

BLOCKED FALLOPIAN TUBES

Damage to the fallopian tubes can affect egg transport and embryo implantation, or even prevent fertilization altogether.

Endometriosis, pelvic inflammatory disease (PID), adhesions from abdominal surgery, and genetic disorders can interfere with the function of the fallopian tubes. These conditions weaken the action of the hairs lining the fallopian tube, which normally brush the egg along its length. If the egg cannot pass down the tube, the sperm will not reach it and conception will not occur. Alternatively, the egg may be fertilized inside the tube and the embryo will grow there. In these "ectopic" pregnancies, the embryo's growth may result in the tube bursting, miscarriage, hemorrhage, and serious risk to the mother. Surgery may be performed to open the tubes, but often IVF, bypassing the need for healthy tubes, may offer a better chance of pregnancy.

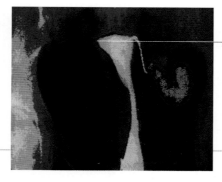

Blocked entrance to fallopian tube

X-ray showing blocked tubes
Dye can be injected through the cervix in a procedure called hysterosalpingography. This can reveal a blockage in the tubes.

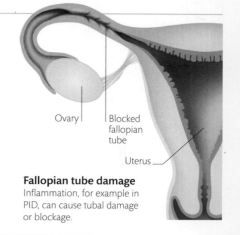

Ovary

Blocked fallopian tube

Uterus

Fallopian tube damage
Inflammation, for example in PID, can cause tubal damage or blockage.

CERVICAL PROBLEMS

The cervix is the gateway to the uterus, through which sperm pass to fertilize the egg, so any defects can reduce fertility and pose a risk of miscarriage.

Cervical cells secrete mucus, which undergoes hormonally influenced cyclical changes to aid fertilization and then protect the uterus. At mid-cycle the mucus becomes clearer, thinner, and more copious, making it easier for sperm to flow up into the uterus. Later, it thickens to provide a barrier to infection, thus protecting the fetus. Cervical problems can be structural or functional. Any congenital (present from birth) abnormal ties or polyps, fibroids (see p.484), or cysts in the cervix may block the passage of sperm. During pregnancy, cervical incompetence, in which the cervical entrance (os) cannot close fully (usually as a result of previous injury or surgery), can cause miscarriage. Functional problems include cervical mucus that stops sperm by being too thick or acidic, or by containing antibodies to sperm. Women may be treated with IVF.

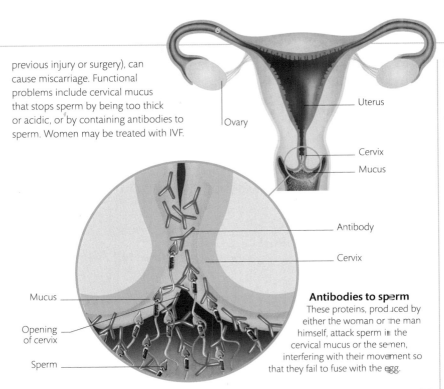

Antibodies to sperm
These proteins, produced by either the woman or the man himself, attack sperm in the cervical mucus or the semen, interfering with their movement so that they fail to fuse with the egg.

Uterus
Cervix
Mucus
Antibody
Cervix
Mucus
Opening of cervix
Sperm
Ovary

PROBLEMS WITH PASSAGE OF SPERM

Fertility may be affected by a blockage of the vas deferens, which transports semen from the testis to the penis, or by the journey of the sperm to the egg.

The sperm, which carry the man's genetic material, are made in each testis and stored in two chambers, called an epididymis. During release (ejaculation), the sperm are combined with seminal fluid from the prostate gland to form semen. This is released from the man's urethra into the woman's vagina, where fewer than 100,000 manage to enter the uterus through the cervix. By the time they get to the egg, somewhere in the fallopian tubes, there may only be 200 left. Even if all else is normal, the majority are lost through wastage as they swim the wrong way, fail to keep moving, or simply become exhausted. In addition, factors such as testicular disease, retrograde ejaculation (when semen is ejaculated, but backward), difficulty of sperm getting through the cervical mucus, uterine abnormalities, or poor fallopian tube function all reduce the chance that the sperm will meet the egg. These problems are hard to treat, but IVF offers an opportunity to bypass them.

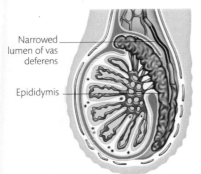

Inflamed vas deferens
Injury or infection can inflame the epididymis and the tube leading from it, the vas deferens, causing blockage that stops the release of sperm.

Narrowed lumen of vas deferens
Epididymis

PROBLEMS WITH SPERM QUALITY AND PRODUCTION

Male factors account for about one half of infertility: in particular, problems with sperm numbers, motility, abnormal shapes, and antibodies to sperm.

Testing for problems relies on semen analysis in the laboratory. Semen volume and pH, sperm numbers and concentration, motility (movement), morphology (shape), and the presence of antibodies (immune system proteins wrongly targeting the man's own sperm as invading organisms) are all assessed. A postcoital test may also be done, to test the sperm's ability to swim in the woman's cervical mucus. Factors affecting sperm quality and quantity include smoking, alcohol, chemical exposure at work, medicines and drug abuse, previous disease such as rubella and STDs, and high testicular temperature. In men with poor sperm counts, the use of ICSI (intracytoplasmic sperm injection), where only a few sperm are needed to be injected into an egg, gives a much better chance of fertilizing the egg.

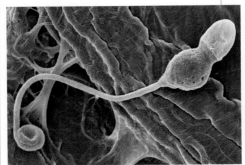

Deformed sperm
Deformed sperm exist in every ejaculate: semen analysis defines a normal sample as having at least 4 percent normal-shaped sperm.

IN-VITRO FERTILIZATION

Commonly abbreviated IVF, in-vitro fertilization is a method of artificially fertilizing an egg outside the body, culturing the embryo in the laboratory, and placing it back inside the uterus. IVF is used for most types of infertility apart from uterine anatomical abnormalities. It starts with the woman having hormone injections to stimulate her ovaries to produce large numbers of eggs which are harvested. Donor eggs or sperm may also be used.
The eggs are incubated with the sperm to achieve fertilization, although ICSI (the direct injection of a sperm into the egg) is used in around half of IVF treatment cycles. The fertilized eggs are cultured for 5–7 days, and then implanted into the uterus. In some cases, assisted hatching is done, in which the shell of the embryo at the eight-cell stage is digested by acid to improve the chances of implantation and pregnancy.

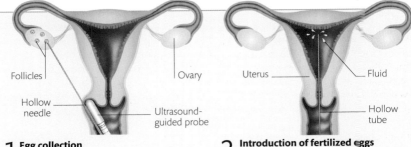

Follicles
Hollow needle
Ovary
Ultrasound-guided probe

1 Egg collection
One eggs has reached a certain maturity, they are retrieved using a needle and probe, and incubated with sperm in the test tube.

Uterus
Fluid
Hollow tube

2 Introduction of fertilized eggs
Three or four of the cultured embryos are inserted via a tube through the cervix into the uterine cavity for implantation to take place.

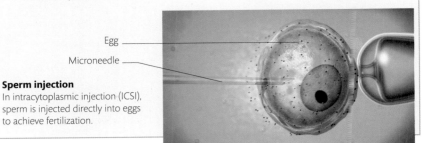

Egg
Microneedle

Sperm injection
In intracytoplasmic injection (ICSI), sperm is injected directly into eggs to achieve fertilization.

EJACULATION PROBLEMS

Sperm delivery occurs by ejaculation: a contraction of the vas deferens, seminal vesicles, ejaculatory ducts, and the muscles around the urethra.

Ejaculation problems range from complete failure to retrograde ejaculation, in which the semen passes back into the bladder rather than down the urethra. These problems can result from many muscular and neurological disorders, such as stroke, spinal injury, or diabetes, and can also occur following prostate or bladder surgery. Investigation includes semen analysis and bladder function studies. Intracytoplasmic insemination (see left) offers hope when ejaculatory failure is not treatable.

DISORDERS OF PREGNANCY AND LABOR

Normal pregnancy lasts about 38 weeks from conception, or 40 weeks from the last menstrual period. Pregnancy and labor (the process of delivering a baby) are usually straightforward. However, problems can affect the mother or baby at any stage.

ECTOPIC PREGNANCY

An ectopic pregnancy is one in which the embryo begins to grow outside the uterus, usually in the fallopian tubes.

Normally, an egg is fertilized and grows into an embryo in the fallopian tube and then implants in the uterine lining. In some cases, however, it implants outside the uterus—most commonly, in the fallopian tube. Most ectopic pregnancies end in miscarriage. If the embryo continues to grow, after 6 to 8 weeks it may cause the tube to rupture, resulting in internal bleeding, shock, and pain, a medical emergency that must be treated with surgery. Ectopic pregnancy is more likely if the fallopian tubes have been damaged due to infection, particularly chlamydia (see p.488), or surgery.

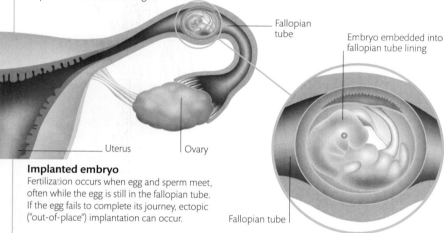

Implanted embryo
Fertilization occurs when egg and sperm meet, often while the egg is still in the fallopian tube. If the egg fails to complete its journey, ectopic ("out-of-place") implantation can occur.

Fallopian tube

Embryo embedded into fallopian tube lining

Uterus

Ovary

Fallopian tube

PREECLAMPSIA

This condition is typified by high blood pressure and edema (tissue swelling); it can be mild or life-threatening.

Preeclampsia can arise any time from 20 weeks of pregnancy to six weeks after the birth. It is more common in first and twin pregnancies.

The disorder is thought to be caused by the placenta not developing properly. The main symptoms are high blood pressure, edema (fluid buildup in the tissues), and protein leakage from the kidneys. In severe cases it can lead to eclampsia, with seizures and possible stroke in the mother, and threat to her life and that of her baby. Delivery is the only cure, and women with preeclampsia may have their baby induced before the pregnancy reaches its full term.

BLEEDING IN EARLY PREGNANCY

Affecting at least 1 in 8 pregnant women, bleeding may occur due to miscarriage or ectopic pregnancy, but in most cases it has a less serious cause.

Bleeding in the first 4 weeks is sometimes thought to result from the embryo implanting itself into the wall of the uterus—so-called implantation bleeding—and may be mistaken for a very light menstrual period. Another common cause is bleeding from the cervix (neck of the uterus) due to the development of an erosion (a raw, red area that bleeds easily) under the influence of pregnancy hormones. Bleeding may also originate from the edge of the growing placenta or be caused by an ectopic pregnancy (see left). Most episodes of bleeding in early pregnancy do not lead to loss of that pregnancy. However, heavier bleeding, with passage of clots or with cramping pain, is more likely to mean the pregnancy is failing.

MISCARRIAGE

Up to 1 in 4 pregnancies can end in miscarriage—the natural loss of a baby before the 24th week.

Pregnancies can fail for a variety of reasons. The embryo may not implant properly, or the fusion of sperm and egg goes slightly wrong so the fertilized egg cannot survive. Occasionally a fetus fails to grow (a "missed" miscarriage), and this may not be discovered until the first scan. Miscarriage can also be caused by a problem in the mother, such as weakness in the cervix (neck of the uterus), infection, or an illness such as diabetes. Often there is no obvious cause. The most common symptoms are bleeding and pain. Many affected women do not even know they were pregnant because the miscarriage occurs at or before the time the period was due. Later miscarriages tend to be more painful and distressing, with greater blood loss and more need for medical attention.

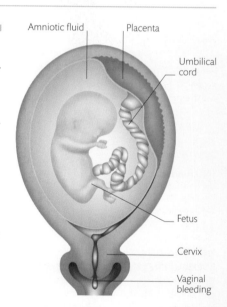

Amniotic fluid

Placenta

Umbilical cord

Fetus

Cervix

Vaginal bleeding

Threatened miscarriage
In this condition, vaginal bleeding occurs but the cervix stays closed and the fetus is alive. In many cases the pregnancy continues to a successful birth, but some may develop into full miscarriage.

PLACENTA PROBLEMS

Some complications in later pregnancy can be due to problems with the placenta, the organ that keeps the fetus alive.

In placenta previa, the placenta lies too low in the uterus, near or over the cervix (neck of the uterus). This can cause painless, bright red bleeding, often at about 29 to 30 weeks—the uterus grows rapidly at this stage. In some cases the growing placenta moves upward and the problem settles. In severe placenta previa (when the placenta covers the opening of the cervix), heavy bleeding may threaten the life of mother and baby. If the placenta is very low, normal birth may not be possible. Placental abruption is the separation of the placenta from the uterus before birth. It can cause vaginal bleeding, or trapped blood may build up behind the placenta. Abruption can cause severe pain in the mother and risk to the baby.

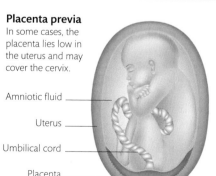

Placenta previa
In some cases, the placenta lies low in the uterus and may cover the cervix.

Amniotic fluid

Uterus

Umbilical cord

Placenta

Placental abruption
As the placenta shears off, blood escapes through the vagina or collects behind the placenta.

Placenta

Blood between uterus and placenta

Uterus

Cervix

PROBLEMS WITH GROWTH AND DEVELOPMENT

The failure of a baby to grow properly in the uterus is called intrauterine growth retardation; this problem can place a baby at risk both before and after birth.

Growth retardation is often caused by a lack of oxygen or nourishment reaching the fetus. It can result from a variety of factors affecting the mother, the fetus, or the placenta (which nourishes the baby in the uterus).

Maternal factors include anemia, which reduces the baby's oxygen supply; preeclampsia, which can reduce blood flow to the placenta; infections such as rubella, which pass to the fetus and affect development; and a prolonged pregnancy, when the placenta becomes less efficient and growth slows. Placental causes of growth retardation include anything that reduces placental blood flow,

such as smoking or alcohol use by the mother, thrombophilia (a disorder that causes a higher risk of blood clots), and preeclampsia. Problems in the fetus include infections such as rubella, blood abnormalities, genetic abnormalities that affect growth, kidney problems, Rh disease (a mismatch between the mother's blood type and that of the fetus), and being one of twins or more. The condition is also often associated

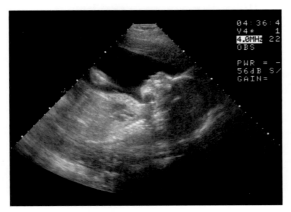

with a reduction in protective fluid around the baby. A fetus whose growth has been severely restricted is at risk of dying in the uterus, likely to have a low birth weight with an increased risk of distress during labor, and liable to complications after birth. Intrauterine growth retardation is usually identified in pregnancy, when doctors measure the growth of the uterus. Ultrasound scans may be done to measure the fetus and assess blood flow through the placenta. If there are signs that the fetus is distressed, or its growth seems to be coming to a halt, the birth may be brought forward.

Growth monitoring
Ultrasound provides an image of the fetus and can be used to monitor growth, check that the baby is developing normally, and measure blood flow in the placental vessels.

PROBLEMS DURING LABOR

An overly long or difficult labor can be stressful, exhausting, and risky for both mother and baby.

There are three stages of normal labor. In the first stage, the muscular walls of the uterus begin to contract and the cervix (neck of the uterus) gradually dilates (opens) to a width of about 4 in (10 cm). In the second stage, the baby is born. The third stage involves delivery of the placenta.

During labor the uterus contractions, dilation of the cervix, and the baby's heartbeat are monitored to detect any problems. If the first stage takes too long and the cervix opens too slowly, the mother can be weakened by pain and exhaustion and this can make the birth more difficult. A long first stage is more common in first-time mothers, whose contractions can be dysfunctional and less

effective in opening the cervix. During the first and second stages, every time the uterus squeezes, the baby's blood supply is briefly reduced. Over time, especially with prolonged labor, the baby will grow tired and stressed, with lowered oxygen levels and increased blood acidity. Small or premature babies,

babies of anemic mothers, or babies who have preexisting problems are more vulnerable. If the contractions are too weak or the baby is showing signs of distress, the mother may be given artificial hormones to strengthen the contractions, or an assisted delivery (see below) may be carried out.

Fetal monitoring
Two sensors are fitted to the mother's abdomen to record uterine contractions and fetal heart rate. Fetal distress may show as sustained rapid beats or as prolonged drops in the heart rate.

ASSISTED DELIVERY

If the mother cannot deliver the baby normally or if the birth needs to be speeded up, an assisted delivery may be carried out. Procedures include the use of forceps or Ventouse, or cesarean section. A Ventouse device is a cap fitted against the baby's head and used to help pull the baby gently as the mother pushes. Forceps are instruments that fit around the baby's head; these also help pull the baby as the mother pushes. Cesarean section is surgical delivery of the baby through the abdomen. The mother may be given a general anesthetic, or she may have a spinal or epidural anesthetic, which numbs the body below the waist.

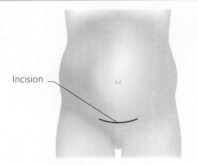

Incision

Cesarean section
In the procedure, a cut is made into the lower part of the uterus, and the baby is removed via this incision. It is used if a vaginal delivery would be too difficult or unsafe for the mother or baby.

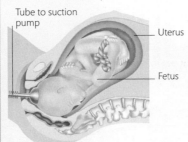

Tube to suction pump

Uterus

Fetus

Vacuum suction delivery
A Ventouse device consists of a cap held onto the baby's head by vacuum suction. The baby may have a swelling on the head afterward, but this soon disappears.

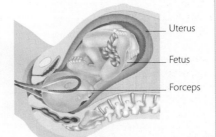

Uterus

Fetus

Forceps

Forceps delivery
Spoon-shaped obstetric forceps are carefully placed around the baby's head. As the mother pushes, the doctor pulls on the forceps until the baby's head reaches the vagina.

ABNORMAL PRESENTATION

The baby has to settle into a particular position ready for birth. Any deviation is known as an abnormal presentation, and this can make the birth more difficult.

Ideally, babies lie with their face toward the mother's back and their head down, over the cervix, ready to push against it as the uterus contracts. In an abnormal presentation the baby may be in a "breech" presentation lying with its bottom first, or it may lie head down but too high in the pelvis to push on the cervix. Occasionally babies lie crosswise, or at an oblique angle, with an arm over the cervix.

A variety of maneuvers are used to deliver babies who present abnormally, but these can cause trauma to the mother or detachment of the placenta. It is possible to turn the baby and then break the waters so that the head comes down against the cervix in a controlled way, but this can be risky because the baby may lie abnormally for a reason (such as the placenta being in the way or the mother's pelvis being too small). If the baby is lying longitudinally, normal delivery may be possible. If the mother's pelvis is too small to let the baby out, or the placenta is in the way, cesarean section (see left) is the only option.

PRETERM LABOR

If labor begins before the 37th week of pregnancy, it is defined as pre-term; if it occurs very early, the baby may suffer health problems or even die.

There are many possible causes of pre-term labor, including abnormalities in the fetus, the placenta, or the mother. The risks are greater for the baby, who may be born before the lungs (and many other organs) are fully mature. If labor begins too soon, drugs may be given to delay or inhibit contractions. Labor may be temporarily halted or delayed long enough for corticosteroids to be given to help the baby's lungs mature, so it is less likely to suffer respiratory problems.

Premature baby
Babies have survived outside the uterus at around 22 weeks, although the risks of lung, brain, and eye damage are extremely high.

ENDOCRINE DISORDERS

The endocrine system is made up of glands and tissues that secrete hormones into the bloodstream to regulate the function of other organs and body systems. Disorders of any gland can affect many other glands and disrupt one or more body systems.

TYPE 1 DIABETES

In this form of diabetes insulin-producing cells in the pancreas are damaged and produce little or no insulin, so the body cannot process glucose properly.

The body takes in glucose from food, uses it to produce energy, and stores any surplus in the liver and muscles. The levels of glucose in the bloodstream (blood sugar levels) are regulated by a hormone, insulin, that is produced in the pancreas in response to food intake. Insulin maintains a steady blood sugar level by helping body cells absorb glucose. If too little insulin is produced, or cells do not take in enough glucose, blood sugar levels become too high, which leads to diabetes mellitus. There are three main types of diabetes: type 1; type 2; and gestational diabetes (see opposite).

Without insulin, cells cannot absorb glucose. Instead, glucose builds up in the blood, and blood sugar levels gradually increase, causing symptoms such as increased thirst, urinating more than normal, nausea, fatigue, weight loss, blurring of vision, and recurrent infections.

Diabetes is diagnosed by testing the urine for sugar and ketones (an acidic by-product of fat breakdown) and blood tests, which show high sugar levels as well as a variety of other chemical changes that occur as the body tries to cope with the metabolic disturbance.

The cause of type 1 diabetes is unclear, but it may be an abnormal reaction by the body's immune system to cells in its own pancreas, triggered by a virus or other infection, and usually occurring in young adult life. As a result, insulin production is reduced or absent. The starving cells try to get energy from fat cells, disrupting the normal metabolism (the chemical reactions that keep the body functioning) and eventually a condition called ketoacidosis, which leads to coma and death.

Type 1 diabetes cannot be cured, but it can be managed by lifelong treatment with insulin to regulate blood sugar levels. The person will be counseled on healthy diet, exercise, and possible complications (see opposite). He or she will be advised to minimize factors that increase the risk of cardiovascular disease (a major risk with diabetes), such as high cholesterol levels, high blood pressure, and unhealthy lifestyle habits including overeating, smoking, and alcohol use.

Injecting insulin
Replacement insulin has to be given, as injections or via a pump, several times a day to control sugar metabolism.

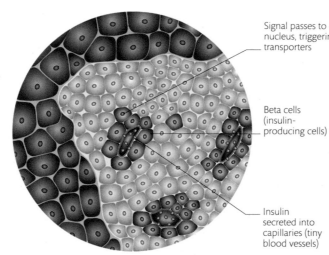

Signal passes to cell nucleus, triggering transporters

Beta cells (insulin-producing cells)

Insulin secreted into capillaries (tiny blood vessels)

Normal beta cell function
Blood sugar is regulated by beta cells, in groups called islets of Langerhans, in the pancreas. These cells secrete the hormones insulin, c-peptide, and amylin during and after eating.

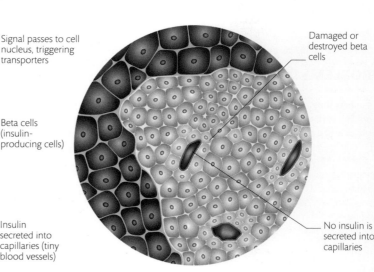

Damaged or destroyed beta cells

No insulin is secreted into capillaries

Damaged beta cells
When the beta cells are damaged, by infection, trauma, or aging, the secretion of hormones such as insulin is reduced and the body's control of blood sugar is impaired.

BLOOD SUGAR REGULATION

The blood sugar must be kept within a narrow range, so that cells have enough glucose but levels do not become too high and thus toxic. The two main hormones involved in glucose regulation are insulin and glucagon, both produced by cells in the islets of Langerhans, inside the pancreas. After food is eaten, high blood glucose levels and the release of gut hormones called incretins, stimulate the beta cells in the islets of Langerhans to produce insulin.

This hormone triggers most of the body cells to increase glucose uptake from the blood; stimulates cells to use more glucose as energy; stimulates the liver and muscle cells to store excess glucose as glycogen; and stimulates fat synthesis from glucose in the liver and adipose (fat) cells. Conversely, when blood sugar falls, such as between meals or during exercise, low sugar levels stimulate another group of pancreatic cells, the alpha cells, to secrete glucagon. This triggers the liver and muscle cells to release previously stored glucose, induces liver and muscle cells to make glucose from other dietary elements, and increases the breakdown of fats to fatty acids and glycerol for use as energy in cells.

Pancreas

Beta cell produces insulin

Alpha cell produces glucagon

Islet of Langerhans
These areas of tissue in the pancreas contain five types of endocrine cells. The islets of Langerhans are responsible for the production of hormones such as insulin, glucagon, and somatostatin, all of which are involved in blood sugar regulation.

TYPE 2 DIABETES

In this condition the pancreas secretes insulin but the body cells do not respond to it as a trigger to take in glucose, so blood sugar levels remain too high.

Body cells obtain energy by taking in glucose, which is released from food during digestion and then carried in the bloodstream to all the tissues. Normally, insulin, a hormone secreted by cells in the pancreas, helps cells absorb glucose. If too little insulin is produced, or cells do not take in enough glucose, the glucose level in the blood becomes too high, which leads to diabetes mellitus.

There are several types of diabetes; in type 2, the disorder results from a combination of decreased insulin secretion, reduced numbers of beta cells in the pancreas, and increased resistance of cells to the effects of insulin. Genetics may play a role, but type 2 diabetes is also strongly linked to obesity. The rapidly increasing incidence in most countries is thought to relate to the rise in weight problems and lack of exercise, resulting in fat storage, especially in the abdomen. The disease may go unnoticed at first, but high sugar levels may cause symptoms such as fatigue, thirst, and recurrent minor infections. If the diabetes is left untreated or poorly controlled, the chronic excess of glucose can damage the blood vessels supplying organs and tissues throughout the body, resulting in retinal damage, vision loss, kidney failure, and nerve damage; it can also increase the risk of cardiovascular diseases such as stroke, heart attack, and peripheral vascular disease (disorders affecting blood vessels in the legs and feet).

Type 2 diabetes is diagnosed using blood and urine tests to detect the excess glucose. Treatment involves regulating blood sugar levels. At first, this may involve simply making lifestyle changes such as adopting a healthy diet, doing regular exercise, and losing weight. Patients are also taught how to monitor their own blood sugar levels. However, as the disease progresses, drugs may be needed to reduce blood sugar levels. Medications may help the pancreas make more insulin or make better use of what is already there, or make the body cells more sensitive to insulin. Some patients eventually need insulin therapy (regular injections of insulin). In addition, it is necessary to control factors such as high blood pressure and high cholesterol to reduce the risk of damage to the kidneys, eyes, nerves, and peripheral blood vessels.

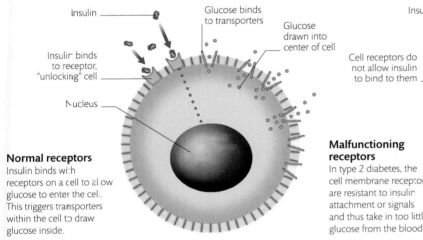

Insulin

Insulin binds to receptor, "unlocking" cell

Glucose binds to transporters

Glucose drawn into center of cell

Nucleus

Normal receptors
Insulin binds with receptors on a cell to allow glucose to enter the cell. This triggers transporters within the cell to draw glucose inside.

Malfunctioning receptors
In type 2 diabetes, the cell membrane receptors are resistant to insulin attachment or signals and thus take in too little glucose from the blood.

Insulin

Transporter inactive

Glucose remains in bloodstream

Cell receptors do not allow insulin to bind to them

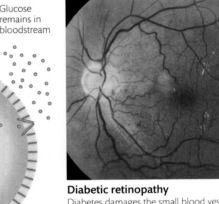

Diabetic retinopathy
Diabetes damages the small blood vessels of the eye through a variety of problems, such as hemorrhages, swelling, and fatty deposits, impairing the light-sensitive cells of the retina.

OBESITY

A growing problem worldwide, obesity can increase the risk of many diseases, including diabetes, heart disease, high blood pressure, arthritis, asthma, infertility, gynecological disorders, and cancers, such as those of the pancreas and colon. The mechanisms by which excess weight increases these risks will vary but it is known that body fat, especially fat that is located centrally in the abdomen, is hormonally active tissue that can have an inflammatory effect on other tissues. Health risks increase as the body mass index (BMI) rises. Many experts believe that waist circumference is a better predictor of future problems: measurements of more than 40 in (102 cm) for men or 35 in (88 cm) for women can indicate excess abdominal fat and an increased risk of diabetes.

Body mass index
The body mass index is found by dividing weight in kilograms by height in meters squared. A BMI of 18.5 to 24.9 is usually taken to be healthy, but BMI can be distorted by age and muscle mass.

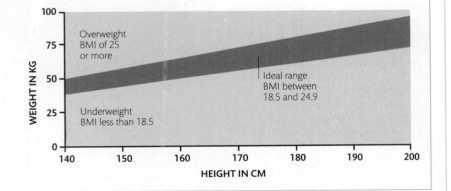

Overweight
BMI of 25 or more

Ideal range
BMI between 18.5 and 24.9

Underweight
BMI less than 18.5

WEIGHT IN KG / HEIGHT IN CM

GESTATIONAL DIABETES

The hormonal changes that occur during pregnancy can cause a form of diabetes called gestational diabetes, which can pose a threat to both mother and baby.

Some pregnancy hormones can counteract the effects of insulin, which normally controls blood sugar levels so sugar levels become too high. Gestational diabetes is more common in overweight women and in those with a family history or personal history of this condition.

Symptoms include thirst, fatigue, and excessive urine production. If left untreated, there are increased risks of the fetus growing too large, congenital heart malformations, miscarriage, stillbirth, or abnormal labor.

Treatment involves controlling blood sugar levels; the woman may receive advice about dietary control and moderate exercise and, if needed, insulin. The fetus will be monitored by ultrasound. With good management, mother and baby will be fine. After delivery, most women's blood sugar levels rapidly return to normal, but for a few this is the start of lifelong diabetes. The risk of recurrence in future pregnancies is high.

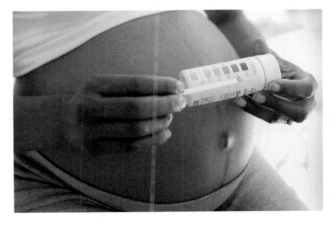

Tests for diabetes
The development of diabetes is relatively common in pregnancy, so pregnant women routinely have their urine dip-tested for sugar; if the test is positive, blood tests may be done to confirm the diagnosis.

HYPOPITUITARISM

The pituitary gland secretes hormones that are vital for major body functions, so hypopituitarism, or low activity of the gland, can give rise to serious disorders.

The pituitary gland helps regulate vital functions such as growth, response to stress or infection, and fertility. It works in conjunction with the hypothalamus and adrenal glands, ovaries, and testes by means of feedback systems acting on other glands. Hypopituitarism can result from a tumor, an infection, a vascular disorder such as stroke, or an autoimmune disease. Symptoms depend on the specific hormone deficiencies and may include loss of sex drive, infertility, or, in children, delayed growth. Treatment involves removing the cause of the disorder and/or correcting deficiencies in "target" hormones such as thyroid hormones.

PITUITARY TUMORS

Accounting for about 15 percent of brain tumors, most pituitary tumors are benign; they usually grow slowly, and gradually secrete excess hormones.

The most common tumors secrete growth hormone and prolactin, producing symptoms such as excessive growth or acromegaly, or excessive breast milk production. Occasionally, the tumor growth has the opposite effect, causing undersecretion of pituitary hormones (see left). Pressure from the tumor causes headaches, partial visual loss if the growth presses on the optic nerves, and palsy (paralysis or spasm) or numbness of the face.

The tumor is diagnosed by skull MRI and CT scans to show the growth and its effects on surrounding tissue, blood tests to show relevant hormone levels, and tests of pituitary function. Treatment depends on the person's age and the size and nature of the tumor. Drugs are given to suppress prolactin and growth hormone secretion. Surgery, chemotherapy, or radiation therapy may be used to remove or shrink the tumor. The person may need replacement hormones afterward.

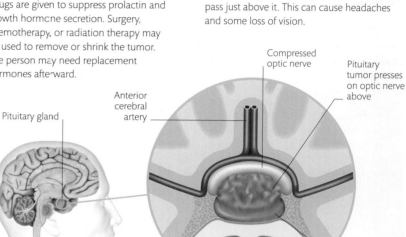

Pituitary tumor
A tumor may press on the optic nerves that pass just above it. This can cause headaches and some loss of vision.

Compressed optic nerve

Pituitary tumor presses on optic nerve above

Anterior cerebral artery

Pituitary gland

Pituitary gland may fail to function normally

LOCATION

HYPOTHYROIDISM

Underproduction of thyroid hormones, or hypothyroidism, causes slowing of the metabolism—the continual chemical reactions that keep the body functioning.

Hypothyroidism is most often seen in adults, due to an autoimmune condition in which the immune system attacks the body's own thyroid tissue, causing thyroiditis (inflammation of the thyroid gland), and is more common in women, especially after menopause. It may also occur in newborn babies due to abnormal development or a genetic disorder of metabolism. Symptoms result from slowing of body functions and include fatigue, weight gain, constipation, dry hair and skin, fluid retention, and mental slowing. Blood tests reveal low levels of thyroxine (T4), secreted by the thyroid, and high levels of thyroid-stimulating hormone (TSH), produced by the pituitary to make the thyroid work. The person will need to take replacement thyroxine for life.

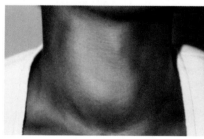

Goiter
This swelling is due to an enlarged thyroid (goiter), which may be visible at the front of the neck. This may result from conditions such as hypothyroidism.

THYROID CARCINOMA

Thyroid cancer, or thyroid carcinoma, is rare but develops slowly and has good survival rates in people who are treated.

There are several types of thyroid carcinoma, each arising from a different cell type: papillary, follicular, and medullary. The most common is papillary carcinoma. Factors such as previous thyroid disease, radiation therapy to the head or neck, or an iodine-deficient diet make this type more likely. Medullary carcinomas are inherited. Thyroid carcinomas grow slowly, causing a lump, swollen glands, or hoarseness.

Ultrasound and biopsy (tissue sampling) are used to confirm the presence of a tumor, and MRI, CT, and radioisotope scans to assess how far it has spread. Treatment includes surgery, radioactive iodine, and radiation therapy to remove or destroy affected thyroid tissue; in some cases, the whole gland is removed. Replacement thyroxine (the hormone normally produced by the thyroid) is usually needed.

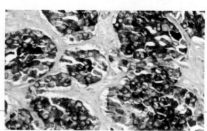

Carcinoma of thyroid
Cancer can develop from any of the main thyroid cell types. Medullary carcinoma (left) spreads at an earlier stage than other types.

HYPERTHYROIDISM

This condition, also called thyrotoxicosis, usually results from excess secretion of thyroid hormones, and causes vital body functions to speed up.

Oversecretion of thyroid hormones can have various causes. The most common is Graves' disease, an autoimmune condition in which the immune system attacks the thyroid gland, stimulating it to make excess hormones; other causes include benign tumors called thyroid nodules, and side effects of medication. Symptoms develop slowly; they reflect the overactivity of the metabolism, and include restlessness, anxiety, irritability, palpitations, weight loss, diarrhea, and breathlessness. People with Graves' disease may develop exophthalmos, or bulging eyes. Complications include heart disease and osteoporosis. Blood tests will show high levels of the thyroid hormone thyroxine, and low levels of thyroid-stimulating hormone (TSH) from the pituitary, as this gland tries to slow down the hormone secretion.

Treatments are designed to reduce levels of circulating thyroxine. Drugs such as carbimazole may be given for 1 to 2 years, until the condition settles. Radioactive iodine may be introduced into the gland to destroy overactive thyroid tissue, or excess thyroid tissue may be removed.

Graves' disease
The autoimmune reaction in Graves' disease causes inflammation and abnormal deposits in the muscles and connective tissue behind the eyes, affecting their shape and function.

NORMAL

ABNORMAL

Eyeball is forced forward; appears unusually prominent (exophthalmos)

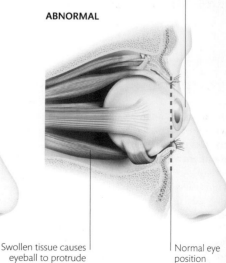

Eyeball sits neatly in socket

Swollen tissue causes eyeball to protrude

Normal eye position

GROWTH PROBLEMS

Growth involves many body systems, so growth disorders may affect not just stature but also organ development, recovery from wounds and disease, and even skin, hair, and nails. Growth hormone, produced by the pituitary gland, plays a major role. In children, excess or deficiency may affect their height. In adults, excess causes acromegaly, while low levels can cause muscle weakness, lack of energy, and depressed mood.

ACROMEGALY

Excessive secretion of growth hormone by the pituitary results in acromegaly: abnormal enlargement of the face, hands, feet, and soft tissues.

Acromegaly is almost always due to a tumor in the pituitary gland (see opposite) that is secreting excessive amounts of growth hormone. The effects can be seen in the bones and soft tissues of the body. In children it may cause gigantism, or excessive growth. Although the bones stop growing after puberty, excess growth hormone can still cause bone enlargement in adults. This process is very gradual, but eventually obvious changes occur: in particular, growth of the hands, feet, lower jaw, and eye sockets. Soft tissue changes include thick lips, large tongue, and leathery, greasy, darkened skin with acne. Internal organs such as the liver, heart, and thyroid also enlarge, causing problems such as heart failure. Excess growth hormone can also induce diabetes and other metabolic disorders, high blood pressure, and nerve and muscle damage.

Blood tests show abnormal hormone and mineral levels, and X-rays and MRI or CT scans can reveal the bone changes. People with a tumor may have surgery or radiation therapy to remove or shrink it. In other cases, drugs may be given to reduce growth hormone levels.

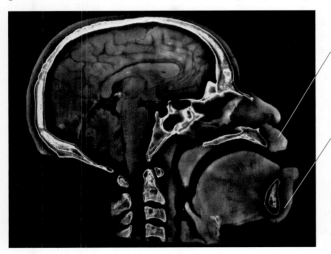

Thick lip

Enlarged, prominent jaw

Effects of acromegaly
This MRI scan shows the enlarged jaw and coarsened facial features of acromegaly.

GROWTH DISORDERS IN CHILDREN

Childhood growth can be affected by abnormalities in genes, hormonal function, nutrition, and general health, as well as growth patterns in the family.

The normal growth of a child is highly complex and affected by every aspect of physical and mental health. Growth abnormalities can be divided into two main types. Abnormal growth patterns causing short stature and/or body disproportion may be due to metabolic disorders or genetic programming, as in achondroplasia (see p.431), a chromosomal abnormality that is one of the most common causes of dwarfism. Some disorders are due to overly high or low levels of hormones, notably growth hormone, produced by the pituitary gland. An excess of growth hormone causes gigantism, or extreme bone growth, while a deficit can cause a child to grow too slowly.

Lack of thyroxine, from the thyroid, can also delay growth and development. By contrast, poor growth (compared to other children the same age) but normal proportions may be due to poor nutrition or chronic disease. To treat a growth disorder, the underlying cause needs to be identified and remedied.

ADDISON'S DISEASE

Damage to the cortex (outer layer) of the adrenal glands may impair hormone production, leading to Addison's disease.

The adrenal cortex produces hormones that help regulate metabolism, control blood pressure, and balance the levels of salt and water in the body. Insufficient corticosteroid levels may be due to an autoimmune reaction in which the immune system attacks the adrenal glands; less common causes are infections, certain drugs, or suddenly stopping

Adrenal anatomy
The adrenal glands sit on the kidneys. The medulla (center) secretes epinephrine and norepinephrine; the cortex produces a variety of hormones.

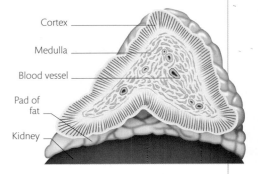

Cortex

Medulla

Blood vessel

Pad of fat

Kidney

CUSHING SYNDROME

If the adrenal glands produce an excess of cortisol (the major corticosteroid in the body), Cushing syndrome may develop.

There are many possible causes of Cushing syndrome; the most common is Cushing's disease, in which the pituitary gland stimulates the adrenals to oversecrete corticosteroids. The symptoms of Cushing syndrome include obesity; excess fat deposits, especially in the face and over the shoulders; excess growth of body hair; high blood pressure; and diabetes. Other symptoms include thinning of the skin and hair, weakness, and osteoporosis leading to fractures. Recurrent infections and sex hormone disturbances, resulting in erectile

DISORDERS OF CALCIUM METABOLISM

An overactive or underactive parathyroid gland can cause levels of calcium in the body to be affected, leading to disorders.

Calcium is needed for bone and tissue growth and muscle and nerve function. The levels are regulated by parathyroid hormone (PTH). If the parathyroid glands are underactive, PTH levels fall too low, leading to low calcium levels; this can cause muscle cramps and nerve problems.

Parathyroid glands
The four glands lie at the back of the thyroid, just under the larynx in the neck. If calcium levels are low they secrete PTH, which draws calcium out of the bones and increases absorption from food.

corticosteroid treatment. Symptoms include fatigue, muscle weakness, nausea, abnormal skin coloring, weight loss, and depression. A sudden illness, injury, or other stress can cause an Addisonian crisis, in which the gland cannot produce enough hormones, causing circulatory collapse; this needs urgent medical attention. Long-term treatment may include replacement corticosteroids and mineralocorticoids.

dysfunction in men and irregular menstrual periods in women, can also occur. Treatment is based on identifying and treating the cause of the syndrome. In Cushing's disease, surgery, with radiation therapy or medication, is used to reduce the pituitary's stimulation of the adrenal glands and thus lower adrenal activity.

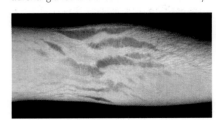

Stretch marks
One of the signs of high corticosteroid levels is the stretching and tearing of skin layers, causing stretch marks, especially where there is underlying fat, as on the torso and upper limbs.

Overactivity, often due to a tumor in a gland, leads to an excess of PTH, which causes calcium to leach from the bones into the blood. This results in thinning of the bones, fractures, and calcium deposits in the kidneys and other tissues. An underactive parathyroid is usually treated with vitamin D and calcium supplements. Surgery may be needed to remove a tumor.

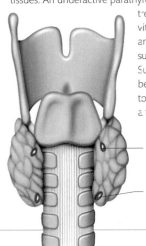

Superior parathyroid gland

Inferior parathyroid gland

Glossary

Terms defined elsewhere in the glossary are in *italics*. All distinct terms are in **bold**.

abduction
The action of moving a limb farther from the midline of the body. In muscle names, **abductor** indicates a muscle that has this action. See also *adduction*.

acetylcholine
A major *neurotransmitter* in the body, conveying signals from *nerves* to muscles as well as between many nerves.

action potential
The electrical *nerve* impulse that travels along the *axon* of a nerve cell (*neuron*).

adduction
The action of moving a limb closer to the midline of the body. In muscle names, **adductor** indicates a muscle that has this action. See also *abduction*.

adipose tissue
Fat-storage *tissue*.

adrenal glands
Also called suprarenal glands. A pair of glands found one on top of each kidney. Each gland consists of an outer **adrenal cortex**, which secretes *corticosteroid hormones*, and an inner **adrenal medulla**, which secretes *epinephrine*. See also corticosteroid.

adrenaline
See *epinephrine*.

afferent
In blood vessels, carrying blood toward an organ, and in *nerves*, conducting impulses toward the *central nervous system*. See also *efferent*.

aldosterone
See *corticosteroid*.

allergy
An unnecessary and sometimes dangerous *immune response* that targets otherwise non-threatening foreign material, such as plant pollen.

alveolus (pl. alveoli)
A small cavity; specifically, one of the millions of tiny air sacs in the lungs where exchange of gases with the blood takes place; also, the technical term for a tooth socket.

amino acid
Proteins are made from up to 20 different types of these small, nitrogen-containing molecules; amino acids also play various other roles in the body. See also *peptide*.

amnion
The *membrane* that encloses the developing *fetus* within the *uterus* (womb). The fluid inside it (**amniotic fluid**) helps cushion and protect the fetus.

anastomosis
An interconnection between two otherwise separate blood vessels (e.g. two *arteries*, or an artery and a *vein*).

androgen
Steroid hormones that tend to promote male body and behavioral characteristics. They are secreted in larger amounts by men than women.

anemia
Damagingly low amounts of *hemoglobin* in the blood. Anemia can have many causes, from undetected bleeding to *vitamin* deficiencies.

angio-
A prefix relating to blood vessels.

angiography
In medical imaging: any technique for obtaining images of blood vessels in the living body.

antagonist
1. A muscle that has the opposite action to another muscle.
2. A drug that interferes with the action of a *hormone*, *neurotransmitter*, etc., by binding to its *receptor*.

anterior
Toward the front of the body, when considered in a standing position. **Anterior to** means in front of. See also *posterior*.

antibiotic
Any of various chemical compounds, natural or synthetic, that destroy or prevent the growth of microorganisms (e.g. *bacteria*, yeasts, and fungi).

antibody
Defensive *proteins* produced by white blood *cells* that recognize and attach to particular "foreign" chemical components (*antigens*), such as the surface of an invading *bacterium* or *virus*. The body is able to produce thousands of different antibodies targeted at different invaders and toxins.

anticoagulant
A substance that prevents blood clotting.

antigen
Any particle or chemical substance that stimulates the *immune system* to produce antibodies against it.

aorta
The body's largest *artery*, conveying blood pumped by the left *ventricle* of the heart. It extends to the lower abdomen, where it divides into the two common iliac arteries.

aponeurosis
A flattened, sheetlike *tendon*.

arteriole
A very small *artery*, leading into *capillaries*.

artery
A vessel carrying blood from the heart to the *tissues* and organs of the body. Arteries have thicker, more muscular walls than *veins*.

articulation
A *joint*, especially one allowing movement; also, a location within a joint where two bones meet in close proximity. A bone in a joint is said to **articulate with** the other bone(s) of the joint.

-ase
A suffix denoting an *enzyme*. For example, sucrase is an enzyme that breaks down *sucrose*.

ATP
Short for **adenosine triphosphate**, an energy-storing *molecule* used by all living *cells*.

atrium (pl. atria)
Either of the two smaller chambers of the heart that receive blood from the *veins* and pass it on to the corresponding *ventricle*.

autoimmunity
A situation where the *immune system* attacks the body's own *tissues*, often leading to disease.

autonomic nervous system
The part of the nervous system that controls non-conscious processes such as the activity of the body's *glands* and the muscles of the gut. It is divided into the **sympathetic nervous system**, the roles of which include preparing the body for "fight or flight," and the **parasympathetic nervous system**, which stimulates movement and secretions in the gut, produces erection of the penis during coitus, and empties the bladder.

axon
A wirelike extension of a *nerve cell* (*neuron*) along which electrical signals are transmitted away from the cell.

bacterium (pl. bacteria)
Any member of a large group of single-celled living organisms, some of which are dangerous *pathogens*. Bacterial *cells* are much smaller than animal and plant cells, and lack *nuclei*.

basal ganglia
Groups of *nerve cells* deep in the *cerebrum*; consists of the caudate nucleus, putamen, globus pallidus, and subthalamic nucleus. Functions include controlling movement.

basophil
A type of *leukocyte* (white blood *cell*).

belly (of muscle)
The widest part of a *skeletal muscle*, which bulges further when it contracts.

bilateral
Concerning or affecting both sides of the body or a body part.

bile
A yellow-green fluid produced by the liver, stored in the *gallbladder*, and discharged into the intestine via the bile duct. It contains excretory products together with bile acids that help with fat digestion.

biopsy
A sample taken from a living body to test for infection, cancerous growth, etc.; also the sampling process.

blood–brain barrier
The arrangements by which the brain is relatively protected from unwanted substances entering it from the blood. It includes *capillaries* that are less permeable to large *molecules* than elsewhere in the body.

brachial
Relating to the arm.

brain stem
The lowest part of the brain, leading down from the rest of the brain to the *spinal cord*. In descending order, it consists of the *midbrain*, pons, and *medulla oblongata*.

bronchus (pl. bronchi)
The air tubes branching from the *trachea* and leading into the lungs; right and left main bronchi enter each lung respectively and divide into lobar bronchi, and eventually into much smaller tubes called **bronchioles**.

calcitonin
See *thyroid gland*.

hemoglobin
The red pigment within *erythrocytes* that gives blood its color and carries oxygen to the *tissues*.

hepatic
Relating to the liver.

histamine
A substance produced by damaged or irritated *tissues* that stimulates an inflammatory response (see *inflammation*).

homeostasis
The maintenance of stable conditions in the body, e.g. in terms of chemical balance or temperature.

hormone
A chemical messenger produced by one part of the body that affects other organs or parts. There also exist **local hormones** that affect only nearby *cells* and *tissues*. Chemically, most hormones are either *steroids*, *peptides*, or small *molecules* related to *amino acids*. See also *neurohormone*, *neurotransmitter*.

hydrocortisone
See *corticosteroid*.

hypothalamus
A small but vital region at the base of the brain, which is the control center for the *autonomic nervous system*, regulating processes such as body temperature and appetite. Also controls the secretion of *hormones* from the *pituitary gland*.

ileum
The last part of the small intestine, ending at the junction with the large intestine (*colon*). N.B: Not the same as **ilium**, one of the bones of the hip

immune response
The body's defensive reactions to invasion by a *bacterium*, *virus*, toxin, etc. It includes general responses such as *inflammation*, as well as specific responses in which an invader is targeted by a particular *antibody* so that it can be recognized and destroyed or disabled.

immune system
The *molecules*, *cells*, organs, and processes that are involved in defending the body against disease.

immunity
Resistance to attack by a *pathogen* (disease-causing organism); specific **immunity** develops as a result of the body's *immune system* being primed to resist a particular *pathogen*.

immunotherapy
Any of various treatments involving either the stimulation or suppression of the activity of the *immune system*.

implantation
The attachment of an early *embryo* to the lining of the *uterus*. It occurs during the first week after *fertilization*, and is followed by the development of the *placenta*.

inferior
Lower down the body, when considered in a standing position (i.e. nearer the feet). See also *superior*.

inflammation
An immediate reaction of body *tissue* to damage, in which the affected area becomes red, hot, swollen, and painful, as white blood *cells* (see *leukocyte*) accumulate at the site to attack potential invaders.

inguinal
Relating to, or in the region of, the groin.

inner ear
The fluid-filled innermost part of the ear, which contains the organs of balance (the *semicircular canals*) and the organs of hearing within the *cochlea*. See also *middle ear*.

insertion
The point of attachment of a muscle to the structure that typically moves when the muscle is contracted. See also *origin*.

insulin
A *hormone* produced by the pancreatic islets (see *pancreas*) that promotes the uptake of *glucose* from the blood, and the conversion of glucose to the storage *molecule*, *glycogen*. See also *diabetes*.

integument
The *external* protective covering of the body.

internal
In anatomy: inside the body, distant from the surface. See also *external*.

interneuron
Any *nerve cell* whose connections are only with other *neurons*, as distinct from a *sensory* or *motor* neuron.

interstitial
Relates to being between things, such as other *cells* or *tissues*, e.g. interstitial fluid surrounds cells.

intra-
Prefix meaning within, as in **intracellular** or **intramuscular**.

intrinsic
Situated within or originating within a particular organ or body part.

ion
An electrically charged atom or *molecule*.

ischemia
Reduction of blood supply to part of the body.

islets of Langerhans
See *pancreas*.

-itis
Suffix meaning "*inflammation*," used in words such as **tonsillitis** and **laryngitis**.

joint
Any junction between two or more bones, whether or not movement is possible between them. See also *articulation*, *suture*, *symphysis*, *synovial joint*.

keratin
A tough *protein* that forms the substance of hair and nails, gives strength to the skin, etc.

labia (sing. labium)
Either of the two paired folds that form part of the *vulva* in females: the outer **labia majora** and the more delicate inner **labia minora**.

labial
Relating to the lips, or to the *labia* of the female genitals.

lactation
Secretion of milk by the breasts.

larynx
The voicebox: a complex structure situated at the top of the *trachea*. It includes the **vocal cords**, structures that function to seal off the trachea when necessary, as well as creating sound when their edges are made to vibrate during breathing.

lateral
Relating to or toward the sides of the body. See also *medial*.

leukocyte
A white blood *cell*. There are several types, acting in different ways to protect the body against disease as part of its *immune response*. Leukocytes are found in *lymph nodes* and other *tissues* generally, as well as in the blood.

levator
Term used in the names of several muscles whose action is to lift up, such as the levator scapula (lifts the shoulder blade). See also *depressor*.

ligament
A tough fibrous band that holds two bones together. Many ligaments are flexible, but they cannot be stretched. The term is also used for bands of *tissue* connecting or supporting some internal organs.

limbic system
Several regions at the base of the brain, involved in memory, behavior, and emotion.

lingual
Relating to the tongue.

lipid
Any of a large variety of fatty or fatlike substances that are found naturally in living things and are relatively insoluble in water.

lumbar
Relating to the lower back and sides of the body between the lowest ribs and the top of the hip bone. The **lumbar vertebrae** are the *vertebrae* that lie within this region.

lumen
The space inside a tubular structure, such as a blood vessel or glandular duct.

lymph node
A small lymphoid organ; lymph nodes serve to filter out and dispose of *bacteria* and debris, such as *cell* fragments.

lymphocyte
A specialized *leukocyte* that produces antibodies including *natural killer cells*, T-cells, and B-cells.

lymphoid tissue
The *tissue* of the lymphatic system, which has an immune function, including *lymph nodes*, the *thymus*, and the *spleen*.

macromolecule
A large *molecule*, especially one that consists of a chain of small similar "building blocks" joined together. *Proteins*, *DNA*, and *starch* are examples of macromolecules.

macrophage
A large type of *leukocyte* that can engulf and dispose of *cell* fragments, *bacteria*, etc.

mammary
Of, or relating to, the breasts.

marrow
In anatomical contexts, usually short for **bone marrow**, the soft material located in the cavities of bones; in some areas this *tissue* is mainly fat; in others, it is blood-forming *tissue*.

matrix
The *extracellular* material in which the *cells* of *connective tissues* are embedded. It may be hard, as in bone; tough, as in *cartilage*; or fluid, as in blood.

meatus
A channel or passage. For example, the *external* **auditory meatus**, the ear canal.

medial
Toward the midline of the body.
See also *lateral*.

medulla
1. Short for **medulla oblongata**,
the elongated lower part of the brain
that connects with the *spinal cord*.
2. The central part or core of some
organs such as the kidneys and
adrenal glands.

melanin
A dark brown naturally occurring pigment
molecule, which occurs in greater amounts
in tanned or darker skin, and protects
deeper *tissues* from ultraviolet radiation.

melatonin
A *hormone* secreted by the pineal *gland* in
the brain, which plays a role in the body's
sleep-wake cycle (see *circadian rhythm*).

membrane
1. A thin sheet of *tissue* covering an
organ, or separating one part of the
body from another.
2. The outer covering of a *cell*
(and similar structures within the cell).
A cell membrane is composed of a
double layer of *phospholipid molecules*
with other molecules such as *proteins*
embedded in it.

meninges
Membranes that enclose the outside of
the brain and *spinal cord*. **Meningitis** is
inflammation of the meninges, usually
resulting from infection.

menopause
The time in a woman's life when *ovulation*
and the *menstrual cycle* permanently
cease.

menstrual cycle
The monthly cycle that takes place in
the *uterus* of a non-pregnant woman of
reproductive age. The *endometrium*
(lining of the uterus) grows thicker in
preparation for possible pregnancy; an
egg is released from the *ovary* (*ovulation*);
then, if the egg is not fertilized, the
endometrium breaks down and is
discharged through the vagina in a
process known as **menstruation**.

mental
1. Relating to the mind (Latin *mens*).
2. Relating to the chin (Latin *mentum*).

mesentery
A folded sheet of *peritoneum*, forming a
connection between the intestines and
the back of the abdominal cavity.

metabolism
The chemical reactions taking place in
the body. The **metabolic rate** is the
overall rate at which these reactions
are occurring.

midbrain
The upper part of the *brain stem*.

middle ear
The air-filled middle chamber of the ear,
between the inner surface of the eardrum
and the *inner ear*. See also *ossicles*.

molecule
The smallest unit of a chemical
compound that can exist, consisting of
two or more atoms joined together by
chemical bonds. The water molecule is
a simple example, consisting of two
hydrogen atoms joined to one oxygen
atom. See also *macromolecule*.

monocyte
A type of *leukocyte* with various roles in
the *immune system*, including giving rise
to *macrophages*.

motor
Adjective relating to the control of muscle
movements, as in **motor neuron, motor
function**, etc. See also *sensory*.

MRI scan
Short for **magnetic resonance
imaging scan**, a medical imaging
technique based on the energy released
when magnetic fields are applied then
removed from the body; it can produce
very detailed images of the soft *tissues* of
the body.

mucosa (pl. mucosae)
A *membrane* that secretes *mucus*.

mucus
A thick fluid produced by some
membranes of the body for protection,
lubrication, etc. (Adjective **mucous**.)

mutation
Any change to the genetic makeup of a
cell, caused for example by accidents or
mistakes during cell division. Mutations in
sex cells (*gametes*) may cause offspring to
have unusual genetic features not present
in their parents.

myelin
Fatty substance forming a layer around
some *nerve axons*, called **myelinated**
axons, insulating them and speeding their
nerve impulses.

myelo-
1. Prefix relating to the *spinal cord*.
2. Prefix relating to bone *marrow*.

myo-
Prefix relating to muscle.

natural killer (NK) cell
A type of *lymphocyte* that can attack and
kill *cancer cells* and *virus*-infected cells.

necrosis
The death of part of an organ or *tissue*.

neocortex
All the *cortex* of the *cerebrum* except
the region concerned with smell
and the hippocampal formation.

nephron
The filtering unit of the kidney, which
regulates the volume and composition
of body fluids by filtering the blood to
produce urine. Waste products, such as
urea and uric acid, are also excreted by
the nephron. There are more than
a million nephrons in each kidney.

nerve
A cablelike structure transmitting
information and control instructions in
the body. A typical nerve consists of *axons*
of many separate nerve *cells* (*neurons*)
running parallel to, but insulated from,
each other; the nerve itself is surrounded
by an overall protective sheath of fibrous
tissue. Nerves may contain nerve fibers
controlling muscles or *glands* (*efferent*
fibers), while others contain fibers
carrying *sensory* information back to
the brain (*afferent* fibers); some nerves
carry both types of nerve fiber.

neurohormone
A *hormone* released by a *nerve cell* rather
than from a *gland*.

neurology
The branch of medicine that specializes
in disorders of the nervous system.
The adjective **neurological** includes
any symptom or disorder that might
fall within the province of neurology.

neuron
A *nerve cell*. A typical neuron consists
of a rounded cell body; branchlike
outgrowths called *dendrites* that carry
incoming electrical signals to the neuron;
and a single, long, wirelike extension,
called an *axon*, which transmits outgoing
messages. There are many variations on
this basic pattern, however.

neurotransmitter
Any of various chemical substances
released at *synapses* by the ends of
nerve cells, where they function to
pass a signal on to another nerve cell or
muscle. Some neurotransmitters act
mainly to stimulate the action of other
cells, others to inhibit them.

neutrophil
The most common type of *leukocyte*
(white blood *cell*). Neutrophils move
quickly toward sites of damage and
engulf invading *bacteria* etc.

nondisjunction
Failure of *chromosomes* to separate from
each other properly during *cell* division,
resulting in daughter cells that have either
too many or too few chromosomes.

noradrenaline
See *norepinephrine*.

norepinephrine
A *neurotransmitter* important in the
sympathetic nervous system.

nucleus (pl. nuclei)
1. The structure within a *cell* that contains
the *chromosomes*.
2. Any of various concentrations of *nerve*
cells within the *central nervous system*.
3. The central part of an atom.

occipital
Relating to the back of the head. The
occipital bone is the skull bone forming
the back of the head. The **occipital lobe**
is the rearmost lobe of each cerebral
hemisphere, lying below the occipital
bone.

olfactory
Relating to the sense of smell.

optic nerve
The *nerve* that transmits visual information
from the *retina* of the eye to the brain.

oral
Relating to the mouth.

orbit
The bony hollow in the skull within which
the eye is contained.

organelle
Any of a variety of small structures inside
a *cell*, usually enclosed within a *membrane*,
which are specialized for functions such
as energy production or secretion.

origin
The point of attachment of a muscle to
the structure that typically remains
stationary when the muscle is contracted.
See also *insertion*.

osmosis
Phenomenon in which water moves from
a less concentrated solution to a more
concentrated one if the two solutions are
separated by a semipermeable *membrane*.

ossi-, osteo-
Prefixes relating to bone.

ossicles
Three small bones of the *middle ear* that
transmit vibrations caused by sound
waves from the eardrum to the *inner ear*.

ovary
Either of the two organs in females that
produce and release egg *cells* (ova).
They also secrete sex *hormones*.

ovulation
The point in the *menstrual cycle* at which an
egg cell (ovum) is released from the *ovary*
and begins to travel toward the *uterus*.

ovum (pl. ova)
An unfertilized egg *cell*.

oxytocin
A *hormone* secreted by the *pituitary gland* involved in dilation of the *cervix* and uterine contractions during childbirth, in *lactation*, and in sexual responses.

palate
The roof of the mouth, comprising the bony **hard palate** in the front and the muscular **soft palate** behind it.

pancreas
A large, elongated *gland* lying behind the stomach, with a dual role in the body. The bulk of its *tissue* secretes digestive *enzymes* into the *duodenum*, but it also contains scattered groups of *cells* called **pancreatic islets** or *islets of Langerhans* that produce important *hormones*, including *insulin* and *glucagon*.

parasympathetic nervous system
See *autonomic nervous system*.

parathyroid glands
Four small *glands* that are often embedded in but are separate from the *thyroid gland*. They produce **parathyroid hormone**, which regulates calcium *metabolism* in the body.

parietal
A term (derived from the Latin word for "wall") with various applications in anatomy. The **parietal bones** form the side walls of the skull, and the **parietal lobes** of the *brain* lie beneath those bones. *Membranes* (such as the *pleura* and *peritoneum*) are described as parietal where they are attached to the body wall.

pathogen
Any disease-causing agent, including *bacteria* and *viruses*.

pathology
The study of disease; also, the physical manifestations of a disease.

pelvic girdle
The hip bones attach to the sacrum to form the pelvic girdle, linking the leg bones to the spine.

pelvis
1. The cavity enclosed by the *pelvic girdle*, or the area of the body containing the pelvic girdle.
2. The renal pelvis is the cavity in the kidney where the urine collects before passing down the *ureter*.

peptide
Any *molecule* consisting of two or more *amino acids* joined together, usually in a short chain. There are many types, some of which are important *hormones*. *Proteins*

are polypeptides: long chains of *amino acids*.

peri-
Prefix meaning round or surrounding.

peripheral
Toward the outside of the body or to the extremities of the body. The term **peripheral nervous system** refers to the whole of the nervous system except for the brain and *spinal cord*. See also central nervous system.

peristalsis
A wavelike contraction of muscles, produced by muscular tubes, such as that which propels digested food through the gut, or urine through the *ureters*, for example.

peritoneum
A thin, lubricated sheet of *tissue* that enfolds and protects most of the organs of the abdomen.

phagocyte
Any *cell* that can engulf and dispose of foreign bodies such as *bacteria*, as well as broken fragments of the body's own cells.

pharynx
The muscular tube behind the nose, mouth, and *larynx*, leading into the esophagus.

phospholipid
A type of *lipid molecule* with a phosphate (phosphorus plus oxygen) group at one end. The phosphate group is attracted to water while the rest of the molecule is not. This property makes phospholipids ideal for forming *cell membranes* if two layers of molecules are situated back-to-back.

physiology
The study of the normal functioning of body processes; also, the body processes themselves.

pituitary gland
Also called the **hypophysis**, a complex pea-sized structure at the base of the brain, sometimes described as the body's "master *gland*." It produces various *hormones*, some affecting the body directly and others controlling the release of hormones by other glands.

placenta
The organ that develops on the inner wall of the *uterus* during pregnancy, allowing the transfer of substances, including nutrients and oxygen, between maternal and fetal blood. See also umbilical cord.

plasma
Blood minus its cellular components (red and white blood *cells*, and *platelets*).

platelets
Specialized fragments of *cells* that circulate in the blood and are involved in blood clotting.

pleura (pl. pleurae)
The lubricated *membrane* that lines the inside of the thoracic cavity and the outside of the lungs.

plexus
A network, usually in reference to *nerves* or blood vessels.

pneum-, pneumo-
1. Prefix relating to air.
2. Prefix relating to the lungs.

portal vein
The large *vein* carrying blood from the intestines to the liver; previously known as the **hepatic portal vein**.

posterior
Toward the back of the body, when considered in a standing position. **Posterior to**, behind. See also *anterior*.

process
In anatomy: a projection or extended part of a bone, *cell*, etc.

progesterone
A *steroid hormone* produced by the *ovaries* and *placenta*, which plays a role in the *menstrual cycle* and in the maintenance and regulation of pregnancy.

prolactin
A *hormone* produced by the *pituitary gland*, the effects of which include stimulating the breasts to produce milk.

pronation
The rotation of the radius around the ulna in the forearm, turning the palms of the hand to face downward or backward. In muscle names, **pronator** indicates a muscle that has this action, e.g. pronator teres. See also *supination*.

prostate gland
A *gland* located below the male bladder; its secretions contribute to *semen*.

proteins
Large *molecules* consisting of long folded chains of small linked units (*amino acids*). There are thousands of different kinds in the body. Nearly all *enzymes* are proteins, as are the tough materials *keratin* and *collagen*. See also *peptide*.

proximal
Relatively closer to the center of the body or from the point of *origin*. See also *distal*.

puberty
The period of sexual maturation between childhood and adulthood.

pulmonary
Relating to the lungs.

pyloric
Relating to the last part of the stomach, or pylorus. The muscle wall of the end of the pylorus is thickened to form the **pyloric sphincter**.

radiation therapy
Cancer treatment using ionizing radiation, carried out by directing beams of radiation at the cancer, or introducing radioactive substances to the body.

receptor
1. Any sense organ, or the part(s) of a sense organ responsible for collecting information.
2. A *molecule* in a *cell*, or on a cell's outer *membrane*, that responds to an outside stimulus, such as a *hormone* molecule attaching to it.

rectum
The short final portion of the large intestine, connecting it to the anal canal.

rectus
In muscle names, a straight muscle.

reflex
An involuntary response in the nervous system to certain stimuli, for example the "knee-jerk" response. Some reflexes, called **conditioned reflexes**, can be modified by learning.

renal
Relating to the kidneys.

respiration
1. Breathing.
2. Also called cellular respiration, the biochemical processes within *cells* that break down fuel *molecules* to provide energy, usually in the presence of oxygen.

retina
The light-sensitive layer that lines the inside of the eye. Light falling onto *cells* in the retina stimulates the production of electrical signals, which are transmitted to the brain via the *optic nerve*.

ribosomes
Particles within *cells* involved in *protein* synthesis.

RNA
Short for **ribonucleic acid**, a long *molecule* similar to *DNA*, but usually single- rather than double-stranded. RNA has many important roles including making copies of the DNA code for *protein* synthesis.

sacral
Relating to or in the region of the **sacrum**, the bony structure made up of fused *vertebrae* at the base of the spine that forms part of the *pelvic girdle*.

sagittal section
A real or imagined section down the body, or part of the body, that divides it into right and left sides.

scrotum
The loose pouch of skin holding the *testes* in males.

sebum
An oily, lubricating substance secreted by sebaceous *glands* in the skin.

semen
The fluid released through the penis when the male ejaculates; it contains *sperm* and a mixture of nutrients and salts. Also called **seminal fluid**.

sensory
Concerned with transmitting information coming from the sense organs of the body.

serotonin
A *neurotransmitter* in the brain that affects many *mental* activities, including mood. It is also active in the gut.

serous membrane
A type of body *membrane* that secretes lubricating fluid and envelops various internal organs and body cavities. The pericardium, *pleura*, and *peritoneum* are all serous membranes.

shock
Medical or circulatory shock: a potentially fatal failure of the blood flow to support the body's needs, as a result of blood loss or other causes. The term is also used more loosely to refer to psychological responses to trauma, etc.

sinus
A cavity; especially:
1. One of the air-filled cavities in the bones of the face that connect to the nasal cavity.
2. An expanded portion of a blood vessel, for example the carotid sinus and coronary sinus.

skeletal muscle
A type of muscle also known as *voluntary* or *striated muscle*, usually under voluntary control. Appears striped under the microscope. Many—but not all—skeletal muscles attach to the skeleton, and are important in movement of the body. See also *smooth muscle*.

smooth muscle
Muscle *tissue* that lacks stripes when viewed under a microscope, in contrast to striated muscle. Smooth muscle is found in the walls of internal organs and structures, including blood vessels, the intestines, and the bladder. It is not under conscious control, but controlled by the autonomic *nervous system*.

somatic
1. Of or relating to the body, e.g. somatic *cells*.
2. Relating to the body wall.
3. Relating to the part of the nervous system involved in voluntary movement and sensing the outside world.

somatosensory
Related to sensations received from the skin and internal organs, including senses such as touch, temperature, pain, and awareness of *joint* position, or proprioception.

sperm
A male sex *cell* (*gamete*), equipped with a long moving "tail" (flagellum) to allow it to swim toward and fertilize an egg in the body of the female. Colloquially the word is also used to mean semen.

sphincter
A ring of muscle that allows a hollow or tubular structure in the body to be drawn closed (e.g. the *pyloric* sphincter and anal sphincter).

spinal cord
The part of the *central nervous system* that extends down from the bottom of the brain through the vertebral column, which protects it. Most *nerves* that supply the body originate in the spinal cord.

spleen
A structure in the abdomen composed of *lymphoid tissue*. It has various roles, including blood storage.

starch
A plant *carbohydrate* made up of long, branched chains of *glucose molecules* linked together.

stem cell
A *cell* in the body that can divide to give rise to more cells. This could be either more stem cells, or a range of more specialized types of cell. Stem cells contrast with highly specialized cells, which play specific roles in the body, and which may have lost the ability to divide completely—such as *nerve* cells.

steroids
Substances that share a basic molecular sturcture, consisting of four rings of carbon atoms fused together. Steroids, which may be naturally occurring or synthetic, are classified as *lipids*. Many of the body's *hormones* are steroids, including estrogen, progesterone, testosterone, and cortisol.

striated muscle
A muscle with *tissue* that presents a striped appearance under a microscope. Striated muscle includes *skeletal muscles* and *cardiac* (heart) muscle. See also *smooth muscle*.

sucrose
See *sugar*.

sugar
1. Commonly used foodstuff, also called *sucrose*.
2. Any of a number of naturally occurring substances that are similar to sucrose. They are all *carbohydrates* with relatively small *molecules*, in contrast to other carbohydrates that are *macromolecules*, such as *starch*.

sulcus (pl. sulci)
One of the grooves on the folded outer surface of the brain. See also *gyrus*.

superficial
Near the surface; **superficial to**, nearer the surface than. (Opposite term: deep.)

superior
Higher up the body, when considered in a standing position. See also *inferior*.

supination
The rotation of the radius around the ulna in the forearm, turning the palms of the hand to face upward or forward. The opposite to *pronation*. In muscle names, **supinator** indicates a muscle having this action, e.g. the supinator of the forearm.

suprarenal glands
See *adrenal glands*.

suture
1. A stitched repair to a wound.
2. A rigid *joint* between two bones, as between the bones of the skull.

sympathetic nervous system
See *autonomic nervous system*.

symphysis
A cartilaginous *joint* between two bones, containing fibrocartilage.

synapse
A close contact between two *nerve cells* (*neurons*) allowing signals to be passed from the end of the first neuron on to the next. Synapses can either be electrical (where the information is transmitted electrically) or chemical (where *neurotransmitters* are released from one neuron to stimulate the next one). Synapses also exist between nerves and muscles.

synovial joint
A lubricated, movable *joint*, such as the knee, elbow, or shoulder. In synovial joints the ends of the bones are covered with smooth *cartilage* and lubricated by a slippery liquid known as **synovial fluid**.

systemic
Relating to or affecting the body as a whole, not just one part of it. The **systemic circulation** is the blood circulation supplying all of the body apart from the lungs.

systole
The part of the heartbeat where the *ventricles* contract to pump blood.

tarsal
1. Relating to the ankle.
2. One of the bones of the tarsus, the part of the foot between the tibia and fibula, and the metatarsals.

temporal
Relating to the temple—the area on either side of the head. The **temporal bones** are two bones, one on each side of the head, that form part of the *cranium*. The **temporal lobes** of the brain are located roughly below the temporal bones.

tendon
A tough fibrous cord that attaches one end of a muscle to a bone or other structure. See also *aponeurosis*.

testis (pl. testes)
Either of the pair of organs in men that produce male sex *cells* (*sperm*). They also secrete the sex *hormone testosterone*.

testosterone
A *steroid hormone* produced mainly in the *testes*, which promotes the development of and maintains male body and behavioral characteristics.

thalamus
Paired structures deep within the brain, forming a relay station for *sensory* and *motor* signals.

thorax
The chest region, which includes the ribs, lungs, heart, etc.

thrombus
A stationary clot in a blood vessel, potentially interfering with circulation. **Thrombosis** is the process by which such a clot is formed.

thymus
A *gland* in the chest composed of *lymphoid tissue*. Largest and most active in childhood, its roles include the maturation of T-lymphocytes.

thyroid gland
An endocrine *gland* located at the front of the throat, close to the *larynx* (voicebox). Thyroid *hormones* such as **thyroxin** are involved in controlling *metabolism*, including regulating overall metabolic rate. The hormone *calcitonin*, which helps regulate the body's calcium, is also secreted by the thyroid.

tissue
Any type of living material in the body that contains distinctive types of *cells*, usually together with *extracellular* material, performing a specific function. Examples of tissues include bone, muscle, *nerve*, and *connective tissue*.

trachea
The windpipe: the tube leading between the *larynx* and the *bronchi*. It is reinforced by rings of *cartilage* to keep it from collapsing.

tract
An elongated structure or connection that runs through a certain part of the body. In the *central nervous system*, the term is used instead of *nerve* for bundles of nerve fibers that connect different body regions.

translocation
1. Transport of material from one part of the body to another. 2. A type of *mutation* in which a *chromosome*, or part of one, becomes physically attached to another chromosome or to a different part of the original chromosome.

transmitter
See *neurotransmitter*.

umbilical cord
The cord that attaches the developing *fetus* to the *placenta* of the mother, within the *uterus*. Blood from the fetus passes through blood vessels inside the cord, transporting nutrients, dissolved gases, and waste products between the placenta and the fetus

urea
A small nitrogen-containing *molecule* formed in the body as a convenient way of getting rid of other nitrogen-containing waste products. It is excreted in the urine.

ureter
Either of two tubes that convey urine from the kidneys to the bladder.

urethra
The tube that conveys urine from the bladder to the outside of the body; in men it also conveys *semen* during ejaculation.

uterus
The womb, in which the *fetus* develops during pregnancy.

vascular system
The network of *arteries*, *veins*, and *capillaries* that conveys blood around the body.

vaso-
Prefix relating to blood vessels.

vein
A vessel carrying blood from the *tissues* and organs of the body back to the heart.

ventral
Relating to the front of the body, or the bottom of the brain.

ventricle
1. Either of the two larger muscular chambers of the heart. The right ventricle pumps blood to the lungs to be oxygenated, while the stronger-muscled left ventricle pumps oxygenated blood to the rest of the body. See also atrium.
2. One of the four cavities in the brain that contain *cerebrospinal fluid*.

venule
A very small *vein*, carrying blood away from *capillaries*.

vertebra (pl. vertebrae)
Any of the individual bones forming the **vertebral column** or spine.

villi (sing. villus)
Small, closely packed, fingerlike protrusions on the lining of the small intestine, giving the surface a velvety appearance and providing a large surface area, which is essential for the absorption of nutrients.

virus
A tiny parasite that lives inside *cells*, often consisting of only a length of *DNA* or *RNA* surrounded by *protein*. Viruses are much smaller than cells, and operate by "hijacking" cells to make copies of themselves. They are unable to replicate by themselves. Many viruses are dangerous *pathogens*.

viscera
Another term for organs. The adjective **visceral** applies to *nerves* or blood vessels, for example, that supply these organs.

vitamin
Any of a variety of naturally occurring substances that are essential to the body in small amounts, but which the body cannot make itself and so must obtain from the diet.

voluntary muscle
See *skeletal muscle*.

vulva
The outer genitalia of females, comprising the entrance to the vagina and surrounding structures.

zygote
A *cell* formed by the union of two *gametes* at *fertilization*.

Index

Acknowledgments

DK Publishing would like to thank the following people for help in the preparation of this book: Hugh Schermuly and Maxine Pedliham, for additional design; Steve Crozier for color work; Nathan Joyce and Laura Palosuo for editorial assistance; Anushka Mody for additional design assistance; Richard Beatty for compiling the glossary. **Medi-Mation** would like to thank: Senior 3D artists: Rajeev Doshi, Arran Lewis, 3D artists: Owen Simons, Gavin Whelan, Gunilla Elam. **Antbits Ltd** would like to thank: Paul Richardson, Martin Woodward, Paul Banville, and Rachael Tremlett.

The publisher would like to thank the following for their kind permission to reproduce their photographs: (Key: a-above; b-below/bottom; c-center; f-far; l-left; r-right; t-top)

Action Plus: 308c, 309cl, 309cr; **Alamy Images:** Dr. Wilfried Bahnmuller 412tr; Alexey Buhantsov 327cl; Kolvenbach 15bl; Gloria-Leigh Logan 394cl; Ross Marks Photography 404cl; Medical-on-Line 483cr; Dr. David E. Scott / Phototake 387cl; Hercules Robinson 459b; Jan Tadeusz 325tr. **Sonia Barbate:** 400cl. **BioMedical Engineering Online:** 2006, 5:30 Sjoerd P Niehof, Frank JPM Huygen, Rick WP van der Weerd, Mirjam Westra, and Freek J Zijlstra, Thermography imaging during static and controlled thermoregulation in complex regional pain syndrome type 1: diagnostic value and involvement of the central sympathetic system, with permission from Elsevier; (doi:10.1186/1475-925X-5-30) 341tr; **Camera Press:** 14bl. **Corbis:** Dr. John D. Cunningham / Visuals Unlimited 390bl; 81A Productions 13br; 402tr, 407bl; Mark Alberhasky 424bc; G. Baden 410tr; Lester V. Bergman 422b, 423tr, 429br, 446bl, 461bl; Biodisc / Visuals Unlimited 344bl; Bernard Bisson / Sygma 417br; Blend Images / ER productions 421br; Markus Botzek 13bc; CNRI 49cl; Dr. John D. Cunningham 298c; Jean-Daniel Sudres/Hemis 310bc; Dennis Kunkel Microscopy, Inc. / Visuals Unlimited 473cr; Dennis Kunkel Microscopy, Inc. / Visuals Unlimited / Terra 441cl; Digital Art 412c; Doc-stock 460br; Eye Ubiquitous / Gavin Wickham 446br; Barbara Galati / Science Faction /Encyclopedia 481tr; Rune Hellestad 405br; Evan Hurd 291tr; Robbie Jack 285tr; Jose Luis Pelaez, Inc. / Blend Images 19cr; Karen Kasmauski 310bl; Peter Lansdorp / Visuals Unlimited 410cl; Lester Lefkowitz 225bl; Dimitri Lundt / TempSport 291br; Lawrence Manning 474cr; Dr. P. Marazzi 424cl; MedicalRF.com 22bl, 477b; Moodboard 310cla; NASA / Roger Ressmeyer 287cr; Sebastian Pfuetze 413bl; Photo Quest LTD 23 (Dense Connective); Photo Quest Ltd. / Science Photo Library 23 (Spongy Bone), 47br; Steve Prezant 447cr; Radius Images 442br; Roger Ressmeyer / Encyclopedia 439t (D55); Martin Ruetschi / Keystone / EPA 442cr; Science Photo Library / Photo Quest Ltd 460cr; Dr. Frederick Skvara / Visuals Unlimited 278br; Howard Sochurek 466br; Gilles Poderins / SPL 429cl; Tom Stewart 479br; Jason Szenes / EPA 308bc; Tetra Images 310clb; Visuals Unlimited 47bc; Visuals Unlimited 424bl, 474bl; Ken Weingart 396bl; Dennis Wilson 398cl; **Lucky Rich Diamond:** 352bl; **Falling Pixel Ltd.:** 13cr. **Fertility and Sterility, Reprinted from:** Vol 90, No 3, September 2008, (doi:10.1016/j.fertnstert.2007.12.049) Jean-Christophe Lousse, MD, and Jacques Donnez, MD, PhD, Department of Gynecology, Université Catholique de Louvain, 1200 Brussels, Belgium, Laparoscopic observation of spontaneous human ovulation; © 2008 American Society for Reproductive Medicine, Published by Elsevier Inc. with permission from Elsevier. 374bl; **Getty Images:** 3D4 Medical.com 460c; 19 (Berber); 297tc, 307br, 407bc, 407br, 407cl, 407cla, 408tr, 426cl; Asia Images Group 407tr; Cristian Baitg 404b, 479bl; Barts Hospital 350tr, 357c; BCC Microimagine 459cra; Alan Boyde 410br; Neil Bromhall 400t; Nancy Brown 19 (Mongolian); Veronika Burmeister 463c; Peter Cade 420cl; Greg Ceo 19crb; Matthias Clamer 19 (Blue Eyed); CMSP / J.L. Carson 420tr; CMSP / J.L. Carson / Collection Mix: Subjects 418bl (D62); Peter Dazeley 44bl, 479t; George Diebold 16br; Digital Vision 14-15 (darker backgrnd), 445bc; f-64 Photo Office / amanaimagesRF 14-15 (light sand); Dr. Kenneth Greer 423tl; Dr. Kenneth Greer / Visuals Unlimited 482c; Jamie Grill 387br; Ian Hooton / Science Photo Library 481br; Dr. Fred Hossler 474c; Image Source 116t, 119tr, 310cb, 312bc, 339br; Jupiterimages 407cra; Kallista Images / Collection Mix: Subjects 438c; Ashley Karyl 19 (Brown Eyed); Dr. Richard Kessel & Dr. Gene Shih 421ca; Scott Kleinman 312bl; Mehau Kulyk / Science Photo Library 439b; PhotoAlto / Teo Lannie 369br; Bruce Laurance 19 (Asian); Wang Leng 296bl; S. Lowry, University of Ulster 420bl, 464tr; National Geographic / Alison Wright 19 (Seychelles); National Geographic / Robert B. Goodman 19 (Maori); Yorgos Nikas 377; Jose Luis Pelaez Inc 405cr; Peres 421crb; Peter Adams 19 (Bolivian), 19ftl; PhotoAlto / Michele Constantini 407crb; Steven Puetzersb 24fbl; Rubberball 408br; John Sann / Riser 417tr; Caroline Schiff / Digital Vision 433cr; Ariel Skelley 307t; AFP 15cla, 281tr, 291cr; SPL 288cra, 289 (Hinge); SPL / Pasieka 6tl, 24cl; Stockbyte 19 (Red Hair); Siqui Sanchez 412b; Michel Tcherevkoff 394-395b; UHB Trust 447b; Alvis Upitis / The Image Bank 418br; Ken Usami 308cl; Nick Veasey 123l, 289 (Saddle); CMSP 18cl, 421tr, 424br, 424tr, 446t; Dr. David Phillips / Visuals Unlimited 421tl; Ami Vitale 19 (Short Beard Indian); Jochem D Wijnands 19 (Indian); Dr. Gladden Willis 23 (smooth tissue), 24bc, 389cl, 421bl, 458c; Dr. G.W. Willis 471cr; G W Willis / Photolibrary 482bl; Brad Wilson 19 (Asian Man); Alison Wright 19 (Bedouin); David Young-Wolff 406t. **Peter Hurst, University of Otago, NZ:** 22t, 23 (Nerve Tissue), 23 (Skeletal Muscle). **iStockphoto.**

com: Johanna Goodyear 312br. **Lennart Nilsson Image Bank:** 398tr. **Dr. Brian McKay / acld. com:** 443cl. **Robert Millard:** Stage Design (c) David Hockney / Photo courtesy LA Music Center Opera, Los Angeles 310br. **The Natural History Museum, London:** 15fcl, 321tr. **Mark Nielsen, University of Utah:** 76bl. **Oregon Brain Aging Study, Portland VAMC and Oregon Health & Science University:** 411br. **Photolibrary:** Peter Arnold Images 49bl. **Reuters:** Eriko Sugita 413cr. **Rex Features:** Granata / Planie 337br. **Dr. Alice Roberts:** 15br, 15tl, 15tr. **Science Photo Library:** David M. Martin, M.D. 355cr, 462cr; Professors P.M. Motta & S. Correr 386cr; 17bl, 45bl, 63br, 321cr, 339cr, 350br, 359bc, 360, 367cr, 379br, 413crb, 422tc, 426bl, 454cr, 455br, 456cr, 463br, 476bl; AJ Photo 430cl; Dr. M.A. Ansary 430c; Apogee 303c; Tom Barrick, Chris Clark, SGHMS 305cr; Alex Bartel 406cl; Dr. Lewis Baxter 442bl; BCC Microimaging 459cr; Juergen Berger 347 (Fungus); PRJ Bernard / CNRI 76-77b; Biophoto Associates 23 (Loose Connective), 187tl, 289bc, 427bl, 434c, 449br, 458cb; Chris Bjornberg 347 (Virus); Neil Borden 71; BSIP VEM 417cr, 452c, 467cl; BSIP, Raguet 419br; Scott Camazine 433bl; Scott Camazine & Sue Trainor 406bl; Cardio-Thoracic Centre, Freeman Hospital, Newcastle-upon-Tyne 453tr; Dr. Isabelle Cartier, ISM 376bl; CC, ISM 444bl, 444tr, 448br; CIMN / ISM 402tl; Hervé Conge, ISM 278bc; E. R. Degginger 13bl; Michelle Del Guercio 457bl; Department of Nuclear Medicine, Charing Cross Hospital 437c; Dept. of Medical Photography, St. Stephen's Hospital, London 475c; Dept. of Clinical Cytogenetics, Addenbrookes Hospital 416c; Du Cane Medical Imaging Ltd 379tc; Edelmann 401tr; Eye of Science 318tc, 346br, 347 (Protazoan), 350tc, 352cl, 354bl, 363c, 377, 468c; Don Fawcett 290cra, 379tl; Mauro Fermariello 288cr, 413c; Simon Fraser / Royal Victoria Infirmary, Newcastle upon Tyne 441br; Gastrolab 462br; GJLP 440cr; Pascal Goetgheluck 457br; Eric Grave 347 (Parasitic Worm); Paul Gunning 287tc; Gustioimages 262bl; Gusto Images 44br, 45br, 284-285cl, 451cl; Dr. M O. Habert, Pitie-Salpetriere, Ism 417cl; Innerspace Imaging 288c, 390bc; Makoto Iwafuji 19cr; Coneyl Jay 394t, 480t; John Radcliffe Hospital 425bc; Kwangshin Kim 475tl; James King-Holmes 72bl; Mehau Kulyk 456bc; Patrick Landmann 430cr; Lawrence Livermore Laboratory 16tr; Jackie Lewin, Royal Free Hospital 458br; Living Art Enterprises 263tr, 304tl; Living Art Enterprises, LCC 445cl; Living Art Enterprises, Llc 289 (Pivot), 432br; Look at Sciences 416bl; Richard Lowenberg 378tr; Lunagrafix 361cr; Dr. P. Marazzi 411c, 416bc, 422c, 422cr, 422tr, 424tc, 425tr, 432bl, 433br, 434bl, 435cr, 435tc, 444tl, 448crb, 457cr, 457t, 460cr, 462tc, 472bl, 475tr; Dr. P. Marazzi 435bc; David M. Martin, M.D. 355cr, 462cr; Arno Massee 366br; Carolyn A. McKeone 431cra; Medimage 20cl, 363cr; Hank Morgan 307c; Dr. G. Moscoso 399tr; Prof. P. Motta / Dept. of Anatomy / University 359br; Prof. P. Motta / Dept. of Anatomy / University "La Sapienza," Rome 352cr; Professor P.

Motta & D. Palermo 362cr; Professor P. Motta & G. Familiari 470tr; Professors P. M. Motta & S. Makabe 476tr; Zephyr 427cr, 429bl, 438tr, 441cr, 465cl, 469tl, 470ca, 471cr; Dr. Gopal Murti 410c; National Cancer Institute 334tl; Susumu Nishinaga 77br, 132bl, 281bl, 335cl, 346tr, 363cl, 372cl, 374c, 409cl; Omikron 347tr; David M. Phillips 347 (Bacterium); Photo Insolite Realite 326cl; Alain Pol / ISM 456cl; K R Porter 20bc; Paul Rapson 449bl; Jean-Claude Revy ISM 363tr; Dave Roberts 286bc; Antoine Rosset 70b; Schleichkorn 336ca; W.W Schultz / British Medical Journal 376tr; Dr. Oliver Schwartz Institute Pasteur 348bl; Astrid & Hans-Frieder Michler 279cl; Martin Dohrn 280tl; Richard Wehr / Custom Medical Stock Photo 279bl; Sovereign, ISM 62bl, 62-63b, 289 (Ball), 302-303c, 432c, 439cr, 473bl, 483cl; SPL 286cl; St. Bartholomew's Hospital, London 434br; Dr. Linda Stannard, UCT 448bl; Volker Steger 12cr, 24bcl; Saturn Stills 416cr; Andrew Syred 396cl; Astrid & Hanns-Frieder Michler 353tr, 468bl; CNRI 285bc, 331bc, 331br, 423b, 428bl, 453cr, 458bl, 462bl; Dee Breger 325bl, 385c; Dr. G. Moscoso 325br; Dr. Gary Settles 331tr; Geoff Bryant 313bc, 313cb; ISM 328bc, 328br, 464cl; Manfred Kage 73br, 320ca; Michael W. Davidson 384bc; Pasieka 354tr, 359cr, 373cl, 380tr, 430bc; Paul Parker 314bl; Richard Wehr / Custom Medical Stock Photop710/226 369cr; Steve Gschmeissner 23 (Adipose Tissue), 23 (Epithelial Tissue), 76bc, 132cl, 287tr, 296br, 296cra, 298bl, 309br, 312cr, 314tc, 335c, 335cr, 336cl, 341cl, 348c, 352crb, 356tr, 362cl, 368tr, 374tc, 375tr, 384clb, 387cr, 391tr, 409clb, 409tr, 469br, 472tr, 477cr; Dr. Harout Tanielian 425br; TEK Image 473br; Javier Trueba / MSF 14cla, 15c, 15cr; David Parker 18cr; M.I. Walker 23 (Cartilage); Garry Watson 466tr; John Wilson 475br; Professor Tony Wright 444br. **SeaPics.com:** Dan Burton 330br; www.skullsunlimited.com <http://www.skullsunlimited.com> 14cl; **Robert Steiner MRI Unit, Imperial College London:** 8-9, 24bl, 34c, 34-35b, 34-35t, 54b, 55b, 134-135 (all), 166-167 (all), 196-197 (all), 234-235 (all), 272-273; **Claire E Stevens, MA PA:** 375b. **Stone Age Institute:** Dr. Scott Simpson (project paleontologist) 14cb. **UNEP/GRID-Arendal:** Emmanuelle Bournay / Sources: GMES, 2006; INTERSUN, 2007. INTERSUN, the Global UV project, is a collaborative project between WHO, UNEP, WMO, the International Agency on Cancer Research (IARC) and the International Commission on Non-Ionizing Radiation Protection (ICNIRP). 280br. **Courtesy of U.S. Navy:** Mass Communication Specialist 2nd Class Jayme Pastoric 327br. **Dr. Katy Vincent, University of Oxford:** 310-311t. **Wellcome Images:** 119br; Joe Mee & Austin Smith 373tr; Dr. Joyce Harper 376clb; Wellcome Photo Library 461br. **Wits University, Johannesburg:** photo by Brett Eloff 14cra

All other images © Dorling Kindersley
For further information see: www.dkimages.com